ELECTRONIC CIRCUITS: DISCRETE AND INTEGRATED

McGraw-Hill Electrical and Electronic Engineering Series

FREDERICK EMMONS TERMAN, *Consulting Editor*
W. W. HARMAN AND J. G. TRUXAL, *Associate Consulting Editors*

AHRENDT AND SAVANT · Servomechanism Practice
ANGELAKOS AND EVERHART · Microwave Communications
ANGELO · Electronic Circuits
ASELTINE · Transform Method in Linear System Analysis
ATWATER · Introduction to Microwave Theory
BERANEK · Acoustics
BRACEWELL · The Fourier Transform and Its Application
BRENNER AND JAVID · Analysis of Electric Circuits
BROWN · Analysis of Linear Time-invariant Systems
BRUNS AND SAUNDERS · Analysis of Feedback Control Systems
CARLSON · Communication Systems: An Introduction to Signals and Noise in Electrical Communication
CHEN · The Analysis of Linear Systems
CHEN · Linear Network Design and Synthesis
CHIRLIAN · Analysis and Design of Electronic Circuits
CHIRLIAN · Basic Network Theory
CHIRLIAN AND ZEMANIAN · Electronics
CLEMENT AND JOHNSON · Electrical Engineering Science
CUNNINGHAM · Introduction to Nonlinear Analysis
D'AZZO AND HOUPIS · Feedback Control System Analysis and Synthesis
EASTMAN · Fundamentals of Vacuum Tubes
ELGERD · Control Systems Theory
EVELEIGH · Adaptive Control and Optimization Techniques
FEINSTEIN · Foundations of Information Theory
FITZGERALD, HIGGINBOTHAM, AND GRABEL · Basic Electrical Engineering
FITZGERALD AND KINGSLEY · Electric Machinery
FRANK · Electrical Measurement Analysis
FRIEDLAND, WING, AND ASH · Principles of Linear Networks
GEHMLICH AND HAMMOND · Electromechanical Systems
GHAUSI · Principles and Design of Linear Active Circuits
GHOSE · Microwave Circuit Theory and Analysis
GLASFORD · Fundamentals of Television Engineering
GREINER · Semiconductor Devices and Applications
HAMMOND · Electrical Engineering
HANCOCK · An Introduction to the Principles of Communication Theory
HARMAN · Fundamentals of Electronic Motion
HARMAN · Principles of the Statistical Theory of Communication
HARMAN AND LYTLE · Electrical and Mechanical Networks
HARRINGTON · Time-harmonic Electromagnetic Fields
HAYASHI · Nonlinear Oscillations in Physical Systems
HAYT · Engineering Electromagnetics
HAYT AND KEMMERLY · Engineering Circuit Analysis
HILL · Electronics in Engineering
JAVID AND BROWN · Field Analysis and Electromagnetics
JOHNSON · Transmission Lines and Networks
KOENIG AND BLACKWELL · Electromechanical System Theory
KOENIG, TOKAD, AND KESAVAN · Analysis of Discrete Physical Systems
KRAUS · Antennas

Kraus · Electromagnetics
Kuh and Pederson · Principles of Circuit Synthesis
Kuo · Linear Networks and Systems
Ledley · Digital Computer and Control Engineering
LePage · Complex Variables and the Laplace Transform for Engineering
LePage and Seely · General Network Analysis
Levi and Panzer · Electromechanical Power Conversion
Ley, Lutz, and Rehberg · Linear Circuit Analysis
Linvill and Gibbons · Transistors and Active Circuits
Littauer · Pulse Electronics
Lynch and Truxal · Introductory System Analysis
Lynch and Truxal · Principles of Electronic Instrumentation
Lynch and Truxal · Signals and Systems in Electrical Engineering
McCluskey · Introduction to the Theory of Switching Circuits
Manning · Electrical Circuits
Meisel · Principles of Electromechanical-energy Conversion
Millman · Vacuum-tube and Semiconductor Electronics
Millman and Halkias · Electronic Devices and Circuits
Millman and Taub · Pulse, Digital, and Switching Waveforms
Minorsky · Theory of Nonlinear Control Systems
Mishkin and Braun · Adaptive Control Systems
Moore · Traveling-wave Engineering
Nanavati · An Introduction to Semiconductor Electronics
Pettit · Electronic Switching, Timing, and Pulse Circuits
Pettit and McWhorter · Electronic Amplifier Circuits
Pfeiffer · Concepts of Probability Theory
Reza · An Introduction to Information Theory
Reza and Seely · Modern Network Analysis
Ruston and Bordogna · Electric Networks: Functions, Filters, Analysis
Ryder · Engineering Electronics
Schilling and Belove · Electronic Circuits: Discrete and Integrated
Schwartz · Information Transmission, Modulation, and Noise
Schwarz and Friedland · Linear Systems
Seely · Electromechanical Energy Conversion
Seely · Electron-tube Circuits
Seely · Introduction to Electromagnetic Fields
Seely · Radio Electronics
Seifert and Steeg · Control Systems Engineering
Shooman · Probabilistic Reliability: An Engineering Approach
Siskind · Direct-current Machinery
Skilling · Electric Transmission Lines
Skilling · Transient Electric Currents
Stevenson · Elements of Power System Analysis
Stewart · Fundamentals of Signal Theory
Strauss · Wave Generation and Shaping
Su · Active Network Synthesis
Terman · Electronic and Radio Engineering
Terman and Pettit · Electronic Measurements
Thaler · Elements of Servomechanism Theory
Thaler and Brown · Analysis and Design of Feedback Control Systems
Thaler and Pastel · Analysis and Design of Nonlinear Feedback Control Systems

Tou · Digital and Sampled-data Control Systems
Tou · Modern Control Theory
Truxal · Automatic Feedback Control System Synthesis
Tuttle · Electric Networks: Analysis and Synthesis
Valdes · The Physical Theory of Transistors
Van Bladel · Electromagnetic Fields
Weeks · Antenna Engineering
Weinberg · Network Analysis and Synthesis

Brooklyn Polytechnic Institute Series

Angelo · Electronic Circuits
Levi and Panzer · Electromechanical Power Conversion
Lynch and Truxal · Signals and Systems in Electrical Engineering
Combining Introductory System Analysis
Principles of Electronic Instrumentation
Mishkin and Braun · Adaptive Control Systems
Schilling and Belove · Electronic Circuits: Discrete and Integrated
Schwartz · Information Transmission, Modulation, and Noise
Shooman · Probabilistic Reliability: An Engineering Approach
Strauss · Wave Generation and Shaping

ELECTRONIC CIRCUITS: DISCRETE AND INTEGRATED

Donald L. Schilling
Associate Professor of Electrical Engineering
Polytechnic Institute of Brooklyn

Charles Belove
Associate Professor of Electrical Engineering
Polytechnic Institute of Brooklyn

McGraw-Hill Book Company
New York St. Louis San Francisco Toronto London Sydney

ELECTRONIC CIRCUITS: DISCRETE AND INTEGRATED

Library of Congress Catalog Card Number 68-19493

55289

1234567890 MAMM 7543210698

to our wives

ANNETTE *and* **GOLDA**

for their assistance and patience

Preface

The field of electronics has undergone a remarkable change over the past two decades. The vacuum tube, which, in some way, influenced every facet of human civilization, has been almost universally supplanted by the transistor. In fact, at the time of this writing it appears that the field of microelectronics and integrated circuits may well make our current thinking obsolete within the next decade.

A revolution such as this creates an immediate need for new teaching philosophies and methods using up-to-date textbooks. This book, intended as a beginning text in the analysis and design of electronic circuits for upper-sophomore- or junior-level engineering and physics students, aims to provide the student with a background which will enable him to handle circuit problems with ease and sufficient depth of understanding to be able to comprehend new devices as they become available. After studying this text, the student should be sufficiently prepared to function competently both in industry and in more sophisticated senior- and graduate-level electronics courses.

The text is based on a particular philosophy of teaching which has been used successfully in the junior electronics courses at the Polytechnic Institute of Brooklyn for several years. It is also being used by the RCA Semiconductor Division in their continuing education program. This philosophy is based on the premise that practical circuit design seldom makes direct use of device physics, but rather uses terminal properties. The authors present physical theory descriptively to enhance the discussion of these terminal properties. The knowledge acquired from this text concerning the practical use of semiconductor devices will prepare the way for a subsequent course in semiconductor physics.

It is assumed that the student has a background in linear passive circuit theory, which includes a thorough grounding in the use of Kirchhoff's laws in dc and ac circuits. This, along with a knowledge of simple power calculations, is all that is required for the first half of the book. For the remainder, some knowledge of the complex-frequency plane and the concept of frequency response is helpful.

The prime objective of this text is to provide insight into the analysis and design of electronic circuits, both discrete and integrated. Chapters 1 to 8 cover transistor circuits at low frequencies. Transistor circuit design procedures and the most useful models are presented. The emphasis in this part is on the graphical approach, in which the authors believe very strongly. The important concepts of dc and ac load lines, along with large-signal and small-signal analyses, are introduced at the outset in connection with diode circuits. When the transistor is introduced in Chapter 3, the student has these techniques firmly established. Bias stability is covered in Chapter 4. The type of bias arrangement employed in integrated circuits is introduced in addition to the standard techniques. Chapter 5 introduces low-frequency power amplifiers. In Chapter 6, the small-signal low-frequency behavior of the various transistor configurations is studied, using the hybrid model. Here the technique of impedance reflection is introduced as a shortcut in the analysis of complicated circuits. The influence of the bias point on the small-signal behavior is also considered. In Chapter 7, multiple-transistor circuits are examined, with emphasis on those circuit configurations (such as the difference amplifier and other dc amplifiers) which lend themselves naturally to integrated circuit fabrication techniques. The principles and advantages of feedback are considered in Chapter 8. Gain, sensitivity, and impedances are studied in detail from both analysis and design points of view. Again, examples stress the type of feedback amplifiers which utilize linear integrated circuits.

Chapter 9 covers fabrication and design techniques for integrated circuits. The operation of the junction and insulated-gate (MOS) field-effect transistor is discussed in Chapter 10. Chapter 11 studies vacuum-tube amplifiers using the triode, tetrode, and pentode.

The four final chapters cover frequency response. Chapters 12 and 13 consider low- and high-frequency response of *RC*-coupled amplifiers, using all the devices previously introduced. The transistor switch is also discussed. Chapter 14 examines tuned narrowband amplifiers and methods for extending the bandwidth of wideband amplifiers. It is unique in that it presents practical VHF and video techniques on a junior undergraduate level. Chapter 15 considers the frequency response of feedback amplifiers. The stability problems associated with integrated circuits are considered, and frequency-compensation techniques are studied. A large section on practical transistor oscillators is also included in this chapter.

Every effort has been made to use practical parameter values in the numerous illustrative examples presented throughout the text. Typical manufacturers' specifications are given, so that the student gains an idea of the practical range of parameter values for the various devices. The circuits presented in the examples have been constructed and tested in the laboratory. More than 450 homework problems are included, ranging from routine drill in analysis to difficult designs. Appendixes II and III include standard resistor and capacitor values and manufacturers' data sheets for various devices, which can be used for the design problems and as future references. A solutions manual is available from the publisher, and the authors will be happy to furnish a set of laboratory experiments which are currently used at P.I.B. in conjunction with the course.

The authors would like to acknowledge gratefully the encouragement and constructive criticism offered by Prof. J. G. Truxal, and the excellent review of the entire manuscript by Prof. G. Anner. We thank our colleagues and students, and in particular, Profs. I. Meth, H. Taub, H. Schachter, and Dr. R. Schilling for their suggestions and criticism of portions of the manuscript. We express our particular appreciation to Mrs. Florence Schiff, who did a fine job of typing clean copy from the penciled notes of two left-handed authors. We also thank R. Werner, one of our students at P.I.B., who checked most of the examples in the laboratory, and J. Oberst for preparing the solutions manual and answer book.

Donald L. Schilling
Charles Belove

Contents

Preface *ix*

Notation *xix*

Chapter 1 Introduction 1

The Transmitter *2*

Chapter 2 Diode-circuit Analysis 8

Introduction to Diodes *8*
2.1 Nonlinear Properties—The Ideal Diode *9*
2.2 An Introduction to Semiconductor Diode Theory *17*
2.3 Analysis of Simple Diode Circuits—The DC Load Line *22*
2.4 Small-signal Analysis—The Concept of Dynamic Resistance *25*
2.5 Small-signal Analysis—The AC Load Line *31*
2.6 Large-signal Analysis—Distortion and Q-point Shift *33*
2.7 Zener Diodes *37*
2.8 Piecewise Linear Analysis and Equivalent Circuits *42*
2.9 Temperature Effects in Diodes *50*
2.10 Manufacturers' Specifications *54*
2.10-1 *The Diode Rectifier* *54*
2.10-2 *The Zener Diode* *56*

Chapter 3 Introduction to Transistor Circuits 73

Introduction *73*
3.1 Current-flow Mechanism in the Junction Transistor *73*
3.1-1 *The Emitter-Base Junction* *75*
3.1-2 *The Collector-Base Junction* *78*
3.2 Current Amplification in the Transistor *80*
3.3 Graphical Analysis of Transistor Circuits *87*
3.4 Power Calculations *96*
3.5 The Infinite Bypass Capacitor *101*
3.6 The Infinite Coupling Capacitor *106*
3.7 The Emitter Follower *110*

Chapter 4 Bias Stability 123

Introduction *123*
4.1 Quiescent-point Variations Due to Uncertainties in β *124*
4.2 The Effect of Temperature on the Q Point *128*
4.3 Stability-factor Analysis *130*
4.4 Temperature Compensation Using Diode Biasing *137*
4.5 Environmental Thermal Considerations in Transistor Amplifiers *142*
4.6 Manufacturers' Specifications for High-power Transistors *145*

Chapter 5 Audio-frequency Linear Power Amplifiers 151

Introduction *151*
5.1 The Class A Common-emitter Power Amplifier *153*
5.1-1 *Q-point Placement* *153*
5.1-2 *Power Calculations* *155*
5.1-3 *The Maximum-dissipation Hyperbola* *158*
5.2 Transformer-coupled Amplifier *164*
5.2-1 *Power Calculations* *165*
5.3 Class B Push-Pull Power Amplifiers *168*
5.3-1 *Load-line Determination* *171*
5.3-2 *Power Calculations* *171*
5.4 Amplifiers Using Complementary Symmetry *177*

Chapter 6 Small-signal Low-frequency Analysis and Design 185

Introduction *185*
6.1 The Hybrid Parameters *186*
6.2 The Common-emitter Configuration *188*
6.3 The Common-base Configuration *198*
6.4 The Common-collector (Emitter-follower) Configuration *202*
6.5 Collection of Significant Parameters for the Three Basic Configurations *212*
6.6 Interpretation of Manufacturers' Specifications for Low-power Transistors *213*

Chapter 7 Multiple-transistor Circuits 220

Introduction *220*
7.1 Cascading of Amplifier Stages *221*

7.2 The Difference Amplifier *233*
7.2-1 *Common-mode Rejection Ratio* *239*
7.2-2 *A Constant-emitter Current Source* *240*
7.3 The Darlington Configuration (Compound Amplifier) *245*
7.4 The Cascode Amplifier *250*

Chapter 8 Feedback-amplifier Fundamentals 262

Introduction *262*
8.1 Basic Concepts of Feedback *263*
8.2 The Gain of a Voltage-feedback Amplifier *266*
8.2-1 *Voltage Feedback with Current Error* *266*
8.2-2 *Voltage Feedback with Voltage Error* *275*
8.3 Feedback Amplifiers and the Sensitivity Function *279*
8.4 Input and Output Impedances *283*
8.4-1 *Input Impedance* *284*
8.4-2 *Output Impedance* *288*
8.5 Examples of Basic Feedback-amplifier Analysis *289*
8.6 Introduction to the Design of Feedback Amplifiers *298*
8.7 Other Applications of Feedback *302*
8.7-1 *An Automatic Volume-control Circuit* *302*
8.7-2 *The Regulated Power Supply* *309*

Chapter 9 Integrated Circuits 326

Introduction *326*
9.1 An Introduction to the Fabrication of an Integrated-circuit Transistor *328*
9.2 The Equivalent Circuit of the Integrated Transistor *331*
9.3 The Integrated Diode *333*
9.4 The Integrated Capacitor *335*
9.4-1 *The Junction Capacitor* *335*
9.4-2 *The Thin-film Capacitor* *336*
9.5 The Integrated Resistor *337*
9.5-1 *The Junction Resistor* *337*
9.5-2 *The Thin-film Resistor* *339*
9.6 The Integrated Inductor *341*
9.7 Design of a Simple Integrated Circuit *341*
9.8 Analysis of a Typical Integrated-circuit Amplifier—The Fairchild μA702 *342*
9.8-1 *Calculation of the Quiescent Operating Points* *345*

9.8-2 *Small-signal Gain* *347*
9.9 Cascading Integrated-circuit Amplifiers *351*

Chapter 10 The Field-effect Transistor 360

Introduction *360*
10.1 Introduction to the Theory of Operation of the JFET *360*
10.2 Introduction to the Theory of Operation of the IGFET (MOSFET) *364*
10.3 Graphical Analysis and Biasing *368*
10.4 Large-signal Analysis—Distortion *372*
10.5 Small-signal Analysis *374*
10.5-1 *The Common-source Voltage Amplifier* *377*
10.5-2 *The Source Follower (The Common-drain Amplifier)* *380*
10.5-3 *The Common-gate Amplifier* *386*
10.6 Typical Manufacturers' Specifications *387*

Chapter 11 The Vacuum Tube 397

Introduction *397*
11.1 Introduction to the Vacuum Tube *398*
11.1-1 *The Diode* *398*
11.1-2 *The Triode* *400*
11.1-3 *The Tetrode* *402*
11.1-4 *The Pentode* *403*
11.2 Graphical Analysis and Biasing *404*
11.3 Distortion *410*
11.4 Small-signal Analysis *413*
11.4-1 *The Grounded-cathode Amplifier* *416*
11.4-2 *The Cathode Follower* *418*
11.4-3 *The Grounded-grid Amplifier* *419*
11.5 Manufacturers' Specifications *420*

Chapter 12 Low-frequency Response of *RC*-coupled Amplifiers 429

Introduction *429*
12.1 The Low-frequency Response of the Transistor Amplifier *431*
12.1-1 *The Emitter Bypass Capacitor* *431*
12.1-2 *Asymptotic (Bode) Plots of Amplifier Transfer Functions* *433*
12.1-3 *The Coupling Capacitor* *440*

12.1-4 *The Base and Collector Coupling Capacitors* *442*
12.1-5 *Combined Effect of Bypass and Coupling Capacitors* *446*
12.2 Low-frequency Response of the FET Amplifier *447*
12.2-1 *The Source Bypass Capacitor* *448*
12.2-2 *The Drain Coupling Capacitor* *450*
12.2-3 *The Gate Coupling Capacitor* *451*
12.3 The Low-frequency Response of the Vacuum Tube *453*

Chapter 13 High-frequency Response of *RC*-coupled Amplifiers 462

13.1 The Transistor Amplifier at High Frequencies *462*
13.1-1 *The Hybrid-pi Equivalent Circuit* *463*
13.1-2 *High-frequency Behavior of the Common-emitter Amplifier—Miller Capacitance* *467*
13.1-3 *The Emitter Follower at High Frequencies* *474*
13.2 The Field-effect Transistor at High Frequencies *480*
13.2-1 *High-frequency Behavior of the Common-source Amplifier—Miller Capacitance* *481*
13.2-2 *High-frequency Behavior of the Source Follower* *483*
13.3 The Vacuum Tube at High Frequencies *487*
13.3-1 *The Cathode Follower at High Frequencies* *490*
13.4 Cascaded *RC* Amplifiers *490*
13.4-1 *Cascading the FET* *495*
13.4-2 *Cascading the Vacuum Tube* *495*
13.5 The Gain-Bandwidth Product *498*
13.5-1 *Gain-Bandwidth Product for a Single-stage Amplifier* *498*
13.5-2 *Gain-Bandwidth Product in a Cascaded Amplifier* *501*
13.6 The Transistor Switch *504*

Chapter 14 Tuned Amplifiers 522

Introduction *522*
14.1 The Single-tuned Amplifier *523*
14.1-1 *The Effect of $r_{bb'}$ on the Response of a Single-tuned Amplifier* *528*
14.1-2 *Impedance Matching to Improve Gain* *530*
14.2 The Cascode Amplifier *539*
14.3 Neutralization *543*
14.4 The Synchronously Tuned Amplifier *549*
14.5 The Stagger- and Double-tuned Amplifiers *553*
14.5-1 *The Stagger-tuned Amplifier* *553*
14.5-2 *The Double-tuned Amplifier* *559*

14.6 Shunt Peaking *561*
14.7 The Distributed Amplifier *565*

Chapter 15 Frequency Response of Feedback Amplifiers 576

Introduction *576*
15.1 Bandwidth and Gain-Bandwidth Product *577*
15.2 The Problem of Stability *580*
15.3 The Nyquist Stability Criterion—Bode Plots *583*
15.4 Stabilizing Networks *586*
15.4-1 *No Frequency Compensation 586*
15.4-2 *Simple Lag Compensation 588*
15.4-3 *More Complicated Lag Compensation 591*
15.4-4 *Lead Compensation 595*
15.5 Examples *598*
15.5-1 *Lag-compensated Feedback Amplifiers 598*
15.5-2 *Lag-compensated IC Amplifier 600*
15.5-3 *Lead Compensation 601*
15.6 Active Filters Using Feedback *605*
15.6-1 *The Integrator 605*
15.6-2 *The Q Multiplier 606*
15.7 Oscillators *609*
15.7-1 *The Phase-shift Oscillator 609*
15.7-2 *The Wien Bridge Oscillator 612*
15.7-3 *The Tuned-circuit Oscillator 614*
15.7-4 *The Colpitts Oscillator 616*
15.7-5 *The Hartley Oscillator 617*

Appendix I Gain Expressed in Logarithmic Units—The Decibel (dB) *623*

Appendix II Standard Values of Resistance and Capacitance *625*

Appendix III Device Characteristics *627*

Index *657*

Notation

The symbols for currents and voltages at the terminals of active devices have subscripts which indicate the pertinent terminal for currents or terminal pair for voltages. In addition, uppercase and lowercase symbols and subscripts are used to distinguish between quiescent values, total values, and incremental values. The International System of Units is used throughout.

EXAMPLES:

I_{BQ}, I_{CQ}, V_{CEQ}	= quiescent-point value
I_B, I_C, V_{CE}	= dc value, with signal
i_B, i_C, i_E, v_{CE}	= total instantaneous value
i_b, i_c, i_e, v_{ce}	= instantaneous value of time-varying component (zero average)
I_b, I_e, V_{ce}	= rms value of sinusoidal component
I_{bm}, I_{em}, V_{cem}	= max (peak) value of time-varying component of variable
V_{BB}, V_{CC}, V_{DD}	= supply voltages

GRAPHICAL ILLUSTRATION OF NOTATION

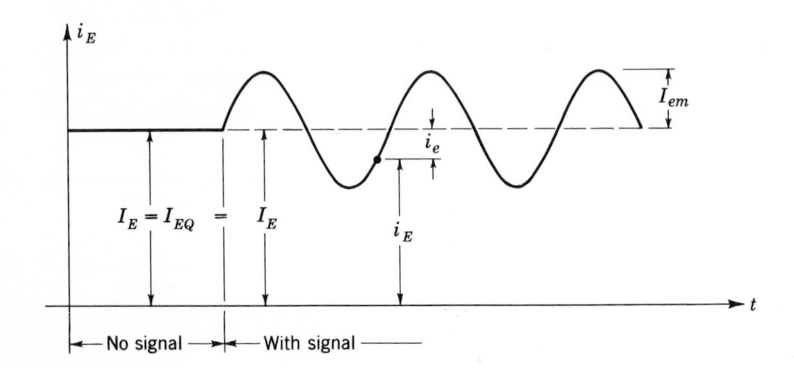

Sinusoidal signal, no distortion

$$i_E(t) = I_E + I_{em} \sin \omega t = I_{EQ} + I_{em} \sin \omega t$$
$$i_e(t) = I_{em} \sin \omega t$$

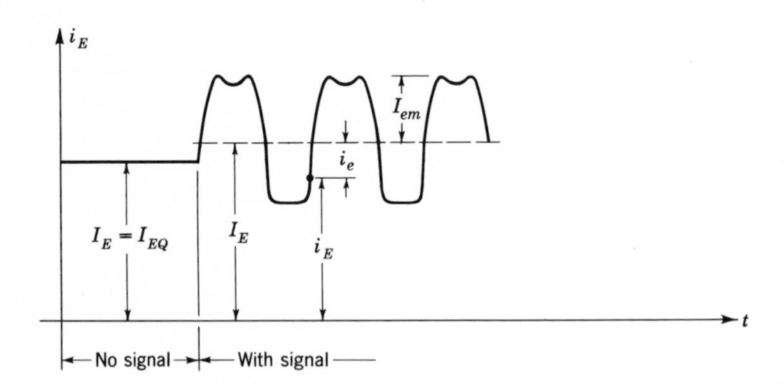

Signal, with distortion

$$i_E(t) = I_E + i_e(t)$$

ELECTRONIC CIRCUITS: DISCRETE AND INTEGRATED

1 Introduction

The title of this text, "Electronic Circuits: Discrete and Integrated," indicates that it will cover one particular facet of that tremendous field of endeavor known somewhat loosely today as *electronics*. Any attempt to subdivide the field leads to a rather large catalog of topics. Broadly speaking, we note that the purpose of most electronics work is the production of a *system* which transfers or transforms energy or information. Radio, television, computers, satellite communication systems, and automatic navigation systems are but a few types of *electronic systems*.

One characteristic of all these systems is that they can be divided into subsystems, or components. Electronic components include inductors, resistors, capacitors, transistors, integrated circuits, etc., and even complete amplifiers. Thus an electronics engineer may be concerned with systems or components, or both.

As the title implies, this text discusses electronic circuits, and in particular, linear amplifiers. In any electronic system, one or more amplifiers are usually required to operate on the *signal* which contains the energy or information. These amplifiers may be linear or nonlinear. A

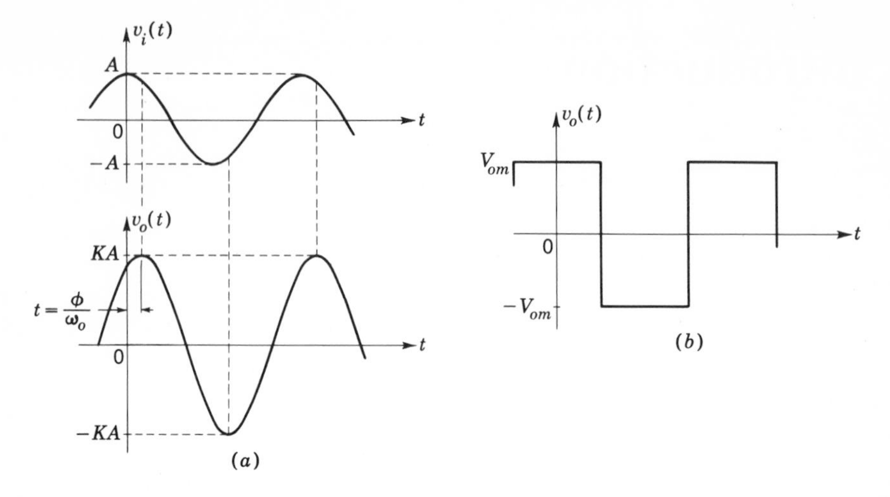

Fig. 1.1 Input and output signals for linear and nonlinear amplifiers. (*a*) Linear amplifier; (*b*) nonlinear amplifier.

linear voltage amplifier, for example, operates on the signal applied to its input terminals,

$$v_i(t) = A \cos \omega_0 t \tag{1.1}$$

to yield an amplified output signal,

$$v_0(t) = KA \cos (\omega_0 t + \phi) \tag{1.2}$$

where K (usually greater than unity) and ϕ are constants. These signals are shown in Fig. 1.1*a*. Any deviation of the output signal from (1.2) constitutes *distortion*, which is usually undesirable. Most linear amplifiers are designed to minimize distortion (Chaps. 10 and 11).

In direct contrast to the linear amplifier are the various kinds of nonlinear amplifiers. For example, in one type, called a *binary amplifier*, the sinusoidal signal (1.1), applied to the input, yields an output which is one constant when $v_i(t)$ is positive, and another constant when $v_i(t)$ is negative, as shown in Fig. 1.1*b*.

As an example of the variety of linear amplifiers which may be required in a typical system, let us consider the various components in an amplitude modulation (AM) radio system, from the announcer's voice in the studio to the loudspeaker in our car or living room.

THE TRANSMITTER

The air-pressure fluctuations caused by the announcer's voice are converted to voltage fluctuations by a microphone, as shown in Fig. 1.2.

These fluctuations are small, often of the order of millivolts or less, and they must be amplified considerably before transmission. The initial stages of amplification usually consist of transistors (Chaps. 3 to 7) or vacuum tubes (Chap. 11).

The frequency content of speech and music varies from about 20 Hz to 20 kHz, the so-called "audio" range. Such low frequencies cannot be transmitted efficiently over long distances by means of electromagnetic radiation through the atmosphere. To accomplish this, they must first be *modulated* onto a high-frequency *carrier* signal, as shown in Fig. 1.3*a*.

Amplitude modulation is one type of modulation technique. It is basically a simple multiplication process. Thus, if the message to be transmitted is written

$$v_m(t) = 1 + m \cos \omega_m t \tag{1.3}$$

where m is the *modulation index* ($m \leq 1$), and the carrier signal is

$$v_c(t) = A \cos \omega_c t \tag{1.4}$$

then the modulated signal is

$$v_T(t) = v_m v_c = A(1 + m \cos \omega_m t) \cos \omega_c t \tag{1.5}$$

Multiplying out and applying the pertinent trigonometric identities, the signal $v_T(t)$ is found to have three frequency components, ω_c, $\omega_c + \omega_m$, and $\omega_c - \omega_m$. These are shown schematically in Fig. 1.3*b*. The modulation process has shifted the message from ω_m to $\omega_c \pm \omega_m$, much higher frequencies (0.55 to 1.65 MHz for standard broadcast AM), which are suitable for transmission over long distances. It should be noted that the message

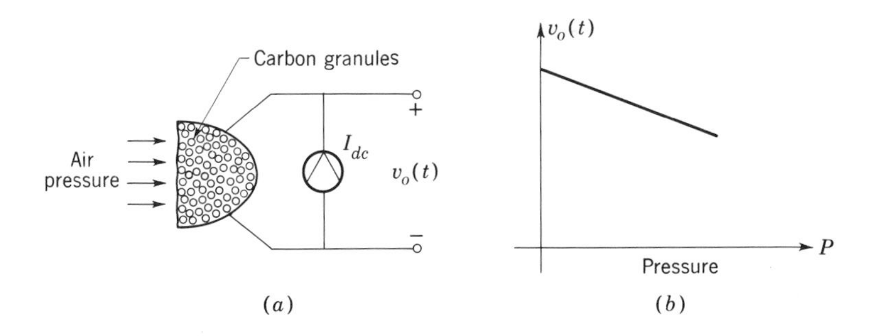

Fig. 1.2 The carbon microphone. (*a*) Air-pressure fluctuations change the effective density of the carbon and therefore its resistance. This resistance change is converted to a fluctuating voltage $v_o(t)$ by passing a dc current through the carbon. (*b*) An increase in air pressure compresses the carbon granules closer together, thereby *decreasing* the resistance. Thus an increase in pressure results in a *decrease* in the output voltage, $P \sim 1/R = I_{dc}/v_o$, $\Delta P \sim -(1/R^2)\,\Delta R$, $\Delta v_o = I_{dc}\,\Delta R \sim -I_{dc}R^2\,\Delta P$.

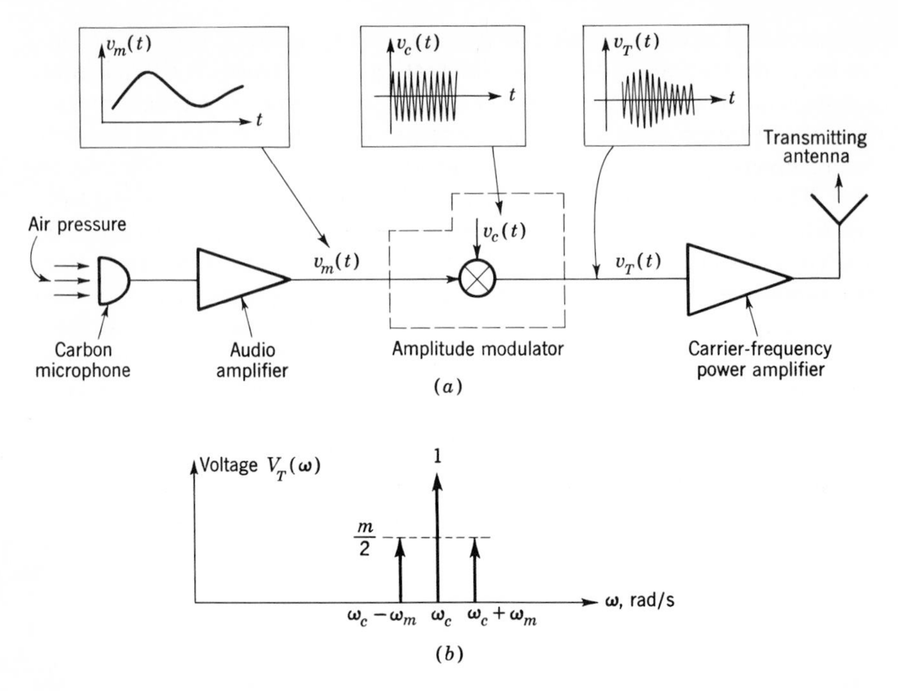

Fig. 1.3 The transmitter. (*a*) AM transmitter; (*b*) frequency spectrum of modulated signal.

will seldom, if ever, consist of a single frequency, but rather of a band of frequencies. We use the single-frequency signal of (1.3) simply to illustrate the process.

The modulated signal is amplified by a *power amplifier* and then transmitted. The power amplification provided depends on the distance over which reception is desired. Several tens of kilowatts are usually transmitted from standard AM broadcast transmitters.

THE RECEIVER

At the receiver, the signal at the antenna terminals will usually have a voltage which lies between 5 and 50 μV. It must be amplified, demodulated, amplified again, and finally converted from a voltage to an audible signal (air-pressure fluctuations) by means of a loudspeaker. This must be done without the introduction of significant distortion.

The first function of the receiver involves separating the carrier frequency of the desired station from all other carriers. The tuning knob of the radio accomplishes this by varying the capacitance in the tuned antenna circuit (Chap. 14). This tuned circuit passes the desired frequencies and attenuates all others so that they are not heard. Often, we hear two stations at the same dial setting. One reason for this is that a station on an adjacent carrier may have so much power that the attenuation of the tuned circuit is not sufficient to keep it from passing through the receiver.

The received signal $v_R(t)$ is amplified in a radio-frequency (RF) amplifier, as shown in Fig. 1.4. This amplifier is designed to amplify low-voltage signals without introducing much noise,* or distortion. It is also sometimes designed to provide additional filtering to further attenuate unwanted stations. For convenience of amplification, the RF carrier frequency is reduced to a fixed intermediate frequency (IF) of 455 kHz by a process called *heterodyning*, or *mixing*. It is this process from which the name *superheterodyne* receiver, often applied to AM radios, is derived. To obtain a *fixed* IF frequency, the carrier signal (ω_c) is multiplied by a *local-oscillator* signal (Chap. 15) of frequency $\omega_c + \omega_0$. As in the modula-

* Broadly speaking, noise is any external disturbance added to the signal. For example, random fluctuation of electrons in a resistor causes a *noise* voltage to appear at the resistor terminals.

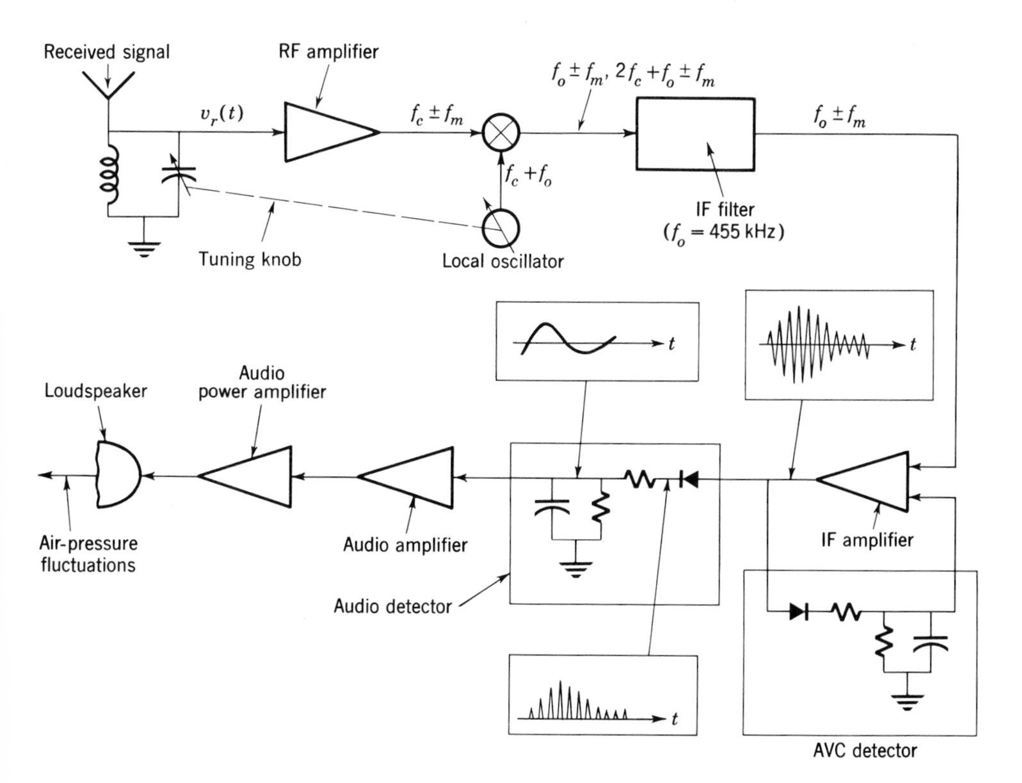

Fig. 1.4 AM receiver.

tion process discussed previously, this multiplication yields sum and difference frequencies, and the difference frequency ($\omega_0 = 455$ kHz) is selected by means of carefully tuned filter circuits. Note that the local-oscillator frequency must vary as the desired carrier varies. This is accomplished in the typical AM radio by mechanically coupling the antenna tuning capacitor to the local-oscillator tuning capacitor. In future receivers the integrated-circuit amplifier (Chap. 9), which is a very high gain wideband device, may make the heterodyne method obsolete. A single integrated amplifier connected to the antenna circuit will provide all the required gain. Tuning will be accomplished electronically, rather than mechanically, by using *varactors,* which are voltage-controlled capacitors.

After the required gain is achieved in the IF amplifier, the signal is demodulated to recover the audio modulation. This is accomplished in a *detector* circuit, as shown in Fig. 1.4. The signal is *rectified* and filtered (Chap. 2) in the detector circuit, the output being proportional to the *envelope* of the IF signal, which has the same waveform as the modulation. This is shown in Fig. 1.5. The audio signal at the output of the detector is then amplified by a power amplifier (Chap. 5). The output of the power amplifier is delivered to the loudspeaker, where the electrical audio signal is converted to an acoustical wave (air-pressure fluctuations).

In order to operate properly, the transistors (or tubes) in the receiver require energy from a dc source. Sometimes batteries are used, but more often the dc voltage is obtained by rectifying and filtering the ac line voltage. This is done in the *power-supply* section of the receiver (Chap. 2). Often, the supply is *regulated* (Chap. 8), so that changes in line voltage or in the load will not affect the dc voltage supplied to the transistors.

A rather sophisticated feature found in most AM radios is called *automatic volume control* (AVC), or *automatic gain control* (AGC) (Chap. 8). This technique automatically compensates for very slow drifts or fluctuations (typically less than 1 Hz) in the received carrier level. Without AVC these very slow changes would be reflected in annoying variations in the loudness of the audio signal from the loudspeaker. These slow changes can occur because of changes taking place over the transmission path, or when the receiver is being tuned from one station to another which may be transmitting at a different power level or be located at a different distance from the receiver. This automatic correction of the volume is done by means of *feedback* (Chap. 8). The AVC system is shown in Fig. 1.4, with typical waveforms in Fig. 1.5. The operation is as follows: The output of the IF amplifier is connected to a detector similar to that used for the audio, the difference being that the filter cutoff frequency is about 1 Hz, so that only very slow changes are recorded. The slowly changing output of this detector is *fed back* to the input stages in such a way that an *increase* in carrier level causes a *decrease* in the gain of the

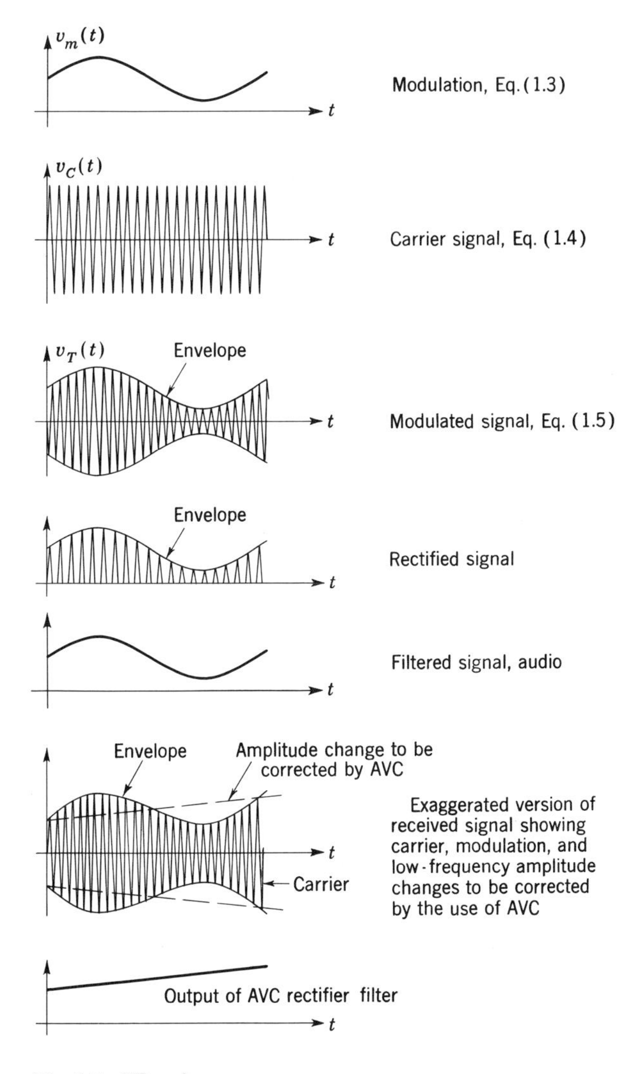

Fig. 1.5 Waveforms.

input stage, thereby maintaining an almost constant level at the output of the IF amplifier.

Amplifiers, rectifiers, feedback, and AVC, as well as many other topics, are discussed in this text. Analysis, as well as the design of these circuits, is studied. The circuits presented to illustrate the principles studied have been constructed and perform as expected. To fully understand the operation of these circuits, the reader should construct and test as many of them as possible.

2
Diode-circuit Analysis

INTRODUCTION TO DIODES

The diode is the simplest of the nonlinear devices with which this text is concerned. It is made in a wide variety of types and used extensively in one form or another in almost every branch of electrical technology. The list includes vacuum diodes, gas diodes, metallic rectifier diodes, semiconductor diodes, tunnel diodes, etc. We concentrate on the semiconductor junction diode; the circuit theory to be developed for this type is almost directly applicable to all the others.

The circuit characteristics of the diode are studied in this chapter. Graphical techniques are emphasized throughout, because they provide a visual picture of circuit operation and often yield insights which are not readily obtained from purely algebraic treatments. These graphical techniques include a thorough treatment of dc and ac load lines as applied to both small and large signals. Although these methods are not often used in the analysis of diode circuits, their introduction at this point serves to establish them firmly in the student's repertoire. When transistors are

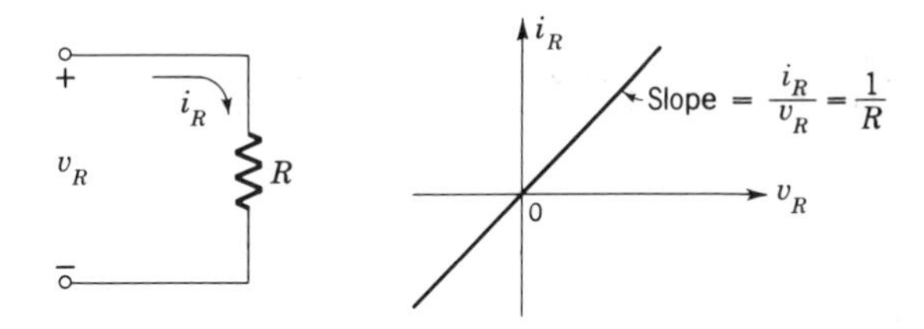

Fig. 2.1-1 The resistance element and its *vi* characteristic.

involved, the additional problems encountered will be more easily solved as a result of the experience gained with diodes.

2.1 NONLINEAR PROPERTIES—THE IDEAL DIODE

Students usually begin their study of circuits by considering models of linear elements, the simplest of these being the resistor. The volt-ampere (*vi*) characteristic of the ideal resistor is described by such a simple relation—Ohm's law—that we sometimes lose sight of its graphical interpretation. The linear character of the resistance is evident in Fig. 2.1-1. The *vi* characteristic of the ideal diode is shown in Fig. 2.1-2. The nonlinear character of the diode is clearly evident here. When the source voltage v_i is positive, i_D is positive, and the diode is a short circuit ($v_D = 0$), while when v_i is negative, i_D is zero, and the diode is an open circuit ($v_D = v_i$). The diode can be thought of as a switch which is controlled by the polarity of the source voltage. The switch is closed for positive source voltages and open for negative source voltages.

Another way to look at this element is to note that the diode conducts current only from p to n (Fig. 2.1-2), and conduction takes place only when the source voltage is positive. The diode does not conduct when the source voltage is negative.

Physical diodes have inherent characteristics and limitations which cause them to differ from the ideal. These are discussed in succeeding sections. For present purposes, the diodes are considered to be ideal.

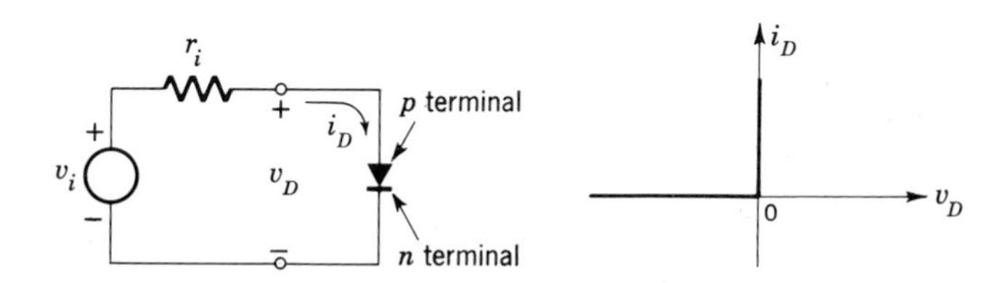

Fig. 2.1-2 The ideal diode and its *vi* characteristic.

The following examples illustrate some of the operations on signals which are often achieved with simple diode circuits.

Example 2.1-1 Half-wave rectifier or clipping circuit One of the principal applications of the diode is in the production of a dc voltage from an ac supply, a process called *rectification*. An often useful by-product of rectification consists of signals at frequencies which are integral multiples of the supply frequency. A typical *half-wave* rectifier circuit is shown in Fig. 2.1-3.

(*a*) The source voltage is sinusoidal, $v_i = V_{im} \cos \omega_0 t$, where $V_{im} = 10$ V. Find and sketch the waveform of the load voltage. Find its average (dc) value.

(*b*) Repeat (*a*) if $v_i = -5 + 10 \cos \omega_0 t$.

Solution (*a*) Kirchhoff's voltage law (KVL) applied to the circuit of Fig. 2.1-3 yields

$$v_i = i_D r_i + v_D + i_D R_L$$

or

$$i_D = \frac{v_i - v_D}{r_i + R_L}$$

This equation contains two unknowns, v_D and i_D. They in turn are related by the diode *vi* characteristic. The solution for i_D or v_D thus requires "substitution" of the *vi* curve into the equation. This can be done in the following way: The diode characteristic indicates that only positive current in the reference direction can flow in this circuit. This requires that $v_i > v_D$. However, when the diode is conducting, $v_D = 0$, so that current flows in the positive direction only when $v_i > 0$.

When v_i is negative, current flow should be opposite to the reference direction; but the diode cannot conduct in this direction; so $i_D = 0$ when $v_i < 0$.

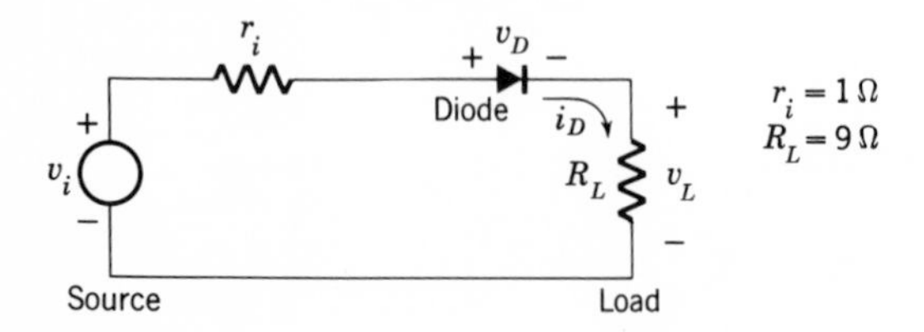

Fig. 2.1-3 Half-wave rectifier circuit for Example 2.1-1.

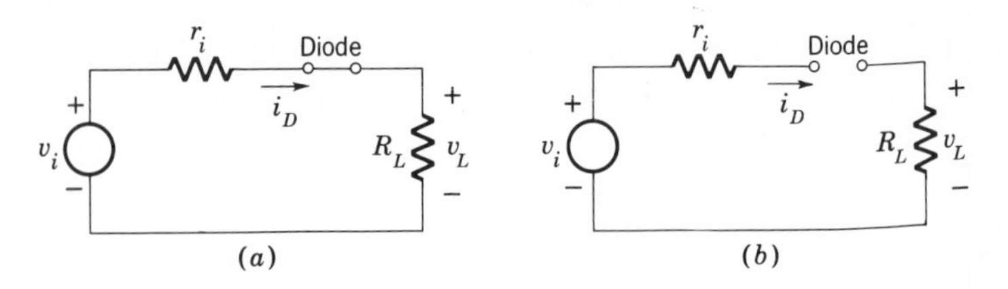

Fig. 2.1-4 Conducting and nonconducting states of the diode rectifier. (a) $v_i > 0$; (b) $v_i < 0$.

This discussion can be summarized by drawing two circuits, one of which holds for $v_i > 0$ and one for $v_i < 0$, as shown in Fig. 2.1-4. Using the circuits shown in this figure, the unknowns v_D and i_D can be found. Thus the diode current i_D is

$$i_D = \begin{cases} \left(\dfrac{V_{im}}{r_i + R_L}\right) \cos \omega_0 t & \text{when } v_i > 0 \\ 0 & \text{when } v_i < 0 \end{cases}$$

and the load voltage v_L is

$$v_L = i_D R_L$$

The load voltage v_L and signal voltage v_i are sketched i 2.1-5. Note that the current waveform has the same shape load voltage v_L. This is a *half-wave-rectified* sine wave. Its value is obtained by dividing the area by the period, 2π.

$$V_{L,\text{dc}} = \left(\frac{1}{2\pi}\right) \int_{-\pi/2}^{\pi/2} (V_{Lm} \cos \omega_0 t)\, d(\omega_0 t) = \frac{V_{Lm}}{\pi} = \frac{9}{\pi} = 2.86 \text{ V}$$

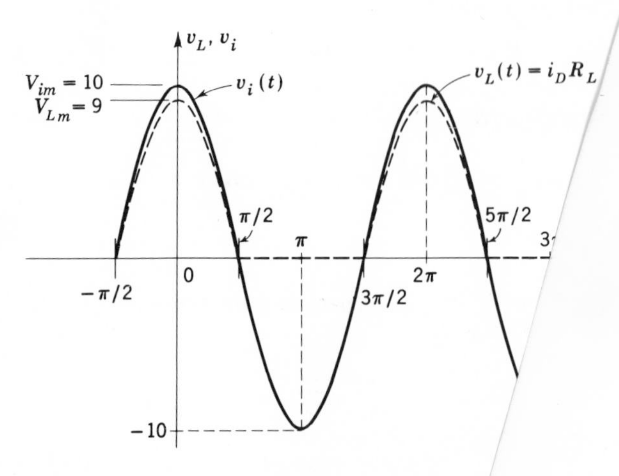

Fig. 2.1-5 Waveforms in the rectifier circuit of

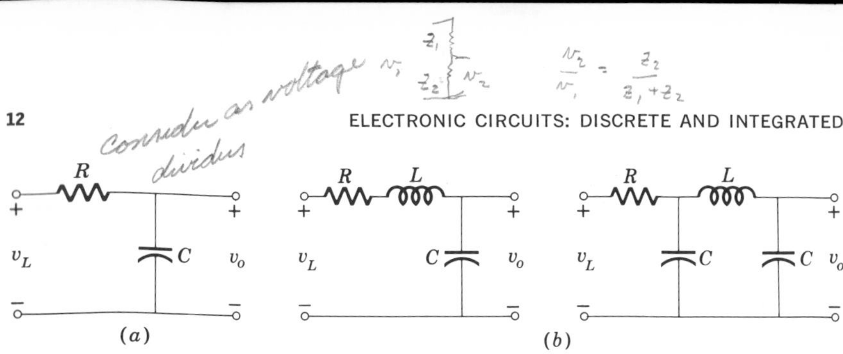

Fig. 2.1-6 Passive power-supply filters.

The Fourier series expansion for $v_L(t)$ is

$$v_L(t) = V_{Lm}\left(\frac{1}{\pi} + \frac{1}{2}\cos\omega_0 t + \frac{2}{3\pi}\cos 2\omega_0 t - \frac{2}{15\pi}\cos 4\omega_0 t + \cdots\right) \tag{2.1-2}$$

This expression shows clearly that the effect of the diode has been to generate not only the dc term and one at the same frequency as the source, but also terms at harmonic frequencies not present in the source voltage.

If the circuit is to produce a dc voltage, the dc component must be separated from the harmonics by filtering $v_L(t)$. This is often done by means of a simple passive filter, as shown in Fig. 2.1-6. The circuit of Fig. 2.1-6a represents a simple RC low-pass filter. If, for example, R and C are adjusted so that $RC = 100/\omega_0$ and if $R \gg R_L$, then the amplitude of the output voltage V_{on}, at the frequency $n\omega_0$, is

$$V_{on} = \frac{V_{Ln}}{\sqrt{1 + (n\omega_0 RC)^2}} \approx \frac{V_{Ln}}{100n} \qquad \text{when } n \geq 1$$

where V_{Ln} is the amplitude of the load voltage at the frequency $n\omega_0$ (for example, $V_{L2} = 2V_{Lm}/3\pi$).

Using superposition, the output voltage is then

$$v_0(t) \approx V_{Lm}\left(\frac{1}{\pi} + \frac{1}{200}\sin\omega_0 t + \frac{1}{300\pi}\sin 2\omega_0 t - \frac{1}{3000\pi}\sin 4\omega_0 t + \cdots\right)$$

Thus the output voltage consists of a dc voltage V_{Lm}/π and a small *ripple* voltage v_r, where

$$v_r = V_{Lm}\left(\frac{1}{200}\sin\omega_0 t + \frac{1}{300\pi}\sin 2\omega_0 t - \cdots\right)$$

The ratio of the rms value of the ripple voltage to the dc voltage is a measure of the effectiveness of the filter in separating the

dc voltage from the harmonics. For the RC filter of this example

$$(v_r)_{\text{rms}} \equiv \left\{ \left(\frac{1}{2\pi}\right) \int_0^{2\pi} [v_r(\omega_0 t)]^2 d(\omega_0 t) \right\}^{1/2}$$

$$= \left(\frac{V_{Lm}}{\sqrt{2}}\right) \sqrt{\frac{1}{(200)^2} + \frac{1}{(300\pi)^2} + \cdots} \approx \frac{V_{Lm}}{280}$$

and

$$\frac{(v_r)_{\text{rms}}}{V_{L,\text{dc}}} \approx \frac{\pi}{280} \approx 0.011$$

Thus the rms ripple is approximately 1 percent of the dc voltage in the output.

More complicated filters, such as the LC or CLC filters shown in Fig. 2.1-6*b*, yield a much smaller rms ripple, which can be calculated approximately using the above method. Exact analyses of rectifier-filter circuits must take into account the nonideal characteristics of the actual diodes used, and are not discussed here.

(*b*) The waveform of v_i is sketched in Fig. 2.1-7. In this case a negative *bias* has been added to the signal. The waveform for v_L is obtained by noting that current will flow only when v_i is positive. The exact time, $\pm t_1$, at which the current flow starts and stops is found by setting $v_i = 0$; then

$$-5 + 10 \cos \omega_0 t_1 = 0$$
$$\cos \omega_0 t_1 = 0.5$$

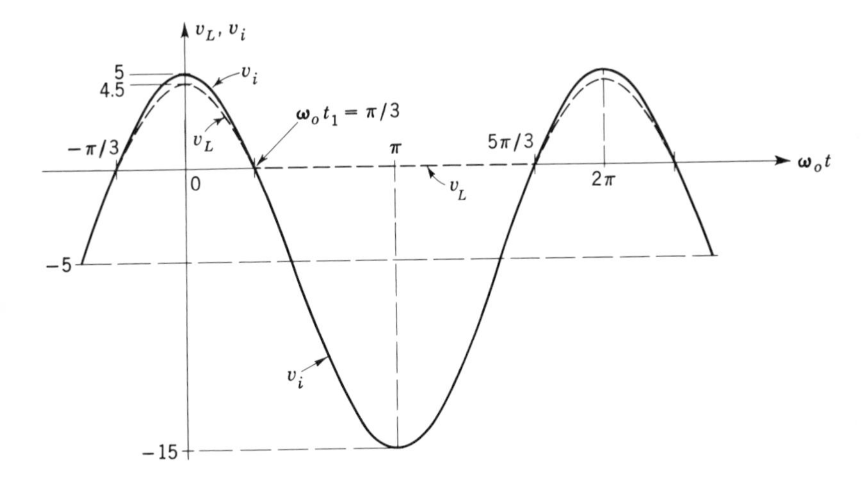

Fig. 2.1-7 Rectifier-circuit waveforms with bias added to signal.

and

$$\omega_0 t_1 = \pm \frac{\pi}{3}$$

From the symmetry of the cosine function, the diode is seen to conduct current when

$$2\pi n - \frac{\pi}{3} \leq \omega_0 t \leq 2\pi n + \frac{\pi}{3}$$

Thus the load voltage is

$$v_L = \begin{cases} -4.5 + 9 \cos \omega_0 t & 2\pi n - \frac{\pi}{3} \leq \omega_0 t \leq 2\pi n + \frac{\pi}{3} \\ 0 & 2\pi n + \frac{\pi}{3} \leq \omega_0 t \leq 2\pi n + \frac{5\pi}{3} \end{cases}$$

The average value of v_L is found as before.

$$\begin{aligned} V_{L,\text{dc}} &= \left(\frac{1}{2\pi}\right) \int_{-\pi/3}^{\pi/3} (-4.5 + 9 \cos \omega_0 t)\, d(\omega_0 t) \\ &= (-4.5)(1/3) + \left(\frac{9}{\pi}\right)\left(\sin \frac{\pi}{3}\right) \\ &= -1.5 + \left(\frac{9}{\pi}\right)\left(\frac{\sqrt{3}}{2}\right) \approx 0.98 \text{ V} \end{aligned}$$

Example 2.1-2 The full-wave rectifier The ripple voltage in the half-wave rectifier is primarily due to the signal component at the fundamental frequency ω_0. The *full-wave rectifier* yields a load voltage with the lowest ripple-frequency term at $2\omega_0$, and in addition, the dc component is doubled. This type of circuit, one form of which is shown in Fig. 2.1-8, is therefore more efficient for the production of a dc voltage with low ripple, and is found in most home radios and

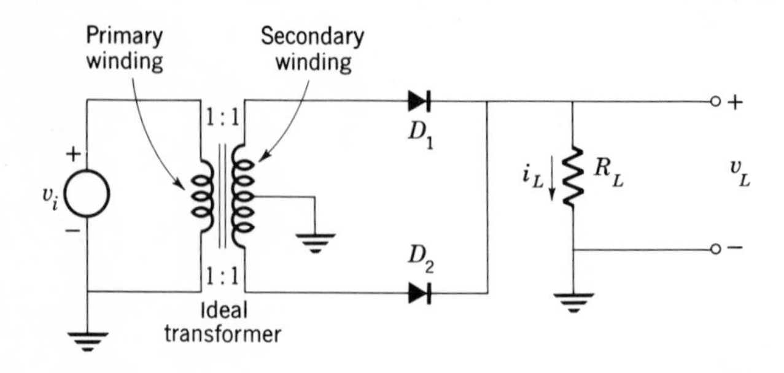

Fig. 2.1-8 Full-wave rectifier.

television sets. It is also the basic rectifier circuit for most dc power supplies.

The circuit operation can be explained qualitatively if the ideal transformer is eliminated by redrawing Fig. 2.1-8 as shown in Fig. 2.1-9*a*. In this figure the transformer is seen to reflect the ac source from the primary into the center-tapped secondary circuit. When v_i is positive, D_1 is a short circuit and D_2 is an open circuit. When v_i is negative, D_1 is an open circuit and D_2 is a short circuit. In each case the load current i_L is in the same positive direction as shown in Fig. 2.1-9*a*, and since one or the other of the diodes D_1 or D_2 is a short circuit on each alternate half-cycle, the load voltage can be written $v_L = |v_i|$. The current and voltage waveforms are shown in Fig. 2.1-9*b*.

The Fourier series for v_L is

$$v_L = V_{Lm}\left(\frac{2}{\pi} + \frac{4}{3\pi}\cos 2\omega_0 t - \frac{4}{15\pi}\cos 4\omega_0 t + \cdots\right) \tag{2.1-3}$$

The dc component is $(2/\pi)V_{Lm}$, which is twice the dc value obtained using the half-wave rectifier. If v_L is passed through the *RC* filter

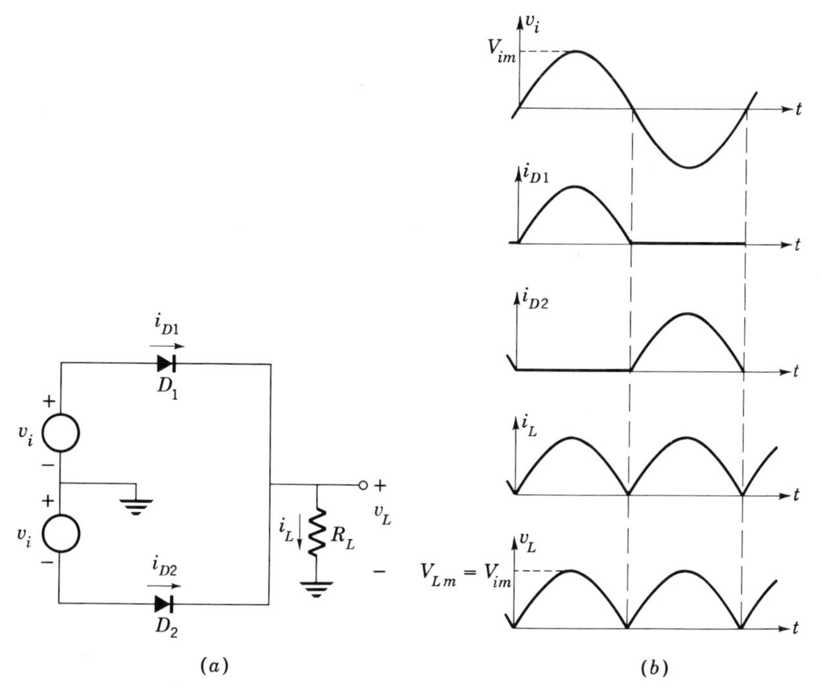

Fig. 2.1-9 Full-wave-rectifier equivalent circuit and waveforms. (*a*) Circuit; (*b*) waveforms.

of Fig. 2.1-6*a*, with $\omega_0 RC = 100$ as before, the output ripple voltage becomes

$$v_r = \left(\frac{4V_{Lm}}{3\pi}\right)\left(\frac{1}{200}\sin 2\omega t - \frac{1}{2000}\sin 4\omega t + \cdots\right)$$

and the rms ripple is

$$(v_r)_{\text{rms}} \approx \frac{V_{Lm}}{210\pi}$$

The ratio of ripple voltage to dc voltage is

$$\frac{(v_r)_{\text{rms}}}{V_{L,\text{dc}}} \approx \frac{1}{420} \approx 0.0024$$

which is considerably less than that obtained using a half-wave rectifier.

Almost all electronic circuits require dc power for their operation, and since ac power is usually available, some sort of rectifier-filter combination will be found in most electronic equipment. Examples 2.1-1 and 2.1-2 have presented the most basic rectifier-filter combinations used to produce a dc voltage from an ac supply. These basic circuits suffer from several drawbacks, which make them unsuitable for many applications. The first of these is the variation of the dc load voltage with load current. This is measured by a quantity called the *regulation*, which is defined as

$$\text{Regulation} = \frac{\text{no-load voltage} - \text{full-load voltage}}{\text{full-load voltage}} \tag{2.1-4}$$

An ideal supply would provide a dc voltage which is constant and independent of the load current, i.e., zero regulation. This is equivalent to saying that the supply would have zero output resistance as seen from the load terminals. However, in practical circuits, the diode resistance, which was neglected in the examples, and the filter-circuit resistance are not negligible and result in a finite output resistance. If the output resistance is equal to the load resistance, the full-load voltage is equal to one-half of the no-load voltage, and the regulation is 100 percent. A second disadvantage of this basic rectifier circuit when used as a dc supply is that the dc output is directly proportional to the magnitude of the ac supply voltage. Since most ac power lines do not maintain an absolutely constant voltage, the dc output will vary proportionally. For many applications, this variation cannot be tolerated, even though it may be relatively small. A third disadvantage of the above circuit is that even the small ripple voltage found in the examples is often more than can be tolerated for proper operation of sophisticated electronic circuits.

Many techniques exist for overcoming the effects mentioned above. Manufacturers of rectifiers provide data and handbooks[1]* which contain complete information on the design of power supplies of many varieties. These handbooks usually contain the latest *state-of-the-art* information and provide an important source of information for the design engineer. Several techniques for improving power-supply performance are discussed in Secs. 2.7 and 8.7.

FREQUENCY MULTIPLICATION

A major application of rectification is frequency multiplication. Equation (2.1-2) indicates that a rectifying circuit generates harmonic frequencies. Thus, to multiply ω_0 by 2, one could simply follow the rectifier by a filter designed to pass only the second harmonic, $2\omega_0$. This is similar, in principle, to the extraction of the dc, or zero-frequency, signal by the use of a low-pass filter.

2.2 AN INTRODUCTION TO SEMICONDUCTOR DIODE THEORY[2]

A brief qualitative discussion of the basic concepts governing current flow in a semiconductor diode is presented in this section. No attempt is made to be rigorous or to derive equations. The interested student should refer to any of the many texts on solid-state physics in order to fill in the details.

The basic material used in the construction of most diodes and transistors today is silicon. Formerly, germanium was used extensively, but it is fast disappearing. Silicon is a semiconductor; that is, at room temperature very few electrons exist in the conduction band of the silicon crystal. Since the current is proportional to the number of electrons in motion, the current is small; hence the material has a high resistance. The conduction and valence bands of pure silicon are shown in Fig. 2.2-1.

At 0°K (absolute zero), all the electrons are at their lowest possible energy levels. At room temperature, an occasional electron has enough

* A superior number refers to a citation in the References at the end of the chapter.

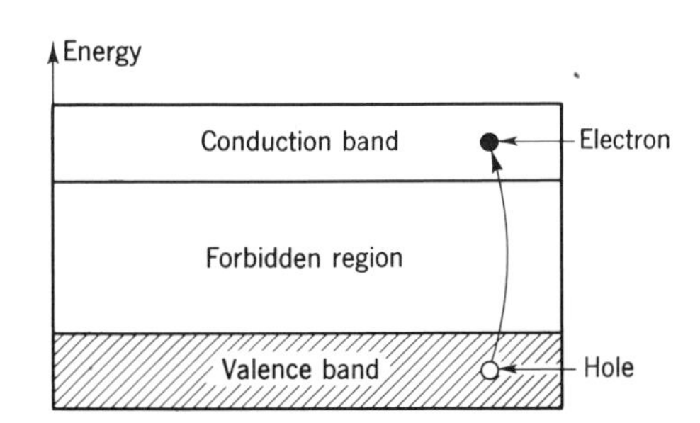

Fig. 2.2-1 Energy bands in silicon at room temperature.

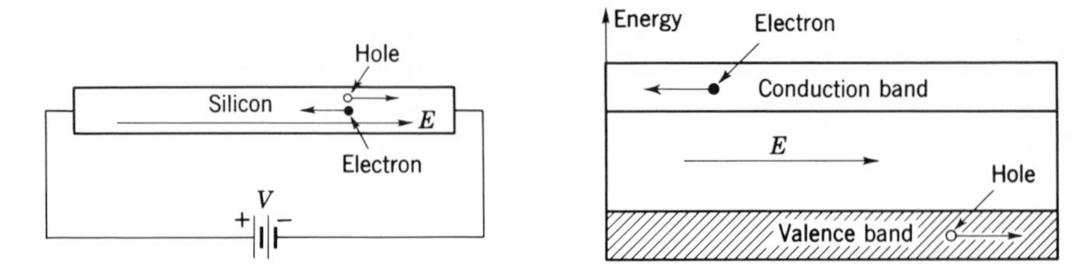

Fig. 2.2-2 Electron and hole motion in silicon with applied electric field.

energy to escape from the valence band and move to the conduction band, as shown by the dot in Fig. 2.2-1. The vacancy left by the electron is shown as a circle, or hole. If an electric field is applied to the material as shown in Fig. 2.2-2, the electron moves toward the positive battery terminal as expected. An electron in the valence band can also move toward the positive battery terminal if it possesses enough energy to take it from its energy level to the energy level of the hole. When this electron does escape into the hole, it leaves a hole behind. Thus it appears that the hole moves to the right, toward the negative battery terminal. The net current is therefore the sum of the current due to the electron motion in the conduction band and the current due to hole motion. We refer to hole, or positive-charge, motion rather than electron motion in the valence band to avoid confusion with the electron motion in the conduction band. The conventional current, due to the flow of electrons, and the hole current are of course in the direction of the electric field.

It should be noted that the electron moves more rapidly to the positive terminal than the hole moves toward the negative terminal since the probability of an electron having the energy required to move to an empty state in the conduction band (which is almost empty) is much greater than the probability of an electron having the energy required to move to an empty state in the valence band (which is almost filled). Thus the current due to electron flow in the conduction band is greater than the hole current in silicon. However, the net current is small, and hence the material is a semiconductor.

To construct a diode, one takes silicon and adds atoms of another element, such as boron. This process is called "doping." Boron is called an *acceptor* material because it is able to accept electrons from the valence band of the silicon. At room temperature, the electrons from the valence band of the silicon fill the acceptor space of the boron, as shown in Fig. 2.2-3, since the probability of the valence electrons having sufficient energy at room temperature to bridge the small gap is very high. The result is that an extremely large number of holes exist. When an electric field is applied across this doped silicon, the hole current is very high, and the

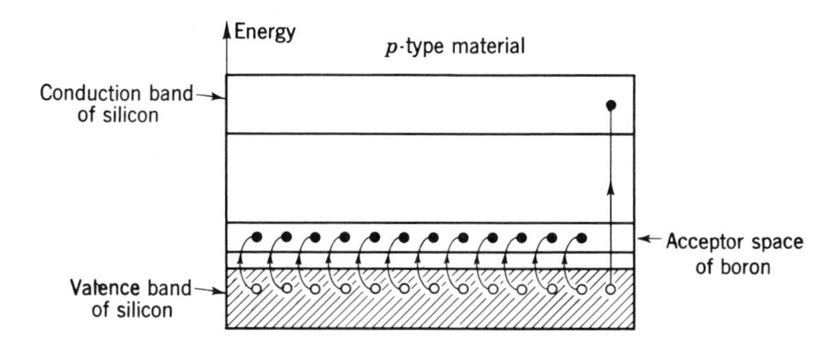

Fig. 2.2-3 Energy bands for silicon with boron added.

material is now a good conductor. This is called *p-type* material. Note that in p-type material conduction is primarily due to the motion of holes.

We then take another piece of silicon and add atoms of another element, such as phosphorus. The phosphorus is called a *donor* material since it is able to donate electrons to the conduction band of the silicon. It therefore "donates" (at room temperature) all its electrons to the conduction band of the silicon, as shown in Fig. 2.2-4. Now the current flow, when an electric field is applied, is due primarily to electron flow. This doped material is called *n-type.*

The diode consists of p- and n-type material joined as shown in Fig. 2.2-5. The *junction* between the n and p materials is the basis for the name *junction diode.* Figure 2.2-5a shows a forward-biased diode and its circuit symbol. Holes from the p region flow into the n region, while electrons from the n region flow into the p region. A small voltage V is sufficient to yield a high current.

Figure 2.2-5b shows a reverse-biased diode. Electrons in the p region now flow into the n region, and holes in the n region flow into the p region. The current flow is hence very small, because of the small number of

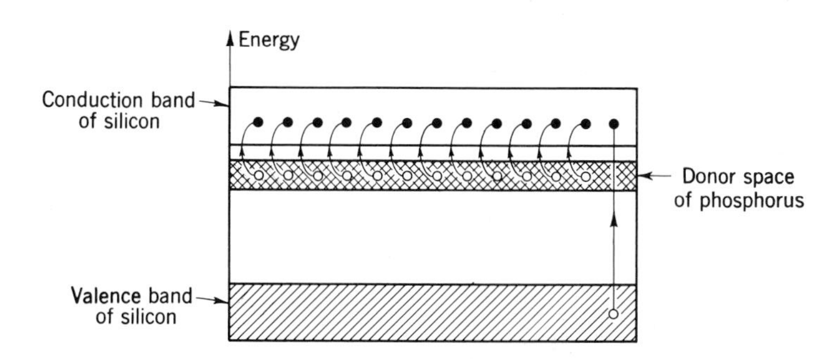

Fig. 2.2-4 Energy bands for silicon with phosphorus added.

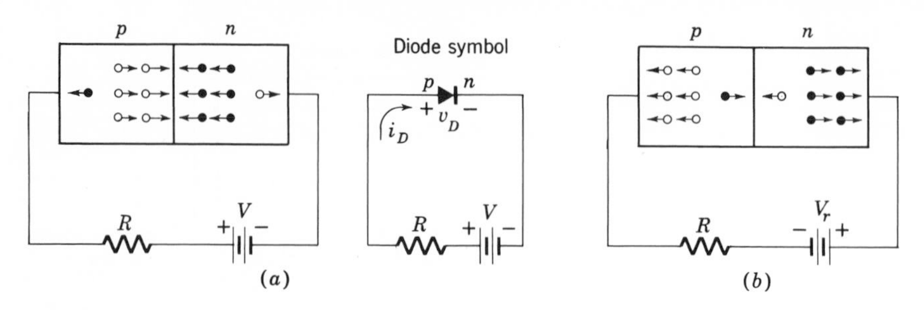

Fig. 2.2-5 The junction diode. (*a*) Forward bias; (*b*) reverse bias.

charges in motion. If V_r is increased beyond the diode's "breakdown voltage," the diode current increases considerably for a very small change in V_r. This is called the *Zener region*, and the breakdown is commonly called a *Zener* or *avalanche* breakdown. The avalanche and Zener breakdown mechanisms differ. The avalanche breakdown occurs at high reverse voltages, and the Zener breakdown occurs at small reverse voltages. However, the effect on the circuit is the same. No distinction is made in this book between these two processes.

The avalanche-breakdown process can be thought of as a moving electron colliding with a fixed electron, knocking it free; these two electrons free two more electrons; etc. This results in a large current flow in this region.

Simple calculations for the diode (neglecting the Zener region) show that the current and voltage are related by

$$i_D = I_o(\epsilon^{qv_D/mkT} - 1) \tag{2.2-1}$$

where i_D = current through diode, A
v_D = voltage across diode, V
I_o = reverse saturation current, A
q = electron charge, 1.6×10^{-19} coulomb
k = Boltzmann's constant, 1.38×10^{-23} joule/°K
T = absolute temperature, °K
m = empirical constant which lies between 1 and 2

At room temperature (300°K),

$$\frac{kT}{q} \approx 25 \text{ mV} \tag{2.2-2}$$

Equation (2.2-1) states that if v_D is negative with magnitude much greater than mkT/q, the current i_D is the reverse saturation current $-I_o$. This reverse current $-I_o$ is a function of material, geometry, and temperature. If, however, v_D is positive and greatly exceeds mkT/q, the forward current is

$$i_D \approx I_o \epsilon^{qv_D/mkT} \tag{2.2-3}$$

Equation (2.2-1) is sketched in Fig. 2.2-6*a*, for germanium and silicon.

The actual characteristic of a typical diode differs from the exponential curve because of various effects. At relatively large forward currents the ohmic resistance of the contacts and the semiconductor material effectively increases the forward resistance. In the reverse direction, surface leakage, which is the current along the surface of the silicon rather than through the junction between the *p*- and *n*-type regions, effectively decreases the reverse resistance. At large reverse voltages the avalanche breakdown takes over. Other effects come into play over various portions of the diode characteristic, but for most practical purposes they are negligible. The curves of Fig. 2.2-6*a*, when plotted to a suitable scale, appear to "turn on" at approximately 0.2 V for germanium and 0.7 V for silicon.

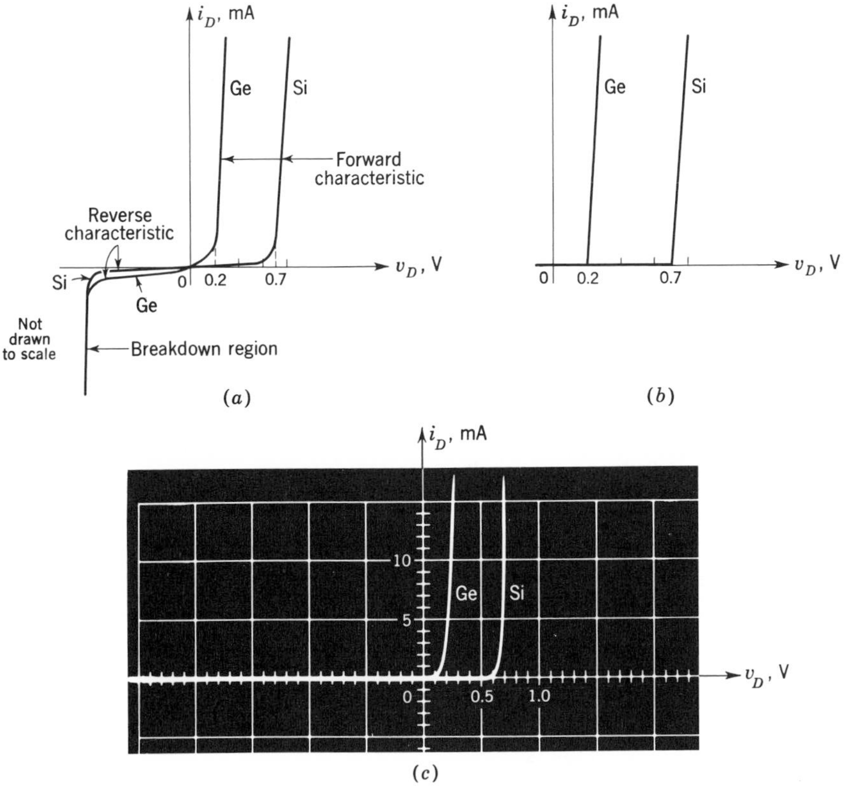

Fig. 2.2-6 Diode characteristics. (*a*) Actual characteristic; (*b*) straight-line (piecewise linear) characteristics; (*c*) diode characteristics.

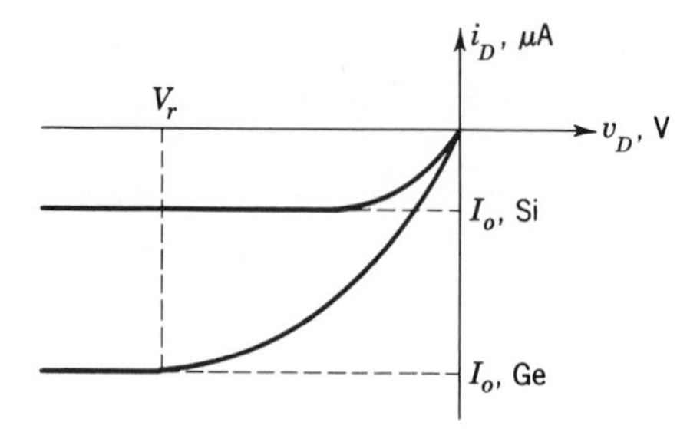

Fig. 2.2-7 Reverse characteristics.

For large-signal applications, the diode is often considered to behave in accordance with the straight-line approximations of Fig. 2.2-6*b*. Such approximations are called *piecewise linear curves.*

Considering the reverse characteristic, it is interesting to note that while germanium has a 0.2 V break in its forward characteristic as compared with 0.7 V for silicon, the reverse characteristics, shown in Fig. 2.2-7, indicate that at the same reverse diode voltage V_r, the silicon diode draws considerably less current than the germanium diode. Calculations indicate that the reverse-current ratio between silicon and germanium diodes should be considerably less than the measured ratios for diodes of comparable current rating. The reason is that a large percentage of the reverse-current flow in silicon is due to surface-leakage current.

2.3 ANALYSIS OF SIMPLE DIODE CIRCUITS—THE DC LOAD LINE

In this section we consider the behavior of simple circuits consisting of independent sources, diodes, and resistors. The circuit to be analyzed is the simple half-wave rectifier shown in Fig. 2.3-1. This circuit was analyzed in Example 2.1-1, assuming an ideal diode. In this section the actual diode characteristics are taken into account.

The philosophy behind the graphical analysis is based on two simple facts:

1. The behavior of the diode is completely characterized at low frequencies by its *vi* characteristic, which is usually available graphically in the manufacturer's specifications or can be easily measured.

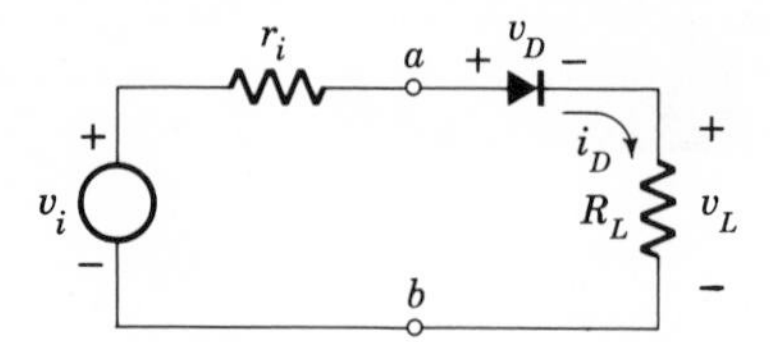

Fig. 2.3-1 Half-wave rectifier.

2. The other elements of the circuit, being linear, can be replaced by a Thévenin equivalent circuit as seen from the diode terminals.

Let us consider the corresponding two parts of the circuit, as shown pictorially in Fig. 2.3-2. The terminal relations for the two parts can be written

Nonlinear element: $$i_D = f(v_D) \tag{2.3-1}$$

Thévenin equivalent: $$v_D = v_T - i_D R_T \tag{2.3-2}$$

We have two equations with two unknowns, v_D and i_D. When the two parts of the circuit are connected, these two relations are satisfied simultaneously, and the circuit will operate at the point given by the solution of the equations. This solution can be arrived at analytically if the functional form of the *vi* characteristic of the nonlinear element is known. If, for example, the element is a silicon diode, (2.2-1) can be used as the nonlinear relation, and the solution found. Because of the exponential nature of (2.2-1), it is clear that this is not going to be a routine calculation, and may, in fact, involve considerable labor. In some cases this may be justifiable, but in most it is not, for two reasons: the accuracy *required* in most cases is not great, so that simpler or approximate methods are justified; and the accuracy *achievable* by a detailed calculation is often meaningless because the behavior of most diodes differs from the theoretical characteristic given by (2.2-1), and in fact large variations will often be encountered in batches of diodes of the same type.

Problems of this type are most often solved graphically by plotting (2.3-1) and (2.3-2) on the same set of axes. The intersection of the two resulting curves gives the operating point of the circuit for the particular conditions which determine v_T and i_D. A typical plot is shown in Fig. 2.3-3. The straight-line characteristic of the Thévenin circuit (usually called the *dc load line*) is drawn for a dc Thévenin voltage of 1.5 V and a

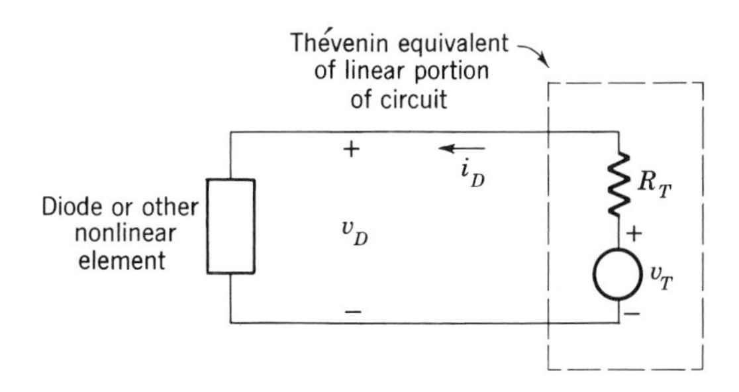

Fig. 2.3-2 General circuit containing a nonlinear element.

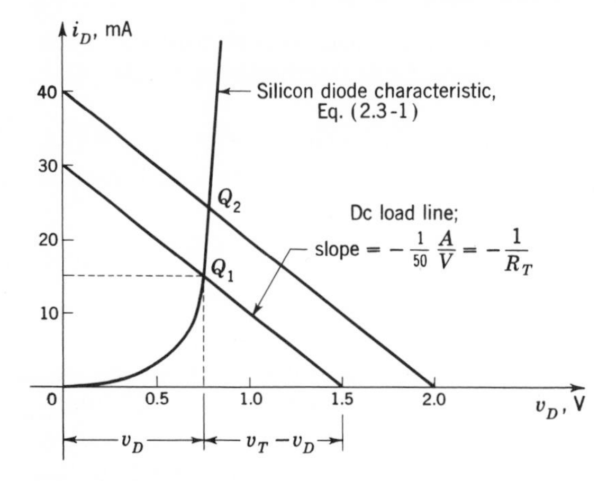

Fig. 2.3-3 Typical diode characteristic and load lines.

Thévenin resistance of 50 Ω. The intersection of the dc load line (2.3-2) and the diode characteristic (2.3-1) gives the operating point (often called the *quiescent*, or Q, point) for these conditions. The intersection of these two curves occurs at the point Q_1, where $v_D \approx 0.7$ V, $i_D \approx 15$ mA. If the Thévenin voltage v_T changes to 2 V, the load line shifts horizontally

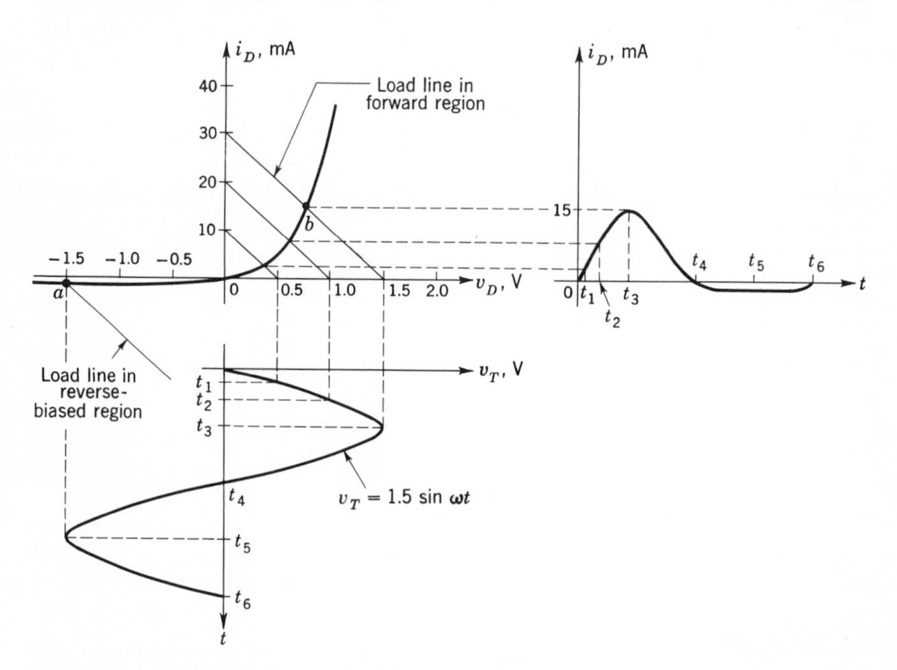

Fig. 2.3-4 Graphical solution for current when sinusoidal voltage is applied.

0.5 V to the right as shown, and the operating point moves to Q_2 on the graph. As long as R_T remains constant, any change in v_T is accounted for by a simple horizontal shift of the load line.

Thus, if v_T is sinusoidal (that is, $V_{Tm} \sin \omega t$) the *current* waveform can be found by choosing several points on the sine wave and drawing the corresponding load lines to find the resulting currents. This technique is illustrated in Fig. 2.3-4 for $V_{Tm} = 1.5$ V. The nonlinear nature of the diode characteristic causes the current waveform to be distorted (i.e., to have a shape different from v_T). This waveform should be compared with that of Fig. 2.1-5, which shows the result when an ideal diode is used.

Returning to the original circuit, Fig. 2.3-1, we observe that for this simple case the Thévenin voltage v_T is the same as the source voltage v_i, and $R_T = r_i + R_L$, so that all one need do is multiply the current in Fig. 2.3-4 by R_L to obtain a plot of the response voltage v_L. Note that the operating path of the diode with this signal applied lies between points a and b, as shown on the diode vi characteristic.

2.4 SMALL-SIGNAL ANALYSIS—THE CONCEPT OF DYNAMIC RESISTANCE

The total peak-to-peak variation (swing) of the ac signal is often a small fraction of the dc; hence the name *small signal*. When this condition occurs, a combined graphical-analytical approach can be employed which greatly simplifies the analysis. This approach is illustrated using the circuit of Fig. 2.3-1. The Thévenin voltage v_T is

$$v_T = V_{\text{dc}} + v_i = V_{\text{dc}} + V_{im} \sin \omega t \tag{2.4-1}$$

where

$$V_{im} \ll V_{\text{dc}}$$

The technique used here is based on the fact that the inequality in (2.4-1) forces the circuit to operate over a very small region of its possible operating range. For most practical purposes the diode characteristic can then be considered linear in this region, and the diode replaced by a resistance. The resulting linear circuit is then amenable to standard circuit-analysis techniques.

Keeping in mind that the circuit is to be linearized, we first determine the operating point for $v_T = V_{\text{dc}}$ (that is, $V_{im} = 0$). This is the quiescent (Q) point. The procedure here is precisely the same as that used in Sec. 2.3, and the pertinent graph is repeated in Fig. 2.4-1 for $V_{\text{dc}} = 1.5$ V and $r_i + R_L = 50\ \Omega$.

The construction necessary to determine the exact current waveform as in Sec. 2.3 (Fig. 2.3-4) is shown on the graph. Only that portion of the diode characteristic lying between points a and b is of importance in

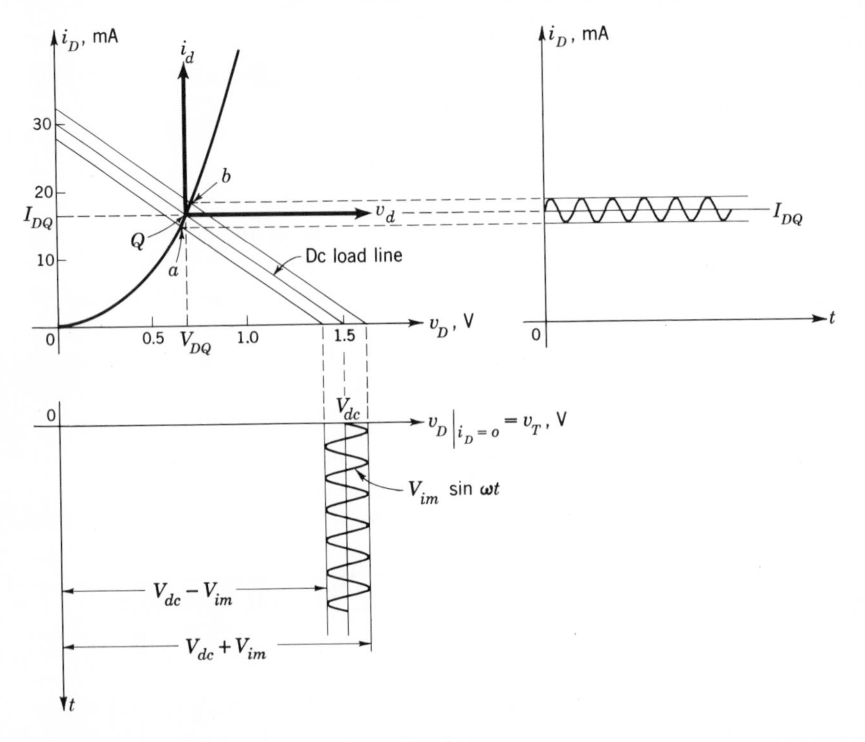

Fig. 2.4-1 Graphical determination of load current.

determining the response. If the characteristic is reasonably linear between these points, it can be replaced by a straight line for purposes of calculating the ac component. In order to focus attention on the ac response, we construct a new set of axes with their origin at the Q point. The variables associated with these axes are then

$$\text{Current:} \qquad i_d = i_D - I_{DQ} \tag{2.4-2a}$$

$$\text{Voltage:} \qquad v_d = v_D - V_{DQ} \tag{2.4-2b}$$

That portion of the graph (Fig. 2.4-1) of interest is shown in Fig. 2.4-2, with the new variables drawn on expanded scales. The operating path ab is assumed to be linear, and passes through the origin. This is equivalent to replacing the diode by a resistance of value equal to the inverse slope of line ab. This is called the *dynamic resistance* r_d of the diode, and can be found by evaluating the inverse slope of the diode characteristic at the Q point. Hence

$$r_d = \frac{\Delta v_D}{\Delta i_D}\bigg|_{Q\text{ point}} \tag{2.4-3}$$

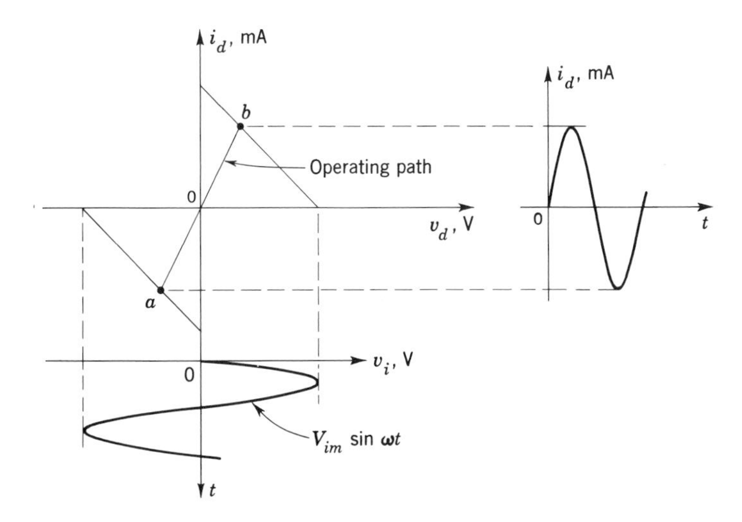

Fig. 2.4-2 Graphical interpretation of auxiliary variables.

Once r_d has been determined, any of the circuit variables (for small-signal ac operation only) can be calculated by simple application of Ohm's law.

The original circuit can be considered to be two separate circuits, as shown in Fig. 2.4-3. Figure 2.4-3*a* is used to find I_{DQ} and V_{DQ} (the quiescent operating point), and Fig. 2.4-3*b* to find i_d and v_d (the small-signal components). The total diode current and voltage can then be found using (2.4-2*a*) and (2.4-2*b*).

The development which led to the equivalent circuits of Fig. 2.4-3 can also be carried through analytically, using a Taylor series expansion of the diode characteristic at the *Q* point. This is an often used engineering approximation. The diode *vi* characteristic is given by

$$i_D = f(v_D) \tag{2.3-1}$$

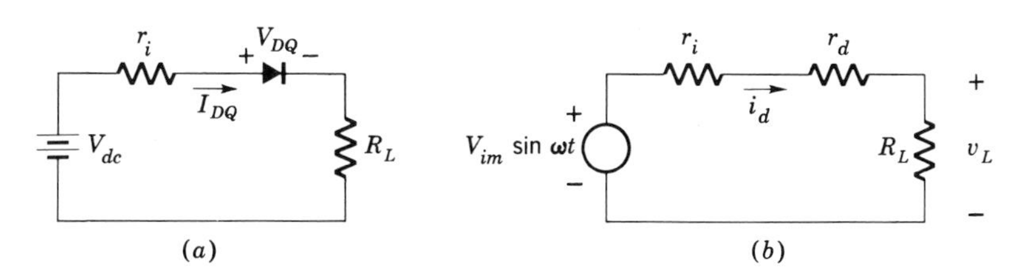

Fig. 2.4-3 Diode circuit considered as two separate circuits. (*a*) Circuit for calculating dc operating point; (*b*) circuit for calculating small-signal ac component.

For small signals and no distortion,

$$i_D = I_{DQ} + i_d \qquad \text{and} \qquad v_D = V_{DQ} + v_d \tag{2.4-4}$$

where

$$|i_d| \ll I_{DQ} \qquad |v_d| \ll V_{DQ}$$

Then (2.3-1) becomes

$$I_{DQ} + i_d = f(V_{DQ} + v_d) \tag{2.4-5}$$

Now Taylor's series, from which $f(x + \Delta x)$ can be found, given $f(x)$, is

$$f(x + \Delta x) = f(x) + \Delta x\, f'(x) + \text{higher-order terms} \tag{2.4-6}$$

Neglecting the higher-order terms, identify x with V_{DQ} and Δx with v_d, so that

$$i_D = I_{DQ} + i_d \approx f(V_{DQ}) + v_d \left(\frac{di_D}{dv_D}\right)\bigg|_Q \tag{2.4-7}$$

Noting that $f(V_{DQ}) = I_{DQ}$, this simplifies to

$$i_d \approx v_d \left(\frac{di_D}{dv_D}\right)\bigg|_Q \tag{2.4-8}$$

and finally,

$$\frac{v_d}{i_d} \approx \frac{dv_D}{di_D}\bigg|_Q \approx \frac{\Delta v_D}{\Delta i_D}\bigg|_Q = r_d \tag{2.4-9}$$

Now Kirchhoff's voltage law (KVL) for the circuit of Fig. 2.3-1 is

$$v_T = v_D + i_D R \tag{2.4-10}$$

where

$$R = r_i + R_L$$

Substituting the small-signal definitions (2.4-1) and (2.4-4) into (2.4-10), this becomes

$$V_{dc} + v_i = V_{DQ} + v_d + I_{DQ}R + i_d R \tag{2.4-11}$$

Since, for the assumed small-signal conditions with no distortion, v_i, v_d, and i_d are all zero-average time-varying signals and V_{dc}, V_{DQ}, and I_{DQ} are constants, (2.4-11) can be separated into a dc and an ac equation as follows:

$$V_{dc} = V_{DQ} + I_{DQ}R \tag{2.4-12}$$

and

$$v_i = v_d + i_d R \tag{2.4-13}$$

Finally, using (2.4-9) in (2.4-13),

$$v_i = i_d(r_d + R) \tag{2.4-14}$$

Equations (2.4-12) and (2.4-14) describe the dc and ac equivalent circuits of Fig. 2.4-3. The dc calculation is performed graphically, using the diode characteristic, while the small-signal analysis is carried out using Ohm's law, with r_d evaluated from the diode characteristic, at the Q point.

CALCULATION OF r_d

An analytic expression for the dynamic resistance of a silicon diode in the forward direction can be found by differentiating the diode equation (2.2-1), inverting the result, and evaluating r_d at the operating point as follows:

$$i_D = I_o(\epsilon^{qv_D/mkT} - 1) \approx I_o\epsilon^{qv_D/mkT} \tag{2.2-1}$$

$$\frac{di_D}{dv_D} = \frac{q}{kT} I_o\epsilon^{qv_D/mkT} = \left(\frac{q}{mkT}\right) i_D$$

and

$$r_d = \frac{dv_D}{di_D}\bigg|_{Q \text{ point}} \approx \frac{mkT/q}{I_{DQ}} \approx m\left(\frac{25 \text{ mV}}{I_{DQ}}\right)$$

$$\text{at } T = 300°\text{K}, I_{DQ} \text{ in amperes} \tag{2.4-15}$$

Typically, the dynamic resistance of a junction diode operating at 1 mA dc current is 25 Ω, assuming $m \approx 1$.

Note that the preceding analysis is valid *only* if the operating portion of the diode characteristic (a to b in Fig. 2.4-1) can be considered a straight line. (If this assumption is not met, the current and voltage waveforms will be distorted.) The final results, the load current and voltage, are found by superimposing the responses of the circuits of Fig. 2.4-3a and b.

$$i_D = I_{DQ} + i_d$$

$$= I_{DQ} + \left(\frac{V_{im}}{r_i + r_d + R_L}\right) \sin \omega t$$

and

$$v_L = R_L i_D$$

REACTIVE ELEMENTS

When small-signal conditions hold, it is a simple matter to take into account reactive elements such as the RC filter shown in Fig. 2.4-4. Clearly, the capacitor can have no effect on the operating point; so the dc calculation is unaltered. Also, the slope of the diode characteristic $1/r_d$ at the Q point does not change, and to determine the ac current and

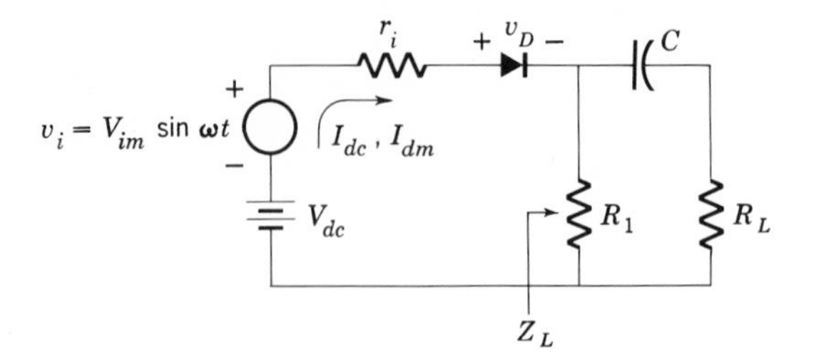

Fig. 2.4-4 Diode circuit with a reactive element.

diode voltage one can use Ohm's law to obtain

$$I_{dm} = \frac{V_{im}}{|r_i + r_d + Z_L|} \tag{2.4-16}$$

where I_{dm} and V_{im} denote the peak current and voltage amplitude, and Z_L is the complex impedance. This is illustrated in the example which follows.

Example 2.4-1 A junction diode is used in the circuit of Fig. 2.4-4 with

$$\begin{aligned} V_{dc} &= 1.5 \text{ V} \\ V_{im} &= 20 \text{ mV} \\ r_i &= 10 \ \Omega \\ R_1 &= 90 \ \Omega \\ R_L &= 200 \ \Omega \\ C &= 100 \ \mu\text{F} \\ \omega &= 10^4 \text{ rad/s} \end{aligned}$$

Find the voltage across R_L.

Solution To find the Q point, the dc load line is drawn through the point $v_D = 1.5$ V, with slope $-1/(r_i + R_1) = -0.01$. The intersection of this load line and the diode characteristic occurs at 7.5 mA

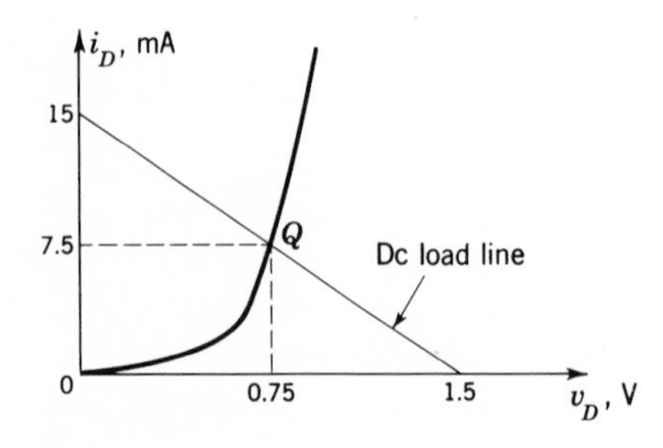

Fig. 2.4-5 Graphical evaluation of Q point for the circuit of Fig. 2.4-4.

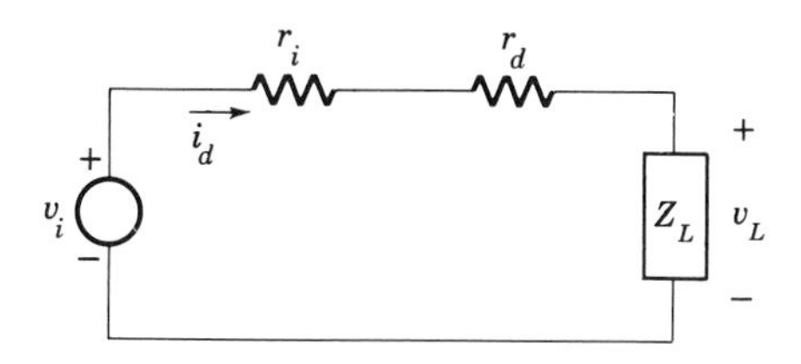

Fig. 2.4-6 Small-signal equivalent circuit.

and 0.75 V (Fig. 2.4-5). The location of the Q point on the curve and the size of the ac signal indicate that small-signal theory will be applicable.

From (2.4-15)

$$r_d = \frac{25 \times 10^{-3}}{7.5 \times 10^{-3}} = 3.3\ \Omega$$

Since $|X_c| = 1/\omega C \approx 1\ \Omega$, the capacitor is seen to have negligible impedance in comparison with R_L. Taking this into account, $Z_L \approx R_1 \| R_L = 62\ \Omega$.* The small-signal circuit analogous to Fig. 2.4-3*b* takes the form shown in Fig. 2.4-6. Using the values obtained for r_d and Z_L,

$$V_{L,\text{dc}} = 0$$

and

$$V_{Lm} = I_{dm}|Z_L| = \frac{V_{im}|Z_L|}{|r_i + r_d + Z_L|} \approx \frac{(20)(62)}{10 + 3.3 + 62} \approx 17 \text{ mV}$$

2.5 SMALL-SIGNAL ANALYSIS—THE AC LOAD LINE

The circuit of Fig. 2.4-4 was described in Sec. 2.4 by a combination of graphical and analytical methods. It is possible, by a simple extension of the dc-load-line concept, to perform both the dc and ac analyses graphically, as long as the reactance of the capacitor is negligible. This procedure leads to the concept of the ac load line, which, while not often used in practice for diode circuits, is often used to analyze and design transistor circuits. The concept is easier to grasp in terms of the simple diode; so we introduce it at this point.

For the circuit of Fig. 2.4-4, the dc load line and Q point are obtained as shown in Fig. 2.5-1. The dc conditions are not influenced by that part of the circuit consisting of the capacitor C and load resistance R_L, because

* The notation $R_1 \| R_L$ is an abbreviation for "R_1 in parallel with R_L," and is used throughout this text.

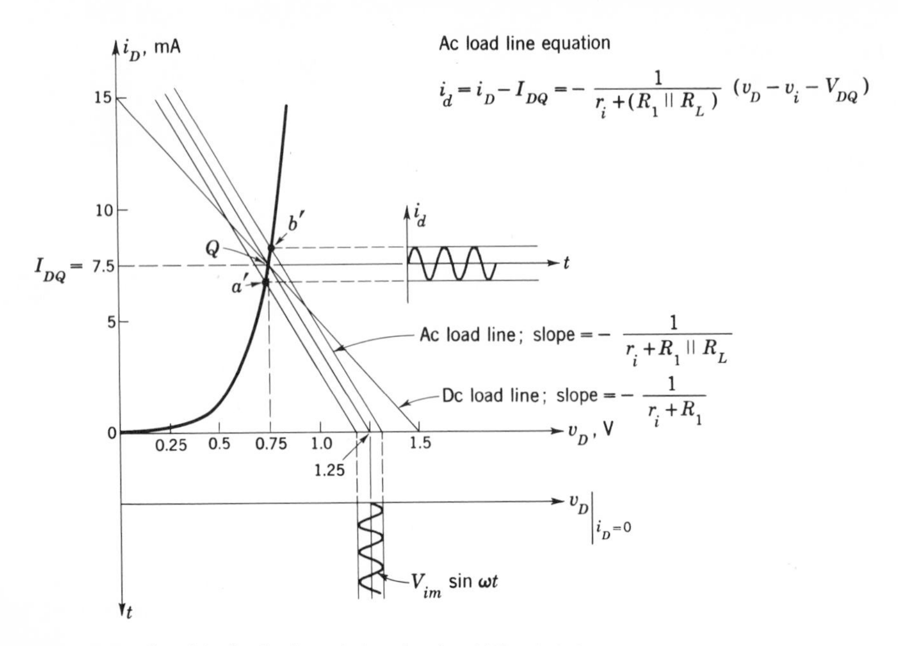

Fig. 2.5-1 Graphical solution of the circuit of Fig. 2.4-4.

of the dc blocking action of the capacitor. The slope of the dc load line is therefore determined by the resistance $r_i + R_1$. When an ac signal is present (assume that the capacitor acts as a short circuit at the frequencies involved), the effective resistance as seen by the diode is $r_i + (R_1 \| R_L)$, which is the negative of the inverse slope of the "ac load line." In order to draw this ac load line, we need only one point, since the slope is known. The point where the ac signal is zero is the easiest to obtain. This is simply the Q point. Thus the ac load line is drawn through the Q point with a slope $-1/[r_i + (R_1 \| R_L)]$, as shown in Fig. 2.5-1.

As the signal varies with time, the ac load line moves back and forth to define the operating path for the diode. Compare this with Fig. 2.4-1, where the dc load line moves back and forth. The difference between these two lies in the fact that the ac impedance is *not* the same as the dc resistance seen by the diode for the circuit being considered in this section.

The amplitude of the ac component of current is found using the graphical construction shown in Fig. 2.5-1. The operating path is along the segment $a'b'$ of the diode characteristic. This procedure will yield results identical with the analytical results obtained in (2.4-16) as long as segment $a'b'$ is approximately linear. As in Sec. 2.4, it is conceptually useful to superimpose a set of i_d-v_d axes on the curves of Fig. 2.5-1. This is left as an exercise.

2.6 LARGE-SIGNAL ANALYSIS—DISTORTION AND Q-POINT SHIFT

When the nonlinearity of the operating path is too great to allow linearization of the diode characteristic about the Q point, waveform distortion results. Circuit calculations now become complicated, and when reactive elements are present, an iterative graphical procedure is often required to determine the waveforms of interest.

Example 2.6-1 Consider the circuit shown in Fig. 2.6-1*a*, with a square-wave input signal v_i, as shown in Fig. 2.6-1*b*. To simplify the calculations further, assume a piecewise linear diode *vi* characteristic given by the equations

$$i_D = \begin{cases} 0 & v_D \leq 0 \\ v_D & 0 < v_D \leq 1 \\ 2v_D - 1 & 1 < v_D \end{cases}$$

This characteristic is plotted in Fig. 2.6-1*c*.

Determine the steady-state diode current i_D.

Solution Using KVL,

$$V_{dc} + v_i = 2 + v_i = v_D + i_D Z_T \tag{2.6-1}$$

Since the dc value of Z_T is R_1 and the ac value is $R_1 \| R_L$,

$$i_D Z_t = R_1 I_D + (R_1 \| R_L) i_d = 2 I_D + i_d \tag{2.6-2}$$

where I_D is the average value of i_D and is *not* the same as I_{DQ} because of the distortion. However, since by definition

$$i_d = i_D - I_D \tag{2.6-3a}$$

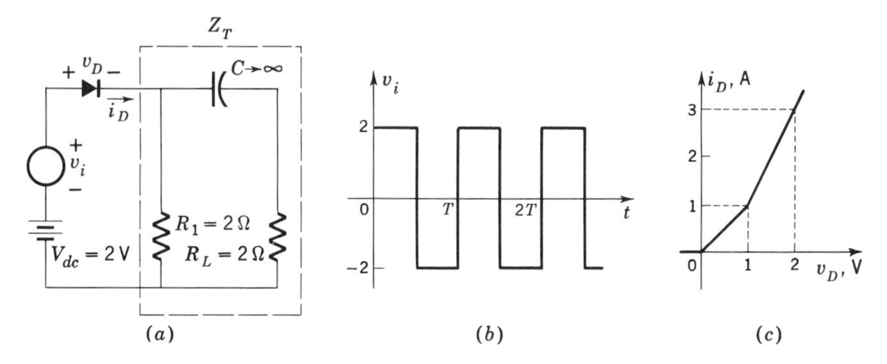

Fig. 2.6-1 Circuit, signal waveform, and piecewise linear diode characteristic for Example 2.6-1. (*a*) Circuit; (*b*) signal waveform; (*c*) piecewise linear characteristic.

(2.6-1) becomes

$$2 + v_i = v_D + I_D + i_D \tag{2.6-3b}$$

Equation (2.6-3*b*) contains three unknowns, v_D, i_D, and I_D. The diode *vi* characteristic is of the form $i_D = f(v_D)$, and can thus be used to eliminate one of the unknowns from (2.6-3*b*). In order to eliminate the other unknown, we use the relation

$$I_D = \left(\frac{1}{T}\right) \int_{-T/2}^{T/2} i_D(t)\, dt \tag{2.6-4}$$

The problem is to simultaneously solve the three equations

$$2 + v_i = v_D + I_D + i_D \tag{2.6-3b}$$

$$i_D = f(v_D) \qquad \text{Fig. 2.6-1}c$$

and

$$I_D = \left(\frac{1}{T}\right) \int_{-T/2}^{T/2} i_D(t)\, dt \tag{2.6-4}$$

A solution is difficult to obtain, even for the simplified conditions of the circuit and signal of this example. An iterative graphical procedure is employed to solve the problem. The sequence of steps used to obtain the solution is as follows:

1. Assume $I_{D1} = I_{DQ}$ (this corresponds to no distortion, and is easily found as before).
2. Using this value for I_D in (2.6-3*b*), find the waveform of i_{D1} from the *vi* characteristic.
3. Perform the integration indicated in (2.6-4) on the waveform of i_{D1} to determine I_{D2}. If this value is sufficiently close to I_{D1}, the solution is i_{D1} as found in step 2. However, this will not be the case when significant distortion is present.
4. Using I_{D2} in (2.6-3*b*), find i_{D2} as in step 2.
5. Average i_{D2} to determine I_{D3}.
6. Continue this process until I_{Dn+1} is sufficiently close to I_{Dn}.

Let us follow this sequence of steps to solve for the diode current in the circuit of Fig. 2.6-1*a*.

1. To find $I_{D1} = I_{DQ}$ let $v_i = 0$ in (2.6-3*b*) and plot the dc load line in Fig. 2.6-2. From the figure $I_{DQ} = \frac{2}{3}$ A; so assume

$$I_{D1} = \tfrac{2}{3}\ \text{A} \tag{2.6-5}$$

2. Now (2.6-3*b*) becomes

$$v_{D1} + i_{D1} = 2 - \tfrac{2}{3} + v_i \tag{2.6-6}$$

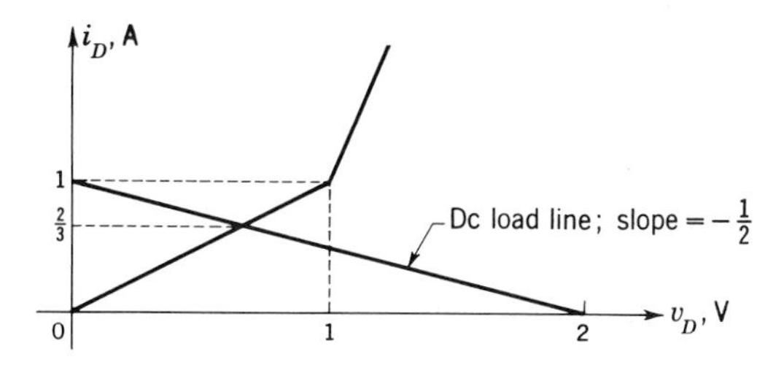

Fig. 2.6-2 First iteration for Example 2.6-1.

This describes the family of ac load lines of slope -1, which correspond to the various values of v_i. The graphical solution for i_{D1} is shown in Fig. 2.6-3.

3. The integration is particularly easy for the square wave:

$$I_{D2} = \left(\frac{1}{T}\right) \int_{-T/2}^{T/2} i_{D1}(t)\, dt = \frac{1.88}{2} = 0.94 \text{ A}$$

This is not sufficiently close to $I_{D1} = 0.67$; so the iteration procedure is continued.

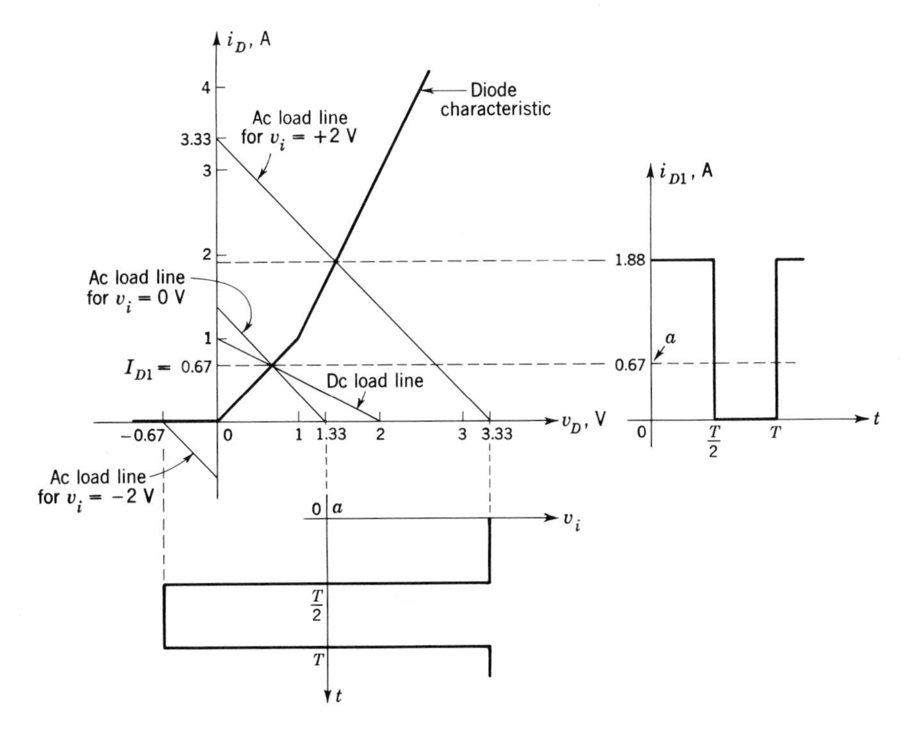

Fig. 2.6-3 Second iteration for Example 2.6-1.

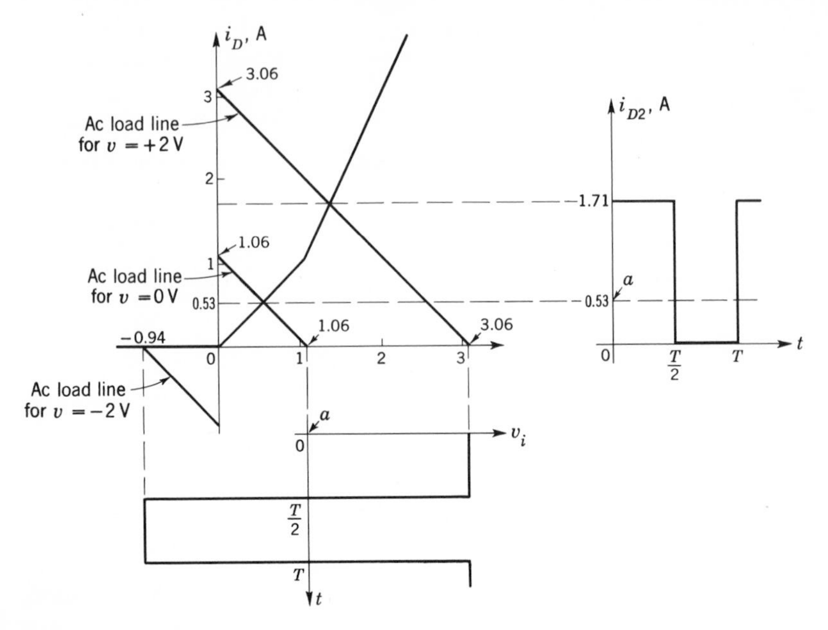

Fig. 2.6-4 Third iteration for Example 2.6-1.

4. Using $I_{D2} = 0.94$ in (2.6-3*b*),

$$v_{D2} + i_{D2} = 2 - 0.94 + v_i \tag{2.6-7}$$

Again, this describes a family of ac load lines, as shown in Fig. 2.6-4. The shifted Q point which corresponds to the point at which $v_i = 0$ (point a in the diagram) is $i_D(v_i = 0) = 1.06/2 = 0.53$ A, and the peak value of $i_{D2}(t)$ is 1.71 A.
5. From Fig. 2.6-4, $I_{D3} = 1.71/2 = 0.855$ A.

Summarizing, we have found

$$I_{D1} = 0.67 \text{ A}$$
$$I_{D2} = 0.94 \text{ A}$$
$$I_{D3} = 0.85 \text{ A}$$

The process seems to be converging to $I_D \approx 0.9$ A (one more iteration actually yields $I_{D4} = 0.88$ A). Thus the diode current in the circuit of Fig. 2.6-1*a* has the waveform shown as $i_{D2}(t)$ in Fig. 2.6-4.

The reader should note carefully the difference between this case and the large-signal analysis of Sec. 2.3. There the load was purely resistive, so that the average value shift, even though still present, did not affect the solution. In other words, there were only two equations, (2.3-1) and (2.3-2), to solve simultaneously. In this problem, the presence of the capacitor led to separate ac and dc conditions, and thus the complication of the extra unknown, I_D. Another point to note here is that the use of a different input-signal waveform would result in a completely different, and probably a much more difficult, solution.

Other graphical approaches can be employed as well. For example, the third equation, (2.6-4), can be replaced by the condition that the average current through the capacitor must be zero in the steady state. These alternative techniques are not pursued here.

Example 2.6-1 indicates how complex analysis of nonlinear devices can become, even in the simplest cases. The digital computer is widely used to solve problems of this type. Using (2.6-3b), (2.6-4), and the nonlinear characteristic, the solution to the problem can be obtained in a very short time.

2.7 ZENER DIODES

Zener, or breakdown, diodes are semiconductor *pn* junction diodes with controlled reverse-biased properties which make them extremely useful in many applications, especially as voltage reference devices. A typical *vi* characteristic is shown in Fig. 2.7-1.

The forward characteristic is similar to that of the standard semiconductor diode. The reverse characteristic, however, exhibits a region in which the terminal voltage is almost independent of the diode current, as discussed in Sec. 2.2. The Zener voltage of any particular diode is

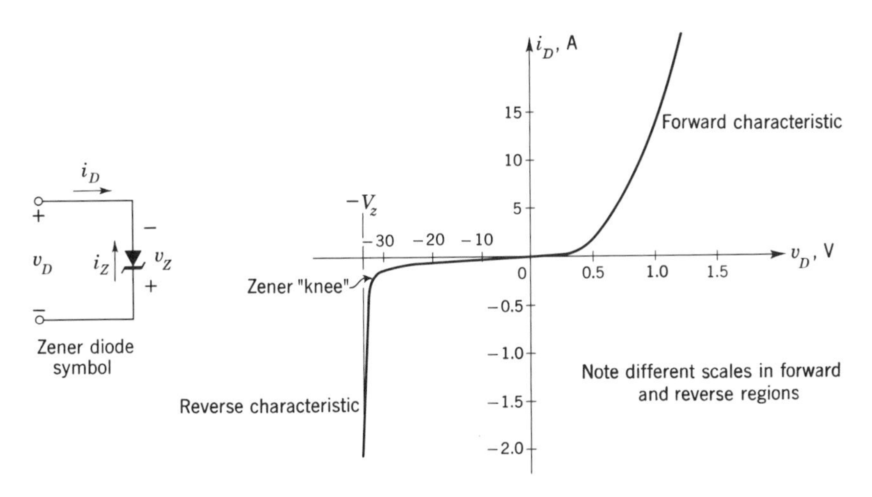

Fig. 2.7-1 Zener-diode circuit symbol and *vi* characteristic.

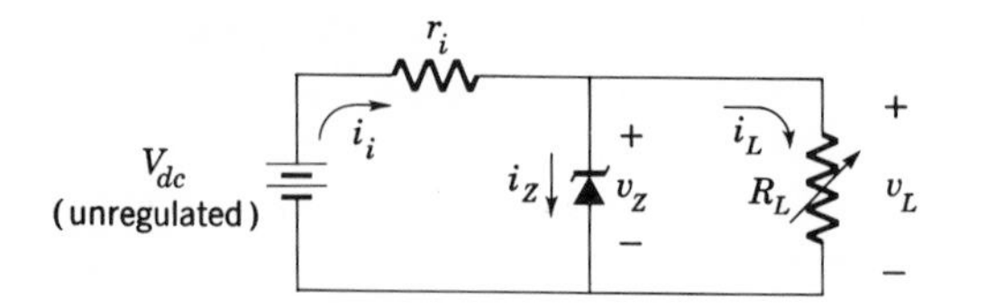

Fig. 2.7-2 Zener-diode voltage regulator.

controlled by the amount of doping applied in the manufacturing process. Typical values range from about 2 to 200 V, with power-handling capabilities up to 100 W.

In most applications, the Zener diode operates in this reverse-biased region. A typical application is the simple voltage-regulator circuit shown in Fig. 2.7-2. When this circuit is properly designed, the load voltage V_L remains at an essentially constant value, equal to the nominal Zener voltage, even though the input voltage V_{dc} and the load resistance R_L vary over a wide range. The operation of the circuit can be explained qualitatively in terms of the *vi* characteristic of Fig. 2.7-1. If the input voltage increases, the diode tends to maintain a constant voltage across the load so that the voltage drop across r_i must increase. The resultant increase in I_i flows through the diode, while the load current remains constant.

Now let the input voltage remain constant, but decrease the load resistance. This requires an increase in load current. This extra current cannot come from the source since the drop in r_i, and therefore the source current, will not change as long as the diode is within its regulating range. The additional load current will, of course, come from a decrease in the Zener-diode current.

Both of these regulating actions depend on the operation of the diode below the knee of the curve, where the diode voltage remains nearly constant, and the range of the regulation will depend on the value of r_i in this particular circuit. (Temperature characteristics will be discussed in Sec. 2.9.)

Example 2.7-1 A 7.2-V Zener diode is used in the circuit of Fig. 2.7-2, and the load current is to vary from 12 to 100 mA. Find the value of r_i required to maintain this load current if the supply voltage $V_{dc} = 12$ V.

Solution For a shunt regulator such as this, an empirical factor of 10 percent of the maximum load current is used as the minimum Zener-diode current. Thus the minimum Zener-diode current for the stated conditions must be at least 10 mA.

Applying Ohm's law to the circuit,

$$r_i = \frac{V_{dc} - V_L}{I_Z + I_L}$$

The voltage across r_i must remain at $12 - 7.2 = 4.8$ V over the regulating range. The minimum Zener current will occur when the load current is maximum, so that

$$r_i = \frac{V_{dc} - V_L}{I_{Z,\min} + I_{L,\max}} = \frac{V_{dc} - V_L}{(1 + 0.1)I_{L,\max}} = \frac{4.8}{0.11} = 43.5\ \Omega$$

As the load current decreases with r_i set at this value, the Zener-diode current will increase, their sum remaining constant at 110 mA. Note that the Zener diode must be capable of dissipating,

$$P = (7.2)(110 \times 10^{-3}) \approx 0.8\ \text{W}$$

to protect it from being destroyed in case the load resistance becomes open-circuited ($I_L = 0$, $I_Z = 110$ mA).

Let us choose a 1-W Zener diode which has a voltage of 7.2 V when drawing 10 mA, and a dynamic resistance r_d of 2 Ω. We shall calculate the output-voltage variation across the load, using graphical methods.

The vi characteristic for this diode is shown in Fig. 2.7-3. Also shown is the maximum power curve,

$$P_Z = \left(\frac{1}{T}\right) \int_{-T/2}^{T/2} v_Z i_Z \, dt = V_{ZQ} I_{ZQ} = 1\ \text{W}$$

This is the equation of the hyberbola sketched in Fig. 2.7-3. Its intersection with the vi characteristic yields the maximum current and voltage that can be sustained by the diode.

The dc-load-line equations can be found by redrawing Fig. 2.7-2 as shown in Fig. 2.7-4. From this circuit we find the load-line equation

$$v_Z = 12 - 43.5(I_L + i_Z) \tag{2.7-1}$$

Thus

$$v_Z + i_Z\,(43.5) \approx \begin{cases} 11.5 & I_L = 12\ \text{mA} \\ 7.65 & I_L = 100\ \text{mA} \end{cases} \tag{2.7-2}$$

The two load lines of (2.7-2) are plotted in Fig. 2.7-3. From the graph, it is evident that the diode reverse voltage v_Z, and hence the load voltage v_L, vary from 7.2 V when $I_L = 100$ mA to 7.37 V when $I_L = 12$ mA. Note that if the Zener diode were not present, the load voltage would vary (keeping $r_i = 43.5$ Ω) from 7.7 V when

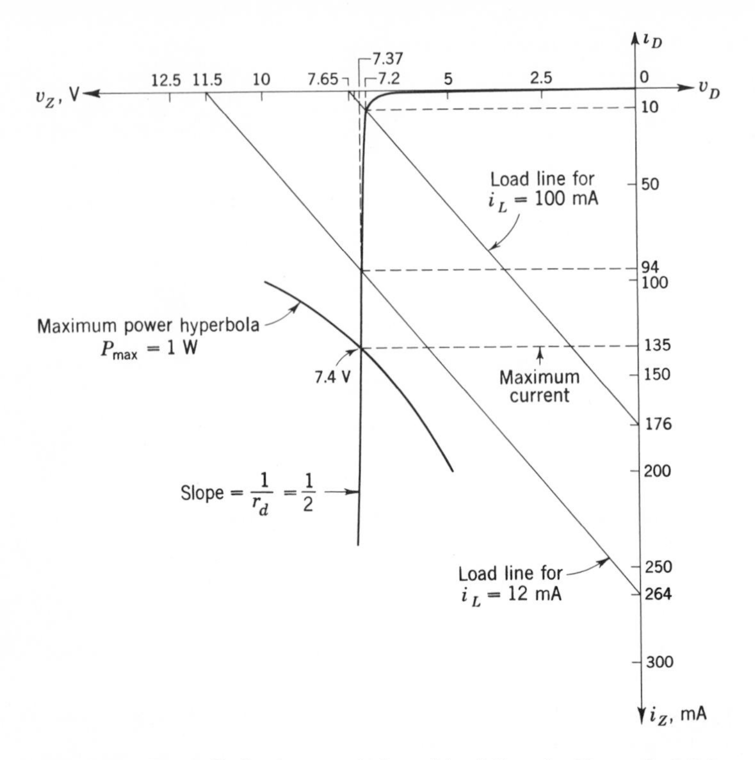

Fig. 2.7-3 Zener-diode characteristic and load lines for Example 2.7-1.

$I_L = 100$ mA to 11.5 V when $I_L = 12$ mA. Thus the Zener diode is seen to have provided voltage regulation with respect to changes in load current. This occurs since the impedance seen by the load is small (2 Ω) compared with the load resistance $R_L > 7.2$ V/100 mA = 72 Ω.

Example 2.7-2 Ripple reduction with the Zener regulator The 7.2-V Zener diode is used in a circuit similar to that shown in Fig. 2.7-2, with an ac ripple voltage added to the unregulated dc voltage. As stated before, these voltages are typical of the output of a dc power

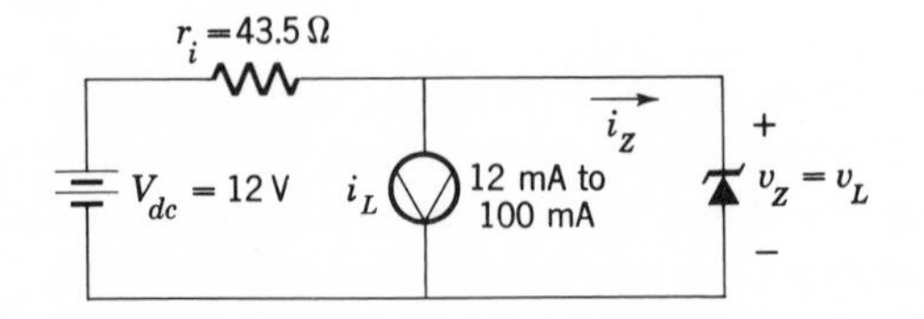

Fig. 2.7-4 The circuit of Fig. 2.7-2 redrawn for clarity.

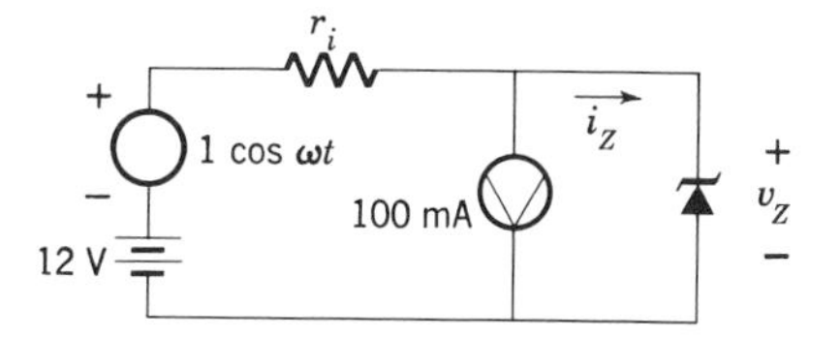

Fig. 2.7-5 Zener regulator for Example 2.7-2.

supply. The load draws a current of 100 mA. The output of the unregulated supply can be represented as

$$v_T = 12 + 1 \cos \omega t \tag{2.7-3}$$

Find the source resistance r_i for proper operation and the peak-to-peak value of the ripple voltage present across the load.

Solution The load-line equation for this problem is easily obtained if we first redraw Fig. 2.7-2 as shown in Fig. 2.7-5. Thus

$$v_Z = 12 + 1 \cos \omega t - r_i(0.1 + i_Z)$$

and

$$v_Z + r_i i_Z = 12 - 0.1 r_i + 1 \cos \omega t \tag{2.7-4}$$

The problem can now be solved graphically, or analytically using the linearized diode characteristic shown in Fig. 2.7-3. We use the graphical technique since it gives insight to the problem. The characteristic shown in Fig. 2.7-3 is redrawn in Fig. 2.7-6.

The range of possible values of source resistance r_i will be bounded because of the minimum and maximum allowable Zener currents. Thus, if $i_{Z,\min} = 10$ mA (which is 10 percent of the maximum current), then $v_{Z,\min} = 7.2$ V, and

$$r_i < \left(\frac{12 - 7.2 + \cos \omega t}{0.1 + 0.01}\right)_{\min} = \frac{3.8}{0.11} = 34.5\ \Omega \tag{2.7-5}$$

A lower bound on r_i is obtained by using the maximum diode voltage and current ratings. Thus, using (2.7-4) and

$$i_{Z,\max} = 135 \text{ mA} \qquad v_{Z,\max} \approx 7.4 \text{ V}$$

the source resistor r_i must exceed,

$$r_i > \left(\frac{12 - 7.4 + \cos \omega t}{0.1 + 0.135}\right)_{\max} = \frac{5.6}{0.235} \approx 24\ \Omega \tag{2.7-6}$$

We arbitrarily choose $r_i = 32\ \Omega$. The load-line equation (2.7-4) can now be plotted on Fig. 2.7-6, and the output ripple

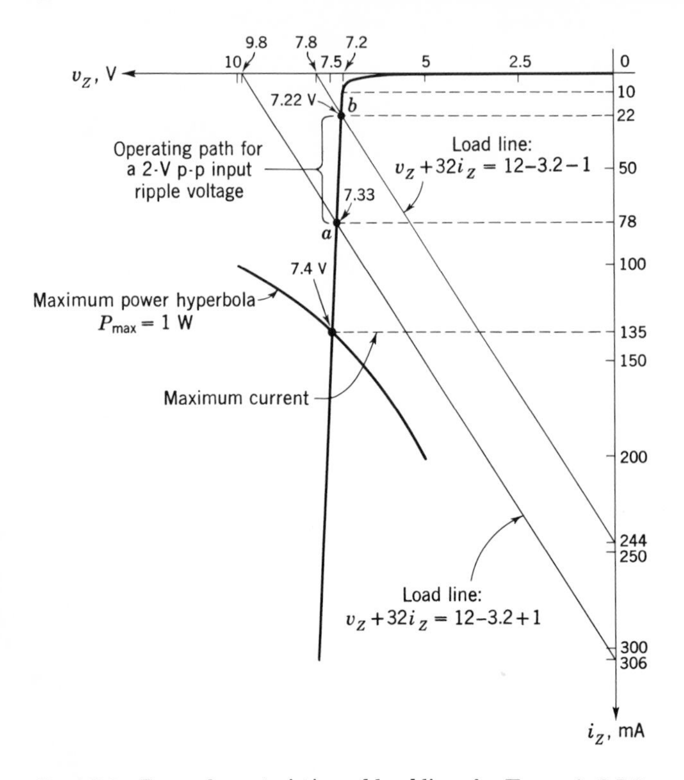

Fig. 2.7-6 Zener characteristic and load lines for Example 2.7-2.

voltage determined. Referring to the figure, the peak-to-peak ripple voltage across the load is $7.33 - 7.22 = 0.11$ V, compared with a 2-V-input peak-to-peak ripple voltage.

2.8 PIECEWISE LINEAR ANALYSIS AND EQUIVALENT CIRCUITS

The technique of piecewise linear analysis involves approximating the characteristic of a nonlinear device with connected straight-line segments, and then utilizing ideal diodes, resistors, and constant-voltage and constant-current sources to construct an equivalent circuit which realizes the piecewise linear curve (it should be noted that this procedure was employed in Sec. 2.6 and again in Sec. 2.7). The equivalent circuit is then amenable to standard circuit-analysis methods, especially for small signals, and graphical procedures are thus avoided.

The piecewise linear technique is widely used in the analysis of electronic circuits. Another application of this technique is to the solution of problems using analog computers. Here a nonlinear differential

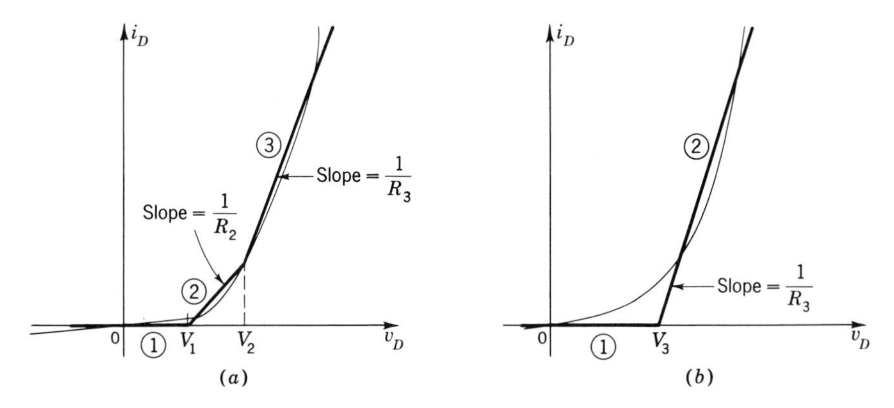

Fig. 2.8-1 Piecewise linear approximations. (*a*) Three segments; (*b*) two segments.

equation representing a physical system is to be solved. Nonlinear terms are approximated by piecewise linear curves and synthesized. The complete problem is then solved electronically, and the result is plotted automatically or displayed on an oscilloscope.

Consider the three-segment approximation to the nonlinear characteristic shown in Fig. 2.8-1*a* as an example of the synthesis of a piecewise linear circuit. Note that any nonlinear curve can be approximated as closely as desired by using one or more straight-line segments. For practical purposes, the number of line segments is kept as small as possible. The curve of Fig. 2.8-1*b*, consisting of only two segments, both semi-infinite, can be used to approximate the curve when the accuracy required permits the large error shown. It is synthesized as the circuit of Fig. 2.8-2*a*. The operation of the circuit is easily explained if one thinks of the input voltage as a dc source which changes slowly from a large negative value to a large positive value. Recalling that the ideal diode will conduct only when the voltage $v_T - V_3$ is positive (forward bias), and is an open circuit when $v_T - V_3$ is negative (reverse bias), no current will

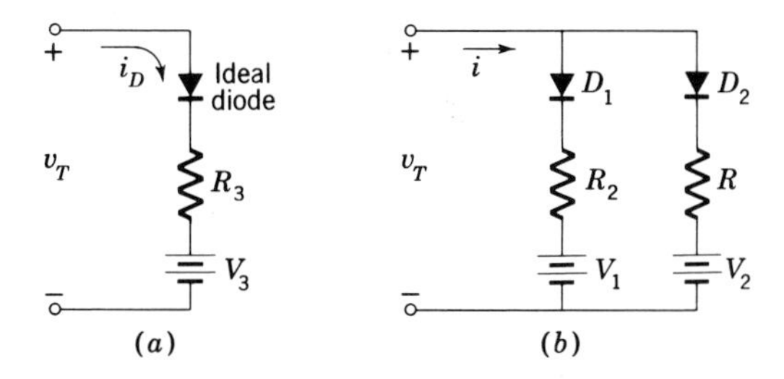

Fig. 2.8-2 Piecewise linear circuits. (*a*) Synthesis of Fig. 2.8-1*b*; (*b*) synthesis of Fig. 2.8-1*a*.

flow unless $v_T \geq V_3$; thus $v_T < V_3$ gives segment 1 of the curve. Now when $v_T = V_3$ (the breakpoint), a change of state occurs. As soon as v_T becomes greater than V_3, the diode conducts, appearing as a short circuit. Thus, for all values of $v_T > V_3$, the circuit consists of resistor R_3 in series with voltage V_3, which furnishes segment 2 of the curve.

This result can be used to synthesize the three-segment curve of Fig. 2.8-1*a*. The circuit of Fig. 2.8-2*a* synthesizes segments 1 and 2 (with R_3 replaced by R_2, and V_3 replaced by V_1). A breakpoint must still be provided at V_2, and the slope of the curve in region 3 must be $1/R_3$. Note that this slope is greater than that of segment 2, so that the equivalent resistance over this range must be less. This suggests that a parallel circuit be added to achieve a reduction in resistance. The circuit of Fig. 2.8-2*b* will produce the desired result. The left-hand branch is the same as the one previously considered, and yields segments 1 and 2 of the curve. The right-hand branch will have no effect until $v_T = V_2$, where D_2 conducts. For $v_T > V_2$, D_2 and D_1 are both short circuits, and the resistance of the circuit is $R \| R_2$. In order to match the desired slope $1/R_3$,

$$\frac{1}{R_3} = \frac{1}{R_2} + \frac{1}{R} \qquad \text{or} \qquad R = \frac{R_2 R_3}{R_2 - R_3}$$

This completes the circuit.

It is important to distinguish between the piecewise linear equivalent circuits under discussion here and the small-signal linear equivalent discussed in Sec. 2.4. The resistance values may differ considerably, because in the small-signal case slopes are measured at a particular operating point, whereas for the piecewise linear case they are values averaged over relatively large ranges. The piecewise linear circuit can be used to calculate *total* currents and voltages, while the small-signal equivalent circuit is restricted to small variations about the operating point.

Example 2.8-1 Synthesize a circuit to yield a piecewise linear approximation to the function $y = \ln x$ for the range $0 < y < 3$.

Solution A plot of $x = \epsilon^y$ for $0 < y < 3$ is shown in Fig. 2.8-3. A three-segment approximation will suffice, with breakpoints at

$$i = \begin{cases} 0 & v_T = 1 \\ 1.61 & v_T = 5 \end{cases}$$

and segment 3 must pass through the point $i = 3$, $v_T = 20$.

One possible synthesis procedure and circuit are described below. Referring to the piecewise linear curve of Fig. 2.8-3, for

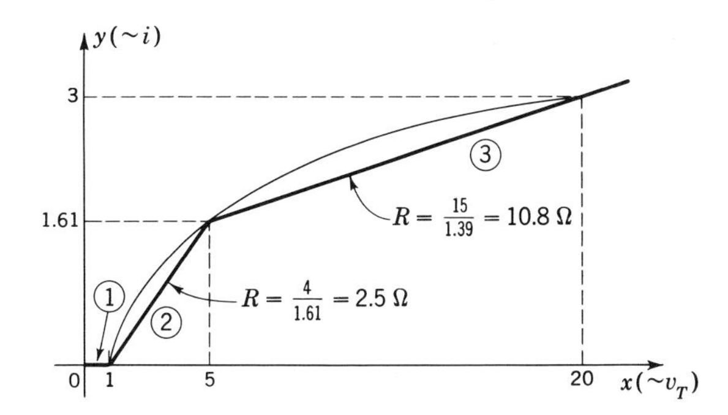

Fig. 2.8-3 The function $x = \epsilon^y$ and its piecewise linear approximation.

v_T less than 1 V, the current i is zero, while from 1 to 5 V the slope is that of a 2.5-Ω resistance. This set of conditions is provided by the circuit of Fig. 2.8-4. Above 5 V the resistance increases to 10.8 Ω. When synthesizing a circuit for the curve of Fig. 2.8-1a, it was seen that if the resistance decreased after a breakpoint, parallel circuits should be added. Reasoning this way, we try a circuit in *series* with the circuit of Fig. 2.8-4 to increase the resistance.

This second circuit must be a short circuit for voltages less than 5 V (currents less than 1.61 A). A possible configuration is shown in Fig. 2.8-5. Observe that for $i < 1.61$ A, D_2 is a short circuit and the 1.61 A from the current source flows through D_2. When $i = 1.61$ A, the current in D_2, i_{D2}, becomes zero, and when i exceeds 1.61 A, D_2 becomes an open circuit. Then, since D_1 is a short circuit, KVL yields

$$v_T = (i - 1.61)(8.3) + 2.5i + 1 \qquad i > 1.61 \text{ A}, v_T > 5 \text{ V}$$

The piecewise linear slope in this range is

$$\frac{\Delta v_T}{\Delta i} = 8.3 + 2.5 = 10.8 \text{ Ω}$$

as required.

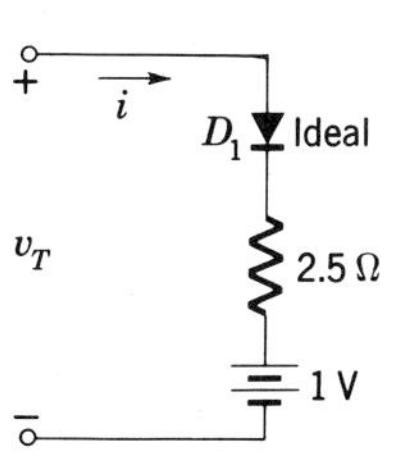

Fig. 2.8-4 Circuit for realizing segments 1 and 2 of Fig. 2.8-3.

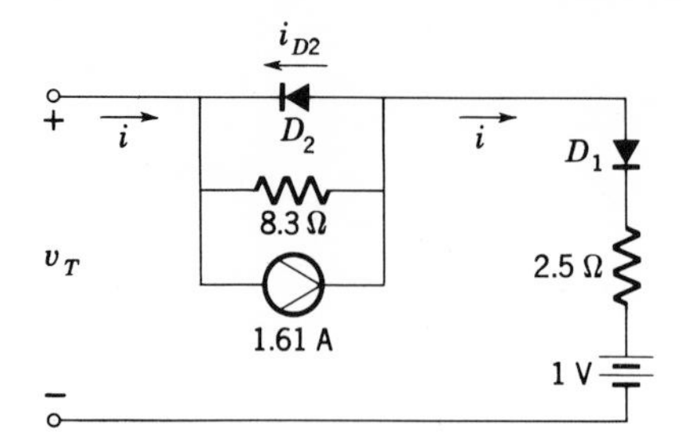

Fig. 2.8-5 Complete circuit for Example 2.8-1.

Several practical problems arise when attempting to construct the circuit of Fig. 2.8-5 using real diodes and resistances, because a real diode has a break voltage and an equivalent series resistance. In addition, the equivalent resistance *decreases* with current, which produces a characteristic with curvature opposite to that desired.

To minimize these problems, we usually *scale* the variables, so that milliamperes, rather than amperes, would flow in the circuit. To see how this transformation is made, refer to the original equation, and rewrite it

$$x = \epsilon^y$$

Now substitute

$$x = av_T$$

and

$$y = bi$$

Then

$$v_T = \left(\frac{1}{a}\right)\epsilon^{bi}$$

If $b = 10^3$, the current is given in milliamperes. By similarly adjusting the constant a, the voltage can be scaled independently. Of course, all resistor values change accordingly. If $a = 1$ and $b = 10^3$, then all resistors in the circuit of Fig. 2.8-5 are multiplied by 10^3, and currents in the new circuit are 10^{-3} times as large as those in the original circuit. Voltages throughout the new circuit are the same as in the original. In general, voltage levels should be chosen high enough so that the diode voltage drops will not cause large errors.

A scaled version of the circuit shown in Fig. 2.8-5 is shown in Fig. 2.8-6. In this circuit the 1-V supply is eliminated and D_1 is replaced by a silicon and germanium diode in series. The sum of the

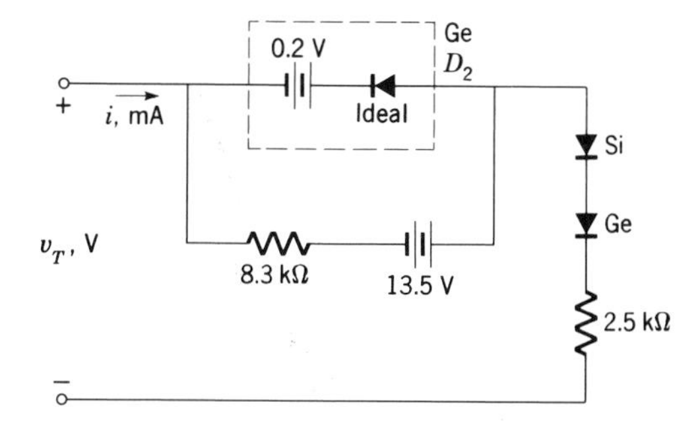

Fig. 2.8-6 Practical version of the circuit of Fig. 2.8-5.

break voltages of these two diodes is approximately equal to 1 V, and their combined internal resistance is much less than 2.5 kΩ at these current levels, so that a reasonably straight-line characteristic would result between 1 and 1.61 mA. The current source is replaced by a 13.5-V battery in series with the 8.3-kΩ resistor.

Example 2.8-2 Find, using piecewise linear circuits, the simultaneous solution of the equations

$$y = x^2 \qquad 0 < x < 2$$

and

$$y = x \qquad 0 < x < 2$$

The solution is, of course, $x = 1$. Let us see how this solution can be obtained electronically.

Solution First represent the equation $y = x^2$ by the three line segments shown in Fig. 2.8-7. A straightforward procedure for choosing the piecewise linear segments is to divide up the interval into equal parts as shown. We start with segments 1 and 2, the slopes being the slopes at the points considered (in this case $x = 0$ and $x = 2$). If the error between the piecewise linear and the true nonlinear equation is too great, add a segment 3 as shown. The slope of segment 3 is the slope at $x = 1$. If the error is still too great, two more segments could be added at $x = \frac{1}{2}$ and $x = \frac{3}{2}$, etc. It must be noted that even with three segments the error could be reduced by simply placing segment 3 as a chord rather than as a tangent to the curve. (Since calculations are simplified by using a tangent, we do not worry about the reduced error here.)

Now let

$$y = i \qquad \text{in mA}$$

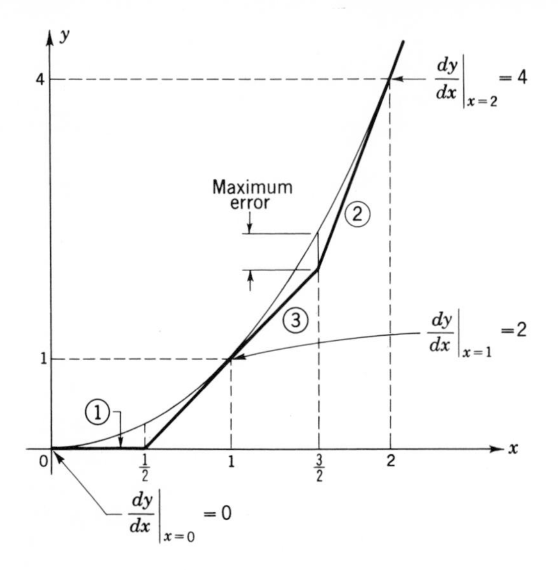

Fig. 2.8-7 Piecewise linear approximation to $y = x^2$.

and

$x = v$ in volts

Then the circuit for $y = x^2$ can be synthesized as shown in Fig. 2.8-8.

The circuit for $y = x$ ($y \sim i$ in mA; $x \sim v$ in volts) is simply a 1-kΩ resistance. To find the simultaneous solution to the equations we force x ($\sim v$) to be the same in both circuits by connecting them in parallel, as shown in Fig. 2.8-9.

The voltage $v = V_1$ is then varied until $i = i_1$. At this point $v = V_1$ ($\sim x$) and $i = i_1$ ($\sim y$) represent the simultaneous solution of the equations. For this simple case $v = V_1 = 1$ V and $i = i_1 = 1$ mA are seen to yield the solution.

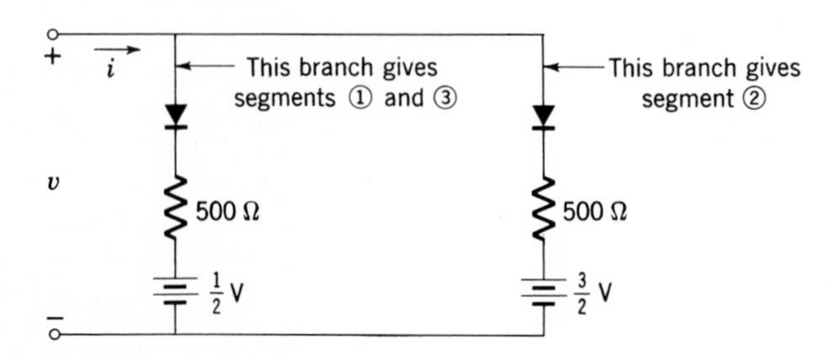

Fig. 2.8-8 Piecewise linear circuit for $y = x^2$.

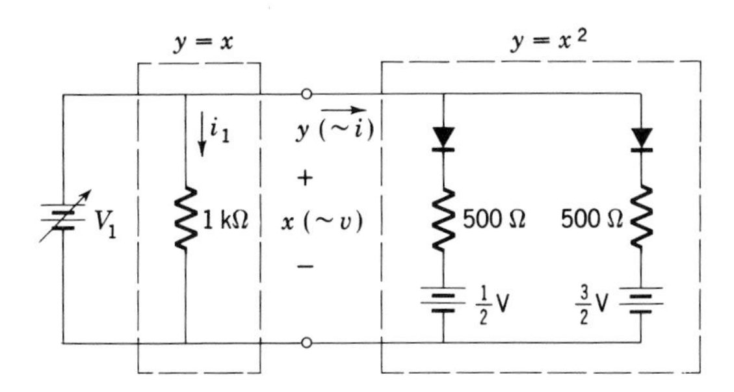

Fig. 2.8-9 Connection for simultaneous solution of $y = x^2$ and $y = x$.

An alternative method involves connecting the 1-kΩ resistance in series with the $y = x^2$ network, thus forcing the current y to be the same in both circuits. The input voltage is then varied until the voltage ($\sim x$) across each circuit is the same.

Example 2.8-3 Use the result of Example 2.8-2 to solve the nonlinear differential equation

$$\frac{dx}{dt} + x^2 = f(t)$$

with

$$x(0) = 0 \qquad \text{and} \qquad f(t) = 4$$

Solution As in Example 2.8-2, we set x to be analogous to v, so that the circuit of Fig. 2.8-8 may be used for the second term of the equation. This yields a current in mA equal to v^2 ($\sim x^2$). The first term can be represented by a 1000-μF capacitor, in which the current is

$$i = C\frac{dv}{dt}$$

or

$$y \sim i \text{ (in mA)} = \frac{dv}{dt} \sim \frac{dx}{dt}$$

Thus we have currents analogous to the two variable terms of the differential equation. This suggests that we use KCL to add the two currents to a 4-mA constant current source, as shown in Fig. 2.8-10.

If the switch is opened at $t = 0$, with the capacitor uncharged, so that $v(0) \sim x(0) = 0$, then an oscilloscope placed across the circuit will record the solution $v(t) \sim x(t)$.

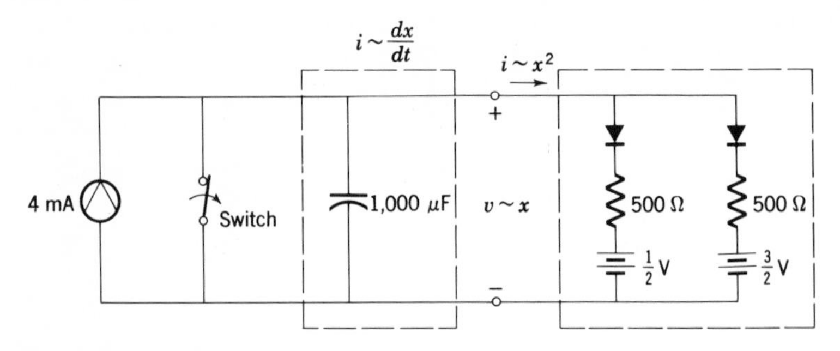

Fig. 2.8-10 Circuit for solving $dx/dt + x^2 = 4$.

The reader will note that a nonzero initial condition can be accounted for by placing an initial charge on the capacitor. A different $f(t)$ would simply require a current source having the same waveform as $f(t)$.

2.9 TEMPERATURE EFFECTS IN DIODES

The practical problem of the maximum allowable heat dissipation in a *pn* junction diode will be considered in this section. The dissipation of electric power as heat in the diode causes the junction temperature to rise. This temperature rise must be kept within acceptable limits, or the diode will suffer physical damage. The thermal problem is easily handled by a simple thermal analog of Ohm's law in which current is replaced by power, voltage by temperature, and electric resistance by thermal resistance θ.

Figure 2.9-1 shows a diode mounted on a heat sink. The diode is

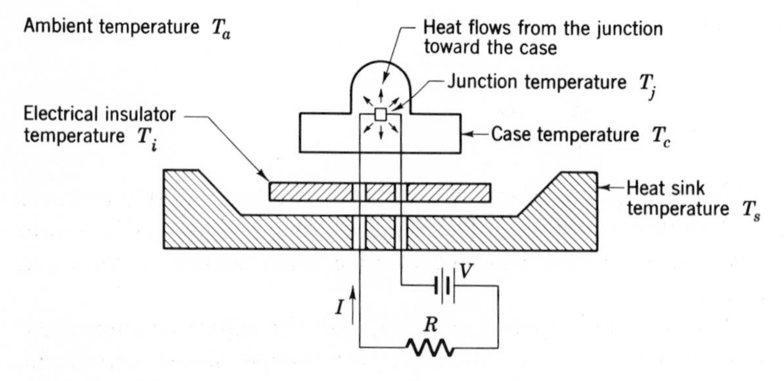

Fig. 2.9-1 Diode mounted on heat sink.

insulated electrically (not thermally) from the large heat sink. The operation of the system is as follows: With no electrical connection made to the diode, the junction temperature T_j will be the same as the ambient temperature T_a. When a signal is applied, power will be dissipated in the diode, causing the junction temperature to rise. The heat produced flows outward to the case, and is then carried by conduction from the diode case to the heat sink. The heat sink has a large surface area from which it radiates the heat to the surrounding environment.

If the power dissipated at the junction is constant and within the power-handling capabilities of the diode, then, after a sufficient time has elapsed, the system will reach a state of thermal equilibrium. Each of the elements will, in general, be at different temperatures. A good approximation is that the temperature rise is in proportion to the power dissipated at the junction. The constant of proportionality is called the *thermal resistance*, and is given the symbol θ.

An increase in the junction temperature above the case temperature is related to the power dissipated by the equation

$$T_j - T_c = \theta_{jc}P_j \tag{2.9-1}$$

where $T_j - T_c$ = rise in junction temperature above case temperature, °C
P_j = electric power dissipated at junction, W
θ_{jc} = thermal resistance between junction and case, °C/W

The thermal resistance is a function of the construction of the diode and case, and is usually specified by the manufacturer.

Consider a diode in a case without any mounting, as shown in Fig. 2.9-2. The case and ambient temperatures differ by an amount equal to the product of P_j and the thermal resistance between the case and ambient, θ_{ca}. Thus the increase in case temperature above ambient temperature is given by

$$T_c - T_a = \theta_{ca}P_j \tag{2.9-2}$$

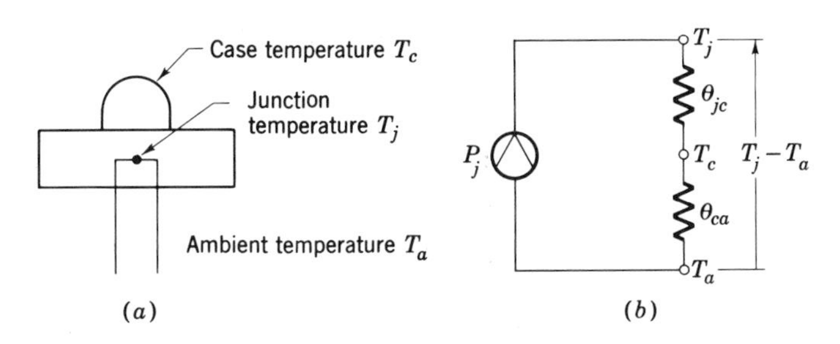

Fig. 2.9-2 Diode and electrical analog of thermal system. (*a*) Diode without mounting; (*b*) electrical analog.

The circuit shown in Fig. 2.9-2*b* represents an electrical analog of the thermal system of Fig. 2.9-2*a*, with the following analogs derived from (2.9-1):

Temperature "drop" $T_j - T_a$	Voltage drop
Power dissipation P_j	Constant-current generator
Thermal resistance $\theta_{jc} + \theta_{ca}$	Electric resistance

It is seen from Fig. 2.9-2*b* and (2.9-1) and (2.9-2) that

$$T_j = P_j\theta_{jc} + P_j\theta_{ca} + T_a \tag{2.9-3}$$

This result can immediately be extended to cover the heat-conduction system of Fig. 2.9-1, which yields

$$T_j = P_j(\theta_{jc} + \theta_{cs} + \theta_{sa}) + T_a \tag{2.9-4}$$

Here the thermal resistances not previously defined are

θ_{cs} = case-to-heat-sink (includes insulator) thermal resistance
θ_{sa} = heat-sink-to-ambient thermal resistance

In a practical design problem, use of this equation is based on the following conditions:

1. The maximum allowable junction temperature is furnished by the manufacturer. Typical values are about 100°C for germanium diodes and 150 to 200°C for silicon diodes.
2. The ambient temperature is an uncontrolled variable depending on the environment in which the equipment will ultimately be operated.
3. The power dissipated at the junction depends on the electric system, and is given, for time-varying currents and voltages, by

$$P_j = \left(\frac{1}{T}\right)\int_0^T v_D(t)i_D(t)\,dt$$

For dc operation this is simply

$$P_j = V_D I_D$$

4. Once a particular diode has been chosen to meet electrical specifications, its thermal resistance θ_{jc} will be fixed. This figure is usually provided by the manufacturer.

Noting these facts and (2.9-3), it is seen that the case-to-ambient thermal resistance is the only variable available to adjust in order to maintain the junction temperature at a safe value.

Solving (2.9-3) yields

$$\theta_{ca} = \frac{T_{j,\max} - T_a}{P_j} - \theta_{jc} \tag{2.9-5}$$

This expression is used to determine the maximum θ_{ca}, given all the factors on the right. Equation (2.9-4) can, of course, be used to determine any of the variables involved.

Example 2.9-1 A 50-W silicon Zener diode is to dissipate 10 W in a particular circuit. The maximum allowable junction temperature is 175°C, the ambient temperature is 50°C, and $\theta_{jc} = 2.4$°C/W. Find the maximum thermal resistance which may be provided between case and ambient so that the junction temperature will not exceed 175°C.

Solution

$$\theta_{ca} = \frac{175 - 50}{10} - 2.4 = 10.1\text{°C/W}$$

A typical heat sink with mica insulator for this type of diode has $\theta_{ca} = 3.2$°C/W, which would be adequate for this application. If this heat sink is used, the actual junction temperature can be calculated from (2.9-4).

$$T_j = (10)(2.4 + 3.2) + 50 = 106\text{°C}$$

DERATING CURVES

The power rating of a diode is usually specified for an ambient temperature of 25°C. The maximum power that the diode can dissipate is determined by the junction temperature. Thus the power rating must be decreased as the ambient temperature rises to keep the junction temperature within a safe limit. The manufacturer usually provides derating curves which can be used to determine the maximum allowable power dissipation for a given case temperature.

A typical derating curve is shown in Fig. 2.9-3. At case temperatures less than T_{co}, the diode can dissipate its maximum allowable power. At case temperatures exceeding T_{co}, the maximum allowable dissipation is decreased until $T_c = T_{j,\max}$. At this maximum temperature the diode cannot dissipate any power without exceeding the maximum allowable junction temperature. The rate of decrease of power with increased case temperature is shown below to be θ_{jc}.

In the region $T_{co} \leq T_c \leq T_{j,\max}$

$$\frac{T_{j,\max} - T_{co}}{P_{j,\max}} = \frac{T_{j,\max} - T_c}{P_j} \tag{2.9-6}$$

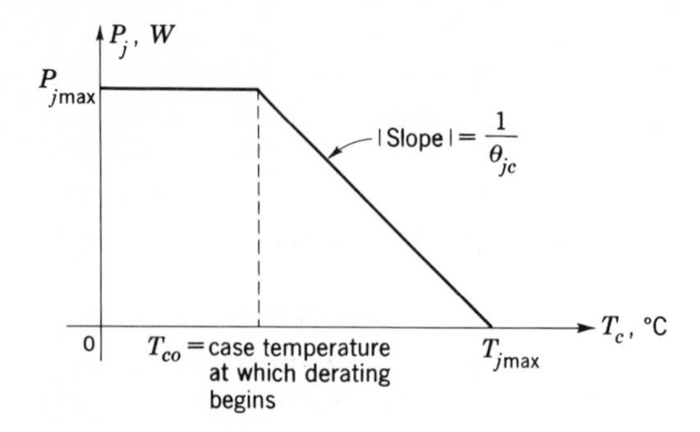

Fig. 2.9-3 Diode derating curve.

Since, by definition,

$$\theta_{jc} \equiv \frac{T_j - T_c}{P_j} \tag{2.9-7}$$

substituting (2.9-7) into (2.9-6) yields

$$\theta_{jc} = \frac{T_{j,\max} - T_{co}}{P_{j,\max}} \tag{2.9-8}$$

2.10 MANUFACTURERS' SPECIFICATIONS

2.10-1 THE DIODE RECTIFIER

The diode characteristics commonly specified by manufacturers are listed as follows (see Figs. 2.10-1 to 2.10-3):

Type: Silicon diode 1N566 (values are given at 25°C)
1. Peak inverse voltage (PIV) = −400 V
2. Maximum reverse current I_0 (at PIV) = 1.5 μA
3. Maximum dc forward voltage = 1.2 V at 500 mA
4. Average half-wave-rectified forward current I_F = 1.5 A
5. Maximum junction temperature $T_{j,\max}$ = 175°C
6. Current derating curve as given in Fig. 2.10-1

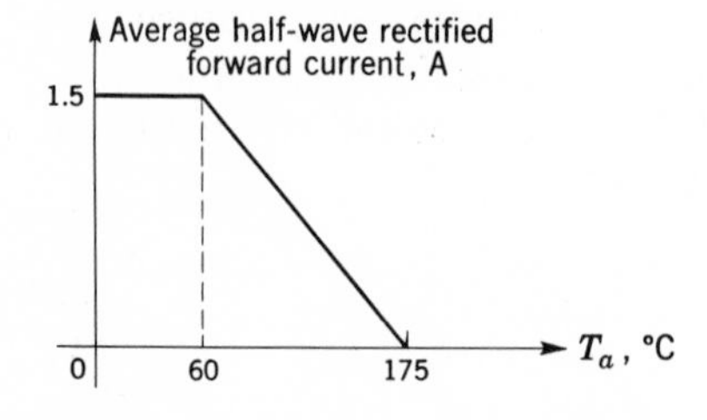

Fig. 2.10-1 Current derating curve for 1N566 silicon diode.

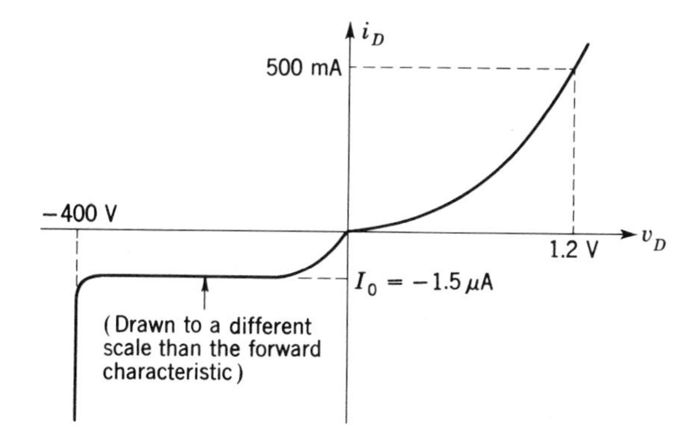

Fig. 2.10-2 Forward and reverse diode characteristics.

Characteristics 1 to 3 are most easily explained by referring to Fig. 2.10-2:

1. The peak inverse voltage (PIV) is the maximum allowable negative voltage that can be applied to the diode before it will break down.
2. The maximum reverse saturation current I_o at this voltage is 1.5 μA. Thus, when the diode is used in a rectifying circuit, the maximum negative current through it is 1.5 μA.
3. If a dc current of 500 mA is passed through the diode, the maximum voltage drop across it will be 1.2 V.

Characteristics 4 to 6 can be explained by referring to Figs. 2.10-3 and 2.10-1.

4. The average half-wave-rectified forward current is

$$I_F = \frac{V_{im}}{\pi R}$$

when the signal is a sinusoid of peak amplitude V_{im}, as shown in Fig. 2.10-3.

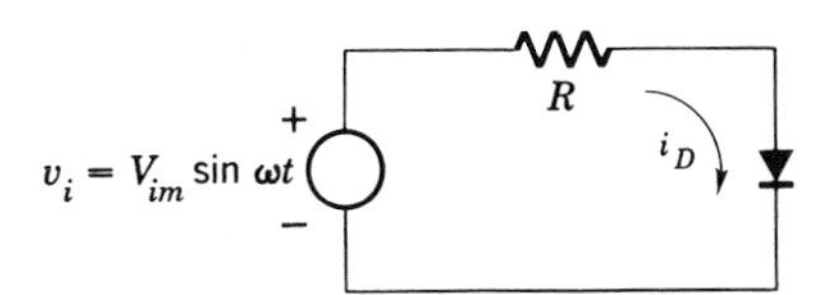

Fig. 2.10-3 Half-wave diode rectifier.

5, 6. Figure 2.10-1 indicates that if the ambient temperature is less than 60°C,

$$I_F = \frac{V_{im}}{\pi R} < 1.5 \text{ A}$$

At temperatures greater than 60°C, this maximum current is linearly derated until the ambient temperature is the same as the maximum junction temperature of 175°C, at which point $I_F = 0$.

When designing diode circuits, good engineering practice dictates the use of a 10 to 20 percent safety factor on all published maximum ratings, to take variations between units into account.

2.10-2 THE ZENER DIODE

Type: 18-V Zener diode 1N2816

1. Nominal reference voltage V_{ZT} = 18 V
2. Tolerance = 5%
3. Maximum dissipation (at 25°C) = 50 W
4. Test current I_{zT} = 700 mA
5. Dynamic impedance at I_{zT}, R_{zT} = 2 Ω
6. Knee current I_{zk} = 5 mA
7. Dynamic impedance at I_{zk}, R_{zk} = 80 Ω
8. Maximum junction temperature = 150°C
9. Temperature coefficient TC = 0.075%/°C

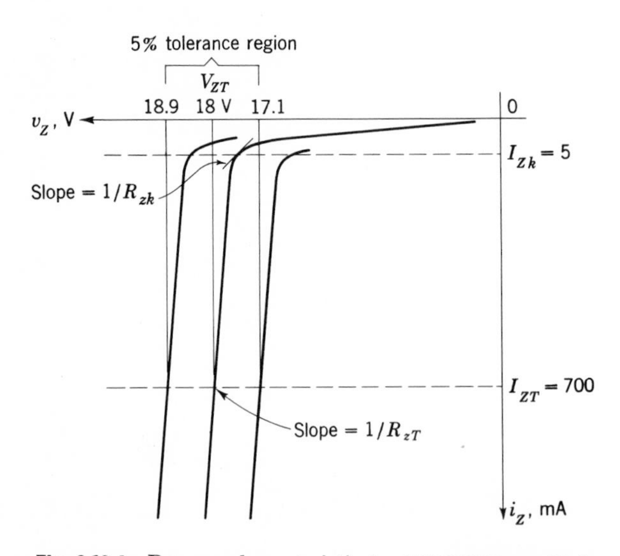

Fig. 2.10-4 Reverse characteristic for 1N2816 Zener diode.

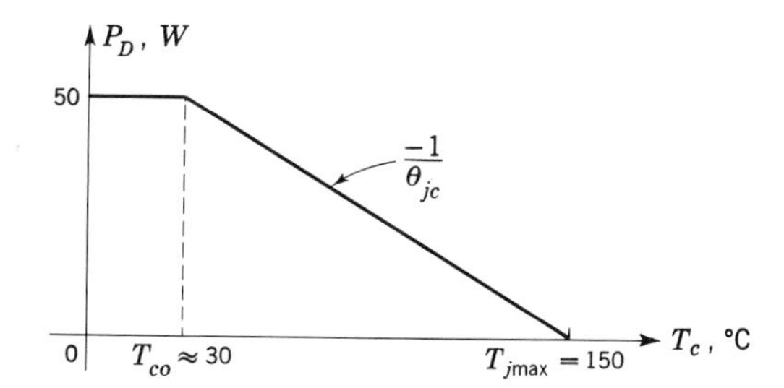

Fig. 2.10-5 Derating curve for 1N2816 Zener diode.

The characteristics are easily explained using Figs. 2.10-4 to 2.10-6. Figure 2.10-4 illustrates items 1, 2, and 4. Notice that an 18-V Zener diode may exhibit a "nominal" voltage anywhere from 17.1 to 18.9 V because of the 5 percent tolerance. The test current indicates the nominal operating region. Item 3, the maximum dissipation, gives the maximum permissible power which the diode can dissipate at room temperature. The dynamic impedance, item 5, is the slope of the reverse-biased diode characteristic measured at the test current I_{zT}. The maximum junction temperature and maximum dissipation are related by the power-derating curve shown in Fig. 2.10-5. θ_{jc} is the thermal resistance ($\approx$2.4°C/W for this diode), and is usually specified by the manufacturer. The temperature T_{co} is the maximum case temperature at which full rated power can be dissipated. The temperature coefficient, item 7, can be explained by reference to Fig. 2.10-6.

At normal operating currents and Zener voltages above about 6 V, an increase in temperature results in an increase in Zener voltage. Manufacturers specify an average or typical temperature coefficient, which is defined as follows:

$$\text{TC}(\%) = \frac{\Delta V_z/V_{zT} \times 100}{T_2 - T_1} \qquad \%/°\text{C}$$

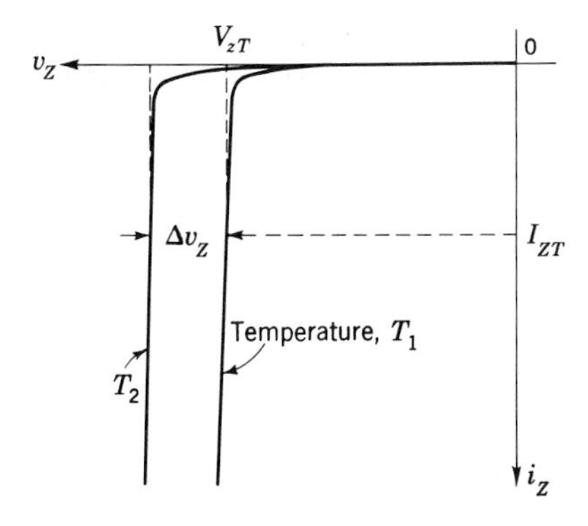

Fig. 2.10-6 Zener-diode characteristic illustrating the temperature coefficient.

For this 18-V, 50-W diode, a temperature rise of 50°C would result in a change in Zener voltage of

$$\frac{\Delta V_z}{V_{zT}} = (0.075\%/°\text{C})(50°\text{C}) = 3.75\%$$

Thus

$$\Delta V_z = (0.0375)(18) = 0.67 \text{ V}$$

and

$$V_{zT} \approx 18.7 \text{ V}$$

when the temperature rises 50°C.

For Zener voltages below about 4 V the temperature coefficient is usually negative. Between about 4 to 6 V, it is possible to obtain a point of zero temperature coefficient.

PROBLEMS

SECTION 2.1

2.1. (*a*) Sketch v_L as a function of time t (in milliseconds).
(*b*) Repeat (*a*) if $v_i(t)$ is sinusoidal, triangular (1-V peak).

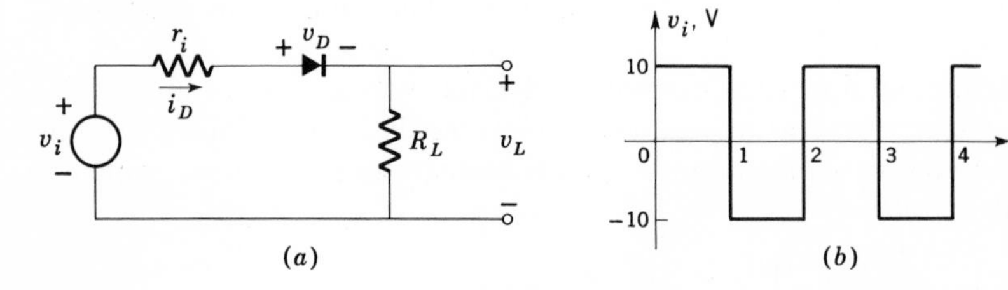

Fig. P2.1

2.2. (*a*) Sketch v_L as a function of time (in milliseconds).
(*b*) Repeat (*a*) if $v_i(t)$ is sinusoidal, triangular (1-V peak).

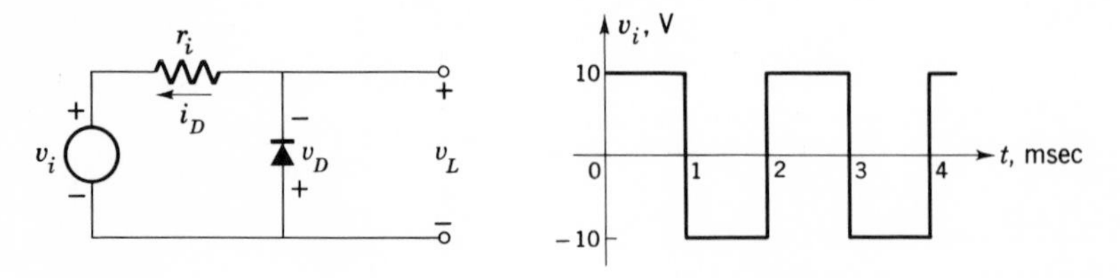

Fig. P2.2

2.3. (*a*) Sketch $v_L(t)$. (Assume $R_b \gg r_i$.)
(*b*) Repeat (*a*) if $v_i(t)$ is sinusoidal, triangular (1-V peak).

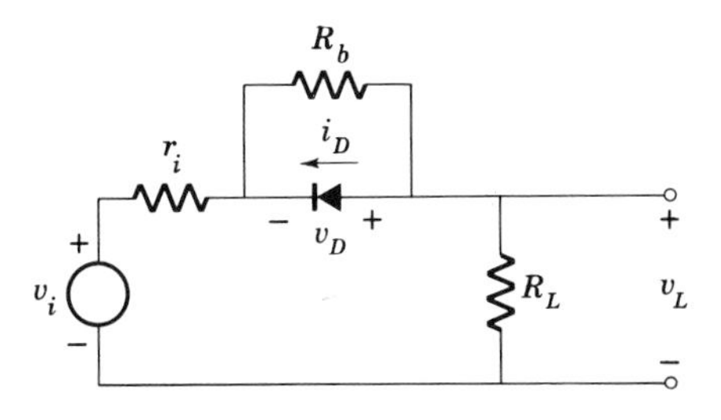

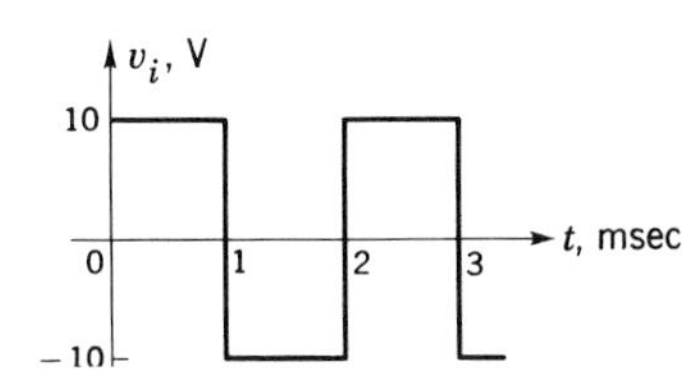

Fig. P2.3

2.4. The signal $v_1(t) = A_0 + A_1 \cos(\omega_0 t + \theta) + A_2 \cos(2\omega_0 t + \phi)$ is passed through a filter with a transfer function

$$H(j\omega) = \frac{1}{1 + j\sqrt{2}\,(\omega/\omega_0) - (\omega/\omega_0)^2}$$

(This is called a second-order *Butterworth* low-pass filter.) Calculate the output voltage $v_2(t)$.

2.5. A periodic function $f(t)$, having a period T, can be represented by a Fourier series

$$f(t) = A_0 + \sum_{n=1}^{\infty} A_n \left(\cos \frac{2\pi nt}{T}\right) + \sum_{n=1}^{\infty} B_n \left(\sin \frac{2\pi nt}{T}\right)$$

where
$$A_0 = \left(\frac{1}{T}\right) \int_{-T/2}^{T/2} f(t)\,dt$$
$$A_n = \left(\frac{2}{T}\right) \int_{-T/2}^{T/2} f(t) \left(\cos \frac{2\pi nt}{T}\right) dt$$
$$B_n = \left(\frac{2}{T}\right) \int_{-T/2}^{T/2} f(t) \left(\sin \frac{2\pi nt}{T}\right) dt$$

Using these equations, derive (2.1-2).

2.6. The output voltage given by (2.1-2) is passed through the LC filter shown. Letting $LC = 10^4/\omega_0^2$ and $RC = 100\sqrt{2}/\omega_0$:
(*a*) Find $v_o(t)$. Compare with the result obtained using the RC filter.
(*b*) Calculate the rms ripple voltage.
(*c*) Calculate the ratio of the rms ripple voltage to the dc voltage.

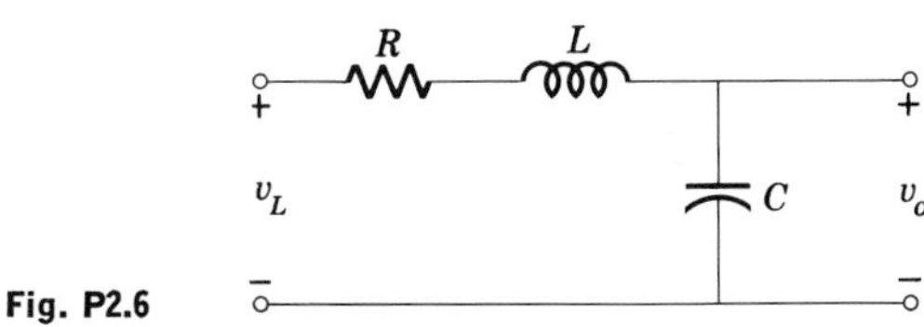

Fig. P2.6

2.7. Repeat Prob. 2.6 using the pi filter shown in Fig. P2.7.

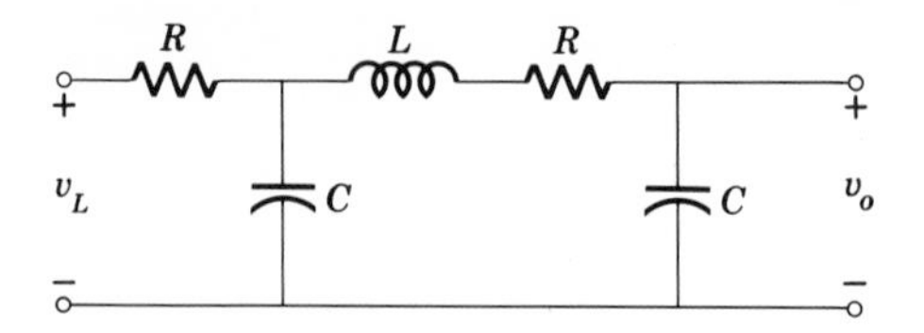

Fig. P2.7

2.8. The circuit shown in Fig. P2.8 is a *clamping* circuit. Find $v_L(t)$ when $v_i(t) = A \cos \omega_0 t$.

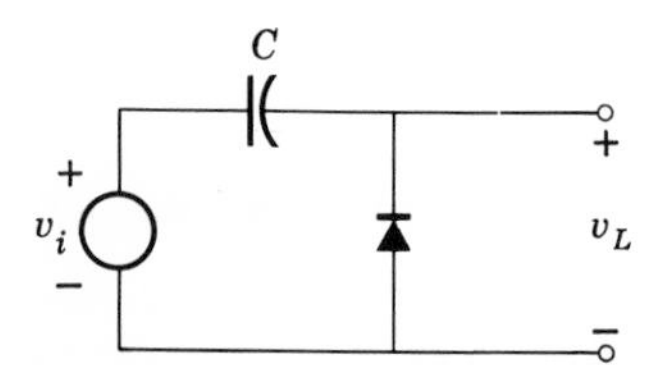

Fig. P2.8

2.9. The circuit shown in Fig. P2.9 is a more practical version of the clamping circuit of Fig. P2.8, since it includes the diode resistances. Sketch $v_L(t)$ when $v_i(t)$ is the square-wave shown. Assume that $R_b \gg r_i$ and that $r_iC \ll T$, while $R_bC \gg T$.

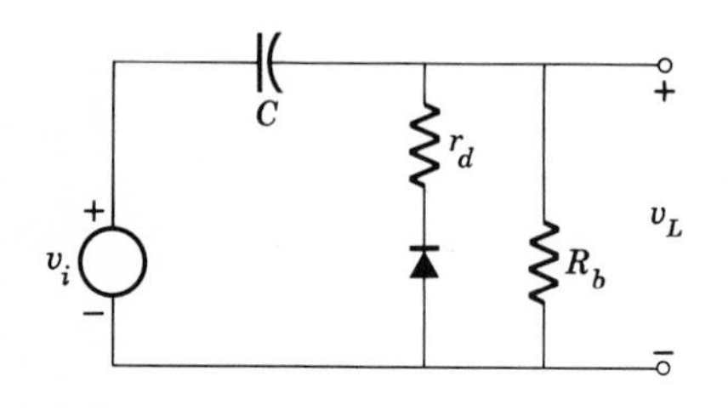

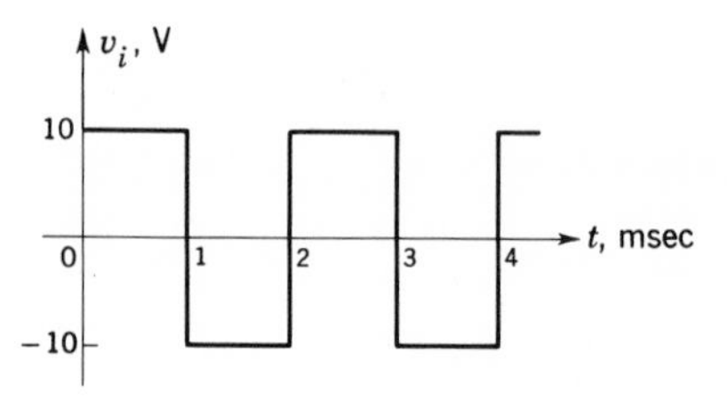

Fig. P2.9

2.10. The circuit shown in Fig. P2.10 is a diode full-wave rectifier bridge. Sketch $v_L(t)$ when $v_i(t)$ is sinusoidal.

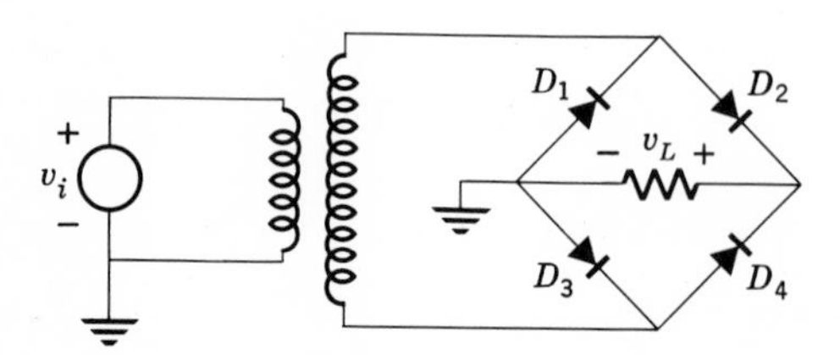

Fig. P2.10

2.11. The full-wave rectifier circuit of Fig. 2.1-8 is complicated because of the presence of resistors in series with D_1 and D_2, as shown in Fig. P2.11. Sketch $v_L(t)$ when v_i is sinusoidal with frequency ω_0.

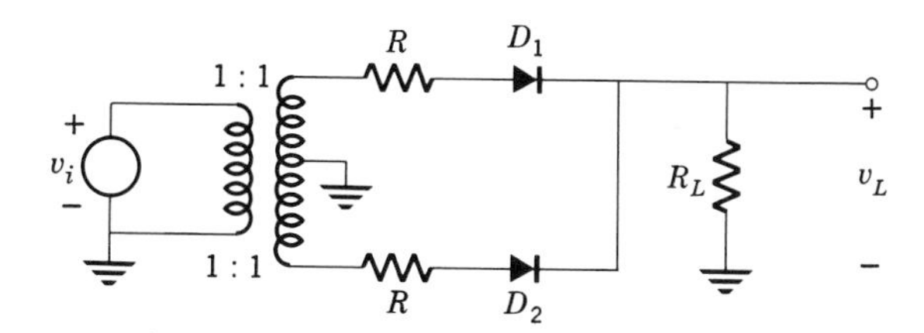

Fig. P2.11

2.12. The voltage v_L of P2.11 is filtered, using the Butterworth filter shown in Fig. P2.6.

(*a*) Calculate and sketch $v_o(t)$.

(*b*) Calculate the rms ripple.

(*c*) Calculate the percent regulation, and plot as a function of R/R_L.

2.13. A tuned circuit is connected to the circuit of Fig. P2.11, as shown in Fig. P2.13. For the tuned circuit, $1/\sqrt{LC} = 2\omega_0$, and the circuit $Q = R/2\omega_0 L = 10$. Neglecting loading, find $v_o(t)$.

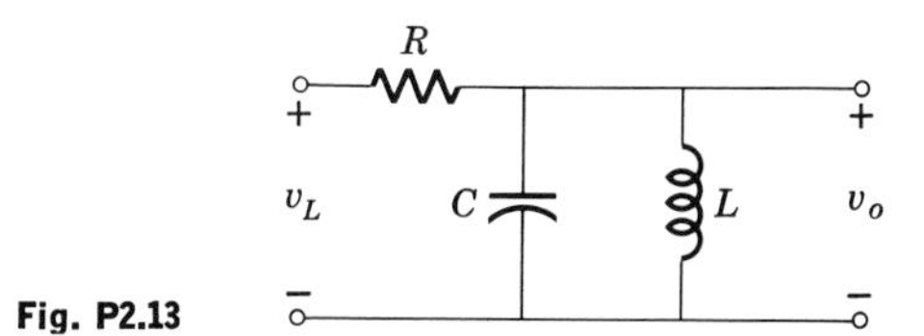

Fig. P2.13

SECTION 2.2

2.14. Sketch i_D/I_o as given by (2.2-1), as a function of the diode voltage v_D, using semilog paper, for $m = 1$, 1.4, and 2. Assume $T = 300°K$ (room temperature).

2.15. Repeat Prob. 2.14 assuming $m = 1$ and $T = 500$ and 200°K. Discuss these results.

2.16. A silicon diode has a reverse saturation current $I_o = 10^{-3}\ \mu A$. Sketch i_D versus v_D. Do not use semilog paper. Assume $m = 1$.

A germanium diode having the same power-dissipation capability as the silicon diode has an $I_o = 100\ \mu A$. Sketch on the same axes i_D versus v_D for the germanium diode.

(*a*) Determine the turn-on voltages.

(*b*) Find approximate straight-line (piecewise linear) representations for the two diodes. Use two straight lines in the forward region to obtain results similar to those shown in Fig. 2.2-6.

2.17. Calculate the $\Delta i_D / i_{D\text{initial}}$ that results from changing the diode voltage by 0.1 V. Assume $T = 300°K$, and repeat the calculation for $m = 1$, 1.4, and 2.

SECTION 2.3

2.18. Sketch i_D when v_T is a square wave having an average value of zero and a peak-to-peak voltage of 2 V. Obtain the answer analytically and also graphically.

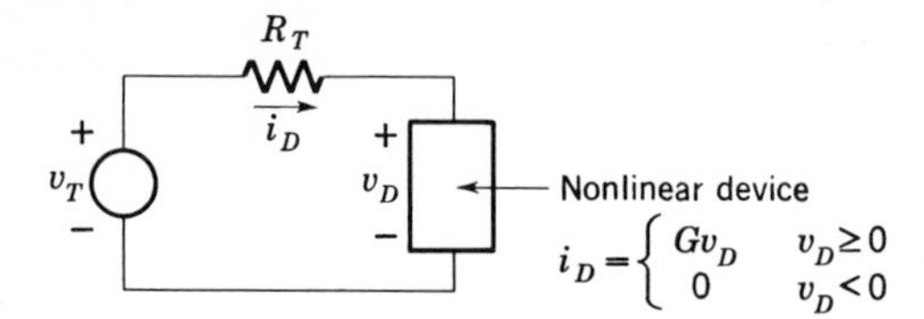

Fig. P2.18

2.19. Repeat Prob. 2.18 for the case when v_T is sinusoidal with a peak value of 1 V (use the graphical technique).

2.20. (*a*) Redraw Fig. P2.20*a* and obtain the Thévenin circuit (v_T and R_T).

(*b*) Let $R = 1\ \text{k}\Omega$, $r_i = 1.5\ \text{k}\Omega$, $R_L = 1.4\ \text{k}\Omega$, $V_{dc} = 5$ V, and $v_i(t) = 10 \sin \omega_0 t$. Plot the dc load line when $\omega_0 t = 0$, $\pi/3$, $\pi/2$, $-\pi/3$, and $-\pi/2$.

(*c*) Sketch $v_L(t)$. *Hint:* $v_L = R_L i_D$.

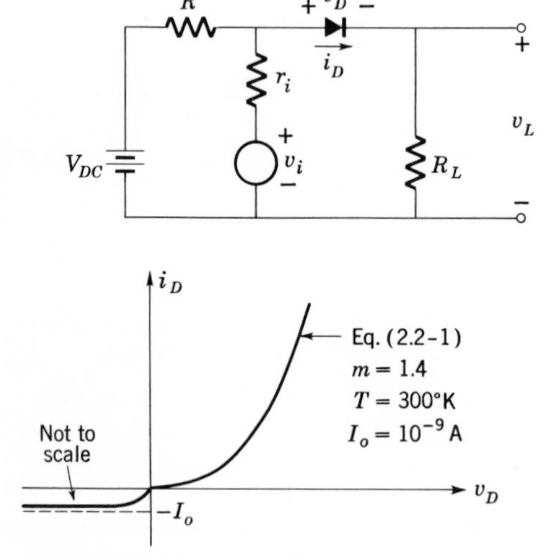

Fig. P2.20

2.21. Repeat Prob. 2.20 with $v_i = 0.1 \sin \omega t$. *Hint:* Expand the diode characteristic given by Fig. P2.20*b* about the quiescent current and voltage obtained when $v_i = 0$. Calculate the diode resistance.

2.22. Calculate $v_L(t)$. Use the characteristic of Fig. P2.20*b*.

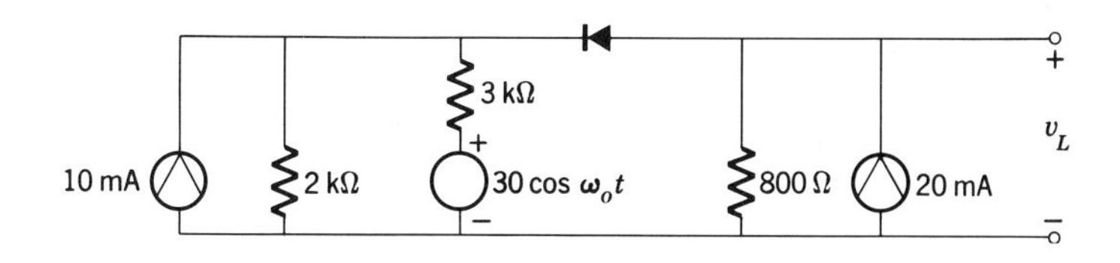

Fig. P2.22

2.23. Diodes D_1 and D_2 are identical. Find v_L.

$$i_{D1} = \begin{cases} 2 \times 10^{-3} v_{D1}^2 & v_{D1} \geq 0 \\ -I_o & v_{D1} < 0 \end{cases}$$

$$i_{D1} = \begin{cases} 2 \times 10^{-3} v_{D1}^2, & v_{D1} \geq 0 \\ -I_o, & v_{D1} < 0 \end{cases}$$

1 kΩ D_1 10 V D_2 + v_L −

Fig. P2.23

2.24. The circuit shown is a "limiter," since it limits the excursions of $v_i(t)$. The diodes D_1 and D_2 are identical, and have the characteristic given in Prob. 2.23 (assume $I_o = 0$).

(*a*) Find $v_L(t)$.

(*b*) Repeat for $v_i(t) = 10 \cos \omega_0 t$.

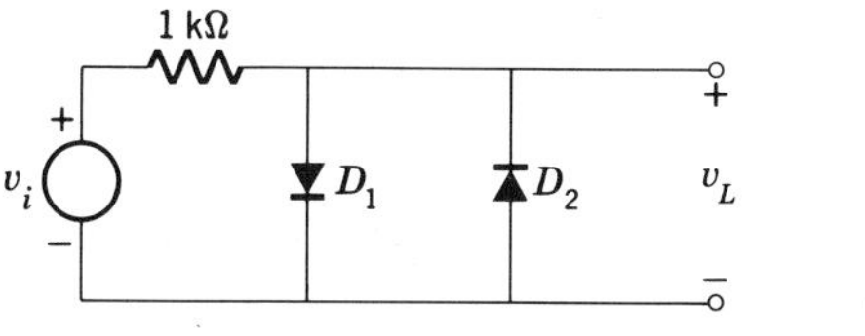

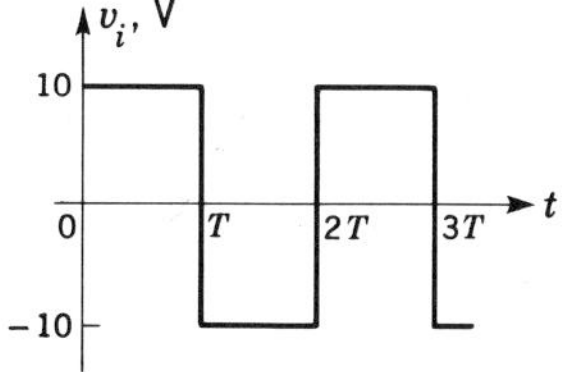

Fig. P2.24

2.25. Repeat Prob. 2.24 using the diode characteristic of Fig. P2.20*b*. *Hint:* $v_{D1} = -v_{D2}$. Redraw the diode characteristics to emphasize this.

2.26. Calculate i_D. Assume D_1 and D_2 are identical and have *vi* characteristics as given by Fig. P2.20*b*. Calculate the voltage drop across each diode.

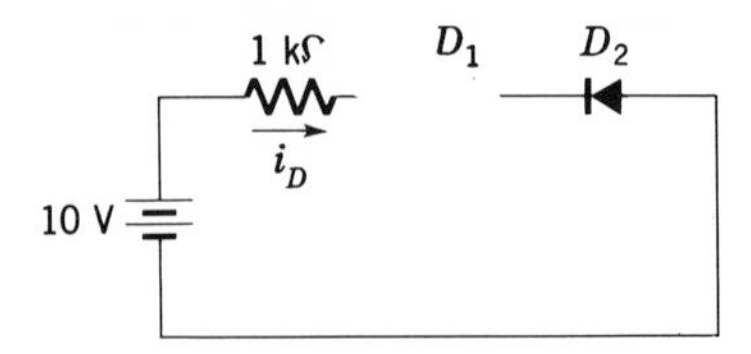

Fig. P2.26

SECTION 2.4

2.27. The diode *vi* characteristic is given by

$$i_D = 10^{-6}(\epsilon^{qv_D/1.4kT} - 1)$$

where $T = 300°K$.

(*a*) Obtain the Thévenin circuit, R_T, and v_T.
(*b*) Determine the quiescent current in the diode.
(*c*) Calculate r_d.
(*d*) Calculate $v_L(t)$.

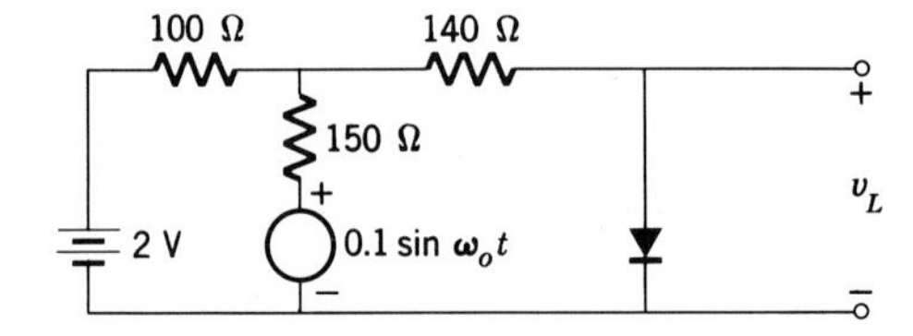

Fig. P2.27

2.28. *Mixing circuit.* (*a*) If the diode is *ideal*, sketch $v_L(t)$.

(*b*) Find the Fourier series for $v_L(t)$.

(*c*) If $v_L(t)$ is passed through a filter which passes frequency components less than ω_0 without attenuation, and completely attenuates frequency components at and above ω_0, find the output voltage of this filter.

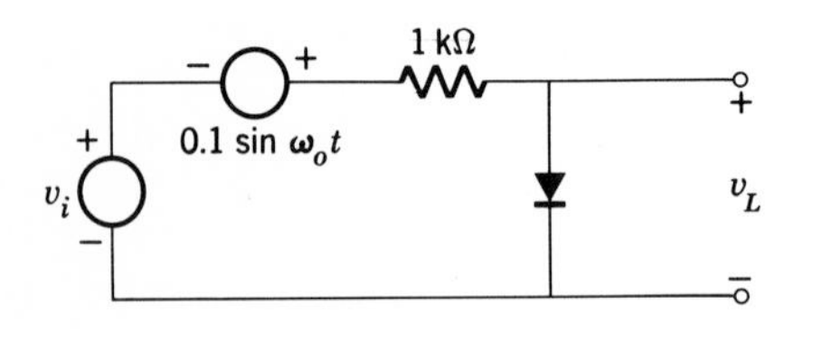

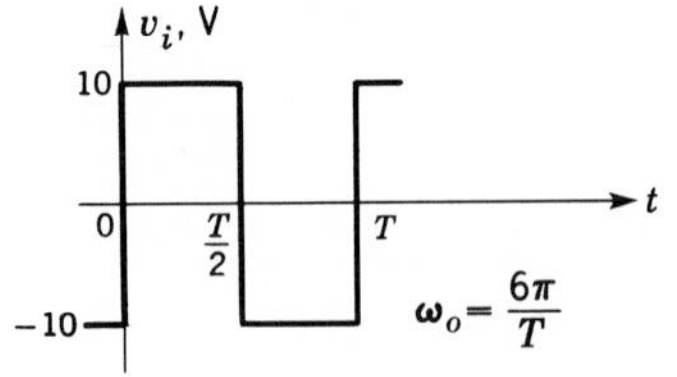

Fig. P2.28

2.29. Repeat Prob. 2.28 using the diode characteristic of Prob. 2.27 to describe the diode.

2.30. Calculate i_D. The diode is characterized by the vi characteristic of Prob. 2.20.

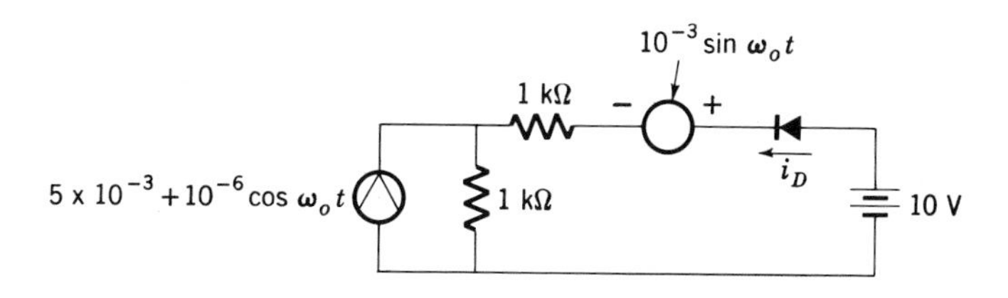

Fig. P2.30

2.31. (*a*) Find the Thévenin equivalent circuit, R_T, and v_T.
(*b*) Sketch $i_D(t)$.
(*c*) Find the average current through the diode.

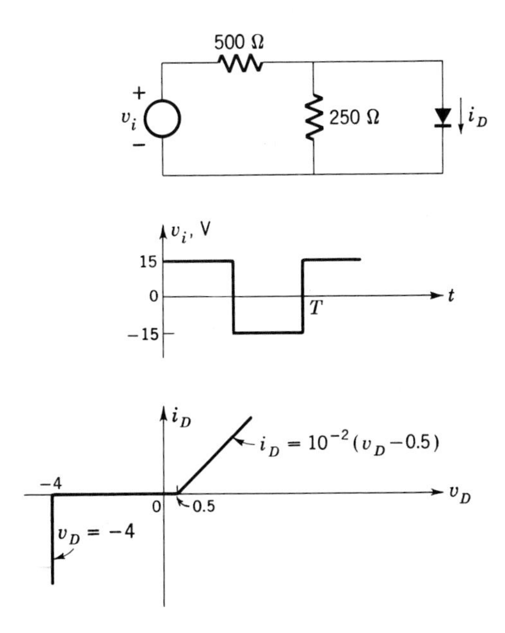

Fig. P2.31

2.32. The diode vi characteristic is given by the equation

$$i_D = I_o\,(\epsilon^{qv_D/mkT} - 1)$$

(*a*) Expand this equation in a Taylor series about the point I_D, V_D.

(*b*) Show that if $v_d \ll mkT/q$, then all terms beyond the first two in the expansion can be dropped (that is, $i_D \approx I_D + v_d/r_d$).

2.33. (*a*) Determine the quiescent diode current. The diode vi characteristic is given in Prob. 2.27.

(*b*) Calculate r_d.
(*c*) Find $v_L(t)$ when $\omega_0 = 10^6, 10^8, 10^{10}$ rad/s.

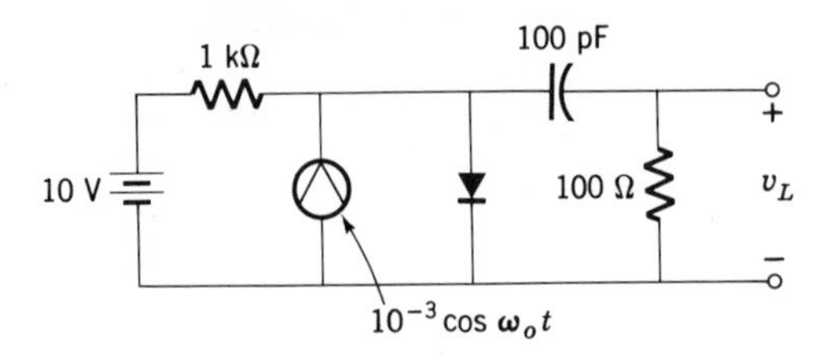

Fig. P2.33

SECTION 2.5

2.34. (*a*) Obtain the equation of the dc load line and plot on the *vi* characteristic. Obtain the quiescent current.
(*b*) Obtain the equation of the ac load line and plot it on the *vi* characteristic.
(*c*) Find and sketch $v_D(t)$ using the graphical technique described in Sec. 2.5.

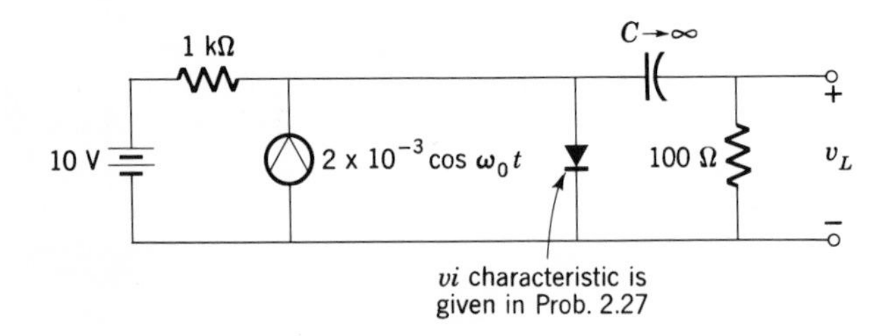

Fig. P2.34

2.35. Find an expression for the silicon diode *vi* characteristic. The diode has a voltage drop of 0.25 V when $i_D = 1$ mA. (*Hint:* One point on the characteristic is known.) Assume $m = 1$ and $T = 300°K$.

SECTION 2.6

2.36. Show that I_{D4} in Example 2.6-1 is 0.88 A.

2.37. Example 2.6-1 can be solved directly, without resorting to an iteration technique. This is accomplished by using Fig. 2.6-1*c*, (2.6-3*b*), and instead of (2.6-4), the fact that the average current through the capacitor must be zero in the steady state; i.e.,

$$I_C = 0$$

or

$$i_{L+} = -i_{L-}$$

where i_{L+} is the current in R_L when $V_{dc} + v_i > 0$ and i_{L-} is the current in R_L when $V_{dc} + v_i < 0$.

Using this result, find I_D.

2.38. The diode circuit of Fig. P2.38 has a signal voltage $v_i = 0.1 \cos \omega_0 t$.
(*a*) Find the quiescent diode voltage and diode current.
(*b*) Construct the ac load line.

(c) Determine the dynamic resistance of the diode.
(d) Calculate v_L.

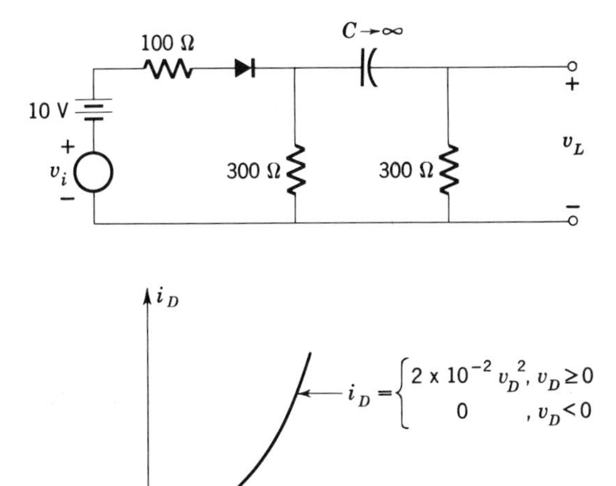

Fig. P2.38

2.39. (a) Obtain an equivalent vi characteristic for the two diodes in series. *Hint:* $i_{D1} = i_{D2} = i_D$ and $v_D = v_{D1} + v_{D2}$.

(b) Calculate v_1 and v_2.

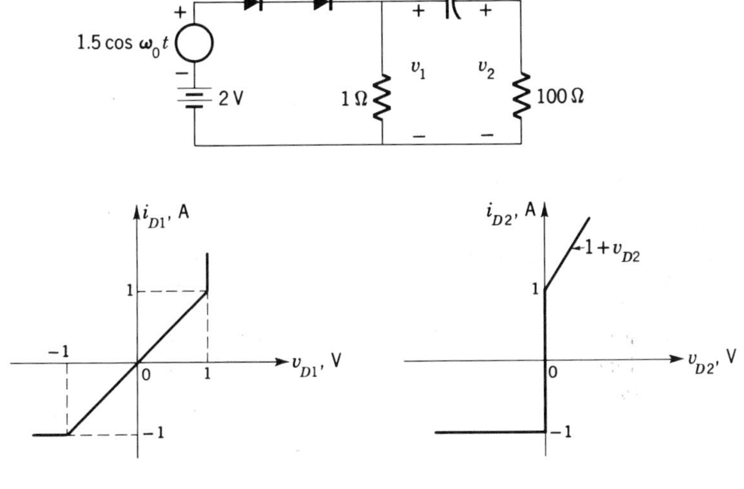

Fig. P2.39

2.40. Repeat Example 2.6-1 letting $v_i = 2 \sin \omega_0 t$.

2.41. Repeat Example 2.6-1 using a diode characteristic given by

$$i_D = v_D^2$$

2.42. Solve Prob. 2.40 by using a digital computer to perform the iteration process.

SECTION 2.7

2.43. The Zener diode has a fixed voltage drop of 18 V across it as long as i_z is maintained between 200 mA and 2 A.

(a) Find r_i so that V_L remains at 18 V while V_{dc} is free to vary from 22 to 28 V.
(b) Find the maximum power dissipated by the diode.

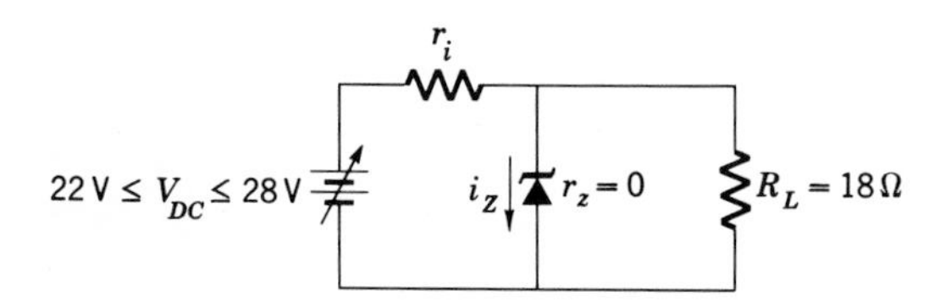

Fig. P2.43

2.44. A 10-V Zener diode is used to regulate the voltage across a variable load resistor. The input voltage v_i varies between 13 and 16 V. The load current i_L varies between 10 and 85 mA. The minimum Zener current is 15 mA.

(a) Calculate the maximum value of r_i.
(b) Calculate the maximum power dissipated by the Zener diode using this value of r_i.

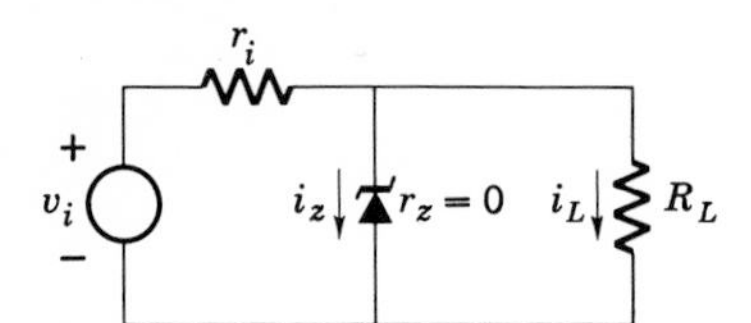

Fig. P2.44

2.45. An unregulated supply varies between 20 and 25 V and has a 10-Ω internal impedance. A 10-V Zener diode is to regulate this voltage for use in a tape recorder. The recorder draws 30 mA while recording, and 50 mA when playing back. The Zener diode has a resistance of 10 Ω when the Zener current is 30 mA. The "knee" of the Zener characteristic occurs at 10 mA. In addition, the Zener diode can dissipate a maximum power of 800 mW.

(a) Find r_i so that the diode regulates continuously.
(b) Find the maximum peak-to-peak output ripple.

2.46. The Zener characteristic is usually due to an avalanche breakdown. This is described approximately by the equation

$$i_Z = \frac{I_o}{(1 - v_z/V_o)^n} \qquad v_z \leq V_o \qquad \frac{i_z}{I_0} \leq 100$$

(*a*) Sketch the $v_z i_z$ characteristic for $n = 4$ and $n = 10$. *Hint:* Plot i_z/I_o versus v_z/V_o to obtain a normalized characteristic.

(*b*) Obtain the "knee" of the curve.

(*c*) Sketch a piecewise linear approximation to the curve using two straight lines.

(*d*) The slope of the straight-line curve below the knee is called the Zener resistance r_z. Calculate r_z. Note that r_z continually changes.

2.47. Find v_L (dc and ripple). Assume D_1 and D_2 are ideal diodes.

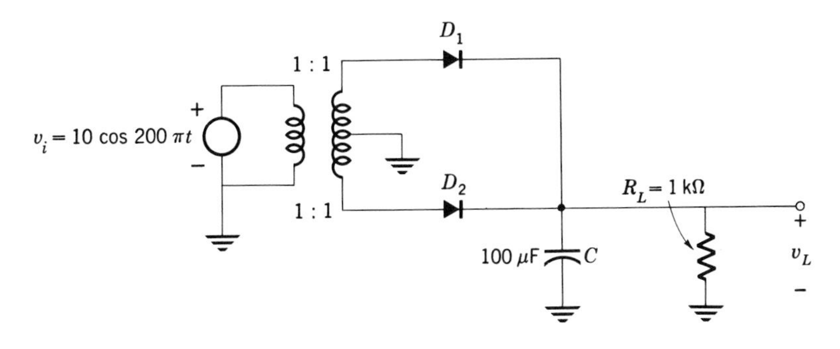

Fig. P2.47

SECTION 2.8

2.48. Obtain a two-segment piecewise linear approximation to the equation

$$y = t^2 \qquad 0 \leq t \leq 3$$

such that the absolute value of the maximum difference between the function y and its approximation $\hat{y}$, at any time t, is always less than 0.5.

2.49. Using one resistor (1 kΩ), one battery (1V), and one ideal diode, synthesize circuits to realize each of the four basic vi characteristics shown in Fig. P2.49.

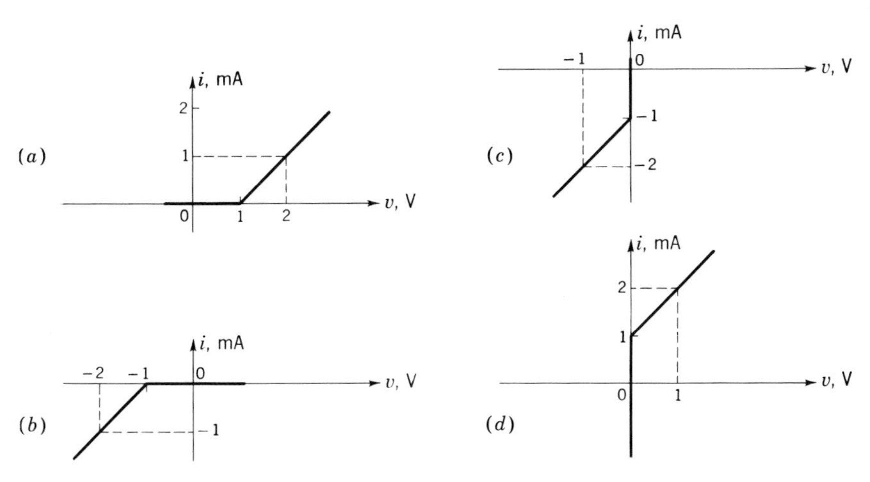

Fig. P2.49

2.50. Obtain a three-segment piecewise linear approximation to the function $y = \ln x$ for the range $0 < y < 3$ that will yield a maximum error $|\hat{y} - y|_{max}$ which is less than that given in Fig. 2.8-3. Use a trial-and-error procedure. How small can you make $|\hat{y} - y|_{max}$?

2.51. Scale the circuit of Example 2.8-1 so that the currents are in μA and the voltages are in volts.

2.52. Design a circuit using resistors and ideal diodes, to synthesize the vi characteristic shown in Fig. P2.52.

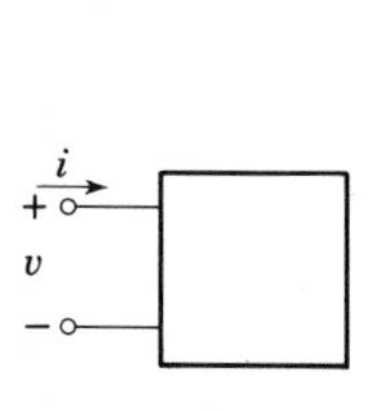

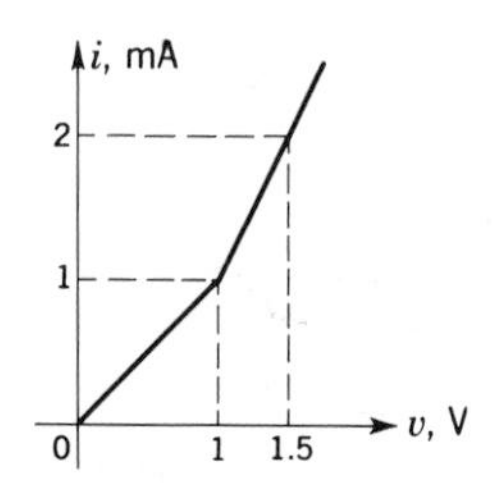

Fig. P2.52

2.53. Design a circuit using resistors and ideal diodes to synthesize the vi characteristic in Fig. P2.53.

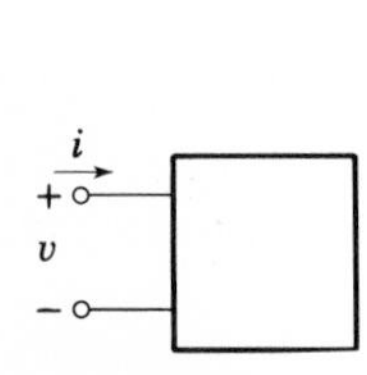

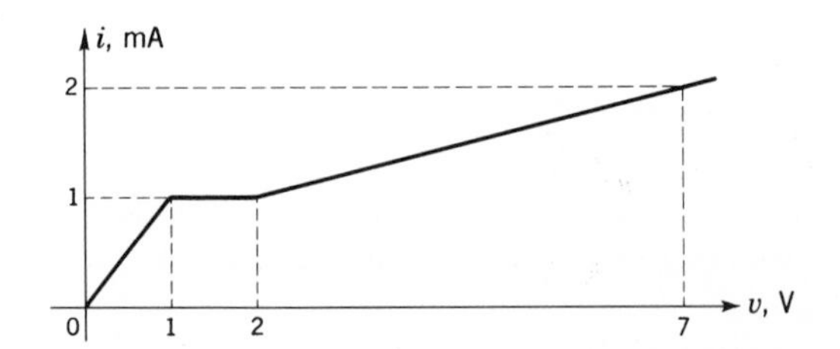

Fig. P2.53

2.54. Design a circuit to solve the nonlinear differential equation

$$\frac{dy}{dt} + \sin y = 0.8 \qquad y(0) = 0 \qquad t \geq 0 \qquad -\frac{\pi}{2} \leq y \leq \frac{\pi}{2}$$

Note: All currents should be in mA, and voltages in volts.

2.55. Solve Example 2.8-3 when $x(0) = 1$.

SECTION 2.9

2.56. Obtain an electrical analog of the thermal system of Fig. 2.9-2a in which temperature is analogous to current. What are P_j and θ_{jc} analogous to?

2.57. A silicon Zener diode is to dissipate 15 W in a particular circuit. The maximum allowable junction temperature is 200°C, the ambient temperature is 25°C, and $\theta_{jc} = 2.4$°C/W.

(a) If an infinite heat sink is provided, find the minimum allowable diode power rating.

(b) What is the junction temperature of the diode?

2.58. Repeat Prob. 2.57 if the heat sink has a thermal resistance θ_{ca} of 3°C/W.

2.59. A diode can dissipate 20 W at temperatures less than 50°C. The maximum junction temperature is 200°C. Find θ_{jc}.

SECTION 2.10

2.60. The reverse saturation current of the diode is 1 μA. Its peak inverse voltage is 500 V. Find r_i so that the peak inverse voltage (PIV) is not exceeded.

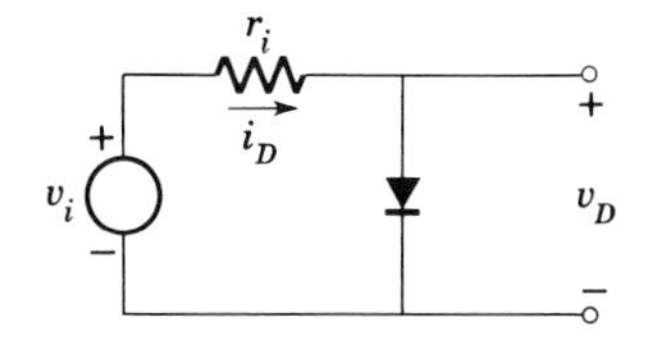

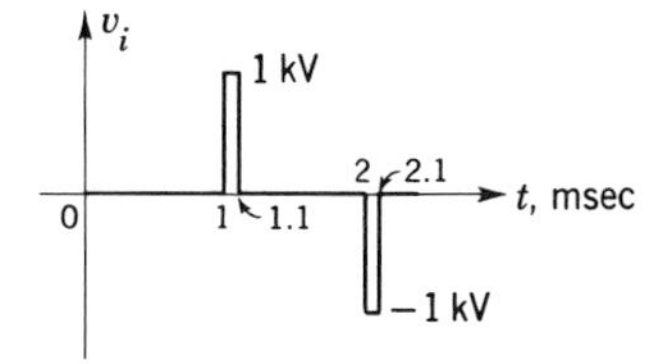

Fig. P2.60

2.61. The maximum average half-wave-rectified forward current through the diode is 1 A.

(*a*) Find R_L so that this value is not exceeded.

(*b*) If $I_o = 1\ \mu$A, find the minimum required PIV to prevent diode breakdown.

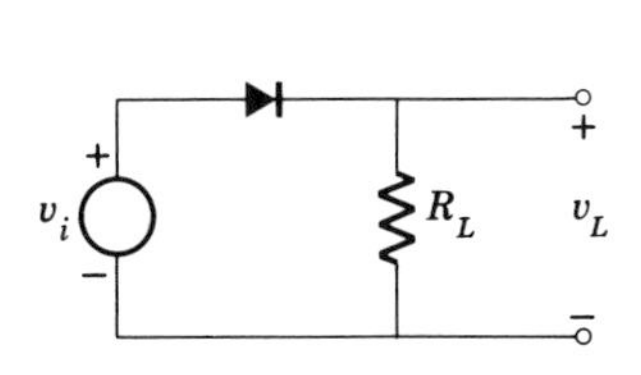

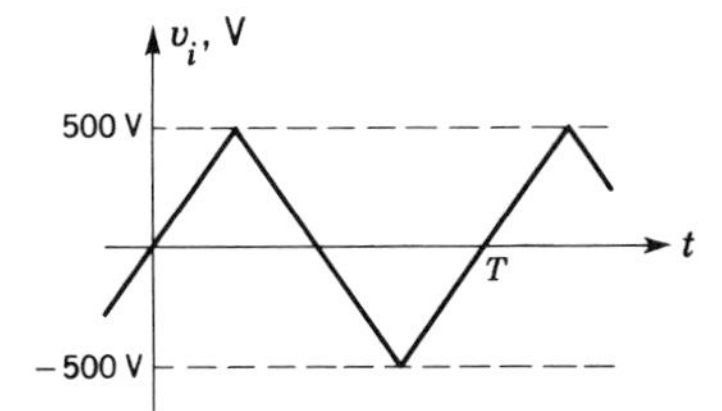

Fig. P2.61

2.62. A Zener diode, at a test current of 100 mA, has a nominal voltage of 20 V. It has a 2 percent tolerance. Calculate the range of operating test voltages.

2.63. The 1N2816 Zener diode is to be used to maintain a fairly constant dc voltage across a load which varies from 10 to 100 Ω. The input voltage varies from 80 to 100 V.

(*a*) Calculate r_i assuming $V_{zT} = 18$ V and $R_{zT} \approx 2\ \Omega$.

(*b*) Calculate v_L taking into account the 5 percent tolerance and the fact that the circuit operates over the temperature range 0 to 25°C.

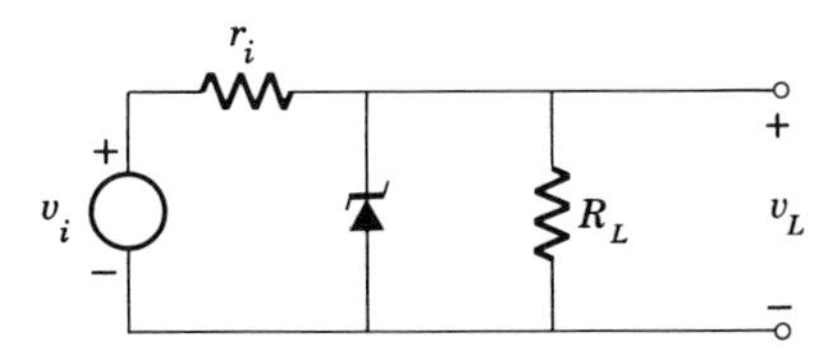

Fig. P2.63

REFERENCES

1. "Silicon Rectifier Handbook" and "Silicon Zener Diode and Rectifier Handbook," Motorola Inc., Semiconductor Products Division, Phoenix, Ariz.
2. Grove, A. S.: "Semiconductor Physics," John Wiley & Sons, Inc., New York, 1967.
3. Motorola Inc., Engineering Staff: "Integrated Circuits," McGraw-Hill Book Company, New York, 1965.
4. Middlebrook, R. D.: "Introduction to Junction Transistor Theory," John Wiley & Sons, Inc., New York, 1957.

3
Introduction to Transistor Circuits

INTRODUCTION

In the preceding chapter the *pn* junction diode and some of its physical and circuit properties were considered. In this chapter, attention is focused on the device which has caused a veritable revolution in the electronics field since the early 1950s. This device is, of course, the transistor. As we shall see, the basic junction transistor consists essentially of two *pn* junctions placed back to back, so that much of the theory of Chap. 2 applies, with minor modifications.

Conceptually, the transistor is a device that acts as a current amplifier, and in this and succeeding chapters its properties, uses, and limitations are studied.

3.1 CURRENT-FLOW MECHANISM IN THE JUNCTION TRANSISTOR[1–3]

Here, the basic mechanism of current flow, as viewed from the terminals of the device, is considered. The junction transistor consists of two

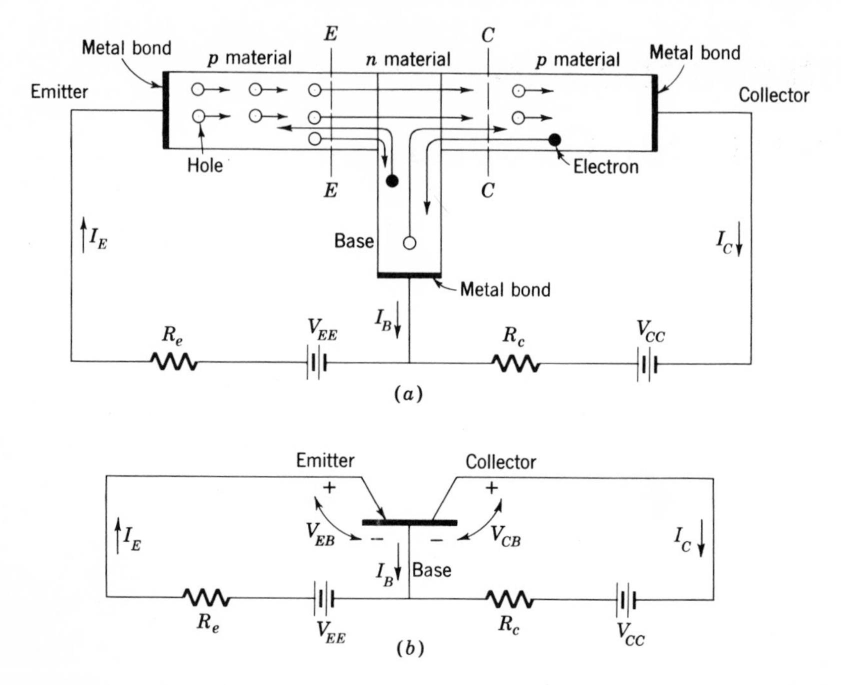

Fig. 3.1-1 The *pnp* junction transistor. (*a*) Pictorial representation; (*b*) circuit symbol and reference directions for the common-base connection.

pn junctions placed back to back, as shown pictorially in Fig. 3.1-1*a*. The circuit symbol used to represent the *pnp* transistor is shown in Fig. 3.1-1*b*. Note that the transistor shown has its semiconductor materials arranged *p-n-p*; hence the name *pnp* transistor. The alternative arrangement, *npn*, is discussed in Sec. 3.2.

Physically, the transistor consists of three parts—emitter, base, and collector, the base region being very thin. Its operation can be explained qualitatively as follows.

The battery V_{EE} forward-biases the emitter-base *pn* junction, causing the emitter to inject holes into the *n* material. Most of the holes travel across the narrow base region, through the second junction, into the right-hand negatively biased *p* region (the collector). A small amount of these holes (approximately 1 percent) is caught in the *n* material and is collected by the base. Electrons in the base material flow into the emitter as shown.

While the emitter-base junction represents a forward-biased diode with its characteristic properties of low impedance and low voltage drop, the collector-base junction is reverse-biased because of the polarity of V_{CC}. This is essentially a back-biased diode, and the collector-base impedance is thus very high.

The current measured in the emitter circuit (the flow of charge across boundary EE per unit time) is called the emitter current I_E, and is positive into the material.* The current measured in the collector circuit (the flow of charge across boundary CC per unit time) is called the collector current I_C. This current consists of two terms, the predominant term representing the percentage of emitter current reaching the collector. The percentage depends almost solely on the construction of the transistor (the size and shape of the material and the doping of the emitter), and can be considered a constant for a particular transistor. The constant of proportionality is defined as α,† so that the major part of the collector current is αI_E. Typical values of α range from 0.90 to 0.99.

The second term represents the current flow through the reverse-biased collector-base junction when $I_E = 0$. This current is called I_{CBO} (it was called I_o in the diode), and, as expected, it is relatively quite small. Since both currents flow out of the collector, this direction is defined as the positive collector-current direction, and

$$I_C = \alpha I_E + I_{CBO} \tag{3.1-1}$$

Applying Kirchhoff's current law (KCL) to the transistor of Fig. 3.1-1, and noting the indicated current directions,

$$I_E = I_B + I_C \tag{3.1-2}$$

Substituting (3.1-1) into (3.1-2), the base current is

$$I_B = (1 - \alpha)I_E - I_{CBO} = \left(\frac{1 - \alpha}{\alpha}\right) I_C - \frac{I_{CBO}}{\alpha} \tag{3.1-3}$$

The symbol β‡ is used to represent the ratio $\alpha/(1 - \alpha)$, which arises continually in the study of transistors. Equations (3.1-1) to (3.1-3) describe the transistor in terms of its terminal currents. In the remainder of this section we show that, in normal operation, the transistor can be considered to be two isolated *pn* junctions, one forward-biased and one reverse-biased. The analysis of transistor circuits can then be carried out using the techniques developed for *pn* junction diodes in Chap. 2.

3.1-1 THE EMITTER-BASE JUNCTION

Applying Kirchhoff's voltage law (KVL) to the emitter-base loop in Fig. 3.1-1, the emitter current is

$$I_E = \frac{V_{EE} - V_{EB}}{R_e} \tag{3.1-4}$$

* Reference directions for the transistor currents are often taken as positive flowing *into* the base, emitter, and collector. In this text, the convention to be used is that currents are positive in the direction in which an ammeter would indicate positive current when the transistor is biased for ordinary linear operation.

† The symbol h_{FB} is often used instead of α.

‡ The symbol h_{FE} is often used instead of β.

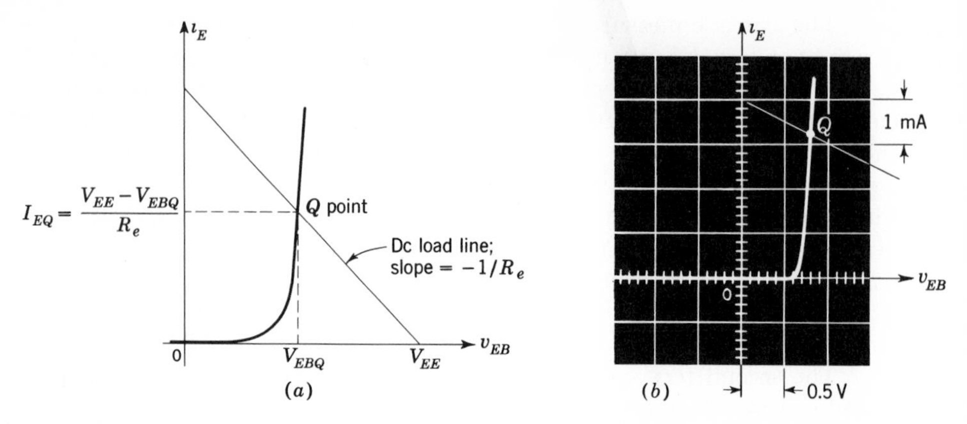

Fig. 3.1-2 Emitter-base *vi* characteristic. (*a*) Emitter-base *vi* characteristic; (*b*) oscillogram of TI 179 6508 silicon transistor emitter-base *vi* characteristic.

where V_{EB} is the voltage across the forward-biased emitter-base junction. The *vi* characteristic of the emitter-base junction is shown in Fig. 3.1-2*a* and *b*. The reader will observe that this characteristic should also depend on the state of the collector junction and the operating temperature. However, in normal operation, the collector-voltage dependence is negligible, and is omitted. Temperature effects are considered in Chap. 4. With this simplification, the analysis of the emitter-base circuit is exactly the same as the analysis of the diode circuits discussed in Secs. 2.3 and 2.4. In particular, the dc load line can be drawn as shown. As with the diode, a voltage "threshold," or "break," exists, where V_{EBQ} is approximately 0.2 V for germanium and 0.7 V for silicon. The transistor with the *vi* characteristic shown in Fig. 3.1-2*b* is operating at $V_{EBQ} = 0.8$ V, when $V_{EE} = 4$ V and $R_e = 1$ kΩ. Notice that V_{EB} varies between 0.7 and 0.9 V, depending on the quiescent emitter (or base) current.

Drawing on our experience with diodes, the emitter-base circuit can be linearized by replacing the emitter-base diode by a piecewise linear equivalent. The corresponding model and *vi* characteristic are shown in Fig. 3.1-3. It must be remembered that when operating near the break voltage, the exact curve of Fig. 3.1-2 should be used.

The resistance r_d represents the slope of the characteristic at the Q point; thus, since the diode equation (2.4-15) applies,

$$r_d = \frac{mV_T}{I_{EQ}} \approx \frac{m(25 \times 10^{-3})}{I_{EQ}} \quad \Omega \text{ (at room temperature)} \tag{3.1-5}$$

where $V_T = kT/q$. This resistance is usually relatively small, and consequently the impedance looking into the emitter-base circuit will be small.

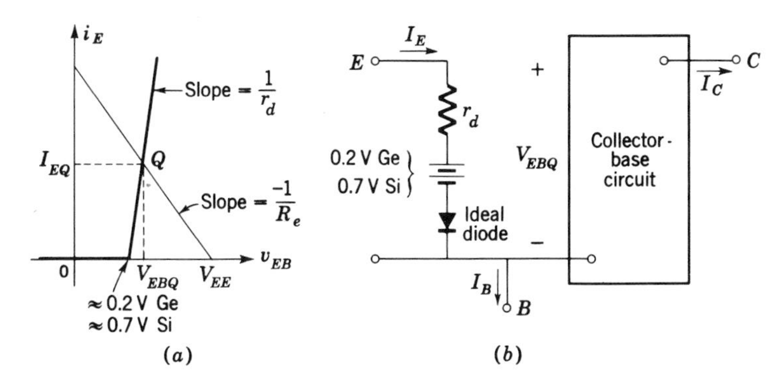

Fig. 3.1-3 Piecewise-linear-transistor input circuit. (*a*) *vi* character istic; (*b*) piecewise linear equivalent.

In Chaps. 3 to 5, which are concerned with "large-signal behavior," this small "input" impedance is neglected, that is, $r_d \approx 0$. The discussion is also restricted to silicon transistors, so that, for all cases, $V_{EBQ} \approx$ 0.7 V. These simplifications permit us to focus more sharply on the central problems considered in these chapters. Later on, when facility has been gained in handling this simplified model, additional effects are considered.

To illustrate the use of the model, assume that an oscillator is inserted in series with V_{EE} as shown in Fig. 3.1-4*a*. The ac signal is

$$v_i = V_{im} \cos \omega t \tag{3.1-6}$$

and from Fig. 3.1-4*b* we must have

$$V_{EE} - V_{im} > V_{EBQ} = 0.7 \text{ V} \tag{3.1-7}$$

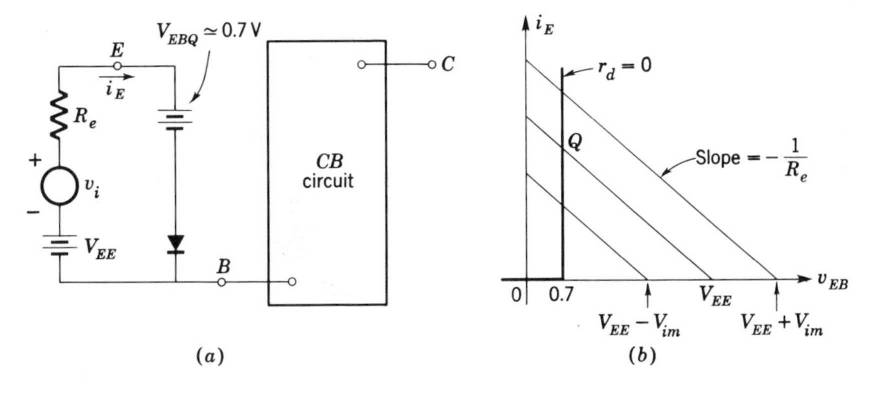

Fig. 3.1-4 Common-base circuit with signal applied. (*a*) Equivalent circuit; (*b*) graphical representation.

so that the junction is always forward-biased and operating past the break. This places an upper bound on the peak signal V_{im} for linear operation, for any given V_{EE}. Thus

$$i_E = \frac{V_{EE} + V_{im}\cos\omega t - V_{EBQ}}{R_e} \tag{3.1-8}$$

and noting that $i_E = I_{EQ} + i_e$,

$$I_{EQ} = \frac{V_{EE} - V_{EBQ}}{R_e} \tag{3.1-9}$$

and

$$i_e = \left(\frac{V_{im}}{R_e}\right)\cos\omega t \tag{3.1-10}$$

The assumptions inherent in (3.1-8) are that the *vi* characteristic of the junction, as shown in Fig. 3.1-4*b*, can be considered a vertical straight line ($r_d \ll R_e$) and that inequality (3.1-7) is satisfied.

3.1-2 THE COLLECTOR-BASE JUNCTION

In order to complete the model of Fig. 3.1-4*a*, an equivalent-circuit model is needed for the collector-base junction. Perhaps the easiest way to find this model is to consider the common-base output *vi* characteristics, as shown in Fig. 3.1-5.

For values of $V_{CB} < 0$ these curves may be considered as a family of straight lines which obey the relation

$$I_C = \alpha I_E + I_{CBO} \tag{3.1-1}$$

This leads to the equivalent circuit of Fig. 3.1-6*a*.

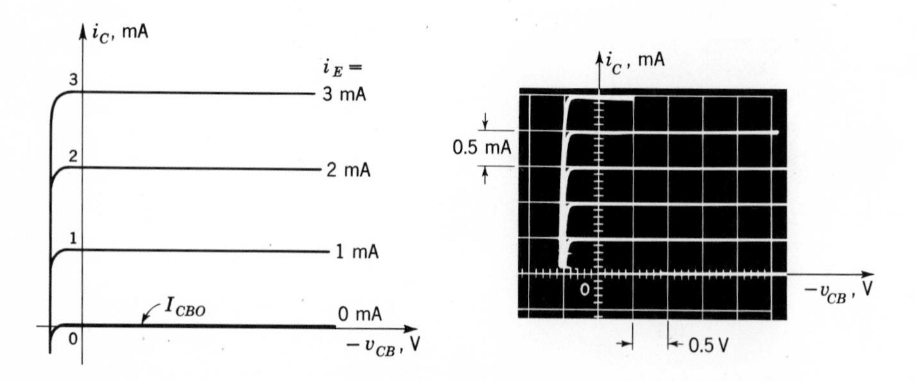

Fig. 3.1-5 Common-base output characteristics.

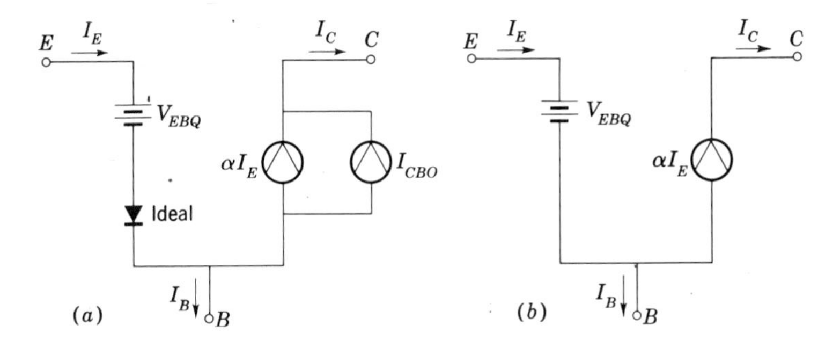

Fig. 3.1-6 Common-base equivalent circuits. (*a*) Basic model; (*b*) simplified model.

Restricting the discussion to silicon transistors, for which $I_{CBO} \ll \alpha I_E$ at normal operating temperatures, the model reduces to that of Fig. 3.1-6*b*, where the ideal diode in the emitter circuit has also been omitted. This is permissible for linear operation if we remember that we are operating past the break at all times, so that the piecewise linear model remains valid. In addition, in order to avoid the nonlinear region to the left of the i_C axis of the output characteristic of Fig. 3.1-5, it is necessary that $v_{CB} < 0$ at all times.

The complete large-signal model of Fig. 3.1-6*b* can be used for most low-frequency large-signal calculations, as illustrated by the example which follows.

Example 3.1-1 In the circuit of Fig. 3.1-1*b*, $\alpha \approx 1$, $I_{CBO} \approx 0$, $V_{EE} = 2$ V, $R_e = 1$ kΩ, $V_{CC} = 50$ V, $R_c = 20$ kΩ, and a 1-V peak sinusoidal source is connected in series with V_{EE}. Find i_E and v_{CB}.

Solution The complete equivalent circuit takes the form shown in Fig. 3.1-7. For the emitter-base circuit, the junction is forward-

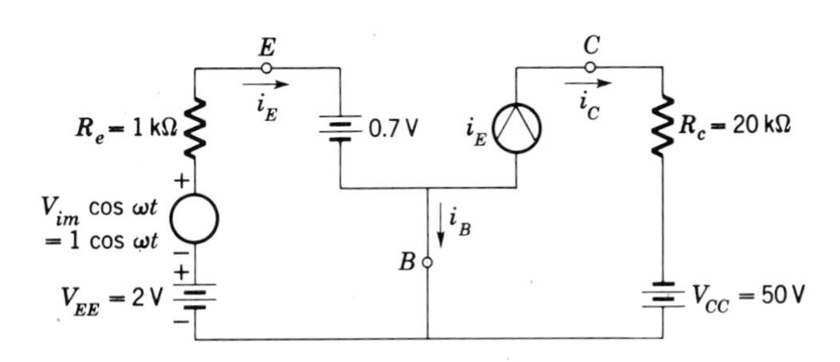

Fig. 3.1-7 Common-base equivalent circuit for Example 3.1-1.

biased as long as $V_{im} < 1.3$ V, and

$$i_E = \frac{V_{EE} - 0.7 + V_{im} \cos \omega t}{R_e} = 1.3 - 1.0 \cos \omega t \qquad \text{mA} \qquad (3.1\text{-}11)$$

For the collector-base circuit KVL yields

$$v_{CB} = -V_{CC} + i_C R_c \approx -V_{CC} + i_E R_c \qquad (3.1\text{-}12)$$

Substituting (3.1-11) into (3.1-12),

$$\begin{aligned} v_{CB} &= -V_{CC} + \left(\frac{V_{EE} - 0.7 + V_{im} \cos \omega t}{R_e}\right) R_c \\ &= -V_{CC} + I_{EQ} R_c + \left(\frac{R_c}{R_e}\right) V_{im} \cos \omega t \qquad (3.1\text{-}13) \end{aligned}$$

Substituting numerical values,

$$\begin{aligned} v_{CB} &= -50 + (1.3)(20) + (20/1) \cos \omega t \\ &= -24 + 20 \cos \omega t \text{ V} \end{aligned}$$

The collector-base junction is always reverse-biased ($v_{CB} < 0$), so that the linear model of Fig. 3.1-7 is valid.

Note that the transistor amplifies the input ac voltage, and the resulting voltage gain A_v is

$$A_v = \frac{V_{cbm}}{V_{im}} = \frac{20}{1} = 20$$

3.2 CURRENT AMPLIFICATION IN THE TRANSISTOR[1]

In the preceding section a much abbreviated description of the mechanism of current flow in the junction transistor was presented. It was just enough to allow us to extract some of the terminal relations for linear operation. In this section the discussion is extended to show how current amplification is achieved. Briefly, the process can be explained this way: Referring to (3.1-1) to (3.1-3), it is seen that if $\alpha \approx 1$ and I_{CBO} is small, a change in emitter current i_E produces a change of approximately the same amount in collector current i_C and a much smaller change in base current (a factor of $1 - \alpha$). To achieve current amplification, the change is initiated in the base current rather than the emitter current. This causes the collector current and the emitter current to change by a factor of approximately $\alpha/(1 - \alpha) = \beta$.

This result assumes that $\beta = \alpha/(1 - \alpha)$ does not vary with base current. Thus, neglecting I_{CBO},

$$i_C = \beta i_B \tag{3.2-1a}$$

A change in i_B results in a change in i_C such that

$$\frac{\Delta i_C}{\Delta i_B} = \beta + \frac{\Delta\beta}{\Delta i_B} i_B \tag{3.2-1b}$$

This ratio is called the small-signal current-amplification factor h_{fe}. Therefore

$$h_{fe} = \beta + \frac{\Delta\beta}{\Delta i_B} i_B \tag{3.2-2}$$

If $(\Delta\beta/\Delta i_B)i_B$ is small compared with β, then

$$h_{fe} \approx \beta \equiv h_{FE} \tag{3.2-3}$$

Figure 3.2-1 shows a typical variation of β and h_{fe} with collector current. Note that in a typical operating range of 1 to 100 mA, h_{fe} is approximately equal to β and is relatively independent of changes in collector current. The assumption that $\beta = h_{fe} =$ constant is often made to simplify analysis. We make this assumption throughout this text.

The explanation presented above is, of course, far from complete. In order to obtain a quantitative picture, consider the circuit of Fig. 3.2-2, which shows a *pnp* transistor in what is known as the *common-emitter* (CE) configuration.

In this circuit V_{BB}, V_{CC}, and R_c are adjusted so that, with no signal present, the base-emitter junction is forward-biased and the collector-base junction is reverse-biased, as was the case for the common-base configuration. Writing KVL around the base-emitter loop yields for the base

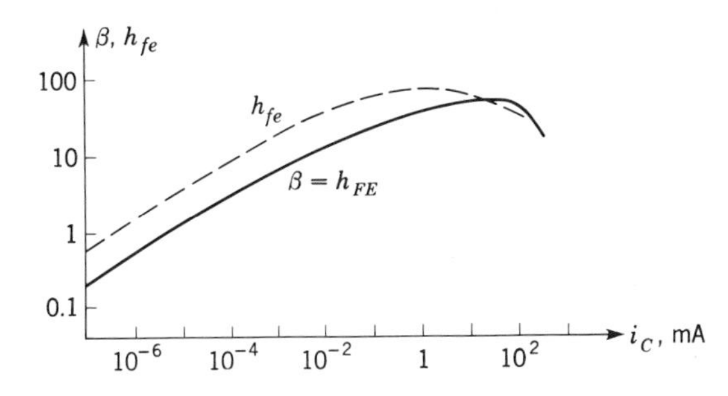

Fig. 3.2-1 Common-emitter large- and small-signal current gains.

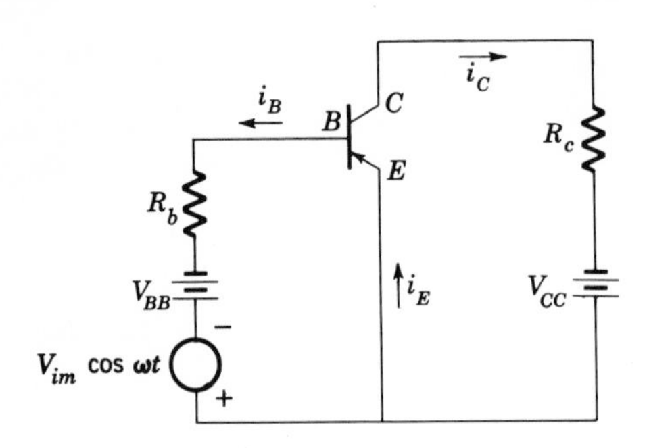

Fig. 3.2-2 Basic transistor amplifier.

current

$$i_B = \frac{V_{BB} + V_{im} \cos \omega t + V_{BEQ}}{R_b} \tag{3.2-4}$$

where the assumption has been made that the base-emitter junction is operating in its linear range with $V_{BB} - V_{im} \gg V_{BEQ}$ (that is, $r_d \approx 0$).

Since linear operation with no distortion is assumed, i_B will consist of both a dc and an ac component,

$$i_B = I_{BQ} + i_b \tag{3.2-5}$$

where

$$I_{BQ} \approx \frac{V_{BB} + V_{BEQ}}{R_b} \tag{3.2-6}$$

$$i_b = \frac{V_{im}}{R_b} \cos \omega t \tag{3.2-7}$$

and the collector current, using (3.2-1*a*), is

$$i_C = I_{CQ} + i_c = \beta(I_{BQ} + i_b) \tag{3.2-8}$$

The small-signal current amplification is then

$$A_i = \frac{i_c}{i_b} \equiv h_{fe} = \beta \tag{3.2-9}$$

This is independent of the external circuit and is a property of the transistor alone, subject to the assumptions made in the derivation.

β is approximately constant for an individual transistor, although it does vary with temperature, and slightly with the collector current. Hence the β and α used are average values. Another transistor of the same type may have a β differing by a factor of 3 or more to 1, from the first transistor. This large variability is caused by very small changes in α. For example, since $\beta = \alpha/(1 - \alpha)$, if α varies from 0.98 to 0.99, β will change from 49 to 99.

THE *npn* TRANSISTOR

Current flow in the *pnp* transistor has been considered in the preceding sections. Now let us briefly investigate current flow in the *npn* transistor, shown in Fig. 3.2-3.

The *npn* transistor has its bias voltages V_{EE} and V_{CC} reversed as compared with the *pnp* transistor. This is necessary to forward-bias the emitter-base junction and reverse-bias the collector-base junction. Since the emitter current is mainly due to electrons moving from the emitter to the collector, the positive direction of emitter current is out of the emitter, as shown in Fig. 3.2-3. The collector and base currents flow into the transistor, and

$$I_C = \alpha I_E + I_{CBO} \tag{3.2-10}$$

$$I_B = (1 - \alpha)I_E - I_{CBO} \tag{3.2-11}$$

as before.

In Figs. 3.1-1 and 3.2-3, the arrow identifies the emitter and indicates the type of transistor. A *pnp* transistor has emitter current flowing into the transistor; hence the arrow points in, as shown in Fig. 3.1-1. In Fig. 3.2-3 the arrow points out, indicating the direction of positive emitter current in an *npn* transistor.

THE COMMON-EMITTER CHARACTERISTIC

For either type of transistor, the actual terminal relations are nonlinear and, of course, have the exponential form of the junction-diode equation. Since there are three terminals to deal with, a variety of ways exist in

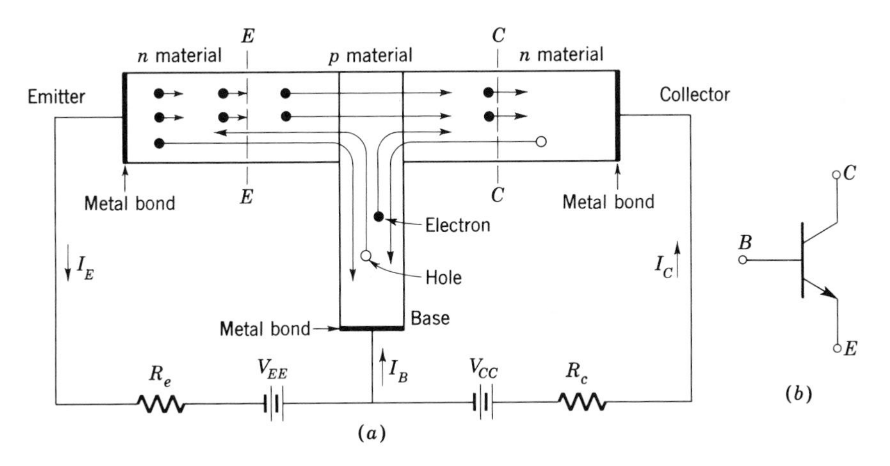

Fig. 3.2-3 The *npn* junction transistor. (*a*) Pictorial representation; (*b*) circuit symbol.

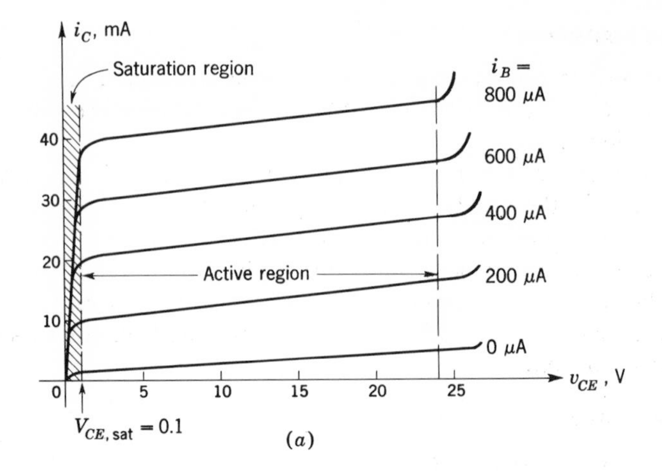

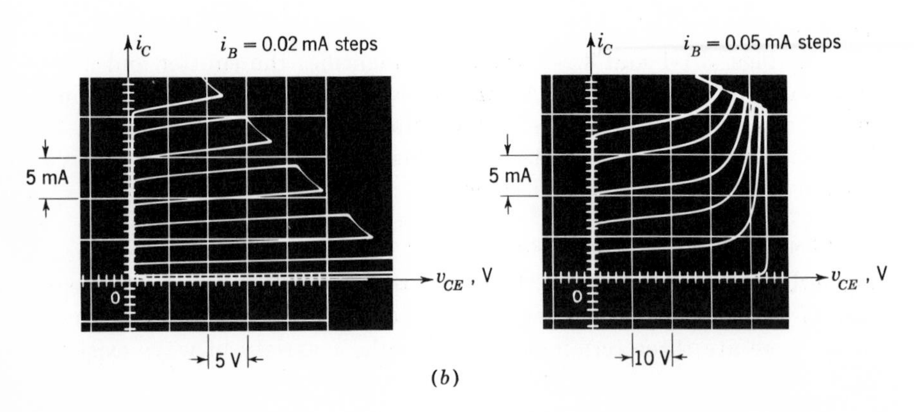

Fig. 3.2-4 Common-emitter output characteristics of *npn* transistor. (*a*) Typical characteristics; (*b*) oscillograms of TI 179 6508 common-emitter output characteristics. Note the different scales and the avalanche breakdown for $V_{CE} > \approx 30$ V.

which the characteristics can be displayed. Probably the most useful curves are the common-emitter static characteristics shown in Fig. 3.2-4 for a typical *npn* low-power silicon transistor.

Figure 3.2-4 is called the common-emitter output (or collector) *vi* characteristic. The shaded region shown at the left of the figure is called the *saturation* region. The collector-emitter saturation voltage which defines this region is typically 0.1 to 0.2 V for low-power (less than 1 W) transistors, and may be as large as 1 to 2 V for high-power transistors. From Fig. 3.1-5 we see that the collector-base saturation voltage is approximately +0.5 V for a *pnp* transistor (and −0.5 V for an *npn* transistor). This result is expected, since $V_{CB} + V_{BE} = V_{CE}$. In this region, an increase in base current does not result in a proportionate increase in collector current. Thus, in the design of linear amplifiers, the saturation region is avoided.

For large values of v_{CE} (about 30 V for the transistor of Fig. 3.2-4b), an avalanche breakdown takes place similar to that described in Sec. 2.2 for the diode. The collector-base breakdown voltage is related to the collector-emitter breakdown voltage by the equation $BV_{CEO} \approx BV_{CBO}/\sqrt[n]{\beta}$, where n varies between 2 and 4 for silicon.

In the region between saturation and breakdown, called the *active* region, the relation between base and collector currents is given by (3.2-9), and it is in this region that linear amplification is obtained.

The active region also has upper and lower bounds on the collector current. The upper bound is the maximum collector current that can flow without causing physical damage to the transistor. This is always specified by the manufacturer. The lower bound is called collector *cutoff*, below which essentially no collector current flows. This is usually taken to be zero.

Example 3.2-1 The behavior of the transistor in the saturation region becomes important in the design of switching circuits. To illustrate this, consider the circuit of Fig. 3.2-5, with $V_{CC} = 10$ V, $R_b = 10$ kΩ, and $R_c = 1$ kΩ. The transistor has $\beta = 100$, $V_{BE} = +0.7$ V, and a saturation voltage $V_{CE,\text{sat}} = 0.1$ V. Find the operating conditions when (*a*) $V_{BB} = 1.5$ V and (*b*) 10.7 V.

Solution (*a*) For $V_{BB} = 1.5$ V application of KVL around the base-emitter loop yields

$$-V_{BB} + I_B R_b + V_{BE} = 0$$

$$I_B = \frac{V_{BB} - V_{BE}}{R_b} = \frac{1.5 - 0.7}{10^4} = 0.08 \text{ mA}$$

$$I_C = \beta I_B = (100)(0.08) = 8 \text{ mA}$$

$$I_E \approx I_C = 8 \text{ mA}$$

$$V_{CE} = V_{CC} - I_C R_c = 10 - (8)(1) = 2 \text{ V}$$

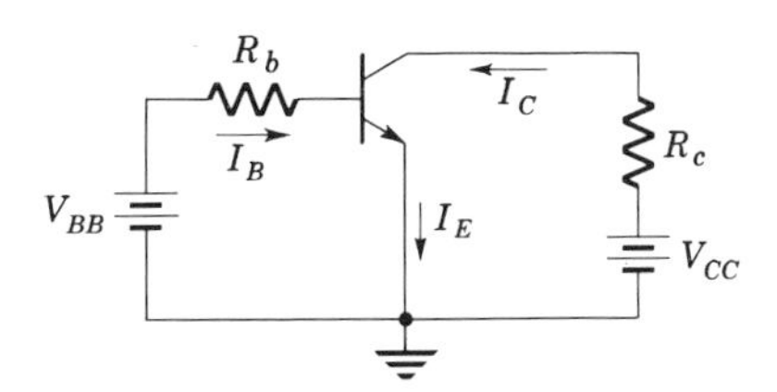

Fig. 3.2-5 Circuit for Example 3.2-1.

Thus the transistor is operating within the active region ($V_{CE} > V_{CE,\text{sat}}$).

(*b*) For $V_{BB} = 10.7$ V,

$$I_B = \frac{10.7 - 0.7}{10^4} = 1 \text{ mA}$$

If the basic relation $I_C = \beta I_B$ were to hold here, we should have $I_C = 100$ mA and $V_{CE} = 10 - 100 = -90$ V, an impossible situation. Thus the transistor is in saturation, and

$$V_{CE} = V_{CE,\text{sat}} = 0.1 \text{ V}$$

The collector current is

$$I_C = \frac{V_{CC} - V_{CE,\text{sat}}}{R_c} = \frac{10 - 0.1}{1 \text{ k}\Omega} = 9.9 \text{ mA}$$

and

$$I_E = I_C + I_B = 10.9 \text{ mA}$$

Note that the effective β in this particular saturation condition is $I_C/I_B = 9.9$.

AN EQUIVALENT CIRCUIT

The vertical distance between adjacent curves in Fig. 3.2-4 obeys the relation $\Delta i_C = h_{fe}\,\Delta i_B$ [Eq. (3.2-1*b*)]. Note that h_{fe} is not absolutely constant over the full range of i_C, as shown in Fig. 3.2-1. For many applications it is sufficient to consider that the output characteristics are a set of uniformly spaced horizontal straight lines, as in Fig. 3.2-6; that is, $h_{fe} = h_{FE} =$ constant (Fig. 3.2-1). It is extremely important to note that while the transistor *vi* characteristic can be used to obtain insight into the operation of the device, it should not be used to derive any quantitative information regarding the variation of i_C with i_B, because the h_{fe} of transistors of the same type may differ considerably.

Note that the assumption of uniformly spaced horizontal lines on

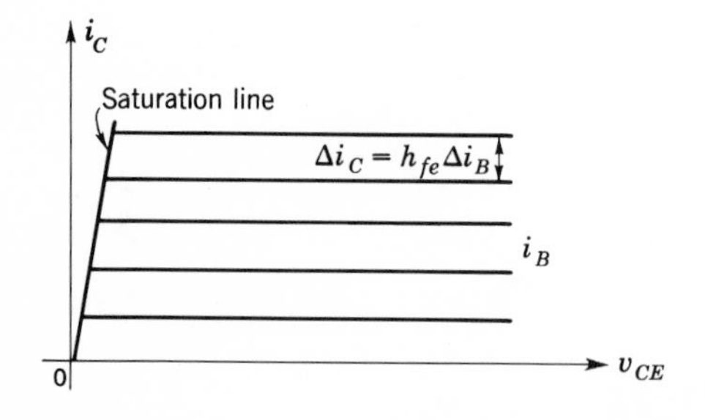

Fig. 3.2-6 Idealized output characteristic for *npn* transistor.

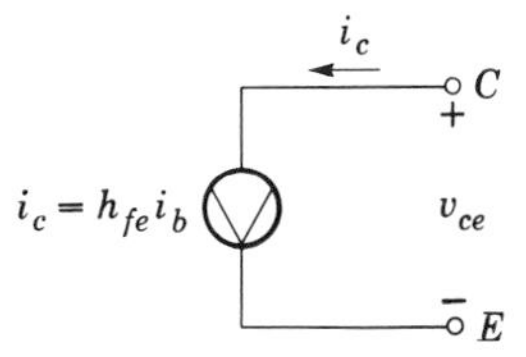

Fig. 3.2-7 Small signal controlled current source model of idealized *npn* transistor valid for the active region.

the output characteristic permits the collector circuit to be replaced by a controlled current source, as shown in Fig. 3.2-7. This model does not implicitly account for the saturation or breakdown regions and is valid only in the active region.

We can go one step further and take into account the nonzero slope of the actual curves by including a resistance R_0 in parallel with the current source i_c, as shown in Fig. 3.2-8*a*. For a silicon transistor this resistor is between 10 and 100 kΩ. Because of its large size, it is often neglected (R_0 is often considered to be infinite). The corresponding output *vi* characteristic is shown in Fig. 3.2-8*b*. It should be noted that the circuit of Fig. 3.2-8 is the small-signal equivalent of the output circuit, and is valid only in the active region (this equivalent circuit will be studied in detail in Chap. 6).

The remainder of this chapter assumes that the transistor is a linear device (some of the nonlinear characteristics of the transistor are studied in subsequent chapters). In addition, unless otherwise stated, it is assumed that silicon transistors are used, that $|V_{BE}| = 0.7$ V, and that I_{CBO} is negligibly small and can be neglected.

3.3 GRAPHICAL ANALYSIS OF TRANSISTOR CIRCUITS

In Chap. 2, the analysis of diode circuits was discussed in terms of three general methods. These were the graphical method, the piecewise linear

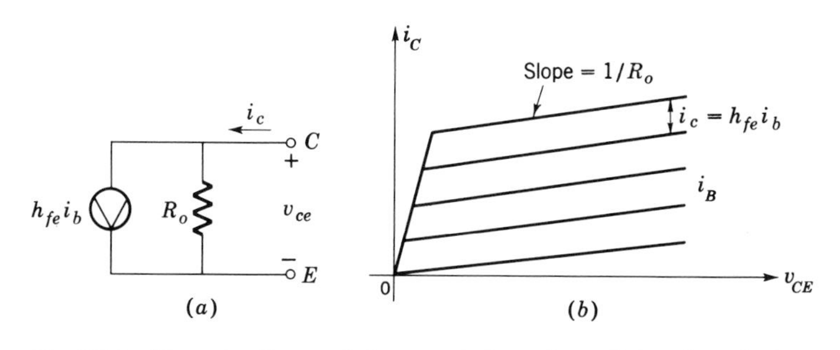

Fig. 3.2-8 Piecewise linear *vi* characteristic and small-signal equivalent circuit of *npn* transistor, including R_0. (*a*) Equivalent circuit; (*b*) piecewise linear *vi* characteristic.

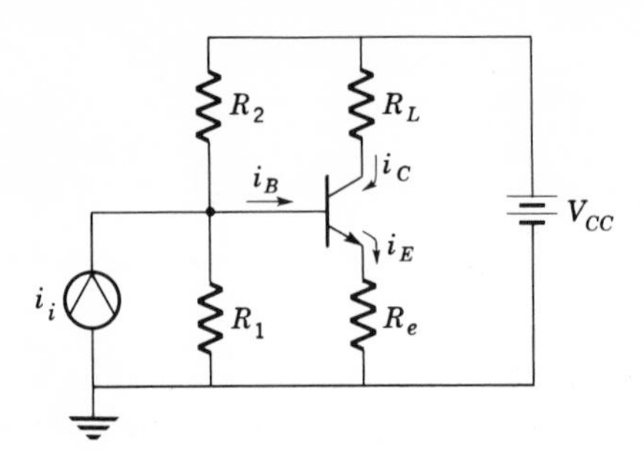

Fig. 3.3-1 Basic common-emitter amplifier.

approximation, and the small-signal linear analysis using *incremental*, or *dynamic*, parameters. All three methods are used for the analysis of transistor circuits. In this and the following sections a combination of the graphical and piecewise linear methods is used to design a common-emitter amplifier so as to obtain maximum symmetrical variation in the collector current. We also determine the maximum power dissipated by the transistor, the maximum power dissipated in the load resistor, and the power furnished by the supply V_{CC}.

THE BASIC AMPLIFIER

A basic transistor amplifier in the common-emitter configuration is shown in Fig. 3.3-1. The resistors R_1, R_2, R_L, and R_e and the supply voltage V_{CC} are chosen so that the transistor operates linearly and a maximum peak-to-peak swing in i_C is possible. Resistors R_1 and R_2 form a voltage divider across the V_{CC} supply. The function of this network is to provide *bias* conditions which ensure that the emitter-base junction is operating in the proper region. Figure 3.3-1 can be simplified by obtaining a Thévenin equivalent circuit for R_1, R_2, and V_{CC}, as shown in Fig. 3.3-2*a*.

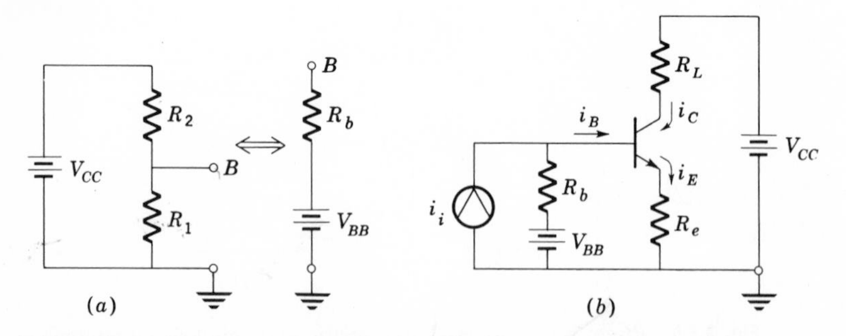

Fig. 3.3-2 A useful simplification of the common-emitter amplifier. (*a*) Thévenin equivalent of bias circuit; (*b*) simplified amplifier.

The Thévenin conversion can be made in either direction, using

$$V_{BB} = \left(\frac{R_1}{R_1 + R_2}\right) V_{CC} \tag{3.3-1a}$$

$$R_b = \frac{R_1 R_2}{R_1 + R_2} \tag{3.3-1b}$$

or

$$R_1 = \frac{R_b}{1 - V_{BB}/V_{CC}} \tag{3.3-1c}$$

$$R_2 = R_b \left(\frac{V_{CC}}{V_{BB}}\right) \tag{3.3-1d}$$

Two equations are needed to describe the operation of the circuit of Fig. 3.3-2*b*. The first equation can be obtained using KVL around the collector-emitter loop. Thus, with $i_i = 0$, so that dc conditions prevail,

$$V_{CC} = v_{CE} + i_C R_L + i_E R_e \tag{3.3-2a}$$

Since $i_C = \alpha i_E \approx i_E$, (3.3-2*a*) becomes

$$V_{CC} \approx v_{CE} + i_C(R_L + R_e) \tag{3.3-2b}$$

The second equation is obtained using KVL around the emitter-base circuit.

$$V_{BB} - i_B R_b = v_{BE} + i_E R_e \tag{3.3-3}$$

Since

$$i_B = i_E(1 - \alpha)$$

(3.3-3) becomes

$$V_{BB} - v_{BE} = i_E[R_e + (1 - \alpha)R_b]$$

and

$$i_E = \frac{V_{BB} - v_{BE}}{R_e + (1 - \alpha)R_b} \tag{3.3-4}$$

Differences in individual transistors can cause $(1 - \alpha) = 1/(\beta + 1)$ to change by a factor of 3 or more to 1, thus changing the dc emitter current. Consequently, the transistor circuit is designed so that

$$R_e \gg (1 - \alpha)R_b \tag{3.3-5}$$

to eliminate variations in i_E due to variations in α.

Using this inequality, (3.3-4) becomes

$$I_{CQ} \approx I_{EQ} \approx \frac{V_{BB} - v_{BE}}{R_e} \qquad (3.3\text{-}6)$$

where $v_{BE} \approx 0.7$ V.

Equations (3.3-2) and (3.3-6) can be solved algebraically for $v_{CE} = V_{CEQ}$.

$$V_{CEQ} = V_{CC} - (V_{BB} - 0.7)\left(1 + \frac{R_L}{R_e}\right) \qquad (3.3\text{-}7)$$

Equations (3.3-6) and (3.3-7) give the dc (quiescent) operating conditions for the transistor circuit of Fig. 3.3-2. We now turn to a graphical analysis in order to determine I_{CQ} (the quiescent collector current) to permit a maximum "swing" in collector current. Before attempting this it will be helpful to discuss the interpretation of some of the equations which have been written, in terms of the diode analysis of Chap. 2 and the transistor theory of Sec. 3.2. Consider (3.3-2*b*), which describes the operation of the collector-emitter loop. This has exactly the same form as that describing the diode circuit of Sec. 2.3, with the important difference that the junction is reverse-biased here. The variables v_{CE} and i_C are also related by the *vi* characteristic given in Fig. 3.2-4. Thus (3.3-2) represents a "dc load line," and we proceed to solve graphically, exactly as for the diode. This load line is plotted in Fig. 3.3-3*a*.

The load line defines the operating path of this circuit. When v_{CE} is less than 0.1 V (Fig. 3.2-4), the transistor is said to be *saturated*, as discussed previously. When the collector current becomes zero ($v_{CE} = V_{CC}$ in this example), the transistor is said to be at *cutoff*, since the collector current cannot become less than zero.

Note that

$$i_C = I_{CQ} + i_c$$

and

$$v_{CE} = V_{CEQ} + v_{ce}$$

Hence, as in Chap. 2, a set of i_c-v_{ce} axes passing through the Q point can be drawn, as shown in the figure.

If the variation of i_b is known, values of v_{ce} and i_c can be found by using the graphical construction of Fig. 3.3-3*b*. This need not be done unless the nonlinearity of the collector characteristics is to be taken into account.

Because of the variability of β, the insertion of the constant i_B curves onto the *vi* characteristic is meaningful quantitatively only if they were measured using the transistor being considered. The results then obtained from Fig. 3.3-3*b* would probably not be valid if we replaced the measured transistor with a different one of the same type but having a different value of β. For this reason the base-current i_B curves are usually omitted when performing a graphical analysis.

Figure 3.3-3*b* is of importance since it illustrates the phase relationships between i_b, i_c, and v_{ce}. Thus, as i_b increases, i_c increases, i_e increases, and v_{ce} decreases. This is clearly seen by referring to points *a* and *b* on the waveforms of Fig. 3.3-3*b*.

Since the collector current can vary from zero to approximately $V_{CC}/(R_L + R_e)$ (neglecting saturation), a quiescent current

$$I_{CQ} = \frac{V_{CC}/2}{R_L + R_e} \tag{3.3-8}$$

will yield the maximum symmetrical collector-current swing. If a sinusoidal base current is applied so as to produce this swing, the total collector current will be

$$i_C = I_{CQ} + i_c \tag{3.3-9}$$

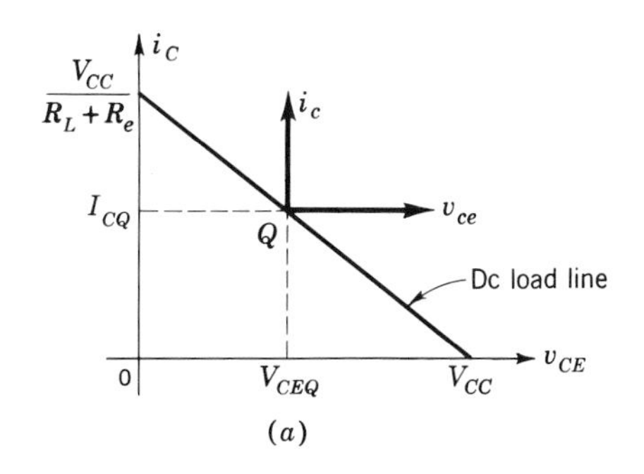

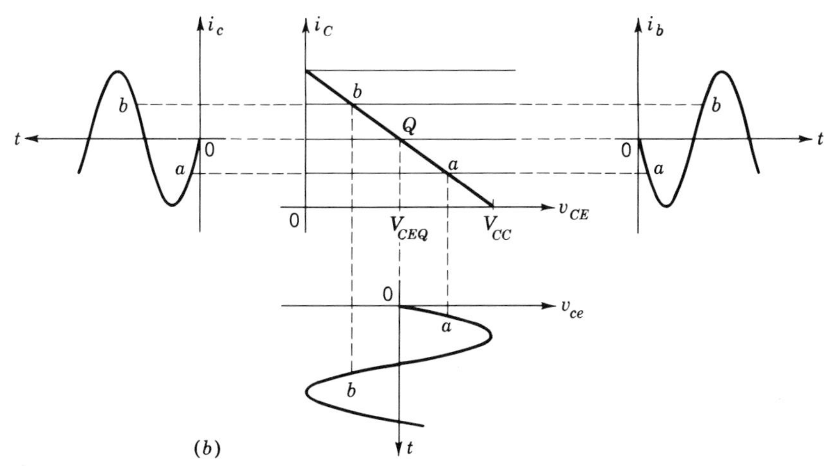

Fig. 3.3-3 Graphical analysis. (*a*) Load line; (*b*) waveforms.

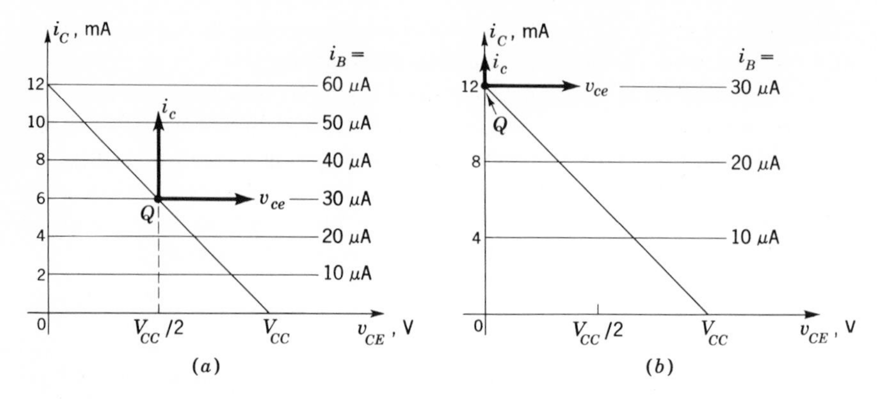

Fig. 3.3-4 Effect of β variation on the Q point. (a) $\beta_1 = 200$; (b) $\beta_2 = 400$.

where

$$I_{CQ} = \frac{V_{CC}/2}{R_L + R_e} \tag{3.3-10}$$

and

$$i_c = \left(\frac{V_{CC}/2}{R_L + R_e}\right) \cos \omega t = I_{cm} \cos \omega t \tag{3.3-11}$$

Limitations on the amount of current that can flow safely, the collector voltage, and the power that can be dissipated in the transistor are discussed in Chap. 5.

It is interesting to observe the effect of variations in β on the response of the amplifier circuit of Fig. 3.3-2*b*. This is accomplished with the aid of Fig. 3.3-4*a*, which shows a nominal quiescent point at

$$V_{CEQ} = \frac{V_{CC}}{2}$$

$$I_{CQ} = 6 \text{ mA}$$

and

$$I_{BQ} = 30\ \mu\text{A}$$

If by appropriately adjusting R_1 and R_2 the quiescent base current were maintained constant at 30 μA, then, by placing a transistor with $\beta_2 = 2\beta_1$ in the circuit, the Q point would shift to (Fig. 3.3-4*b*)

$$V_{CEQ} = 0$$

$$I_{CQ} = 12 \text{ mA}$$

$$I_{BQ} = 30\ \mu\text{A}$$

Thus, with one transistor, we have a maximum collector-voltage swing, while the other transistor is in saturation, and hence there is no swing. This is the basic reason for requiring the inequality in (3.3-5) to be met so that the transistor is biased with a constant-emitter rather than a constant-base current. The important question of how to arrange the input circuit so as to achieve the desired Q point is postponed until later.

Example 3.3-1 In the circuit of Fig. 3.3-2, $V_{CC} = 15$ V, $R_L = 1$ kΩ, and $R_e = 500$ Ω. Determine the maximum symmetrical swing in collector current and the Q point.

Solution The dc load line is plotted in Fig. 3.3-5. In order to obtain the maximum symmetrical swing, choose the quiescent collector current in the center of the load line. Thus

$$I_{CQ} = 5 \text{ mA}$$

and

$$V_{CEQ} = 7.5 \text{ V}$$

The peak-to-peak collector current can reach 10 mA.

Example 3.3-2 Find the Q point for maximum symmetrical collector-current swing in the circuit of Fig. 3.3-6.

Solution The dc-load-line equation for the circuit is

$$9 \approx V_{CEQ} + I_{CQ}(1000 + 200)$$

where it has been assumed that $200I_{EQ} \approx 200I_{CQ}$, that is, $\alpha \approx 1$. The dc-load-line equation is then plotted as shown. When the collector-emitter voltage is zero (saturation), maximum collector current flows (7.5 mA). Thus, to achieve a maximum symmetrical swing, the Q point should be set at 3.75 mA, so that a peak current of 3.75 mA results.

Note that if the Q point were set at 3.5 mA, the maximum

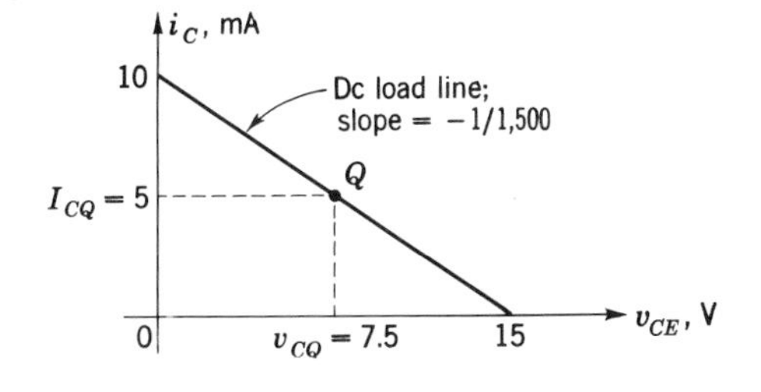

Fig. 3.3-5 Load line for Example 3.3-1.

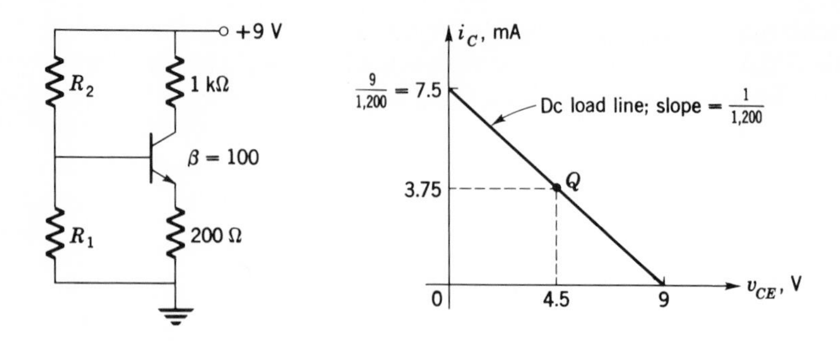

Fig. 3.3-6 Circuit and load line for Example 3.3-2.

symmetrical swing would be 3.5 mA; if the Q point were set at 4 mA, the maximum symmetrical swing would again be 3.5 mA.

Example 3.3-3 Find R_1 and R_2 in Example 3.3-2 to set the Q point at $I_{CQ} = 3.75$ mA, $V_{CEQ} = 4.5$ V.

Solution The Thévenin equivalent of the bias network is shown in Fig. 3.3-7. The quiescent emitter voltage V_{EQ} is

$$V_{EQ} = I_{EQ}(200) \approx (3.75 \times 10^{-3})(200)$$
$$= 0.75\ V$$

The quiescent base voltage V_{BQ} is then

$$V_{BQ} = V_{BE} + 0.75 \approx 0.7 + 0.75 = 1.45 \text{ V}$$

Note from (3.3-6) that if the voltage drop across R_b is small,

$$V_{BB} \approx V_{BE} + V_{EQ}$$

Therefore

$$V_{BB} \approx V_{BQ} = 1.45 \text{ V}$$

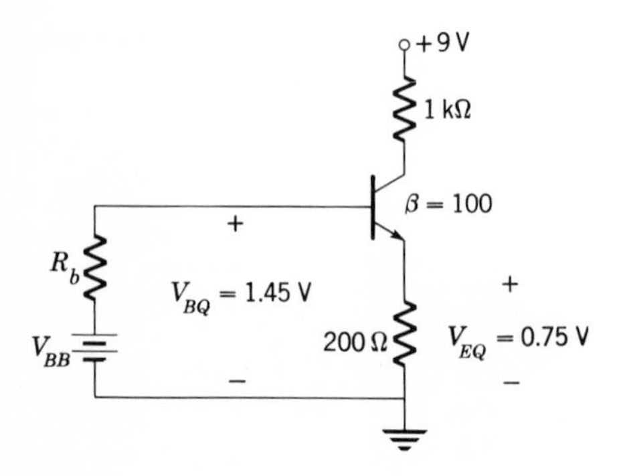

Fig. 3.3-7 Circuit for Example 3.3-3.

From (3.3-5), the inequality $R_e \gg (1 - \alpha)R_b$ must be satisfied if the quiescent current is to be stabilized against β variations. Thus, rearranging terms, the inequality can be written

$$R_b \ll \frac{R_e}{1 - \alpha}$$

But $1/(1 - \alpha) = \beta + 1$, so that

$$R_b \ll (\beta + 1)R_e \approx \beta R_e = 100R_e = 20 \text{ k}\Omega$$

and R_b can safely be chosen as 2 kΩ.

Knowing V_{BB} and R_b, R_1 and R_2 can be determined from (3.3-1c) and (3.3-1d).

$$R_1 = \frac{R_b}{1 - V_{BB}/V_{CC}} = \frac{2 \text{ k}\Omega}{1 - 1.45/9} \approx 2.4 \text{ k}\Omega$$

and

$$R_2 = \left(\frac{V_{CC}}{V_{BB}}\right) R_b = \left(\frac{9}{1.45}\right) (2 \text{ k}\Omega) \approx 12.4 \text{ k}\Omega$$

As a practical matter, standard resistors would be used. Thus

$$R_1 = 2.2 \text{ k}\Omega$$

and

$$R_2 = 12 \text{ k}\Omega$$

These standard values for R_1 and R_2 in turn yield

$$V_{BB} = \left(\frac{2.2}{14.2}\right) (9) \approx 1.4 \text{ V}$$

and

$$R_b = \frac{(2.2)(12)}{14.2} \approx 1.9 \text{ k}\Omega$$

The resulting quiescent current is then [Eq. (3.3-6)]

$$I_{CQ} \approx \frac{1.4 - 0.7}{200} = 3.5 \text{ mA}$$

Thus, because of the use of standard resistors, the maximum swing is 3.5 mA, rather than 3.75 mA. Since saturation and the drop in R_b have been neglected, we should expect the maximum swing to be, actually, somewhat less than 3.5 mA.

3.4 POWER CALCULATIONS[2]

Power calculations are extremely important for the following reasons: The transistor may be destroyed if its maximum allowable dissipation is exceeded; the power supply is capable of supplying only a finite amount of power; and resistors have a power rating (such as 0.1, 1, 2, 10 W), which, if exceeded, will cause them to burn out. We always try to use the smallest resistor possible to save space. Wattage, size, and cost vary proportionally. In this section we calculate the power supplied by the V_{CC} supply, the power dissipated in R_L and R_e, and the power dissipated in the transistor.

The average power supplied to or dissipated by any linear or nonlinear device is, simply,

$$P = \left(\frac{1}{T}\right) \int_0^T V(t)I(t)\,dt \tag{3.4-1}$$

where V is the total voltage across the device, I is the total current through it, and T is the period of any periodic time-varying part of V (or I). If

$$V = V_{av} + v(t) \tag{3.4-2a}$$

and

$$I = I_{av} + i(t) \tag{3.4-2b}$$

where V_{av} and I_{av} are average values and $v(t)$ and $i(t)$ are time-varying components having an average value of zero, then

$$\begin{aligned} P &= \left(\frac{1}{T}\right) \int_0^T [V_{av} + v(t)][I_{av} + i(t)]\,dt \\ &= \left(\frac{1}{T}\right) \int_0^T V_{av}I_{av}\,dt + \left(\frac{1}{T}\right) \int_0^T V_{av}i(t)\,dt \\ &\quad + \left(\frac{1}{T}\right) \int_0^T I_{av}v(t)\,dt + \left(\frac{1}{T}\right) \int_0^T v(t)i(t)\,dt \\ &= V_{av}I_{av} + \left(\frac{1}{T}\right) \int_0^T v(t)i(t)\,dt \end{aligned} \tag{3.4-3a}$$

since

$$\left(\frac{1}{T}\right) \int_0^T i(t)\,dt = \left(\frac{1}{T}\right) \int_0^T v(t)\,dt = 0 \tag{3.4-3b}$$

by definition of $i(t)$ and $v(t)$. Equation (3.4-3a) shows that the average power supplied (or dissipated) by a device consists of the sum of the power in the dc (average) terms and the power in the ac terms.

AVERAGE POWER DISSIPATED IN LOAD

Let us now turn our attention to the transistor circuit of Fig. 3.3-2. The ac power dissipated in the load, $P_{L,\text{ac}}$, is

$$P_{L,\text{ac}} = \left(\frac{1}{T}\right) \int_0^T i_c^2 R_L \, dt \tag{3.4-4a}$$

Assuming i_c sinusoidal,

$$i_c = I_{cm} \cos \omega t \tag{3.4-4b}$$

the ac power dissipated in the load becomes

$$\begin{aligned} P_{L,\text{ac}} &= \left(\frac{1}{T}\right) \int_0^T R_L I_{cm}^2 \cos^2 \omega t \, dt \\ &= R_L \left(\frac{1}{T}\right) \int_0^T \frac{I_{cm}^2}{2} (1 + \cos 2\omega t) \, dt \\ &= \frac{I_{cm}^2 R_L}{2} \end{aligned} \tag{3.4-4c}$$

Equation (3.4-4*c*) is plotted in Fig. 3.4-1 as a function of I_{cm}. Since the power increases parabolically with I_{cm}, maximum ac power is dissipated in the load when I_{cm} is a maximum. If the quiescent collector current is chosen for maximum swing,

$$\max I_{cm} = I_{CQ}$$

The maximum average power dissipated in the load is then

$$\max P_{L,\text{ac}} = \frac{I_{CQ}^2 R_L}{2} \tag{3.4-5a}$$

Using (3.3-10),

$$\max P_{L,\text{ac}} = \frac{V_{CC}^2 R_L}{8(R_L + R_e)^2} \tag{3.4-5b}$$

To maximize the power dissipated in the load resistor, the emitter resistor is made much smaller than the load resistor; that is, $R_L \gg R_e$. Then the peak-to-peak ac voltage swing across R_L is approximately V_{CC}, and

$$\max P_{L,\text{ac}} \approx \frac{V_{CC}^2}{8R_L} \tag{3.4-5c}$$

It must be noted that decreasing R_e requires a decrease in R_b [Eq. (3.3-5)], which in turn results in a decrease in current gain (Chap. 6). There is a practical limit to how small R_e can be made for a given degree of stability against β and temperature variations. This is discussed in Chap. 4.

AVERAGE POWER DELIVERED BY THE SUPPLY

The average power delivered by the supply is

$$P_{CC} = \left(\frac{1}{T}\right) \int_0^T V_{CC} i_C \, dt = \left(\frac{1}{T}\right) \int_0^T V_{CC}[I_{CQ} + i_c(t)] \, dt$$
$$= V_{CC} I_{CQ} \tag{3.4-6a}$$

The supplied power is seen to be a constant, independent of signal power for the distortionless conditions that have been assumed. Under the above conditions, when I_{CQ} is chosen for maximum swing,

$$I_{CQ} = \frac{V_{CC}}{2(R_L + R_e)} \tag{3.4-6b}$$

Hence

$$P_{CC} = \frac{V_{CC}^2}{2(R_L + R_e)} \approx \frac{V_{CC}^2}{2R_L} \qquad \text{when } R_L \gg R_e \tag{3.4-6c}$$

AVERAGE POWER DISSIPATED IN THE COLLECTOR

The power dissipated in the collector P_C is

$$P_C = \left(\frac{1}{T}\right) \int_0^T v_{CE} i_C \, dt = \left(\frac{1}{T}\right) \int_0^T [V_{CC} - (R_L + R_e) i_C] i_C \, dt$$
$$= \left(\frac{1}{T}\right) \int_0^T V_{CC} i_C \, dt - (R_L + R_e) \left(\frac{1}{T}\right) \int_0^T i_C^2 \, dt \tag{3.4-7a}$$

The first term is recognized as the power delivered by the supply, P_{CC}. The second integral represents the dc and ac power dissipated in the load, P_L, and in the emitter resistor, P_E. This result should be intuitively obvious, since the power delivered by the supply must equal the sum of all the other power components:

$$P_{CC} = P_C + P_L + P_E \tag{3.4-7b}$$

Equation (3.4-7a) can be evaluated after performing the integration

$$\left(\frac{1}{T}\right) \int_0^T i_C^2 \, dt = \left(\frac{1}{T}\right) \int_0^T (I_{CQ} + I_{cm} \cos \omega t)^2 \, dt$$
$$= I_{CQ}^2 + \frac{I_{cm}^2}{2} \tag{3.4-7c}$$

Thus

$$P_C = P_{CC} - (R_L + R_e) I_{CQ}^2 - (R_L + R_e) \left(\frac{I_{cm}^2}{2}\right) \tag{3.4-7d}$$

The power dissipated in the collector [Eq. (3.4-7d)] is plotted in Fig. 3.4-1. It is seen that the collector dissipation is a maximum when no signal is present.

$$\max P_C = P_{CC} - (R_L + R_e) I_{CQ}^2 = \frac{V_{CC}^2}{4(R_L + R_e)} \approx \frac{V_{CC}^2}{4R_L} \tag{3.4-8}$$

In most low-power transistor circuits the power dissipated in the input circuit is small, so that P_C represents the total dissipation internal to the transistor. The maximum value of P_C is always specified by the manufacturer, and this value must not be exceeded if the junction temperature is to be kept within safe limits. This is discussed in detail in Chap. 4.

It is interesting and important to note that the power supply furnishes only dc power. However, the useful power dissipated in the load is ac power, i.e., dissipation resulting from an ac signal present in the load. This ac load power is generated in the transistor amplifier, and appears as the third term of (3.4-7*d*). Thus increasing the ac current increases the ac generated power. This decreases the collector dissipation and increases the power delivered to the load (and to the emitter resistor, in this case).

The maximum collector dissipation is twice the maximum power that can be delivered to the load. Thus this device is a very inefficient power amplifier. Ideally, one would like no power to be dissipated if there were no signal present. For example, consider using this amplifier in the receiver of an intercom system. The receiver is always on, yet we do not want to dissipate power in the receiver unless a voice signal is present. Use of the amplifier above would not be economical in this application. The class B amplifier studied in Chap. 5 is a circuit which dissipates almost no power unless a signal is present.

EFFICIENCY

The ratio of the ac power dissipated in the load resistor to the power delivered by the supply is defined as the efficiency η of the amplifier.

$$\eta = \frac{P_{L,\text{ac}}}{P_{CC}} = \frac{I_{cm}^2(R_L/2)}{V_{CC}^2/2R_L} \tag{3.4-9a}$$

The maximum efficiency occurs when the signal is maximum, since P_{CC} is constant and $P_{L,\text{ac}}$ increases with increasing current. Then

$$\max \eta = \frac{V_{CC}^2/8R_L}{V_{CC}^2/2R_L} = 0.25 \qquad R_L \gg R_e \tag{3.4-9b}$$

This type of amplifier is extremely inefficient, and is not often used to amplify or deliver large amounts of power. It is, however, used extensively, as will be seen in Chap. 6, to amplify current at low power levels (under 500 mW).

The ratio of the maximum collector power to the maximum ac load power is

$$\frac{\max P_C}{\max P_{L,\text{ac}}} \approx \frac{V_{CC}^2/4R_L}{V_{CC}^2/8R_L} = 2 \tag{3.4-10}$$

Thus, using this type of amplifier, to obtain 1 W dissipation in the load requires a transistor capable of handling 2 W of collector dissipation. This is extremely wasteful, and this ratio can be significantly reduced (to ⅕), using a class B amplifier. This configuration is discussed in detail in Chap. 5.

Example 3.4-1 For the circuit of Example 3.3-1 calculate the power supplied by the collector supply, the power dissipated in the load and emitter resistors, the power dissipated in the transistor, and the efficiency of operation.

Solution The power furnished by the collector supply is

$$P_{CC} = V_{CC}I_{CQ} = (15)(5 \times 10^{-3}) = 75 \text{ mW}$$

The power dissipated in the load and emitter resistors (assuming a sinusoidal signal) is

$$\begin{aligned} P_L + P_E &\approx I_{CQ}^2(R_L + R_e) + \left(\frac{1}{T}\right)\int_0^T i_c^2(t)(R_L + R_e)\,dt \\ &= (R_L + R_e)\left(I_{CQ}^2 + \frac{I_{cm}^2}{2}\right) \end{aligned}$$

and

$$\begin{aligned} \max\,(P_L + P_E) &= (1.5 \times 10^3)\left[(5 \times 10^{-3})^2 + \frac{(5 \times 10^{-3})^2}{2}\right] \\ &= 56.25 \text{ mW} \end{aligned}$$

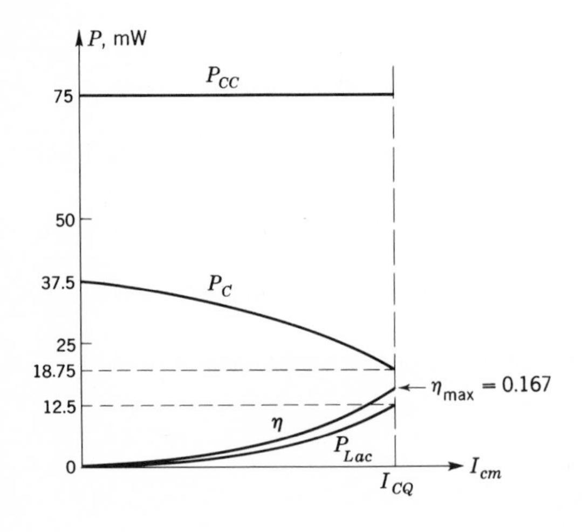

Fig. 3.4-1 Power relations in the CE amplifier of Example 3.4-1.

The power dissipated in the transistor varies with ac collector current.

$$P_C = V_{CEQ}I_{CQ} + \left(\frac{1}{T}\right)\int_0^T v_{ce}(t)i_c(t)\,dt$$

$$= (7.5)(5 \times 10^{-3}) - \left(\frac{1}{T}\right)\int_0^T (R_L + R_e)i_c^2(t)\,dt$$

$$= (37.5 \times 10^{-3}) - \left(\frac{1500}{2}\right)I_{cm}^2$$

Notice that maximum collector dissipation occurs when there is no ac signal.

$$P_{C,\max} = 37.5 \text{ mW}$$

The efficiency is

$$\eta = \frac{P_{L,\text{ac}}}{P_{CC}} = \frac{\frac{1}{2}I_{cm}^2R_L}{V_{CC}I_{CQ}} = \frac{I_{cm}^2(10^3)}{(2)(15)(5 \times 10^{-3})} = \frac{10^6}{150}I_{cm}^2$$

Note that the efficiency is 16.7 percent at max I_{cm}. This value is less than 25 percent [Eq. (3.4-9)] because R_e is not negligible for this amplifier.

The results are plotted as a function of I_{cm} in Fig. 3.4-1.

3.5 THE INFINITE BYPASS CAPACITOR

The emitter resistor R_e is inserted to obtain the desired quiescent emitter current. However, the inclusion of R_e causes a decrease in amplification. Thus, in many applications the emitter resistor is "bypassed" by a capacitor, as shown in Fig. 3.5-1. In this section infinite capacitance is

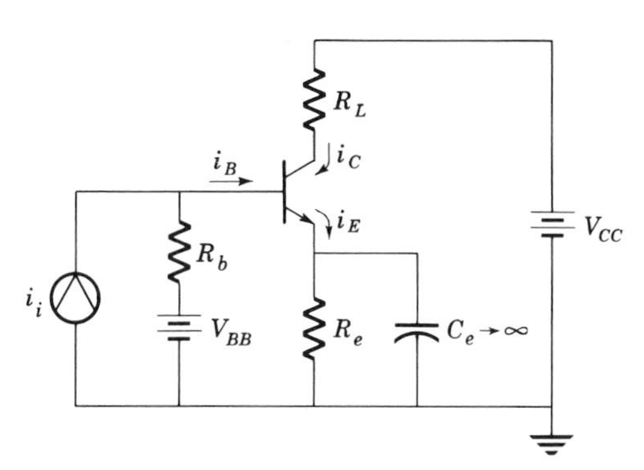

Fig. 3.5-1 CE amplifier with emitter bypass capacitor.

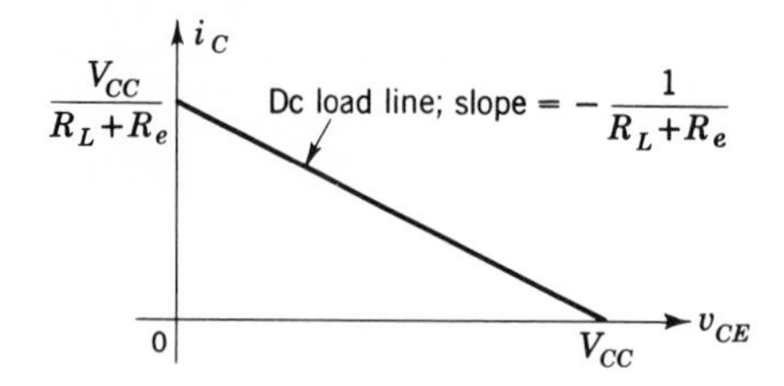

Fig. 3.5-2 Dc load line for the CE amplifier.

assumed, to avoid frequency effects. The effect of a finite capacitor on the response of the amplifier is determined in Chap. 12.

The quiescent point for this circuit is found as before, since at dc the capacitor acts as an open circuit. The dc load line is shown in Fig. 3.5-2.

Now we are faced with a situation similar to that of Sec. 2.5, where both dc and ac conditions had to be satisfied. For ac signals the collector-emitter circuit impedance is not $R_L + R_e$ as in Sec. 3.3, but simply R_L, because the capacitor effectively short-circuits R_e at all signal frequencies. Thus we must construct an ac load line with a slope $-1/R_L$. The ac load line will then be the operating path for ac signals, and must pass through the Q point, because when the ac signal is reduced to zero, the operating path must reduce to the Q point. Let us locate the Q point so as to obtain a maximum symmetrical swing. The situation is illustrated graphically in Fig. 3.5-3.

The equation of the ac load line is

$$i_c + \frac{v_{ce}}{R_L} = 0$$

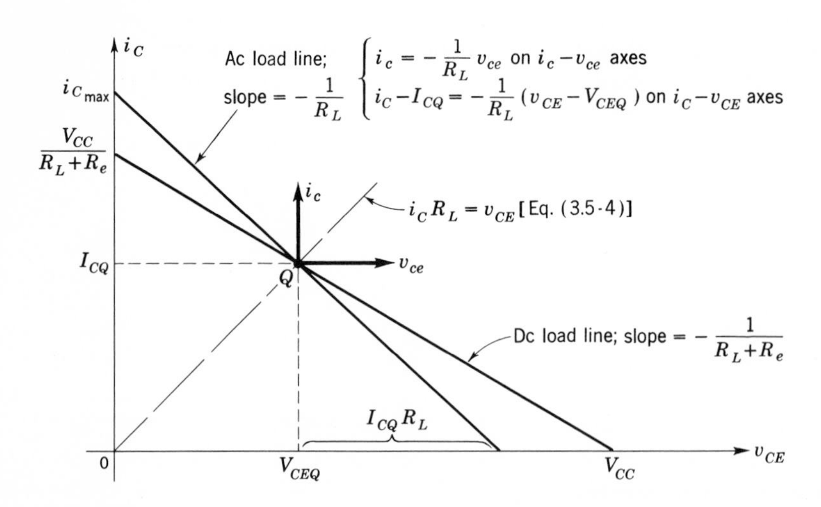

Fig. 3.5-3 Dc and ac load lines for the amplifier of Fig. 3.5-1.

which can be written

$$i_C - I_{CQ} = -\left(\frac{1}{R_L}\right)(v_{CE} - V_{CEQ}) \tag{3.5-1}$$

The maximum value of i_C occurs when $v_{CE} = 0$.

$$i_{C,\max} = I_{CQ} + \frac{V_{CEQ}}{R_L} \tag{3.5-2}$$

To obtain maximum symmetrical swing the Q point should bisect the ac load line so that

$$i_{C,\max} = 2I_{CQ} \tag{3.5-3}$$

Substituting in (3.5-2),

$$2I_{CQ} = I_{CQ} + \frac{V_{CEQ}}{R_L}$$

and

$$I_{CQ} = \frac{V_{CEQ}}{R_L} \tag{3.5-4}$$

The Q point [Eq. (3.5-4)] lies on the line $i_C R_L = v_{CE}$, which passes through the origin. Its intersection with the dc load line yields the Q point for maximum symmetrical swing, as shown in Fig. 3.5-3. The design procedure is to first construct the dc load line, then the line $i_C R_L = v_{CE}$ through the origin, to determine the Q point. The ac load line is then drawn through the Q point with slope $-1/R_L$. The operating conditions must be chosen so that the maximum collector dissipation is not exceeded.

It should be noted that the optimum Q point can also be obtained analytically by substituting (3.5-4) into the dc-load-line equation.

$$V_{CC} = V_{CEQ} + I_{CQ}(R_L + R_e) \tag{3.5-5}$$

Then

$$I_{CQ} \text{ (for maximum symmetrical swing)} = \frac{V_{CC}}{2R_L + R_e} \tag{3.5-6a}$$

and

$$V_{CEQ} = \frac{V_{CC}}{2 + R_e/R_L} \tag{3.5-6b}$$

Now that the Q point and ac load line are specified, the current and voltage waveforms can be sketched. Figure 3.5-4 shows that, as i_B increases, i_C increases and v_{CE} decreases. Thus i_b is in phase with i_c, and v_{ce} is 180° out of phase with both i_b and i_c.

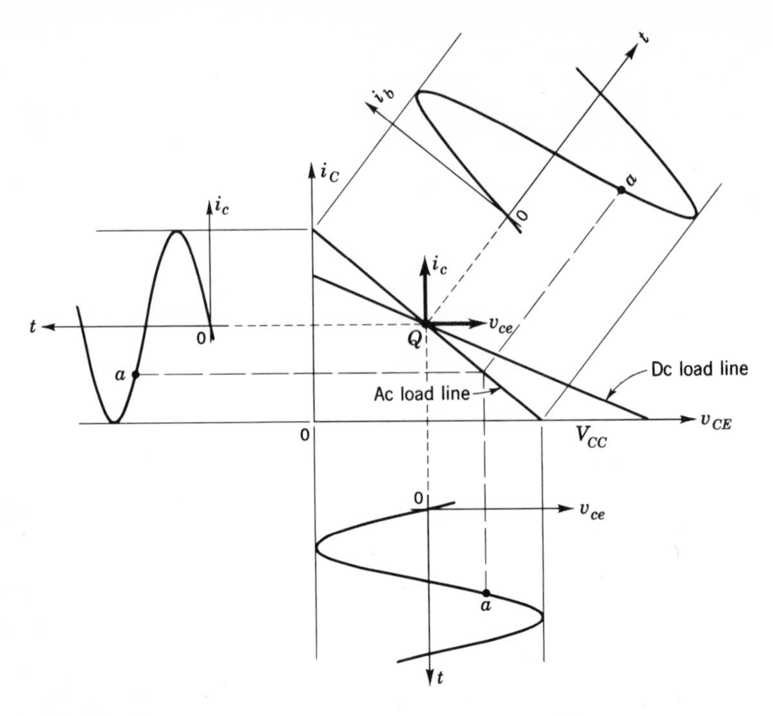

Fig. 3.5-4 Waveforms in the CE amplifier of Fig. 3.5-1 when adjusted for maximum swing.

Example 3.5-1 In Fig. 3.5-1, $V_{CC} = 15$ V, $R_L = 1$ kΩ, and $R_e = 500$ Ω. Find the maximum possible collector swing and the Q point.

Solution The dc and ac load lines are shown in Fig. 3.5-5. This result can also be obtained analytically. The dc equation

$$V_{CC} = V_{CEQ} + I_{CQ}(R_L + R_e)$$

$$15 = V_{CEQ} + I_{CQ}(1500)$$

when combined with (3.5-4),

$$V_{CEQ} = 1000I_{CQ}$$

yields

$$I_{CQ} = 15/2500 = 6 \text{ mA}$$

and hence

$$V_{CEQ} = 6 \text{ V}$$

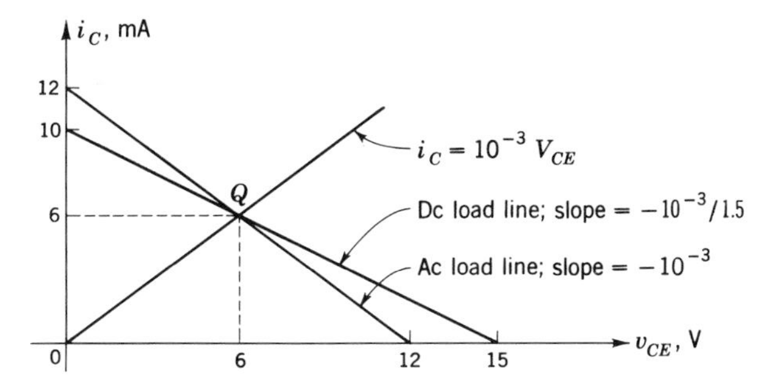

Fig. 3.5-5 Load lines for Example 3.5-1.

The maximum peak collector swing is therefore

$$I_{cm} = 6 \text{ mA}$$

Example 3.5-2 Using the optimum Q point found in Example 3.5-1, calculate R_1 and R_2.

Solution Since $I_{CQ} = 6$ mA,

$$V_{EQ} = 3 \text{ V}$$

and

$$V_{BQ} = 3.7 \text{ V}$$

Thus

$$V_{BB} \approx 3.7 \text{ V}$$

and, for stability against β variation, we require that [Eq. (3.3-5)]

$$R_b \ll \frac{R_e}{1-\alpha} \approx \frac{\alpha R_e}{1-\alpha} = \beta R_e$$

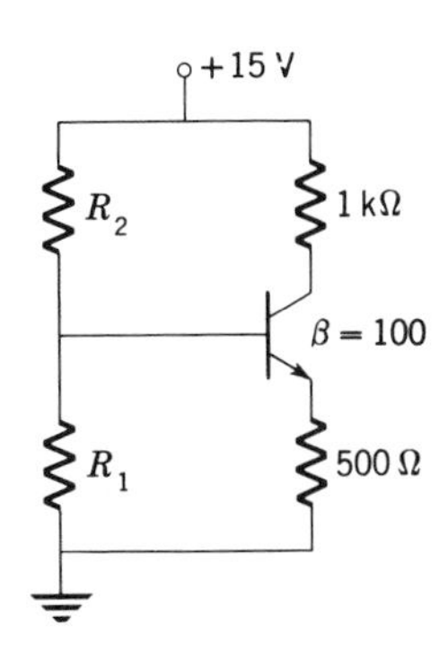

Fig. 3.5-6 Dc circuit for Example 3.5-2.

Thus

$$R_b \ll (100)(500) = 50\ \text{k}\Omega$$

Letting $R_b = 5\ \text{k}\Omega$,

$$R_2 = R_b\left(\frac{V_{CC}}{V_{BB}}\right) = (15/3.7)(5000) = 20.3\ \text{k}\Omega$$

and

$$R_1 = \frac{R_b}{1 - V_{BB}/V_{CC}} = \frac{5000}{1 - 3.7/15} = 6.63\ \text{k}\Omega$$

Using standard resistance values,

$$R_2 = 20\ \text{k}\Omega$$

and

$$R_1 = 6.8\ \text{k}\Omega$$

3.6 THE INFINITE COUPLING CAPACITOR

Quite often the load resistor must be ac-coupled to the transistor so that dc current will not flow through the load. This is usually accomplished by inserting a *coupling* capacitor between the collector and load, as shown in Fig. 3.6-1. This capacitor serves to *block* dc currents while permitting currents at signal frequencies to pass. The effect of this capacitor on the frequency characteristics of the amplifier is studied in Chap. 12. In this section maximum swing conditions are found when C_c is assumed to be infinite.

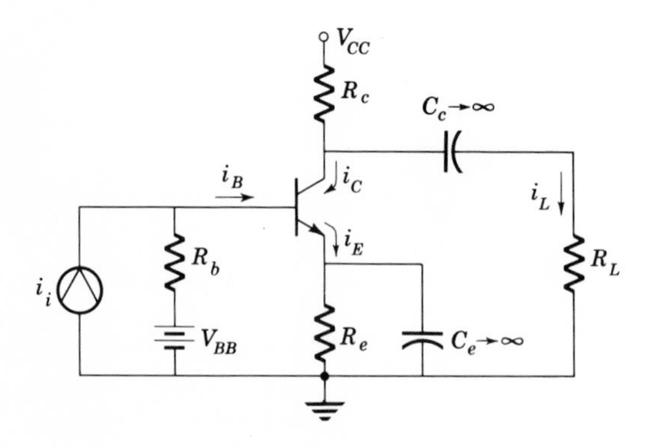

Fig. 3.6-1 Amplifier with ac-coupled load.

The dc-load-line equation is

$$V_{CC} = i_C(R_c + R_e) + v_{CE} \tag{3.6-1}$$

and the ac-load-line equation

$$i_C - I_{CQ} = -\left(\frac{R_L + R_c}{R_L R_c}\right)(v_{CE} - V_{CEQ}) \tag{3.6-2}$$

Note that the resistance which determines the slope of the ac load line is the resistance seen by the collector at signal frequencies.

Since (3.6-2) is similar to (3.5-1), the same procedure can be used to determine the Q point for maximum swing. Thus the intersection of (3.6-1) and the straight line

$$i_C = \left(\frac{R_c + R_L}{R_c R_L}\right) v_{CE} \tag{3.6-3}$$

determines the Q point for maximum symmetrical swing. (The graphical interpretation of these three equations is the same as the curves shown in Fig. 3.5-3.) The quiescent current can be calculated by combining (3.6-1) and (3.6-3). This yields

$$V_{CC} = i_C\left(R_c + R_e + \frac{R_L R_c}{R_L + R_c}\right)$$

and at the Q point

$$I_{CQ} = \frac{V_{CC}}{R_c + R_e + R_L R_c/(R_L + R_c)} \tag{3.6-4}$$

The maximum sinusoidal ac collector current with these bias conditions is

$$i_c = \frac{V_{CC}}{R_c + R_e + R_L R_c/(R_L + R_c)} \cos \omega t \tag{3.6-5}$$

and the maximum current in the load R_L is

$$i_L = \left(\frac{R_c}{R_c + R_L}\right)\left[\frac{V_{CC}}{R_c + R_e + R_L R_c/(R_L + R_c)}\right] \cos \omega t \tag{3.6-6}$$

Note that the analysis above was performed algebraically, without reference to the previously described graphical technique. This is possible because of our assumption that the transistor is a *linear* amplifier over the range of voltages and currents of interest. The graphical technique is recommended in conjunction with the analysis because of the insight it provides. For example, in a design problem, a glance at the appropriate diagram will often immediately indicate the effects of parameter changes, which often become obscured in an equation.

Example 3.6-1 In Fig. 3.6-1, $V_{CC} = 15$ V, $R_c = 1$ kΩ, $R_e = 500$ Ω, and $R_L = 1$ kΩ. Find the Q point and the maximum symmetrical collector current swing.

Solution The dc and ac load lines are plotted in Fig. 3.6-2. The Q point can also be obtained analytically by combining the dc-load-line equation,

$$15 = v_{CE} + i_C(1500)$$

with (3.6-3),

$$v_{CE} = 500 i_C$$

Thus

$$I_{CQ} = 7.5 \text{ mA}$$

and

$$V_{CEQ} = 3.75 \text{ V}$$

The maximum peak sinusoidal ac collector current is

$$I_{cm} = 7.5 \text{ mA}$$

Example 3.6-2 If in Example 3.6-1 the emitter resistor is unbypassed, find the Q point, the maximum peak ac collector-current swing, and R_1 and R_2 (Fig. 3.6-3*a*).

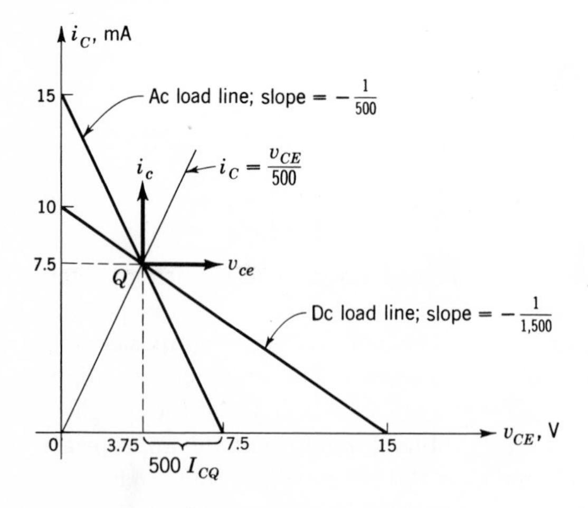

Fig. 3.6-2 Load lines for Example 3.6-1.

Solution The quiescent current can be obtained either graphically or by solving (3.6-1) and (3.6-3) simultaneously.

$$15 = v_{CE} + 1500i_C$$

Since the ac load seen by the collector is $(R_c \| R_L) + R_e$,

$$v_{CE} = 1000i_C$$

Thus

$$I_{CQ} = 6 \text{ mA}$$

and the maximum peak ac collector-current swing is

$$I_{cm} = 6 \text{ mA}$$

The bias resistors R_1 and R_2 are calculated as before. Note that

$$V_{EQ} = 3.0 \text{ V} \qquad V_{BB} \approx V_{BQ} = 3.7 \text{ V}$$

Also, using $\beta = 50$,

$$R_b \ll (50)(500) = 25 \text{ k}\Omega$$

We therefore choose

$$R_b = 2.5 \text{ k}\Omega$$

Then

$$R_2 = (2500)(15/3.7) = 10.1 \text{ k}\Omega$$

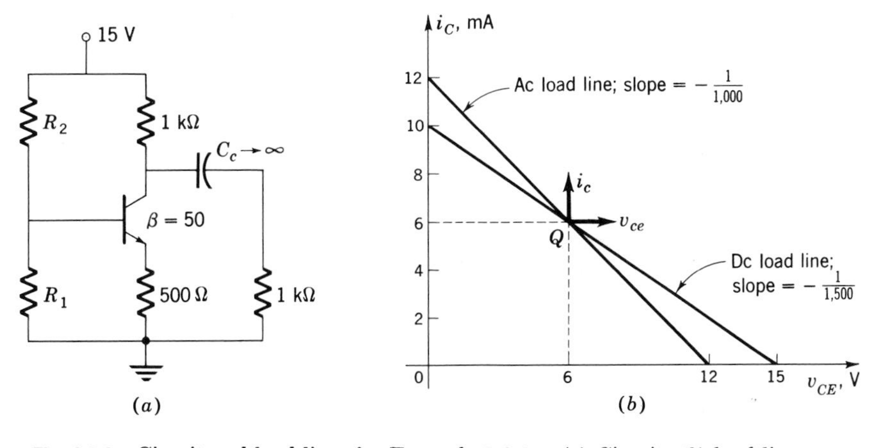

Fig. 3.6-3 Circuit and load lines for Example 3.6-2. (*a*) Circuit; (*b*) load lines.

and

$$R_1 = \frac{2500}{1 - 3.7/5} = 3.32 \text{ k}\Omega$$

Using standard resistors, $R_2 = 10 \text{ k}\Omega$ and $R_1 = 3.3 \text{ k}\Omega$.

3.7 THE EMITTER FOLLOWER

The circuit shown in Fig. 3.7-1*a* represents the common-collector, or emitter-follower (EF), configuration. The term *follower* refers to the fact that the output voltage "follows" the signal voltage quite closely, as will be seen.

As before, the bias network is replaced by its Thévenin equivalent to obtain the circuit of Fig. 3.7-1*b*. The dc load line for this circuit has a slope $-1/R_e$ and is shown in Fig. 3.7-2.

Note that, for this circuit, application of KVL to the collector-emitter loop yields the simple equation

$$V_{CC} = v_{CE} + v_E$$

Therefore

$$v_E = V_{CC} - v_{CE} \tag{3.7-1}$$

Since V_{CC} is the point at which the dc load line intersects the v_{CE} axis, we can easily construct a v_E scale as shown in Fig. 3.7-2. Now, if the Q point is placed in the center of the load line (at $I_{CQ} = V_{CC}/2\,R_e$), the output swing will be symmetrical, varying from $v_E = 0$ to $v_E = V_{CC}$.

Writing KVL around the base-emitter loop,

$$v_B = v_{BE} + v_E \tag{3.7-2}$$

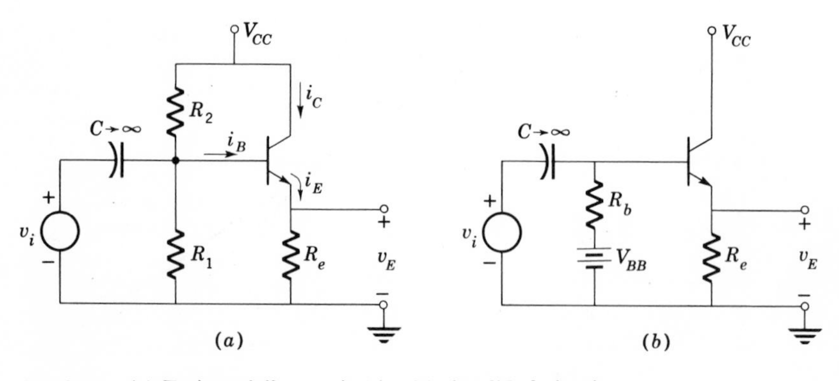

Fig. 3.7-1 (*a*) Emitter-follower circuit; (*b*) simplified circuit.

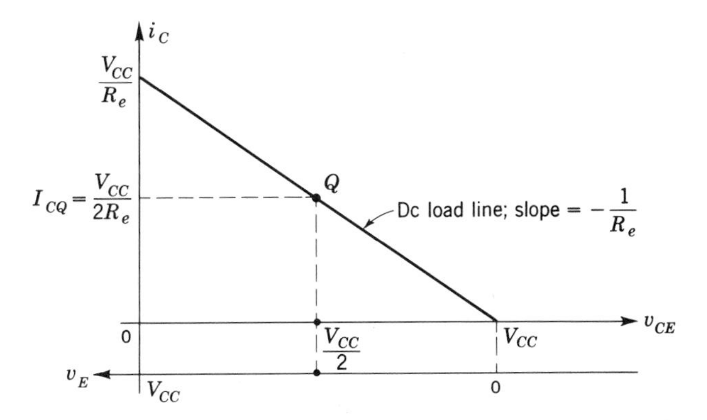

Fig. 3.7-2 Load line and auxiliary output-voltage scale for emitter follower.

If the time variation of v_{BE} is negligible, we have

$$V_{BB} \approx V_E + 0.7$$

$$v_i \approx v_e \tag{3.7-3}$$

so that the output voltage "follows" the signal.

Usually, the load resistor is ac-coupled to the emitter, as shown in Fig. 3.7-3*a*. The corresponding dc and ac load lines are sketched in Fig. 3.7-3*b*. As before, the maximum symmetrical output swing occurs when the Q point is placed at the point where the equation

$$i_C = \left(\frac{R_e + R_L}{R_e R_L}\right) v_{CE} \tag{3.7-4}$$

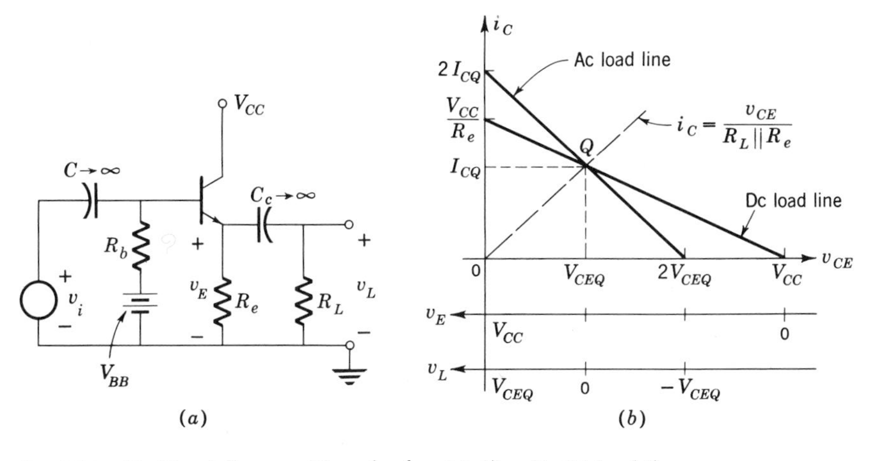

Fig. 3.7-3 Emitter follower with ac load. (*a*) Circuit; (*b*) load lines.

intersects the dc load line. The ac load line is drawn through this point, with slope $-(R_e + R_L)/R_eR_L$. The peak ac voltage swing across the load is V_{CEQ}. The peak collector-current swing is I_{CQ}.

Notice that we have performed the same calculations when considering the common-emitter amplifier circuits of Secs. 3.5 and 3.6 as when handling the emitter follower. The reason is that, approximately, the same current flows in the collector and emitter circuits. Thus, as far as the collector-emitter circuit is concerned, R_e and R_c are interchangeable. The difference is, of course, that when the output is taken from the emitter, $v_e \approx v_i$, while when the output is taken from the collector, $i_L \approx h_{fe}i_i$. The small-signal amplification and impedance characteristics of this device are discussed in detail in Chap. 6.

Example 3.7-1 Consider the circuit of Fig. 3.7-4. Find the values of R_1 and R_2 which will permit a maximum possible swing in the output.

Solution The quiescent point is the intersection of the dc load line $V_{CC} = v_{CE} + 1500i_C$ and the equation $v_{CE} = 600i_C$. Thus

$$I_{CQ} = {}^{21}\!/_{2100} = 10 \text{ mA}$$

The Q point and the load lines are shown in Fig. 3.7-5. From the curves, v_L is seen to have a maximum peak-to-peak ac swing of 12 V.

Since $V_{EQ} = 15$ V, $V_{BQ} = 15.7$ V. R_1 and R_2 are easily found if we assume I_{BQ} very small compared with $V_{CC}/(R_1 + R_2)$. With this assumption,

$$\frac{V_{CC} - V_{BQ}}{R_2} = \frac{V_{BQ}}{R_1}$$

and

$$\frac{21 - 15.7}{R_2} = \frac{15.7}{R_1}$$

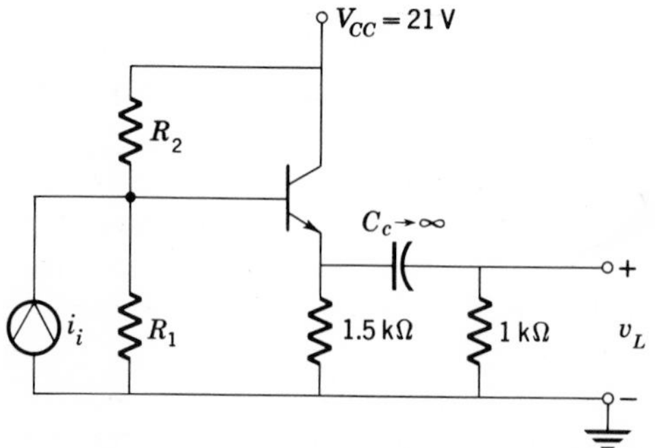

Fig. 3.7-4 Emitter follower with ac load and bias network.

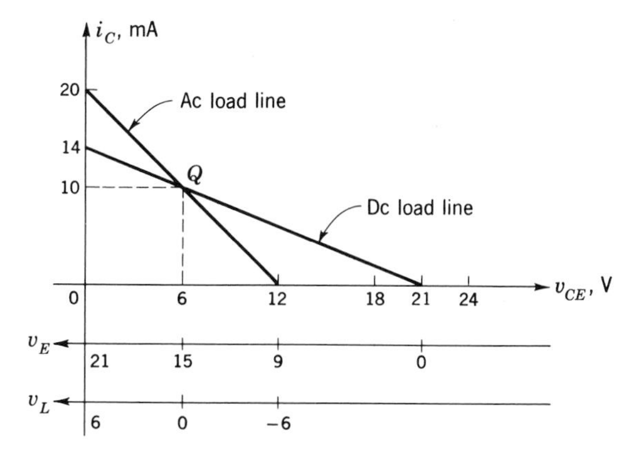

Fig. 3.7-5 Load lines for Example 3.7-1.

Thus

$$\frac{R_2}{R_1} = \frac{5.3}{15.7} = 0.34$$

Let $R_1 = 10\ \text{k}\Omega$ and $R_2 = 3.3\ \text{k}\Omega$, the closest standard resistance value. Checking the base-current assumption,

$$\frac{V_{CC}}{(R_1 + R_2)} = \frac{21}{13.3} \approx 1.6\ \text{mA}$$

so I_{BQ} must be much less than 1.6 mA. If $\beta = 100$,

$$I_{BQ} \approx \tfrac{1}{100} I_{CQ} = 0.1\ \text{mA}$$

The procedure employed to determine R_1 and R_2 is therefore satisfactory.

If this circuit were built using a randomly selected transistor and 10 percent resistors, the actual output swing would probably differ from the design value of 12 V peak to peak. The difference might be as much as 20 percent, and is due to the approximations made in the design and the use of standard resistors having ± 10 percent tolerance.

One often is required to build a single amplifier, and is not concerned with Q-point changes with β variation due to transistor replacement. In this case R_1 can usually be eliminated. Figure 3.7-6 shows the emitter follower of Fig. 3.7-4 with R_1 removed. This is called *base-injection* bias.

Let us find R_2 so that v_L can undergo a maximum symmetrical swing. To do this we first determine the Thévenin equivalent

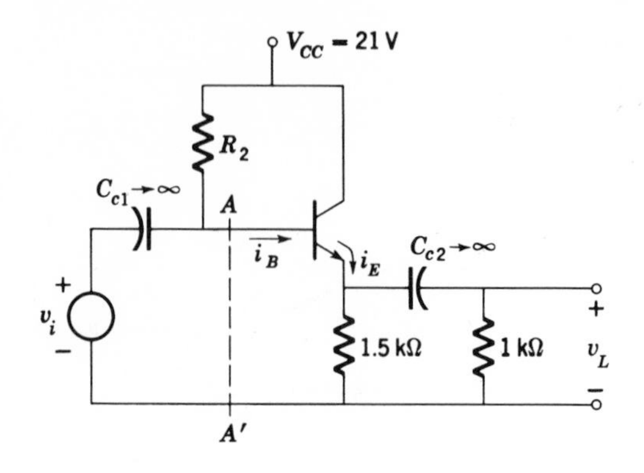

Fig. 3.7-6 Emitter follower with base-injection bias.

circuit looking into the base at AA'. The dc portion of the circuit as seen from AA' is shown in Fig. 3.7-7.

Application of KVL yields

$$v_B = V_{BE} + i_E R_e$$

or

$$v_B = 0.7 + \left(\frac{i_B}{1 - \alpha}\right) R_e \tag{3.7-5}$$

The circuit described by (3.7-5) is shown in Fig. 3.7-8. This is the desired Thévenin equivalent piecewise linear circuit looking into the base of the transistor.

Now return to Fig. 3.7-6. Making use of the Thévenin equivalent, insert R_2 and V_{CC} as shown in Fig. 3.7-9.

In this example $V_{CC} = 21$ V, $V_{BQ} = 15.7$ V, $R_e = 1500\ \Omega$, and $\alpha = 0.99$. Thus

$$I_{BQ} = \frac{V_E}{R_e/(1 - \alpha)} = \frac{15}{1500/(1 - 0.99)} = 100\ \mu\text{A}$$

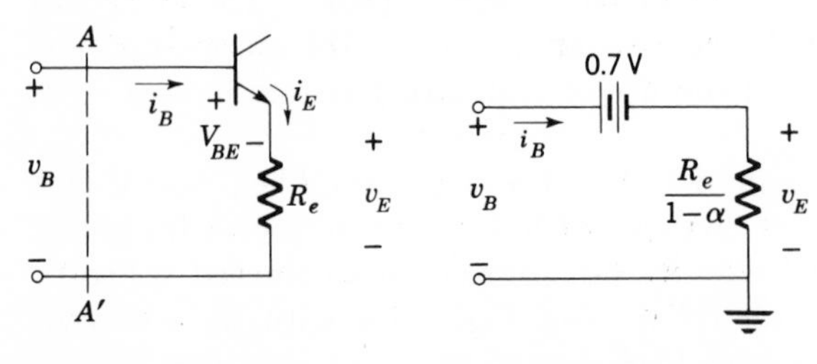

Fig. 3.7-7 Dc portion of EF of Fig. 3.7-6.

Fig. 3.7-8 Equivalent circuit looking into the base terminal.

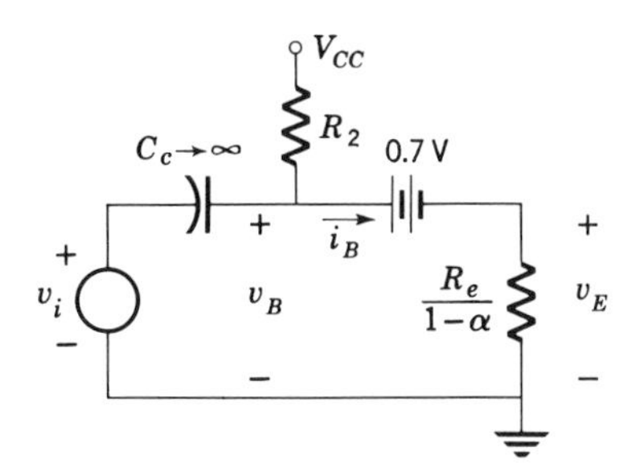

Fig. 3.7-9 Complete dc equivalent circuit for base-injection bias.

and

$$R_2 = \frac{V_{CC} - V_{BQ}}{I_{BQ}} = \frac{21 - 15.7}{10^{-4}} = 53\ \text{k}\Omega$$

(A standard value of 56 kΩ would be used in practice.)

Note that if $R_2 = 53$ kΩ, and the transistor is replaced with another, where $\alpha = 0.98$ instead of 0.99, so that $R_e/(1 - \alpha) = 75$ kΩ, then

$$V_{EQ} = \frac{(V_{CC} - 0.7)[R_e/(1 - \alpha)]}{R_2 + R_e/(1 - \alpha)} \approx 12\ \text{V}$$

instead of 15 V. The maximum peak-to-peak symmetrical swing under these conditions will be reduced accordingly.

An interesting result of this analysis is that the emitter resistor as viewed from the base terminal appears as $R_e/(1 - \alpha) = (1 + \beta)R_e$, a much larger resistance. This point is discussed in Chap. 6, where small-signal equivalent circuits are considered.

PROBLEMS

In all cases, a sketch of the pertinent load lines should be included and the problems should be solved graphically, wherever possible.

SECTION 3.1

3.1. The emitter-base junction of a *pnp* silicon transistor in the common-base configuration can be represented approximately as a 0.5-V battery in series with a 10-Ω resistance and an ideal diode (Figs. 3.1-1 and 3.1-3). Find V_{EBQ} for $R_e = 1000$ Ω and 10 kΩ and $V_{EE} = 6$ V.

3.2. The emitter-base junction of a *pnp* transistor is characterized by the equation

$$i_E = 10^{-7}(\epsilon^{v_{EB}/m(25\times10^{-3})} - 1)$$

Find V_{EBQ} when $R_e = 100$ Ω and $V_{EE} = 6$ V. Let $m = 1$ and 1.6.

SECTION 3.2

3.3. The dc current-amplification factor β for a hypothetical transistor is given by the equation

$$\beta = 100 i_C(\epsilon^{-6(i_C - 0.1)^2})$$

(*a*) Calculate h_{fe}.
(*b*) Plot h_{fe} and β as a function of i_C.

3.4. The Ebers-Moll equations are often used to describe analytically the transistor characteristics (*pnp* transistor).

$$i_E = a_{11}(\epsilon^{v_{EB}/V_T} - 1) + a_{12}(\epsilon^{v_{CB}/V_T} - 1)$$
$$i_C = a_{12}(\epsilon^{v_{EB}/V_T} - 1) + a_{22}(\epsilon^{v_{CB}/V_T} - 1)$$

where

$$\frac{a_{12}}{a_{11}} = \alpha$$
$$I_{CBO} = \frac{a_{12}{}^2}{a_{11}} - a_{22}$$

Letting $a_{11} = 10^{-3}$, $\alpha = 0.99$, and $I_{CBO} = 1\ \mu\text{A}$,
(*a*) Find i_B.
(*b*) Plot i_C versus v_{CB} with i_E as a parameter.

3.5. Using the values of Prob. 3.4, plot i_E versus v_{EB} with v_{CB} as a parameter.

3.6. (*a*) Show, using (3.1-3), that

$$I_C = \beta I_B + \frac{I_{CBO}}{1 - \alpha}$$

(*b*) The above equation is valid only in the active region. At high collector-emitter voltages the equation becomes

$$I_C = \beta I_B + \frac{I_{CBO}}{1 - \dfrac{\alpha}{[1 - (V_{CE}/BV_{CBO})^n]}}$$

(1) Find the value of V_{CE}, where I_C becomes infinite. This voltage is called the common-emitter breakdown voltage BV_{CEO}. n varies between 2 and 4 in silicon.
(2) Find the breakdown region, i.e., the region where I_C greatly increases. Let $I_B = 1$ mA, $\beta = 100$, $I_{CBO} = 0.1\ \mu\text{A}$, $BV_{CBO} = 20$ V, and $n = 3$.

3.7. Find V_{BB} to saturate the transistor.

$$V_{CE,\text{sat}} = 0.1 \text{ V}$$
$$V_{BE} \approx 0.7 \text{ V}$$
$$\beta = 10$$

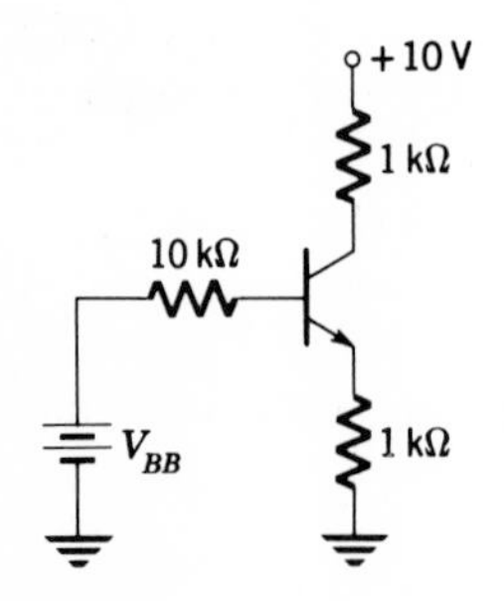

Fig. P3.7

3.8. Using the Ebers-Moll equations with $\alpha = 0.99$, $I_{CBO} = 1\ \mu A$, and $a_{11} = 1\ \mu A$, find the value of V_{BB} required to cut off the transistor.

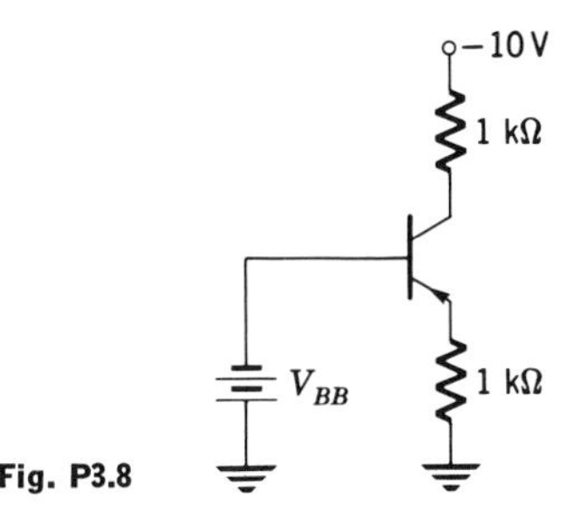

Fig. P3.8

SECTION 3.3

3.9. Find the Q point of the amplifier of Fig. P3.9 for

(a) $R_b = 1\ k\Omega$
(b) $R_b = 10\ k\Omega$

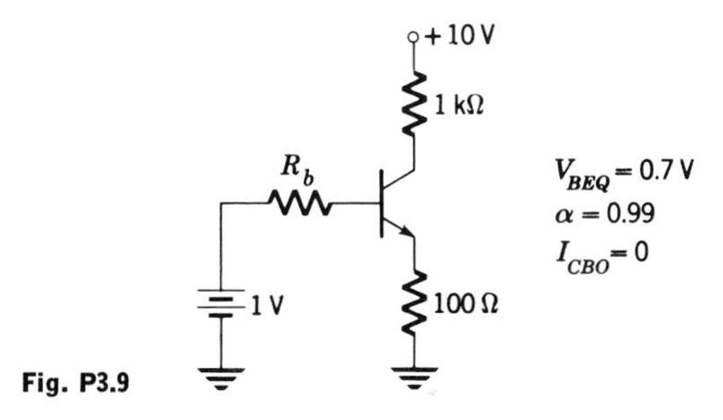

Fig. P3.9

3.10. Find R_1 and R_2 so that $V_{CEQ} \approx 5$ V. The quiescent current I_{CQ} must not vary by more than 10 percent as β varies from 20 to 60.

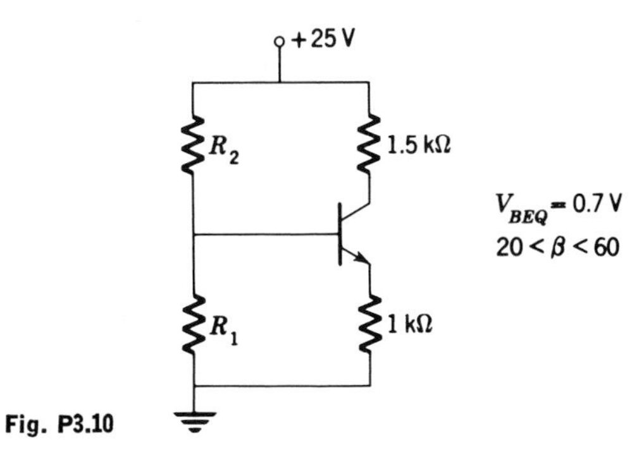

Fig. P3.10

3.11. For the amplifier of Fig. P3.10 ($\beta = 100$) find new values of R_1 and R_2 which will permit a maximum symmetrical swing in i_C.

SECTION 3.4

3.12. For the amplifier of Prob. 3.11 calculate the quiescent power

(*a*) Supplied by the battery.

(*b*) Dissipated in R_1, R_2, R_e, R_c, and at the collector junction.

3.13. (*a*) Find V_{BB} for maximum symmetrical collector swing. Calculate the efficiency under this condition.

(*b*) Repeat part *a*, assuming that the transistor saturates at $v_{CE} = 2$ V.

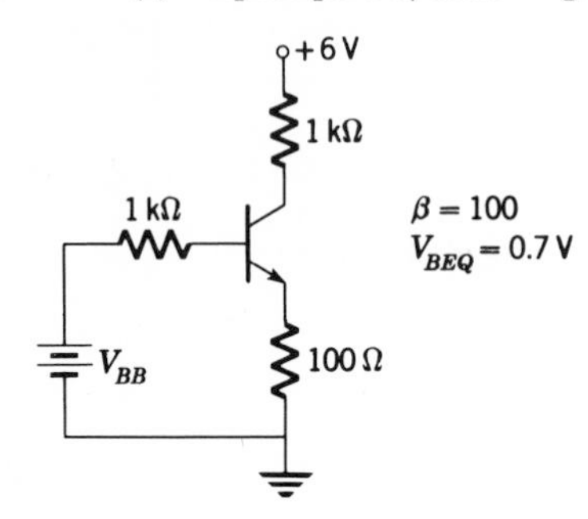

Fig. P3.13

3.14. In the circuit of Fig. P3.13, $V_{BB} = 1.2$ V and $\beta = 20$. Find the maximum possible symmetrical collector swing and the efficiency.

SECTION 3.5

3.15. (*a*) Find R_1 and R_2 so that $I_{CQ} = 10$ mA. (Be sure that $R_b \ll \beta R_e$.)

(*b*) Find the maximum symmetrical collector swing possible with these values of R_1 and R_2.

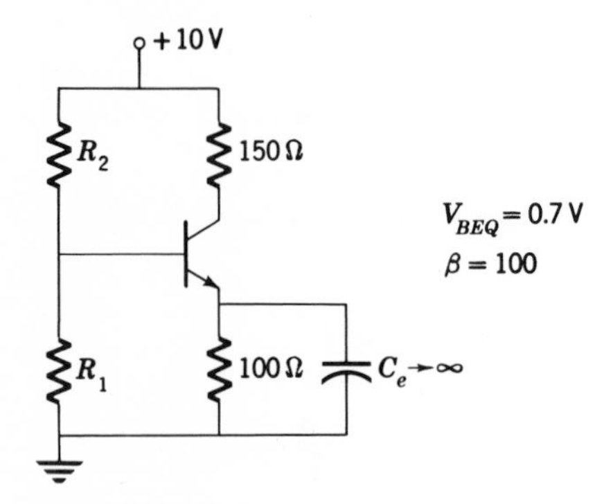

Fig. P3.15

3.16. In Fig. P3.15 find R_1 and R_2 for maximum symmetrical collector-current swing. Specify the Q point for this condition.

3.17. In Fig. P3.15 the maximum required collector-current swing is 10 mA peak-to-peak. In order to reduce the current demand on the power supply, I_{CQ} is to be as small as possible. Specify the Q point and the required values of R_1 and R_2, assuming that the transistor is cut off at $i_C = 0$.

SECTION 3.6

3.18. Find R_1, R_2, R_c, and R_e so that 40 mA peak ac current can flow in the 100-Ω load. Note that the solution is not unique.

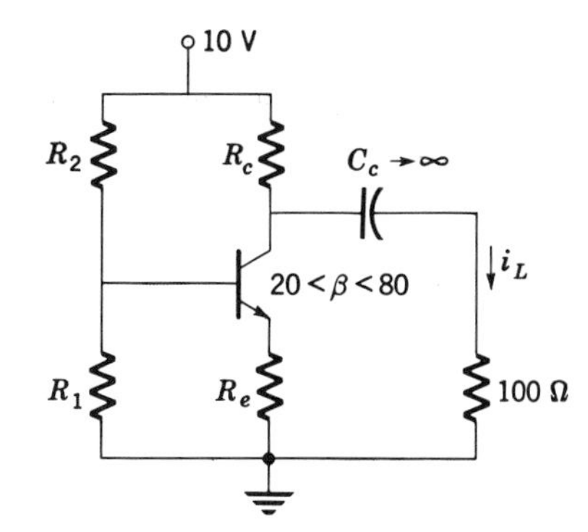

Fig. P3.18

3.19. For the final circuit of Prob. 3.18, find the maximum peak ac current that can flow in the load if $V_{CE,\text{sat}} = 0.5$ V.

3.20. Find R_1 and R_2 for maximum symmetrical load-current swing. Find the collector- and load-current swings for this condition.

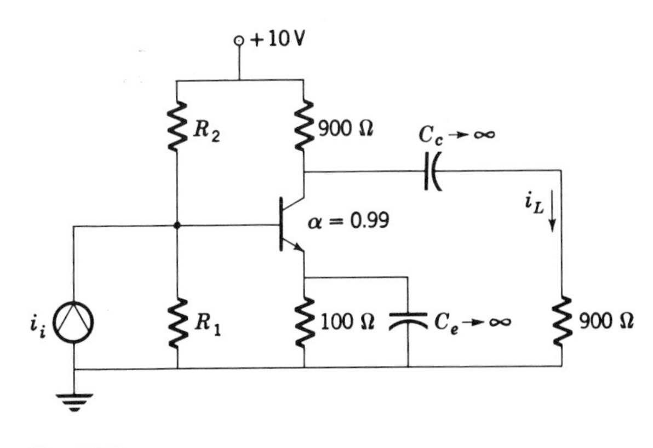

Fig. P3.20

3.21. Find R_c for a maximum symmetrical output voltage v_L.

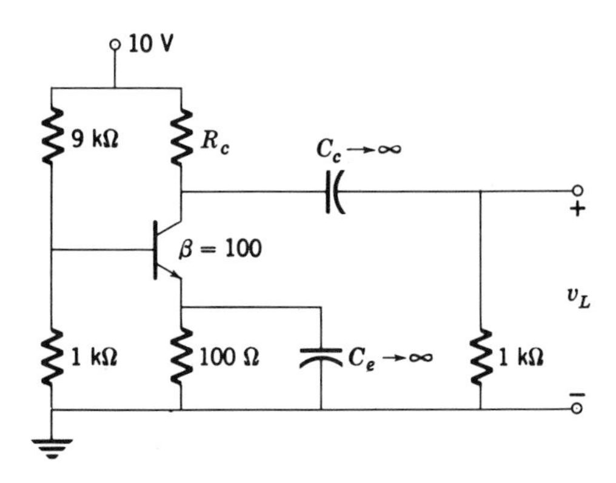

Fig. P3.21

SECTION 3.7

3.22. Find the maximum possible symmetrical swing in v_L.

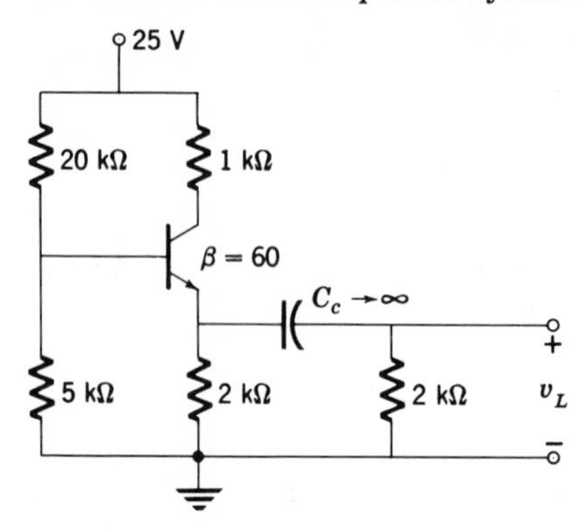

Fig. P3.22

3.23. (*a*) Find the Q point.

(*b*) Another transistor of the same type is to be plugged into the same circuit. What is the minimum β that the new transistor may have if the quiescent collector current is not to change by more than 10 per cent?

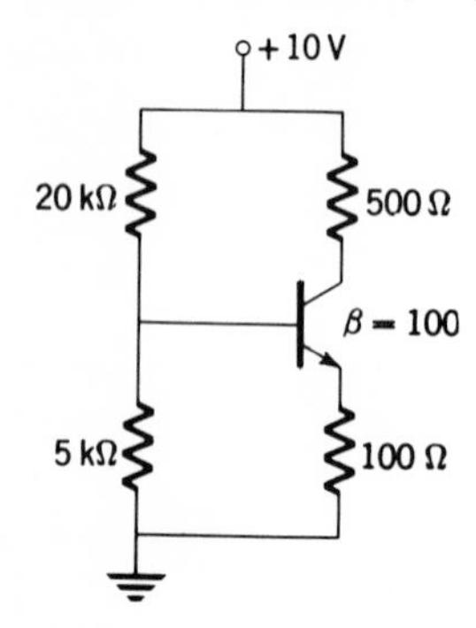

Fig. P3.23

3.24. Find the Q point and the maximum symmetrical v_L.

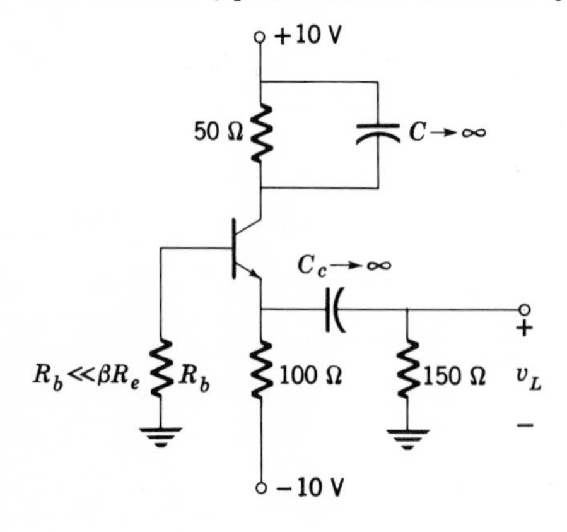

Fig. P3.24

3.25. (*a*) Find R_2 so that $I_{CQ} = 5$ mA.
(*b*) Find the maximum undistorted value of v_L with this value of I_{CQ}.

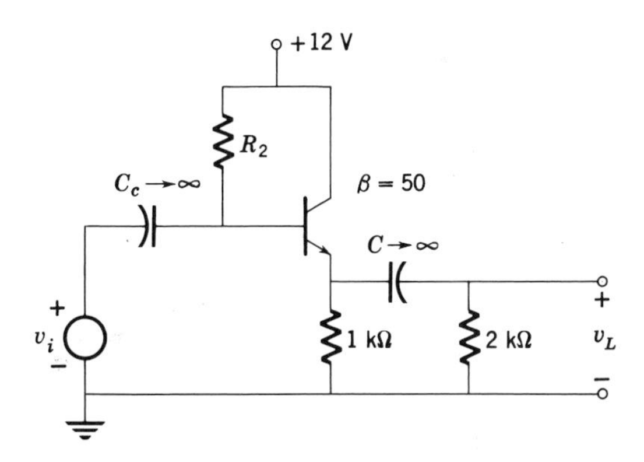

Fig. P3.25

3.26. (*a*) Find V_{BB} so that $V_{EQ} = 0$.
(*b*) Find the maximum v_L.

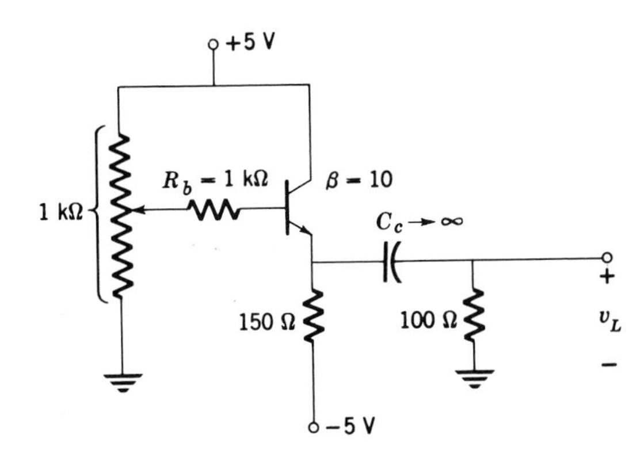

Fig. P3.26

3.27. (*a*) If $v_i = 0$, find v_C.
(*b*) If $v_i = -3$ V dc, find v_C.
(*c*) Find v_i for $v_C = 2.5$ V.
(*d*) Find v_C and v_i for cutoff and saturation.

(*e*) Plot v_C versus v_i for $-6 < v_i < 6$. This is the *transfer* characteristic of the amplifier.

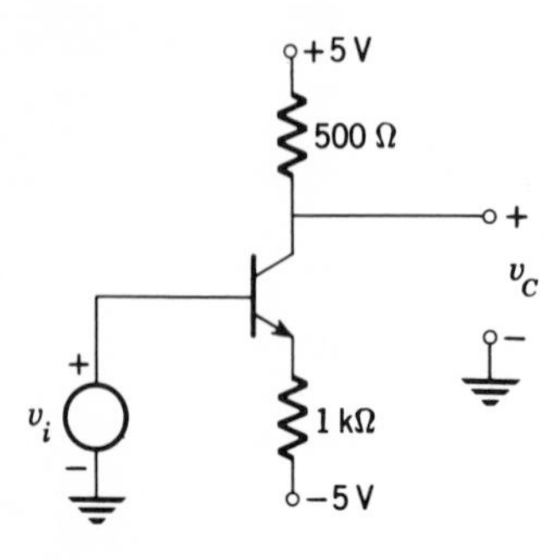

Fig. P3.27

3.28. (*a*) Plot the CB characteristics i_C versus v_{CB} for the transistor.

(*b*) Find the Q point and plot with dc and ac load lines on the CB characteristics.

(*c*) Find the maximum symmetrical v_L.

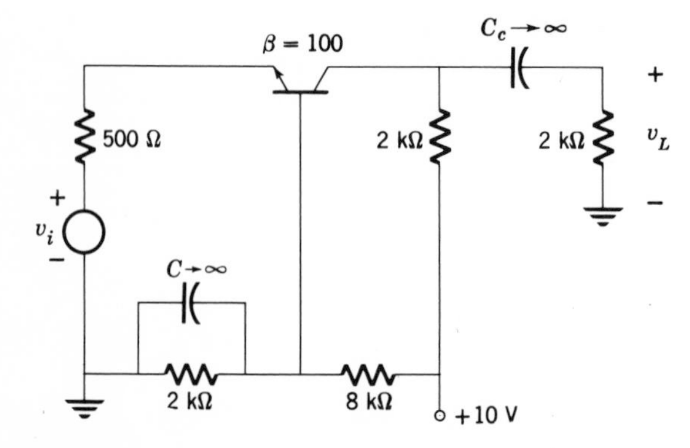

Fig. P3.28

3.29. In the amplifier of Prob. 3.20 a sinusoidal signal is applied which is sufficient to produce maximum symmetrical load-current swing. Under this condition, calculate the power dissipated in each resistance (ac and dc) and in the transistor. Compare with the power drawn from the V_{CC} supply. Neglect the power drawn from the signal source and the ac power dissipated in the bias network.

REFERENCES

1. Grove, A. S.: "Semiconductor Physics," John Wiley & Sons, Inc., New York, 1967. Motorola Inc. Engineering Staff: "Integrated Circuits," McGraw-Hill Book Company, New York, 1965.
2. Brenner, E., and M. Javid: "Analysis of Electric Circuits," McGraw-Hill Book Company, New York, 1959.
3. Searle, C. L., et al.: "Elementary Circuit Properties of Transistors," SEEC, vol. 3, John Wiley & Sons, Inc., New York, 1964.

4
Bias Stability

INTRODUCTION

In the practical design of transistor circuits, the quiescent operating (Q) point is carefully established in order that the transistor operate over a specified range and to ensure that linearity (and perhaps a maximum linear swing) is achieved and that $P_{C,\max}$ is not exceeded. Once a design has been completed, it is necessary to check for quiescent-point variations due to temperature changes and possible unit-to-unit amplifier-parameter variations. These variations must be kept within acceptable limits as set by the specifications.

Among the parameters which cause a shift of the Q point are variations in the collector cutoff current I_{CBO}, the current amplification factor β, the quiescent base-emitter voltage V_{BEQ}, the supply voltages, and the circuit resistances. All these are independent parameters. The current I_{CBO} varies with temperature; V_{BEQ}, for which we have assumed a nominal value of 0.7 V for silicon units, can vary from 0.5 to 0.9 V, depending on the transistor employed and the quiescent current, and is also a function of temperature. As pointed out previously, β often varies by 3 (or more) to 1 for a particular transistor type. Resistors used in transistor circuits often have a 10 percent tolerance, and power-supply voltages may fluctuate if they are not adequately regulated.

In this chapter we discuss all these factors, along with methods for minimizing the effect of their changes on the Q point.

4.1 QUIESCENT-POINT VARIATION DUE TO UNCERTAINTY IN β[1]

When transistors first appeared, engineers used common-emitter characteristics such as those shown in Fig. 4.1-1 to establish the Q point for an amplifier. A suitable dc load line was drawn as shown, and the Q point was established at some base current I_{BQ}. The input circuit was designed to maintain the quiescent *base* current at this value. A problem arose, however, when the amplifier was mass-produced. Since the β of various transistors of the same type may be subject to a typical variation of 3:1, the quiescent collector current is subject to the same variation (if I_{BQ} is held constant). This has the effect of changing the scale of i_C so that the Q point, with I_{BQ} fixed, could be at saturation in one circuit, where the β was high, or at cutoff in another, where the β was low.

When we studied the common-emitter amplifier in Sec. 3.3 we found that the quiescent collector current could be stabilized against unit-to-unit β variation by using an emitter resistor and maintaining a certain relation between the base- and emitter-circuit resistances. This relation is given by the inequality

$$R_e \gg R_b(1 - \alpha) \tag{4.1-1}$$

We shall now show that when this inequality is satisfied, the Q point is essentially independent of the transistor characteristics. The circuit

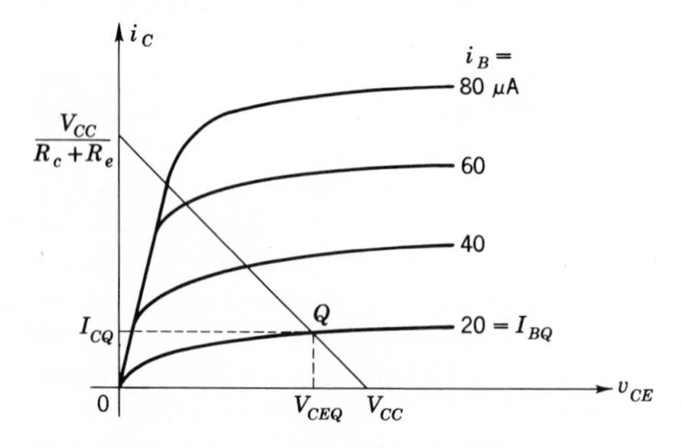

Fig. 4.1-1 Common-emitter characteristics.

employed is shown in Fig. 4.1-2. It should be noted that while Fig. 4.1-2 can be thought of as a common-emitter circuit, it also represents the dc circuit for the common-base configuration, and if we let R_c equal zero, the circuit represents the emitter-follower configuration. Thus Fig. 4.1-2 and all the results which follow apply equally well to all three configurations.

Using results obtained in Sec. 3.3, we have

For the collector current:

$$I_C = \beta I_B + (\beta + 1)I_{CBO} \tag{4.1-2}$$

For the collector circuit using KVL:

$$V_{CC} = I_C R_c + I_E R_e + V_{CE} \tag{4.1-3}$$

For the base circuit using KVL:

$$V_{BB} = I_B R_b + V_{BE} + I_E R_e \tag{4.1-4}$$

From (3.1-3):

$$I_B = (1 - \alpha)I_E - I_{CBO} \tag{4.1-5}$$

Combining (4.1-4) and (4.1-5), we obtain

$$V_{BB} = V_{BE} - I_{CBO}R_b + I_E[R_e + (1 - \alpha)R_b] \tag{4.1-6}$$

Now using

$$I_C = \alpha I_E + I_{CBO} \tag{4.1-7}$$

in (4.1-6), the quiescent collector current becomes

$$I_{CQ} = \frac{\alpha(V_{BB} - V_{BE}) + I_{CBO}(R_e + R_b)}{R_e + (1 - \alpha)R_b} \tag{4.1-8}$$

The quiescent collector-emitter voltage V_{CEQ} can be obtained from (4.1-3) and (4.1-8). These equations can be simplified considerably by making two practical assumptions:

1. $\alpha \approx 1$ (4.1-9a)

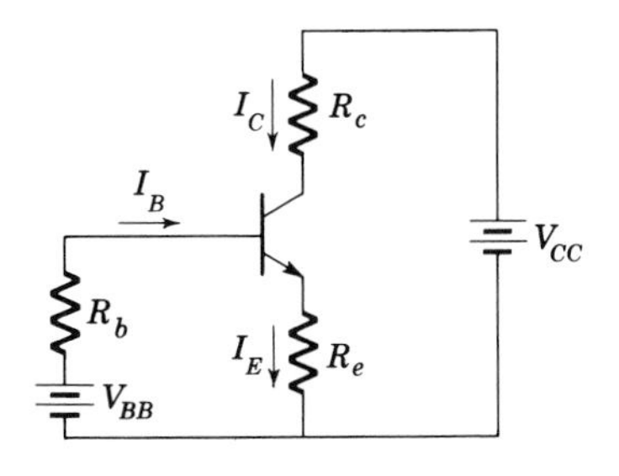

Fig. 4.1-2 Common-emitter circuit.

2. I_{CBO} is very small in silicon transistors, so that (4.1-7) becomes

$$I_C \approx \alpha I_E \approx I_E \tag{4.1-9b}$$

$$I_{CBO}(R_e + R_b) \ll \alpha(V_{BB} - V_{BE}) \approx V_{BB} - V_{BE} \tag{4.1-9c}$$

With these assumptions, (4.1-3) becomes

$$V_{CC} \approx V_{CE} + I_C(R_c + R_e) \tag{4.1-10}$$

and (4.1-8) becomes

$$I_{CQ} \approx \frac{V_{BB} - V_{BE}}{R_e + (1 - \alpha)R_b} \tag{4.1-11a}$$

If the inequality (4.1-1) is satisfied, (4.1-11a) simplifies to

$$I_{CQ} \approx \frac{V_{BB} - V_{BE}}{R_e} \approx \frac{V_{BB} - 0.7}{R_e} \tag{4.1-11b}$$

since V_{BE} is assumed equal to 0.7 V for silicon units.

The location of the Q point as given by (4.1-11b) is seen to be independent of β when the inequality of (4.1-1) is satisfied.

A design procedure can now be prescribed.

1. Choose a suitable dc load line and Q point on the basis of considerations such as available supply voltage V_{CC}, desired current swing, desired quiescent power dissipation, etc.

   The slope of the dc load line then fixes $R_c + R_e$, and the intercept on the v_{CE} axis fixes V_{CC}.
2. V_{BB}, R_b, and either R_c or R_e remain to be determined, and considerable latitude is available. Equation (4.1-11b) can be used to determine V_{BB} once a suitable value of R_e has been chosen. For a given value of R_e, (4.1-1) represents an upper bound on R_b. In Chap. 6 we show that R_b should be as large as possible so that current gain is not lost through attenuation in the input circuit. With this requirement in mind, we assume that a factor of 10 will satisfy the inequality, and write

$$R_b(1 - \alpha_{\min}) = \frac{R_e}{10} \tag{4.1-12a}$$

and since $1/(1 - \alpha) = 1 + \beta \approx \beta$,

$$R_b \approx \frac{\beta_{\min} R_e}{10} \tag{4.1-12b}$$

This fixes R_b, and since V_{BB} and V_{CC} are known, the practical bias circuit of Fig. 3.3-1 can be determined using (3.3-1c) and (3.3-1d).

Example 4.1-1 In the circuit of Fig. 4.1-2, let $V_{CC} = 10$ V, $V_{BB} = 1.75$ V, $I_{CQ} = 10$ mA, $V_{CEQ} = 5$ V, $R_c = 400\ \Omega$, and $40 \leq \beta \leq 120$.

Find suitable values for (*a*) R_e and (*b*) R_b. (*c*) Calculate the variation in Q point as β varies over its total indicated range.

Solution The specifications give enough information to determine the load line and locate the Q point shown in Fig. 4.1-3.

(*a*) From the load line

$$R_c + R_e = \frac{10}{20 \times 10^{-3}} = 500\ \Omega$$

and since $R_c = 400\ \Omega$,

$$R_e = 100\ \Omega$$

(*b*) From (4.1-12*b*)

$$R_b \approx \frac{\beta_{\min} R_e}{10} = \frac{(40)(100)}{10} = 400\ \Omega$$

(*c*) To calculate the effect of β on the Q point, we use (4.1-11*a*), evaluated for the quiescent conditions

$$V_{BB} = V_{BEQ} + I_{EQ}[R_e + (1 - \alpha)R_b]$$

Assuming $\alpha \approx 1$, $I_{CQ} \approx I_{EQ}$, and $1 - \alpha \approx 1/\beta$, this becomes

$$I_{CQ} \approx \frac{V_{BB} - 0.7}{R_e + R_b/\beta}$$

When $\beta = 40$,

$$I_{CQ} = \frac{1.75 - 0.7}{100 + 400/40} = \frac{1.05}{110} \approx 9.5 \text{ mA}$$

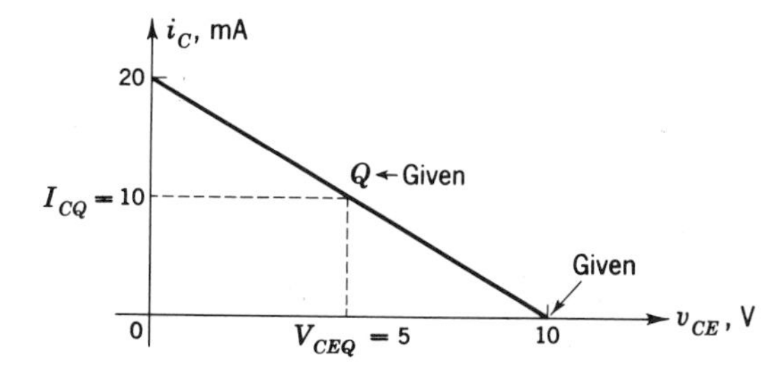

Fig. 4.1-3 Load lines for Example 4.1-1.

When $\beta = 120$,

$$I_{CQ} = \frac{1.75 - 0.7}{100 + 400/120} \approx 10.2 \text{ mA}$$

Thus a 3:1 variation in β produces a negligible shift in the Q point.

4.2 THE EFFECT OF TEMPERATURE ON THE Q POINT[1]

In the preceding section the variation of Q-point position with respect to unit-to-unit variations in β, and a biasing arrangement which minimized these variations, were discussed. Another important cause of Q-point variation is the transistor operating temperature. As we shall see, the effects of temperature variations are much reduced, and are usually negligible when silicon, rather than germanium, transistors are used in low-power ac-coupled amplifiers. However, even with silicon, temperature compensation must often be used in power amplifiers and dc-coupled amplifiers, to avoid significant Q-point shifts. In this section we study the variation of the Q point due to the dependence of I_{CBO} and V_{BE} on temperature.

We begin the analysis with the exact expression for the quiescent collector current [Eq. (4.1-8)]. This expression can be simplified by noting that $\alpha \approx 1$ and usually $R_e + (1 - \alpha)R_b \approx R_e$, because we have stabilized against variations of β as in Sec. 4.1. Thus (4.1-8) becomes

$$I_{CQ} \approx \frac{V_{BB} - V_{BE}}{R_e} + I_{CBO}\left(1 + \frac{R_b}{R_e}\right) \tag{4.2-1}$$

This is the desired relation between collector current and the two temperature-dependent variables I_{CBO} and V_{BE}.

In the preceding section we neglected I_{CBO} and used (4.1-11b) to find the quiescent collector current. The assumptions leading to (4.1-11b) are valid at room temperature. Equation (4.2-1) is now used to investigate the validity of (4.1-11b) at elevated temperatures for both silicon and germanium.

The base-emitter voltage V_{BE} is found to decrease linearly with temperature according to the relation

$$\Delta V_{BE} = V_{BE2} - V_{BE1} = -k(T_2 - T_1) \tag{4.2-2}$$

where

$$k \approx 2.5 \text{ mV/°C} \qquad T \text{ in °C}$$

The reverse saturation current I_{CBO} approximately doubles for every 10°C temperature rise. This implies that

$$I_{CBO2} = I_{CBO1}(\epsilon^{K(T_2 - T_1)}) \tag{4.2-3}$$

where

$$K \approx 0.07/°\text{C} \qquad T \text{ in } °\text{C}$$

Typical values for the I_{CBO} of low-power transistors at room temperatures are

Silicon: $1\ \mu\text{A}$ or less

Germanium: $100\ \mu\text{A}$

For germanium, the effect of changes in V_{BE} is insignificant compared with the effect of changes in I_{CBO}. The opposite is true when dealing with silicon, because I_{CBO} is extremely small at room temperature. We demonstrate this by a calculation of the collector-current change due to a temperature change.

The variation of I_{CQ} with temperature can be found from (4.2-1). Assuming that only V_{BE} and I_{CBO} vary,

$$\frac{\Delta I_{CQ}}{\Delta T} = -\left(\frac{1}{R_e}\right)\left(\frac{\Delta V_{BE}}{\Delta T}\right) + \left(1 + \frac{R_b}{R_e}\right)\left(\frac{\Delta I_{CBO}}{\Delta T}\right) \tag{4.2-4}$$

From (4.2-2), with $\Delta T = T_2 - T_1$,

$$\frac{\Delta V_{BE}}{\Delta T} = -k \tag{4.2-5}$$

and from (4.2-3),

$$\frac{\Delta I_{CBO}}{\Delta T} = \frac{I_{CBO2} - I_{CBO1}}{\Delta T} = \frac{I_{CBO1}(\epsilon^{K\,\Delta T} - 1)}{\Delta T} \tag{4.2-6}$$

Substituting (4.2-5) and (4.2-6) into (4.2-4), we get

$$\frac{\Delta I_{CQ}}{\Delta T} = \frac{k}{R_e} + \left(1 + \frac{R_b}{R_e}\right) I_{CBO1}\left(\frac{\epsilon^{K\,\Delta T} - 1}{\Delta T}\right) \tag{4.2-7}$$

from which

$$\Delta I_{CQ} = \frac{k\,\Delta T}{R_e} + \left(1 + \frac{R_b}{R_e}\right) I_{CBO1}(\epsilon^{K\,\Delta T} - 1) \tag{4.2-8}$$

In the example to follow, some typical values will be calculated.

Example 4.2-1 Consider the circuit of Fig. 4.1-2, with $R_b = 400\ \Omega$, $R_e = 100\ \Omega$, and $I_{CQ} = 10$ mA, at room temperature (25°C). Calculate the change in I_{CQ} if the temperature increases to 55°C, for (*a*) silicon and (*b*) germanium transistors.

Solution Substituting the given values in (4.2-8) with $\Delta T = 30°\text{C}$,

$$\Delta I_{CQ} = \frac{(2.5)(10^{-3})(30)}{100} + (1 + 4)I_{CBO1}(7.2) = 0.75 \times 10^{-3} + 36I_{CBO1}$$

(*a*) A typical value for a low-power silicon transistor is

$$I_{CBO1} = 1\ \mu\text{A}$$

Therefore

$$\Delta I_{CQ} = (0.75 + 0.036) \times 10^{-3} = 0.786\ \text{mA}$$

(*b*) A typical value for a low-power germanium transistor is

$$I_{CBO1} = 100\ \mu\text{A}$$

Therefore

$$\Delta I_{CQ} = (0.75 + 3.6) \times 10^{-3} = 4.35\ \text{mA}$$

We see from this example that the shift in I_{CQ} for the germanium transistor is almost 50 percent of the design current value, and might well make the amplifier operation unsatisfactory. On the other hand, the shift in I_{CQ} for the silicon transistor would be negligible for most applications. Thus the use of silicon transistors greatly reduces the problem of Q-point stabilization against changes in ambient temperature. Note also that 95 percent of the current change for the silicon unit is due to the change in base-emitter voltage V_{BE}.

4.3 STABILITY-FACTOR ANALYSIS

The method of analysis presented in this section is one that is often used in engineering practice. Briefly stated, the problem is the following: Given a physical variable (in our case, I_{CQ}), what change will it undergo when the variables on which it depends (in our case, I_{CBO}, V_{BE}, β, V_{CC}, etc.) change by prescribed (usually small) amounts? This type of analysis goes under various names: sensitivity analysis, variability analysis, and stability-factor analysis, for example. All these methods are based on the assumption that, for *small* changes, the variable of interest is a linear function of the other variables and can be expressed in the form of a total differential. For our case, we write

$$I_{CQ} = I_{CQ}(I_{CBO}, V_{BE}, \beta, \ldots) \tag{4.3-1}$$

Then the total differential is

$$dI_{CQ} = \left(\frac{\partial I_{CQ}}{\partial I_{CBO}}\right) dI_{CBO} + \left(\frac{\partial I_{CQ}}{\partial V_{BE}}\right) dV_{BE} + \left(\frac{\partial I_{CQ}}{\partial \beta}\right) d\beta + \cdots \tag{4.3-2}$$

Now we define *stability factors*

$$S_I = \frac{\Delta I_{CQ}}{\Delta I_{CBO}} \approx \frac{\partial I_{CQ}}{\partial I_{CBO}} \tag{4.3-3a}$$

$$S_V = \frac{\Delta I_{CQ}}{\Delta V_{BE}} \approx \frac{\partial I_{CQ}}{\partial V_{BE}} \tag{4.3-3b}$$

$$S_\beta = \frac{\Delta I_{CQ}}{\Delta \beta} \approx \frac{\partial I_{CQ}}{\partial \beta} \tag{4.3-3c}$$

where the partial derivatives must be evaluated at the nominal (design) Q point. If the changes are reasonably small,

$$\Delta I_{CQ} \approx dI_{CQ} \qquad \Delta V_{BE} \approx dV_{BE} \qquad \Delta\beta \approx d\beta \tag{4.3-4}$$

and finally,

$$\Delta I_{CQ} \approx S_I\,\Delta I_{CBO} + S_V\,\Delta V_{BE} + S_\beta\,\Delta\beta + \cdots \tag{4.3-5}$$

From this relation we can easily find ΔI_{CQ} for changes in any of the independent variables, provided the changes are small enough so that our assumption that the increment ΔI is approximately equal to the differential dI is valid. If the assumption is not valid, we must calculate the actual increment. (This will usually be the case when variations in β are considered.*)

Equation (4.1-8) is the complete relation between I_{CQ} and the variables of interest for the conventional CE amplifier. Taking the partial derivatives as indicated by (4.3-3*a*) and (4.3-3*b*), we find, assuming $R_e \gg (1 - \alpha)R_b$,

$$S_I = \frac{\partial I_{CQ}}{\partial I_{CBO}} = \frac{R_e + R_b}{R_e + (1-\alpha)R_b} \approx 1 + \frac{R_b}{R_e} \tag{4.3-6a}$$

$$S_V = \frac{\partial I_{CQ}}{\partial V_{BE}} = \frac{-\alpha}{R_e + (1-\alpha)R_b} \approx -\frac{1}{R_e} \tag{4.3-6b}$$

To find S_β we must calculate the actual increment because of the large change involved. The first step is to rewrite (4.1-8), using $\alpha = \beta/(\beta + 1)$. Since $V_{BB} - V_{BE} \gg I_{CBO}(R_e + R_b)$ in the active region, the I_{CBO} terms will be neglected. Thus

$$I_{CQ} \approx \frac{\beta(V_{BB} - V_{BE})}{R_b + (\beta + 1)R_e} \tag{4.3-7}$$

Now let β_2 and β_1 represent the upper and lower limits, respectively, on β, with I_{CQ2} and I_{CQ1} the corresponding collector currents. Next, we

* If I_{CQ} is a linear function of a variable x, then $\Delta I_{CQ}/\Delta x = dI_{CQ}/dx$, even when Δx is large (since the slope of a straight line is a constant). However, if I_{CQ} is a nonlinear function of x, $\Delta I_{CQ}/\Delta x$ does not represent the slope of the curve except in the limit where $\Delta x \rightarrow dx$.

form the ratio

$$\frac{I_{CQ2}}{I_{CQ1}} = \left(\frac{\beta_2}{\beta_1}\right)\left[\frac{R_b + (\beta_1 + 1)R_e}{R_b + (\beta_2 + 1)R_e}\right] \tag{4.3-8}$$

Now unity is subtracted from both sides of (4.3-8), and the result manipulated to yield

$$\frac{I_{CQ2} - I_{CQ1}}{I_{CQ1}} = \frac{\Delta I_{CQ}}{I_{CQ1}} = \frac{\Delta\beta(R_b + R_e)}{\beta_1[R_b + (\beta_2 + 1)R_e]} \tag{4.3-9}$$

where

$$\Delta I_{CQ} = I_{CQ2} - I_{CQ1} \qquad \text{and} \qquad \Delta\beta = \beta_2 - \beta_1$$

Therefore

$$S_\beta \equiv \frac{\Delta I_{CQ}}{\Delta\beta} = \left(\frac{I_{CQ1}}{\beta_1}\right)\left[\frac{R_b + R_e}{R_b + (\beta_2 + 1)R_e}\right] \tag{4.3-10}$$

Thus, finally,

$$\Delta I_{CQ} \approx \left(1 + \frac{R_b}{R_e}\right)\Delta I_{CBO} - \left(\frac{1}{R_e}\right)\Delta V_{BE} + \left(\frac{I_{CQ1}}{\beta_1}\right)\left[\frac{R_b + R_e}{R_b + (\beta_2 + 1)R_e}\right]\Delta\beta + \cdots \tag{4.3-11}$$

The increments in I_{CBO} and V_{BE} can be related directly to temperature, by making use of (4.2-2) and (4.2-3). Thus

$$\Delta V_{BE} = -k\,\Delta T \tag{4.2-2}$$

and

$$\Delta I_{CBO} = I_{CBO1}(\epsilon^{K\,\Delta T} - 1) \tag{4.2-3}$$

Note that it is relatively easy to take into account other factors which might affect I_{CQ}. For example, if changes in I_{CQ} due to changes in V_{CC} and R_e are also to be found, we write

$$\Delta I_{CQ} = S_I\,\Delta I_{CBO} + S_V\,\Delta V_{BE} + S_\beta\,\Delta\beta + S_{V_{CC}}\,\Delta V_{CC} + S_{R_e}\,\Delta R_e \tag{4.3-12}$$

where

$$S_{V_{CC}} \approx \frac{\partial I_{CQ}}{\partial V_{CC}} \qquad \text{and} \qquad S_{R_e} \approx \frac{\partial I_{CQ}}{\partial R_e} \tag{4.3-13}$$

for small changes of I_{CQ}.

Equation (4.3-12) shows that the total change in the quiescent current, ΔI_{CQ}, is proportional to the changes in each of the independent variables and to their stability factors. Thus, to design for small ΔI_{CQ}, we design to *minimize* the stability factors. In the usual application of

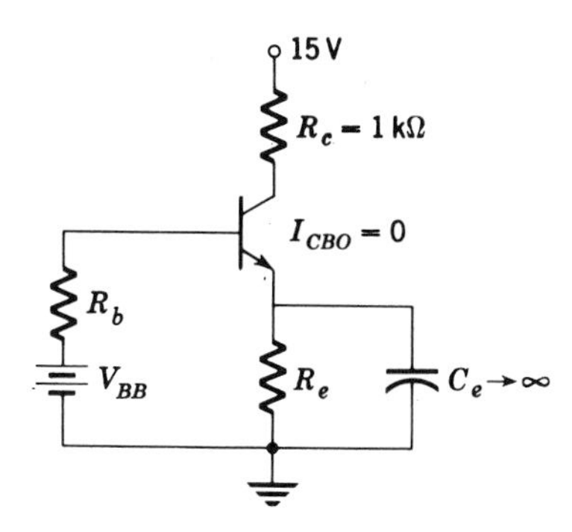

Fig. 4.3-1 Circuit for Example 4.3-1.

(4.3-12), extreme accuracy is not required, and it is customary to use this type of analysis for relatively large changes (20 percent or more) and to exercise engineering judgment in the interpretation of the results. The examples which follow illustrate typical orders of magnitude for these quantities for a silicon transistor.

Example 4.3-1 Find V_{BB}, R_b, and R_e for the amplifier shown in Fig. 4.3-1, so that i_C can swing by at least ± 5 mA. β varies from 40 to 120, and V_{BEQ} is between 0.6 and 0.8 V. The collector-emitter saturation voltage $V_{CE,\text{sat}}$ is 0.1 V.

Solution To have a peak swing of ± 5 mA requires that the quiescent current I_{CQ} be bounded by I_{CQ1} and I_{CQ2}, as shown on Fig. 4.3-2. The point Q_1 ensures that i_C can swing -5 mA before cutoff, while

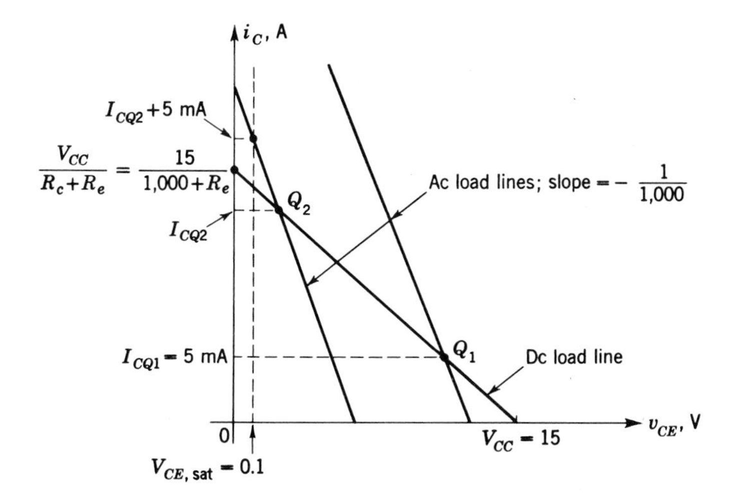

Fig. 4.3-2 Load lines for Example 4.3-1.

Q_2 ensures that i_C can swing $+5$ mA before saturation. Assuming cutoff at $i_C = 0$, $I_{CQ1} = 5$ mA. From Fig. 4.3-2, I_{CQ2} is seen to be given by the intersection of the dc and ac load lines.

Dc load line: $\quad V_{CC} = V_{CEQ2} + I_{CQ2}(R_c + R_e)$

Ac load line: $\quad V_{CEQ2} - V_{CE,\text{sat}} = R_c(I_{CQ2} + 5 \times 10^{-3} - I_{CQ2})$

Eliminating V_{CEQ2},

$$I_{CQ2} = \frac{V_{CC} - 5 \times 10^{-3}R_c - V_{CE,\text{sat}}}{R_c + R_e}$$

$$= \frac{(15) - (5) - (0.1)}{1000 + R_e} \approx \frac{10}{1000 + R_e}$$

Therefore the quiescent collector current must satisfy the inequality

$$I_{CQ1} \leq I_{CQ} \leq I_{CQ2}$$

$$5 \text{ mA} \leq I_{CQ} \leq \frac{10}{1000 + R_e} \qquad \text{mA}$$

Solving, we find that R_e must be less than 1 kΩ in order to ensure that the Q point will be between Q_1 and Q_2.

The next step in the solution is to investigate the effect of the expected β and V_{BEQ} variations. The pertinent relation is (4.3-7), which is repeated for convenience:

$$I_{CQ} = \frac{\beta(V_{BB} - V_{BE})}{R_b + (\beta + 1)R_e} \approx \frac{V_{BB} - V_{BE}}{R_e} \qquad (\beta + 1)R_e \gg R_b$$

From (4.3-11)

$$\Delta I_{CQ} \approx -\frac{1}{R_e}\Delta V_{BE} + \left(\frac{I_{CQ1}}{\beta_1}\right)\left[\frac{R_b + R_e}{R_b + (\beta_2 + 1)R_e}\right]\Delta\beta$$

The worst case will occur when $\Delta V_{BE} = -0.2$ V and

$$\Delta\beta = 120 - 40 = 80$$

Then

$$\Delta I_{CQ} \approx \frac{0.2}{R_e} + \frac{I_{CQ1}}{40}\left[\frac{(R_b + R_e)(80)}{R_b + 120R_e}\right]$$

$$\approx \frac{0.2}{R_e} + \left(\frac{I_{CQ1}}{60}\right)\left(1 + \frac{R_b}{R_e}\right) \qquad 120R_e \gg R_b$$

Clearly, a wide range of values of R_b, R_e, and V_{BB} will maintain the Q point between the limits Q_1 and Q_2. If we choose

$$R_e = 500\ \Omega$$
$$R_b = 2000\ \Omega$$
$$V_{BB} = 3.7\text{ V}$$

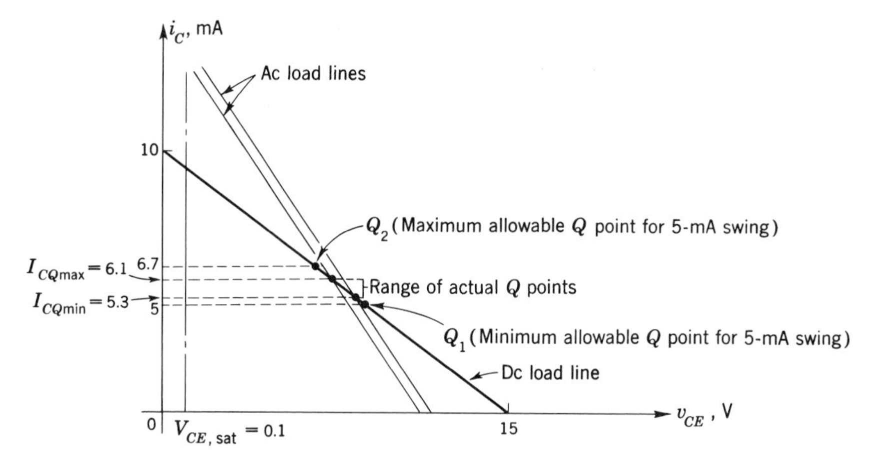

Fig. 4.3-3 Q-point variation for Example 4.3-1.

then the lower limit on I_{CQ} occurs when $V_{BE} = 0.8$ V and $\beta = 40$. From (4.3-7)

$$I_{CQ,\text{min}} = \frac{(40)(3.7 - 0.8)}{2000 + (40)(500)} \approx 5.3 \text{ mA}$$

and

$$\Delta I_{CQ} = \frac{0.2}{500} + \frac{5.3}{60} \times 10^{-3}\left(1 + \frac{2000}{500}\right) \approx 0.8 \text{ mA}$$

Thus

$$I_{CQ,\text{max}} = 5.3 + 0.8 = 6.1 \text{ mA}$$

The load lines shown in Fig. 4.3-3 illustrate the possible Q-point variation under these conditions. The collector current is seen to be capable of ± 5-mA swing for the specified range of β and V_{BE}.

Example 4.3-2 (*a*) Find I_{CQ} at room temperature, using the nominal values given.

(*b*) Find ΔI_{CQ} for the indicated tolerances on V_{CC}, R_e, and β. The ambient temperature ranges from 25 to +125°C.

Solution (*a*) From (4.1-8), the nominal quiescent current is

$$I_{CQ} = \frac{\alpha(V_{BB} - V_{BEQ}) + I_{CBO}(R_e + R_b)}{R_e + (1 - \alpha)R_b}$$

$$= \frac{\alpha\{[R_1/(R_1 + R_2)]V_{CC} - V_{BEQ}\} + I_{CBO}(R_e + R_b)}{R_e + (1 - \alpha)R_b}$$

$$\approx \frac{(0.5/5.5)(20) - 0.7}{100 + 455/76} = \frac{1.12}{106} = 10.6 \text{ mA}$$

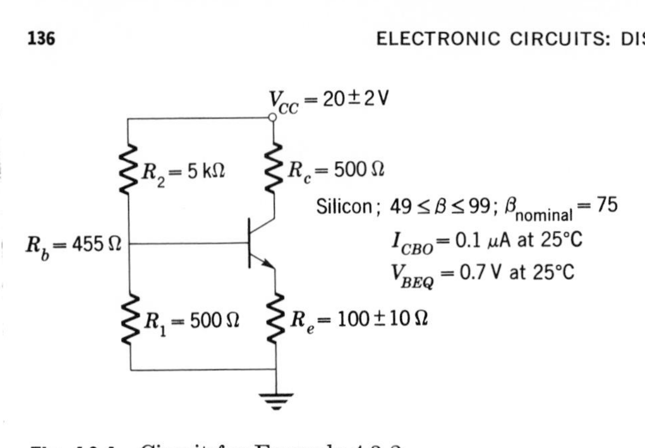

Fig. 4.3-4 Circuit for Example 4.3-2.

(*b*) To find ΔI_{CQ} we use (4.3-5). The required stability factors are S_I, S_V, S_β, $S_{V_{CC}}$, and S_{R_e}.

From (4.3-6*a*)

$$S_I = \frac{R_e + R_b}{R_e + (1 - \alpha)R_b} = \frac{100 + 455}{100 + 6} = 5.25 \text{ mA/mA}$$

and since $R_e \gg (1 - \alpha)R_b$, the approximations in (4.3-6) can be used. From (4.3-6*b*)

$$S_V \approx -\frac{1}{R_e} = -0.01 \text{ A/V} = -10 \text{ mA/V}$$

From (4.3-10)

$$S_\beta \approx \frac{10.6}{49} \times 10^{-3} \left(\frac{455 + 100}{455 + 10{,}000}\right) = 0.0116 \text{ mA per unit change in } \beta$$

To find $S_{V_{CC}}$, we refer to (*a*). Then

$$S_{V_{CC}} = \frac{\partial I_{CQ}}{\partial V_{CC}} = \left[\frac{\alpha}{R_e + (1 - \alpha)R_b}\right]\left(\frac{R_1}{R_1 + R_2}\right) \approx \frac{R_1}{(R_1 + R_2)R_e}$$

$$= \frac{500}{(5500)(100)} = +0.91 \text{ mA/V}$$

To find S_{R_e}, we apply the definition to (4.1-8) (neglecting I_{CBO})

$$S_{R_e} = \left(\frac{\partial I_{CQ}}{\partial R_e}\right) \approx \left\{\frac{-\alpha(V_{BB} - V_{BE})}{[R_e + (1 - \alpha)R_b]^2}\right\} \approx \frac{-1.1}{106^2}$$

$$\approx -10^{-4} \text{ A/Ω} = -0.1 \text{ mA/Ω}$$

Thus all the stability factors are determined. The next step in the solution is to find ΔI_{CBO}, ΔV_{BE}, $\Delta\beta$, ΔV_{CC}, and ΔR_e.

Using (4.2-3),

$$\Delta I_{CBO} = I_{CBO1}(\epsilon^{K\,\Delta T} - 1) = 0.1 \times 10^{-6}(\epsilon^{(0.07)(100)} - 1) \approx 0.11 \text{ mA}$$

and from (4.2-2),

$$\Delta V_{BE} = -k\,\Delta T = -(2.5)(10^{-3})(100) = -250 \text{ mV}$$

From the specifications

$$\Delta\beta = 50$$
$$\Delta V_{CC} = 4 \text{ V}$$
$$\Delta R_e = 20\ \Omega$$

The *worst* possible Q-point shift from the *minimum* value will be

$$\Delta I_{CQ} = |S_I\,\Delta I_{CBO}| + |S_V\,\Delta V_{BE}| + |S_\beta\,\Delta\beta| + |S_{V_{CC}}\,\Delta V_{CC}| + |S_{R_e}\,\Delta R_e|$$
$$= (5.25)(0.11) + (10)(0.25) + (0.0116)(50) + (0.91)(4) + (0.1)(20)$$
$$= (0.58 + 2.5 + 0.56 + 3.64 + 2) \text{ mA}$$

and

$$\Delta I_{CQ} \le 9.3 \text{ mA}$$

The maximum possible variation from the *nominal* value of 10.6 mA will then be about one-half of this figure, or $\approx \pm 4.6$ mA.

The large variation in temperature resulted in approximately 3.1 mA of the total 9.3-mA quiescent-current variation. Note that this variation is due primarily to variations in V_{BE}. To reduce the variations due to V_{CC} and R_e, a regulated supply could be used so that $\Delta V_{CC} = 0.1$ V instead of 4 V, and a 1 rather than a 10 percent resistor employed so that $\Delta R_e = \pm 1\ \Omega$. Then

$$\Delta I_{CQ} \le \pm 2 \text{ mA}$$

Even this change, approximately 20 percent of the nominal I_{CQ}, is not insignificant. This analysis procedure is therefore extremely important since the information enables the designer to determine if the required peak collector-current swing can be obtained. If it cannot be obtained, a redesign is required.

4.4 TEMPERATURE COMPENSATION USING DIODE BIASING[2]

In preceding sections we saw that changes in ambient temperature could result in significant variation in quiescent collector current. With silicon transistors this variation is due primarily to the base-emitter voltage V_{BE}, which is a function of temperature (the effect of temperature on I_{CBO} is usually negligible when using silicon transistors).

One method of reducing this current variation becomes obvious when one considers the stability factor S_V of (4.3-6*b*). If R_e is increased, S_V

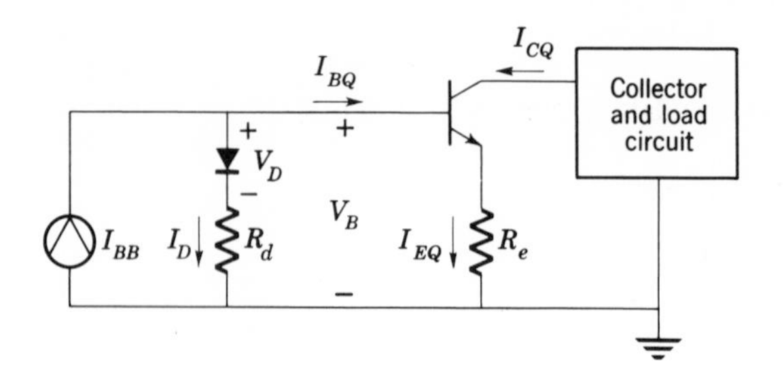

Fig. 4.4-1 Diode-biasing: simplified circuit.

decreases and ΔI_{CQ} is decreased. However, this also reduces the quiescent current. Thus this scheme of minimizing S_V cannot be carried very far.

An alternative method for reducing the base-emitter voltage variation is to use *diode compensation*. To understand this technique, consider the circuit of Fig. 4.4-1.

In this circuit, the bias is supplied by a constant *current* source, I_{BB}. If the diode is chosen to *match* the base-emitter junction characteristic of the transistor,

$$\frac{\Delta V_D}{\Delta T} = \frac{\Delta V_{BE}}{\Delta T} \tag{4.4-1}$$

The bias supply current is constant, so that

$$I_{BB} = I_D + I_{BQ} = I_D + \frac{I_{EQ}}{\beta + 1} = \text{constant} \tag{4.4-2}$$

The base voltage is found from Fig. 4.4-1.

$$V_B = V_D + I_D R_d = V_{BEQ} + I_{EQ} R_e \tag{4.4-3}$$

Thus, using (4.4-2) and (4.4-3), the quiescent emitter current is

$$I_{EQ} = \frac{V_D - V_{BEQ} + I_{BB} R_d}{R_e + [R_d/(\beta + 1)]} \tag{4.4-4}$$

Since I_{BB} is constant,

$$\frac{\Delta I_{EQ}}{\Delta T} = \frac{\Delta V_D/\Delta T - \Delta V_{BE}/\Delta T}{R_e + [R_d/(\beta + 1)]} = 0 \tag{4.4-5}$$

Thus I_{EQ} is insensitive to temperature variations.

Note also from (4.4-4) that the emitter current I_{EQ} is relatively independent of variations in β, if

$$R_e \gg \frac{R_d}{\beta + 1} \tag{4.4-6}$$

(See Prob. 4.17.)

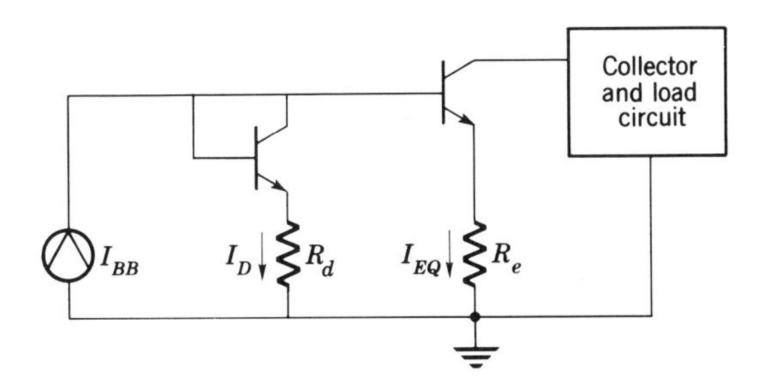

Fig. 4.4-2 Diode biasing with a transistor.

Clearly, the degree of temperature stabilization depends on the matching of the external diode to the base-emitter diode of the transistor. When using discrete components care should be used in selecting the diode.

An interesting way of almost eliminating this matching problem is to use a transistor (of the same type as the transistor being biased) connected as a diode, as shown in Fig. 4.4-2. In ordinary transistor-amplifier design the use of a second transistor is costly. In integrated-circuit design, however (Chap. 9), diodes are formed from transistors by connecting the base to the collector, as shown in the figure. In integrated circuits, the transistor is the most inexpensive component, and this biasing procedure is quite common. It is interesting to note at this point that in integrated circuits resistors are also "formed" from transistors, and hence it is less expensive to use transistors than resistors. In addition, in manufacture of integrated circuits, transistors are easily made to be almost identical.

The current source I_{BB} is obtained by connecting the supply voltage V_{CC}, in series with a large resistor R_b, as shown in Example 4.4-1.

Example 4.4-1 A practical diode biasing circuit, used in many integrated circuits, is shown in Fig. 4.4-3. Determine the effect of temperature changes on the quiescent current.

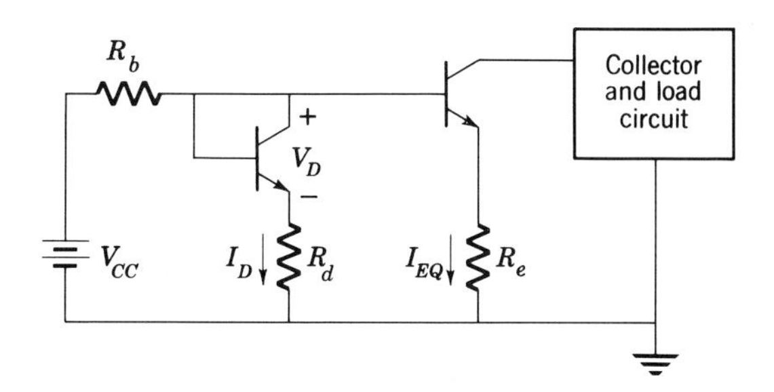

Fig. 4.4-3 Practical diode-biasing circuit.

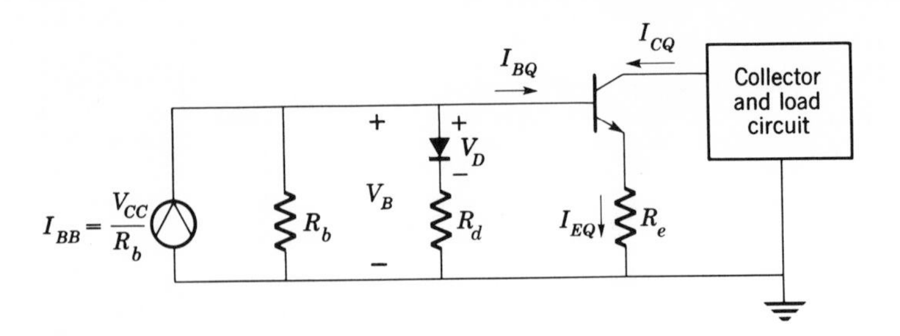

Fig. 4.4-4 Equivalent circuit.

Solution We first redraw the circuit as shown in Fig. 4.4-4. Then, applying KCL, we have

$$I_{BB} = \frac{V_B}{R_b} + \frac{V_B - V_D}{R_d} + I_{BQ}$$

Let us choose R_b and R_d so that

$$I_{BQ} \ll \frac{V_B}{R_b}$$

and

$$I_{BQ} \ll \frac{V_B - V_D}{R_e}$$

(that is, $\beta R_e \gg R_b$ and R_d). Then

$$I_{BB} \approx V_B \left(\frac{R_b + R_d}{R_b R_d}\right) - \frac{V_D}{R_d}$$

and

$$V_B = \left(I_{BB} + \frac{V_D}{R_d}\right)\left(\frac{R_b R_d}{R_b + R_d}\right)$$

The quiescent emitter current is therefore

$$I_{EQ} = \frac{V_B - V_{BEQ}}{R_e} = \left(\frac{1}{R_e}\right)\left(\frac{V_{CC}R_d}{R_b + R_d} + \frac{V_D R_b}{R_b + R_d} - V_{BEQ}\right)$$

This current can be set at the desired value by adjusting V_{BB} and R_b (or R_d), as in the standard bias circuit.

The change in I_{EQ} due to a change in temperature is

$$\frac{\partial I_{EQ}}{\partial T} = \left(\frac{1}{R_e}\right)\left(\frac{R_b}{R_b + R_d}\frac{\partial V_D}{\partial T} - \frac{\partial V_{BEQ}}{\partial T}\right)$$

But

$$\frac{\partial V_D}{\partial T} = \frac{\partial V_{BEQ}}{\partial T} = -k \qquad k \approx 2.5 \text{ mV/°C}$$

Therefore

$$\frac{\partial I_{EQ}}{\partial T} = \left(+\frac{k}{R_e}\right)\left(\frac{1}{1 + R_b/R_d}\right)$$

We saw in Sec. 4.3 that without diode compensation

$$\frac{\partial I_{EC}}{\partial T} \approx +\frac{k}{R_e} \qquad \text{neglecting } I_{CBO}$$

Thus diode compensation, when employed as above, reduces sensitivity to temperature. If, for example, $R_b = 2.5\ \text{k}\Omega$ and $R_d = 250\ \Omega$, then

$$\frac{\partial I_{EQ}}{\partial T} = \frac{k}{11R_e}$$

and temperature effects are reduced by a factor of 11, as compared with an amplifier stage without diode stabilization.

EXPERIMENTAL RESULTS

The two amplifier circuits shown in Fig. 4.4-5 were built and tested. Both amplifiers were tested at room temperature (25°C), and V_{BB} of the diode-compensated amplifier adjusted so that the emitter current of the compensated amplifier, I_{EC}, was the same as the emitter current of the uncompensated amplifier, I_{ENC}. The emitter-to-ground voltage V_E was measured and found to be 0.720 V for each amplifier.

The temperature was then raised to 90°C ($\Delta T = 65$°C), and V_E again measured. The emitter voltage of the uncompensated amplifier

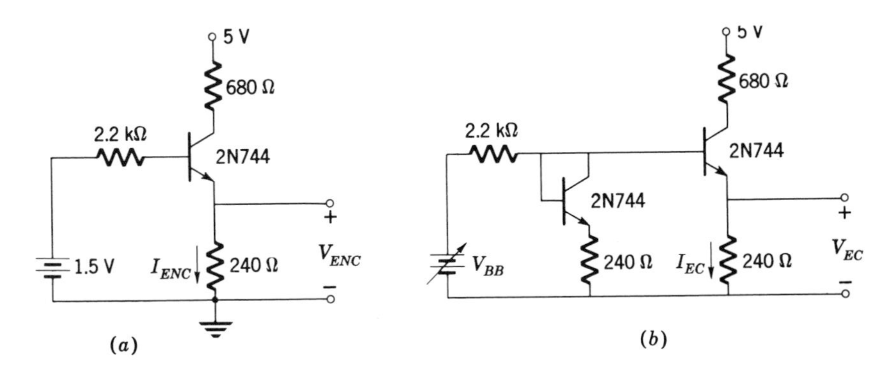

Fig. 4.4-5 Experimental models. (*a*) The uncompensated amplifier; (*b*) the compensated amplifier.

V_{ENC} was found to be 0.856 V ($\Delta V_{ENC} = 0.136$ V), while for the compensated amplifier, V_{EC} was measured as 0.736 V ($\Delta V_{EC} = 0.016$ V).

The results were also calculated using the following relations.

With no compensation:

$$\frac{\Delta I_{ENC}}{\Delta T} = \frac{k}{R_e}$$

Hence

$$\Delta V_{ENC} = \Delta(I_{ENC}R_e) = k\,\Delta T = 0.162\text{ V}$$

as compared with

$$\Delta V_{ENC}\text{ (measured)} = 0.136\text{ V}$$

With diode compensation:

$$\Delta V_{EC} = \Delta(I_{EC}R_e) = \frac{k\,\Delta T}{1 + (R_b/R_d)} = \frac{(2.5)(10^{-3})(65)}{1 + 2200/240} \approx 0.016\text{ V}$$

which is exactly the same as the measured ΔV_{EC}.

These results indicate the usefulness of diode compensation, and also that the theory accurately predicts the degree of compensation.

4.5 ENVIRONMENTAL THERMAL CONSIDERATIONS IN TRANSISTOR AMPLIFIERS

The practical design of transistor circuits almost always involves thermal as well as electrical considerations, because the maximum average power that the transistor can dissipate is limited by the temperature that the collector-base junction can withstand. Thus all circuit designs should include or be followed by a calculation of thermal conditions, to ensure that the maximum allowable junction temperature is not exceeded. The maximum allowable operating tempereture for germanium is in the range 90 to 120°C, and for silicon, 150 to 200°C. At higher temperatures the transistor will suffer physical damage. The average power dissipated in the collector circuit, P_C, is equal to the average of the product of the collector-current and the collector-base voltage. The maximum allowable average collector power is specified by the manufacturer. This rating can be exceeded momentarily provided the transistor does not have sufficient time to heat up to the point where it burns out.

The analysis of the thermal situation in the case of a transistor is the same as for the junction diode considered in Sec. 2.9. The typical physical configuration shown in Fig. 4.5-1 is described exactly by (2.9-3), and the discussion in connection with that equation carries over directly. Information on the thermal resistances θ_{jc} and θ_{ca} is usually provided by the transistor manufacturer.

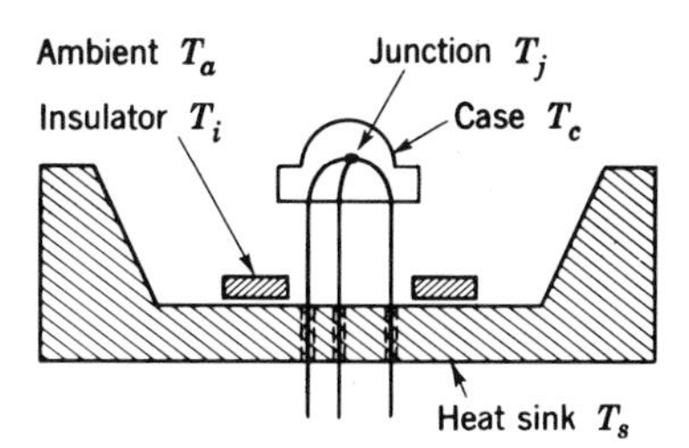

Fig. 4.5-1 Transistor and heat sink.

Example 4.5-1 A silicon transistor has the following thermal ratings:

$T_{j,\max} = 150°C$
$\theta_{jc} = 0.7°C/W$

Find (*a*) the power this transistor could dissipate if the case could be maintained at 50°C, regardless of the junction temperature; (*b*) the power that could be dissipated with an ambient temperature of 50°C and a heat sink having $\theta_{ca} = 1°C/W$.

Solution (*a*) For these conditions

$$P_j = \frac{T_j - T_c}{\theta_{jc}} = \frac{150 - 50}{0.7} \approx 143 \text{ W}$$

(*b*) Using (2.9-3),

$$P_j = \frac{T_j - T_a}{\theta_{jc} + \theta_{ca}} = \frac{150 - 50}{0.7 + 1} \approx 59 \text{ W}$$

Thus an "infinite" heat sink as in (*a*) permits the transistor to dissipate over twice the power permitted when using the "real" heat sink specified in (*b*).

DERATING CURVES

The variation of maximum collector dissipation with case temperature is an important characteristic supplied by the transistor manufacturer. A typical curve is shown in Fig. 4.5-2. At case temperatures less than T_{co}, the transistor can dissipate its maximum allowable power. At case tem-

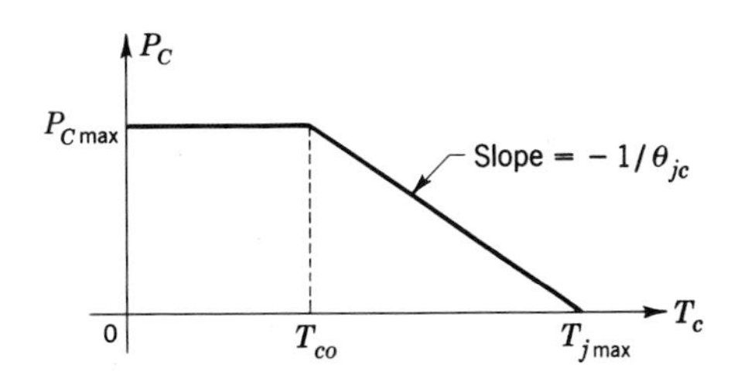

Fig. 4.5-2 Derating curve.

peratures exceeding T_{co}, the maximum allowable collector dissipation is decreased, as shown.

The collector dissipation in this region is given by the equation

$$\frac{T_{j,\max} - T_{co}}{P_{C,\max}} = \frac{T_{j,\max} - T_c}{P_C} \tag{4.5-1}$$

Since

$$T_j - T_c = P_C \theta_{jc} \tag{4.5-2}$$

we see that

$$\theta_{jc} = \frac{T_{j,\max} - T_{co}}{P_{C,\max}} \tag{4.5-3}$$

Example 4.5-2 A high-power silicon transistor can dissipate 150 W as long as the case temperature is less than 45°C. Above this temperature the collector power is linearly derated, as shown in Fig. 4.5-2. The maximum junction temperature is 120°C. The amplifier is to be capable of operating at very high ambient temperatures, reaching 80°C. Determine the maximum power this transistor can dissipate, and the thermal resistance of the heat sink and insulator required, to avoid having the junction temperature exceed its maximum allowable value.

Solution Let

$$T_j = T_{j,\max} = 120°\text{C}$$

Then, from (4.5-2),

$$T_j - T_c = P_C \theta_{jc}$$

where θ_{jc} can be found using (4.5-3):

$$\theta_{jc} = \frac{120 - 45}{150} = 0.5°\text{C/W}$$

In order that the maximum junction temperature not be exceeded,

$$T_{j,\max} = T_{a,\max} + P_C \theta_{ja}$$

Thus

$$120 - 80 = 40 = P_C \theta_{ja} = P_C(\theta_{jc} + \theta_{ca}) = P_C(0.5 + \theta_{ca})$$

At this point engineering judgment must be used. It is clear that, with an infinite heat sink, $P_C = 80$ W. However, infinite heat sinks cannot be bought. A very good heat sink (with insulator) has a thermal resistance of 0.5°C/W. Thus

$P_{C,\max} = 40$ W

Observe that the 150-W transistor could only dissipate 40 W because of the high-ambient-temperature requirement.

4.6 MANUFACTURERS' SPECIFICATIONS FOR HIGH-POWER ($P_{C,\max} > 1$ W) TRANSISTORS

In this section, some common specifications provided by transistor manufacturers are discussed. The specifications given below are for an *npn* silicon transistor.

Transistor type: 2N1016, silicon *npn*

1. Maximum thermal resistance, $\theta_{jc} = 0.7$°C/W
2. Maximum collector dissipation with infinite heat sink at 25°C, $P_C = 150$ W
3. Maximum junction temperature, $T_{j,\max} = 140$°C
4. Absolute maximum ratings at 25°C:
   (*a*) $I_C = 7.5$ A
   (*b*) $I_B = 5$ A
   (*c*) Breakdown voltage
      (1) Collector-base (BV_{CBO}) = 30 V
      (2) Emitter-base (BV_{EBO}) = 25 V
      (3) Collector-emitter (BV_{CEO}) = 30 V
5. Max I_{CBO} at maximum V_{CB} at 25°C = 10 mA
6. Current amplification β, at $V_{CE} = 4$ V and $I_C = 5$ A, $10 < \beta \leq 18$
7. Common-emitter cutoff frequency, $f_\beta = 30$ kHz

EXPLANATION OF SYMBOLS

The maximum thermal resistance θ_{jc}, the maximum collector dissipation $P_{C,\max}$, and the maximum junction temperature $T_{j,\max}$ have been discussed in Secs. 2.9 and 4.5. They can be summarized using the power-derating curve shown in Fig. 4.6-1. Thus, if an infinite heat sink is

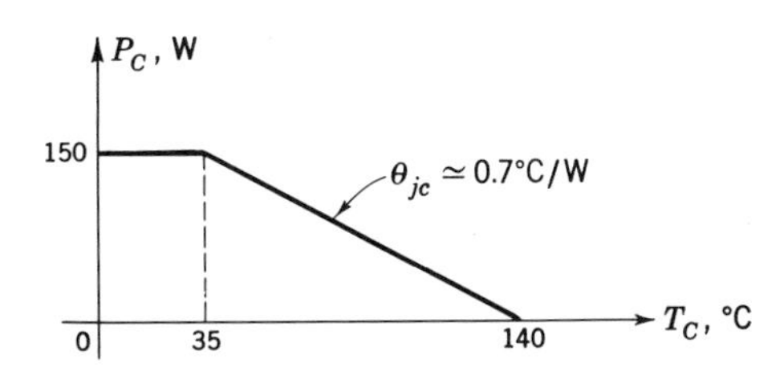

Fig. 4.6-1 Derating curve.

employed ($\theta_{ca} = 0$), 150 W can be dissipated as long as the case temperature (which equals the ambient temperature for an infinite heat sink) is less than 35°C. If the case temperature increases above this value, the allowable power dissipation in the transistor decreases, as shown.

The absolute maximum ratings specified indicate upper bounds on the current and voltage capability of the transistor. Thus a collector current of 7.5 A should never be exceeded. A base current of 5 A (making sure that $I_C < 7.5$ A) should never be exceeded. The collector-base and collector-emitter voltages should not exceed 30 V. This is called the *breakdown voltage*. When this voltage is exceeded, the junction breaks down (as in a Zener diode), and avalanche current multiplication results in a *vi* characteristic, as shown in Fig. 3.2-4*b*. Notice the large I_{CBO} (=10 mA) that can occur. This is not very large if we note that 7 A can flow in the collector. High-power transistors normally have higher relative collector-base currents than low-power transistors. This is also true of the emitter-to-base breakdown voltage, which for this transistor is 25 V.

We see that the current-amplification factor β is only 10 to 18. This is due to construction problems. In a high-power transistor the base region is widened to increase the breakdown voltage. As a result, β decreases, and the maximum usable frequency decreases. (This frequency is inversely proportional to the time required for an electron leaving the emitter in an *npn* unit to pass through the base, which has been widened, and reach the collector.) This is discussed in Chap. 13.

PROBLEMS

SECTION 4.1

4.1. The silicon transistor to be used in the circuit of Fig. P4.1 has a β that varies from 50 to 200.

If $V_{BB} = 3$ V and $R_e = 200\ \Omega$, find the variation in Q point for $R_b = 1$ kΩ and $R_b = 10$ kΩ.

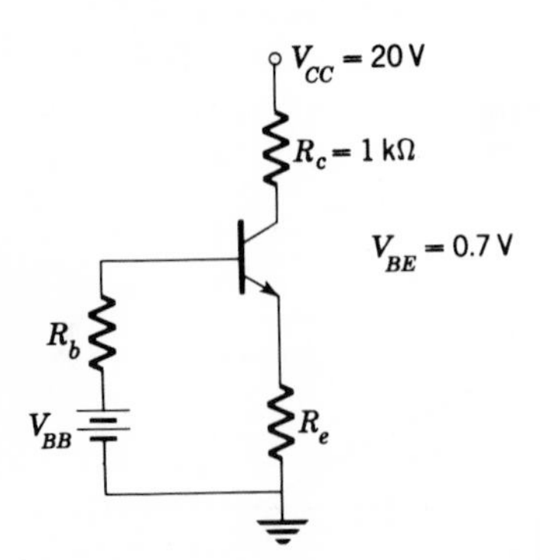

Fig. P4.1

4.2. The transistor of Prob. 4.1 is to be used in the circuit of Fig. P4.2. Find the variation in quiescent current as β varies from 50 to 200.

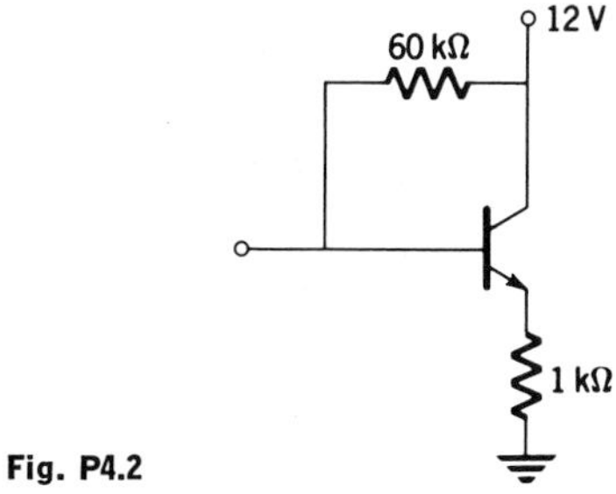

Fig. P4.2

4.3. The amplifier shown in Fig. P4.3 is to be designed to have a maximum symmetrical swing. If β varies from 50 to 150 for this type of transistor, find V_{BB}, R_e, and the maximum swing.

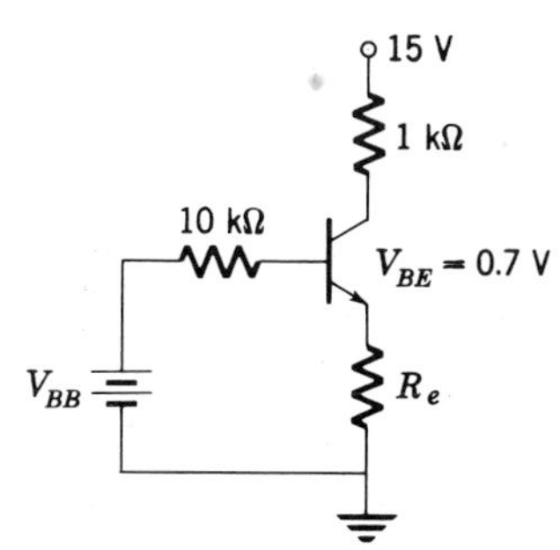

Fig. P4.3

SECTION 4.2

4.4. In the circuit of Fig. P4.1, $V_{BB} = 3$ V, $I_{CBO} = 0.1\ \mu$A, $R_b = 1$ kΩ, and $R_e = 200\ \Omega$. Find the variation in quiescent current as the temperature varies from 25 to 175°C. Assume $\beta = 100$.

4.5. In the circuit of Fig. P4.2, $I_{CBO} = 10\ \mu$A and $\beta = 100$. Find the variation in quiescent current as the temperature varies from 25 to 175°C and from 25 to −55°C.

4.6. Calculate V_C as the temperature varies from 25 to 100°C. Assume T_1 and T_2 are identical silicon transistors with $I_{CBO} = 1\ \mu$A and $\beta = 20$.

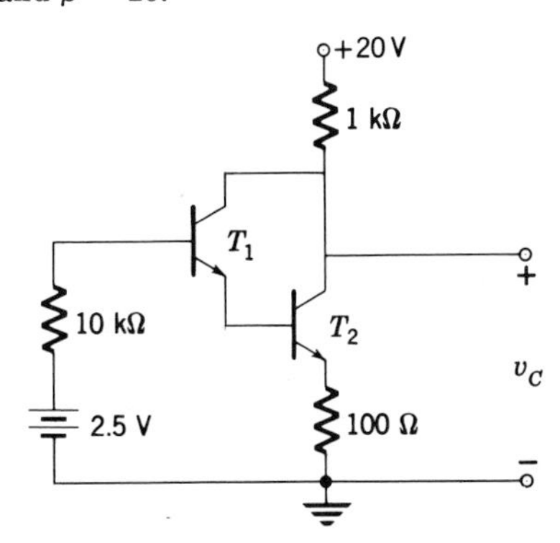

Fig. P4.6

4.7. The amplifier shown in Fig. P4.7 is to be designed to have a maximum symmetrical swing. The temperature range lies between −55 and +125°C.

(*a*) Find V_{BB} and R_e.

(*b*) Find the maximum swing.

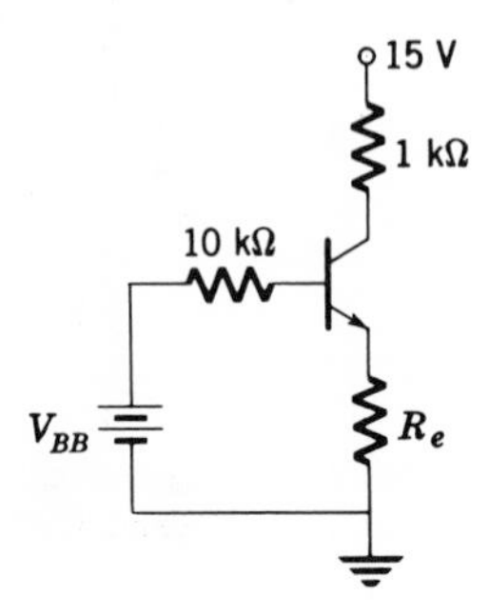

Fig. P4.7

SECTION 4.3

4.8. Using the stability factor S_β, find values for R_e and R_b in the circuit of Fig. P4.1 such that the voltage across R_c will not vary by more than ±0.5 V as β varies from 50 to 200. The quiescent current is to be ≈10 mA.

4.9. In the circuit of Fig. P4.9

$$50 < \beta < 200$$
$$25°\text{C} < T < 75°\text{C}$$
$$V_{CC} = 6 \text{ V} \pm 0.2 \text{ V}$$
$$I_{CBO} = 0.01\ \mu\text{A at } 25°\text{C}$$

Find the quiescent current, all the pertinent stability factors, and the worst case quiescent-current shift.

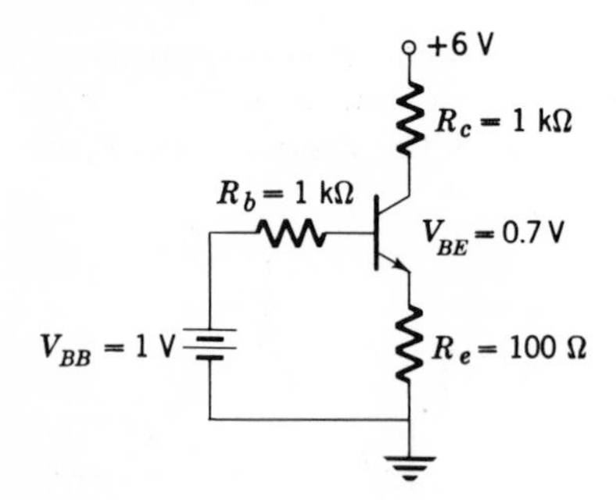

Fig. P4.9

4.10. The circuit of Fig. P4.10 illustrates the use of *feedback* (Chap. 8) to bias a transistor. Without using feedback techniques, show that

$$I_{CQ} \approx \frac{\beta[V_{CC} - V_{BEQ} + (R_c + R_b)I_{CBO}]}{\beta R_c + R_b}$$

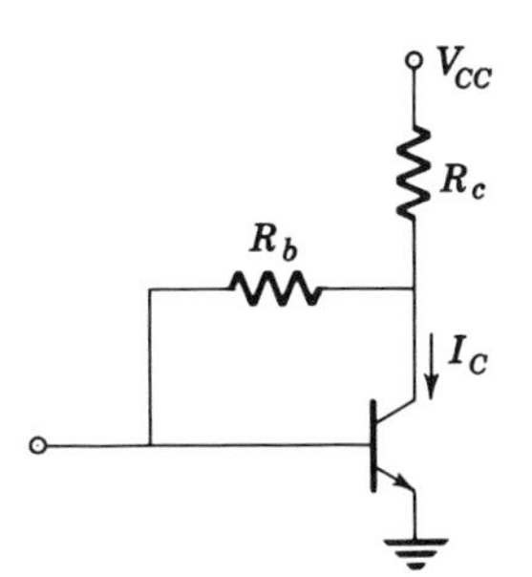

Fig. P4.10

4.11. For the circuit of Fig. P4.10 find S_I, S_V, and S_β.

4.12. In Fig. P4.10, $V_{CC} = 20$ V and $R_c = 1$ kΩ. Find R_b so that $I_{CQ} \approx 10$ mA. Calculate S_I, S_V, and S_β. Determine the Q-point shift for $50 < \beta < 200$ and $25°\text{C} < T < 100°\text{C}$, with $I_{CBO}(25°\text{C}) = 1\ \mu\text{A}$.

4.13. An emitter resistor R_e is added to the circuit of Fig. P4.10. Show that

$$I_{CQ} \approx \frac{\beta(R_e + R_b + R_c)I_{CBO} + \beta(V_{CC} - V_{BEQ})}{R_b + \beta(R_e + R_c)}$$

4.14. Find S_I, S_V, and S_β for the circuit of Prob. 4.13.

4.15. Using the results of Probs. 4.13 and 4.14, find the Q-point shift when $50 < \beta < 200$, $25°\text{C} < T < 100°\text{C}$, and $I_{CBO} = 1\ \mu\text{A}$ at room temperature. Find R_b such that $I_{CQ} = 10$ mA, $V_{CC} = 20$ V, $R_c = 800\ \Omega$, and $R_e = 200\ \Omega$.

4.16. The amplifier shown in Fig. P4.16 is to be operated over the temperature range -25 to 75°C. The range of β, for the transistor used, lies between 100 and 300. The transistor has $I_{CBO} = 0.1\ \mu\text{A}$ and $V_{BE} = 0.7$ V at room temperature. If $R_1 \| R_2 \geq 1$ kΩ, find the maximum possible symmetrical swing. Specify R_1, R_2, and R_e.

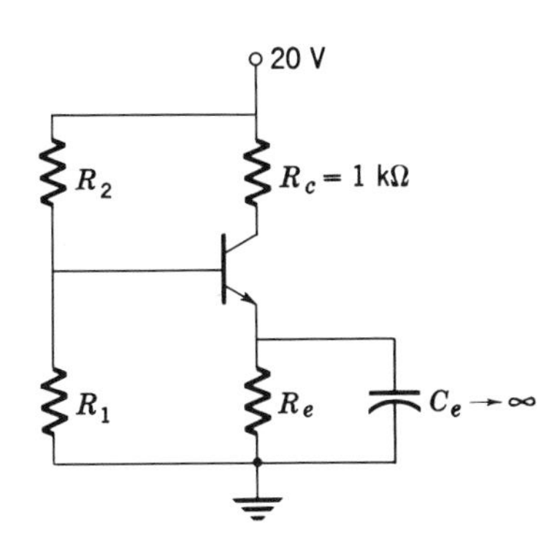

Fig. P4.16

SECTION 4.4

4.17. For the diode biasing circuit of Fig. 4.4-5*b* find S_I, S_V, and S_β.

4.18. Find S_I, S_V, and S_β for the uncompensated circuit of Fig. 4.4-5*a* and compare with the results of Prob. 4.17.

4.19. The amplifiers shown in Fig. 4.4-5 are to operate over the temperature range 25 to 90°C. Calculate the change in quiescent current for the uncompensated and compensated amplifiers, and compare.

SECTION 4.5

4.20. A 50-W transistor (derating curve shown in Fig. P4.20) is to dissipate 6 W in a particular circuit. The ambient temperature is 85°C.

(*a*) Find θ_{ca} so that the transistor will not overheat.

(*b*) Find T_c for these conditions.

(*c*) If an infinite heat sink were available, how much power could this transistor dissipate?

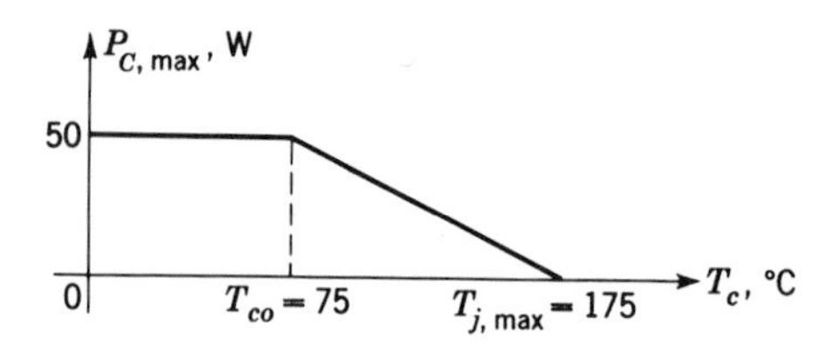

Fig. P4.20

4.21. A silicon power transistor has the following thermal ratings:

$$P_{C,\max} = 200 \text{ W}$$
$$T_{j,\max} = 175°\text{C}$$
$$\theta_{jc} = 0.7°\text{C/W}$$

Find the power that this transistor could dissipate if the case could be maintained at room temperature (25°C).

4.22. The transistor of Prob. 4.21 is mounted directly to a flat aluminum heat sink which has $\theta_{sa} = 8°\text{C/W}$. The direct mounting results in $\theta_{cs} = 0.2°\text{C/W}$. Find the maximum allowable dissipation.

4.23. The heat sink of Prob. 4.22 is used with the transistor of Prob. 4.21, but the transistor is electrically insulated from the sink with a mica washer, so that θ_{cs} increases to 2°C/W. Find the maximum allowable dissipation.

4.24. The transistor of Prob. 4.22 is used with a large finned heat sink which has $\theta_{sa} \approx 0.9°\text{C/W}$. Find the maximum allowable dissipation if the transistor is mounted directly, or if it is electrically insulated by the mica washer, as in Prob. 4.23.

4.25. Find the temperature of the case and of the heat sink for the circuits of Probs. 4.22 and 4.23.

SECTION 4.6

4.26. Sketch a set of typical common-emitter output characteristics for the 2N1016 silicon transistor, at temperatures of 25 and 110°C. On the curves, indicate the area of safe operation, and carefully specify its boundaries. Comment on your results.

REFERENCES

1. Searle, C. L., et al.: "Elementary Circuit Properties of Transistors," SEEC, vol. 3, chap. 5, John Wiley & Sons, Inc., New York, 1964.
2. Widlar, R. J.: Circuit Design Techniques for Linear Integrated Circuits, *IEEE Trans. on Circuit Theory,* December, 1965.

5
Audio-frequency Linear Power Amplifiers

INTRODUCTION

Several major problems associated with amplifiers which are required to furnish large amounts of power are discussed in this chapter. The aim in most applications is to furnish the required power as economically as possible while meeting other specifications, which may include limitations on size, weight, dc supply voltage, distortion, etc. The designer often has to make several compromises along the way in order to achieve an optimum design. Often, the transistors are driven to the limits of their useful operating range, and careful design is required to ensure that physical damage, due to excessive heating, does not occur.

Power amplifiers are classified according to the portion of the input sine-wave cycle during which load current flows:

Class A: Current flows for 360° (full-cycle operation).

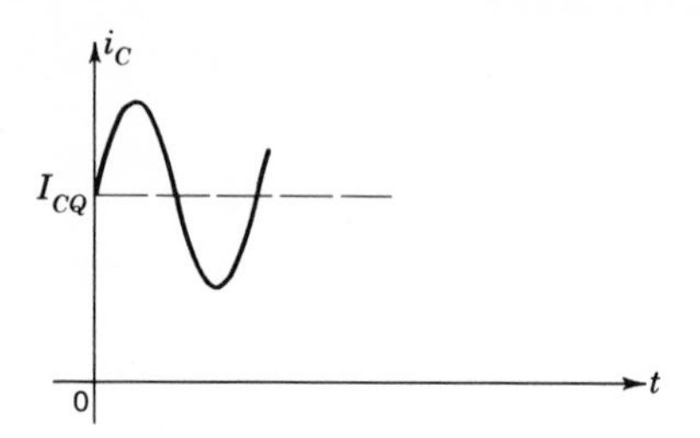

Class AB: Current flows for more than one half-cycle but less than full cycle.

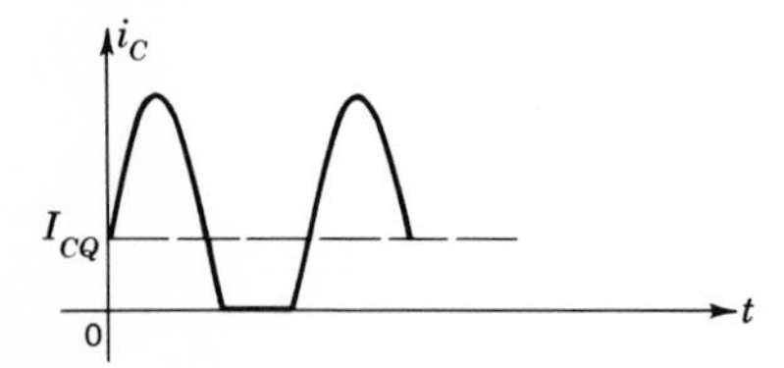

Class B: Current flows for one half-cycle.

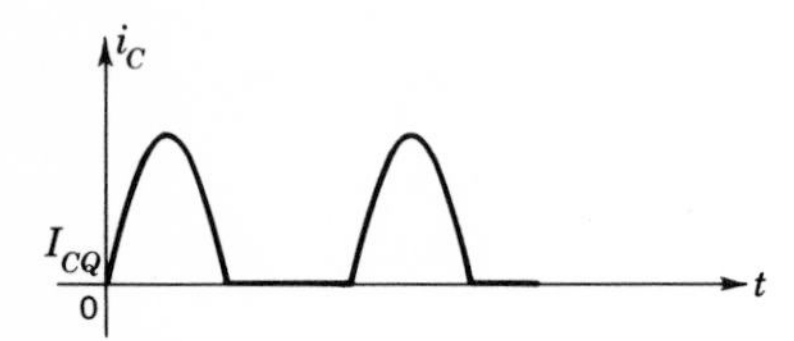

Class C: Current flows for less than one half-cycle.

To achieve low-distortion amplification of audio-frequency signals, only class A would seem to apply. However, if one uses a *push-pull* arrangement, described in Secs. 5.3 and 5.4, class AB and class B amplifiers can also yield essentially linear amplification. Class C power amplifiers are used extensively at radio frequencies where tuned circuits remove the distortion resulting from the nonlinear operation of the circuit.

In this chapter, Q-point placement and power relations in the most commonly used classes A and B audio-frequency circuits are studied. Distortion is neglected. Distortion in a vacuum-tube amplifier is discussed in Sec. 11.3. That analysis is directly applicable to the transistor.

5.1 THE CLASS A COMMON-EMITTER POWER AMPLIFIER

In Chap. 3 it was observed that a large amount of power was dissipated in the collector resistor R_c because of the quiescent collector current I_{CQ}. This resulted in a maximum operational efficiency of only 25 percent. Thus, when 1 W of signal power is to be dissipated in the load under maximum signal conditions, 4 W must be furnished continuously by the dc power supply. In this section it is shown that replacing R_c with a large inductor (often called a "choke") eliminates some of this dissipation and increases the maximum efficiency to 50 percent.

5.1-1 Q-POINT PLACEMENT

A class A circuit with choke coupling is shown in Fig. 5.1-1a. The circuit is designed so that all capacitors are essentially short circuits and the inductor is essentially an open circuit at signal frequencies. At dc, the capacitors are, of course, open circuits, while the inductor is a short circuit. The inductor is assumed, for simplicity, to have no internal resistance.

To determine the Q point, apply KVL around the collector circuit, including only dc voltage drops. The dc-load-line equation for this ampli-

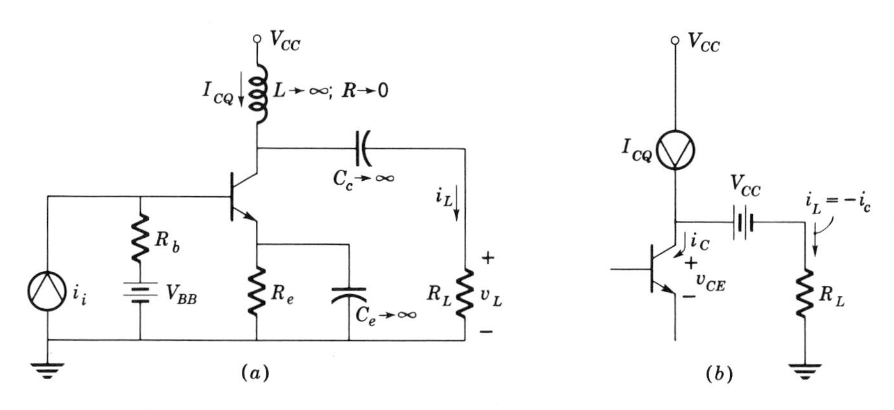

Fig. 5.1-1 Inductor-coupled power amplifier. (a) Circuit; (b) equivalent circuit.

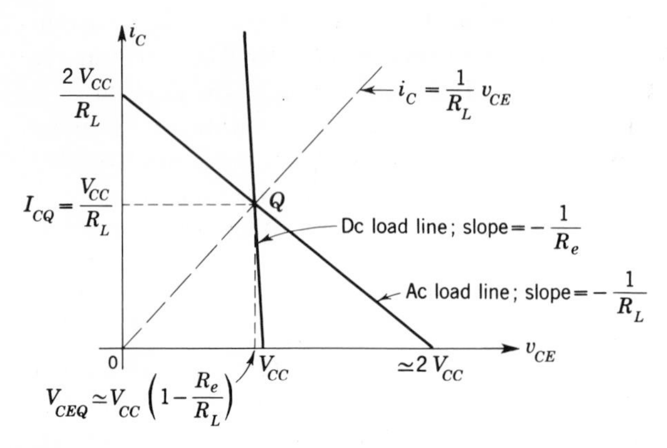

Fig. 5.1-2 Power-amplifier load lines.

fier is then

$$V_{CC} = v_{CE} + i_C R_e \tag{5.1-1}$$

The emitter resistor is kept as small as possible in order to minimize bias-circuit power loss while maintaining adequate Q-point stability. Thus the slope of the dc load line is almost vertical, as shown in Fig. 5.1-2.

Applying KVL around the collector circuit (Fig. 5.1-1a), including only ac voltage drops, yields the ac-load-line equation

$$v_{ce} = -i_c R_L = i_L R_L \tag{5.1-2a}$$

which can be written

$$i_C - I_{CQ} = \left(-\frac{1}{R_L}\right)(v_{CE} - V_{CEQ}) \tag{5.1-2b}$$

To place the Q point for maximum symmetrical swing, we can, as before, plot the equation

$$i_C = \left(\frac{1}{R_L}\right) v_{CE} \tag{3.5-4}$$

and determine the intersection of this line with the dc load line, as shown in Fig. 5.1-2. However, because R_e is small, the dc load line is almost vertical.

Neglecting the voltage drop across R_e,

$$V_{CEQ} \approx V_{CC}$$

which is independent of I_{CQ}. Then, using (3.5-4),

$$I_{CQ} = \left(\frac{1}{R_L}\right) V_{CC} \tag{5.1-3}$$

The ac load line passes through the Q point with slope $-1/R_L$, as shown in Fig. 5.1-2. Note that the maximum collector-current swing is from 0 to $2I_{CQ}$ as v_{CE} swings from $2V_{CC}$ to 0. Note, too, that v_{CE} is limited by the saturation voltage of the transistor to a minimum voltage of $V_{CE,\text{sat}}$. To simplify calculations, the saturation voltage is assumed to be zero. Its effect, however, is considered in the examples.

It is interesting to consider how the collector-emitter voltage can become twice the supply voltage. Since the inductance is very large, no ac current will flow through it, and for purposes of analysis, it may be replaced by a constant current source of strength I_{CQ}. Since the capacitive reactance is very small, no ac voltage will appear across the capacitor, and it can be replaced by a battery of voltage V_{CC}, the voltage to which it is charged when no signal is present. With these two substitutions, the collector-load circuit takes the equivalent form shown in Fig. 5.1-1*b*. Assume that a sinusoidal signal is present, and consider an instant of time when $i_C = 0$. At that instant, $i_L = I_{CQ}$, so that $v_{CE} = V_{CC} + i_L R_L$. From (5.1-3), $I_{CQ}R_L = i_L R_L = V_{CC}$, so that $v_{CE} = 2V_{CC}$. This establishes the upper limit on v_{CE}. When the signal reverses polarity, $i_C = 2I_{CQ}$. Then i_L must equal $-I_{CQ}$ so that $v_{CE} = 0$, establishing the lower limit.

5.1-2 POWER CALCULATIONS

With the amplifier biased as in (5.1-3), the currents and voltages of interest are (the power dissipated in the emitter resistor is neglected for simplicity)

$$i_C = I_{CQ} + i_c = \frac{V_{CC}}{R_L} + i_c \tag{5.1-4a}$$

$$i_L = -i_c \tag{5.1-4b}$$

$$i_{\text{supply}} = i_L + i_C = I_{CQ} = \frac{V_{CC}}{R_L} \tag{5.1-4c}$$

$$v_{CE} = V_{CC} - i_c R_L \tag{5.1-5a}$$

(see Fig 5.1-1*b*), and

$$v_L = +i_L R_L = -i_c R_L \tag{5.1-5b}$$

If the signal current is sinusoidal,

$$i_i = I_{im} \sin \omega t \tag{5.1-6a}$$

then

$$i_c = I_{cm} \sin \omega t \tag{5.1-6b}$$

It should be noted that the *maximum* peak value of ac collector current is I_{CQ}, so that

$$\max i_c = I_{CQ} \sin \omega t \tag{5.1-7a}$$

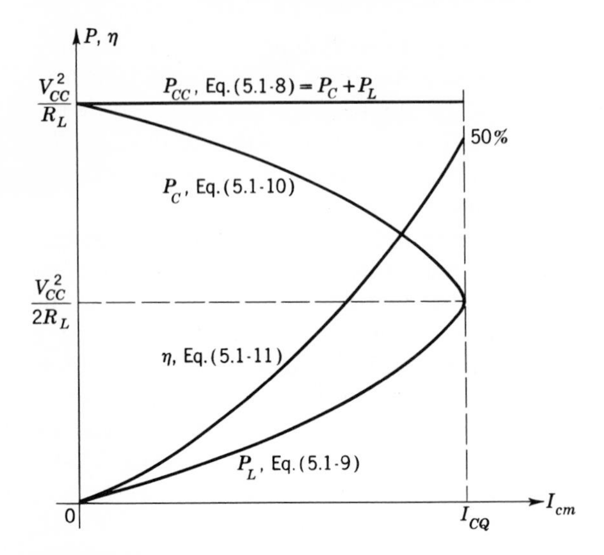

Fig. 5.1-3 Variation of power and efficiency with collector current.

Therefore

$$I_{cm} \leq I_{CQ} \tag{5.1-7b}$$

The power supplied, the power dissipated in the collector and the load, and the efficiency are found in the same manner as in Sec. 3.4. The results are given in (5.1-8) to (5.1-11) and are plotted in Fig. 5.1-3.

Supplied power

$$P_{CC} = V_{CC} I_{CQ} \approx \frac{V_{CC}^2}{R_L} \tag{5.1-8}$$

which is constant and essentially independent of signal current as long as the distortion is negligible.

Power transferred to load

$$P_L = \frac{I_{Lm}^2 R_L}{2} = \frac{I_{cm}^2 R_L}{2} \tag{5.1-9a}$$

since $i_L = -i_c$, $I_{Lm} = -I_{cm}$.

The *maximum* average power dissipated by the load occurs when

$$I_{cm} = I_{CQ}$$

Thus

$$P_{L,\max} = \frac{I_{CQ}^2 R_L}{2} = \frac{V_{CC}^2}{2R_L} \tag{5.1-9b}$$

Collector dissipation

$$P_C = P_{CC} - P_L = \frac{V_{CC}^2}{R_L} - \frac{I_{cm}^2 R_L}{2} \tag{5.1-10a}$$

Thus the minimum power dissipated in the collector is

$$P_{C,\min} = \frac{V_{CC}^2}{2R_L} \tag{5.1-10b}$$

which occurs when the maximum power is dissipated by the load. The maximum power dissipated by the collector is

$$P_{C,\max} = \frac{V_{CC}^2}{R_L} = V_{CEQ} I_{CQ} \tag{5.1-10c}$$

which occurs when no signal is present.

Efficiency The efficiency of operation of the inductor-coupled amplifier for a sinusoidal signal is

$$\eta = \frac{P_L}{P_{CC}} = \frac{I_{cm}^2 (R_L/2)}{V_{CC} I_{CQ}} = \left(\frac{1}{2}\right)\left(\frac{I_{cm}}{I_{CQ}}\right)^2 \tag{5.1-11a}$$

Thus maximum efficiency occurs at maximum signal current. Then

$$\eta_{\max} = \tfrac{1}{2} = 50\% \tag{5.1-11b}$$

The efficiency of operation has been doubled by using an inductor rather than a resistor R_c in the dc collector circuit.

The variations of supply, load, and collector power and efficiency are plotted in Fig. 5.1-3 as a function of collector current, for sinusoidal signals. Note that as the load power increases, the collector dissipation decreases, their sum remaining constant ($P_{CC} = P_C + P_L$). Also note that P_C is a maximum when there is no signal present.

Figure of merit A useful figure of merit for a power amplifier is the ratio of maximum transistor collector dissipation to maximum power dissipated in the load. Using (5.1-10*c*) and (5.1-9*b*) or Fig. 5.1-3,

$$\frac{P_{C,\max}}{P_{L,\max}} = 2 \tag{5.1-12}$$

This is the same result obtained in Chap. 3. Thus, if max $P_L = 25$ W, the collector junction must be capable of dissipating at least 50 W. In Chap.

4 it was found that in order to operate at high ambient temperatures, transistors often must be derated. Thus, to dissipate 25 W might require a transistor with an allowable collector dissipation of 100 W, if the ambient temperature is high.

5.1-3 THE MAXIMUM-DISSIPATION HYPERBOLA

Once the maximum power to be delivered to the load and the anticipated temperature range are known, the power rating of the transistor can be determined and the transistor can be selected. The maximum allowable collector dissipation is, as stated above, usually less than the maximum rating specified for the transistor.

In addition to having the specified power rating, the transistor must be able to handle currents as high as $2I_{CQ}$ (Fig. 5.1-2) and a collector-emitter voltage up to $2V_{CC}$. It must also have an operating frequency at least as high as the signal frequency. These ratings are generally supplied by the transistor manufacturer (Sec. 4.6).

Design and transistor specifications will, in general, include the following items:

$\max i_C$

BV_{CEO}

and

$$P_{C,\max} = V_{CEQ}I_{CQ} \tag{5.1-10c}$$

These maximum ratings place bounds on the permissible operating region of the transistor, as in Fig. 5.1-4. The figure shows that for safe operation the Q point must lie on or below the hyperbola

$$v_{CE}i_C = P_{C,\max}$$

This hyperbola represents the locus of all operating points at which the collector dissipation is exactly $P_{C,\max}$.

The ac load line, having a slope $-1/R_L$, must pass through the Q

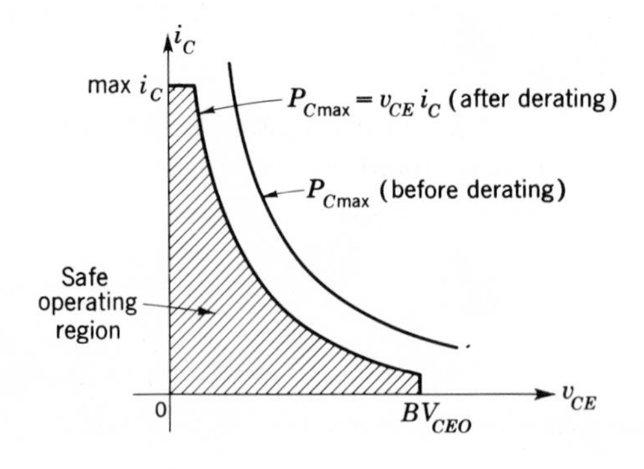

Fig. 5.1-4 Maximum-collector-dissipation hyperbola.

point and intersect the v_{CE} axis at a voltage less than BV_{CEO}, and must intersect the i_C axis at a current less than max i_C; that is,

$$2V_{CC} \leq BV_{CEO} \tag{5.1-13a}$$

$$2I_{CQ} \leq \max i_C \tag{5.1-13b}$$

In order to obtain a maximum symmetrical swing we must have

$$I_{CQ} = \left(\frac{1}{R_L}\right) V_{CEQ} \tag{3.5-4}$$

Combining (3.5-4) and (5.1-10c), the Q point is at

$$I_{CQ} = \sqrt{\frac{P_{C,\max}}{R_L}} \tag{5.1-14a}$$

and

$$V_{CEQ} = \sqrt{P_{C,\max} R_L} \tag{5.1-14b}$$

It is interesting to note that at the Q point the slope of the hyperbola is

$$\frac{\partial i_C}{\partial v_{CE}} = -\frac{I_{CQ}}{V_{CEQ}} = -\frac{1}{R_L} \tag{5.1-15}$$

Thus the slope of the ac load line is the same as the slope of the hyperbola, and the ac load line is tangent to the hyperbola at the Q point when maximum symmetrical swing is obtained.

Example 5.1-1 A transistor having the following specifications is used in the circuit of Fig. 5.1-1:

$P_{C,\max}$ (after derating) $= 4$ W

BV_{CEO} $= 40$ V

max i_C $= 2$ A

The load resistance R_L is 10 Ω. Determine the Q point so that maximum power is dissipated by the load resistor. Specify the required supply voltage V_{CC}.

Solution The permissible operating range is shown in Fig. 5.1-5. The quiescent point is obtained by plotting the equation

$$i_C = \left(\frac{1}{R_L}\right) v_{CE} = \frac{v_{CE}}{10} \tag{3.5-4}$$

on the vi characteristic and finding the intersection of this equation with the $P_{C,\max}$ hyperbola. The intersection is the desired Q point, and the ac load line is tangent to the hyperbola at this point.

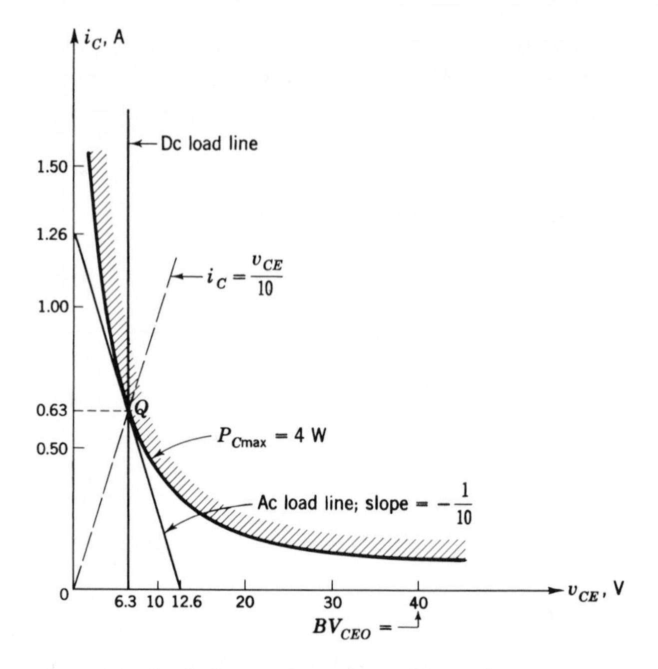

Fig. 5.1-5 Load lines and maximum-dissipation hyperbola for Example 5.1-1.

The Q point is obtained graphically as in Fig. 5.1-5, or analytically, using (5.1-14a) and (5.1-14b).

$$I_{CQ} = \sqrt{4/10} \approx 0.63 \text{ A}$$

and

$$V_{CEQ} = \sqrt{(4)(10)} \approx 6.3 \text{ V}$$

These results are seen to verify those obtained graphically.

The supply voltage V_{CC} is equal to V_{CEQ} (if we neglect the voltage drop in R_e). Thus

$$V_{CC} = 6.3 \text{ V}$$

and the maximum v_{CE} is equal to 12.6 V and is less than the breakdown voltage. The maximum collector current i_C is equal to 1.26 A and is less than the maximum allowable current.

The maximum power delivered to the load is then

$$P_{L,\max} = \frac{I_{CQ}^2 R_L}{2} = \frac{(0.63)^2 \times 10}{2} = 2 \text{ W} \qquad (5.1\text{-}9b)$$

SELECTING R_e, R_b, AND V_{BB} It should be noted that the quiescent operating point is obtained using the standard techniques of Chap. 3. Referring to Fig. 5.1-1a, R_b is chosen so that

$$R_e \gg \frac{R_b}{\beta + 1}$$

In addition, R_e is chosen to be small, so that its power dissipation is negligible. For example, we could choose $R_e = 1\ \Omega$, so that

$$P_{R_e} = I_{CQ}^2(1) = 0.4\text{ W} \ll P_{C,\max} = 4\text{ W}$$

Then R_b is chosen so that

$$R_b = \frac{(\beta + 1)R_e}{10}$$

If $\beta = 40$,

$$R_b \approx 4\ \Omega$$

The base supply voltage is

$$V_{BB} \approx 0.7 + (0.63)(1) = 1.33\text{ V}$$

The quiescent point is found to be shifted slightly if the effect of R_e is included in the design. This results in an increase in the V_{CC} required.

Let us consider the same problem with the maximum collector-current rating changed to max $i_C = 1$ A. The previous solution results in a max i_C equal to 1.26 A, which exceeds the new maximum allowable current. If the Q point is left unchanged, the peak safe ac current swing is reduced to 0.37 A. If the signal is sinusoidal, the maximum ac current is then

$$i_C = 0.37 \sin \omega t \qquad \text{A}$$

and the maximum power dissipated in the load now becomes

$$P_{L,\max} = (\tfrac{1}{2})(0.37)^2 \times 10 \approx 0.69\text{ W}$$

Since the 10-Ω load is fixed, the slope of the ac load line is fixed. However, if the load line is moved so that it intersects the i_C axis at max $i_C = 1$ A, and the Q point placed at

$$I_{CQ} = 0.5\text{ A} \qquad V_{CEQ} = V_{CC} = 5\text{ V}$$

the maximum ac collector current is

$$i_C = 0.5 \sin \omega t \qquad \text{A}$$

and the maximum power in the load becomes

$$P_{L,\max} = (\tfrac{1}{2})(0.5)^2 \times 10 = 1.25 \text{ W}$$

In either case the power delivered to the load is far below the available power of 2 W. This is because we are unable to compensate for the reduced maximum allowable collector current. In the next section the transformer-coupled power amplifier is considered. Using this circuit, the Q point can be placed without regard to the actual load resistance by making use of the impedance-transforming property of the transformer. The problem faced above, of not being able to deliver maximum available power to the load, is thus alleviated.

Example 5.1-2 The 2N1724 *npn* silicon transistor is used in the circuit of Fig. 5.1-6*a* with a 10-Ω load resistor. The maximum ratings of this transistor are

$$P_{C,\max} = 2.5 \text{ W (after derating)}$$

$$BV_{CEO} = 80 \text{ V}$$

$$V_{CE,\text{sat}} = 2 \text{ V}$$

Determine the maximum attainable swing and the maximum power dissipated by the load.

Solution The amplifier circuit and the *vi* characteristic are shown in Fig. 5.1-6.

Maximum power is dissipated by the load when the collector-current swing is a maximum. The ac-load-line equation is

$$R_L(i_C - I_{CQ}) = -(v_{CE} - V_{CEQ})$$

When i_C is a maximum, $i_C = 2I_{CQ}$ (Sec. 3.5) and $v_{CE} = V_{CE,\text{sat}}$. Then

$$R_L I_{CQ} = V_{CEQ} - V_{CE,\text{sat}} \tag{5.1-16}$$

To avoid exceeding the maximum average collector dissipation, we set

$$I_{CQ}V_{CEQ} = P_{C,\max} \tag{5.1-17}$$

Combining (5.1-16) and (5.1-17) and solving the resulting equation for I_{CQ} and V_{CEQ} yields

$$I_{CQ} = -\frac{V_{CE,\text{sat}}}{2R_L} + \sqrt{\frac{P_{C,\max}}{R_L} + \left(\frac{V_{CE,\text{sat}}}{2R_L}\right)^2} \tag{5.1-18a}$$

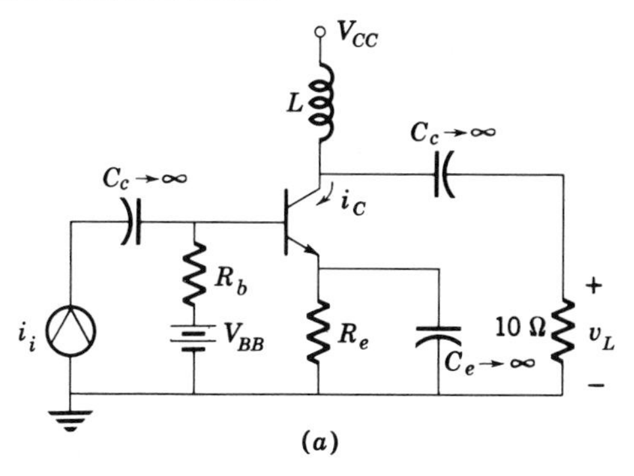

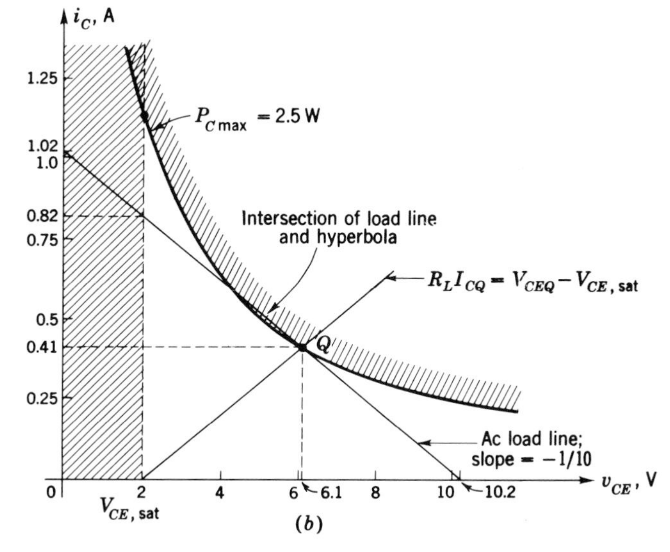

Fig. 5.1-6 Circuit for Example 5.1-2. (*a*) Circuit; (*b*) operating conditions.

and

$$V_{CEQ} = \frac{V_{CE,\text{sat}}}{2} + \sqrt{P_{C,\max}R_L + \left(\frac{V_{CE,\text{sat}}}{2}\right)^2} \tag{5.1-18b}$$

The quiescent point can, of course, also be obtained graphically from the intersection of (5.1-16) and (5.1-17), as shown in Fig. 5.1-6*b*.

The Q point is found to be

$I_{CQ} = 0.41$ A

and

$V_{CEQ} = 6.1$ V

Thus the collector supply voltage V_{CC} is 6.1 V.

Note that the ac load line *crosses* the maximum power hyperbola. This does *not* mean that the average power dissipated in the collector circuit exceeds $P_{C,\max}$. Since maximum collector dissipation occurs when there is no signal, and under this condition the collector dissipation is $P_{C,\max}$ [Eq. (5.1-17)], maximum collector dissipation is not exceeded.

The maximum peak ac collector current is 0.41 A, and the maximum average power dissipated by the load resistor is

$$P_{L,\max} = \tfrac{1}{2} I_{CQ}^2 R_L = (\tfrac{1}{2})(0.41)^2(10) = 0.84 \text{ W}$$

Note that the maximum efficiency, neglecting the loss in R_e, is only

$$\eta_{\max} = \frac{P_{L,\max}}{P_{CC}} = \frac{0.84}{(6.1)(0.41)} = \frac{0.84}{2.5} = 33.6\%$$

5.2 TRANSFORMER-COUPLED AMPLIFIER

A class A transformer-coupled amplifier is shown in Fig. 5.2-1. In the analysis presented below the transformer is assumed to be "ideal." This implies that

$$v_c = N v_L \tag{5.2-1}$$

and

$$N i_c = -i_L \tag{5.2-2}$$

Multiplying (5.2-1) and (5.2-2),

$$v_c(-i_c) = v_L i_L \tag{5.2-3}$$

In addition, dividing (5.2-1) by (5.2-2), we get

$$\frac{v_c}{-i_c} = N^2 \left(\frac{v_L}{i_L}\right) \equiv R_L' \tag{5.2-4}$$

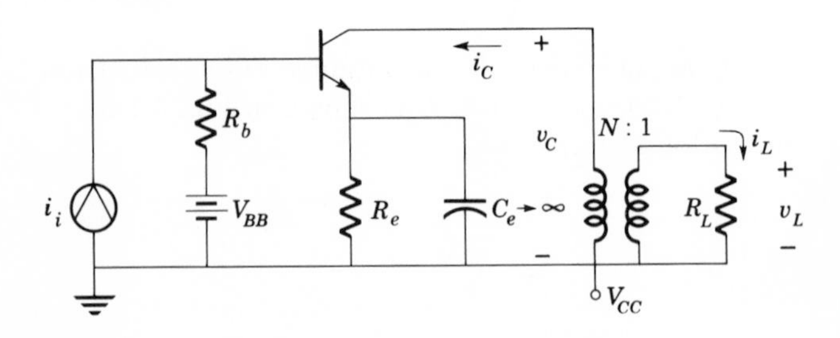

Fig. 5.2-1 Transformer-coupled power amplifier.

Therefore the ac impedance R'_L, seen looking into the transformer, is N^2 times the load resistance R_L.

The dc-load-line equation for this amplifier is the same as for the inductor-coupled amplifier.

$$V_{CC} = v_{CE} + i_E R_e \approx v_{CE} + i_C R_e \tag{5.2-5}$$

R_e is again made small. The ac load line can be obtained directly from (5.2-4), noting that $v_{ce} = v_c$. Thus the slope of the ac load line is

$$\frac{i_c}{v_{ce}} = -\frac{1}{R'_L} \tag{5.2-6}$$

The dc and ac load lines are plotted in Fig. 5.2-2. If R_e is chosen so that

$$R_e \ll R'_L$$

the quiescent current for maximum symmetrical swing is

$$I_{CQ} \approx \frac{V_{CC}}{R'_L} \tag{5.2-7}$$

Thus the impedance-transformation ratio provides the extra freedom required to set the Q point for maximum power transfer to the load.

5.2-1 POWER CALCULATIONS

The power calculations presented below are identical with those made in Sec. 5.1, with R_L replaced by R'_L. The signal i_i is sinusoidal; thus

$$i_c = I_{cm} \sin \omega t \tag{5.2-8}$$

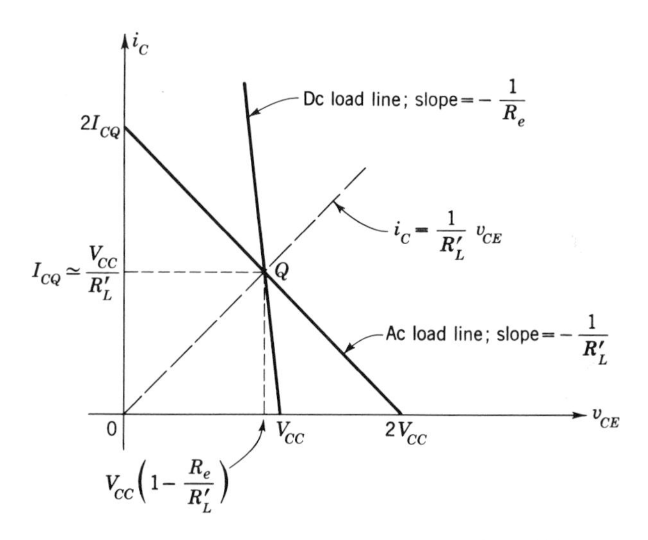

Fig. 5.2-2 Load lines for transformer-coupled amplifier.

Supplied power

$$P_{CC} = V_{CC}I_{CQ} = \frac{V_{CC}^2}{R'_L} \tag{5.2-9}$$

Power transferred to load When the signal is sinusoidal, the load current is also sinusoidal (neglecting distortion).

$$i_L = I_{Lm} \sin \omega t \tag{5.2-10}$$

Then

$$P_L = \left(\frac{I_{Lm}^2}{2}\right) R_L \tag{5.2-11}$$

Using (5.2-2),

$$I_{Lm} = NI_{cm} \tag{5.2-12}$$

Therefore

$$P_L = \left(\frac{I_{cm}^2}{2}\right) R'_L \tag{5.2-13a}$$

and

$$P_{L,\max} = \left(\frac{I_{CQ}^2}{2}\right) R'_L = \frac{V_{CC}^2}{2R'_L} \tag{5.2-13b}$$

Collector dissipation

$$P_C = \frac{V_{CC}^2}{R'_L} - \left(\frac{I_{cm}^2}{2}\right) R'_L \tag{5.2-14a}$$

and the maximum collector dissipation is, with no signal present,

$$P_{C,\max} = \frac{V_{CC}^2}{R'_L} = V_{CEQ}I_{CQ} \tag{5.2-14b}$$

Efficiency The efficiency of operation is also unchanged.

$$\eta = \left(\frac{1}{2}\right)\left(\frac{I_{cm}}{I_{CQ}}\right)^2 \tag{5.2-15a}$$

and

$$\eta_{\max} = 50\% \tag{5.2-15b}$$

Figure of merit The transistor figure of merit remains

$$\frac{P_{C,\max}}{P_{L,\max}} = 2 \tag{5.2-16}$$

From the above equations it is seen that the transformer performs a single function over and above that of the inductor-capacitor coupling circuit shown in Fig. 5.1-1a, i.e., of transforming the load impedance. This factor provides the flexibility needed to place the ac operating path in the optimum position. It has been assumed in the above analysis that this transformation takes place without any power loss. In practice, however, the transformer is not ideal, and power losses occur in it, reducing the load power and the efficiency of operation. In addition, audio-frequency transformers are always of the iron-core variety, and hence may be heavy and space-consuming and often introduce distortion.

Example 5.2-1 Using the transistor of Example 5.1-1, where max $i_C = 1$ A, with transformer coupling to the 10-Ω load, redesign the amplifier for maximum power transfer to the load. Specify the required supply voltage, the power dissipated in the load, and the transformer turns ratio N.

Solution The vi characteristic showing the permissible operating region is presented in Fig. 5.2-3. The quiescent point which will provide maximum power transfer to the load can be obtained graphically or analytically using (5.1-14).

$$I_{CQ} = \sqrt{\frac{P_{C,\max}}{N^2 R_L}} = \sqrt{\frac{0.4}{N^2}} = \frac{0.63}{N} \qquad \text{A} \tag{5.2-17a}$$

and

$$V_{CEQ} = \sqrt{P_{C,\max} N^2 R_L} = 6.3N \qquad \text{V} \tag{5.2-17b}$$

Thus, using a transformer, the Q point may be chosen almost arbitrarily, as long as

$$2I_{CQ} = \frac{1.26}{N} < 1 = \max i_C \tag{5.2-18a}$$

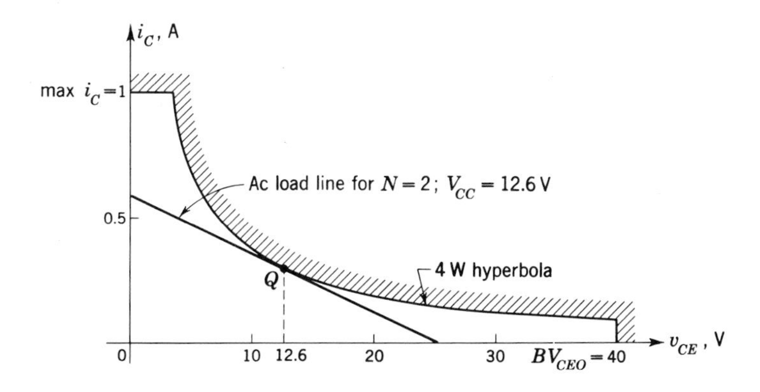

Fig. 5.2-3 Operating region and optimum load line for Example 5.2-1.

and

$$2V_{CEQ} = 12.6N < 40 = BV_{CEO} \tag{5.2-18b}$$

These inequalities set bounds on the turns ratio; i.e.,

$$1.26 < N < 3.17$$

When faced with the problem of choosing from a range of Q points, other considerations must be taken into account. For example, it is good practice to use as little current as possible, since the greater the current capability of the dc supply, the greater the size and cost. Often a specified supply voltage must be used, and this determines V_{CC} (and hence V_{CEQ}). A very important consideration is the availability of a transformer with the proper turns ratio. One common turns ratio is $N = 2$. If a transformer having this turns ratio is chosen,

$$I_{CQ} \approx 0.32 \text{ A}$$

and

$$V_{CEQ} = 12.6 \text{ V} \approx V_{CC}$$

Then

$$P_{L,\max} = (\tfrac{1}{2})(0.32)^2(2)^2(10) = 2 \text{ W}$$

The load line is shown in Fig. 5.2-3 for these conditions.

5.3 CLASS B PUSH–PULL POWER AMPLIFIERS

As found in the previous sections, the maximum attainable efficiency in class A operation is 50 percent, because the peak ac collector current never exceeds the quiescent collector current. In the class B amplifier, the dc collector current is less than the peak ac current. Thus less collector dissipation results, and the efficiency increases. The class B push-pull amplifier shown in Fig. 5.3-1 has a maximum efficiency of 78.5 percent, an improvement of 28.5 percent over the class A amplifiers discussed in Secs. 5.1 and 5.2.

Let us first examine the operation of this circuit, assuming ideal transistors, to determine the upper limits on efficiency and power output. Circuit operation can be explained in terms of the waveforms shown in Fig. 5.3-2. The center-tapped input transformer supplies two base currents of equal amplitudes but 180° out of phase (Fig. 5.3-2*b* and *c*). On the first half-cycle, i_{B1} is zero, and because T_1 is biased at cutoff, i_{C1} is

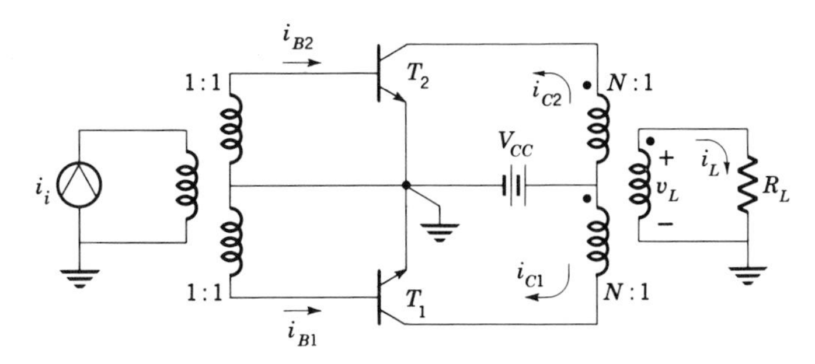

Fig. 5.3-1 Push-pull amplifier.

zero as in (*d*) of the figure. However, in this same interval, i_{B2} is positive, T_2 conducts, and the collector current i_{C2} is shown in (*e*). Thus one transistor is cut off, while the other is conducting. On the second half-cycle, the roles are reversed: T_2 is cut off, and T_1 conducts. When T_2 conducts, the current shown in (*e*) flows through the upper half of the primary winding, and the resulting time-varying flux in the transformer core induces a voltage in the secondary winding. This voltage, in turn, produces the first half-cycle of current through the load (*f*). When T_1 conducts, the current i_{C1} induces a flux in the core in a direction opposite to the flux of the previous half-cycle, resulting in the second half-cycle of load current. The final load current under these ideal conditions is thus directly proportional to the signal current i_i. It is seen from Fig. 5.3-1 that the load current i_L is related to the individual currents by

$$i_L = N(i_{C1} - i_{C2}) \tag{5.3-1}$$

If the circuit of Fig. 5.3-1 were used in practice, the load current would be extremely distorted near the zero crossing, as seen from the oscillogram of the load current shown in Fig. 5.3-2*g*. This effect is called *crossover distortion*, and is due to the base-emitter voltage v_{BE} being zero when no signal is applied. However, linear operation of the transistor begins only when i_B is positive enough for v_{BE} to exceed the break voltage, which is assumed to be 0.7 V for silicon. This distortion is shown in dotted lines in Fig. 5.3-2*d* to *f*, and is clearly visible in the oscillogram of Fig. 5.3-2*g*.

To eliminate this distortion, base-emitter junctions are biased at approximately 0.7 V. The result is then class AB rather than class B operation, although it is so nearly class B that it is usually called, simply, class B. This bias is called the *turn-on* bias. In practice, one often allows crossover distortion and relies on the transformer and internal and stray capacitances to filter it out.

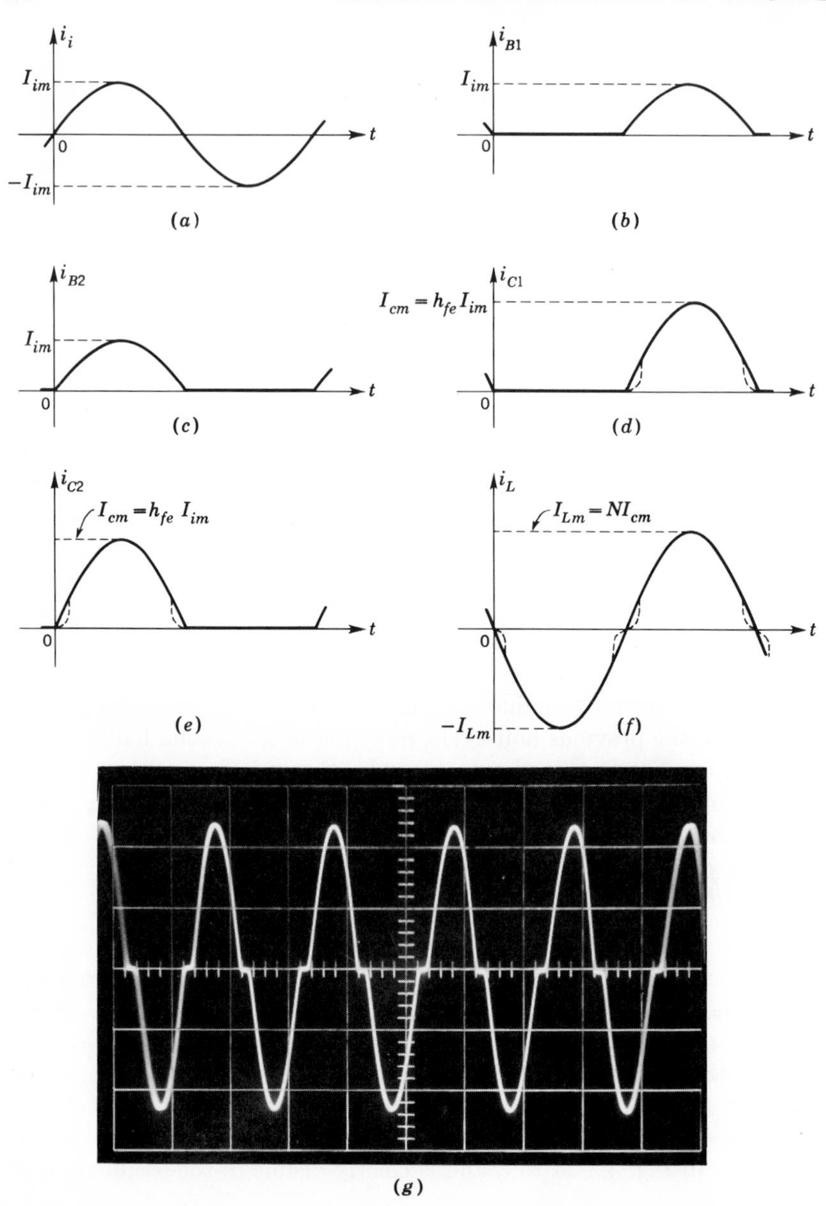

Fig. 5.3-2 Waveforms in the push-pull amplifier. (*a*) Input current; (*b*) base current in T_1; (*c*) base current in T_2; (*d*) collector current in T_1; (*e*) collector current in T_2; (*f*) load current; (*g*) oscillogram of load-current waveform, illustrating crossover distortion.

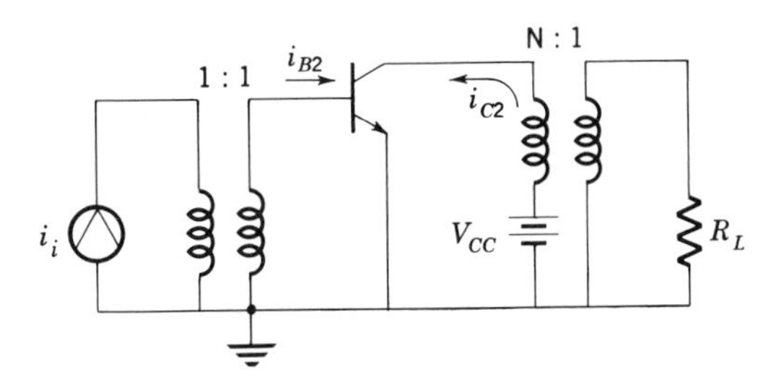

Fig. 5.3-3 One-half of class B push-pull stage.

5.3-1 LOAD–LINE DETERMINATION

Since each transistor operates in a symmetrical fashion and only half the time, we need study the operation of only one of the transistors. Consider T_2 as shown in Fig. 5.3-3. This circuit enables us to describe the operation of the amplifier. The dc load line is a vertical line, $v_{CE} = V_{CC}$, and the ac load line has a slope

$$\frac{i_C}{v_{CE}} = -\frac{1}{R'_L} \tag{5.3-2}$$

as shown in Fig. 5.3-4. During the time that T_2 is off, $i_{C2} = 0$ and

$$v_{CE2} = V_{CC} + Nv_L$$

varies from V_{CC} (which occurs when $v_{CE1} = V_{CC}$, and hence $Nv_L = 0$) to $2V_{CC}$ (which occurs when $v_{CE1} = 0$, and hence $Nv_L = V_{CC}$). Thus, while the transistor is off, the ac load line is horizontal, with $i_{C2} = 0$.

The maximum value of both i_{C1} and i_{C2} (Figs. 5.3-2*d* and *e* and 5.3-4) is

$$I_{cm} = \frac{V_{CC}}{R'_L} \tag{5.3-3}$$

5.3-2 POWER CALCULATIONS

Assume that the signal current is sinusoidal,

$$i_i = I_{im} \sin \omega t$$

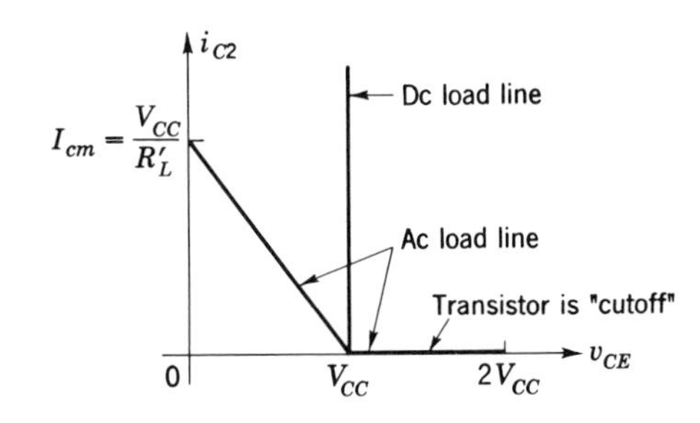

Fig. 5.3-4 Load lines for class B stage.

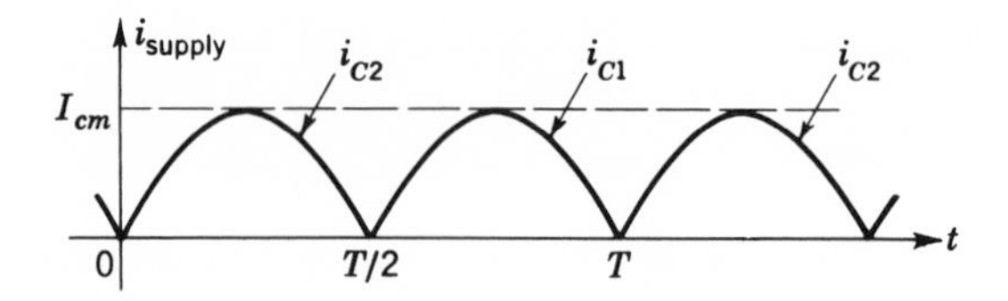

Fig. 5.3-5 Power-supply current waveform.

Supplied power The power delivered by the supply is

$$P_{CC} = V_{CC}\left(\frac{1}{T}\right)\int_{-T/2}^{T/2}[i_{C1}(t) + i_{C2}(t)]\,dt \tag{5.3-4a}$$

The current $i_{C1} + i_{C2}$ is the current flowing through the supply. From Fig. 5.3-2*d* and *e*, this is a full-wave-rectified current, as shown in Fig. 5.3-5. The average value of a full-wave-rectified sine wave is $2/\pi$ times its peak value. Thus

$$\left(\frac{1}{T}\right)\int_{-T/2}^{T/2}(i_{C1} + i_{C2})\,dt = \left(\frac{2}{\pi}\right)I_{cm}$$

The supplied power P_{CC} is then

$$P_{CC} = \left(\frac{2}{\pi}\right)V_{CC}I_{cm} \tag{5.3-4b}$$

Its maximum value is (Fig. 5.3-4)

$$P_{CC,\max} = \left(\frac{2}{\pi}\right)V_{CC}\left(\frac{V_{CC}}{R_L'}\right) = \frac{2V_{CC}^2}{\pi R_L'} \tag{5.3-4c}$$

Power transferred to load The power transferred to the load is

$$P_L = \tfrac{1}{2}I_{Lm}^2R_L = \tfrac{1}{2}I_{cm}^2N^2R_L = \tfrac{1}{2}I_{cm}^2R_L' \tag{5.3-5a}$$

Its maximum value is

$$P_{L,\max} = \frac{V_{CC}^2}{2R_L'} \tag{5.3-5b}$$

Power dissipated in the collector The power dissipated in the collectors of transistors T_1 and T_2 totals

$$2P_C = P_{CC} - P_L$$

Using (5.3-4*b*) and (5.3-5*a*),

$$2P_C = \left(\frac{2}{\pi}\right)V_{CC}I_{cm} - \frac{R_L'I_{cm}^2}{2} \tag{5.3-6}$$

The maximum value of collector dissipation $P_{C,\max}$ is found by differentiating P_C with respect to I_{cm} and setting the result equal to zero.

$$2\left(\frac{dP_C}{dI_{cm}}\right) = \left(\frac{2}{\pi}\right) V_{CC} - R'_L I_{cm} = 0 \tag{5.3-7}$$

The collector current at which the collector dissipation is a maximum is then

$$I_{cm} = \left(\frac{2}{\pi}\right)\left(\frac{V_{CC}}{R'_L}\right) \tag{5.3-8}$$

and combining (5.3-6) and (5.3-8), the maximum collector dissipation is

$$2P_{C,\max} = \left(\frac{2}{\pi}\right) V_{CC}\left(\frac{2V_{CC}}{\pi R'_L}\right) - \left(\frac{R'_L}{2}\right)\left[\left(\frac{2}{\pi}\right)\left(\frac{V_{CC}}{R'_L}\right)\right]^2 = \left(\frac{2}{\pi^2}\right)\left(\frac{V_{CC}^2}{R'_L}\right) \tag{5.3-9}$$

The power dissipated in each collector is then

$$P_{C,\max} = \left(\frac{1}{\pi^2}\right)\left(\frac{V_{CC}^2}{R'_L}\right) \approx 0.1\left(\frac{V_{CC}^2}{R'_L}\right) \tag{5.3-10}$$

Efficiency The efficiency of operation η is calculated from (5.3-4*b*) and (5.3-5*a*).

$$\eta = \frac{P_L}{P_{CC}} = \frac{\frac{1}{2}R'_L I_{cm}^2}{(2/\pi)V_{CC}I_{cm}} = \left(\frac{\pi}{4}\right)\left(\frac{I_{cm}}{V_{CC}/R'_L}\right) \tag{5.3-11}$$

Since the maximum attainable collector current is V_{CC}/R'_L, the maximum attainable efficiency is

$$\eta_{\max} = \frac{\pi}{4} \approx 78.5\% \tag{5.3-12}$$

The supplied power, the load power, the collector dissipation, and the efficiency are plotted in Fig. 5.3-6. These results should be compared with those obtained using the class A amplifier of Sec. 5.1, which are shown in Fig. 5.1-3.

Figure of merit The transistor utilization figure of merit for the class B push-pull amplifier is

$$\frac{P_{C,\max}}{P_{L,\max}} = \frac{V_{CC}^2/\pi^2 R'_L}{V_{CC}^2/2R'_L} = \frac{2}{\pi^2} \approx \frac{1}{5} \tag{5.3-13}$$

Note the improvement by a factor of 10 in the figure of merit achieved over the class A amplifier. Thus, if $P_{L,\max}$ is to be 25 W, each collector must dissipate only 5 W. Although it is true that the circuit requires two transistors and two transformers, the lower power rating of each of the transistors means that they will take up less space and require significantly less heat sinking than will one high-power transistor. In this case a single-transistor class A amplifier would require 50 W of power-dissipation capability.

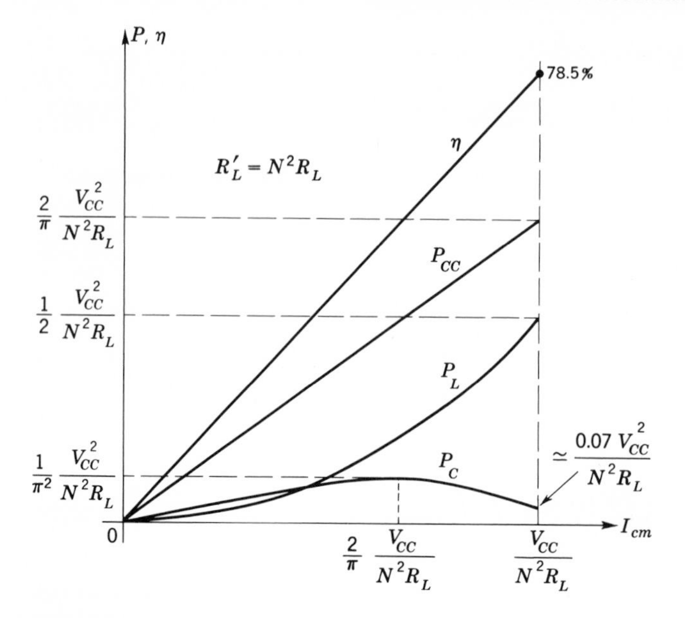

Fig. 5.3-6 Power and efficiency variation in the class B push-pull amplifier.

Another major advantage of class B operation is that the no-signal current drain from the battery is zero, while in class A operation, the no-signal current is the same as the full-load current.

It is important to keep in mind that the efficiency and collector dissipation ratings are derived for sinusoidal signals, and are theoretical maxima, which can only be approached in actual practice.

Example 5.3-1 Design a push-pull class B amplifier to achieve maximum power output to a 10-Ω load. Use two transistors with ratings as in Example 5.1-1 (max $i_C = 1$ A). Specify V_{CC}, N, and a bias network to eliminate crossover distortion. Calculate the power output and efficiency.

Solution For convenience, the transistor ratings are repeated.

$$P_{C,\max} = 4 \text{ W}$$

$$BV_{CEO} = 40 \text{ V}$$

$$\max i_C = 1 \text{ A}$$

The maximum power output is given by (5.3-5*b*).

$$P_{L,\max} = \frac{V_{CC}^2}{2R'_L} = \frac{V_{CC} I_{cm}}{2}$$

Thus the power output can be increased by increasing V_{CC} and I_{cm}. However, V_{CC} and I_{cm} cannot be increased indefinitely. The transistor ratings establish upper bounds on V_{CC} and I_{cm} as follows:

$$V_{CC} \leq \tfrac{1}{2}BV_{CEO} = 20 \text{ V}$$

$$I_{cm} \leq \max i_C = 1 \text{ A}$$

and

$$P_{L,\max} = \frac{V_{CC}I_{cm}}{2} \leq 5P_{C,\max} = 20 \text{ W}$$

The Q point is chosen to drive the transistor to its rated maximum i_C and BV_{CEO}. Thus

$$V_{CC} = 20 \text{ V}$$

and

$$I_{cm} = 1 \text{ A}$$

Then

$$P_{L,\max} = 10 \text{ W}$$

The turns ratio N is found as follows: Since

$$I_{cm} = \frac{V_{CC}}{N^2R_L}$$

$$N^2 = 2$$

and

$$N = 1.414$$

Note that the two transistors in push-pull provide *five times* the power that a single transistor can supply to the given load resistance without exceeding maximum ratings (see Example 5.2-1).

Figure 5.3-7 shows the *vi* characteristic of each transistor and the ac load line. Note that the ac load line intersects the maximum average collector-dissipation curve. This means that the instantaneous power *can* exceed the maximum average power. Our power restriction is simply that the *average* collector dissipation be less than the maximum allowable average collector dissipation.* Note

* If a very low frequency signal is being amplified, the time during which the maximum dissipation curve is exceeded may be long enough to result in the transistor overheating. The lowest frequency which can be amplified depends on the thermal time constant of the transistor.

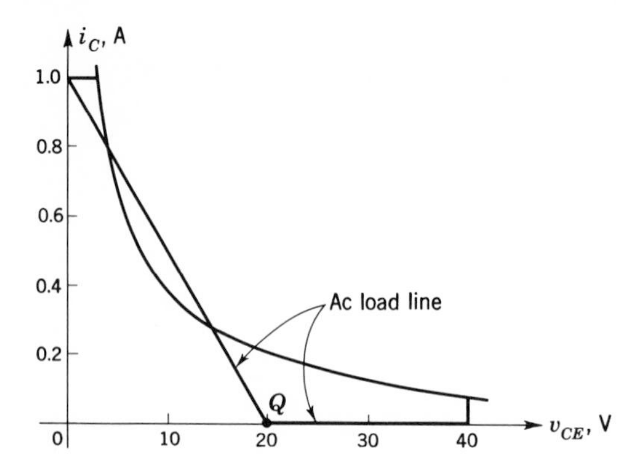

Fig. 5.3-7 Load line for Example 5.3-1.

that we do not optimize the design of a class B push-pull amplifier by making the ac load line tangent to the $P_{C,\max}$ hyperbola.

To complete the design, a bias network must be selected. The circuit to be used, Fig. 5.3-8, shows the push-pull amplifier with a separate bias supply. V_{BB} is adjusted to the "turn-on" voltage of the transistor, which, for silicon, is approximately 0.7 V.

In practice, the V_{CC} supply with a suitable voltage divider is used rather than a separate supply as shown in Fig. 5.3-8*b*. R_1 and R_2 are chosen so that the base-emitter drop is about 0.7 V (for silicon transistors). The parallel combination of R_1 and R_2 is kept as small as practicable. A silicon diode is often used in place of R_2, since the voltage drop across a silicon diode operating above its break voltage is similar to the quiescent base-emitter voltage required to turn on the transistor.

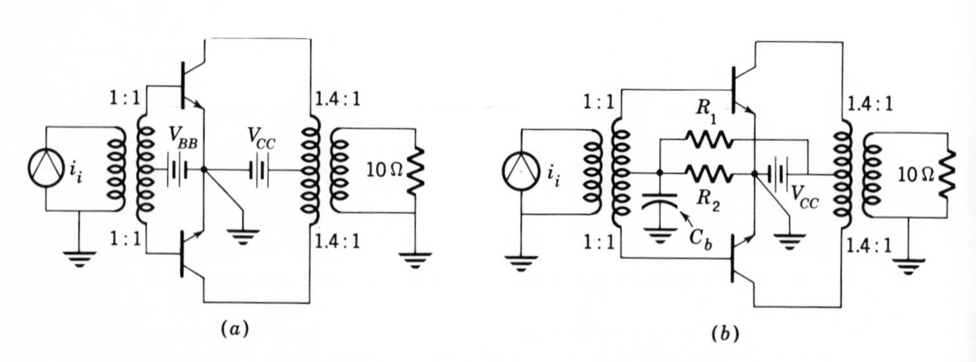

Fig. 5.3-8 Push-pull amplifier with bias supply. (*a*) Battery supply; (*b*) resistance divider supply.

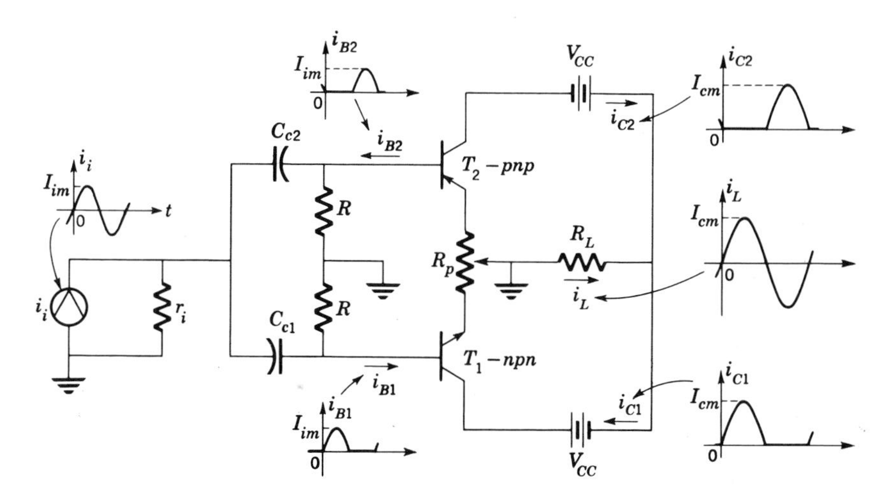

Fig. 5.4-1 Complementary-symmetry amplifier.

5.4 AMPLIFIERS USING COMPLEMENTARY SYMMETRY

Figure 5.4-1 illustrates a type of push-pull class B amplifier which employs one *pnp* and one *npn* transistor and requires *no* transformers. This type of amplifier uses *complementary symmetry*. Its operation can be explained by referring to the figure. When the signal current is positive, T_1 (the *npn* transistor) conducts, while T_2 (the *pnp* transistor) is cut off. When the signal current is negative, T_2 conducts while T_1 is cut off. The load current is

$$i_L = i_{C1} - i_{C2} \tag{5.4-1}$$

The load-line and output-circuit power relations for this amplifier are the same as for the conventional class B amplifier of Sec. 5.3. Some advantages of the circuit are that the transformerless operation saves on weight and cost, and "balanced" push-pull input signals are not required. Disadvantages are the need for both positive and negative supply voltages, and the problem of obtaining pairs of transistors that are matched closely enough to achieve low distortion. To help balance the circuit, a potentiometer is often inserted in the emitter circuit, as shown in Fig. 5.4-1.

Example 5.4-1 When using a complementary-symmetry push-pull amplifier, the center tap of the power supply is often grounded, rather than the emitters of T_1 and T_2. This results in an emitter-follower type of configuration, as shown in Fig. 5.4-2. Using transis-

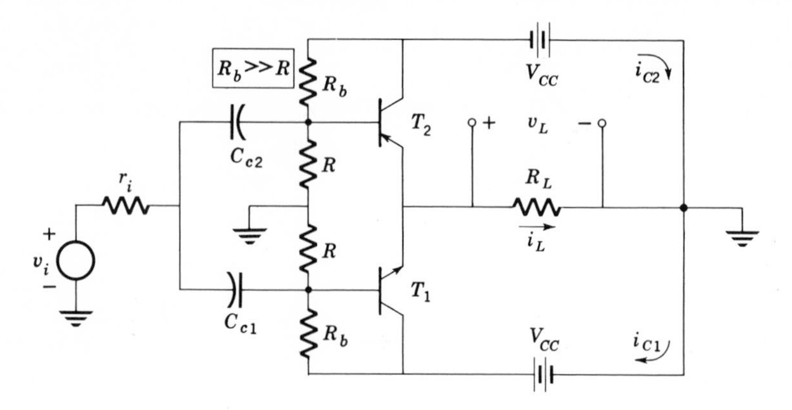

Fig. 5.4-2 Complementary-symmetry emitter follower.

tors having the same specifications as before,

$P_{C,\max} = 4$ W

$BV_{CEO} = 40$ V

$\max i_C = 1$ A

Find V_{CC} and the maximum power which can be delivered to a 10-Ω load.

Solution Each transistor is essentially a class B emitter follower. Consider T_2. Its equivalent circuit during conduction and the corresponding load line and operating path are shown in Fig. 5.4-3. Since the peak value of i_{C2} cannot exceed 1 A,

$$\frac{V_{CC}}{R_L} = 1 \text{ A}$$

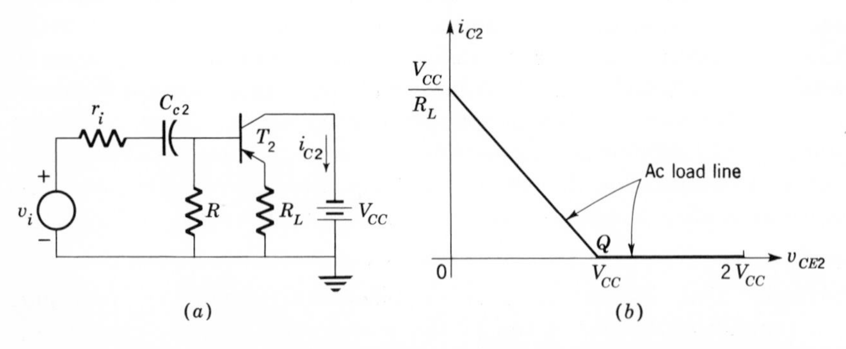

Fig. 5.4-3 Circuit and load line for T_2 of the complementary-symmetry emitter-follower circuit. (*a*) Circuit; (*b*) load line.

Therefore

$$V_{CC} = 10 \text{ V}$$

Note that the $2V_{CC}$ swing is less than the collector-emitter breakdown voltage, as required.

The load current is

$$i_L = i_{C1} - i_{C2}$$

If v_i is sinusoidal, i_L is also sinusoidal, with a maximum peak current

$$I_{Lm} = \frac{V_{CC}}{R_L}$$

Thus

$$I_L = I_{Lm} \sin \omega t$$

$$= \frac{V_{CC}}{R_L} \sin \omega t$$

The maximum power into the load is

$$P_{L,\max} = \frac{V_{CC}^2}{2R_L} = 5 \text{ W}$$

Note that the push-pull amplifier of Sec. 5.3 could deliver 10 W because the transformer doubled the effective load impedance.

We can get some idea of the signal voltage required to drive this amplifier by assuming that r_i is zero and R infinite. Then, because the circuit is basically an emitter follower, the maximum signal $v_i = V_{im} \sin \omega t$ must have a peak value of V_{CC} (that is, $V_{im} = V_{CC}$) in order to transfer the full 5 W to the load.

SUMMARY

The basic class A and push-pull class B power amplifiers have been considered, and procedures have been presented for establishing preliminary circuit designs. In all cases, the circuit design of a power amplifier must be accompanied by a thermal design which will ensure that the junction temperature is maintained within safe limits. It was found that the push-pull class B configuration yielded significantly higher efficiency and lower power-supply drain. Thus, for a given power output, much smaller and lighter transistors are required. In particular, the complementary-symmetry arrangement is very attractive because of the simplicity of the circuit and the availability of matched pairs of transistors.

In practice, a preliminary power-amplifier design is usually done on paper, and final adjustments made in the laboratory where power output, distortion, efficiency, and temperature stability are easily measured.

PROBLEMS

SECTION 5.1

5.1. For the circuit of Fig. P5.1, calculate

(*a*) The maximum power dissipated in the load.
(*b*) The total power delivered by the V_{CC} supply.
(*c*) The power dissipated at the collector.
(*d*) The efficiency.

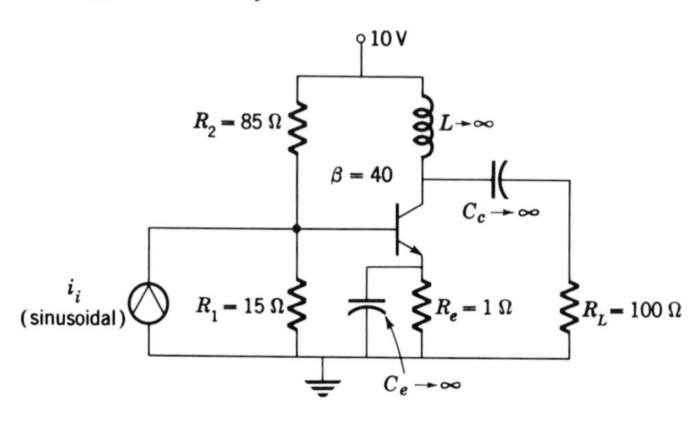

Fig. P5.1

5.2. In the circuit of Prob. 5.1, the transistor has $20 < \beta < 60$. Find new values for R_1 and R_2 which will permit maximum power to be dissipated in the load. Recalculate Prob. 5.1 for these new values.

5.3. The circuit shown is a class A power amplifier which must supply a *maximum* undistorted power of 2 W to the 10-Ω load. The *minimum* necessary transistor ratings are to be specified. Find I_{CQ}, P_{CC}, and η. Also specify $P_{C,\max}$, $v_{CE,\max}$, and $i_{C,\max}$ for the transistor. Neglect R_e and bias circuit losses.

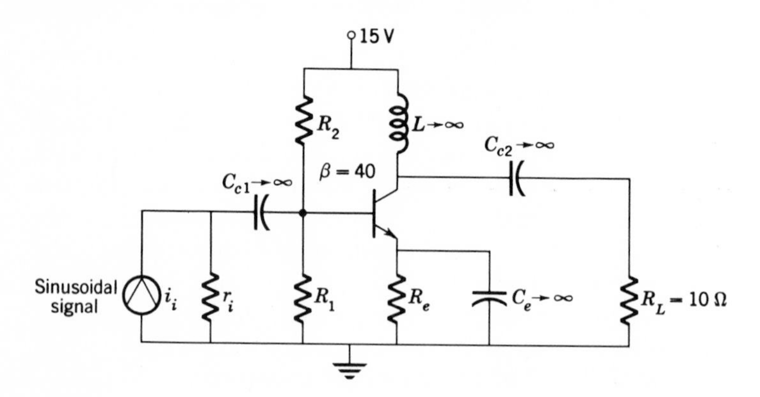

Fig. P5.3

5.4. Verify Eqs. (5.1-18*a*) and (5.1-18*b*).

5.5. In Example 5.1-2 assume that $i_{C,\max} = 0.75$ A. Determine the maximum attainable swing and the maximum power dissipated in the load.

5.6. Repeat Prob. 5.1 if the transistor collector-emitter saturation voltage is 1 V.

SECTION 5.2

5.7. The transistor ratings are to be specified for class A operation. The maximum required load power is 2 W. Neglect R_e and bias-circuit losses.

(*a*) Find P_{CC}, assuming that the amplifier is designed for maximum efficiency.
(*b*) Find I_{CQ}.
(*c*) Specify the $i_{C,\max}$, $v_{CE,\max}$, and $P_{C,\max}$ ratings for the transistor.
(*d*) If $R_L = 6.25\ \Omega$, find the turns ratio N.

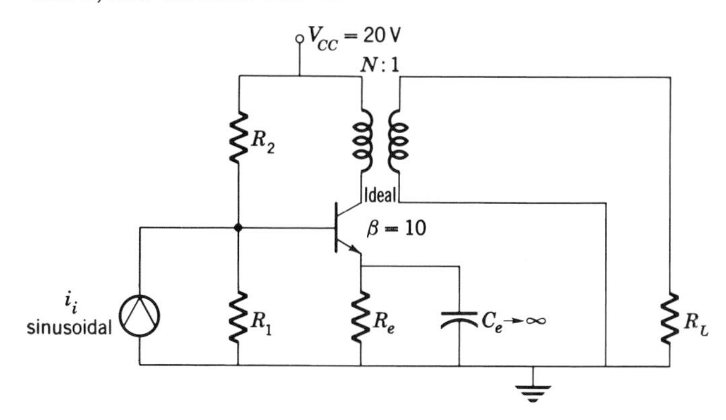

Fig. P5.7

5.8. Repeat Prob. 5.7, assuming that the transformer efficiency is 75 percent.

5.9. In Prob. 5.7, the maximum power of $P_L = 2$ W is obtained when $i_i = I_{im} \sin \omega t$. Find P_L, when

$$i_i = \left(\frac{I_{im}}{2}\right) \sin \omega t + \left(\frac{I_{im}}{2}\right) \sin 3\omega t$$

5.10. Repeat Prob. 5.7, assuming $V_{CE,\text{sat}} = 1$ V. Include the effect of losses in the emitter and bias circuits. Assume $R_e = 1\ \Omega$ and $R_1 = 10\ \Omega$.

5.11. In the emitter-coupled class A power amplifier shown, $P_{C,\max} = 100$ W. Find R_b, V_{BB}, and N, so that maximum power can be transmitted to the load. Also find P_{CC}, $P_{L,\max}$, P_C, and η.

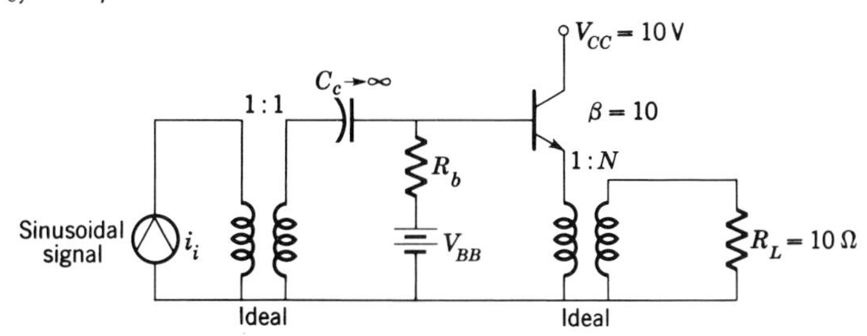

Fig. P5.11

5.12. Find N so that maximum power can be dissipated in the load. Calculate $P_{L,\max}$, $P_{C,\max}$, and P_{CC}. Include the effect of losses in the bias and emitter circuits.

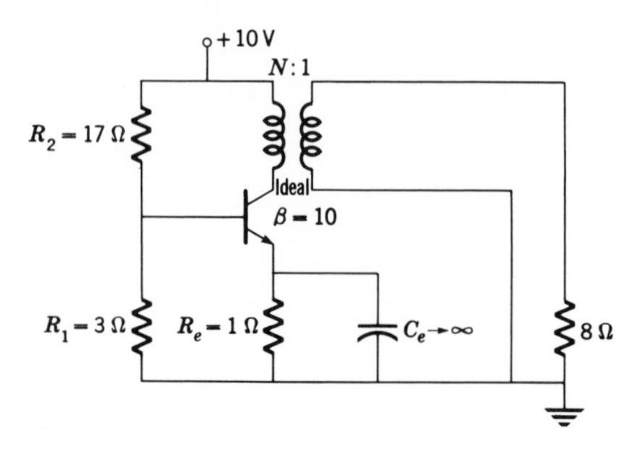

Fig. P5.12

5.13. In Example 5.2-1 calculate N and P_{Lm} if $V_{CC} = 9$ V. Compare with the result in the example, where $V_{CC} = 12.6$ V.

5.14. Repeat Prob. 5.13 if $V_{CC} = 18$ V.

SECTION 5.3

5.15. For the class B push-pull amplifier shown, calculate the maximum values of i_C, i_L, v_{CE}, P_L, P_C, and P_{CC}. Plot P_{CC}, P_L, and P_C versus i_C for the range $0 < i_C < \max i_C$.

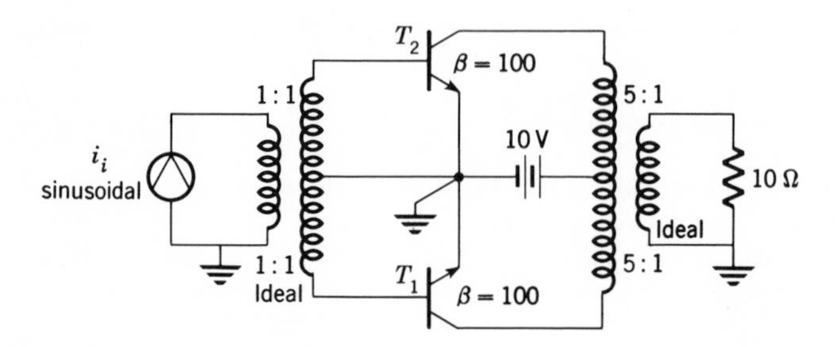

Fig. P5.15

5.16. Design a class B push-pull amplifier to deliver 10 W to a 10-Ω load, using transistors which have $BV_{CEO} = 40$ V. Specify $P_{C,\max}$ for each transistor, V_{CC}, and the minimum N required.

5.17. Repeat Prob. 5.15 with the transformer in the emitter circuit.

5.18. Transistors having $BV_{CEO} = 50$ V and $P_{C,\max} = 1$ W are available. Design a push-pull class B amplifier using these transistors, if $V_{CC} = 22.5$ V.

(a) Specify the reflected load resistance, and find the maximum power output.

(b) Find the input-current swing required if $\beta \approx 50$.

5.19. A loudspeaker rated at 8 Ω and 500 mW is to be driven by a class B push-pull amplifier. The power supply is 9 V. The transistors to be used have $V_{CE,\text{sat}} = 1$ V. Select a suitable value of N, and find P_{CC} and P_C when 500 mW is being dissipated in the load.

5.20. Transistors T_1 and T_2 are nonlinear, so that

$$i_{C2} = 10i_{B2} + i_{B2}^2$$

and

$$i_{C1} = 10i_{B1} + i_{B1}^2$$

If

$$i_i = \cos \omega_0 t$$

find i_L. Note the distortion resulting from the nonlinearities.

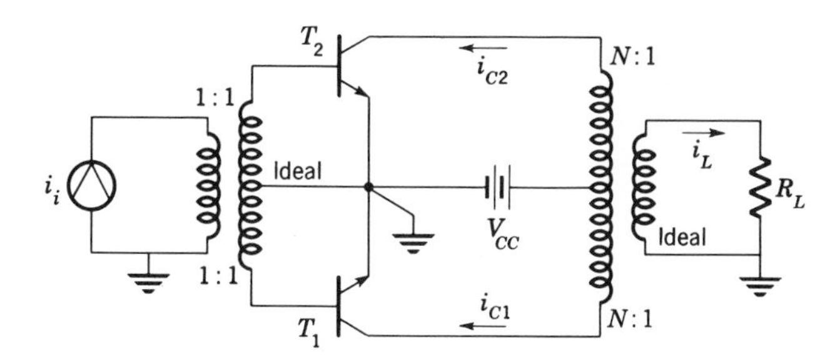

Fig. P5.20

5.21. The push-pull amplifier shown is operating in the class A mode. The quiescent current is I_{CQ}. Derive expressions for

(*a*) P_{CC}
(*b*) P_L
(*c*) P_C
(*d*) η
(*e*) Plot these terms as a function of the peak collector current I_{cm}.
(*f*) Calculate the figure of merit $P_{C,\max}/P_{L,\max}$.
Hint: $V_{CC} = 2N^2R_LI_{CQ}$

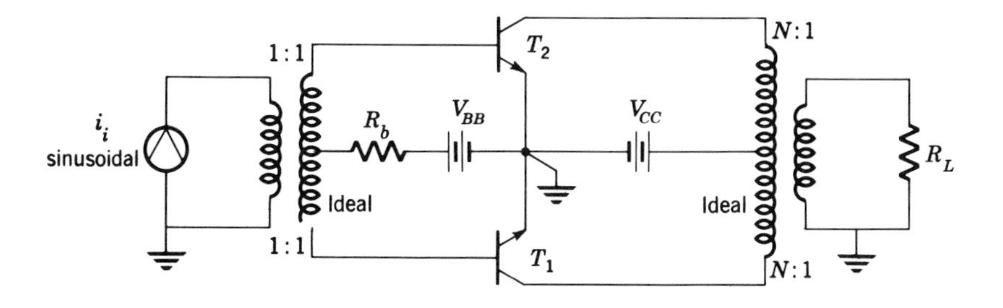

Fig. P5.21

5.22. The push-pull amplifier shown in Fig. P5.21 is operating in the class A mode. If $V_{CC} = 20$ V, $R_L = 100\ \Omega$, $P_{C,\max}$ of each transistor is 4 W, $BV_{CEO} = 40$ V, find

(*a*) I_{CQ} for maximum power dissipated in the load.
(*b*) The power furnished by the supply.
(*c*) The power dissipated in the load.
(*d*) The power dissipated in the collector.
(*e*) The maximum efficiency.
(*f*) The ratio $P_{C,\max}/P_{L,\max}$.

5.23. Repeat Prob. 5.22. However, plot P_{CC}, P_L, P_C, and η as a function of the peak collector-current swing I_{cm}.

5.24. The class A push-pull amplifier shown in Fig. P5.21 has nonlinear transistors. The nature of the nonlinearity is given in Prob. 5.20. If $i_i = \cos \omega_0 t$, calculate i_L.

Note that a result of using the class A push-pull amplifier is significant reduction of distortion as compared with the class B push-pull amplifier.

SECTION 5.4

5.25. For the complementary symmetry amplifier shown,

(*a*) Calculate the maximum power dissipated in the load.
(*b*) Calculate the maximum power furnished by both the V_{CC} and $-V_{EE}$ supplies.
(*c*) Calculate the maximum power dissipated by each transistor.
(*d*) Plot P_L, P_{CC}, and P_C versus I_{cm} in the range $0 < I_{cm} < 2$ A.

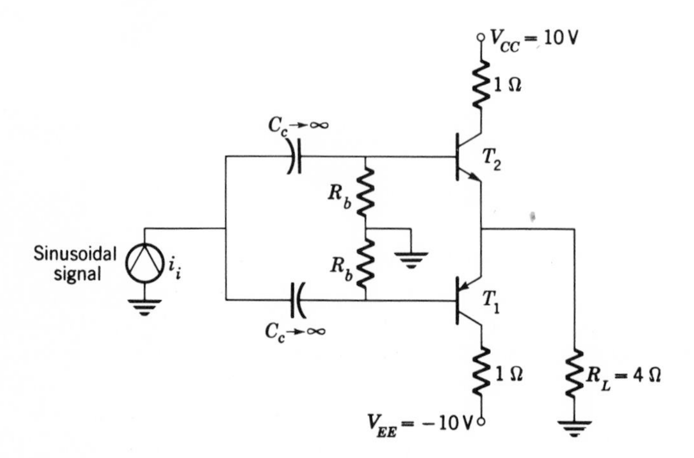

Fig. P5.25

6
Small-signal Low-frequency Analysis and Design

INTRODUCTION

Up to this point, the graphical approach has been emphasized as a conceptual aid in the solution of problems in analysis and design. This approach was found useful when considering dc biasing and power amplifiers, where large signals are encountered. In this chapter, the response of transistor circuits to *small signals* is studied.

When the collector swing is very small, the transistor is considered to be linear and can be replaced, for purposes of analysis, by a small-signal *equivalent-circuit model.* The graphical approach is abandoned, and this linear model can be analyzed using standard network analysis techniques (e.g., loop or node equations) in order to determine the small-signal response. To simplify response calculations, the technique of "impedance reflection" is introduced. Making use of this technique, complicated circuits can often be simplified to the point where they can be solved by inspection.

The small-signal equivalent circuits developed in this chapter are assumed to be independent of frequency. Frequency response is considered in detail in Chaps. 12 to 15.

In practice, the design of small-signal current or voltage amplifiers is broken down into two essentially separate parts. The first consists of setting the dc bias, i.e., finding a suitable Q point; here the graphical method is used. The second part involves gain and impedance calculations at signal frequencies; here the small-signal equivalent circuit is used. These two parts are not completely independent because, as will be seen, the values of some of the components in the equivalent circuit depend on the Q point.

6.1 THE HYBRID PARAMETERS

The elements of the equivalent circuit for the transistor can be developed from the internal physics of the device or from its terminal properties. The latter approach is employed here since it is more general and has many conceptual advantages.

When analyzing or designing a transistor amplifier, attention is focused on two pairs of terminals, input and output. Thus use can be made of the applicable results of two-port network theory. There are six possible pairs of equations relating input and output quantities, which can be used to define completely the terminal behavior of the two-port network shown in Fig. 6.1-1*a*. These six pairs of equations involve impedance, admittance, hybrid, and chain parameters, all of which are interrelated. For most transistor work, the *hybrid*, or h, parameters are the most useful, because they are easily measured and therefore most often specified in manufacturers' data sheets, and they provide quick estimates of circuit performance.

The standard form for the hybrid equation is

$$v_1 = h_{11}i_1 + h_{12}v_2 \tag{6.1-1a}$$

$$i_2 = h_{21}i_1 + h_{22}v_2 \tag{6.1-1b}$$

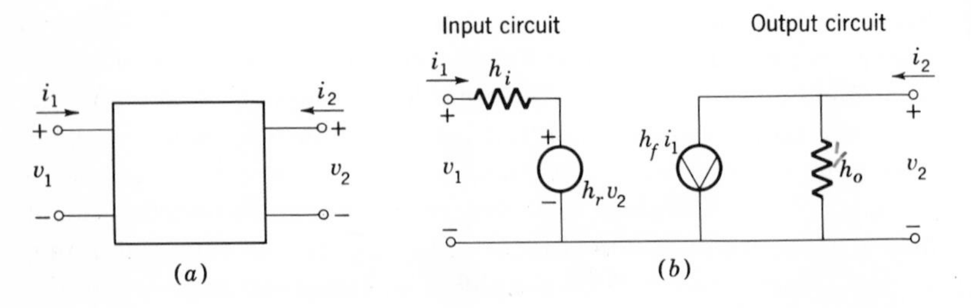

Fig. 6.1-1 The two-port network. (*a*) General two-port; (*b*) h-parameter equivalent circuit.

In these equations, the independent variables are the input current i_1 and the output voltage v_2. The voltage and current variables are understood to represent small variations about the quiescent operating point. Note that the small-signal-current directions shown in Fig. 6.1-1*b* are into the network. This differs from the convention employed to describe dc and total currents in the transistor.

In transistor circuit theory, the numerical subscripts are exchanged for letters which identify the physical nature of the parameter.

$$v_1 = h_i i_1 + h_r v_2 \tag{6.1-2}$$

$$i_2 = h_f i_1 + h_o v_2 \tag{6.1-3}$$

Referring to the equivalent circuit of Fig. 6.1-1*b*, the *input* circuit is derived from (6.1-2), using KVL, and the *output* circuit from (6.1-3), using KCL. The physical meaning of the h parameters can be obtained from the defining equations or from the circuit. For example, (6.1-2) indicates that h_i is dimensionally an impedance. Referring to the circuit of Fig. 6.1-1*b*, it is seen to be the input impedance with the output short-circuited ($v_2 = 0$). The subscript i thus stands for *input*. Similarly, h_r is dimensionless and represents the *reverse* open-circuit voltage ratio.

Terminal definitions for all four parameters are as follows:

$$h_i = \left.\frac{v_1}{i_1}\right|_{v_2=0} = \text{short-circuit input impedance} \tag{6.1-4}$$

$$h_r = \left.\frac{v_1}{v_2}\right|_{i_1=0} = \text{open-circuit reverse voltage gain} \tag{6.1-5}$$

$$h_f = \left.\frac{i_2}{i_1}\right|_{v_2=0} = \text{short-circuit forward current gain} \tag{6.1-6}$$

$$h_o = \left.\frac{i_2}{v_2}\right|_{i_1=0} = \text{open-circuit output admittance} \tag{6.1-7}$$

The equivalent circuit of Fig. 6.1-1*b* is extremely useful for several reasons: First, it isolates the input and output circuits, their interaction being accounted for by the two controlled sources; second, the two parts of the circuit are in a form which makes it simple to take into account source and load circuits. The input circuit is a Thévenin equivalent, and the output circuit a Norton equivalent.

The circuit and definitions of (6.1-4) to (6.1-7) also suggest small-signal methods of measurement for the various parameters. For example, (6.1-6) indicates that h_f can be measured by placing an ac short circuit (a large capacitor) across the output (so that $v_2 = 0$), applying a small ac current to the input and then measuring the current ratio. Note that the dc conditions must be maintained so that the h parameters can be determined with respect to a specified Q point. We show below that the h parameters are each a function of the Q point.

The objective of the rest of this chapter is to find the equivalent circuits for the common-base, common-emitter, and common-collector (emitter-follower) configurations, using the h parameters wherever practicable. It should be noted that these parameters must all be evaluated at the Q point. The parameters are in general different for each configuration, and are distinguished by adding an identifying letter as a second subscript. Thus, for example, h_{ob} is the output admittance for the CB configuration.

Often, the manufacturer will give the CB h parameters, and the designer may need the CE parameters. The conversion from one set to another can be made using simple circuit-analysis methods, to be described later.

6.2 THE COMMON-EMITTER CONFIGURATION

In this section we find the small-signal equivalent-circuit model for the common-emitter configuration shown in Fig. 6.2-1*a*. In Chaps. 3 and 5, where large-signal behavior was considered, it was assumed that the time variation of v_{BE} was negligible, compared with the signal. Thus the base-emitter circuit was represented by a battery ($\approx$0.7 V for silicon transistors). In this chapter, which is concerned with *small* signals, this assumption, which would imply that h_{ie} and h_{re} are zero, is *not* made.

The load current i_L, in the circuit of Fig. 6.2-1*a*, contains a dc, as well as a small-signal ac, component. Since linear operation is assumed, the ac and dc components can be treated separately, using superposition.

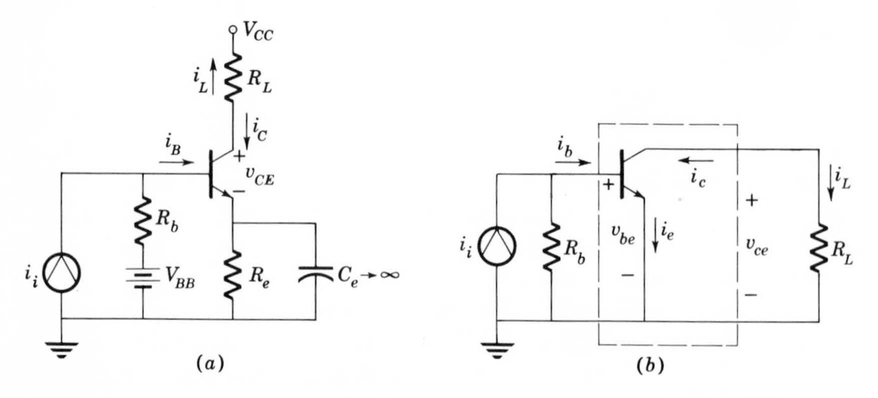

Fig. 6.2-1 The common-emitter configuration. (*a*) Complete circuit; (*b*) small-signal circuit.

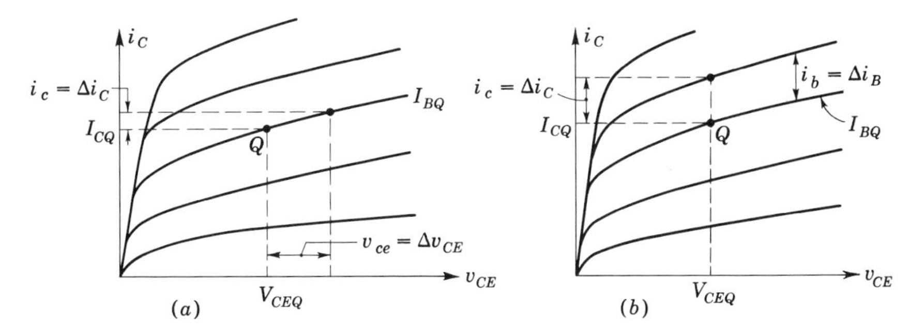

Fig. 6.2-2 Estimating h_{oe} and h_{fe} from the vi characteristic. (a) h_{oe}; (b) h_{fe}.

Thus the batteries and the capacitor can be replaced by short circuits, yielding the small-signal ac circuit of Fig. 6.2-1b.

Now, of the four hybrid parameters, three can be disposed of rather quickly. The reverse voltage gain h_{re} is usually negligible, and is omitted. The output admittance h_{oe} can be written using (6.1-7).

$$h_{oe} = \frac{i_c}{v_{ce}}\bigg|_{i_b=0} \tag{6.2-1a}$$

i_c and v_{ce} are defined as small variations about the nominal operating point. Thus the parameter h_{oe} is simply the slope of the collector characteristic at the Q point, as shown in Fig. 6.2-2a (see also Fig. 3.2-4).

$$h_{oe} = \frac{\Delta i_C}{\Delta v_{CE}}\bigg|_{Q\text{ point}} \tag{6.2-1b}$$

Numerical values for h_{oe} can be obtained from the vi characteristic, if it is available. For most silicon transistors, h_{oe} has a value less than 10^{-4} mho, and since it is in parallel with a load resistance R_L, it can be neglected as long as R_L is less than 1 or 2 kΩ, which is very often the case.

The short-circuit current gain h_{fe} is obtained by setting $R_L = 0$. Then, from (6.1-6),

$$h_{fe} = \frac{i_c}{i_b}\bigg|_{Q\text{ point}} = \frac{\Delta i_C}{\Delta i_B}\bigg|_{Q\text{ point}} \tag{6.2-1c}$$

This parameter can also be obtained from the vi characteristic, as shown in Fig. 6.2-2b. Figure 3.2-1 shows that h_{fe} is approximately equal to h_{FE} and is a function of the quiescent current.

Finally, h_{ie} is calculated using (6.1-4).

$$h_{ie} = \frac{v_{be}}{i_b}\bigg|_{v_{ce}=0} \tag{6.2-2}$$

Refer to the forward-biased junction diode, which is seen looking into the base-emitter terminals. The small-signal ratio v_{be}/i_b represents the dynamic resistance of that junction, evaluated at the Q point. This resistance was determined in Sec. 3.1-1. Making use of that result, we write (assuming $m = 1$)

$$h_{ie} = \left.\frac{v_{be}}{i_b}\right|_{Q\text{ point}} = \frac{V_T}{I_{BQ}} \approx h_{FE}\left(\frac{V_T}{I_{EQ}}\right) \approx h_{fe}\left(\frac{V_T}{I_{EQ}}\right) \tag{6.2-3a}$$

Experimentally, we find that [Eq. (3.1-5)]

$$h_{ie} \approx mh_{fe}\left(\frac{V_T}{I_{EQ}}\right) \tag{6.2-3b}$$

The value of m varies from 1 to 2, and for silicon transistors is often given as 1.4. At room temperature $V_T \approx 25$ mV, so that a transistor with $h_{fe} = 100$ and $I_{CQ} = 10$ mA would have an input impedance

$$250\ \Omega < h_{ie} < 500\ \Omega \tag{6.2-3c}$$

Note that h_{fe} may vary by 3:1 for the same transistor. If the h_{fe} of this transistor varied from 50 to 150,

$$125\ \Omega < h_{ie} < 750\ \Omega \tag{6.2-3d}$$

When designing transistor circuits we most often use the nominal value of h_{fe} and assume that $m = 1$. (In this example $h_{ie} = 250\ \Omega$.) However, we always keep in mind the possible variation of this resistance.

To sum up the results for the common-emitter configuration, the equivalent circuit is shown in three successively simplified versions in Fig. 6.2-3. The simple version of Fig. 6.2-3*c* is easily remembered, and serves adequately for most calculations.

Let us return to the amplifier circuit of Fig. 6.2-1*b* and insert the equivalent circuit in place of the transistor, as shown in Fig. 6.2-4.

The important quantities are the input and output impedances and the current and voltage gain. These are easily calculated directly from the circuit.

For the current gain

$$i_b = i_i\left(\frac{R_b}{R_b + h_{ie}}\right) = i_i\left(\frac{1}{1 + h_{ie}/R_b}\right) \tag{6.2-4}$$

and

$$i_L = -i_c = -h_{fe}i_b \tag{6.2-5}$$

Thus

$$A_i = \frac{i_L}{i_i} = \left(\frac{i_L}{i_b}\right)\left(\frac{i_b}{i_i}\right) = \frac{-h_{fe}}{1 + h_{ie}/R_b} = \frac{-h_{fe}}{1 + h_{fe} \times [(25 \times 10^{-3})/I_{EQ}]} \tag{6.2-6}$$

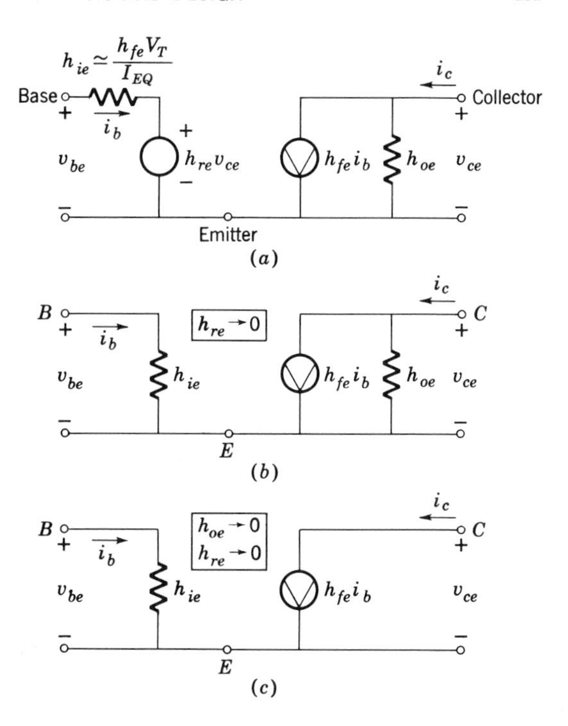

Fig. 6.2-3 Small-signal equivalent circuits for the transistor in the CE configuration. (a) Complete hybrid circuit; (b) circuit neglecting h_{re}; (c) circuit neglecting h_{re} and h_{oe}.

In order for the current gain to approach the theoretical maximum value of h_{fe}, h_{ie}/R_b should be as small as possible, that is, $R_b \gg h_{ie}$. This result implies that for large current gain most of the signal current must flow into the base of the transistor, and a minimal amount may be lost in the bias network. In Sec. 4.1 it was found that for good stability against h_{FE} variation and temperature effects, the inequality $R_b \ll h_{FE}R_e$ should

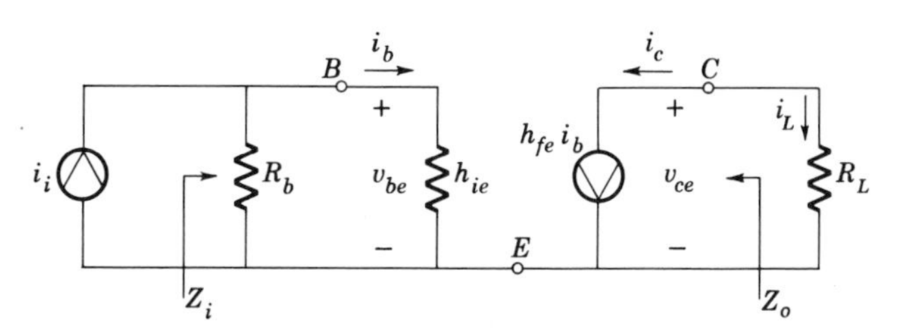

Fig. 6.2-4 Complete common-emitter amplifier equivalent circuit.

be satisfied. Thus, to meet requirements of high gain and stability simultaneously, we should design so that

$$h_{ie} \ll R_b \ll h_{FE}R_e \approx h_{fe}R_e \tag{6.2-7}$$

If this inequality can be satisfied, the amplifier will have high current gain and good stability. Otherwise a compromise must be made in one requirement or the other.

We next calculate the input and output impedances. Looking to the right from the current source i_i, the input impedance Z_i is

$$Z_i = \frac{R_b h_{ie}}{R_b + h_{ie}} \approx h_{ie} \qquad \text{if } R_b \gg h_{ie} \tag{6.2-8}$$

This simple expression is the result of the assumption that h_{re} is negligible. The calculation of output impedance is even simpler. If h_{oe} is taken into account, then

$$Z_o = \left.\frac{v_{ce}}{i_c}\right|_{i_i=0} = \frac{1}{h_{oe}} \tag{6.2-9}$$

If we neglect h_{oe}, then $Z_o \to \infty$.

The parameters h_{re} and h_{oe} are almost never specified, and are usually neglected in calculations (see Sec. 6.6 for a detailed discussion of manufacturers' specifications).

Example 6.2-1 In Fig. 6.2-5, the silicon transistor has $h_{fe} = h_{FE} = 50$. All bypass and coupling capacitors are assumed to have zero reactance at the signal frequency. Find

(*a*) Quiescent conditions.

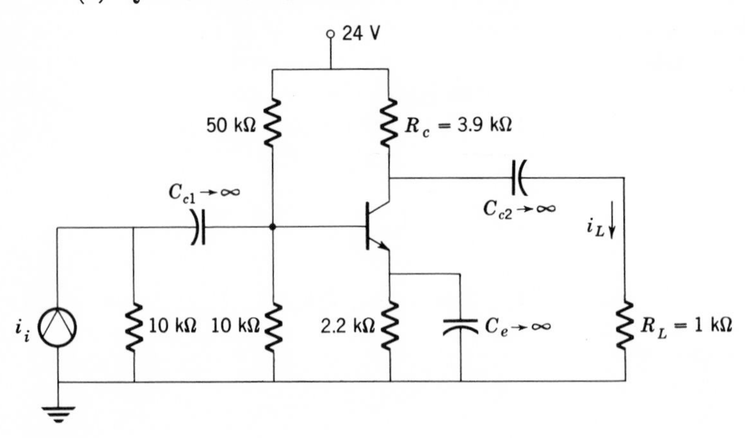

Fig. 6.2-5 Circuit for Example 6.2-1.

(*b*) The small-signal equivalent circuit, neglecting h_{oe} and h_{re}.
(*c*) The current gain, $A_i = i_L/i_i$.
(*d*) The input impedance seen by the signal current source, i_i.
(*e*) The output impedance seen by the 1-kΩ load.

Solution (*a*) $V_{BB} = \left(\frac{10}{10 + 50}\right)(24) = 4 \text{ V}$

$$R_b = \frac{(10)(50)}{10 + 50} \text{ k}\Omega = 8.3 \text{ k}\Omega$$

Note that $R_b \ll h_{FE}R_e$, so that the resulting bias is insensitive to Q-point variations. Neglecting I_{BQ}, the emitter voltage V_{EQ} is

$$V_{EQ} \approx V_{BB} - V_{BEQ} = 4 - 0.7 = 3.3 \text{ V}$$

and

$$I_{EQ} = \frac{3.3 \text{ V}}{2.2 \text{ k}\Omega} \approx 1.5 \text{ mA}$$

and

$$V_{CEQ} \approx 24 - I_{EQ}(R_c + R_e) = 15 \text{ V}$$

(*b*) $h_{ie} \approx \frac{25h_{fe} \times 10^{-3}}{I_{EQ}} = \frac{(25)(50)(10^{-3})}{(1.5)(10^{-3})} = 833 \ \Omega$

Thus $h_{ie} \ll R_b$. The resulting small-signal equivalent circuit is shown in Fig. 6.2-6.

(*c*) $A_i = \frac{i_L}{i_i} = \left(\frac{i_b}{i_i}\right)\left(\frac{i_L}{i_b}\right)$

$$\frac{i_b}{i_i} = \frac{(4.5)(10^3)}{(4.5 + 0.83)(10^3)} = 0.85$$

$$\frac{i_L}{i_b} = (-50)\frac{(3.8)(10^3)}{(3.8 + 1)(10^3)} = -39.6$$

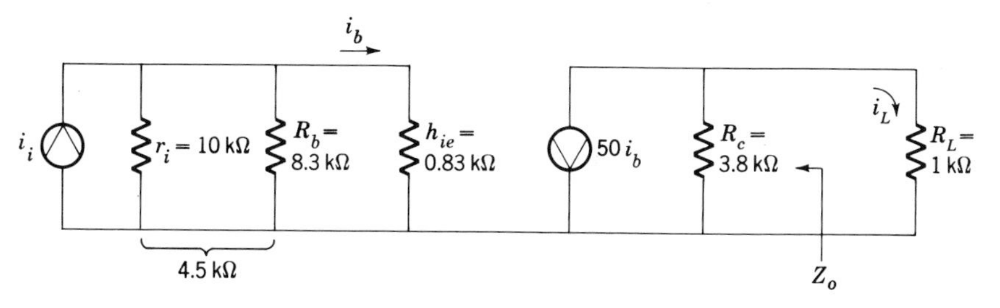

Fig. 6.2-6 Small-signal equivalent circuit for Example 6.2-1.

Thus

$$A_i = (0.85)(-39.6) = -33.7 \approx -34$$

Note that the minus sign occurs in A_i because the positive direction for i_L is opposite to that of i_c.

(*d*) $Z_i = 10 \text{ k}\Omega \| 8.3 \text{ k}\Omega \| 0.83 \text{ k}\Omega \approx 700\ \Omega$

(*e*) $Z_o = 3.8 \text{ k}\Omega$ neglecting h_{oe}

Example 6.2-2 Find the current gain of the amplifier of Example 6.2-1 if $h_{re} = 10^{-4}$ and $h_{oe} = 10^{-4}$ mho.

Solution The equivalent circuit of the amplifier is shown in Fig. 6.2-7. Referring to the output circuit

$$i_L = (-50)\left(\frac{2.75}{2.75 + 1}\right) i_b = -36.7 i_b$$

Compare with Example 6.2-1, where $i_L/i_b = -39.6$.

Note that

$$v_{ce} = -36.7 \times 10^3 i_b$$

Turning to the input circuit and applying KVL,

$$v_b = 830 i_b - (10^{-4})(36.7 \times 10^3) i_b$$
$$= (830 - 3.67) i_b \approx 830 i_b$$

The effect of h_{re} is therefore negligible in this (and most) examples.

The current gain is

$$A_i = \frac{i_L}{i_i} = \left(\frac{i_L}{i_b}\right)\left(\frac{i_b}{v_b}\right)\left(\frac{v_b}{i_i}\right)$$

where

$$\frac{i_L}{i_b} = -36.7 \qquad \frac{i_b}{v_b} = \frac{1}{830}$$

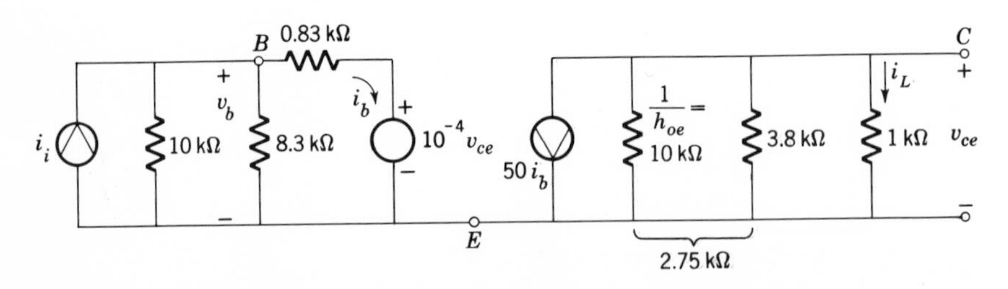

Fig. 6.2-7 Small-signal equivalent circuit for Example 6.2-2.

and

$$i_i = v_b\left(\frac{1}{10\text{ k}\Omega} + \frac{1}{8.3\text{ k}\Omega}\right) + i_b \approx v_b\left(\frac{1}{10\text{ k}\Omega} + \frac{1}{8.3\text{ k}\Omega} + \frac{1}{830\ \Omega}\right) = \frac{v_b}{700}$$

Thus

$$A_i \approx (-36.7)(1/830)(700) = -31$$

The effect of h_{oe} is to reduce the gain from $A_i \approx -34$ to $A_i \approx -31$. This small reduction in gain is usually insignificant.

Example 6.2-3 A silicon *npn* transistor has $h_{fe} = 120$.* Design a single-stage amplifier (Fig. 6.2-8) to achieve a small-signal current gain of 60. The load resistor is 470 Ω and is capacitively coupled to the collector. The supply voltage V_{CC} is 9 V, and the signal-source impedance is 10 kΩ. A peak load current of 0.1 mA is required.

Solution The small-signal equivalent circuit is shown in Fig. 6.2-9 (it is assumed that $h_{oe} = h_{re} = 0$ and $R_1 \| R_2 = R_b$).

The current gain A_i is

$$A_i = \frac{i_L}{i_i} = \left(\frac{i_L}{i_b}\right)\left(\frac{i_b}{i_i}\right) = -120\left(\frac{R_c}{470 + R_c}\right)\left[\frac{10^4\|R_b}{h_{ie} + (10^4\|R_b)}\right]$$

Since $A_i = -60$, this becomes

$$\frac{1}{2} = \left(\frac{R_c}{470 + R_c}\right)\left[\frac{10^4\|R_b}{h_{ie} + (10^4\|R_b)}\right]$$

* Throughout the remainder of this text, it will be assumed, unless otherwise stated, that $h_{fe} = h_{FE}$. The symbol h_{fe} will be used.

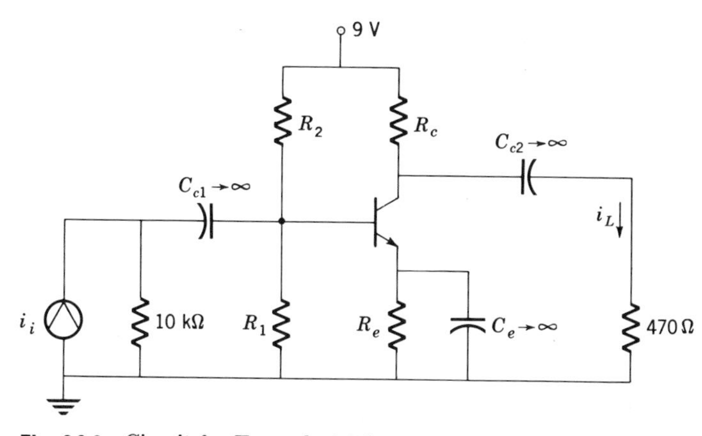

Fig. 6.2-8 Circuit for Example 6.2-3.

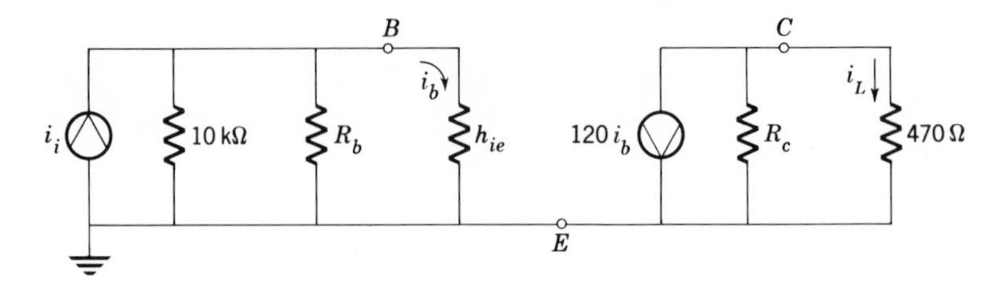

Fig. 6.2-9 Equivalent circuit for Example 6.2-3.

Clearly, many combinations of R_c, R_b, and h_{ie} will satisfy this requirement. In the absence of other requirements, we *arbitrarily* choose to let each of the above factors be equal. (This simplifies calculations and usually provides a good design.) Then, to obtain a current gain greater than or equal to 60,

$$\frac{R_c}{470 + R_c} = \frac{10^4 \| R_b}{h_{ie} + (10^4 \| R_b)} \geq \sqrt{\frac{1}{2}} = 0.707$$

and

$$R_c \geq 1.13\ \text{k}\Omega$$

The choice of R_b and h_{ie} will determine the quiescent current and stability of the stage. Keeping this in mind, R_b is chosen to be 10 kΩ. Then

$$h_{ie} \leq 2.1\ \text{k}\Omega$$

(Note that if R_b were infinite, $h_{ie} = 4.14$ kΩ; therefore a large change in R_b would not result in a large change in h_{ie}.)

If $h_{ie} = 2.1$ kΩ, the quiescent current I_{CQ} is [Eq. (6.2-3)]

$$I_{CQ} \approx \frac{120 \times 25 \times 10^{-3}}{2100} = 1.43\ \text{mA}$$

Now R_e can be chosen to complete the design. The dc and ac load lines are shown in Fig. 6.2-10.

The emitter resistor R_e is chosen so that

1. $I_{CQ} = 1.43\ \text{mA} < \dfrac{9}{1130 + R_e}$

2. $R_b \ll h_{fe}R_e = 120R_e$ to achieve Q-point stability

In this example Q-point stability is not critical, since a peak swing of only 0.1 mA is required and the quiescent current is 1.43 mA. Thus a value of $R_e = 1$ kΩ will satisfy conditions 1 and 2. The

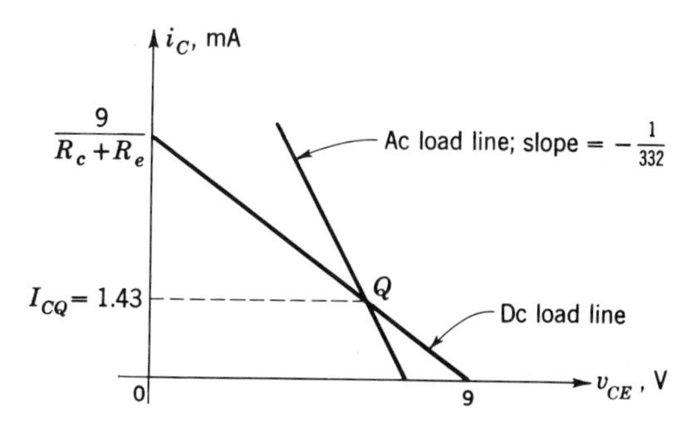

Fig. 6.2-10 Load lines for Example 6.2-3.

peak swing available is approximately 1.43 mA, which more than meets the specifications.

The final values of the resistors are then

$R_c = 1.2\ \text{k}\Omega$ the nearest standard value to 1.13 kΩ

$R_e = 1\ \text{k}\Omega$

$R_b = 10\ \text{k}\Omega$ $h_{ie} \ll R_b \ll h_{fe}R_e$

To find R_1 and R_2 note that

$$I_{CQ} \approx \frac{V_{BB} - V_{BE}}{R_e + R_b/h_{fe}}$$

Thus

$$V_{BB} = (1.43 \times 10^{-3})(10^3 + 83.3) + 0.7 = 2.25\ \text{V}$$

Knowing R_b and V_{BB}, R_1 and R_2 can be found from (3.3-1c) and (3.3-1d).

$R_1 \approx 13\ \text{k}\Omega$ (use 12 kΩ)

$R_2 \approx 40\ \text{k}\Omega$ (use 39 kΩ)

This completes the design. It is important to note that many designs are possible because of the "loose" specifications. In particular, the arbitrary division of the current-gain formula into equal factors might not be desirable if the specifications were more stringent. For example, if the gain requirement were increased from $A_i = 60$ to $A_i = 90$, R_c would have to be increased and h_{ie} decreased. In order to decrease h_{ie}, the quiescent current would have to be increased. This increase is in turn limited by the increased collector resistance, so that a compromise would have to be made.

6.3 THE COMMON-BASE CONFIGURATION

The circuit of the common-base amplifier is shown in Fig. 6.3-1.

This configuration does not provide current gain, but does provide some voltage gain. It has properties which make it useful at high frequencies. The hybrid small-signal equivalent circuit is shown in Fig. 6.3-1*c*. The hybrid equations using the notation and reference directions in the figure are

$$v_{eb} = h_{ib}(-i_e) + h_{rb}v_{cb} \tag{6.3-1}$$

$$i_c = +h_{fb}i_e + h_{ob}v_{cb} \tag{6.3-2}$$

The input resistance h_{ib} is defined as

$$h_{ib} = \frac{v_{eb}}{i_i} = \left.\frac{v_{eb}}{-i_e}\right|_{v_{cb}=0} = \frac{V_T}{I_{EQ}} \approx \frac{h_{ie}}{h_{fe}+1} \tag{6.3-3a}$$

[Eq. (6.2-3)]. Thus, if $h_{ie} = 1\ \text{k}\Omega$ and $h_{fe} = 100$, $h_{ib} = 10\ \Omega$. The input resistance of the common-base amplifier is usually significantly smaller than that of the common-emitter amplifier.

The reverse voltage gain h_{rb} is of the order of 10^{-4}, and can usually be neglected.

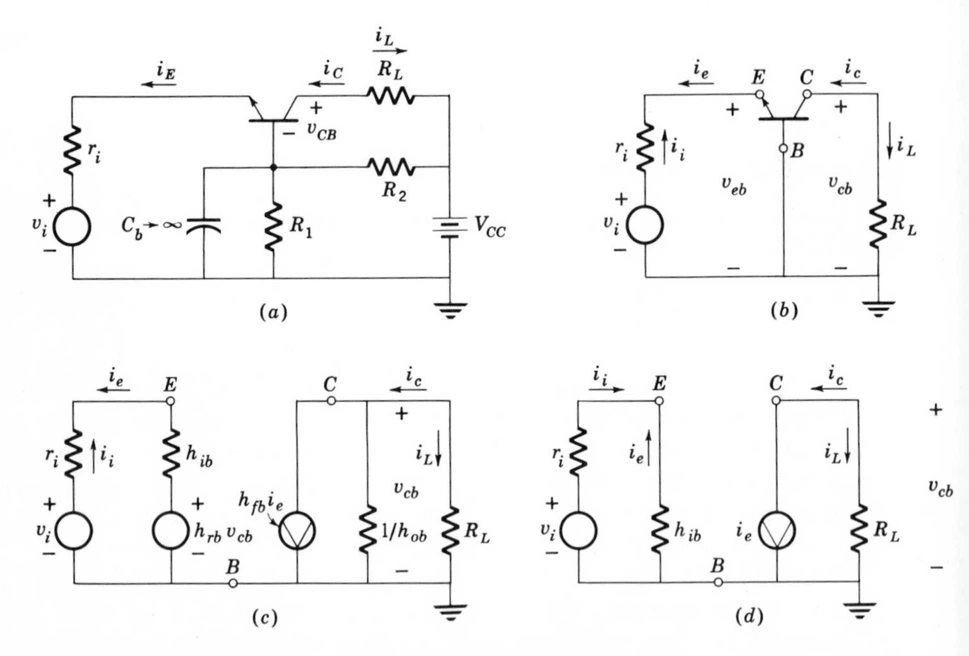

Fig. 6.3-1 Common-base amplifier. (*a*) Complete circuit; (*b*) ac circuit; (*c*) equivalent circuit using hybrid model; (*d*) simplified equivalent circuit.

The forward current amplification factor h_{fb} is defined as

$$h_{fb} = \frac{i_c}{i_e}\bigg|_{v_{cb}=0} \tag{6.3-3b}$$

Thus h_{fb} is approximately equal to $+1$ (note the current directions of i_c and i_e).

The output admittance h_{ob} is

$$h_{ob} = \frac{i_c}{v_{cb}}\bigg|_{i_e=0} \tag{6.3-3c}$$

This value is typically 10^{-6} mho, and is often neglected.

An approximate CB equivalent circuit is shown in Fig. 6.3-1*d*, which neglects h_{rb} and h_{ob} and assumes h_{fb} is unity. Another method for obtaining Fig. 6.3-1*c* involves rearranging the equivalent circuit of Fig. 6.2-3*c* so that the emitter is the input terminal and the base the common terminal, as shown in Fig. 6.3-2. Now the definitions of (6.3-3) are applied to the rearranged circuit by writing KCL at the emitter terminal, with terminals CB shorted.

$$-i_e + i_b + h_{fe}i_b = 0 \tag{6.3-4}$$

$$i_e = (1 + h_{fe})i_b = (1 + h_{fe})\left(\frac{-v_{eb}}{h_{ie}}\right) \tag{6.3-5}$$

Therefore

$$-\frac{v_{eb}}{i_e}\bigg|_{v_{cb}=0} \equiv h_{ib} = \frac{h_{ie}}{1 + h_{fe}} \approx \frac{h_{ie}}{h_{fe}} \tag{6.3-6}$$

The short-circuit current gain is simply

$$h_{fb} = \frac{i_c}{i_e}\bigg|_{v_{cb}=0} = \left(\frac{i_c}{i_b}\right)\left(\frac{i_b}{i_e}\right) = \frac{h_{fe}}{h_{fe} + 1} \tag{6.3-7}$$

To find the output admittance of the common-base configuration, rearrange Fig. 6.2-3*b* as shown in Fig. 6.3-3, where h_{oe} has been included.

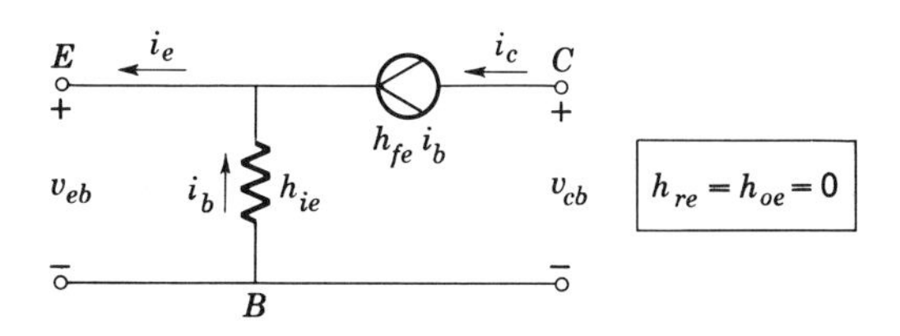

Fig. 6.3-2 Rearranged CE circuit for finding h parameters of common-base configuration.

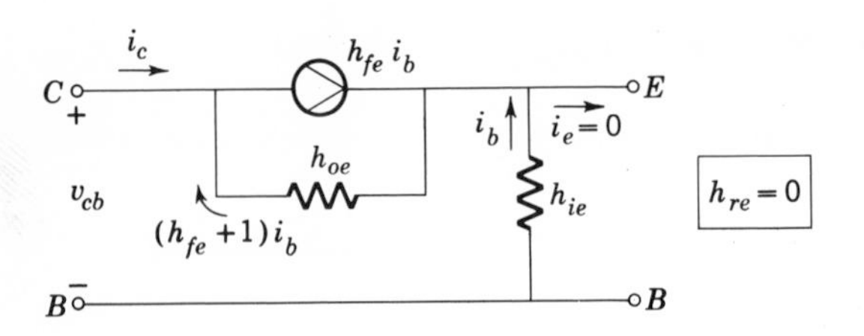

Fig. 6.3-3 Common-base equivalent for finding h_{ob}.

Using (6.3-3c), we note that $i_e = 0$. Then the current through h_{oe} is $(h_{fe} + 1)i_b$.

Applying KVL around the loop, we have

$$v_{cb} + (h_{fe} + 1)\left(\frac{i_b}{h_{oe}}\right) + i_b h_{ie} = 0 \qquad (6.3\text{-}8)$$

Since $i_e = 0$, $i_b = -i_c$. Also $(h_{fe} + 1)/h_{oe} \gg h_{ie}$, so that (6.3-8) becomes

$$v_{cb} \approx \left(\frac{h_{fe} + 1}{h_{oe}}\right) i_c$$

and

$$h_{ob} = \left.\frac{i_c}{v_{cb}}\right|_{i_e=0} = \frac{h_{oe}}{h_{fe} + 1} \qquad (6.3\text{-}9)$$

This will be of the order of several megohms for most transistors. As in the common-emitter circuit, the reverse transmission h_{rb} and, usually, the output admittance h_{ob} are neglected for low-frequency calculations.

Note that, to find the CB parameters h_{ob}, h_{fb}, and h_{ib}, one simply divides the corresponding CE parameter by $1 + h_{fe}$. Thus, if the CE parameters of a particular transistor are $1/h_{oe} = 10$ kΩ, $h_{fe} = 100$, and $h_{ie} = 250$ Ω, the same transistor in the CB configuration will have $1/h_{ob} = 1$ mΩ, $h_{fb} = 100/101 = 0.99$, and $h_{ib} = 2.5$ Ω. The CB stage thus has lower input impedance and higher output impedance than the CE stage.

Example 6.3-1 (*a*) Find the CB parameters for the transistor of Example 6.2-1. Use

$$\frac{1}{h_{oe}} = 10 \text{ k}\Omega$$

(*b*) The transistor is used in a CB configuration with $r_i = 100$ Ω and $R_L = 5$ kΩ (Fig. 6.3-4). Find the current gain, voltage gain, and input and output impedances.

Solution (*a*) From Example 6.2-1 we have

$$h_{fe} = 50 \qquad h_{oe} \approx 10^{-4} \text{ mho}$$
$$h_{ie} = 0.83 \text{ k}\Omega \qquad h_{re} \approx 0$$

Using (6.3-7) and (6.3-3),

$$h_{fb} = \frac{h_{fe}}{h_{fe}+1} = \frac{50}{51} = 0.98$$
$$h_{ib} = \frac{h_{ie}}{h_{fe}+1} \approx \frac{830}{51} \approx 16\ \Omega$$
$$h_{ob} = \frac{h_{oe}}{h_{fe}+1} \qquad \frac{1}{h_{ob}} \approx 500 \text{ k}\Omega$$
$$h_{rb} \approx 0$$

(*b*) The current gain (from current-source equivalent) is

$$A_i = \frac{i_L}{i_i} = \left(\frac{100}{100+16}\right)(0.98)\left(\frac{500}{500+5}\right) \approx 0.83$$

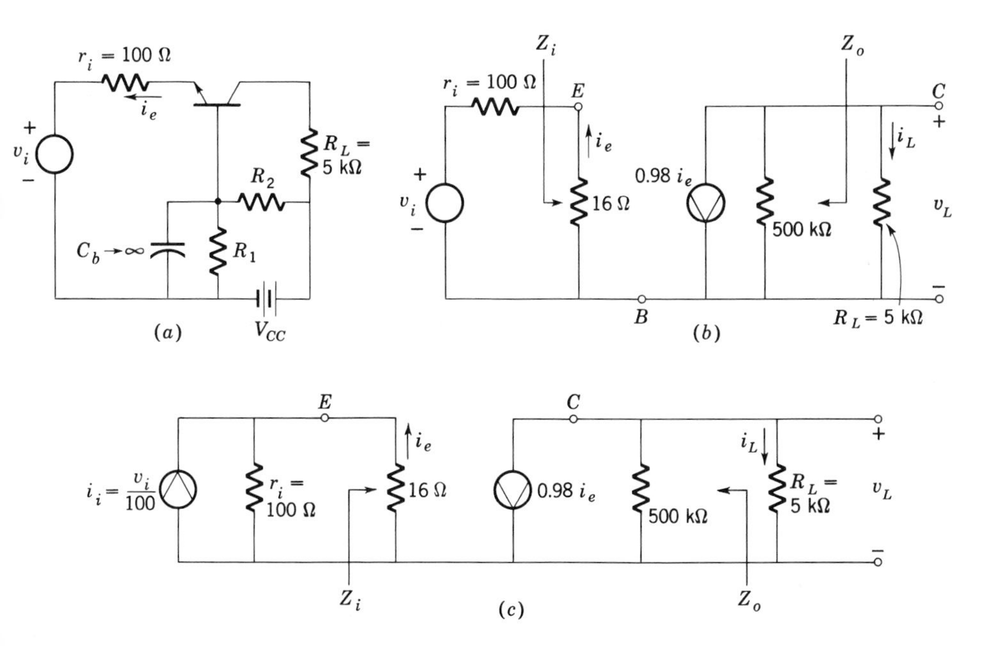

Fig. 6.3-4 Circuit for Example 6.3-1. (*a*) Complete circuit; (*b*) ac circuit; (*c*) current-source equivalent.

The voltage gain is

$$\frac{v_L}{v_i} = \left(\frac{v_L}{i_e}\right)\left(\frac{i_e}{v_i}\right)$$

$$i_e = \left(\frac{v_i}{100}\right)\left(\frac{-100}{100 + 16}\right) = \frac{-v_i}{116}$$

$$v_L \approx -0.98 i_e (5000) \approx -5000 i_e$$

$$A_v = \frac{v_L}{v_i} \approx 43$$

The input and output impedances, by inspection, are

$$Z_i = 16\ \Omega$$

$$Z_o = 500\ \text{k}\Omega$$

6.4 THE COMMON-COLLECTOR (EMITTER-FOLLOWER) CONFIGURATION

The emitter-follower (EF) configuration is characterized by a voltage gain slightly less than unity, high input impedance, and low output impedance. It is generally used as an *impedance transformer* in the input and output circuits of amplifier systems. When placed in the input circuit, its high input impedance reduces the loading on the source. When placed in the output circuit, it serves to isolate the preceding stage of the amplifier from the load and, in addition, provides a low output impedance.

The emitter follower and its ac circuit are shown in Fig. 6.4-1. Following the procedure outlined in preceding sections, a set of common-collector (CC) hybrid parameters can be defined, and an equivalent circuit drawn. This, however, results in a circuit not often used for design

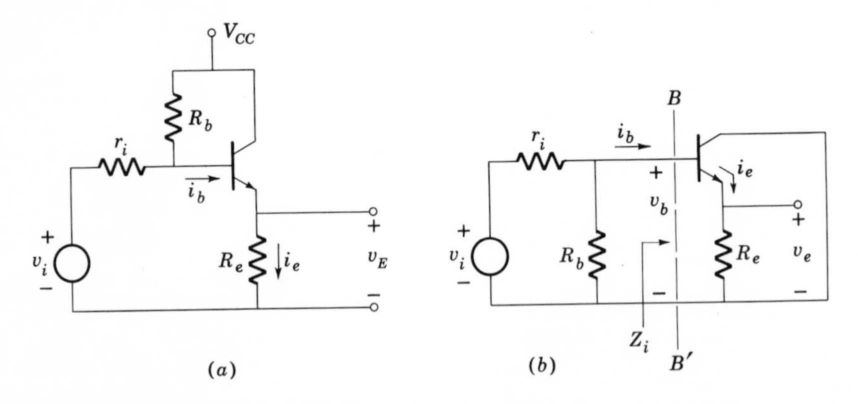

Fig. 6.4-1 (*a*) Emitter-follower circuit; (*b*) ac circuit.

purposes. Instead, an equivalent circuit is obtained directly from Fig. 6.4-1*b*, using KVL.

At terminals BB', KVL yields

$$v_b = v_{be} + i_e R_e \tag{6.4-1}$$

where, from Fig. 6.2-3*c*,

$$v_{be} = i_b h_{ie} \tag{6.4-2}$$

In addition, note that

$$i_e R_e = i_b[(h_{fe} + 1)R_e] \tag{6.4-3}$$

Substituting (6.4-2) and (6.4-3) into (6.4-1),

$$v_b = i_b h_{ie} + i_b[(h_{fe} + 1)R_e] \tag{6.4-4}$$

Equation (6.4-4) indicates that the equivalent circuit of the EF as seen looking into terminals BB' is a series combination of h_{ie} and $(1 + h_{fe})R_e$, as shown in Fig. 6.4-2.

From this circuit the voltage gain A_v is found by simple voltage division.

$$A_v = \frac{v_e}{v_i} = \left[\frac{(1 + h_{fe})R_e}{h_{ie} + (1 + h_{fe})R_e}\right]\left[\frac{R_b \| [h_{ie} + (1 + h_{fe})R_e]}{r_i + \{R_b \| [h_{ie} + (1 + h_{fe})R_e]\}}\right]$$

After some manipulation this can be put in the form

$$A_v = \left(\frac{R_b}{r_i + R_b}\right)\left\{\frac{1}{1 + [h_{ie} + (r_i \| R_b)]/[(1 + h_{fe})R_e]}\right\} \tag{6.4-5}$$

Thus, if $(1 + h_{fe})R_e$ is much greater than the sum of h_{ie} and $r_i \| R_b$, as is often the case, the quantity in braces will be close to unity, and the voltage gain determined by the $r_i - R_b$ voltage divider.

Note that the input impedance of the emitter follower, defined as the impedance seen looking into terminals BB', is simply

$$Z_i = h_{ie} + (h_{fe} + 1)R_e \tag{6.4-6}$$

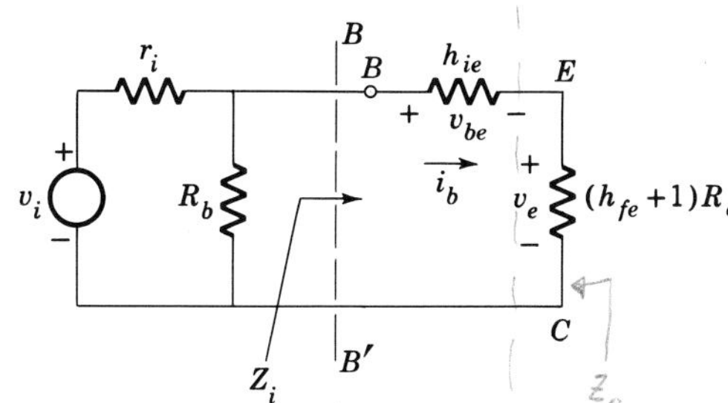

Fig. 6.4-2 Equivalent circuit for emitter follower.

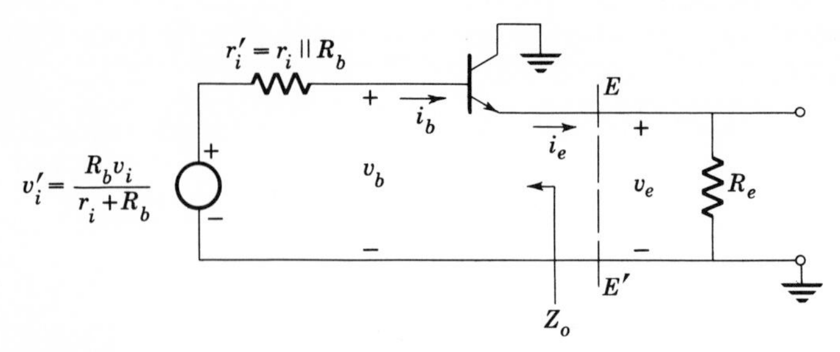

Fig. 6.4-3 Emitter-follower ac circuit.

In a similar way, the equivalent circuit looking into the emitter (output) is easily obtained by redrawing Fig. 6.4-1b as shown in Fig. 6.4-3.

Writing KVL around the emitter-base loop,

$$v_i' = r_i' i_b + v_{be} + v_e \tag{6.4-7}$$

where

$$v_{be} = h_{ib} i_e \tag{6.4-8}$$

and

$$r_i' i_b = \left(\frac{r_i'}{h_{fe} + 1}\right) i_e \tag{6.4-9}$$

Substituting (6.4-8) and (6.4-9) into (6.4-7),

$$v_i' = \left(\frac{r_i'}{h_{fe} + 1}\right) i_e + h_{ib} i_e + v_e \tag{6.4-10}$$

Equation (6.4-10) yields the equivalent circuit of the EF as seen looking into terminals EE'. This circuit is shown in Fig. 6.4-4.

The voltage gain A_v as calculated from this circuit is, of course, the same as that obtained from the equivalent circuit of Fig. 6.4-2.

The output impedance as seen from terminals EE' is

$$Z_o = h_{ib} + \frac{r_i'}{h_{fe} + 1} \tag{6.4-11}$$

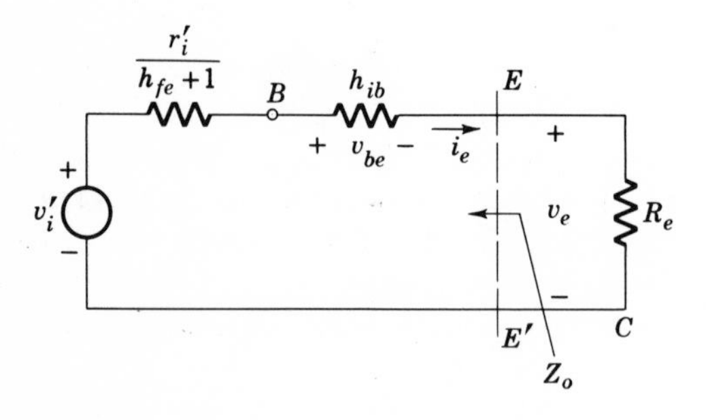

Fig. 6.4-4 Another equivalent circuit for the emitter follower.

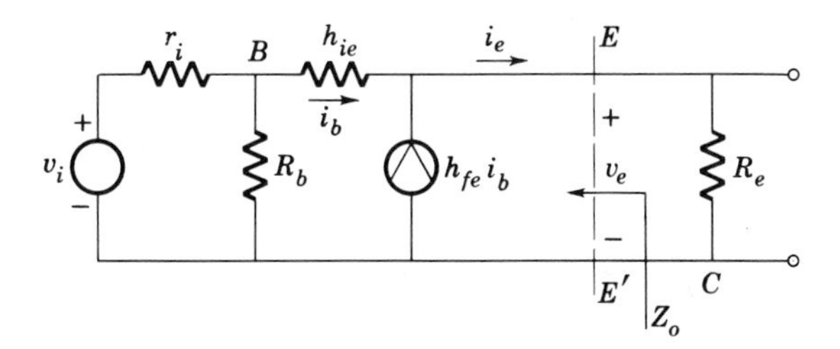

Fig. 6.4-5 Emitter follower using CE equivalent circuit.

The equivalent circuits of Figs. 6.4-2 and 6.4-4 can also be obtained by replacing the transistor in Fig. 6.4-1*b* by the CE equivalent circuit, as shown in Fig. 6.4-5.

The Thévenin output impedance at EE', with R_e removed, is

$$Z_o = \frac{v_e}{-i_e}\bigg|_{v_i=0} = \frac{v_e}{-(h_{fe}+1)i_b} = \frac{-i_b[h_{ie} + (r_i\|R_b)]}{-(h_{fe}+1)i_b}$$

$$= \frac{h_{ie} + (r_i\|R_b)}{h_{fe}+1} = h_{ib} + \frac{r_i'}{h_{fe}+1} \tag{6.4-12}$$

To obtain the Thévenin open-circuit voltage v_i' at EE', remove R_e. Then $i_e = 0$ and

$$i_b = \frac{i_e}{h_{fe}+1} = 0$$

Thus

$$v_i' = \left(\frac{R_b}{R_b + r_i}\right) v_i \tag{6.4-13}$$

which leads directly to the circuit of Fig. 6.4-4.

IMPEDANCE REFLECTION IN THE TRANSISTOR

Figures 6.4-2 and 6.4-4 illustrate an extremely useful small-signal property of the base-emitter circuit. For example, consider Fig. 6.4-2. Looking between terminal B (the base terminal) and ground, one sees h_{ie} in series with the emitter-to-ground impedance multiplied by $h_{fe} + 1$. All currents in the circuit are at "base-current level." The current through the resistor $(1 + h_{fe})R_e$ is *not* the actual ac emitter current i_e, but rather $i_e/(1 + h_{fe})$. The output voltage v_e, in Fig. 6.4-2, is the same as the output voltage in the actual EF of Fig. 6.4-1. Thus, when drawing an equivalent circuit, one can *reflect* the emitter circuit *through* the junction by simply multiplying the emitter-circuit impedance by $h_{fe} + 1$. Voltages throughout are approximately preserved, while currents in the reflected impedance are *reduced* by $h_{fe} + 1$.

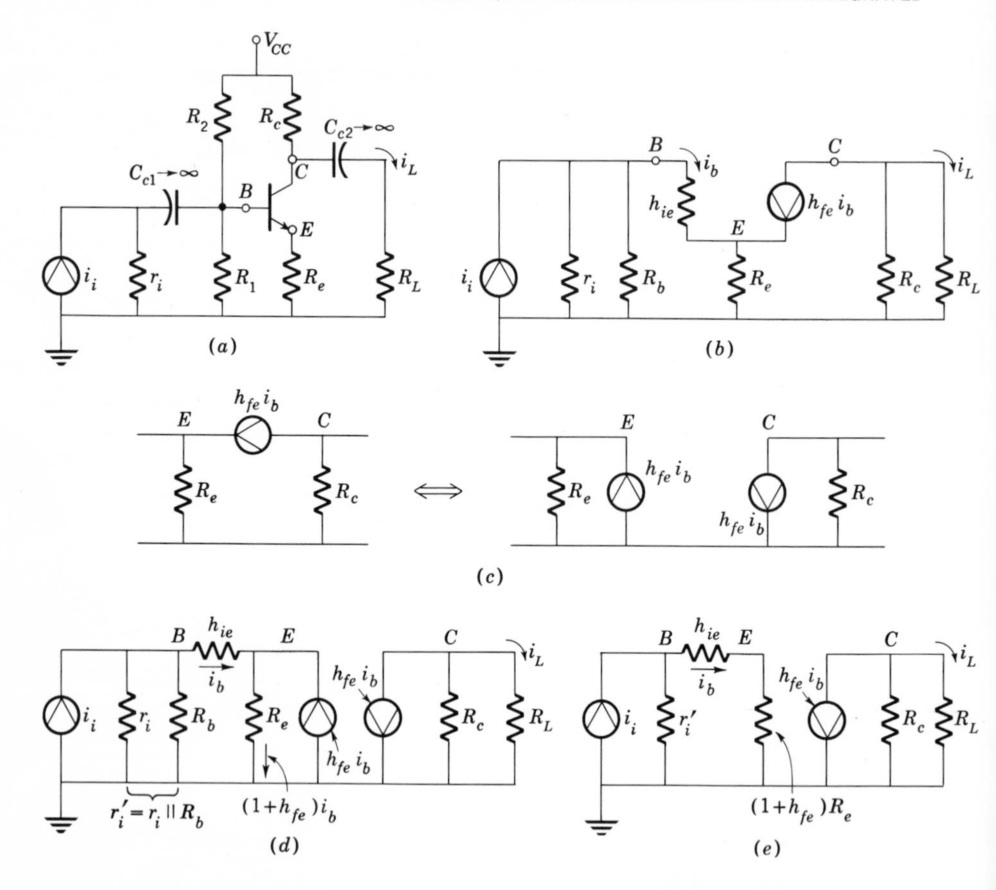

Fig. 6.4-6 Impedance reflection applied to the common-emitter amplifier. (*a*) Amplifier; (*b*) small-signal equivalent circuit ($h_{oe} = 0$, $h_{re} = 0$); (*c*) splitting the current source; (*d*) current splitting applied to the amplifier; (*e*) final equivalent circuit with "reflected" emitter resistance.

Consideration of Fig. 6.4-4 indicates that when reflecting in the other direction, i.e., looking into the emitter, one *divides* the base-circuit impedance by $h_{fe} + 1$.* Again, voltages are approximately preserved, while currents in the reflected impedances are at "emitter-current level," that is, $h_{fe} + 1$ times as large.

It should be pointed out that these results are approximate, since h_{oe} and h_{re} have been neglected. The errors introduced are negligible for most practical circuits.

As an example of the use of this property, consider the CE amplifier with unbypassed emitter resistance, shown in Fig. 6.4-6*a*.

* Note that $h_{ib} = h_{ie}/(h_{fe} + 1)$.

The small-signal equivalent circuit is shown in Fig. 6.4-6*b*. The current source can be "split" by employing KCL, as shown in Fig. 6.4-6*c*. This manipulation results in the circuit of Fig. 6.4-6*d*. This is then equivalent to the circuit of Fig. 6.4-6*e*, where the parallel combination of R_e and the $h_{fe}i_b$ source have been replaced by the *reflected* resistance $(1 + h_{fe})R_e$. The equivalence is established by noting that, in both circuits, $v_e = (1 + h_{fe})i_bR_e$. The current gain can then be found by inspection from the circuit of Fig. 6.4-6*e*.

$$A_i = -h_{fe}\left(\frac{R_c}{R_c + R_L}\right)\left[\frac{r_i'}{r_i' + h_{ie} + (h_{fe} + 1)R_e}\right] \tag{6.4-14}$$

This technique is used extensively in Chap. 7, where multiple-transistor circuits are considered.

Example 6.4-1 Using the EF shown in Fig. 6.4-1, plot

(*a*) Z_i versus R_e

(*b*) Z_o versus r_i

(*c*) A_v versus R_e

Solution Equations (6.4-6), (6.4-11), and (6.4-5) are the required relations. They are most easily plotted on log-log coordinates. For example, consider Z_i as a function of R_e.

$$Z_i = h_{ie} + (h_{fe} + 1)R_e \tag{6.4-6}$$

The asymptotic values of Z_i are

$$Z_i \approx \begin{cases} h_{ie} & \text{for } h_{ie} \gg (h_{fe} + 1)R_e \\ (h_{fe} + 1)R_e & \text{for } h_{ie} \ll (h_{fe} + 1)R_e \end{cases}$$

The first of these is a constant (assuming the transistor parameters do not change if R_e is changed). The second is a straight line of slope $h_{fe} + 1$ on log-log coordinates. The two asymptotes intersect where $h_{ie} = (h_{fe} + 1)R_e$. At this point, the actual value of Z_i is $2h_{ie}$. This is shown in Fig. 6.4-7*a*.

Equation (6.4-6) implies that $Z_i \to \infty$ as $R_e \to \infty$. However, when R_e becomes very large, the common-base output admittance must be included in the equivalent circuit. The equivalent circuit can be obtained from Fig. 6.4-2 by inserting the resistance $1/h_{ob}$ between base and collector. The input impedance is then seen to approach $1/h_{ob}$ as R_e approaches infinity. This is shown in Fig. 6.4-7*a*.

Z_o and A_v are plotted in a similar fashion in Fig. 6.4-7*b* and *c*.

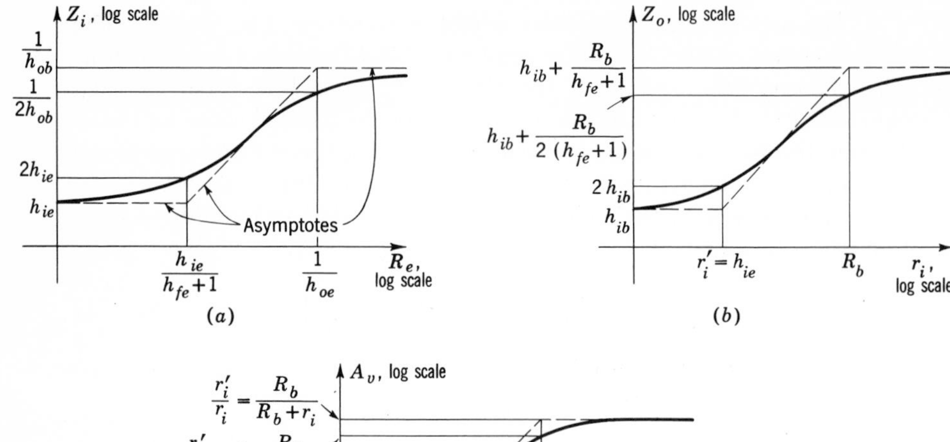

Fig. 6.4-7 Variation of EF parameters. (*a*) $Z_i = h_{ie} + (1 + h_{fe})Re$ [Eq. (6.4-6)]; (*b*) $Z_o = h_{ib} + \dfrac{r_i \| R_b}{h_{fe}+1}$ [Eq. (6.4-11)]; (*c*) $A_v = \left(\dfrac{r_i'}{r_i}\right)\left[\dfrac{1}{1 + \dfrac{h_{ie} + r_i'}{(h_{fe}+1)\,Re}}\right]$.

Example 6.4-2 Design an EF to meet the following specifications:

(*a*) $A_v \geq 0.9$ for small signals
(*b*) $V_{im} \leq 4$ V
(*c*) $r_i = 100\ \Omega$
(*d*) R_L (ac-coupled) $= 50\ \Omega$
(*e*) $100 \leq h_{fe} \leq 200$
(*f*) $V_{CC} = 15$ V
(*g*) $V_{CE,\text{sat}} = 1$ V

The circuit is shown in Fig. 6.4-8.

Solution As in previous design examples, many solutions are possible. In this example, it can be shown that the maximum value that R_e can have and still meet the specifications of a 4-V peak swing with a 1-V saturation voltage is $R_e = 75\ \Omega$. Using this value of R_e, $I_{CQ} = 133$ mA and $V_{CEQ} = 5$ V. To use this value for R_e would require that h_{fe} be known exactly. Thus a smaller value of R_e is required to accommodate the variation of h_{fe}. To simplify the calculation, we let

$$R_e = R_L = 50\ \Omega$$

The dc and ac load lines are shown in Fig. 6.4-9. Two operating points are shown. The first, Q_1, is chosen on the basis of the minimum current which will satisfy the specifications: A peak signal-voltage swing of 4 V implies that the output voltage must be capable of swinging approximately 4 V. Since the ac load is 25 Ω, a peak current swing of 160 mA is required. Thus Q_1 is placed at $I_{CQ1} = 160$ mA, and the maximum signal will drive the transistor almost to cutoff.

The second operating point, Q_2, is placed to avoid the saturation voltage of 1 V. Thus $V_{CEQ2} = 5$ V and $I_{CQ2} = 200$ mA.

The base resistance R_b is chosen so that the Q point will not shift outside the range from Q_1 to Q_2, when h_{fe} varies from 100 to 200. This can be done by finding the limits on I_{CQ} as a function of h_{fe}.

The quiescent current is given by

$$I_{CQ} \approx \frac{V_{BB} - V_{BE}}{R_e + R_b/h_{fe}} = \frac{V_{BB} - 0.7}{50 + R_b/h_{fe}}$$

This will be a minimum when $h_{fe} = 100$.

Then, since the minimum allowable current is 160 mA,

$$160 \times 10^{-3} \leq \frac{V_{BB} - 0.7}{50 + R_b/100}$$

and when $h_{fe} = 200$,

$$200 \times 10^{-3} \geq \frac{V_{BB} - 0.7}{50 + R_b/200}$$

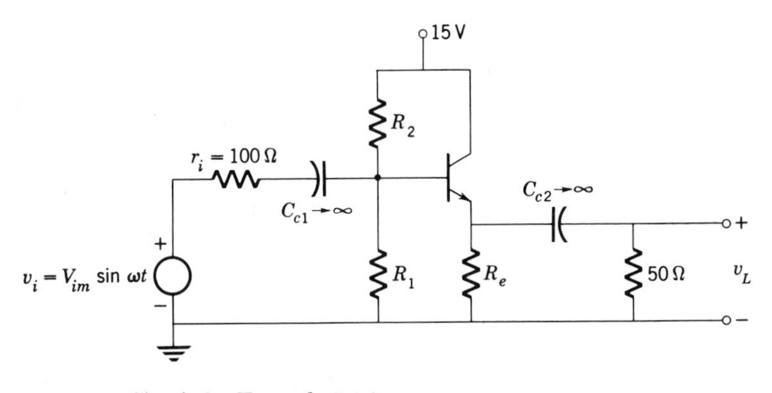

Fig. 6.4-8 Circuit for Example 6.4-2.

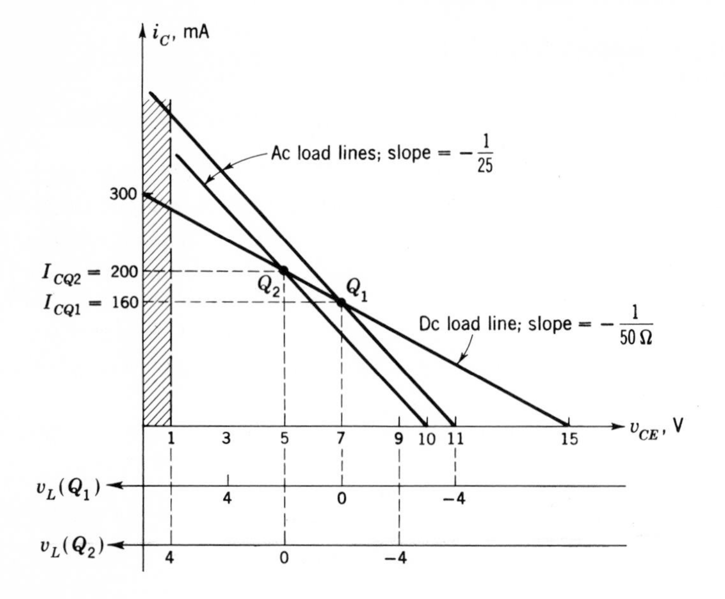

Fig. 6.4-9 Load lines for Example 6.4-2.

Combining these inequalities,

$$(0.16)\left(50 + \frac{R_b}{100}\right) \leq V_{BB} - 0.7 \leq (0.2)\left(50 + \frac{R_b}{200}\right)$$

Simplifying,

$$8 + 1.6 \times 10^{-3}R_b \leq 10 + 10^{-3}R_b$$

$$0.6 \times 10^{-3}R_b \leq 2$$

and

$$R_b \leq 3.3\ \text{k}\Omega$$

To achieve a gain exceeding 0.9 requires that

$$A_v \approx \left(\frac{R_b}{r_i + R_b}\right)\left[\frac{R_L \| R_e}{(R_L \| R_e) + h_{ib} + (r_i \| R_b)/h_{fe}}\right] \geq 0.9$$

Thus, with $h_{ib} = (h_{ib})_{\text{av}} = \dfrac{V_T}{I_{EQ,\text{av}}} \approx \dfrac{25 \times 10^{-3}}{180 \times 10^{-3}} \approx 0.14\ \Omega$

$$0.9 \leq \left(\frac{1}{1 + 100/R_b}\right)\left[\frac{25}{25 + 0.14 + \left(\dfrac{100R_b}{100 + R_b}\right)\left(\dfrac{1}{h_{fe,\text{min}}}\right)}\right]$$

Note that $h_{fe,\text{min}}$ is employed here to ensure that a gain of at least 0.9 results for any h_{fe}.

If R_b is chosen to be 2.5 kΩ,

$$A_v = \left(\frac{1}{1 + 100/2500}\right)\left[\frac{25}{25 + 0.14 + \left(\frac{250{,}000}{2600}\right)\left(\frac{1}{100}\right)}\right] \approx 0.92$$

and $R_b = 2.5$ kΩ satisfies the gain specification. To find V_{BB} return to the original inequality:

$$(0.16)(50 + 25) \leq V_{BB} - 0.7 \leq (0.2)(50 + 12.5)$$

$$12.7 \leq V_{BB} \leq 13.2$$

If we choose $V_{BB} = 13$ V, then, from (3.3-1), $R_2 \approx 2.9$ kΩ and $R_1 \approx 19$ kΩ. We would use standard resistors so that $R_2 = 2.7$ kΩ and $R_1 = 18$ kΩ.

Example 6.4-3 The circuit in Fig. 6.4-10 is a *phase inverter*. Calculate v_2 and v_1.

Solution The emitter voltage v_1 is found as if the circuit were an EF, as shown in Fig. 6.4-10*b*.

Using (6.4-5),

$$v_1 = v_i\left(\frac{R_b}{r_i + R_b}\right)\left\{\frac{1}{1 + (h_{ie} + r_i')/[(h_{fe} + 1)R_e]}\right\}$$

The emitter current i_e is

$$i_e = \frac{v_1}{R_e}$$

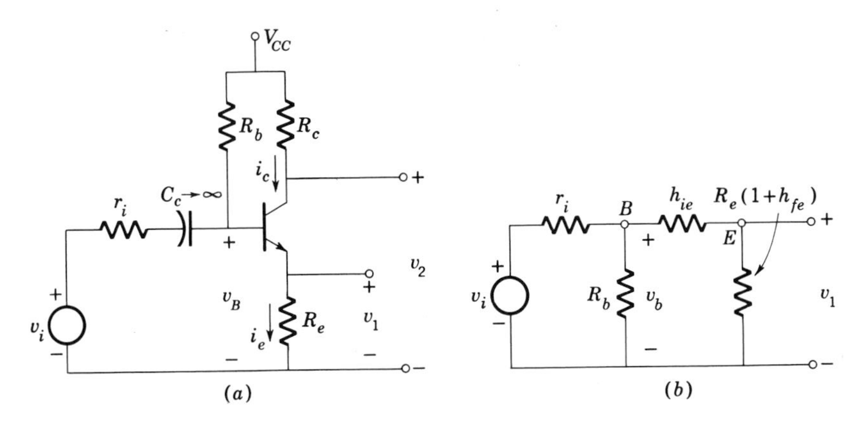

Fig. 6.4-10 Phase inverter for Example 6.4-3. (*a*) Complete circuit; (*b*) base-emitter equivalent.

and

$$v_2 = -R_c i_c = -R_c h_{fb} i_e = -h_{fb}\left(\frac{R_c}{R_e}\right) v_1$$

If $h_{fb}R_c = R_e$,* then $v_1 = -v_2$. Thus a phase inverter gives two outputs that can be made equal in amplitude, and are 180° out of phase. It is often used to provide out-of-phase input signals for the push-pull amplifier, which was discussed in Sec. 5.3.

6.5 COLLECTION OF SIGNIFICANT PARAMETERS FOR THE THREE BASIC CONFIGURATIONS

The analyses of Secs. 6.2 to 6.4 yielded approximate formulas for the h parameters of the CE and CB stages and the input impedance, output impedance, and voltage gain of the EF (CC) stage. The results are summarized in Table 6.5-1.

* Since $h_{fb} \approx 1$, R_c is usually made (to within component tolerances) equal to R_e. If necessary, the outputs are made equal by adjusting R_c or R_e experimentally.

Table 6.5-1

	Configuration		
	CE	*EF(CC)*	*CB*
Gain	$A_i \approx -h_{fe}$	$A_v \approx 1$	$A_i \approx h_{fb} = \dfrac{h_{fe}}{1 + h_{fe}}$
Input impedance	$h_{ie} = \dfrac{m(25 \times 10^{-3})h_{fe}}{I_{EQ}}$	$Z_i = h_{ie} + (h_{fe} + 1)R_e$	$h_{ib} = \dfrac{h_{ie}}{1 + h_{fe}}$
Output impedance	$\dfrac{1}{h_{oe}} > 10^4\ \Omega$	$Z_o \approx h_{ib} + \dfrac{r_i'}{h_{fe} + 1}$	$\dfrac{1}{h_{ob}} = \dfrac{1 + h_{fe}}{h_{oe}}$
Simplest equivalent circuit	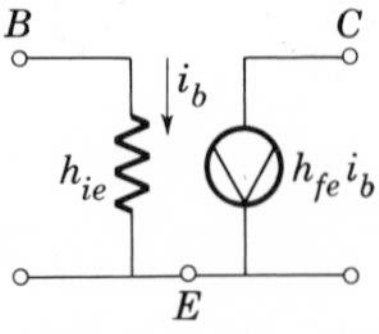 B, C, i_b, h_{ie}, $h_{fe}i_b$, E	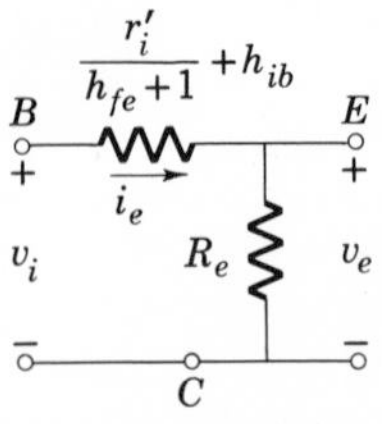 $\dfrac{r_i'}{h_{fe}+1} + h_{ib}$, B, E, +, i_e, v_i, R_e, v_e, −, C	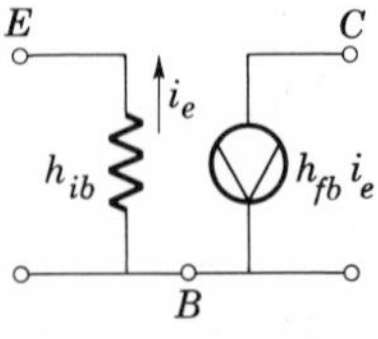 E, C, i_e, h_{ib}, $h_{fb}i_e$, B

6.6 INTERPRETATION OF MANUFACTURERS' SPECIFICATIONS FOR LOW-POWER TRANSISTORS ($P_C < 1$ W)

In this section we discuss some common specifications given by manufacturers.

To illustrate the manufacturers' specifications, consider the 2N3647 silicon *npn* transistor.

1. Maximum collector dissipation in free air at 25°C, $P_{C,\max} = 400$ mW
2. Derating factor in free air, $\theta_{jc} = 0.4$°C/mW
3. Maximum junction temperature, $T_{j,\max} = 200$°C
4. Absolute maximum ratings at 25°C
   a. $BV_{CBO} = 40$ V
   b. $BV_{CEO} = 10$ V
   c. $BV_{EBO} = 6$ V
   d. $I_{C,\max} = 500$ mA
   e. $I_{CBO} = 25$ nA
5. Typical h parameters at 25°C
   a. $h_{fe} = 150$ (typical maximum value)
   b. $h_{oe} = 10^{-4}$ mho (maximum)
   c. $h_{ie} = 4.5$ kΩ
   d. $h_{re} = 10^{-4}$
6. $C_{ob} = 4$ pF (maximum)
7. Common-base cutoff frequency, $f_\alpha \geq 350$ MHz

This transistor is capable of dissipating 400 mW at room temperature using an infinite heat sink. It is derated linearly at the rate of 0.4°C/mW, as shown in Fig. 6.6-1.

The breakdown voltages differ considerably from those of the high-power transistor of Sec. 4.6. Note, for example, that the emitter-base-junction breakdown voltage is only 6 V, compared with 25 V for the power transistor. In addition, I_{CBO} is 0.025 μA, an extremely small value.

Manufacturers often list typical hybrid parameters; for this transistor $h_{oe} = 10^{-4}$ mho and $h_{re} = 10^{-4}$. Referring to Fig. 6.2-3, we see that h_{oe} can be neglected when this transistor is to be used as a CE amplifier if

$$R_L \ll \frac{1}{h_{oe}} = 10 \text{ k}\Omega$$

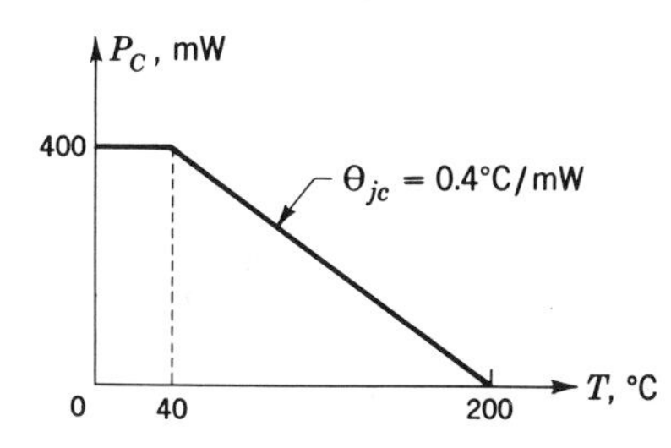

Fig. 6.6-1 Derating curve for 2N3647.

In addition, h_{re} can be neglected if

$$h_{ie}i_b \gg h_{re}v_{ce} = 10^{-4}v_{ce}$$

or, using $|v_{ce}| \approx h_{fe}i_bR_L$,

$$h_{ie}i_b \gg h_{re}h_{fe}R_Li_b$$

Using the typical parameters given, this becomes

$$h_{ib} \gg h_{re}R_L = 10^{-4}R_L$$

Thus h_{re} can be neglected as long as h_{ib} is much larger than $10^{-4}R_L$.

The capacitance C_{ob} and the cutoff frequency are discussed in Chap. 13.

PROBLEMS

In all cases a complete small-signal equivalent circuit should accompany the solution.

SECTION 6.1

6.1. For the amplifier shown in Fig. P6.1, $h_i = 2$ kΩ, $h_r = 0$, $h_f = 200$, and $1/h_o =$ 10 kΩ. Find $A_i = i_2/i_i$ and $A_v = v_2/v_i$, where $v_i = i_ir_i$.

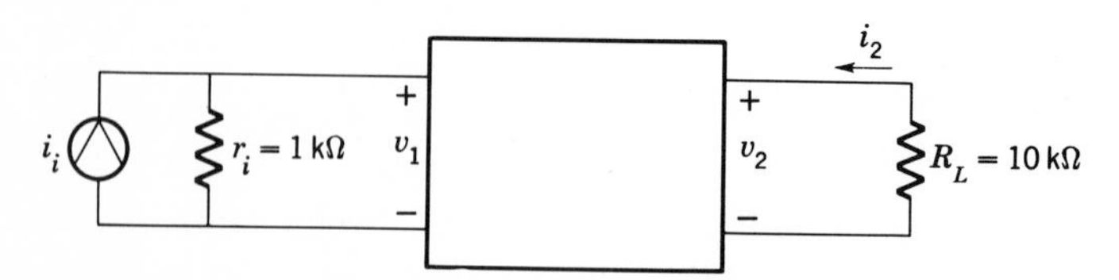

Fig. P6.1

6.2. High-frequency transistors are often specified in terms of the y parameters as defined by the equations

$$i_1 = y_{11}v_1 + y_{12}v_2$$
$$i_2 = y_{21}v_1 + y_{22}v_2$$

(a) Draw the equivalent circuit similar to Fig. 6.1-1b using the y parameters.
(b) Give terminal definitions for the y parameters as in (6.1-4) to (6.1-7).

6.3. Find the relations from which one can calculate the y parameters, given the h parameters.

6.4. Two-port parameters are often defined in terms of partial derivatives evaluated at the operating point. For the hybrid parameters, (6.1-2) and (6.1-3) are written

$$v_{1T} = V(i_{1T},v_{2T})$$
$$i_{2T} = I(i_{1T},v_{2T})$$

where $i_{1T} = I_{1Q} + i_1$, etc.

Expand v_{1T} and i_{2T} in a Taylor's series in two variables about the operating point I_{1Q}, V_{1Q}; neglect high-order terms; and find the definitions of the h parameters in terms of the partial derivatives of the functions V and I.

SECTION 6.2

6.5. Using the characteristics for the 2N3904 silicon *npn* transistor given in Appendix III, estimate h_{oe} and h_{fe} at $I_{CQ} = 1$ and 5 mA, $V_{CEQ} = 10$ V.

6.6. Sketch two circuits which may be used to measure each of the h parameters. *Hint:* Refer to Fig. 6.1-1*b* and recall that an ac short circuit can be obtained by using a capacitor which has a very small reactance at the frequency being used.

6.7. For the silicon transistor, $h_{fe} = 100$ and $h_{re} = h_{oe} = 0$. Find h_{ie}, A_i, Z_i, Z_o.

$r_i = 2$ kΩ
$R_1 = 3.5$ kΩ
$R_2 = 20$ kΩ
$R_c = 1500$ Ω
$R_e = 500$ Ω
$R_L = 1500$ Ω
All C's $\rightarrow \infty$

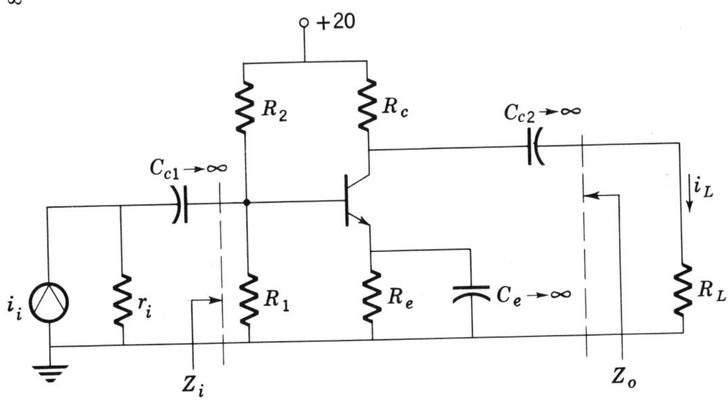

Fig. P6.7

6.8. For the transistor, $h_{ie} = 1$ kΩ, $h_{re} = 10^{-4}$, $h_{oe} = 10^{-5}$ mho, $h_{fe} = 50$.

(*a*) Assume that $R_b \gg h_{ie}$. Plot $A_i = i_L/i_i$ as a function of R_L.

(*b*) Assume that $h_{re} = h_{oe} = 0$. Plot A_i versus R_L on the same axes as in part *a*. Compare.

(*c*) Assume $R_L = 10$ kΩ. Plot A_i versus R_b. On the same axes, plot A_i versus R_b if $h_{re} = 0$. Compare.

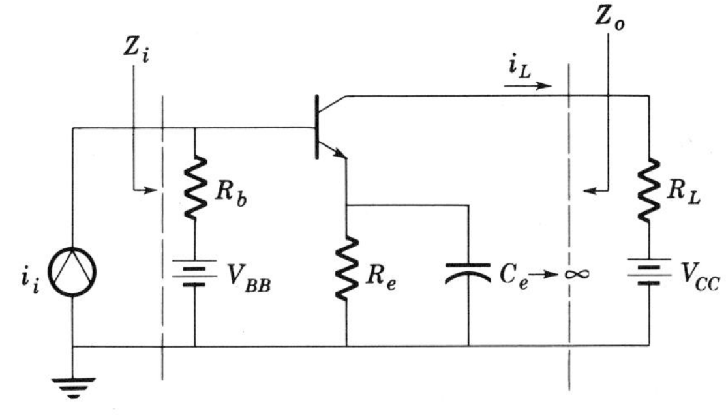

Fig. P6.8

6.9. Repeat Prob. 6.8 for Z_i instead of A_i.

6.10. Repeat Prob. 6.8 for Z_o instead of A_i.

6.11. For this transistor, $h_{re} = h_{oe} = 0$; $50 < h_{fe} < 150$; and $m = 1.4$. Calculate the range of A_i and Z_i to be expected.

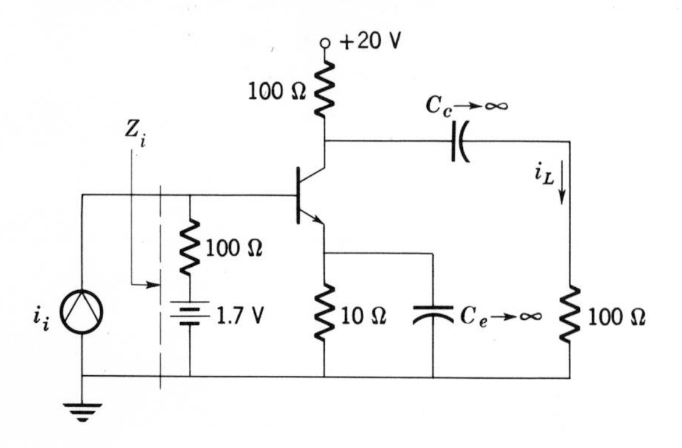

Fig. P6.11

6.12. The emitter resistor is not bypassed. Find A_i and Z_i if $h_{oe} = h_{re} = 0$.

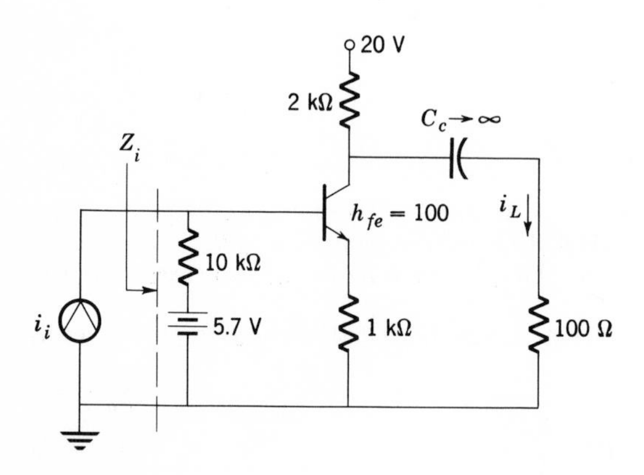

Fig. P6.12

6.13. Repeat Prob. 6.12 with $h_{oe} = 10^{-4}$ mho and $h_{re} = 10^{-5}$.

6.14. Find the voltage gain $A_v = v_c/v_i$ and Z_i. Assume $h_{fe} = 100$ and $h_{oe} = h_{re} = 0$.

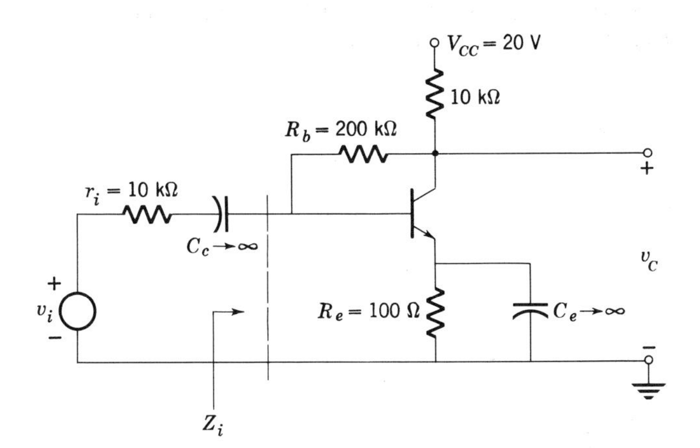

Fig. P6.14

6.15. Repeat Prob. 6.14 with $C_e = 0$.

6.16. Repeat the design problem of Example 6.2-3 to achieve as high a current gain as possible.

SECTION 6.3

6.17. Find the four common-base hybrid parameters in terms of the four common-emitter hybrid parameters.

6.18. The collector and emitter currents are given by the Ebers-Moll equations (Prob. 3.4).

$$i_E = a_{11}(\epsilon^{v_{EB}/V_T} - 1) + a_{12}(\epsilon^{v_{CB}/V_T} - 1)$$
$$i_C = a_{12}(\epsilon^{v_{EB}/V_T} - 1) + a_{22}(\epsilon^{v_{CB}/V_T} - 1)$$

Calculate h_{ib}, h_{rb}, h_{fb}, and h_{ob}. *Hint:* Use the result of Prob. 6.4.

6.19. Find the four CB hybrid parameters for the transistor of Prob. 6.8.

6.20. The transistor of Prob. 6.19 is used in the circuit of Fig. 6.3-4*a*. Find A_i, A_v, Z_i, and Z_o.

6.21. A transistor has $h_{fe} = 10$, $h_{oe} = 10^{-4}$ at 1 mA, and $h_{re} = 0$. Design a CB amplifier as in Fig. 6.3-4*a* for a maximum voltage gain if $r_i = 50\ \Omega$ and $R_L = 10$ kΩ.

SECTION 6.4

6.22. The emitter follower (common-collector) can be described by a set of hybrid parameters. Find h_{ic}, h_{rc}, h_{oc}, and h_{fc} in terms of the CE h parameters.

6.23. Find h_{ie}, A_v, Z_i, and Z_o.

$$h_{fe} = 100$$
$$h_{oe} = h_{re} = 0$$
$$r_i = 500\ \Omega$$
$$R_b = 100\ \text{k}\Omega$$
$$R_e = 1\ \text{k}\Omega$$
$$R_L = 1\ \text{k}\Omega$$

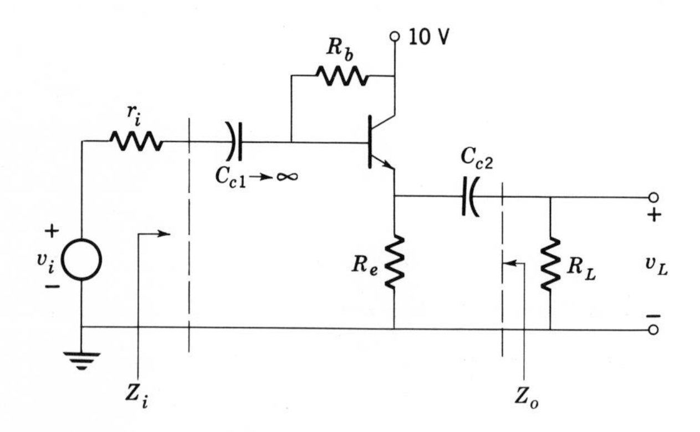

Fig. P6.23

6.24. (a) Plot A_v versus r_i for $0 < r_i < \infty$ ($R_L = 1$ kΩ).
(b) Plot A_v versus R_L for $0 < R_L < \infty$ ($r_i = 1$ kΩ).
(c) Plot Z_i versus R_L for $0 < R_L < \infty$.

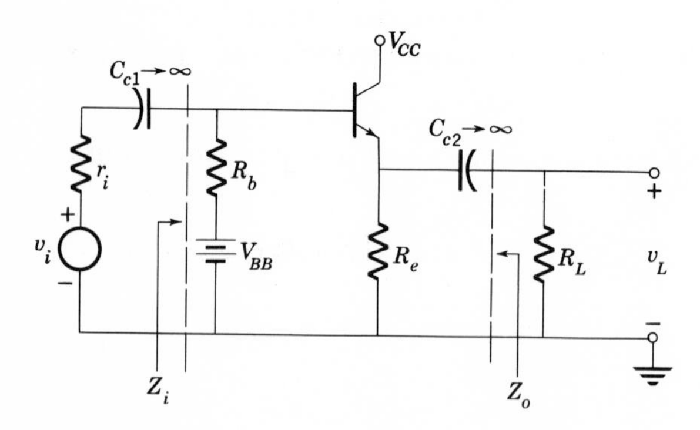

Fig. P6.24

(*d*) Plot Z_o versus r_i for $0 < r_i < \infty$.

$h_{fe} = 100$
$h_{ie} = 1\ \text{k}\Omega$
$h_{oe} = h_{re} = 0$
$R_b \to \infty$
$R_e = 1\ \text{k}\Omega$

6.25. Repeat Prob. 6.24*a* and *b*, if $h_{re} = 10^{-4}$ and $h_{oe} = 10^{-4}$ mho.

6.26. In Example 6.4-2 show that the maximum R_e for a 4-V peak swing is 75 Ω.

6.27. A 2N118 *npn* silicon transistor is used in an emitter-follower circuit, with $I_{EQ} = 1$ mA, $V_{CEQ} = 5$ V. At this *Q* point, $h_{ib} = 42\ \Omega$ and $h_{fb} = +0.96$. In the circuit $r_i = 10\ \text{k}\Omega$, $R_b = 100\ \text{k}\Omega$, and $R_e \| R_L = 500\ \Omega$. Find A_v, A_i, Z_o, and Z_i.

6.28. In the circuit shown, $R_1 = 10\ \text{k}\Omega$, $R_2 = 20\ \text{k}\Omega$, $R_3 = 100\ \text{k}\Omega$, $V_{CC} = 20$ V, $R_e = 1\ \text{k}\Omega$, and $h_{fe} = h_{FE} = 50$.

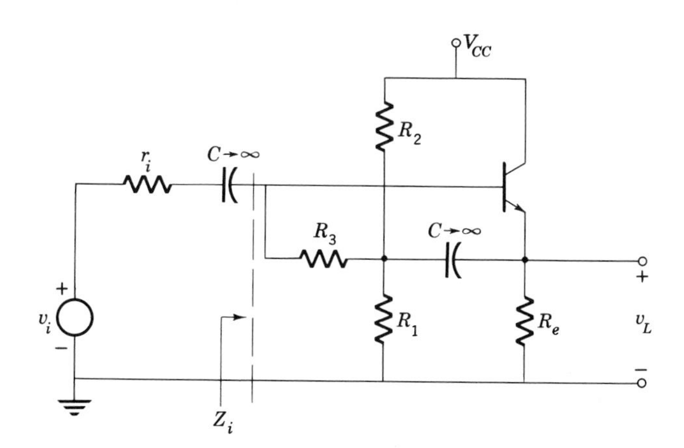

Fig. P6.28

(*a*) Find the quiescent conditions throughout the circuit. Be sure to take R_3 into account.

(*b*) Show that

$$Z_i \approx \frac{R_3 \| h_{ie}}{1 - A_v}$$

where $A_v \approx 1$ is the voltage gain v_L/v_b. This is an example of a technique called "bootstrapping," which is sometimes used to raise input-impedance levels.

REFERENCE

1. Searle, C. L., et al.: "Elementary Circuit Properties of Transistors," vol. 3, John Wiley & Sons, Inc., New York, 1964.

7
Multiple-transistor Circuits

INTRODUCTION

In the usual design problem, the specifications of gain, power output, and frequency response are such that a single transistor is seldom adequate. The designer then has to use more than one transistor in order to satisfy the requirements of the design. In this chapter several of the more commonly used multiple-transistor circuits are studied from the dc and small-signal points of view. These circuits fall into essentially two categories. Those employing *direct* coupling between stages are called *dc amplifiers* because they can amplify signals from dc (zero frequency) to the upper limit of their range. The second category, called *ac amplifiers*, utilizes capacitive coupling between stages, and as a result, the lower limit of their useful range is usually of the order of a few Hertz or so. For many applications the range of frequencies from dc to this lower limit is unimportant, and in these cases it is advantageous to use ac coupling.

A disadvantage of the dc amplifier is that any Q-point shifts (due to temperature variation, for example) are amplified in succeeding stages because of the direct coupling. Thus a small dc shift in the first stage can cause the final stage to be either saturated or cut off. All integrated circuit amplifiers are direct-coupled because of the difficulty of fabricating large integrated capacitors. This leads to special problems in their design.

In this chapter both analysis and design are considered, and special attention is given to circuits which are currently being used in integrated-circuit amplifiers.

7.1 CASCADING OF AMPLIFIER STAGES

When the gain required from an amplifier exceeds that obtainable from a single transistor, two or more individual transistors can be connected in *cascade*. This means that a portion of the output of the first stage is used as the input to the second stage, and so on. The cascade connection of two ac-coupled common-emitter stages is shown in Fig. 7.1-1. One can, of course, connect different types of stages in cascade (for example, a CE-CC cascaded pair is frequently used).

In Chaps. 4 and 6 the biasing, small-signal analysis, and design of single-stage amplifiers were discussed. It was found that in all cases the design procedure consisted of a definite series of steps involving, first, the determination of a suitable quiescent point, and then adjustment of the parameters to achieve the required gain. Similar procedures apply to the cascade amplifier.

Consider the small-signal analysis of the two-stage amplifier shown

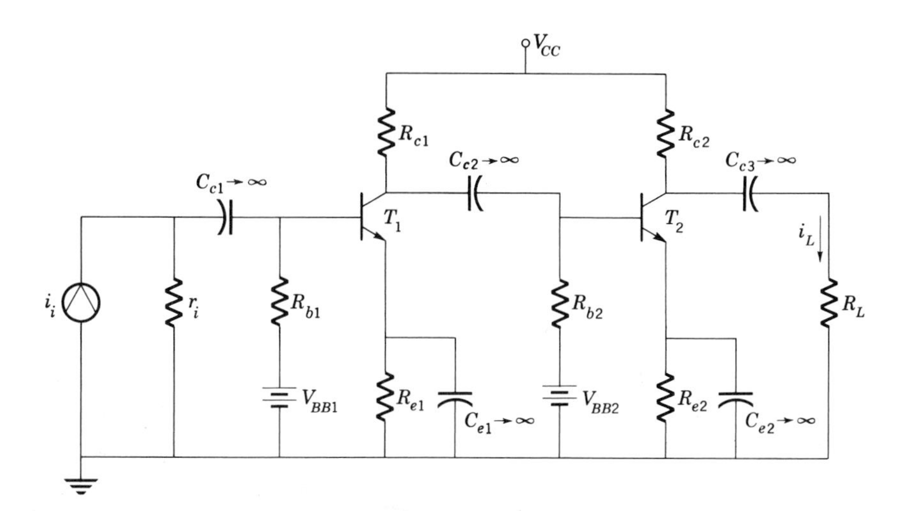

Fig. 7.1-1 Cascaded ac-coupled CE stages.

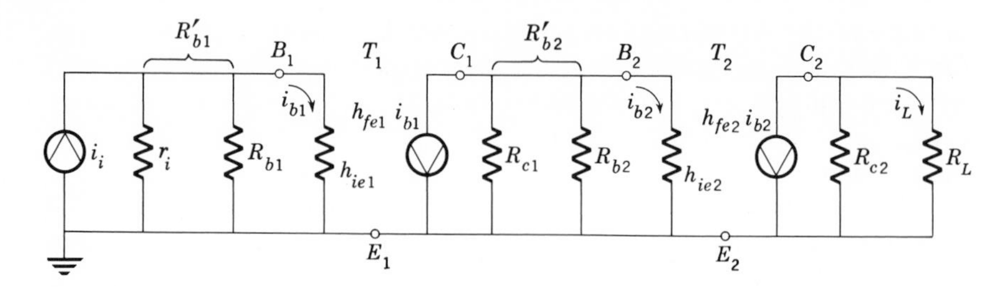

Fig. 7.1-2 Small-signal equivalent circuit.

in Fig. 7.1-1. The small-signal equivalent circuit is shown in Fig. 7.1-2. In this chapter all coupling and bypass capacitors are assumed to be short circuits, and transistor capacitances are assumed to be open circuits.

The overall current gain $A_i = i_L/i_i$ can be calculated by successive application of the current-division principle. However, it is instructive to consider the case where

$$h_{ie1} \ll R_{b1}\|r_i \equiv R'_{b1} \tag{7.1-1a}$$

$$h_{ie2} \ll R_{c1}\|R_{b2} \equiv R'_{b2} \tag{7.1-1b}$$

$$R_L \ll R_{c2} \tag{7.1-1c}$$

Maximum current gain is obtained when (7.1-1) is satisfied. For this case

$$A_i = \frac{i_L}{i_i} = \left(\frac{i_L}{i_{b2}}\right)\left(\frac{i_{b2}}{i_{b1}}\right)\left(\frac{i_{b1}}{i_i}\right) \approx (-h_{fe2})(-h_{fe1})(1) = h_{fe2}h_{fe1} \tag{7.1-2}$$

Clearly, if N identical stages are cascaded, the theoretical maximum current gain will be

$$A_i = (-h_{fe})^N \tag{7.1-3}$$

From Fig. 7.1-2 the actual current gain is

$$A_i = \left(\frac{i_L}{i_{b2}}\right)\left(\frac{i_{b2}}{i_{b1}}\right)\left(\frac{i_{b1}}{i_i}\right) = \left(\frac{-h_{fe2}R_{c2}}{R_{c2} + R_L}\right)\left(\frac{-h_{fe1}R'_{b2}}{R'_{b2} + h_{ie2}}\right)\left(\frac{R'_{b1}}{R'_{b1} + h_{ie1}}\right) \tag{7.1-4}$$

Because the inequalities of (7.1-1) are not always satisfied, the actual gain given by (7.1-4) is often considerably less than the theoretical maximum.

In practice, an amplifier design problem might include specifications on gain, input and output impedance, output current swing, and other factors. Constraints such as available transistor types, available power-supply voltages, source and load resistances, size and weight limitations, etc., are usually present. In view of all these constraints, it should be clear that there is seldom, if ever, a unique design which will meet a given set of specifications. It is the ingenuity of the designer that will ultimately determine if the design is an optimum one for the intended application.

The examples which follow illustrate typical analysis-and-design problems.

Example 7.1-1 Calculate the voltage gain and the maximum symmetrical output-voltage swing for the CE-CC amplifier of Fig. 7.1-3. Assume identical transistors with $h_{fe} = 100$.

Solution In order to determine the voltage gain, h_{ie1} and h_{ie2} must first be calculated. The gain can then be obtained from the small-signal equivalent circuit. However, to find h_{ie1} and h_{ie2} and to calculate the maximum symmetrical swing, the quiescent currents I_{CQ1} and I_{CQ2} must be determined.

The circuit of the emitter-follower output stage is shown in Fig. 7.1-4*a*. To calculate the quiescent current I_{CQ2}, KVL is used around the base-emitter circuit.

$$V_{CC} = R_{b2}I_{BQ2} + V_{BE2} + R_eI_{EQ2}$$
$$20 = 10^5 I_{BQ2} + V_{BE2} + 10^3 I_{EQ2}$$

Substituting $V_{BE2} = 0.7$ V and $I_{BQ2} \approx I_{EQ2}/100$, one finds

$$I_{EQ2} = 9.65 \text{ mA}$$

The Q point and the ac and dc load lines are shown in Fig. 7.1-4*b*. Since the ac load line has a slope equal to $-1/500$, a maximum output-voltage swing of ± 5 V can be obtained.

The circuit and *vi* characteristic for the common-emitter first

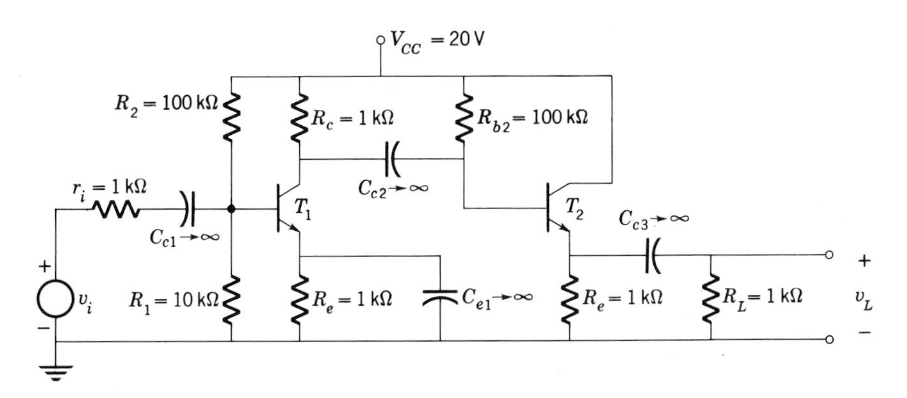

Fig. 7.1-3 Amplifier of Example 7.1-1.

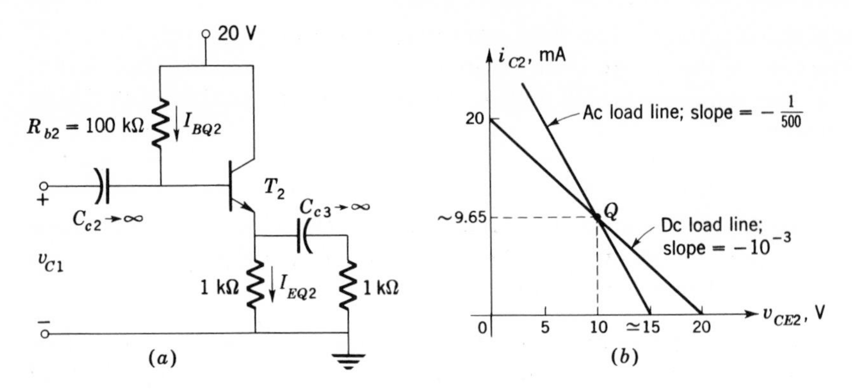

Fig. 7.1-4 Output stage for Example 7.1-1. (*a*) Second stage; (*b*) load lines.

stage are shown in Fig. 7.1-5. The Q point is obtained as in Chap. 3. The input impedance h_{ie1} can now be calculated, using (6.2-3).

$$h_{ie1} \approx \frac{25 \times 10^{-3} h_{fe1}}{I_{CQ1}} = \frac{25 \times 10^{-3} \times 100}{1.3 \times 10^{-3}} \approx 1920\ \Omega$$

From Fig. 7.1-5*b*, the maximum symmetrical collector swing is 2.6 V peak-to-peak. Since this represents the maximum undistorted voltage that can be applied to the emitter follower, the maximum undistorted voltage swing at the output terminals will be only about 2.6 V peak-to-peak, the EF always having a voltage gain of less than unity. (Note that the maximum swing is limited by T_1. The swing provided by T_1 could have been increased simply by increasing I_{CQ1}. For example, if $I_{CQ1} = 6.5$ mA, then $V_{CEQ1} = 7$ V and a peak-

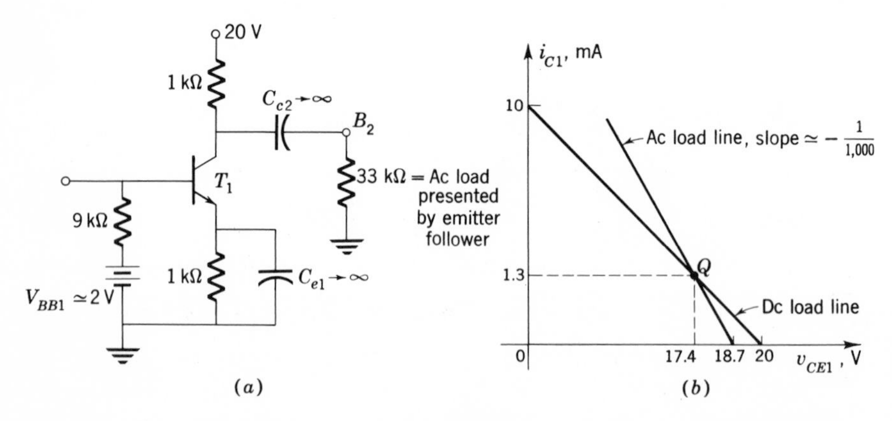

Fig. 7.1-5 Common-emitter stage for Example 7.1-1. (*a*) Circuit; (*b*) load lines.

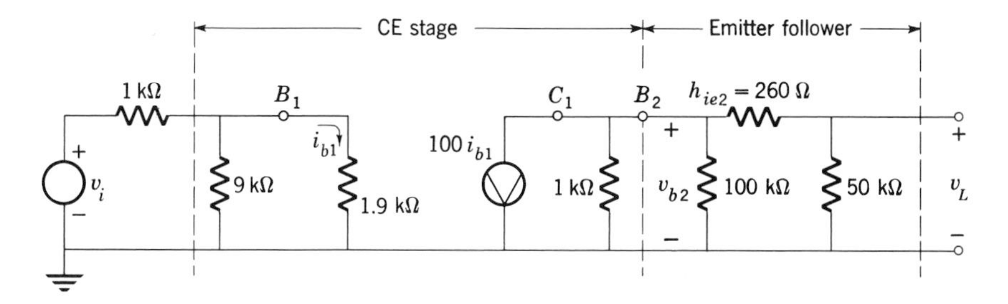

Fig. 7.1-6 Small-signal equivalent circuit for Example 7.1-1.

to-peak swing of 13 V would be possible with T_1. To achieve this increased current, R_1 would be increased to ≈ 120 kΩ.)

The voltage gain $A_v = v_L/v_i$ can now be calculated using the small-signal equivalent circuit shown in Fig. 7.1-6. Note that in the emitter follower all components have been reflected into the base circuit.

The voltage gain is

$$A_v = \frac{v_L}{v_i} = \left(\frac{v_L}{v_{b2}}\right)\left(\frac{v_{b2}}{i_{b1}}\right)\left(\frac{i_{b1}}{v_i}\right)$$

$$\approx (1)(-10^5)(0.32 \times 10^{-3}) = -32\ (= 30\text{ dB})^*$$

Thus, to obtain an output voltage of 2.6 V peak-to-peak, a signal voltage of 2.6/32 = 81 mV peak-to-peak is required.

Example 7.1-2 The circuit shown in Fig. 7.1-7 is a direct-coupled amplifier used for wideband high-gain applications. It is currently available as a self-contained integrated circuit.

Find the quiescent conditions, voltage gain, and maximum output-voltage swing, assuming $h_{fe1} = h_{fe2} = 100$.

Solution

QUIESCENT OPERATING POINT AND MAXIMUM OUTPUT SWING
At first glance, the calculation of quiescent conditions appears to be a formidable task. However, if certain simplifying assumptions are made at the outset, approximate results can be obtained quickly, without recourse to simultaneous equations or graphical procedures. The key assumptions are that $V_{BE} = 0.7$ V and $I_B \approx 0$ for all transistors. The approximate results can then be used to find more accurate values for I_B by iterating the initial approximate result. However, parameter variations from unit-to-unit and temperature

* See Appendix I for a definition of the decibel (dB).

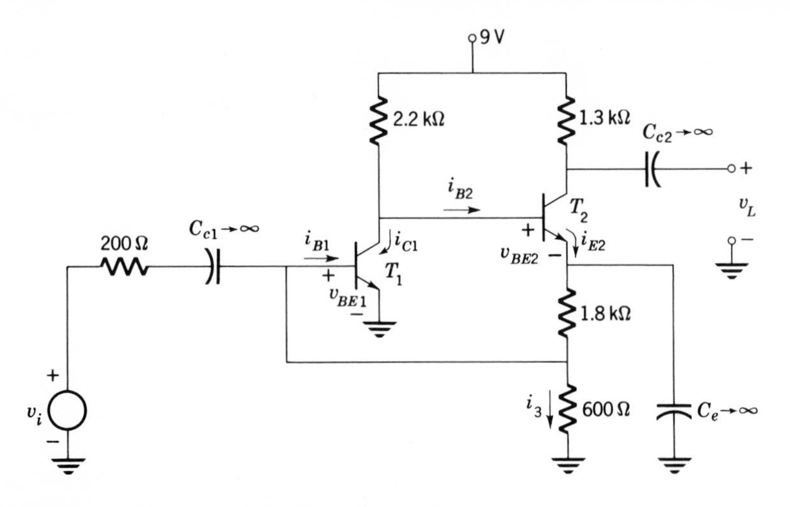

Fig. 7.1-7 Integrated-circuit amplifier.

effects will usually produce greater errors than those of the approximate calculation.

In the circuit of Fig. 7.1-7 it is convenient to begin by assuming that $V_{BE1} = 0.7$ V. Then $I_3 = 0.7/600 = 1.17$ mA. Now we assume that $I_{B1} \ll I_3$, so that $I_{E2} \approx I_3 = 1.17$ mA.

The quiescent voltages for T_2 are then

$$V_{E2} \approx (2.4 \text{ k}\Omega)(1.17 \text{ mA}) \approx 2.8 \text{ V}$$

$$V_{C2} \approx 9 - (1.3)(1.17) \approx 7.5 \text{ V}$$

$$V_{CE2} \approx 7.5 - 2.8 = 4.7 \text{ V}$$

The collector voltage of T_1 and the base voltage of T_2 are the same.

$$V_{C1} = V_{B2} \approx 0.7 + V_{E2} = 3.5 \text{ V}$$

Assuming that $I_{B2} \ll I_{C1}$, the collector current in T_1 is

$$I_{C1} \approx \frac{9 - 3.5}{2.2 \times 10^3} = 2.5 \text{ mA}$$

This completes the calculation of approximate quiescent conditions.

At this point a second approximation to the base currents I_{B1} and I_{B2} can be made:

$$I_{B1} = \frac{I_{C1}}{h_{FE1}} \approx \frac{2.5 \text{ mA}}{100} = 25\ \mu\text{A}$$

$$I_{B2} = \frac{I_{C2}}{h_{FE2}} \approx \frac{1.17 \text{ mA}}{100} \approx 12\ \mu\text{A}$$

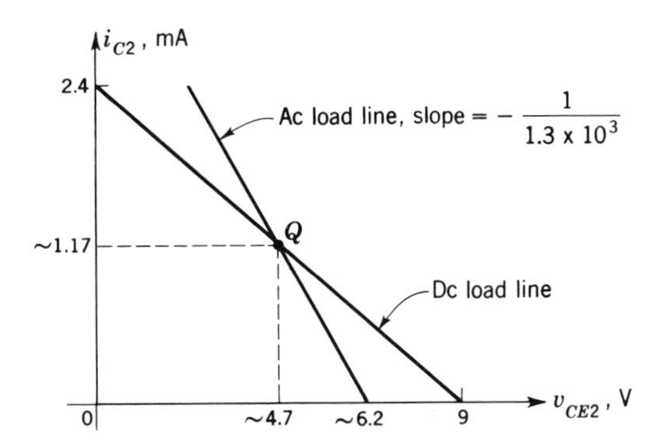

Fig. 7.1-8 Load lines for Example 7.1-2.

These values are easily shown to be small enough so that a second iteration is not necessary (that is, $I_{B1} \ll I_3$ and $I_{B2} \ll I_{C1}$).

To find the maximum output swing, the load lines for T_2 can be drawn as shown in Fig. 7.1-8. From the ac load line it is seen that the collector current can swing ± 1.17 mA. Thus the output voltage v_L can swing $\pm(1.17)(1.3) \approx \pm 1.5$ V.

SMALL-SIGNAL ANALYSIS The small-signal equivalent circuit is shown in Fig. 7.1-9. Using this circuit, the open-circuit voltage

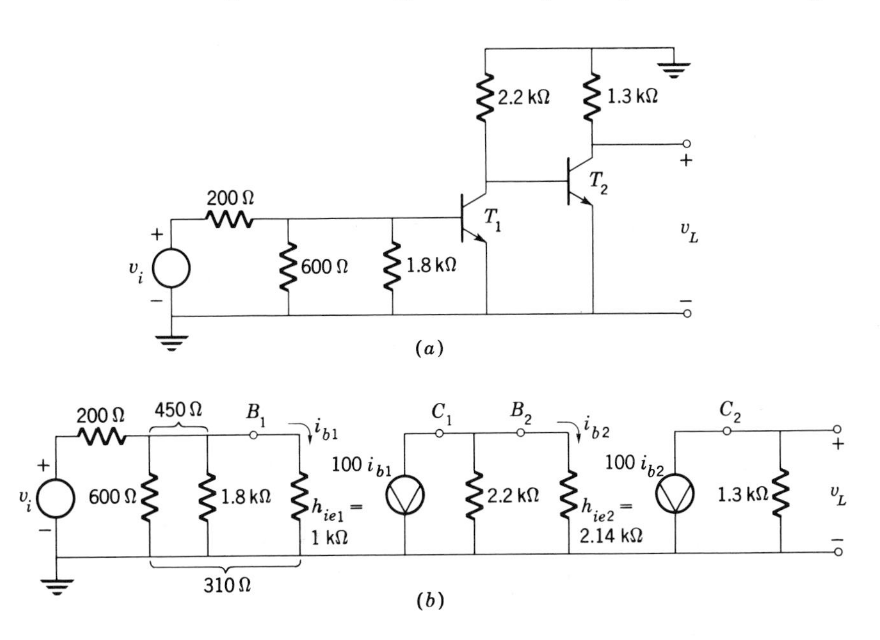

Fig. 7.1-9 Small-signal equivalent circuits for Example 7.1-2. (*a*) Ac circuit; (*b*) small-signal model, $h_{ie1} = (100)(25/2.5) = 1$ kΩ, $h_{ie2} = (100)(25/1.17) = 2.14$ kΩ.

gain is

$$A_v = \frac{v_L}{v_i} = \left(\frac{v_L}{i_{b2}}\right)\left(\frac{i_{b2}}{i_{b1}}\right)\left(\frac{i_{b1}}{v_i}\right)$$

$$= (-1.3 \times 10^5)\left(-\frac{100 \times 2.2}{4.34}\right)\left(\frac{310}{510} \times 10^{-3}\right)$$

$$\approx 4000 \; (= 72 \text{ dB})$$

BIAS STABILITY

When using direct coupling we must exercise extra care in order to avoid operating-point shifts, because any change in dc conditions, especially in the low-level stages, is amplified in succeeding stages. This problem is not as acute in ac-coupled amplifiers, because the operating point of each stage is independent of the other stages.

Consider the problem of quiescent-current shift due to temperature variation of V_{BE} in the circuit of Fig. 7.1-7. Since $R_e = 0$ in the emitter circuit of T_1, one might expect the dc stability of this stage to be poor. However, this is not the case, because of the connection between the emitter circuit of T_2 and the base of T_1. This connection provides a *feedback* path which results in excellent bias stability. The feedback action can be described qualitatively as follows: Suppose V_{BE1} decreases because of a rise in temperature. Then I_{C1} will decrease and V_{CE1} will increase. This results in an increase in V_{BE2}, and hence V_{E2}. This in turn causes I_{E2}, and hence I_{B1}, to increase. The increase in I_{B1} is accompanied by an increase in I_{C1}, which partly counteracts the original decrease. Thus the feedback tends to maintain I_{C1} at a nearly constant value.

Quantitatively, the stability is measured by the ratios $\Delta I_{C1}/\Delta T$ and $\Delta I_{C2}/\Delta T$. These can be found by expressing the currents I_{C1} and I_{C2} in terms of the temperature-dependent variables V_{BE1} and V_{BE2}. The pertinent circuit equations are

$$V_{B2} = V_{C1} = 9 - 2200\left(\frac{I_{E2}}{h_{fe2}} + I_{C1}\right) \tag{7.1-5}$$

$$V_{E2} = V_{B2} - V_{BE2} = 1800 I_{E2} + V_{BE1} \tag{7.1-6}$$

$$I_{E2} = I_{B1} + \frac{V_{BE1}}{600} = \frac{I_{C1}}{h_{fe1}} + \frac{V_{BE1}}{600} \tag{7.1-7}$$

Solving these three equations for I_{C1} in terms of V_{BE1} and V_{BE2} yields

$$I_{C1}\left(\frac{1800}{h_{fe1}} + \frac{2200}{h_{fe1}h_{fe2}} + 2200\right) = 9 - V_{BE1}\left(\frac{11}{3h_{fe2}} + 4\right) - V_{BE2}$$

$$I_{C1} \approx \frac{9 - 4V_{BE1} - V_{BE2}}{2200} \tag{7.1-8}$$

Assuming $\Delta V_{BE}/\Delta T = -k \approx -2.5$ mV/°C (Sec. 4.2) for both transistors,

$$\frac{\Delta I_{C1}}{\Delta T} = \left(\frac{\Delta I_{C1}}{\Delta V_{BE}}\right)\left(\frac{\Delta V_{BE}}{\Delta T}\right) = \frac{(-5)(-2.5)(10^{-3})}{2200} = 5.7\ \mu\text{A/°C} \tag{7.1-9}$$

Using (7.1-7), with $I_{C2} \approx I_{E2}$,

$$\frac{\Delta I_{C2}}{\Delta T} \approx \left(\frac{1}{h_{fe1}}\right)\left(\frac{\Delta I_{C1}}{\Delta T}\right) + \left(\frac{1}{600}\right)\left(\frac{\Delta V_{BE1}}{\Delta T}\right) \tag{7.1-10}$$

$$\approx \frac{(-2.5)(10^{-3})}{600} = -4.2\ \mu\text{A/°C} \tag{7.1-11}$$

If the temperature varies by as much as ± 50°C, the quiescent current for T_1 varies only ± 0.3 mA and for T_2 only ∓ 0.2 mA.

Example 7.1-3 Design an amplifier to have a current gain of 100. The source resistance is 1 kΩ, and the load resistance is 1 kΩ (ac-coupled). The peak-to-peak output-current swing must be at least 10 mA.

Silicon *npn* transistors having $h_{fe} = 20$ are available. The power supply is +15 V.

Solution We note in advance that the solution given below is only one of many possible solutions. Considering the specifications, either a CE-CC or CE-CE configuration may be used. The CE-CE circuit of Fig. 7.1-1 is chosen for convenience. A reasonable way to start the design is at the second, or load, stage. This is designed to yield the desired output swing consistent with low distortion. The first stage is then designed to make up the required gain, taking into account the loading between stages.

Design of the Second Stage The load current swing is to be 10 mA peak-to-peak. Therefore, since the load is in parallel with R_{c2}, the collector voltage of T_2 must be capable of swinging.

$$\begin{aligned} v_{CE} &= i_L R_L \\ &= 10 \times 10^{-3} \times 1000 \\ &= 10\text{V peak-to-peak} \end{aligned}$$

The dc and ac load lines are sketched in Fig. 7.1-10.

The Q point is chosen at $V_{CEQ2} = 6$ V, so that a swing of ± 5 V will not cause saturation. Since ± 5-mA load current is required, I_{CQ2} must be greater than 5 mA. I_{CQ2} is found as follows: Neglecting R_{e2} for the moment, the dc-load-line equation is

$$V_{CC} = 15 = V_{CE2} + I_{c2}R_{c2} \tag{7.1-12}$$

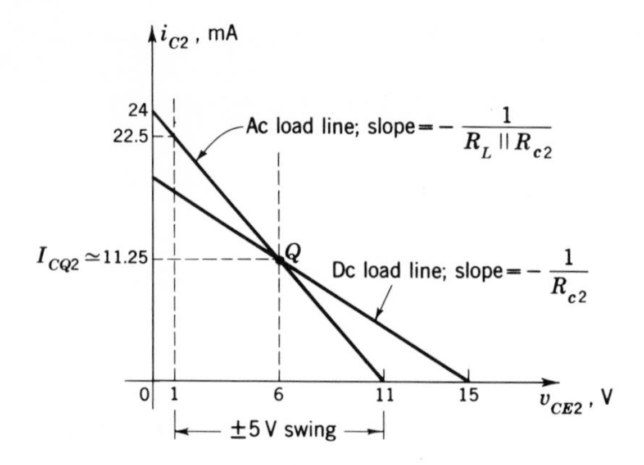

Fig. 7.1-10 Load lines for Example 7.1-3.

Thus, at the Q point,

$$I_{CQ2}R_{c2} = 9 \tag{7.1-13}$$

The equation of the ac load line is

$$i_{C2} - I_{CQ2} = -\left(\frac{R_L + R_{c2}}{R_L R_{c2}}\right)(v_{CE2} - 6) \tag{7.1-14}$$

Now, when $i_{C2} = 0$, $v_{CE2} = 11$ V. Hence

$$\left(\frac{R_L R_{c2}}{R_L + R_{c2}}\right) I_{CQ2} = 5 \tag{7.1-15}$$

where $R_L = 1$ kΩ.
Equations (7.1-13) and (7.1-15) are solved for R_{c2} and I_{CQ2}. The results are

$$R_{c2} = 800\ \Omega \tag{7.1-16}$$

$$I_{CQ2} = 11.25 \text{ mA} \tag{7.1-17}$$

Note that to have the required ± 5 mA swing in the 1-kΩ load requires $(\pm 5)[(1 + 0.8)/0.8] = \pm 11.25$ mA swing in the collector. Thus (when $v_{CE2} = 1$ V) i_{C2} should be 22.5 mA. From (7.1-14)

$$i_{C2} = 11.25 - (1.8/0.8)(-5) = 22.5 \text{ mA}$$

which verifies the result.

To complete the design of T_2, R_{e2} and the bias network consisting of R_{b2} and V_{BB2} must be specified. These steps are postponed until the design of T_1 is completed, since they affect the gain of T_1. We must also remember to check that $R_{c2} \gg R_{e2}$, as was assumed in (7.1-12) and (7.1-13).

DESIGN OF FIRST STAGE As noted before, the first stage is designed to make up the required gain. The overall gain is given by (7.1-4). The contribution of the collector circuit of T_2 is

$$\frac{i_L}{i_{b2}} = -h_{fe2}\left(\frac{R_{c2}}{R_{c2} + R_L}\right) = (-20)\left(\frac{800}{1800}\right) \approx -9 \tag{7.1-18}$$

Thus the remaining gain to be supplied by T_1 is

$$\left|\frac{i_{b2}}{i_i}\right| = \left|\left(\frac{i_{b2}}{i_{b1}}\right)\left(\frac{i_{b1}}{i_i}\right)\right| \geq \frac{100}{|i_L/i_{b2}|} = \frac{100}{9} \tag{7.1-19}$$

$$\frac{i_{b2}}{i_{b1}} = -h_{fe1}\left(\frac{R'_{b2}}{R'_{b2} + h_{ie2}}\right) = -20\left[\frac{R'_{b2}}{R'_{b2} + (25 \times 10^{-3}h_{fe2})/I_{CQ2}}\right]$$

$$= \frac{-20R'_{b2}}{R'_{b2} + (20 \times 25 \times 10^{-3})/(11.25 \times 10^{-3})} \approx -\frac{20}{1 + (44/R'_{b2})} \tag{7.1-20}$$

Also,

$$\frac{i_{b1}}{i_i} = \frac{R'_{b1}}{R'_{b1} + h_{ie1}} = \frac{R'_{b1}}{R'_{b1} + (20 \times 25 \times 10^{-3})/I_{CQ1}}$$

$$= \frac{1}{1 + (0.5/R'_{b1}I_{CQ1})} \tag{7.1-21}$$

Substituting (7.1-20) and (7.1-21) into (7.1-19),

$$\left(\frac{20}{1 + 44/R'_{b2}}\right)\left(\frac{1}{1 + 0.5/R'_{b1}I_{CQ1}}\right) \geq \frac{100}{9} \tag{7.1-22}$$

Equation (7.1-22) can be satisfied easily. The single major restriction is the minimum value possible for I_{CQ1}. I_{CQ1} must be chosen to provide sufficient current to T_2 so the 10-mA peak-to-peak current can be delivered to the load. From (7.1-18) one finds that

$$i_{b2} = 10/9 \approx 1.11 \text{ mA peak-to-peak}$$

The quiescent collector current I_{CQ1} must be chosen so that this value of i_{b2} is obtained.

For convenience, choose $I_{CQ1} = 5$ mA. Then (7.1-22) becomes

$$\left(\frac{20}{1 + 44/R'_{b2}}\right)\left(\frac{1}{1 + 100/R'_{b1}}\right) \geq \frac{100}{9} \tag{7.1-23}$$

If R_{c1} is chosen, for simplicity, to be 1 kΩ, then

$$R'_{b2} = R_{b2}\|1 \text{ k}\Omega \tag{7.1-24a}$$

$$R'_{b1} = R_{b1}\|1 \text{ k}\Omega \tag{7.1-24b}$$

From (7.1-23) it is seen that the required gain will be achieved if

$$R'_{b2} > 100\ \Omega \tag{7.1-25a}$$

$$R'_{b1} > 250\ \Omega \tag{7.1-25b}$$

[Note that this is not the only combination that satisfies (7.1-23).]

The design is not yet completed since R_{b1}, R_{b2}, V_{BB1}, V_{BB2}, R_{e1}, and R_{e2} must be chosen. For stability, we have the additional inequalities

$$R_{e1} \gg \frac{R_{b1}}{h_{fe1}} \tag{7.1-26a}$$

$$R_{e2} \gg \frac{R_{b2}}{h_{fe2}} \tag{7.1-26b}$$

and for the design of T_2,

$$R_{c2} \gg R_{e2} \tag{7.1-27}$$

Thus

$$R_{c2} \gg R_{e2} \gg \frac{R_{b2}}{20}$$

and since $R_{c2} = 800\ \Omega$,

$$800 \gg R_{e2} \gg \frac{R_{b2}}{20}$$

Choosing $R_{b2} = 400\ \Omega$,

$$800 \gg R_{e2} \gg 20 \tag{7.1-28}$$

and $R_{e2} = 100\ \Omega$ will be satisfactory. Note that $R'_{b2} = 286\ \Omega > 100\ \Omega$.

To select R_{e1}, note that

$$R_{c1} = 1000 > R_{e1} \gg \frac{R_{b1}}{20}$$

Choosing $R_{b1} = 1\ \text{k}\Omega$ [$R'_{b1} = 500\ \Omega$, which satisfies (7.1-25*b*)],

$$1000 > R_{e1} \gg 50 \tag{7.1-29}$$

Consider choosing

$$R_{e1} = 500\ \Omega$$

Then the dc and ac load lines for T_1 can be sketched as shown in Fig. 7.1-11.

The important aspect of the *vi* characteristic for T_1 is that the desired quiescent current of 5 mA can be achieved. If this were

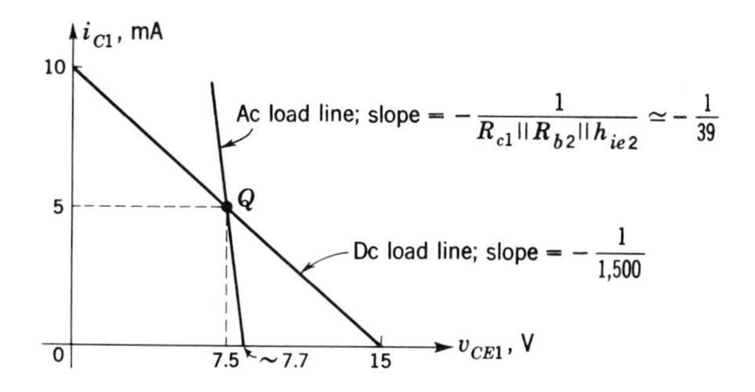

Fig. 7.1-11 Load lines for input stage of Example 7.1-3.

not possible, a redesign would be necessary. Since I_{CQ1} was chosen so that i_{b2} could swing 1.1 mA peak-to-peak, an I_{CQ1} as low as 1 mA would also satisfy the input-current requirements of the second stage. However, different resistance values would have to be selected.

The bias voltages are found using KVL as before.

$$V_{BB1} \approx 0.7 + I_{CQ1}R_{e1} = 3.2 \text{ V}$$

$$V_{BB2} \approx 0.7 + I_{CQ2}R_{e2} = 1.8 \text{ V}$$

Summarizing,

$$R_{b1} = 1 \text{ k}\Omega \qquad R_{b2} = 400 \ \Omega$$

$$R_{e1} = 500 \ \Omega \qquad R_{e2} = 100 \ \Omega$$

$$R_{c1} = 1 \text{ k}\Omega \qquad R_{c2} = 800 \ \Omega$$

$$V_{BB1} = 3.2 \text{ V} \qquad V_{BB2} = 1.8 \text{ V}$$

The standard bias networks utilizing a voltage divider across the collector supply are easily designed using the given values of V_{BB1}, V_{BB2}, R_{b1}, and R_{b2}.

7.2 THE DIFFERENCE AMPLIFIER

The difference-amplifier configuration is extremely important because it is used as a basic building block in many feedback amplifiers and most integrated-circuit linear amplifiers. The basic difference-amplifier circuit is shown in Fig. 7.2-1. It is shown below that the load current (and hence the load voltage) is proportional to the *difference* between the two input signals.

The circuit will be analyzed initially under the assumption that

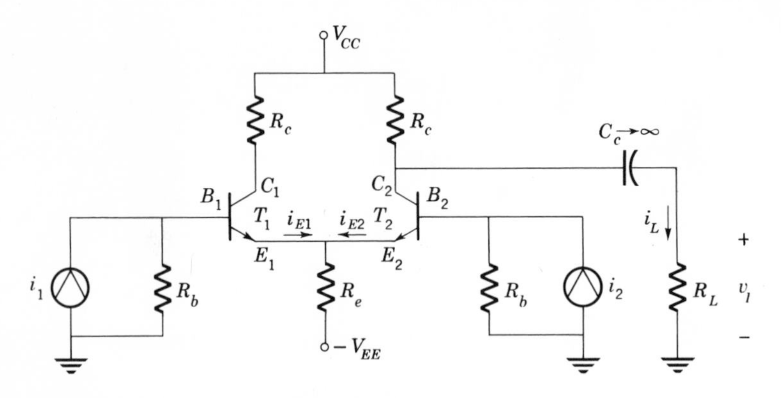

Fig. 7.2-1 Basic difference amplifier.

perfect symmetry exists; i.e., the transistors and the external base circuits are identical.

QUIESCENT-POINT ANALYSIS

Since perfect symmetry is assumed, we can, for dc analysis, separate the emitters, placing a resistance of $2R_e$ in each emitter circuit as shown in Fig. 7.2-2. This can be verified by using KVL:

$$V_{E1} = V_{E2} = (I_{E1} + I_{E2})R_e - V_{EE}$$
$$= I_{E1}(2R_e) - V_{EE} = I_{E2}(2R_e) - V_{EE}$$

since $I_{E1} = I_{E2}$, because of symmetry. Applying KVL to the base-emitter loop,

$$I_{EQ1} = I_{EQ2} \approx \frac{V_{EE} - 0.7}{2R_e + R_b/h_{fe}} \tag{7.2-1}$$

For the collector loop

$$V_{CEQ1} = V_{CEQ2} = V_{CC} - I_{CQ}R_c - I_{EQ}(2R_e) + V_{EE}$$
$$\approx V_{CC} + V_{EE} - I_{CQ}(R_c + 2R_e) \tag{7.2-2}$$

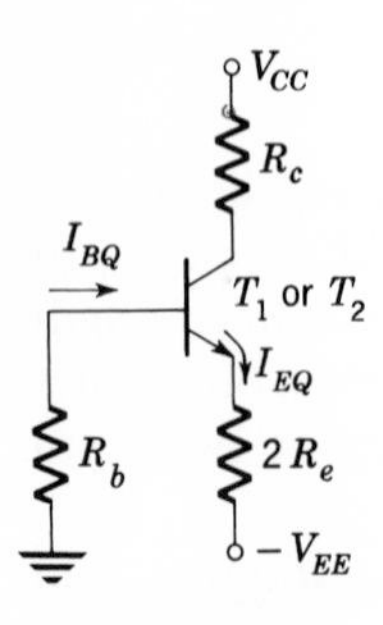

Fig. 7.2-2 Circuit for either transistor for quiescent-point analysis.

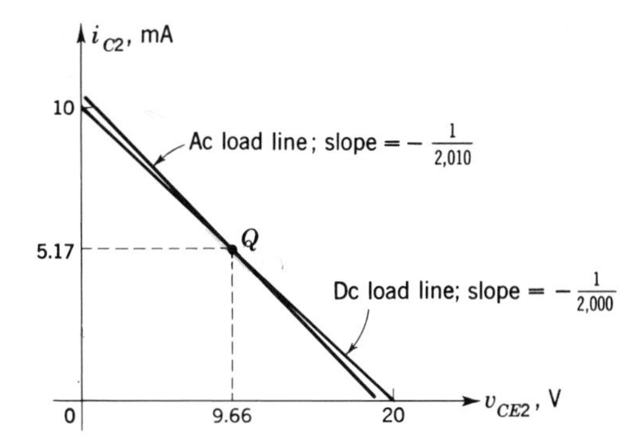

Fig. 7.2-3 Load lines for Example 7.2-1.

These two equations can be used to determine the Q point, or alternatively, to design for a specified Q point. Once the Q point is determined, the ac load line can be constructed, and the load-current swing found. This is illustrated in the following example.

Example 7.2-1 In the circuit of Fig. 7.2-1,

$$V_{CC} = V_{EE} = 10 \text{ V} \qquad R_b = 200\ \Omega \qquad R_e = 900\ \Omega$$
$$R_c = 200\ \Omega \qquad R_L = 10\ \Omega$$

Find the maximum possible load-current swing.

Solution From (7.2-1), assuming $2R_e \gg R_b/h_{fe}$,

$$I_{CQ} \approx I_{EQ} \approx \frac{10 - 0.7}{1800} = 5.17 \text{ mA}$$

From (7.2-2)

$$V_{CEQ} = 10 + 10 - (5.17 \times 10^{-3})(200 + 1800) = 9.66 \text{ V}$$

The dc and ac load lines as seen by transistor T_2 are shown in Fig. 7.2-3. The maximum available swing into the load is seen to be ± 5.17 mA.

SMALL-SIGNAL ANALYSIS

The utility of the difference amplifier lies in its behavior when small signals are applied. To investigate this a small-signal equivalent circuit is drawn by reflecting the sources in Fig. 7.2-1 into the emitter circuits (Sec. 6.4). This yields the ac equivalent circuit of Fig. 7.2-4. In order to provide insight into the operation of the circuit, we shall analyze it using a technique which emphasizes the fact that the amplifier output is approximately proportional to the *difference* current $i_2 - i_1$, and which

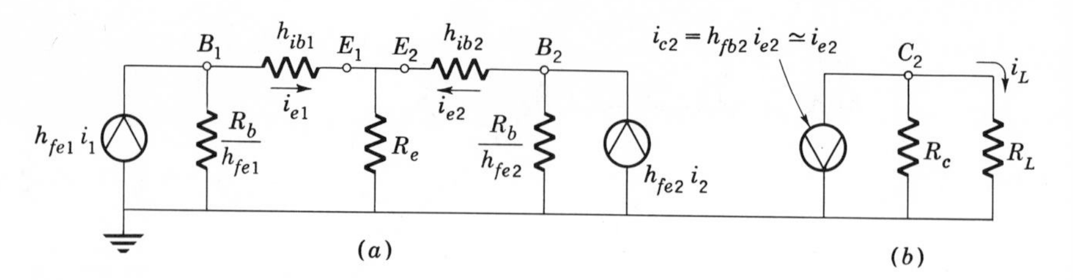

Fig. 7.2-4 Small-signal equivalent circuit of the difference amplifier. (a) Emitter circuit; (b) output circuit.

points out the conditions required for optimum design of the difference amplifier.

The analysis is begun by expressing the input currents in terms of *common-mode* and *differential-mode* components. A graphical interpretation of this process is shown in Fig. 7.2-5, from which i_1 and i_2 can be written as

$$i_2 = i_0 + \frac{\Delta i}{2} \tag{7.2-3a}$$

and

$$i_1 = i_0 - \frac{\Delta i}{2} \tag{7.2-3b}$$

where

$$\Delta i = i_2 - i_1 \tag{7.2-4a}$$

and

$$i_0 = \frac{i_2 + i_1}{2} \tag{7.2-4b}$$

Equation (7.2-3) and Fig. 7.2-5 illustrate that i_1 and i_2 can be represented in terms of the average current i_0, called the *common-mode* current, and the difference current Δi, called the *differential-mode* current.*

* Note that (7.2-3) and (7.2-4) are general definitions, and apply when i_1 and i_2 are arbitrary time functions or constants.

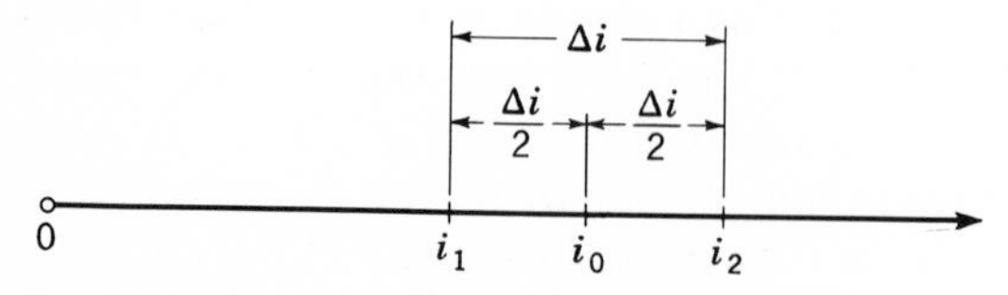

Fig. 7.2-5 Decomposition of input currents.

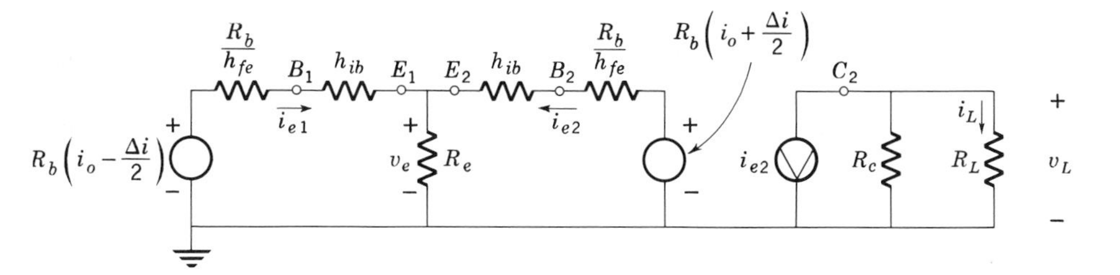

Fig. 7.2-6 Equivalent circuit using the common- and differential-mode currents.

If, in Fig. 7.2-4, the current sources $h_{fe1}i_1$ and $h_{fe2}i_2$ are converted to voltage sources, and (7.2-3) is used to express i_1 and i_2 in terms of i_0 and Δi, Fig. 7.2-4 can be redrawn as shown in Fig. 7.2-6. The circuit has been further simplified from that shown in Fig. 7.2-4 by assuming that transistors T_1 and T_2 are identical, so that $h_{fe1} = h_{fe2}$. Since the Q points of T_1 and T_2 are set so that $I_{EQ1} = I_{EQ2}$ [see (7.2-1)], we have also set $h_{ib1} = h_{ib2}$.

The emitter circuit of Fig. 7.2-6 is completely symmetrical, except for the input currents. Thus, if we set $i_1 = i_2 = i_0$, then $\Delta i = 0$, and the input sources are equal. Then, by symmetry, $i_{e1} = i_{e2}$. Similarly if $i_2 = -i_1$, then $\Delta i/2 = i_2$ and $i_0 = 0$. For this condition $i_{e2} = -i_{e1}$. Making use of these two conditions and superposition, we shall find i_{e2} and hence i_L. We do this by first finding the emitter current due to the common-mode current i_0, and then finding the emitter current due to the differential-mode current Δi. The total emitter current is then, by superposition, $i_{e2} = i_{e2c} + i_{e2d}$, where i_{e2c} is due to the current i_0, and i_{e2d} is due to the current Δi.

The circuits used to determine i_{e2c} and i_{e2d} are shown in Fig. 7.2-7. Fig. 7.2-7*a* is the circuit employed to calculate i_{e2c} with the differential-mode current $\Delta i = 0$. By symmetry $i_{e1c} = i_{e2c}$, and the voltage drop across R_e is $v_e = R_e(2i_{e2c})$. Thus, to calculate i_{e2c}, the circuit can be reduced to that shown in Fig. 7.2-7*b* and the common-mode emitter current is

$$i_{e2c} = \frac{R_b i_0}{2R_e + h_{ib} + R_b/h_{fe}} \tag{7.2-5a}$$

The circuit used to determine i_{e2d} is shown in Fig. 7.2-7*c*. In this circuit $i_0 = 0$. We note that here the voltage drop across R_e is

$$v_e = R_e(i_{e2d} + i_{e1d}) = 0$$

since $i_{e1d} = -i_{e2d}$. Thus, R_e can be replaced by a short circuit. The reduced equivalent circuit is shown in Fig. 7.2-7*d* and the differential-mode

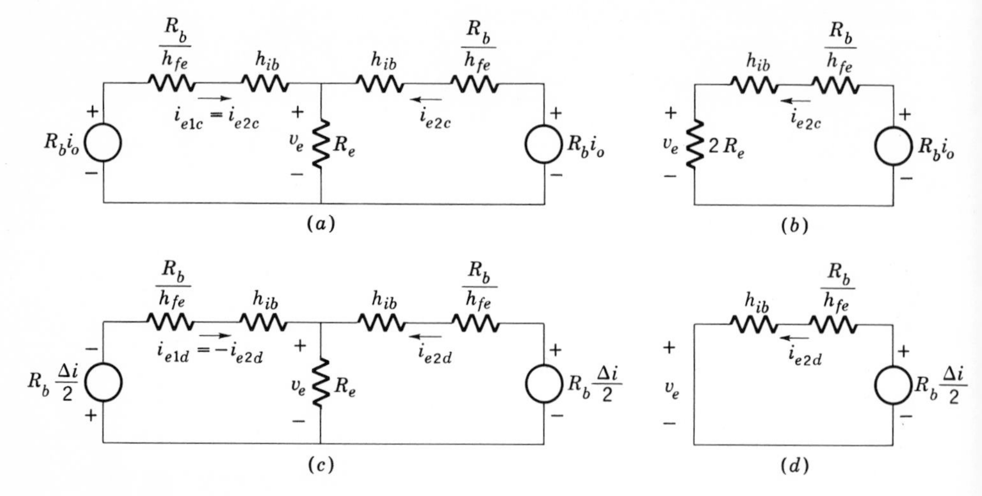

Fig. 7.2-7 Equivalent emitter circuits used to calculate the emitter current $i_{e2} = i_{e2c} + i_{e2d}$. (a) Circuit used to calculate common-mode emitter current i_{e2c}; (b) reduced equivalent circuit; (c) circuit used to calculate differential-mode current i_{e2d}; (d) reduced equivalent circuit.

current is

$$i_{e2d} = \frac{R_b \, \Delta i}{2(h_{ib} + R_b/h_{fe})} \tag{7.2-5b}$$

The total emitter current is found by adding the results of (7.2-5a) and (7.2-5b):

$$i_{e2} = i_{e2c} + i_{e2d} = \left[\frac{R_b}{2R_e + h_{ib} + R_b/h_{fe}}\right] i_0 + \left[\frac{R_b}{2(h_{ib} + R_b/h_{fe})}\right] \Delta i \tag{7.2-6}$$

The load current i_L can now be calculated from the output circuit shown in Fig. 7.2-6. Thus,

$$i_L = \frac{-R_c}{R_c + R_L} i_{e2} = \left(\frac{-R_c}{R_c + R_L}\right)\left(\frac{R_b}{2R_e + h_{ib} + R_b/h_{fe}}\right) i_0 + \left(\frac{-R_c}{R_c + R_L}\right)\left[\frac{R_b}{2(h_{ib} + R_b/h_{fe})}\right] \Delta i \tag{7.2-7}$$

At this point it is useful to define two current gains—the common-mode current gain A_c due to i_0 which is

$$A_c = \left(\frac{-R_c}{R_c + R_L}\right)\left(\frac{R_b}{2R_e + h_{ib} + R_b/h_{fe}}\right) \tag{7.2-8a}$$

and the differential-mode current gain A_d due to Δi

$$A_d = \left(\frac{-R_c}{R_c + R_L}\right)\left[\frac{R_b}{2(h_{ib} + R_b/h_{fe})}\right] \tag{7.2-8b}$$

Thus,

$$i_L = A_c i_0 + A_d\,\Delta i \tag{7.2-8c}$$

7.2-1 COMMON-MODE REJECTION RATIO

In an ideal differential amplifier, the load current is proportional to Δi, and does not depend on the common-mode current i_0. Thus, in an ideal differential amplifier $A_c = 0$. This condition cannot be realized in practice, since to make $A_c = 0$, R_e would have to be infinite. In order to measure the departure from the ideal, a quantity called the *common-mode rejection ratio* (CMRR) is used. The CMRR is defined as the ratio of the differential-mode gain to the common-mode gain,

$$\text{CMRR} = \frac{A_d}{A_c} \tag{7.2-9a}$$

Using (7.2-8*a*) and (7.2-8*b*), the CMRR becomes

$$\text{CMRR} = \frac{2R_e + h_{ib} + R_b/h_{fe}}{2(h_{ib} + R_b/h_{fe})} \tag{7.2-9b}$$

If, as is often true in practice, $2R_e \gg \dfrac{R_b}{h_{fe}} + h_{ib}$, then

$$\text{CMRR} \approx \frac{R_e}{h_{ib} + R_b/h_{fe}} \tag{7.2-9c}$$

To illustrate the usefulness of the CMRR consider that $i_2 = 10.5$ mA and $i_1 = 9.5$ mA. Then $i_0 = 10$ mA and $\Delta i = 1$ mA. The load current i_L is, then, from (7.2-8*c*),

$$i_L = A_d\,\Delta i\left(1 + \frac{A_c}{A_d}\frac{i_0}{\Delta i}\right) = A_d\,\Delta i\left(1 + \frac{10}{\text{CMRR}}\right) \tag{7.2-10}$$

In order that i_L be proportional to Δi, the CMRR should be much greater than 10. In general, we see that the CMRR should be chosen so that

$$\text{CMRR} \gg \frac{i_0}{\Delta i} \tag{7.2-11}$$

if the load current is to be proportional to Δi.

Example 7.2-2 In Example 7.2-1 the common-mode signal current is 1 mA. Find the common-mode rejection ratio. Assume

$$h_{fe1} = h_{fe2} = 100$$

Find the differential-mode signal for which the differential output is at least 100 times the common-mode output.

Solution From (7.2-8*a*)

$$A_c = -\left(\frac{200}{200+10}\right)\left[\frac{200}{(2)(900) + \dfrac{25 \times 10^{-3}}{5.17 \times 10^{-3}} + \dfrac{200}{100}}\right] \approx -0.1$$

and, from (7.2-8*b*),

$$A_d = -\left(\frac{200}{200+10}\right)\left[\frac{200}{(2)\left(\dfrac{25 \times 10^{-3}}{5.17 \times 10^{-3}} + \dfrac{200}{100}\right)}\right] \approx -14$$

Thus

$$i_L \approx -0.1 i_0 - 14\Delta i$$

The common-mode rejection ratio is

$$\text{CMRR} = \frac{A_d}{A_c} = 140 \ (\approx 43 \ dB)$$

For good differential-amplifier operation

$$\Delta i \gg \frac{i_0}{\text{CMRR}} = \frac{i_0}{140}$$

If $i_0 = 1$ mA and i_{Ld} is to be at least 100 i_{Lc} (1 percent departure from a pure difference signal in the output), then

$$\Delta i \gg (100)(1/140) \approx 0.7 \text{ mA}$$

If Δi is less than 0.7 mA, the departure from a pure difference signal in the output will be greater than 1 percent.

7.2-2 A CONSTANT-EMITTER CURRENT SOURCE

In order to increase the CMRR given by (7.2-9*b*), a third transistor is often used, as shown in Fig. 7.2-8.

The third transistor, T_3, supplies a constant current I_{EQ3} to the emitter circuits of T_1 and T_2. The impedance looking into the collector of T_3 is $1/h_{ob3}$. Thus, T_3 is basically equivalent to a very large resistance in series with a large negative supply. Since i_{C3} is approximately constant because the base is biased at a constant voltage, $-V_{BB}$, T_3 can be considered a constant-current source.

Quiescent operation In the circuit shown the quiescent emitter currents are determined by T_3 and are, for identical T_1 and T_2,

$$I_{EQ1} = I_{EQ2} = \tfrac{1}{2} I_{EQ3} = \left(\frac{1}{2}\right)\left(\frac{V_{EE} - V_{BB} - 0.7}{R_e}\right) \qquad (7.2\text{-}12)$$

Then, since

$$V_{CC} = I_{CQ1}R_c + V_{CEQ1} - V_{BE1} - \frac{R_b I_{CQ1}}{h_{fe}} \tag{7.2-13}$$

the quiescent collector-to-emitter voltages are

$$V_{CEQ1} = V_{CEQ2} \approx V_{CC} - R_c I_{CQ1} - 0.7 - \frac{R_b I_{CQ1}}{h_{fe}} \tag{7.2-14}$$

If $V_{CC} \gg 0.7$ V and $R_c \gg R_b/h_{fe}$, the collector-emitter voltages are approximately independent of V_{BE} and h_{fe}:

$$V_{CEQ1} = V_{CEQ2} \approx V_{CC} - R_c I_{CQ1} \tag{7.2-15}$$

Small-signal operation The small-signal equivalent circuit for the amplifier shown in Fig. 7.2-8 is identical to that shown in Fig. 7.2-6. The load current i_L is given by (7.2-8). The important difference between the two-transistor amplifier shown in Fig. 7.2-1 and the three-transistor amplifier shown in Fig. 7.2-8 is the value of R_e and hence the CMRR. When using a constant current transistor, $R_e \approx 1/h_{ob}$ which is much larger than the value of R_e attainable using a passive resistance. To illustrate this consider that $1/h_{ob} = 1$ MΩ. If a passive resistor is to be used instead of T_3, and the quiescent current supplied to each transistor is to be 1 mA, then 2 mA must flow in R_e. In order to be able to use $R_e = 1$ MΩ, V_{EE} must be approximately 2000 V.

The balance control When T_1 and T_2 have different characteristics, a variable resistor R_v is often connected between the emitters of T_1 and T_2

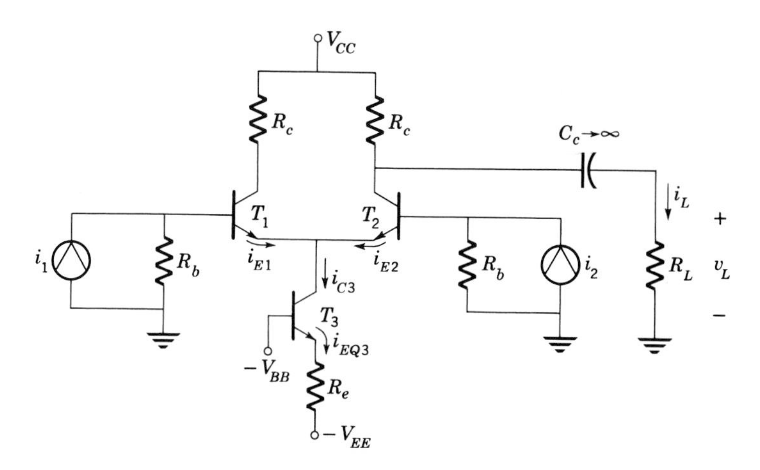

Fig. 7.2-8 Difference amplifier with constant current source.

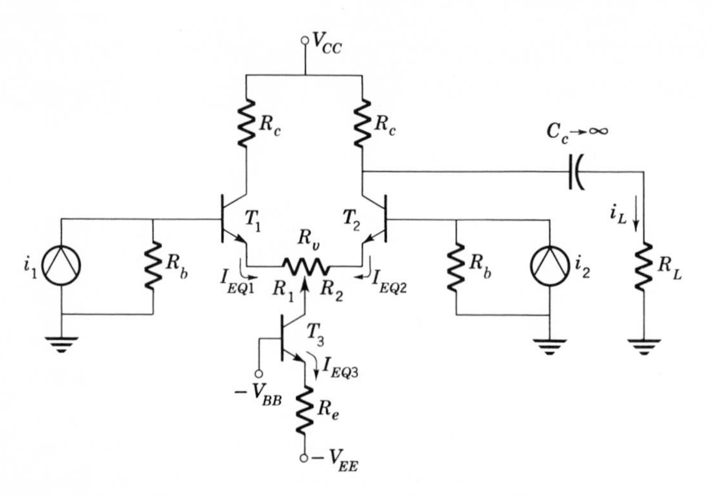

Fig. 7.2-9 Difference amplifier with balance control R_v.

to serve as a balancing control, as shown in Fig. 7.2-9. The slider of the variable resistance is adjusted so that $I_{EQ1} = I_{EQ2}$. A small value of R_v, typically 100 Ω, is usually sufficient to compensate for large differences between h_{fe1} and h_{fe2}.

To evaluate the effect of the balance control note that KVL, applied from the arm of the control to ground, must yield the same voltage when taken around the base-emitter loop of either T_1 or T_2. Thus

$$\left(\frac{R_b}{h_{fe1}} + R_1\right) I_{EQ1} + V_{BE1} = \left(\frac{R_b}{h_{fe2}} + R_2\right) I_{EQ2} + V_{BE2} \tag{7.2-16}$$

If $V_{BE1} = V_{BE2}$, the condition which ensures that the emitter currents of T_1 and T_2 are the same is obtained by setting $I_{EQ1} = I_{EQ2}$ in (7.2-16). This yields

$$R_2 - R_1 = R_b\left(\frac{1}{h_{fe1}} - \frac{1}{h_{fe2}}\right) \tag{7.2-17}$$

and since

$$R_2 + R_1 = R_v = \text{constant} \tag{7.2-18}$$

balance is obtained when

$$R_2 = \frac{R_v}{2} + \left(\frac{R_b}{2}\right)\left(\frac{1}{h_{fe1}} - \frac{1}{h_{fe2}}\right) \tag{7.2-19a}$$

and

$$R_1 = \frac{R_v}{2} - \left(\frac{R_b}{2}\right)\left(\frac{1}{h_{fe1}} - \frac{1}{h_{fe2}}\right) \tag{7.2-19b}$$

For example, if $h_{fe1} = 50$, $h_{fe2} = 150$, and $R_b = 1.5$ kΩ, then R_v must be at least equal to 20 Ω. For this value of R_v, we find $R_2 = 20$ Ω and $R_1 = 0$.

Inclusion of R_v results in symmetrical operation, but also causes a loss in current gain. If R_1 and R_2 are adjusted as in (7.2-19) and the effective emitter resistance is very large so that

$$(R_e)_{\text{eff}} \approx \frac{1}{h_{ob3}} \gg h_{ib1} + \frac{R_b}{h_{fe1}} + R_1 \tag{7.2-20}$$

then the differential-mode gain of (7.2-8*b*) becomes

$$A_d = \frac{i_L}{\Delta i} \approx \frac{-R_b R_c/(R_c + R_L)}{R_b(1/h_{fe1} + 1/h_{fe2}) + 2h_{ib} + R_v} \tag{7.2-21}$$

This should be compared with (7.2-8*b*) which gives the gain for an ideal circuit.

Example 7.2-3 The difference amplifier shown in Fig. 7.2-10 makes use of an integrated circuit package (shown inside the dashed lines). Design the amplifier to obtain a voltage common-mode rejection ratio of 100 (40 dB). The load resistance is 1 kΩ ac-coupled. The signal sources v_1 and v_2 have internal resistances of 1 kΩ each, and the transistors have $h_{fe} = 100$.

Solution The difference-amplifier CMRR is obtained from (7.2-9). However, because of the two 50-Ω resistors, the expressions are altered slightly. Taking account of the fact that the *voltages* are

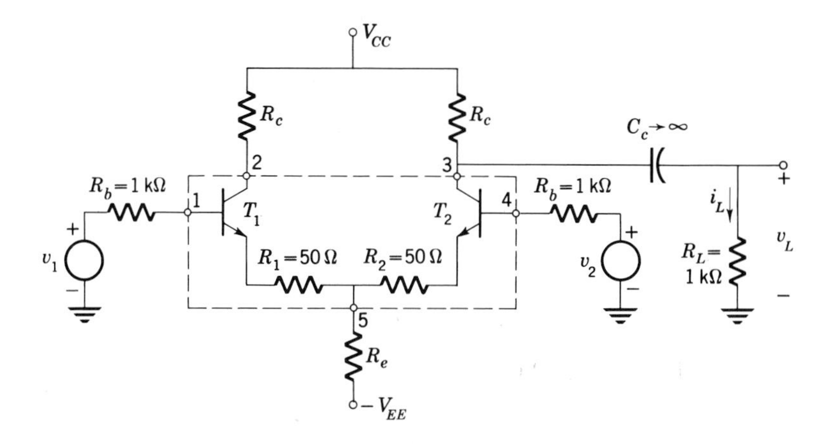

Fig. 7.2-10 Circuit for Example 7.2-3.

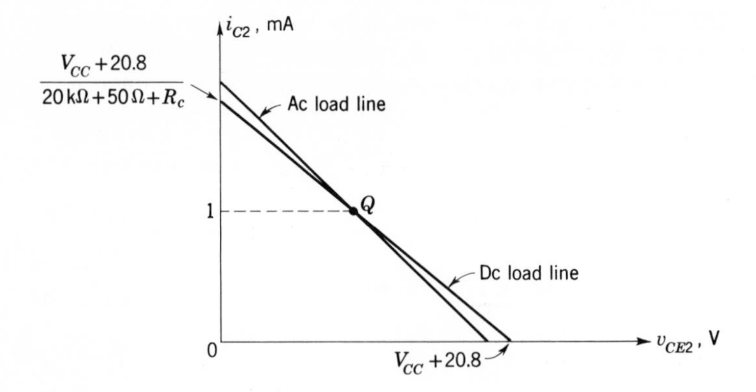

Fig. 7.2-11 Load lines for output stage of Example 7.2-3.

the variables of interest in this example,

$$\left(\frac{R_b}{R_L}\right) v_L = -\left(\frac{R_c}{R_c + R_L}\right)\left(\frac{R_b}{2R_e + h_{ib} + R_1 + R_b/h_{fe}}\right) v_0$$
$$- \left(\frac{R_c}{R_c + R_L}\right)\left(\frac{R_b}{2h_{ib} + R_1 + R_2 + 2R_b/h_{fe}}\right) \Delta v$$
$$= A_{cv} v_0 + A_{dv}\,\Delta v$$

where the common- and differential-mode voltage signals are

$$v_0 = \frac{v_1 + v_2}{2}$$

and

$$\Delta v = v_2 - v_1$$

The CMRR is

$$\frac{A_{dv}}{A_{cv}} = \frac{2R_e + h_{ib} + 50 + 1000/100}{2h_{ib} + 50 + 50 + (2)\,1000/100} \approx \frac{R_e}{60 + h_{ib}}$$

In order to obtain a CMRR of 40 dB,

$$R_e \geq 100(60 + h_{ib})$$

If $I_{EQ1} = I_{EQ2} = 1$ mA, then $h_{ib} = 25\ \Omega$, and $R_e \geq 8.5$ kΩ. Let $R_e = 10$ kΩ. Then V_{EE} is found using KVL (refer to Fig. 7.2-2 and add R_1 to the circuit shown):

$$\left(\frac{R_b}{h_{fe}}\right) I_{EQ1} + V_{BE1} + R_1 I_{EQ1} + 2I_{EQ1}R_e = V_{EE}$$

Thus

$$V_{EE} = 10 \times 10^{-3} + 0.7 + 50 \times 10^{-3} + 20 \approx 20.8 \text{ V}$$

The collector resistance R_c and supply voltage V_{CC} are chosen on the basis of output swing and gain. This can be seen from the load lines shown in Fig. 7.2-11.

It should be noted that the slopes of the dc and ac load lines are almost the same if $2R_e \gg R_c$, as is usually the case. Thus, if $R_c = 1$ kΩ, the slope of the dc load line is $-1/21{,}050$, while the slope of the ac load line is $-1/20{,}550$. If, in addition, $V_{CC} \approx 20$ V, the Q point will be approximately in the center of the load line. However, because $I_{CQ2} = 1$ mA, the maximum possible current swing in the parallel combination of R_c and R_L will be 1 mA peak. Since $R_c = R_L$, the maximum output voltage will then be 0.5 V peak. More output-voltage swing can be obtained by increasing R_c.

7.3 THE DARLINGTON CONFIGURATION (COMPOUND AMPLIFIER)

This amplifier configuration joins N transistors, so that the resultant circuit behaves like a single transistor with a short-circuit current gain approaching $h_{fe}{}^N$. It is often used in direct-coupled integrated amplifiers. A two-stage compound amplifier is shown in Fig. 7.3-1. The operation of this amplifier can be explained qualitatively as follows: An increase in the input current i_i causes an increase in i_{B1}. The emitter current i_{E1} ($= i_{B2}$) increases, resulting in an increase in i_{E2} (and i_{C2}) of approximately $h_{fe1}h_{fe2}i_i$. If the load resistance R_L is much smaller than R_c, the current gain is approximately $h_{fe1}h_{fe2}$.

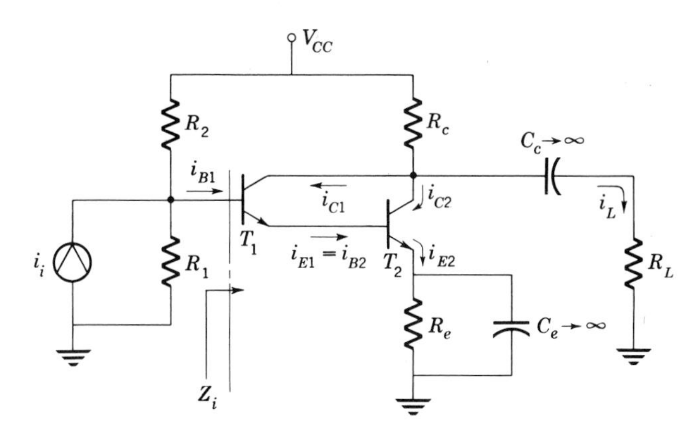

Fig. 7.3-1 Compound amplifier (Darlington).

SETTING THE QUIESCENT POINT

The Q points for each transistor T_1 and T_2 are set in the standard way. Note that

$$V_{CE2} = V_{CE1} + V_{BE2} \approx V_{CE1} + 0.7 \tag{7.3-1}$$

In addition, $I_{C2} \gg I_{C1}$ because $I_{C1} \approx I_{B2}$. The dc load line for T_2 is obtained from

$$\begin{aligned} V_{CC} &\approx V_{CE2} + R_c\left(I_{C2} + \frac{I_{C2}}{h_{fe}}\right) + I_{C2}R_e \\ &\approx V_{CE2} + I_{C2}(R_c + R_e) \end{aligned} \tag{7.3-2}$$

The base voltage

$$V_{B1} \approx 1.4 + I_{C2}R_e \tag{7.3-3}$$

The resistors R_1 and R_2 are chosen so that

$$\frac{R_1V_{CC}}{R_1 + R_2} = V_{B1} \tag{7.3-4}$$

and

$$R_b = \frac{R_1R_2}{R_1 + R_2} \ll h_{fe1}h_{fe2}R_e \tag{7.3-5}$$

The ac load line for T_2 has a slope $\approx 1/(R_L \| R_c)$. This is shown in Fig. 7.3-2.

Q-point placement will depend, as usual, on maximum load swing and gain requirements. Since

$$I_{C1} \approx \frac{I_{C2}}{h_{fe2}}$$

and

$$V_{CE1} \approx V_{CE2} - 0.7$$

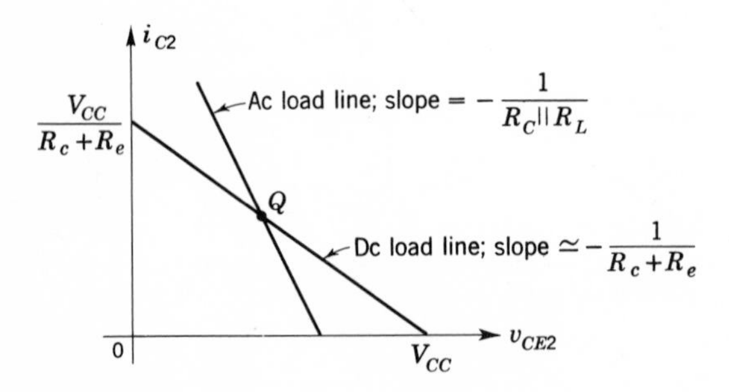

Fig. 7.3-2 Load lines for compound amplifier.

the Q point for T_1 is at

$$V_{CEQ1} \approx V_{CEQ2} - 0.7 \tag{7.3-6}$$

and

$$I_{CQ1} \approx \frac{I_{CQ2}}{h_{fe2}} \tag{7.3-7}$$

SMALL-SIGNAL OPERATION

The small-signal equivalent circuit for the Darlington amplifier is obtained by reflecting the base circuit of T_1 into the emitter circuit, and the emitter circuit of T_2 into its base circuit, as shown in Fig. 7.3-3. The current gain of the amplifier as calculated from this figure is

$$A_i = \frac{i_L}{i_i} = \left(\frac{i_L}{i_{b2}}\right)\left(\frac{i_{b2}}{i_i}\right)$$
$$\approx -\left(\frac{R_c}{R_c + R_L}\right) h_{fe2} \left(\frac{R_b}{R_b/h_{fe1} + h_{ib1} + h_{ie2}}\right) \tag{7.3-8}$$

Note that, since [Eq. (7.3-7)]

$$I_{EQ2} = h_{fe2} I_{EQ1}$$

the transistor impedances are

$$h_{ib1} = \frac{25 \times 10^{-3}}{I_{EQ1}} \tag{7.3-9a}$$

and

$$h_{ie2} = h_{fe2}\left(\frac{25 \times 10^{-3}}{I_{EQ2}}\right) \approx \frac{25 \times 10^{-3}}{I_{EQ1}} = h_{ib1} \tag{7.3-9b}$$

The current gain (7.3-8) therefore becomes

$$A_i = \frac{i_L}{i_i} \approx -h_{fe1} h_{fe2} \left(\frac{R_c}{R_c + R_L}\right)\left(\frac{1}{1 + 2h_{ie1}/R_b}\right) \tag{7.3-10a}$$

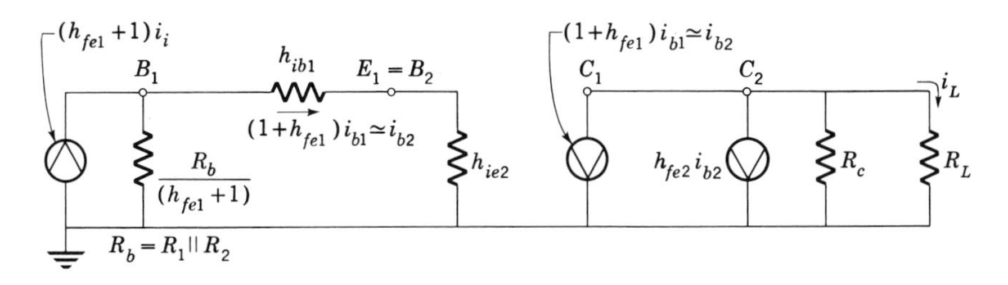

Fig. 7.3-3 Small-signal equivalent of Darlington amplifier.

If the collector resistor R_c is much larger than the load resistor R_L, and if R_b is much greater than $2h_{ie}$, then maximum current gain is obtained. This gain is

$$A_{i,\max} \approx -h_{fe1}h_{fe2} \tag{7.3-10b}$$

and is not often achieved in practice because of the attenuation resulting from practical values of R_b (the attenuation caused by R_c is usually small).

The Darlington configuration can be thought of as a common-emitter amplifier preceded by an emitter follower, so that the input impedance seen looking directly into the base of T_1 is [Eq. (7.3-9b)]

$$Z_i \approx h_{ie1} + h_{fe1}h_{ie2} \approx 2h_{ie1} \tag{7.3-11}$$

Example 7.3-1 The Darlington differential amplifier shown in Fig. 7.3-4 is available as a self-contained integrated circuit. Determine the quiescent operating conditions and the maximum possible output-voltage swing. Assume $h_{fe} = 100$ for all transistors.

Solution Transistors T_3, T_4, and T_5 form a difference amplifier with a constant emitter-current supply. The quiescent current supplied by T_5 is calculated from the circuit of Fig. 7.3-5*a*. Neglecting I_{B5},

$$V_{B5} = (-6)\left(\frac{2.9}{2.9 + 1.3}\right) = -4.14 \text{ V}$$

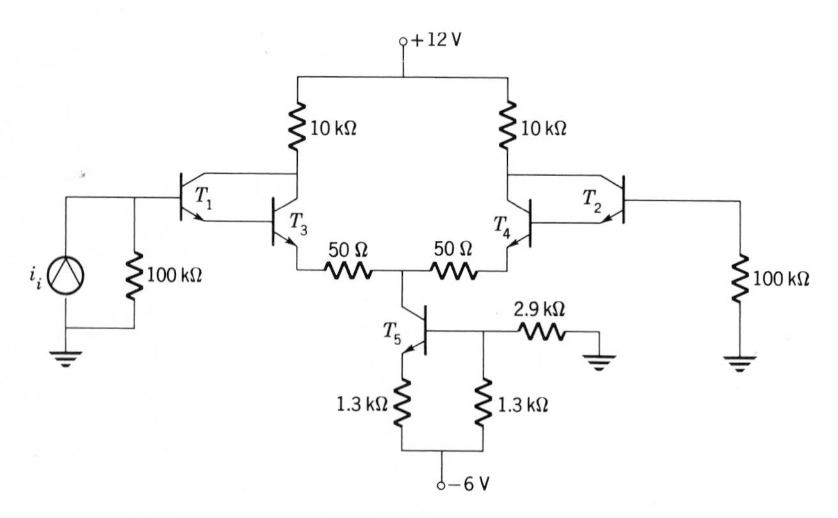

Fig. 7.3-4 Darlington differential amplifier.

The emitter voltage V_{E5} is

$$V_{E5} = V_{B5} - 0.7 = -4.84 \text{ V}$$

Thus

$$I_{CQ5} \approx I_{EQ5} = \frac{6 - 4.84}{1.3 \; k\Omega} \approx 0.9 \text{ mA}$$

It is assumed that this current divides evenly between T_3 and T_4, so that

$$I_{EQ3} = I_{EQ4} \approx 0.45 \text{ mA}$$

To determine the quiescent operating point of T_1 and T_3 (and T_2 and T_4), consider the circuit of Fig. 7.3-5*b*. From the figure

$$I_{EQ1} \approx \frac{I_{EQ3}}{h_{fe}} = \frac{0.45 \text{ mA}}{100} = 4.5 \; \mu\text{A}$$

and

$$I_{BQ1} = \frac{I_{EQ1}}{h_{fe}} = \frac{4.5 \; \mu\text{A}}{100} = 45 \times 10^{-9} \text{ A}$$

Next the collector voltages can be found:

$$V_{CQ1} = V_{CQ3} \approx 12 - (10^4)[(0.45 \times 10^{-3}) + (4.5 \times 10^{-6})] \approx 7.5 \text{ V}$$

$$V_{EQ1} \approx (-10^5)(45 \times 10^{-9}) - 0.7 \approx -0.7 \text{ V}$$

and

$$V_{EQ3} \approx -1.4 \text{ V}$$

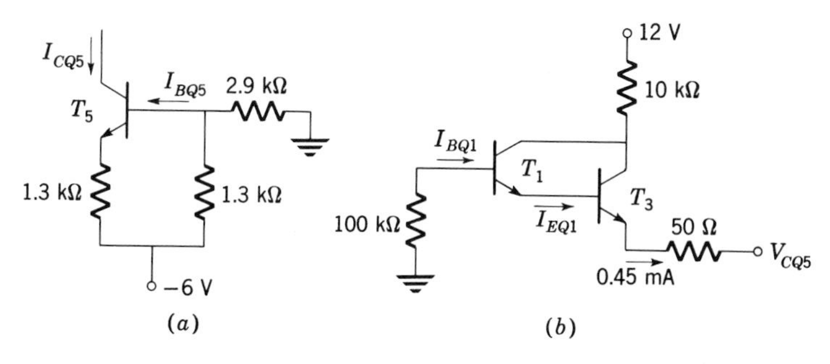

Fig. 7.3-5 Circuits for Example 7.3-1. (*a*) Constant current source for Darlington differential amplifier of Fig. 7.3-4; (*b*) portion of Darlington amplifier of Example 7.3-1.

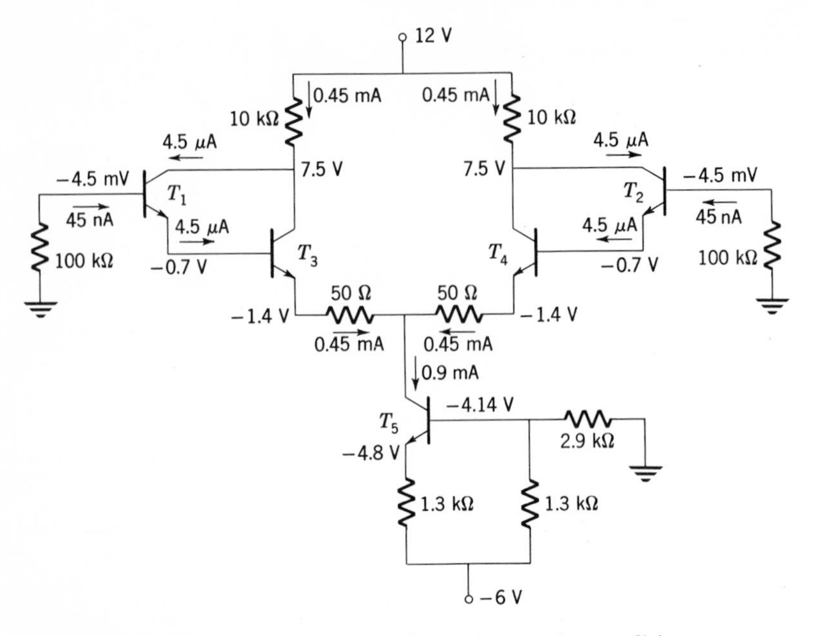

Fig. 7.3-6 Compound difference amplifier with operating conditions.

Hence

$$V_{CQ5} \approx -1.4 - (50)(0.45 \times 10^{-3}) \approx -1.4 \text{ V}$$

The complete circuit with all currents and voltages is shown in Fig. 7.3-6.

The collector can swing symmetrically from 3 to 12 V. Thus the maximum symmetrical output swing is 9 V peak-to-peak if the load resistance (not shown in Fig. 7.3-6) is much greater than 10 kΩ.

It should be noted that the quiescent currents and voltages are essentially *independent* of h_{fe}, and depend primarily on the values of the resistances external to the transistors. This is extremely important, since h_{fe} can vary by 3:1 among transistors of the same type (Chaps. 3 and 4).

This example should be compared with Example 7.1-2, where a similar step-by-step procedure was used to calculate quiescent conditions in a direct-coupled amplifier.

7.4 THE CASCODE AMPLIFIER

Another compound type of transistor connection leads to the *cascode* amplifier, one version of which is shown in Fig. 7.4-1.

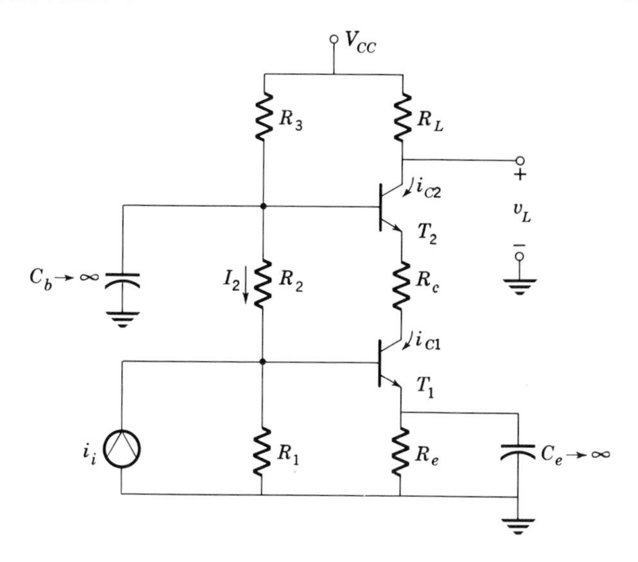

Fig. 7.4-1 Cascode amplifier.

In this version a common-emitter stage is followed by a common-base stage. This configuration has several desirable high-frequency characteristics, which are discussed in Chap. 14.

SMALL-SIGNAL ANALYSIS

The small-signal equivalent circuit is shown in Fig. 7.4-2. The transfer gain $A_T = v_L/i_i$ is often of interest in applications of this circuit. From the figure

$$A_T = \frac{v_L}{i_i} = \left(\frac{v_L}{i_{e2}}\right)\left(\frac{i_{e2}}{i_{b1}}\right)\left(\frac{i_{b1}}{i_i}\right)$$

$$= -h_{fb2}R_L(h_{fe1})\left(\frac{R_1\|R_2}{h_{ie1} + R_1\|R_2}\right) \tag{7.4-1}$$

When

$$h_{ie1} \ll R_1\|R_2$$

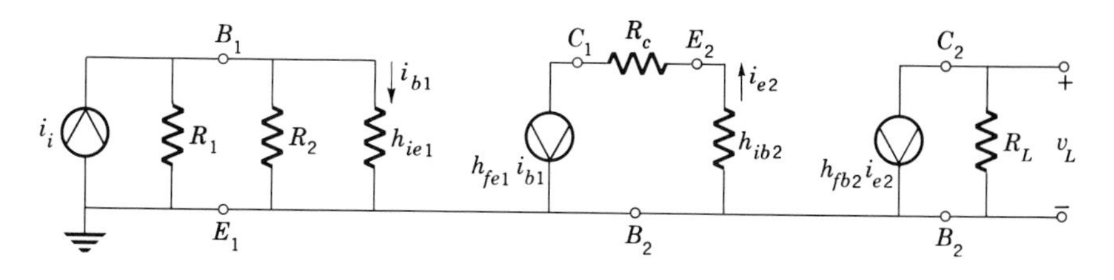

Fig. 7.4-2 Small-signal equivalent circuit of the cascode amplifier.

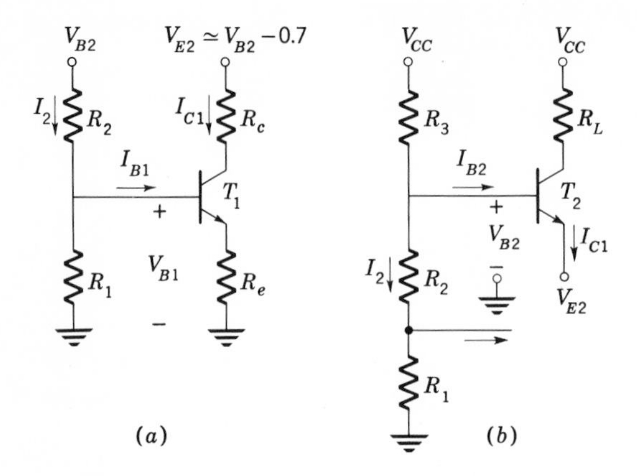

Fig. 7.4-3 Cascode amplifier separated for dc analysis. (*a*) Dc circuit for T_1; (*b*) dc circuit for T_2.

the transfer gain of the amplifier becomes

$$A_T \approx -h_{fe1}R_L \tag{7.4-2}$$

Thus the gain of the amplifier is approximately the same as the gain of the CE stage. An advantage of cascode connection becomes apparent at high frequencies, where the capacitance present between base and collector reduces the gain of the amplifier. Use of the cascode connection significantly reduces the effect of this capacitance and results in a wider-band high-gain amplifier than provided by a CE stage alone (see Chap. 14 for a detailed discussion of this point).

DC ANALYSIS

To determine the Q points of T_1 and T_2, we first separate the stages as shown in Fig. 7.4-3*a* and *b*.

Referring to Fig. 7.4-3*a* and assuming that $I_2 \gg I_{B1}$,

$$V_{B1} \approx \frac{V_{B2}R_1}{R_1 + R_2} \tag{7.4-3}$$

and

$$I_{C1} \approx I_{E1} \approx \frac{V_{B1} - 0.7}{R_e} = \frac{V_{B2}R_1/(R_1 + R_2) - 0.7}{R_e} \tag{7.4-4}$$

The collector-to-emitter voltage of T_1 is

$$\begin{aligned} V_{CE1} &\approx V_{B2} - 0.7 - R_cI_{C1} - R_eI_{C1} \\ &= V_{B2} - 0.7 - \left(\frac{R_c + R_e}{R_e}\right)\left[V_{B2}\left(\frac{R_1}{R_1 + R_2}\right) - 0.7\right] \end{aligned} \tag{7.4-5}$$

The quiescent point of T_2 is found using Fig. 7.4-3*b*. Again assuming $I_2 \gg I_{B1}$ and $I_2 \gg I_{B2}$,

$$V_{B2} \approx \frac{V_{CC}(R_1 + R_2)}{R_1 + R_2 + R_3} \tag{7.4-6}$$

and

$$\begin{aligned} V_{CE2} &\approx V_{CC} - I_{C1}R_L - V_{E2} \\ &\approx V_{CC} - I_{C1}R_L - V_{B2} + 0.7 \end{aligned} \tag{7.4-7}$$

Example 7.4-1 In the circuit of Fig. 7.4-1, let $V_{CC} = 20$ V, $R_e = 1$ kΩ, $R_c = 500$ Ω, $R_L = 1$ kΩ, $R_1 = 10$ kΩ, $R_2 = 10$ kΩ, and $R_3 = 10$ kΩ. Find V_{CEQ1}, V_{CEQ2}, I_{CQ1}, and I_{CQ2}.

Solution Using (7.4-6),

$$V_{BQ2} \approx \frac{(20)(20 \times 10^3)}{30 \times 10^3} \approx 13.3 \text{ V}$$

From (7.4-3)

$$V_{BQ1} \approx \frac{(13.3)(10^4)}{2 \times 10^4} \approx 6.7 \text{ V}$$

The collector current I_{CQ1} is then found from (7.4-4).

$$I_{CQ1} = \frac{6.7 - 0.7}{1\ k\Omega} = 6 \text{ mA} \approx I_{CQ2}$$

The collector-to-emitter voltages can now be calculated. Using (7.4-5),

$$V_{CEQ1} \approx 13.3 - 0.7 - (1.5 \times 10^3)(6 \times 10^{-3}) = 3.6 \text{ V}$$

and using (7.4-7),

$$V_{CEQ2} \approx 20 - (10^3)(6 \times 10^{-3}) - (13.3 - 0.7) \approx 1.4 \text{ V}$$

Example 7.4-2 Another version of the cascode amplifier is used to provide *level shifting* in direct-coupled amplifiers. The circuit shown in Fig. 7.4-4 is a level-shifting configuration commonly used in integrated-circuit amplifiers. T_1 is an ordinary common-emitter amplifier. Transistors T_2 and T_3 form a cascode amplifier which shifts the dc signal voltage from the collector of T_1 to ground without introducing much attenuation. Thus the dc level at the output is zero, and the peak value of v_L is approximately equal to the peak value of v_{c1}. To demonstrate this, we must calculate the quiescent points of T_2 and T_3 and the small-signal voltage gain from the collector of T_1 to the output.

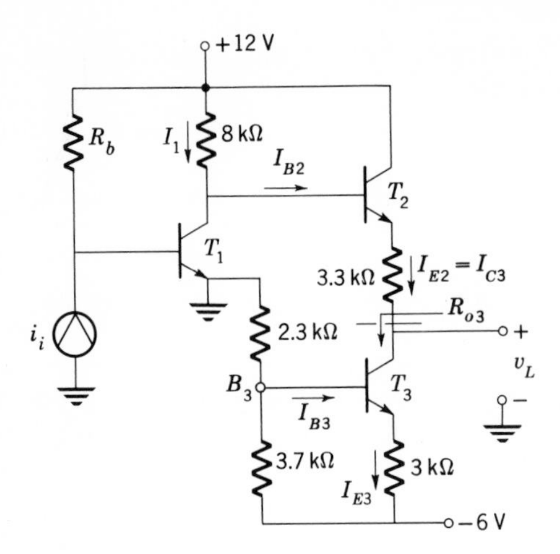

Fig. 7.4-4 Cascode amplifier used for level shifting.

Solution

QUIESCENT-POINT ANALYSIS The calculation is begun by determining V_{B3} and then V_{E3}. Neglecting I_{B3},

$$V_{B3} = \left(-6\right)\left(\frac{2.3}{2.3 + 3.7}\right) = -2.3 \text{ V}$$

and

$$V_{E3} \approx -2.3 - 0.7 = -3 \text{ V}$$

Hence

$$I_{E3} = \frac{3}{3 \times 10^{+3}} = 1 \text{ mA} \approx I_{C3}$$

Now the purpose of the circuit is to make $V_L = 0$. With this condition

$$V_{E2} = (3.3 \text{ k}\Omega)I_{C3} + 0 = 3.3 \text{ V}$$

and

$$V_{B2} = V_{BE2} + V_{E2} = 0.7 + 3.3 = 4 \text{ V}$$

Thus the current I_1 in the 8-kΩ load of T_1 must be adjusted to 1 mA. Since I_{B2} is much less than 1 mA, $I_{C1} \approx 1$ mA. R_b is then chosen to provide a base current of

$$I_{B1} = \frac{1 \text{ mA}}{h_{fe1}}$$

SMALL-SIGNAL ANALYSIS The small-signal equivalent circuit from which the gain v_L/v_{c1} can be obtained is shown in Fig. 7.4-5.

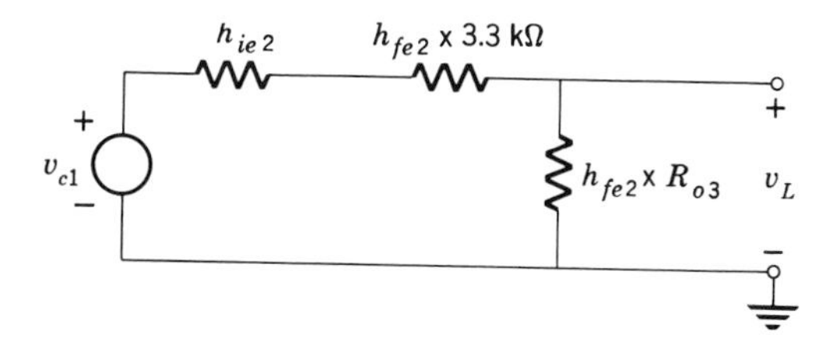

Fig. 7.4-5 Small-signal equivalent for level-shifting amplifier.

In this circuit R_{o3} is the impedance seen looking into the collector of T_3, and will be of the order of $1/h_{ob3}$. Thus

$$h_{fe2}R_{o3} \approx \frac{h_{fe2}}{h_{ob3}} \gg h_{ie2} + 3300\ h_{fe2}$$

The voltage gain A_v is

$$A_v = \frac{v_L}{v_{c1}} \approx 1$$

Thus, using the cascode configuration enabled us to shift the dc voltage from V_{C1} $(= V_{B2} = 4\text{ V})$ to V_L $(= 0\text{ V})$ with negligible loss in gain.

PROBLEMS

SECTION 7.1

7.1. (a) Find $A_i = i_L/i_i$.
(b) Find the maximum possible symmetrical swing in i_L.

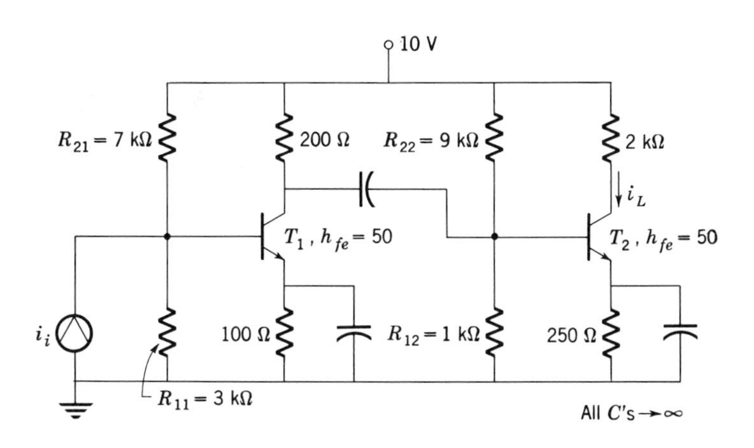

Fig. P7.1

7.2. In the circuit of Fig. P7.1 find new values for R_{11}, R_{21}, R_{12}, and R_{22} so as to permit maximum possible swing in i_L. Find A_i under these conditions.

7.3. (*a*) Find R_{11}, R_{21}, R_{12}, and R_{22} so that v_L can undergo maximum swing.
(*b*) Find v_L/i_i.

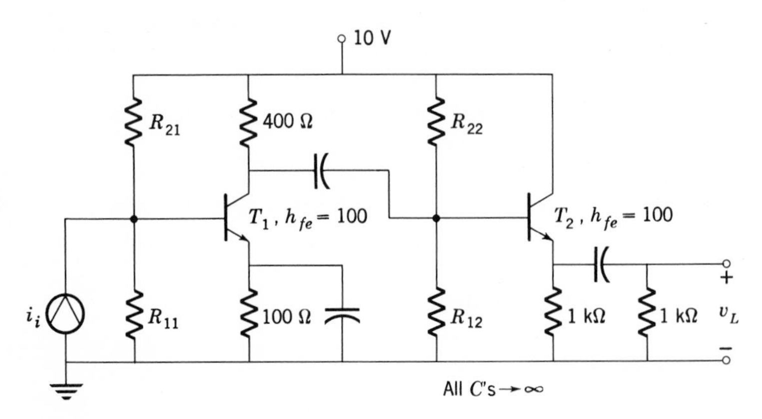

Fig. P7.3

7.4. (*a*) Find R so that $v_{o1}/i_i = -v_{o2}/i_i$.
(*b*) Find v_{o1}/i_i.

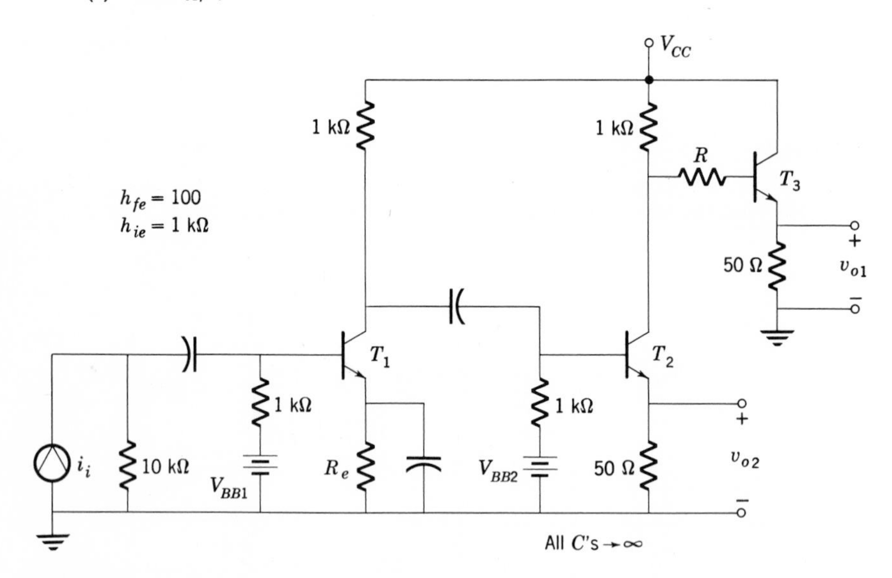

Fig. P7.4

7.5. Design a two-stage amplifier to meet the following specifications:
(*a*) $A_i = 200$
(*b*) $Z_i \geq 1$ kΩ
(*c*) $R_L = 10$ Ω ac-coupled

(*d*) Maximum peak swing in $R_L = 50$ mA
(*e*) $V_{CC} = 10$ V

Transistors with $h_{fe} = 50$ are to be used.

Be sure to specify all resistors and transistor ratings.

7.6. Verify Eq. (7.1-8) by solving (7.1-5) to (7.1-7) for I_{C1}.

7.7. For the circuit of Fig. 7.1-7 find $\Delta I_{C1}/\Delta T$ and $\Delta I_{C2}/\Delta T$ if the feedback connection is open-circuited and T_1 is biased at the same current as in Example 7.1-2. Compare with the result of Example 7.1-2.

7.8. Availability of both *npn* and *pnp* transistors opens up many interesting possibilities for cascaded dc amplifiers.

Calculate the small-signal gain of the amplifier shown in Fig. P7.8.

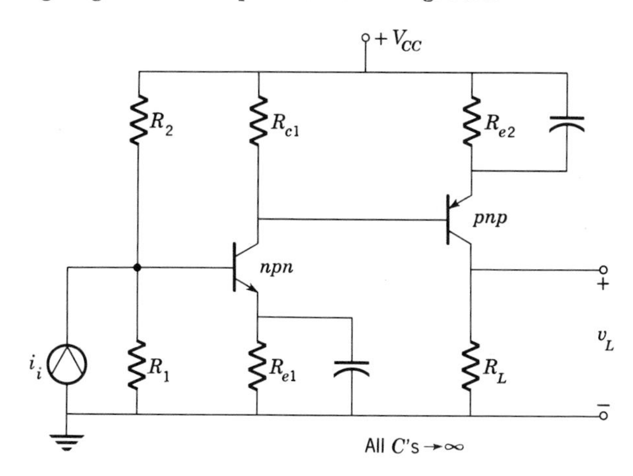

Fig. P7.8

7.9. (*a*) Find all quiescent conditions in the circuit of Fig. P7.9. $h_{fe1} = h_{fe2} = 100$.
(*b*) Calculate $\Delta v_L/\Delta T$.

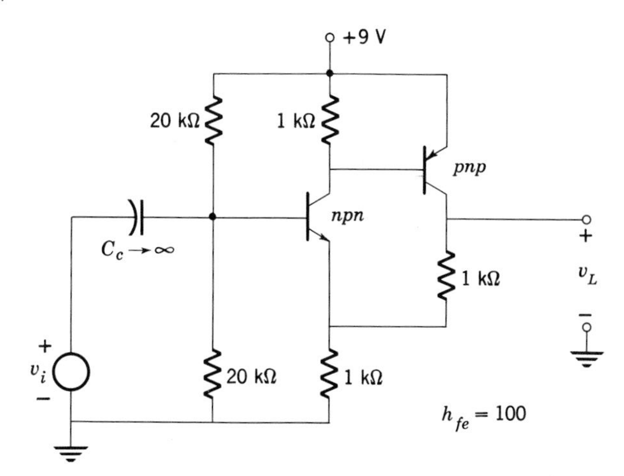

Fig. P7.9

SECTION 7.2

7.10. Verify Eq. (7.2-6) by applying standard network analysis techniques to the circuit of Fig. 7.2-6.

7.11. Find the small-signal equivalent circuit for the difference amplifier of Fig. 7.2-1 by reflecting the appropriate elements into the base circuits. Analyze the resulting circuit to verify (7.2-6).

7.12. Verify Eq. (7.2-21).

7.13. Find i_L in terms of the common- and differential-mode signals.

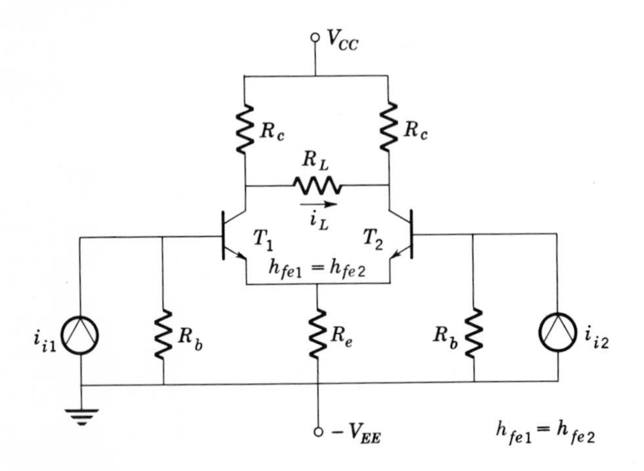

Fig. P7.13

7.14. T_1 and T_2 are not identical. T_1 has h_{fe1} and h_{ie1}, and T_2 has h_{fe2} and h_{ie2}.

(*a*) Find i_L.

(*b*) Find values for R_1 and R_2 which will balance the amplifier and maximize the CMRR.

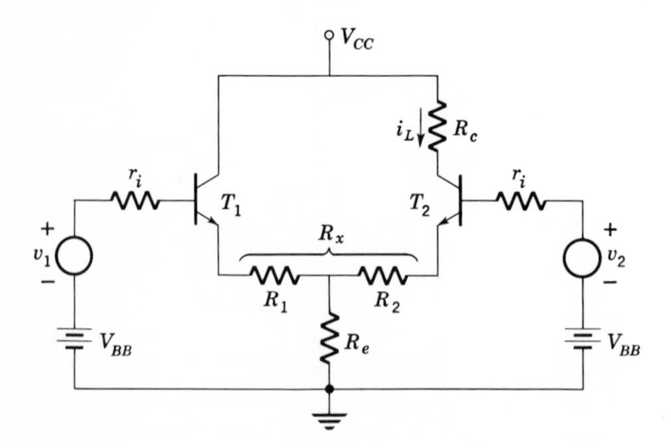

Fig. P7.14

7.15. Design a difference amplifier as in Fig. 7.2-1 to have a CMRR of 40 dB. The load resistance is 100 Ω ac-coupled, and the voltage sources each have 1-kΩ internal resistance. Use 2N3904 transistors (Appendix III) and two power supplies, one positive and one negative.

7.16. Find $\Delta I_{c1}/\Delta T$ and $\Delta I_{c2}/\Delta T$ for the basic difference amplifier of Fig. 7.2-1. Assume that $I_{CBO} = 0$ and only V_{BE} varies with temperature ($\Delta V_{BE}/\Delta T = -2.5$ mV/°C).

7.17. Repeat Prob. 7.16 if $R_{c1} = 0$.

7.18. Find the common-mode gain for the difference amplifier with a balance control [see Eq. (7.2-21) and Fig. 7.2-9].

7.19. In Example 7.2-3 draw the small-signal equivalent circuit and verify the first equation in the solution.

SECTION 7.3

7.20. Find the hybrid equivalent (h_i, h_o, and g) circuit for the compound transistor connection of Fig. P7.20. Use the simplified hybrid equivalent for each transistor, but do not assume identical parameters.

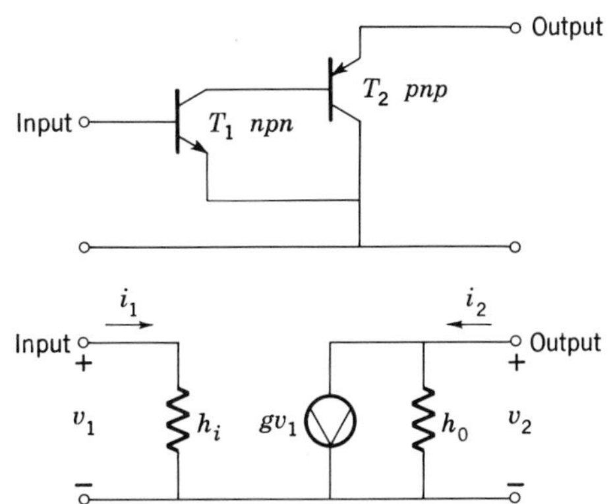

Fig. P7.20

7.21. Repeat Prob. 7.20 for the parallel connection of Fig. P7.21. Assume identical transistors, and comment on the possible advantages of this circuit.

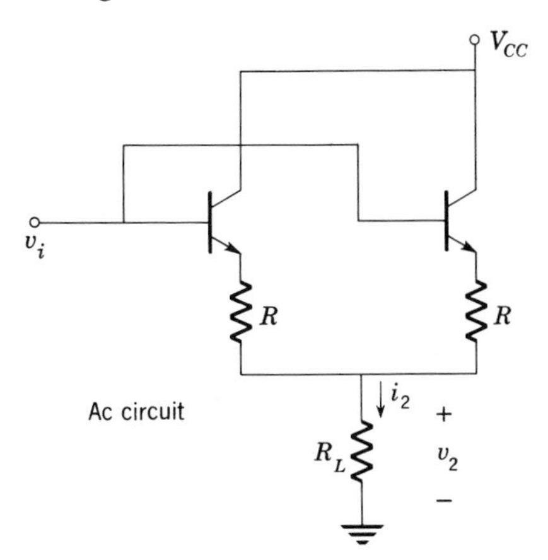

Fig. P7.21

7.22. Repeat Prob. 7.21 for the series connection of Fig. P7.22. Assume $h_{oe} = 0$.

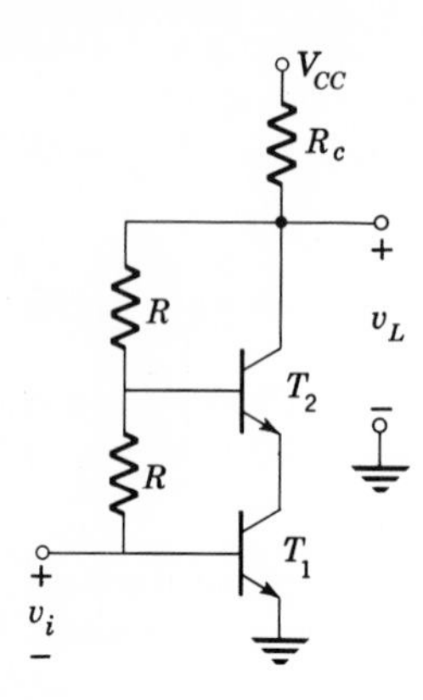

Fig. P7.22

7.23. (*a*) Find quiescent conditions throughout the circuit.
(*b*) Find $A_i = i_L/i_i$.

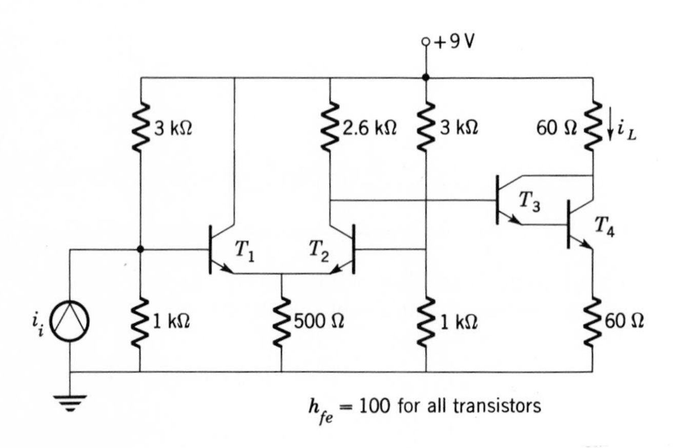

Fig. P7.23

7.24. Find
(*a*) Quiescent conditions.
(*b*) v_L/i_i
(*c*) Z_i
(*d*) Z_o

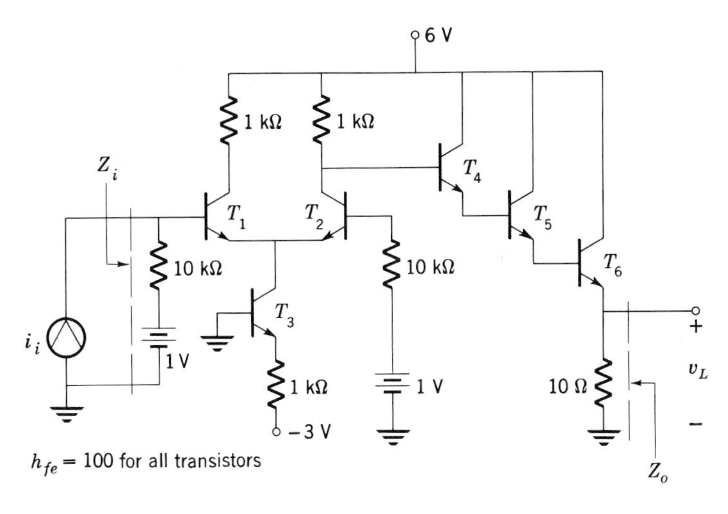

Fig. P7.24

SECTION 7.4

7.25. (*a*) Find R_1, R_2, R_c, R_e, R_b, and R_L for maximum swing in v_L and maximum gain v_L/i_i.

(*b*) Calculate v_L/i_i.

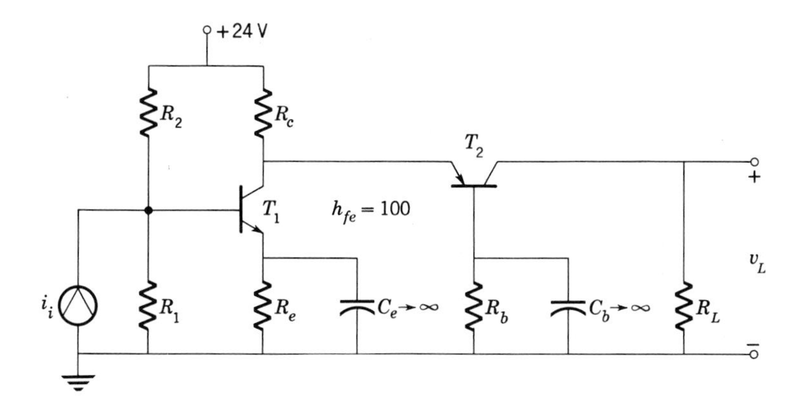

Fig. P7.25

REFERENCE

1. Thornton, R. D., et al.: "Multistage Transistor Circuits," vol. 5, John Wiley & Sons, Inc., New York, 1965.

8
Feedback-amplifier Fundamentals

INTRODUCTION

An ideal linear amplifier in the midfrequency region, where all capacitance effects can be neglected, provides an output signal which is an exact replica of the input signal, except for a scale factor, representing the degree of amplification, and a phase shift. Practical amplifiers depart from the ideal for many reasons, among them nonlinearity of transistor characteristics, parameter variations, temperature effects, etc. Some of these imperfections can be minimized by improving the basic devices involved, but often the point of diminishing returns is rapidly reached. Usually, the error-correcting properties of feedback are utilized to achieve a desired degree of improvement. In a feedback amplifier, a signal, derived from the output signal, is added to (or subtracted from) the input signal in such a way that the operation of the amplifier automatically corrects, to some degree, departures from the ideal.

The study of feedback is recognized as being applicable to such diverse fields as economics and biology, and the literature on the subject is vast. In this chapter we study the application of feedback to transistor amplifiers and its effect on gain, impedance levels, and sensitivity to parameter variation.

8.1 BASIC CONCEPTS OF FEEDBACK

Feedback can be applied to amplifiers in many ways. The resulting circuits can be analyzed using standard loop and node equations, without even utilizing the fact that feedback is present. However, we are analyzing circuits mainly to provide an insight into their design. Thus the properties of feedback are of great importance to us, and the analysis techniques to be presented place in evidence the actual feedback mechanism. This chapter considers the midfrequency response of feedback amplifiers. The frequency response of feedback amplifiers is postponed until Chap. 15.

There are four basic types of feedback circuits, as shown in Fig. 8.1-1. These circuits each contain

1. An amplifier having an input resistance R_i, an output resistance r_o, and a current gain A_1' (or a voltage gain A_v').
2. A feedback network β.* This network may contain active elements (transistors, etc.) or may be completely passive (resistors, capacitors, inductors). It is represented as a simplified hybrid equivalent circuit for purposes of calculation; i.e., it has an input impedance R_β, an output impedance R_f, and a controlled current or voltage source. It is thus, essentially, the same as the hybrid transistor equivalent-circuit model of Sec. 6.1.
3. A difference or summing circuit. This circuit will subtract (or add) voltages or currents, depending on the design of the amplifier.

To compare the feedback amplifier with the amplifier without feedback, we define the following types of gain:

1. The external, or overall, gain of the feedback amplifier
   a. Current gain, $A_{if} = i_L/i_i$ (Fig. 8.1-1*a* and *c*)
   b. Voltage gain, $A_{vf} = v_L/v_i$ (Fig. 8.1-1*b* and *d*)
2. The forward, or internal, amplifier gain, i.e., the gain of the amplifier without feedback
   a. Current gain, $A_i = A_{if}\Big|_{G_i=0 \text{ or } K_i=0}$ (Fig. 8.1-1*a* and *c*)
   b. Voltage gain, $A_v = A_{vf}\Big|_{K_v=0 \text{ or } R_\beta=0}$ (Fig. 8.1-1*b* and *d*)

* This β should not be confused with the $\beta = h_{FE}$ of the transistor.

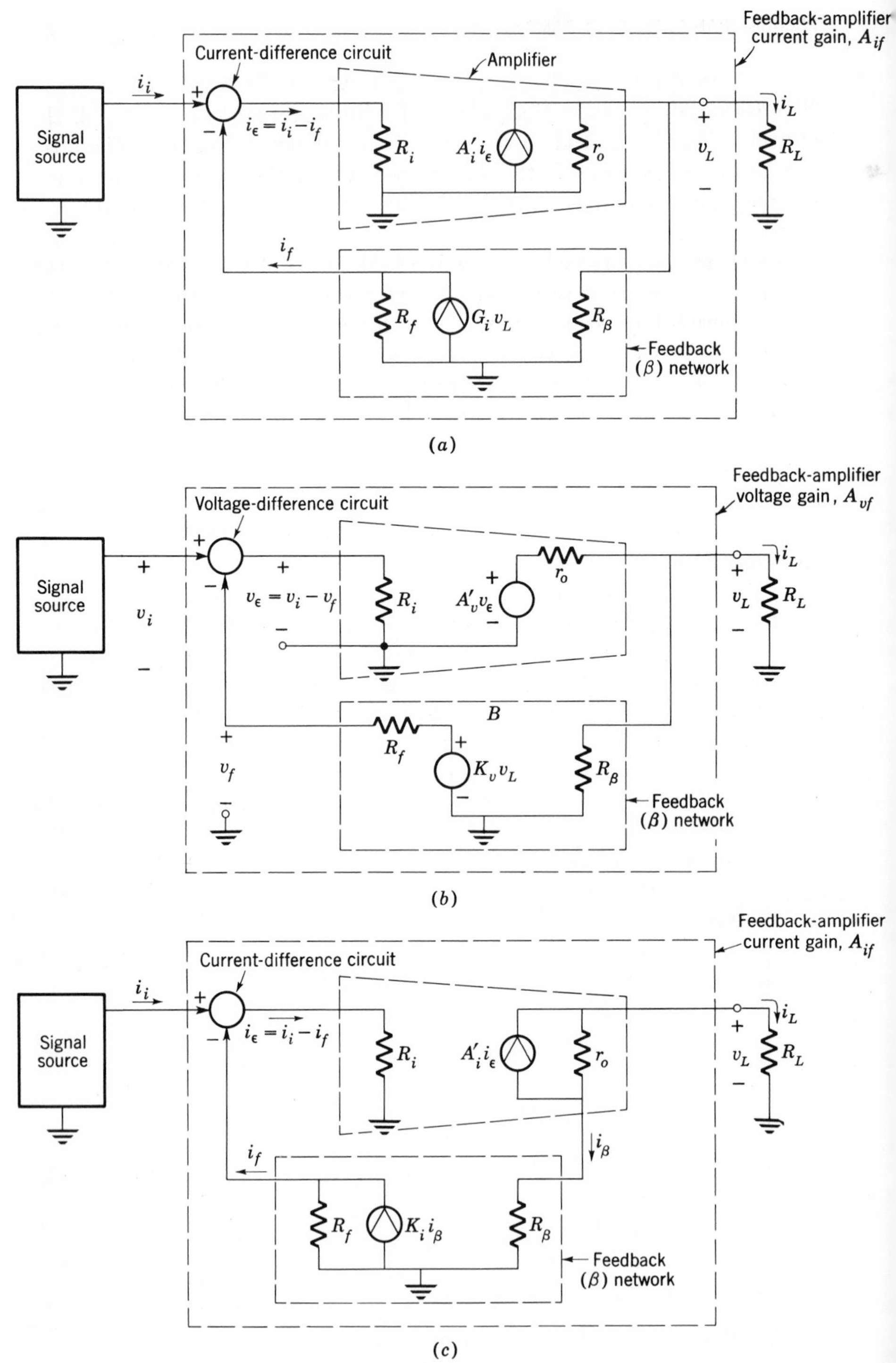
Feedback-amplifier current gain, A_{if}
Current-difference circuit
Amplifier
i_i
Signal source
$i_\epsilon = i_i - i_f$
R_i
$A'_i i_\epsilon$
r_o
v_L
i_L
R_L
i_f
R_f
$G_i v_L$
R_β
Feedback (β) network
(a)
Feedback-amplifier voltage gain, A_{vf}
Voltage-difference circuit
Signal source
v_i
$v_\epsilon = v_i - v_f$
R_i
$A'_v v_\epsilon$
r_o
v_L
i_L
R_L
B
R_f
$K_v v_L$
R_β
v_f
Feedback (β) network
(b)
Feedback-amplifier current gain, A_{if}
Current-difference circuit
i_i
Signal source
$i_\epsilon = i_i - i_f$
R_i
$A'_i i_\epsilon$
r_o
v_L
i_L
R_L
i_β
i_f
R_f
$K_i i_\beta$
R_β
Feedback (β) network
(c)

Figure 8.1-1*a* is a circuit which has *voltage feedback* and *current error*. This terminology means that, via the feedback network, a fraction of the output *voltage* v_L is used to generate the feedback *current* i_f. The *error* (or *difference*) *current* i_ϵ is the difference between the input current i_i and the feedback current i_f. As the amount of feedback (the proportion of v_L used to generate i_f) is increased, the error current i_ϵ is decreased. This is called *negative feedback*. Negative voltage feedback implies that the output *voltage* is used to reduce the error (in this case, a *current error*).

Figure 8.1-1*b* is similar, in that a fraction of v_L is used to generate a feedback *voltage* v_f. However, in this case, the error is a *voltage error*, and v_ϵ is the difference between the input signal v_i and the feedback signal v_f. Increasing the fraction of v_L fed back to the input decreases the error voltage v_ϵ.

In the circuit of Fig. 8.1-1*c*, a part of the load current i_L is fed back to the input as a feedback current i_f. Thus this amplifier is said to have *current feedback* with a *current error*. Similarly, Fig. 8.1-1*d* shows an amplifier with *current feedback* and *voltage error*.

The remainder of this chapter deals with the gain, the sensitivity, and the impedance levels of voltage-feedback amplifiers, the type most often used in practice. The techniques used in the analysis and design of voltage-feedback amplifiers can also be used for current feedback amplifiers.[1]

When feedback is used to reduce the sensitivity of the gain of the

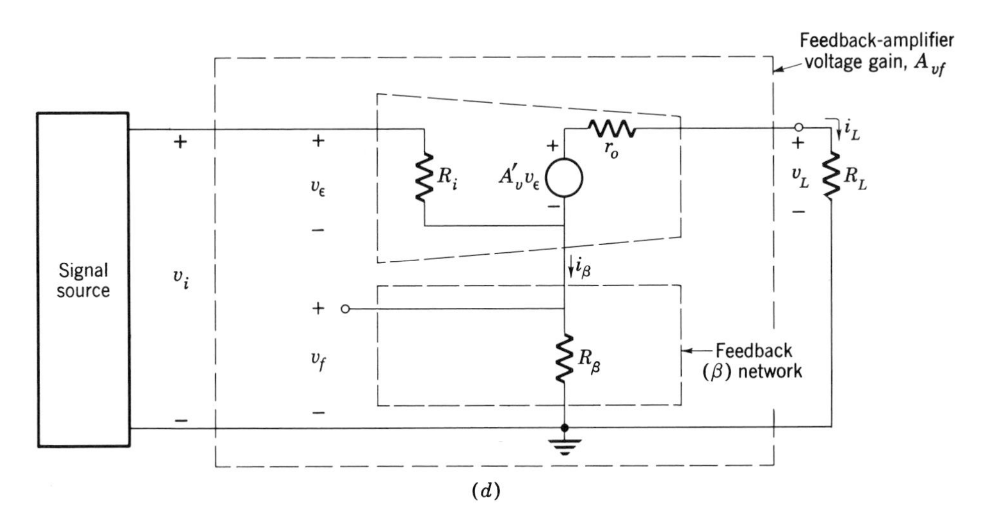

Fig. 8.1-1 Basic feedback configurations. (*a*) Voltage feedback–current error; (*b*) voltage feedback–voltage error; (*c*) current feedback–current error; (*d*) current feedback–voltage error.

feedback amplifier (A_{vf} or A_{if}) to amplifier parameter variations, such as h_{fe} or supply voltage variations, the *gain* of the feedback amplifier is *decreased*, and the feedback is said to be negative. If an increased gain is desired, *positive* feedback can be used. This results in an error signal which is larger than the input signal, i_i (or v_i) Positive feedback increases the sensitivity of the gain of the feedback amplifier to disturbances.*

8.2 THE GAIN OF A VOLTAGE–FEEDBACK AMPLIFIER

8.2-1 VOLTAGE FEEDBACK WITH CURRENT ERROR

In order to determine the current gain A_{if} of this type of feedback amplifier, redraw Fig. 8.1-1*a* as shown in Fig. 8.2-1. Note that the input impedance of the feedback network is in parallel with R_L. Usually, the feedback network is designed so that it does not load the output, that is, $R_\beta \gg R_L$. This represents proper design procedure. In this text, R_β is always assumed much larger than R_L, and hence neglected. Its effect is considered in the problems.

It is also useful to designate the input voltage to the feedback network as v'_L rather than v_L, for reasons which will become clear shortly. When calculating the amplifier response with the feedback network connected, we simply set $v'_L = v_L$.

The Norton equivalent circuit of the feedback network and amplifier are shown in Fig. 8.2-2. If the impedance seen by the feedback

* Because of the presence of capacitive elements, a feedback amplifier can have negative feedback at some frequencies and positive feedback at other frequencies. This is discussed in Chap. 15.

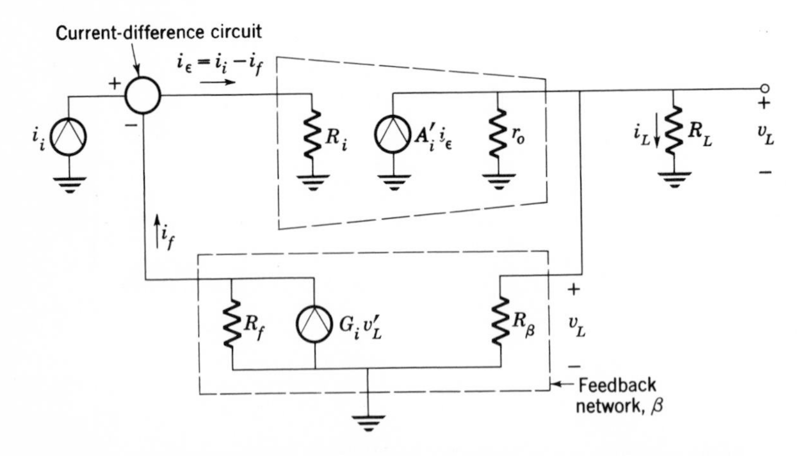

Fig. 8.2-1 Voltage-feedback–current-error configuration with the feedback network isolated from the output.

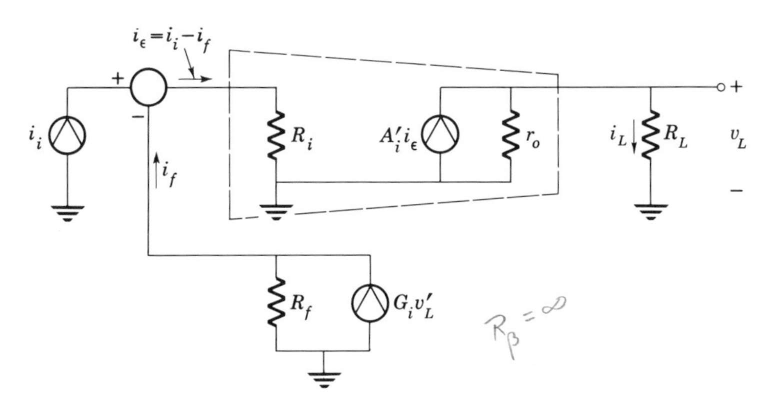

Fig. 8.2-2 The feedback network of Fig. 8.2-1 replaced by its Norton equivalent, assuming $R_\beta \to \infty$.

network, when looking into the difference circuit, is much less than R_f, then the feedback current i_f is

$$i_f = G_i v'_L \tag{8.2-1}$$

The error current (assuming $R_f \gg R_i$) is

$$i_\epsilon = i_i - i_f = i_i - G_i v'_L \tag{8.2-2}$$

Hence the load current is (for $R_L \ll r_o$)

$$i_L = A'_i i_\epsilon = A'_i(i_i - G_i v'_L) \tag{8.2-3a}$$

Letting $v'_L = v_L$, in order to find the overall current gain,

$$v'_L = v_L = i_L R_L \tag{8.2-3b}$$

and the overall current gain A_{if} is obtained by substituting (8.2-3*b*) into (8.2-3*a*).

$$A_{if} = \frac{i_L}{i_i} = \frac{A'_i}{1 + A'_i G_i R_L} \tag{8.2-4a}$$

In this example, the forward current gain (the current gain without feedback) is

$$A_i = A_{if}\Big|_{G_i=0} = A'_i \tag{8.2-4b}$$

Thus

$$A_{if} = \frac{A_i}{1 + A_i G_i R_L} = \left(\frac{1}{G_i R_L}\right)\left(\frac{1}{1 + 1/A_i G_i R_L}\right) \tag{8.2-4c}$$

As long as $A_i G_i R_L > 0$, the forward current gain is reduced by the use of feedback. In addition, if

$$A_i G_i R_L \gg 1$$

the gain with feedback becomes

$$A_{if} \approx \frac{1}{G_i R_L} \tag{8.2-5}$$

When (8.2-5) is valid, the gain with feedback A_{if} is *independent* of the forward current gain A_i, and hence independent of variations in A_i. Feedback amplifiers are usually designed so that this relation holds.

Loop gain At this point it is useful to introduce the concept of *loop gain.* The loop gain T is the voltage gain measured around the complete feedback circuit—from v'_L, through the feedback network, then through the amplifier to v_L (Fig. 8.2-2). The loop gain is defined as

$$T = \frac{v_L}{v'_L}\bigg|_{i_i=0} \tag{8.2-6}$$

For the circuit of Fig. 8.2-2, with $i_i = 0$, and assuming R_f is infinite,

$$i_\epsilon = -i_f \approx -G_i v'_L$$

and

$$v_L = i_L R_L = A'_i i_\epsilon R_L = -A'_i G_i R_L v'_L \tag{8.2-7}$$

Therefore

$$T = \frac{v_L}{v'_L}\bigg|_{i_i=0} = -A'_i G_i R_L = -A_i G_i R_L \tag{8.2-8}$$

since $A_i = A'_i$ for this amplifier.

Comparing (8.2-8) and (8.2-4), the current gain with feedback A_{if} can be expressed as

$$A_{if} = \frac{A_i}{1 - T} \tag{8.2-9}$$

Equation (8.2-9) is a general expression for the current gain with feedback. The "magic number" $1 - T$ will be seen to alter, not only the gain, but also the input and output impedances and the sensitivity.

Stability When the loop gain T is negative, we say that the feedback is *negative,* and when the loop gain is positive, the feedback is *positive.* Equation (8.2-9) indicates that A_{if} becomes infinite when the feedback is positive and $T = +1$. Under these conditions, the feedback amplifier will be unstable and will oscillate, since any small input signal (or disturbance) would be amplified until limited by the nonlinear characteristics (cutoff and saturation, for example) of the device. A loop gain $T \geq +1$ is avoided in amplifier design. However, a loop gain of about $+\frac{1}{2}$ is sometimes used to obtain increased amplifier gain.* A discussion of stability and instability (oscillations) is presented in Chap. 15.

Example 8.2-1 Find A_{if} and A_{vf} for the amplifier of Fig. 8.2-3. This is a voltage-feedback–current-error circuit. The signal current and feedback current are not subtracted as in Fig. 8.2-2, but are added, since they are connected together at the amplifier input. To obtain the difference signal, an inversion is provided in the high-gain amplifier.

Solution This circuit can be redrawn as shown in Fig. 8.2-4*a*. The β network in this figure is slightly more complicated than the β network shown in Fig. 8.2-1, due to the presence of the voltage source $v_1 = i_\epsilon R_i$. It is convenient to convert this voltage source to a current source $(R_i/R_f)i_\epsilon$ as shown in Fig. 8.2-4*b*. In practical feedback amplifiers, it is usually true that $A_i' \gg R_i/R_f$ and $R_f \gg R_L$. These assumptions will be made throughout this chapter.

With these assumptions, and using Fig. 8.2-4*b*, the error current is found to be

$$i_\epsilon \approx \left[\frac{r_i \| R_f}{R_i + (r_i \| R_f)}\right]\left(i_i + \frac{v_L'}{R_f}\right)$$

Then

$$i_L \approx -A_i' i_\epsilon = -A_i' \left[\frac{r_i \| R_f}{R_i + (r_i \| R_f)}\right]\left(i_i + \frac{v_L'}{R_f}\right)$$

* See, for example, the "bootstrap" circuit of Sec. 9.8.

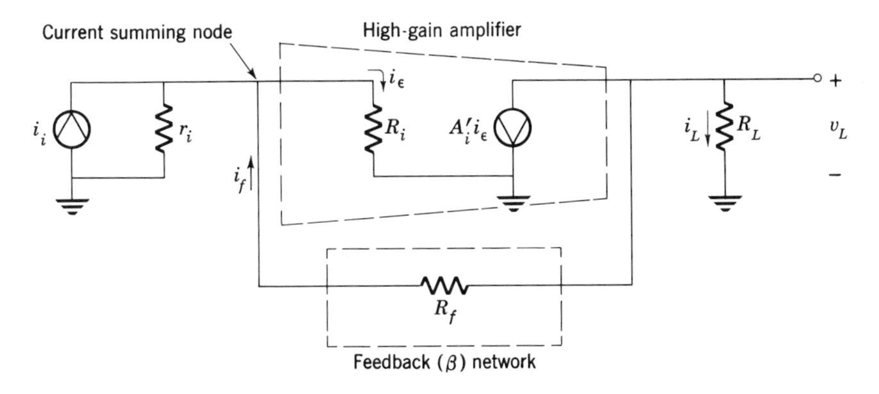

Fig. 8.2-3 Voltage-feedback–current-error amplifier.

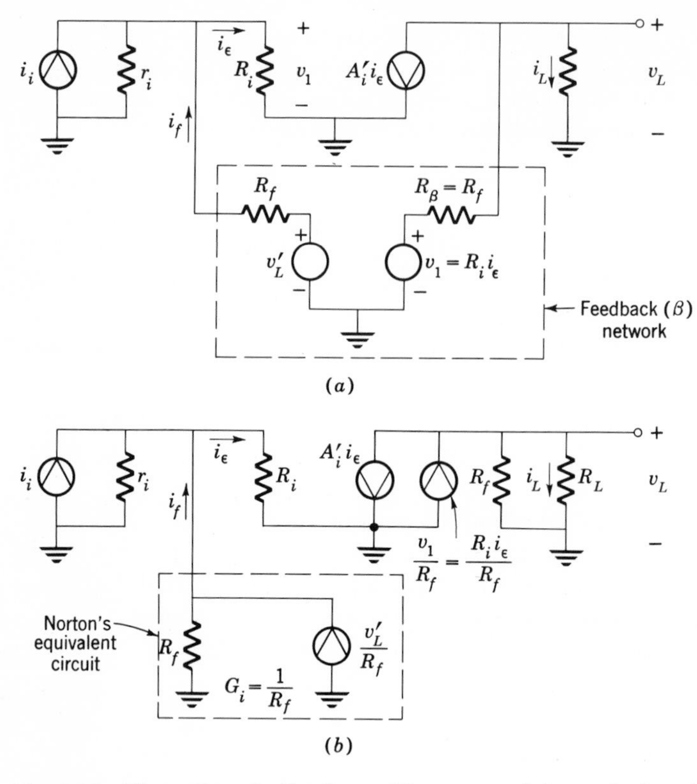

Fig. 8.2-4 The voltage-feedback amplifier prepared for analysis. (*a*) Complete equivalent circuit; (*b*) current-source equivalent. Typically: $A_i' \gg R_i/R_f$, $R_f \gg R_L$.

Since

$$v_L' = v_L = i_L R_L$$

the gain with feedback is

$$A_{if} = \frac{i_L}{i_i} = \frac{-A_i'\left[\dfrac{r_i\|R_f}{R_i + (r_i\|R_f)}\right]}{1 + A_i'\left[\dfrac{r_i\|R_f}{R_i + (r_i\|R_f)}\right]\left(\dfrac{R_L}{R_f}\right)}$$

Comparing this equation with (8.2-9) we find

$$A_i = -A_i'\left[\frac{r_i\|R_f}{R_i + (r_i\|R_f)}\right]$$

and

$$T = -A_i'\left[\frac{r_i\|R_f}{R_i + (r_i\|R_f)}\right]\left(\frac{R_L}{R_f}\right)$$

Let us calculate A_i and T directly. From the definition of T, given by (8.2-6), and Fig. 8.2-4b with $i_i = 0$,

$$i_\epsilon = \left[\frac{r_i \| R_f}{R_i + (r_i \| R_f)}\right]\left(\frac{v'_L}{R_f}\right)$$

and

$$v_L = -A'_i i_\epsilon R_L = -A'_i \left[\frac{r_i \| R_f}{R_i + (r_i \| R_f)}\right]\left(\frac{R_L}{R_f}\right) v'_L$$

The loop gain T becomes

$$T = \frac{v_L}{v'_L}\bigg|_{i_i=0} = -A'_i \left[\frac{r_i \| R_f}{R_i + (r_i \| R_f)}\right]\left(\frac{R_L}{R_f}\right)$$

as expected.

In order to calculate the current gain without feedback A_i, refer to (8.2-4), from which it is seen that A_i equals the overall current gain with feedback A_{if} when the feedback transconductance G_i is zero. Thus A_i should be thought of as the gain *without* feedback, but with the feedback network impedance R_f in place. This can also be obtained by setting $v'_L = 0$ (see Fig. 8.2-2). In either case the feedback network current $G_i v'_L$ is reduced to zero, the output resistance R_f of the feedback network is left unchanged, and A_i is

$$A_i = A_{if}\bigg|_{v'_L=0}$$

In this example, when the feedback-current source is open-circuited so as to supply no feedback current (Fig. 8.2-4), the current gain without feedback is

$$A_i = A_{if}\bigg|_{v'_L=0} = \frac{i_L}{i_i}\bigg|_{v'_L=0} = \left(\frac{i_\epsilon}{i_i}\right)\left(\frac{i_L}{i_\epsilon}\right)\bigg|_{v'_L=0} = -A'_i \left[\frac{r_i \| R_f}{R_i + (r_i \| R_f)}\right]$$

Notice that although the output voltage is prevented from feeding back to the input, the loading effect of the feedback network on the input circuit is included.

If $-T \gg 1$, the current gain with feedback is

$$A_{if} = \frac{i_L}{i_i} \approx \frac{-R_f}{R_L}$$

If the voltage gain A_{vf} is defined to be

$$A_{vf} = \frac{v_L}{v_i}$$

where $v_L = i_L R_L$

$$v_i = i_i r_i$$

(note that v_i is the Thévenin source voltage, *not* the voltage across R_i), then

$$A_{vf} \approx -\frac{R_f}{r_i}$$

Thus, with large loop gain, the voltage gain with feedback depends solely on the *ratio of feedback to source resistance,* and the current gain on the *ratio of feedback to load resistance.*

Example 8.2-2 In Fig. 8.2-5 a difference amplifier drives a high-gain current amplifier. Voltage feedback is employed between the input and the output. Because the feedback is returned to the base of T_1 (providing current error), the circuit is basically the same as that of Fig. 8.2-3.

Calculate the gain with feedback by first calculating the current gain without feedback A_i, and then calculating the loop gain T. Assume $h_{re} = h_{oe} = 0$. We also assume, as in Example 8.2-1, that $A_i' i_\epsilon \gg v_{b1}/R_f$, and that $R_f \gg R_L$.

Solution The gain A_i is obtained by disconnecting the feedback network from the load and grounding it (this is the same as setting $v_L' = 0$ and replacing the feedback network by its Norton impedance R_f). The resulting circuit is shown in Fig. 8.2-6*a*.

From the equivalent circuit of Fig. 8.2-6*b*, assuming identical transistors, the collector current in T_2 is

$$h_{fe}i_{b2} = i_{c2} \approx -h_{fe}\left[\frac{r_i \| R_f}{(r_i \| R_f) + 2h_{ie} + R_T}\right] i_i$$

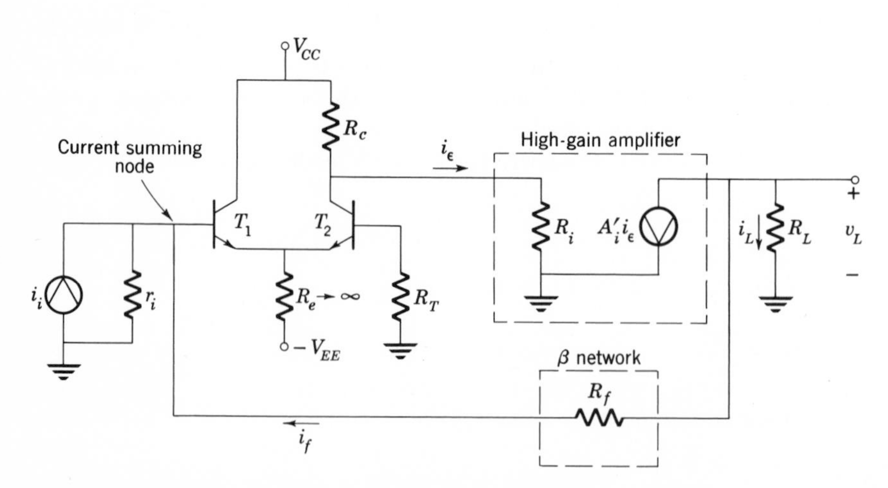

Fig. 8.2-5 Feedback amplifier for Example 8.2-2.

Thus the error current is

$$i_\epsilon = -\left(\frac{R_c}{R_c + R_i}\right) i_{c2} = i_i \left(\frac{h_{fe}R_c}{R_c + R_i}\right) \left[\frac{r_i \| R_f}{(r_i \| R_f) + 2h_{ie} + R_T}\right]$$

Therefore

$$A_i = \left.\frac{i_L}{i_i}\right|_{v'_L = 0} = \left(\frac{i_L}{i_\epsilon}\right)\left(\frac{i_\epsilon}{i_{c2}}\right)\left(\frac{i_{c2}}{i_i}\right)$$

$$= -h_{fe}\left(\frac{A_i'}{1 + R_i/R_c}\right)\left[\frac{r_i \| R_f}{(r_i \| R_f) + 2h_{ie} + R_T}\right]$$

The loop gain T is found by setting $i_i = 0$, removing the feedback network from the load, and calling the voltage to be fed back v_L', as shown in Fig. 8.2-7*a*.

The collector current i_{c2} is then

$$i_{c2} = -\left(\frac{h_{fe}v_L'}{R_f}\right)\left[\frac{r_i \| R_f}{(r_i \| R_f) + 2h_{ie} + R_T}\right]$$

and since

$$v_L = i_L R_L = -A_i' R_L i_\epsilon$$

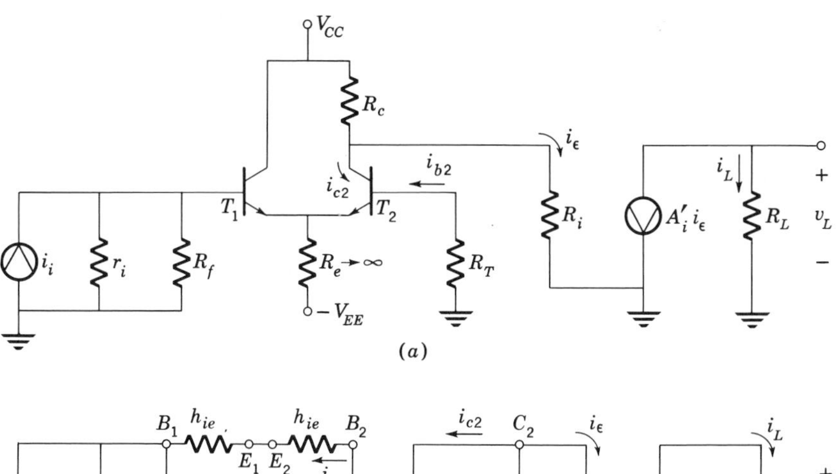

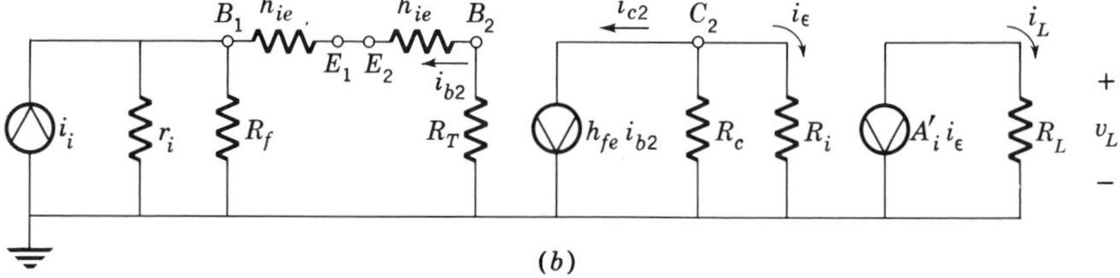

Fig. 8.2-6 Circuit for finding A_i ($R_f \gg R_L$ and $A_i' i_\epsilon \gg v_{b1}/R_f$). (*a*) Circuit with loop opened ($v_L' = 0$); (*b*) small-signal equivalent circuit (reflected into base of T_2).

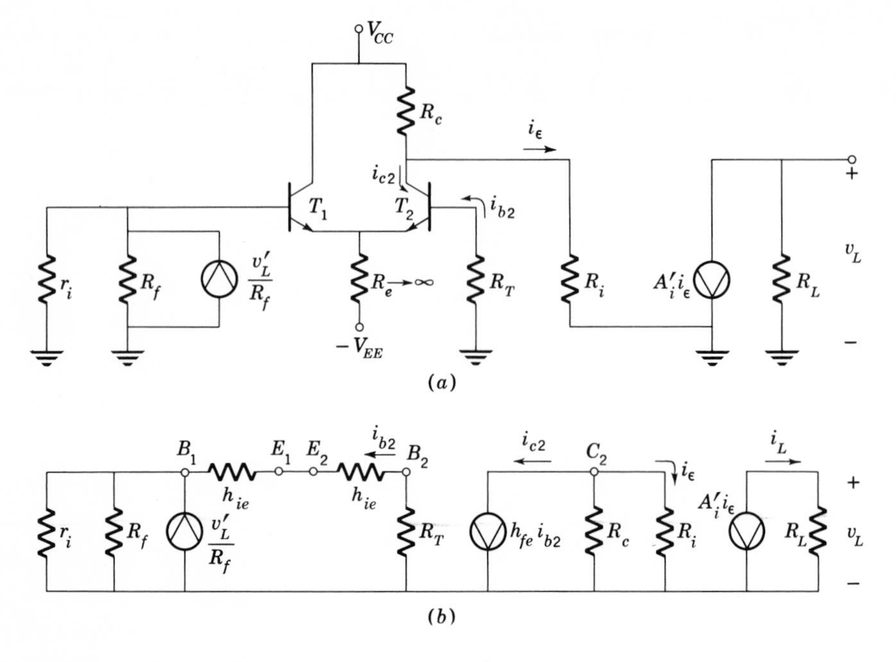

Fig. 8.2-7 Circuit for finding T ($R_f \gg R_L$, $A_i' i_\epsilon \gg v_{b1}/R_f$). (a) Circuit with loop open ($i_i = 0$); (b) small-signal equivalent circuit (reflected into base of T_2).

where

$$i_\epsilon = -\left(\frac{R_c}{R_c + R_i}\right) i_{c2}$$

the loop gain T is

$$T = \frac{v_L}{v_L'}\bigg|_{i_i=0} = -h_{fe} A_i' R_L \left(\frac{R_c}{R_c + R_i}\right)\left[\frac{r_i \| R_f}{(r_i \| R_f) + 2h_{ie} + \mathrm{R}_T}\right]\left(\frac{1}{R_f}\right)$$

The current gain with feedback is then

$$A_{if} = \frac{A_i}{1 - T} = \frac{\left[-\dfrac{h_{fe} A_i'}{1 + (R_i/R_c)}\right]\left[\dfrac{r_i \| R_f}{(r_i \| R_f) + 2h_{ie} + R_T}\right]}{1 + A_i' R_L \left(\dfrac{h_{fe} R_f}{1 + R_i/R_c}\right)\left[\dfrac{r_i \| R_f}{(r_i \| R_f) + 2h_{ie} + R_T}\right]}$$

If $-T \gg 1$, then

$$A_{if} \approx \frac{-R_f}{R_L}$$

as in Example 8.2-1. Note that the gain is independent of h_{fe} and A_i'.

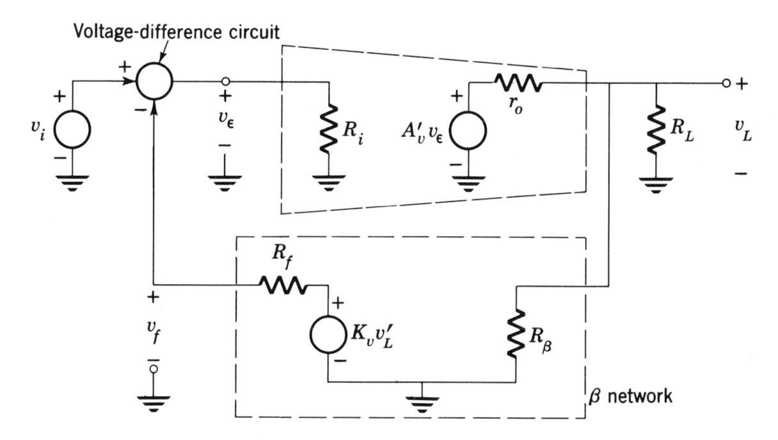

Fig. 8.2-8 Voltage-feedback–voltage-error configuration.

The gain with feedback A_{if} can be calculated directly to verify these results.

8.2-2 VOLTAGE FEEDBACK WITH VOLTAGE ERROR

In order to determine the gain for this configuration, redraw Fig. 8.1-1*b* as shown in Fig. 8.2-8. Again assume that the feedback network does not load the output circuit ($R_\beta \gg R_L$) and replace the β network by its Thévenin equivalent. The resulting circuit is shown in Fig. 8.2-9.

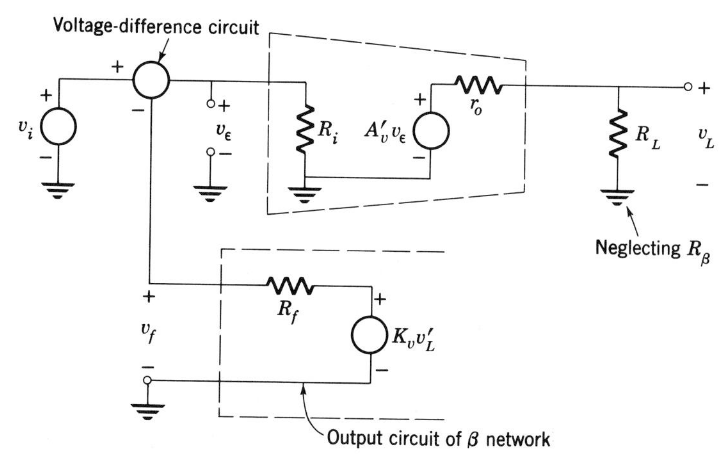

Fig. 8.2-9 Circuit of Fig. 8.2-8 prepared for analysis.

If the effect of R_f on the difference circuit is neglected, the error voltage v_ϵ is (setting $v'_L = v_L$ to calculate A_{vf} directly)

$$v_\epsilon = v_i - K_v v'_L = v_i - K_v v_L \tag{8.2-10}$$

The output voltage is, for $R_L \gg r_o$,

$$v_L = A'_v v_\epsilon \tag{8.2-11}$$

Thus the voltage gain with feedback A_{vf} is

$$A_{vf} = \frac{v_L}{v_i} = \frac{A'_v}{1 + K_v A'_v} \tag{8.2-12a}$$

The voltage gain without feedback A_v is (for this example)

$$A_v = A_{vf}\Big|_{K_v=0} = A'_v \tag{8.2-12b}$$

Thus

$$A_{vf} = \frac{A_v}{1 + K_v A_v} = \left(\frac{1}{K_v}\right)\left(\frac{1}{1 + 1/K_v A_v}\right) \tag{8.2-12c}$$

The loop gain for this configuration is

$$T = \frac{v_L}{v'_L}\Big|_{v_i=0} = -K_v A'_v = -K_v A_v \tag{8.2-13}$$

so that

$$A_{vf} = \frac{A_v}{1 - T} \tag{8.2-14}$$

Note that the basic forms of these expressions are the same as (8.2-8) and (8.2-9). In the above derivation, loading of the input circuit was neglected. This assumption is not made in the following example. The reader will observe that input loading reduces the loop gain and A_v.

Example 8.2-3 Find A_{vf} for the amplifier of Fig. 8.2-10. It consists of a difference amplifier followed by a high-gain voltage amplifier. The input and output stages of this high-gain amplifier are assumed to be composed of emitter followers, so that R_i is very large (compared with R_c) and r_o very small (compared with R_L). The feedback network is assumed not to load the output ($r_o \ll R_2$). The loop gain T can be adjusted by varying R_1 and/or R_2. The emitter follower T_3 is used as the output stage of the feedback circuit, so that the Thévenin impedance R_f is small compared with the input impedance to the difference amplifier.

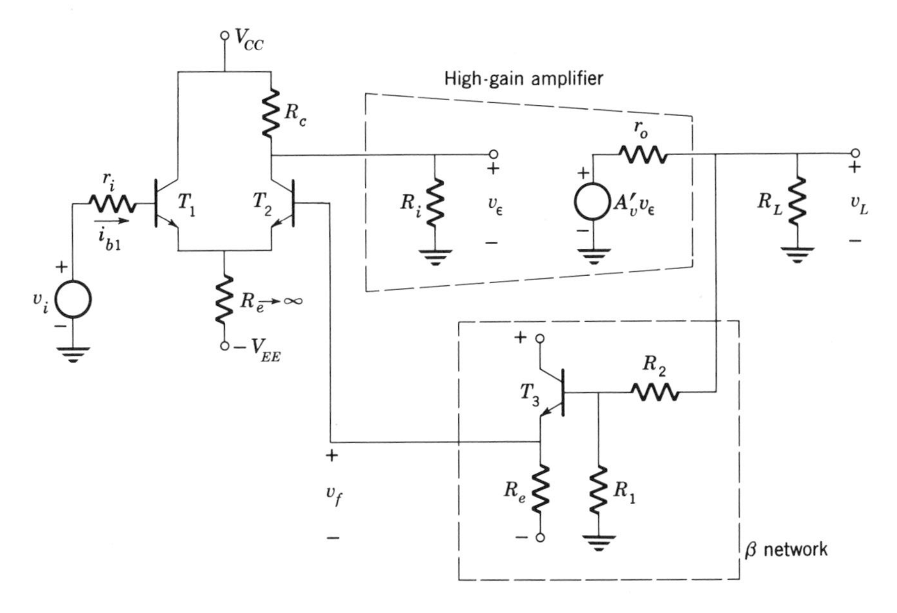

Fig. 8.2-10 Circuit for Example 8.2-3.

Solution Figure 8.2-10 is redrawn as shown in Fig. 8.2-11. To calculate the voltage gain directly, let $v'_L = v_L$. Then

$$i_{c2} = h_{fe} \left\{ \frac{[R_1/(R_1 + R_2)]v_L - v_i}{r_i + R_f + 2h_{ie}} \right\}$$

In addition,

$$v_\epsilon = -i_{c2} \left(\frac{R_i R_c}{R_i + R_c} \right) = h_{fe} \left(\frac{R_i R_c}{R_i + R_c} \right) \left\{ \frac{[-R_1/(R_1 + R_2)]v_L + v_i}{r_i + R_f + 2h_{ie}} \right\}$$

and

$$v_L = A'_v \left(\frac{R_L}{r_o + R_L} \right) v_\epsilon$$

Hence

$$A_{vf} = \frac{v_L}{v_i} = \frac{h_{fe} A'_v \left(\dfrac{R_L}{r_o + R_L} \right) [(R_i \| R_c)/(r_i + R_f + 2h_{ie})]}{1 + (h_{fe} A'_v) \left(\dfrac{R_L}{r_o + R_L} \right) (R_i \| R_c) \left[\dfrac{R_1/(R_1 + R_2)}{r_i + R_f + 2h_{ie}} \right]}$$

$$= \frac{A_v}{1 - T}$$

When $-T \gg 1$ the overall voltage gain becomes

$$A_{vf} \approx \frac{R_1 + R_2}{R_1} = 1 + \frac{R_2}{R_1}$$

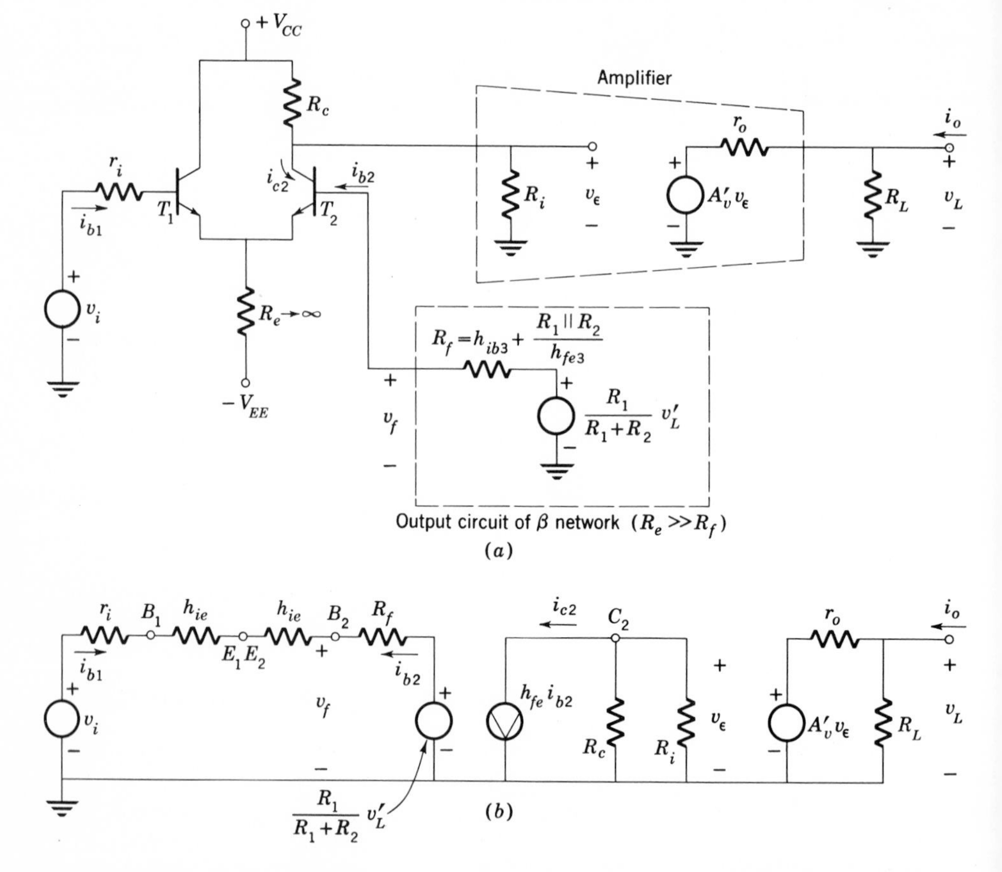

Fig. 8.2-11 Voltage-feedback–voltage-error configuration. (*a*) Circuit; (*b*) small-signal equivalent circuit.

It should be noted that there is no significant difference in the method of calculating the gain with feedback, whether a current error or a voltage error is present. Note, also, that by shifting the position of the output terminal of the feedback network from the base of T_2 to the base of T_1, current rather than voltage error can be obtained.

The voltage gain A_{vf} can also be obtained by calculating T and A_v. Referring to Fig. 8.2-11*b*,

$$T = \left.\frac{v_L}{v'_L}\right|_{v_i=0} = \left.\left(\frac{v_L}{v_\epsilon}\right)\left(\frac{v_\epsilon}{i_b}\right)\left(\frac{i_b}{v'_L}\right)\right|_{v_i=0}$$

$$= \frac{-h_{fe}A'_v[R_L/(r_o + R_L)](R_i \| R_c)[R_1/(R_1 + R_2)]}{r_i + R_f + 2h_{ie}}$$

and

$$A_v = \frac{v_L}{v_i}\bigg|_{v'_L=0} = \left(\frac{v_L}{v_\epsilon}\right)\left(\frac{v_\epsilon}{i_b}\right)\left(\frac{i_b}{v_i}\right)$$

$$= \frac{h_{fe}A'_v[R_L/(r_o + R_L)](R_i\|R_c)}{r_i + R_f + 2h_{ie}}$$

8.3 FEEDBACK AMPLIFIERS AND THE SENSITIVITY FUNCTION

Practical feedback amplifiers are designed so that the open-loop gain A_i (or A_v) is extremely large (40 to 120 dB or more). The resulting closed-loop gain is then primarily a function of the feedback network. The gain is almost totally independent of variations in h_{fe}, supply voltage, temperature, etc.

SENSITIVITY TO GAIN VARIATIONS

A quantitative measure of the effectiveness of the feedback in making A_{if} independent of the internal amplifier gain A_i is the sensitivity function $S_{A_i}^{A_{if}}$. This is defined as

$$S_{A_i}^{A_{if}} = \frac{dA_{if}/A_{if}}{dA_i/A_i} = \left(\frac{A_i}{A_{if}}\right)\left(\frac{dA_{if}}{dA_i}\right) \tag{8.3-1}$$

The sensitivity of A_{if} with respect to variations in A_i is the ratio of the fractional (or percentage) change in A_{if} to the fractional (or percentage) change in A_i. Clearly, one would like to achieve zero sensitivity, because this means that changes in the forward gain A_i would not cause any change in the gain with feedback A_{if}.

We can calculate $S_{A_i}^{A_{if}}$ for the basic feedback amplifier of Fig. 8.1-1*a* by differentiating (8.2-9). This yields

$$\frac{dA_{if}}{dA_i} = \frac{1}{1-T} - \left[\frac{A_i}{(1-T)^2}\right]\left[\frac{d(1-T)}{dA_i}\right] \tag{8.3-2a}$$

Using (8.2-8),

$$A_i\left[\frac{d(1-T)}{dA_i}\right] = -T \tag{8.3-2b}$$

Hence

$$\frac{dA_{if}}{dA_i} = \frac{1}{(1-T)^2} \tag{8.3-2c}$$

Substituting (8.3-2) in (8.3-1) yields

$$S_{A_i}^{A_{if}} = \frac{1}{1-T} \tag{8.3-3}$$

Equation (8.3-3) states that if $-T \gg 1$, a 10 percent change in the forward current gain A_i due to transistor replacement, temperature changes, etc., will appear as a change in overall gain A_{if} of approximately $10/T$ percent. Equations (8.3-3) and (8.2-9) can be used to establish preliminary design figures, as shown by the following example.

Example 8.3-1 A feedback amplifier is to be designed to have an overall gain A_{if} of 40 dB and a sensitivity of 5 percent to internal amplifier gain variations.

Find the required loop gain and forward-current gain.

Solution From (8.2-9)

$$A_{if} = \frac{A_i}{1 - T} \approx \frac{A_i}{-T} = 100 \ (= 40 \text{ dB})$$

From (8.3-3)

$$S_{A_i}^{A_{if}} \approx \frac{1}{-T} = 0.05$$

so that

$$|T| \approx 20 \ (= 26 \text{ dB})$$

and

$$A_i \approx (20)(100) = 2000 \ (= 66 \text{ dB})$$

Thus, in order to achieve an overall gain of 100 (40 dB), we must start with an amplifier having a gain of 2000 (66 dB). This sacrifice of gain has resulted in a considerable degree of stability.

REDUCTION OF INTERNAL DISTURBANCES

In order to illustrate this aspect of feedback, consider the block diagram of Fig. 8.3-1. In this diagram, the disturbance is represented by a signal i_d, introduced at a suitable point in the internal current amplifier. Small amounts of nonlinear distortion (Secs. 10.4 and 11.3), drift due to temperature effects, and power-supply variation can often be accounted for by this method. Assume that loading effects can be ignored. Under these conditions (Sec. 8.2),

$$v_L = R_L A_1 A_2 (i_i - G_i v_L) + A_2 R_L i_d \tag{8.3-4}$$

Solving,

$$v_L = \left(\frac{A_i R_L}{1 - T}\right) i_i + \left(\frac{A_2 R_L}{1 - T}\right) i_d \tag{8.3-5}$$

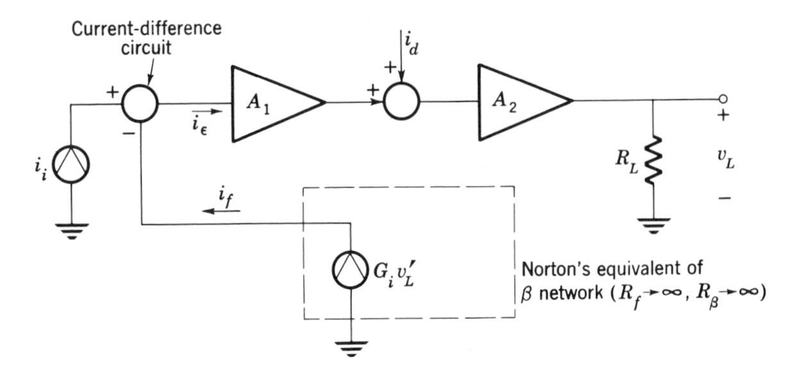

Fig. 8.3-1 Feedback amplifier with disturbing signal.

where $A_i = A_1 A_2$, and $T = -A_i G_i R_L$. Thus

$$v_L = \left(\frac{A_i R_L}{1 - T}\right)\left(i_i + \frac{i_d}{A_1}\right) \tag{8.3-6}$$

Equation (8.3-6) implies that the distortion source can be replaced by an equivalent generator at the input, as shown in Fig. 8.3-2. This new source represents the distorting signal i_d, referred to the input, and has the same effect on the output as the actual distortion generator.

Clearly, the effect of the disturbance at the output depends on the point inside the amplifier at which it is introduced. If we set $i_i = 0$ in (8.3-6), then

$$v_L \bigg|_{i_i=0} = \left(\frac{A_i R_L}{1 - T}\right)\left(\frac{i_d}{A_1}\right) = \left(\frac{A_2 R_L}{1 - T}\right) i_d \tag{8.3-7}$$

If the disturbance is, for example, power-supply drift in the final stage of the amplifier, having a gain A_{io}, then $A_1 = A_{io}$, and $A_2 = 1$ (since a disturbance in the collector supply is not amplified by the transistor—a demonstration of this point is reserved for the problems). The output

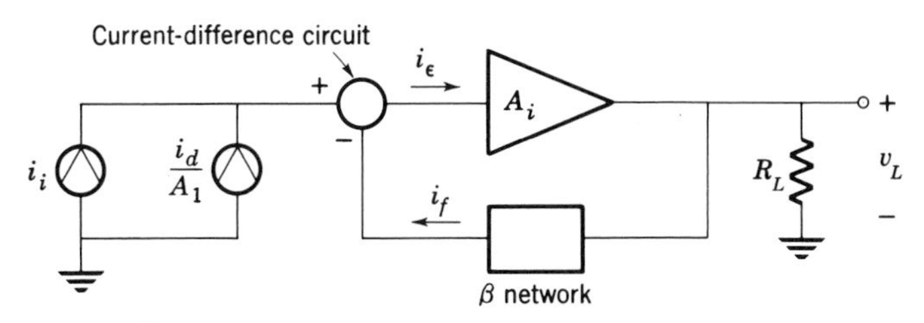

Fig. 8.3-2 Alternative representation of amplifier of Fig. 8.3-1 with disturbing signal.

voltage is then

$$v_L \Big|_{i_i=0} \approx \left(\frac{R_L}{1 - T}\right) i_d$$

and the drift is considerably reduced in the output. If, on the other hand, the disturbance is noise at the input to the amplifier, then $A_1 = 1$, and A_2 becomes A_{io}. The noise then appears in the output amplified to the same degree as the input signal, and *is not reduced at all.*

From the foregoing discussion it is seen that feedback is definitely not a cure-all, and the design of the internal amplifier must be carried through with care; in particular, the low-level stages must be distortion-free and contribute little noise.

Another problem arises when signal levels are such that the internal amplifier is driven near the saturation or the cutoff levels. When this happens, the feedback, which tries to maintain a linear relation between input and output waveforms, results in an increase in the distortion within the amplifier. In some situations the increased internal distortion can be so large that serious overloading can occur.[1]

Example 8.3-2 The single-stage amplifier of Fig. 8.3-3 has a direct feedback path from collector to base. This connection provides voltage feedback at signal frequencies and, in addition, supplies bias current. In the circuit, power-supply ripple is represented as a voltage v_{cc} in series with the supply. Find the ripple voltage present in the output voltage v_L. Assume $r_i \gg R_f \gg R_L$.

Solution We begin by obtaining an equivalent circuit as shown in Fig. 8.3-4. The output voltage v_L is

$$v_L = -h_{fe}R_L i_b + v_{cc}$$

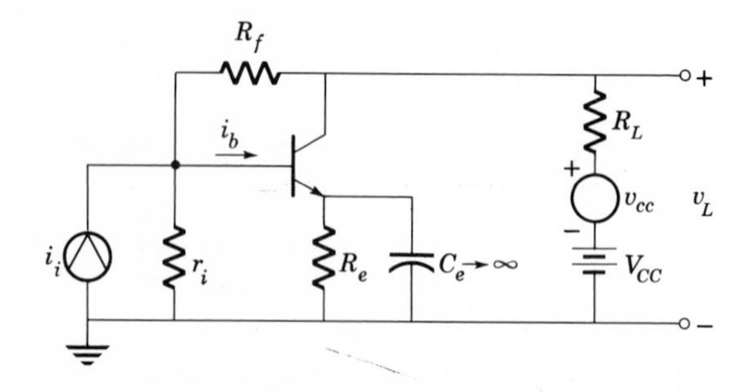

Fig. 8.3-3 Circuit for Example 8.3-2.

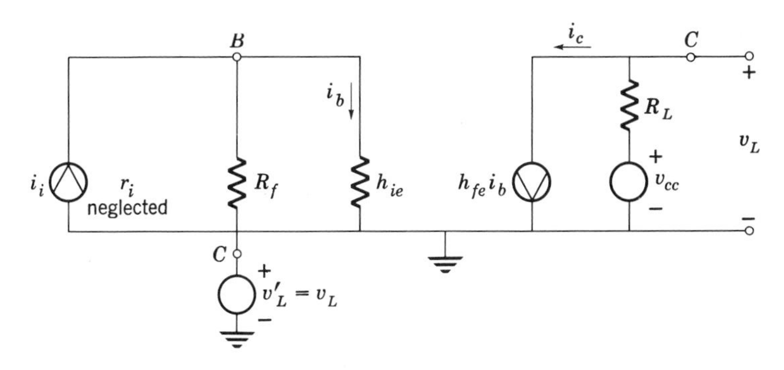

Fig. 8.3-4 Small-signal equivalent of amplifier of Example 8.3-2.

and

$$i_b \approx i_i \left(\frac{R_f}{R_f + h_{ie}}\right) + \frac{v_L}{R_f + h_{ie}}$$

Thus

$$v_L = v_{cc} - \left(\frac{h_{fe}R_LR_f}{R_f + h_{ie}}\right) i_i - \left(\frac{h_{fe}R_L}{R_f + h_{ie}}\right) v_L$$

and finally,

$$v_L = \frac{v_{cc}}{1 + h_{fe}R_L/(R_f + h_{ie})} - \left[\frac{h_{fe}R_LR_f/(R_f + h_{ie})}{1 + h_{fe}R_L/(R_f + h_{ie})}\right] i_i$$

The loop gain is

$$T \approx -\frac{h_{fe}R_L}{R_f + h_{ie}} \approx -\frac{h_{fe}R_L}{R_f} \qquad \text{if } R_f \gg h_{ie}$$

If $-T \gg 1$, the output is, approximately,

$$v_L \approx \frac{v_{cc}}{-T} - R_f i_i$$

As expected, the ripple in the output is reduced by the loop gain.

8.4 INPUT AND OUTPUT IMPEDANCES

The application of feedback to an amplifier produces impedance-level changes, as well as the gain and sensitivity changes discussed in the preceding sections. In this section, the effects of various types of feedback on input and output impedance are investigated.

In order to calculate (or measure) the input or output impedance of an amplifier, one can connect a test current or voltage source to the appropriate terminals. Impedance is then defined as the ratio of impressed voltage to current (or voltage to impressed current), with all other independent sources reduced to zero (impedances throughout the circuit are, of course, left unchanged). With this definition in mind, refer to Fig. 8.1-1.

Figure 8.1-1*a* utilizes feedback to force the current error toward zero. If the input impedance of the internal amplifier is R_i (and is much less then the feedback-network impedance), then the voltage drop caused by the impressed current i_i is $i_\epsilon R_i$, which tends toward zero. Hence the input impedance with feedback Z_{if} is

$$Z_{if} \approx \frac{i_\epsilon R_i}{i_i} \rightarrow 0$$

The converse is true when dealing with voltage error, as in Fig. 8.1-1*b*. In this case the error voltage v_ϵ is made small by the use of feedback. Hence the resulting current, which is proportional to v_ϵ/R_i, decreases, and the input impedance is increased by the use of negative feedback.

Figure 8.1-1*a* and *b* employs voltage feedback to reduce the error. To determine the output impedance, we let i_i in Fig. 8.1-1*a* be zero, connect a test voltage source v_L across the output terminals, and calculate the resulting current. If there were no feedback, i_ϵ would be zero because i_i is zero, the amplifier would amplify no current, and the only current flowing would flow through R_L and r_o. The output impedance would then, of course, be $R_L \| r_o$. With negative feedback, i_ϵ is equal to $-i_f$, and the resulting output current is the current in R_L and r_o and the current $A_i i_\epsilon$. Thus the output current has increased, and so the output impedance has decreased. The same result is observed in Fig. 8.1-1*b*. Without feedback, the output impedance is the parallel combination of R_L and the output impedance of the amplifier (see Fig. 8.2-10, for example). With negative feedback, v_ϵ is not zero, and additional current flows, resulting in a decrease in output impedance.

In a similar manner one can show that, when using negative feedback, the output impedance in a current-feedback system is increased.

A very interesting and useful result, which is demonstrated below, is that the impedance levels are increased or decreased by the factor $1 - T$, the same factor by which the gain is changed.

8.4-1 INPUT IMPEDANCE

Current error To demonstrate that when negative feedback is used to reduce the current error the input impedance is reduced, refer to Fig. 8.2-3, which is redrawn for convenience in Fig. 8.4-1. The input impedance Z_{if}

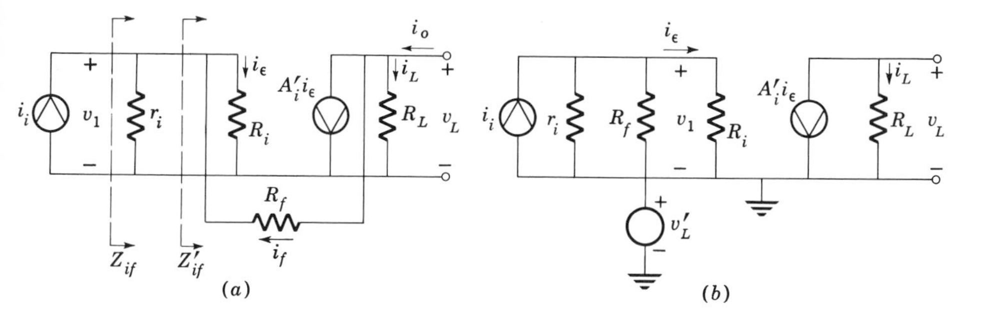

Fig. 8.4-1 Circuit for calculating input and output impedance. (*a*) Amplifier; (*b*) amplifier with feedback loop open.

is defined as

$$Z_{if} = \frac{v_1}{i_i} \tag{8.4-1}$$

where

$$v_1 = i_\epsilon R_i \tag{8.4-2}$$

and

$$i_\epsilon = \left(i_i + \frac{v_L}{R_f}\right)\left[\frac{r_i \| R_f}{R_i + (r_i \| R_f)}\right] \tag{8.4-3}$$

Also, for $R_L \ll R_f$ (which is usually the case),

$$v_L = -A_i' i_\epsilon R_L \tag{8.4-4}$$

After some algebra we obtain

$$Z_{if} = \frac{R_i\left[\dfrac{r_i \| R_f}{R_i + (r_i \| R_f)}\right]}{1 + A_i'\left(\dfrac{R_L}{R_f}\right)\left[\dfrac{r_i \| R_f}{R_i + (r_i \| R_f)}\right]} = \frac{R_i \| r_i \| R_f}{1 - T} \tag{8.4-5a}$$

where

$$T = -A_i'\left(\frac{R_L}{R_f}\right)\left[\frac{r_i \| R_f}{R_i + (r_i \| R_f)}\right] \tag{8.4-5b}$$

(The expression for the loop gain T was obtained in Example 8.2-1.)

The parallel combination of R_i, r_i, and R_f is the input impedance without feedback Z_i. This value of Z_i can be obtained from the equation

$$Z_i = \frac{v_1}{i_i}\bigg|_{v'_L = 0} \tag{8.4-5c}$$

Thus

$$Z_{if} = \frac{Z_i}{1 - T} \tag{8.4-5d}$$

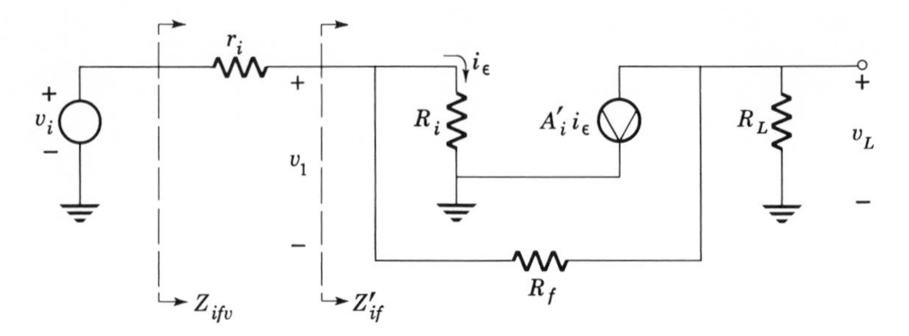

Fig. 8.4-2 Feedback amplifier of Fig. 8.4-1 driven by a voltage source.

The input impedance when current error is used is reduced from its value without feedback by the factor $1 - T$.

Usually, interest would center on the impedance Z'_{if}, shown in Fig. 8.4-1. This is the impedance seen by the source (considering r_i as part of the source). To find this input impedance, we set $r_i = \infty$ in (8.4-5*a*). Then

$$Z'_{if} = Z_{if} \qquad r_i = \infty \tag{8.4-6a}$$

Since the limit of $r_i \| R_f$ as r_i becomes infinite is R_f,

$$Z'_{if} = \frac{R_i \| R_f}{1 + A'_i[R_L/(R_i + R_f)]} \tag{8.4-6b}$$

Note that as $A'_i \to \infty$, $Z'_{if} \to 0$. The impedance Z'_{if} of most feedback amplifiers using current error is negligibly small.

This result can also be obtained in a straightforward manner by simply noting that Z_{if} is equal to Z'_{if} in parallel with r_i.

$$\frac{1}{Z_{if}} = \frac{1}{r_i} + \frac{1}{Z'_{if}}$$

Then

$$Y'_{if} = \frac{1}{Z'_{if}} = Y_{if} - \frac{1}{r_i} \tag{8.4-6c}$$

We can use this result to find the input impedance when the amplifier is driven by a voltage source rather than a current source, as shown in Fig. 8.4-2. The impedance Z'_{if} measured at the *amplifier input* is the same in both Figs. 8.4-1 and 8.4-2 because the amplifiers are the same. However, the impedance seen by the *voltage* source of Fig. 8.4-2 alone is

$$Z_{ifv} = r_i + Z'_{if} \tag{8.4-7a}$$

while that seen by the current source of Fig. 8.4-1 is

$$Z_{if} = r_i \| Z'_{if} \tag{8.4-7b}$$

Since Z'_{if} is usually small, the impedance $Z_{ifv} \approx r_i$.

Voltage error To illustrate that negative voltage feedback with a voltage error increases the input impedance, refer to the amplifier of Example 8.2-3 (Figs. 8.2-10 and 8.2-11).

The input impedance is Z_{ifv}, where

$$Z_{ifv} = \frac{v_i}{i_{b1}} \tag{8.4-8a}$$

From Fig. 8.2-11*b* the current i_{b1v} is

$$i_{b1} = \frac{v_i - [R_1/(R_1 + R_2)]v_L}{r_i + R_f + 2h_{ie}} \tag{8.4-8b}$$

Since $A_{vf} = v_L/v_i$, the input impedance is

$$Z_{ifv} = \frac{r_i + R_f + 2h_{ie}}{1 - [R_1/(R_1 + R_2)]A_{vf}} \tag{8.4-9}$$

We note from Example 8.2-3 that

$$A_{vf} = \frac{A_v}{1 - T} \tag{8.4-10a}$$

where

$$T = -\left(\frac{R_1}{R_1 + R_2}\right) A_v \tag{8.4-10b}$$

Thus

$$A_{vf} = -\frac{[(R_1 + R_2)/R_1]T}{1 - T} \tag{8.4-11}$$

Substituting (8.4-11) into (8.4-9),

$$Z_{ifv} = \frac{r_i + R_f + 2h_{ie}}{1 + [T/(1 - T)]} = (1 - T)(r_i + R_f + 2h_{ie}) \tag{8.4-12}$$

The input impedance without feedback is calculated with $v'_L = 0$. From Fig. 8.2-11*b*, we see by inspection that, with $v'_L = 0$,

$$Z_i = r_i + R_f + 2h_{ie} \tag{8.4-13}$$

It follows that

$$Z_{ifv} = Z_i(1 - T) \tag{8.4-14}$$

and voltage error results in an increase in input impedance. For example, if $r_i = 20\ \Omega$, $R_f = 1\ \text{k}\Omega$, and $h_{ie} = 240\ \Omega$, then $Z_i = 1500\ \Omega$. If we required that the input impedance be 600 kΩ, a loop gain of 52 dB would be required.

8.4-2 OUTPUT IMPEDANCE

Voltage feedback–current error Let us again refer to the amplifier of Fig. 8.4-1 to calculate Z_{of}. With $i_i = 0$, we apply a voltage source of value v_L to the output terminals. Neglecting loading due to R_f, the current i_o is

$$i_o = i_L + A_i' i_\epsilon \tag{8.4-15}$$

Note that $i_L = v_L/R_L$, and from (8.4-3)

$$i_\epsilon = \left(\frac{v_L}{R_f}\right)\left[\frac{r_i \| R_f}{R_i + (r_i \| R_f)}\right] \tag{8.4-16}$$

Using (8.4-5*b*), (8.4-15), and (8.4-16), the output admittance $Y_{of} = 1/Z_{of}$ becomes

$$Y_{of} = \frac{i_o}{v_L} = \frac{i_L + A_i' i_\epsilon}{v_L} = \frac{1}{R_L} + \left(\frac{A_i'}{R_f}\right)\left[\frac{r_i \| R_f}{R_i + (r_i \| R_f)}\right]$$

$$= \left(\frac{1}{R_L}\right)\left\{1 + \left(\frac{A_i' R_L}{R_f}\right)\left[\frac{r_i \| R_f}{R_i + (r_i \| R_f)}\right]\right\} = \left(\frac{1}{R_L}\right)(1 - T) \tag{8.4-17}$$

Thus

$$Z_{of} = \frac{R_L}{1 - T} \tag{8.4-18}$$

and the output impedance has been *reduced* because of the voltage feedback. If, for example, $R_L = 100\ \Omega$ and $T = 60$ dB, the output impedance with feedback becomes 0.1 Ω.

Voltage feedback–voltage error If we now refer to Fig. 8.2-11*b* with $v_i = 0$, the output impedance Z_{of} is again defined as

$$Z_{of} = \frac{v_L}{i_o}$$

where

$$i_o = \frac{v_L}{R_L} + \frac{v_L - A_v' v_\epsilon}{r_o} \tag{8.4-19}$$

The error voltage v_ϵ and output voltage v_L are related by the feedback loop, so that

$$v_\epsilon = -\left[\left(\frac{h_{fe} R_1}{R_1 + R_2}\right) v_L\right]\left(\frac{1}{R_f + r_i + 2h_{ie}}\right)(R_c \| R_i) \tag{8.4-20}$$

In addition, the loop gain is

$$T = -\left(\frac{h_{fe}R_1}{R_1 + R_2}\right)\left(\frac{1}{R_f + r_i + 2h_{ie}}\right)(R_c \| R_i)\left[A_v'\left(\frac{R_L}{r_0 + R_L}\right)\right] \tag{8.4-21}$$

Therefore

$$v_\epsilon = \left\{\frac{T}{A_v'[R_L/(r_o + R_L)]}\right\} v_L \tag{8.4-22}$$

Substituting (8.4-22) into (8.4-19) yields

$$Y_{of} = \frac{1}{Z_{of}} = \frac{1}{R_L} + \frac{1}{r_o} - \frac{T}{r_oR_L/(r_o + R_L)} = \left(\frac{1}{R_L} + \frac{1}{r_o}\right)(1 - T) \tag{8.4-23}$$

The output impedance without feedback is $Z_o = r_o \| R_L$; so the output impedance with feedback is

$$Z_{of} = \frac{Z_o}{1 - T} \tag{8.4-24}$$

Thus, for voltage feedback, using either current or voltage error, the output impedance with feedback is reduced from the value without feedback by the factor $1 - T$.

8.5 EXAMPLES OF BASIC FEEDBACK-AMPLIFIER ANALYSIS

In this section, several popular feedback amplifiers are analyzed to determine overall gain, loop gain (which determines sensitivity), and input and output impedance. The difference amplifier is emphasized as an input stage because of its versatility for feedback applications. Most integrated-circuit linear amplifiers utilize a differential amplifier for the input stage.

Example 8.5-1 In the circuit of Fig. 8.5-1, a differential stage is followed by a Darlington emitter follower. Voltage feedback is used with voltage error. Find A_{vf}, A_{if}, T, Z_{if}, and Z_{of}. Assume that all transistors are identical, with $h_{re} = h_{oe} = 0$.

Solution As a first step, we redraw the amplifier as shown in Fig. 8.5-2*a* to *c*. The small-signal equivalent circuits are obtained as in Secs. 7.2 and 7.3.

The loop gain T is calculated using Fig. 8.5-2*b*.

$$T = \left.\frac{v_L}{v_L'}\right|_{v_i=0} = \left(\frac{v_L}{i_{b3}}\right)\left(\frac{i_{b3}}{i_{b2}}\right)\left(\frac{i_{b2}}{v_L'}\right)$$

$$= [(h_{fe} + 1)^2R_L]\left(\frac{-h_{fe}R_c}{R_c + R_A}\right)\left(\frac{1}{R_f + r_i + 2h_{ie1}}\right)$$

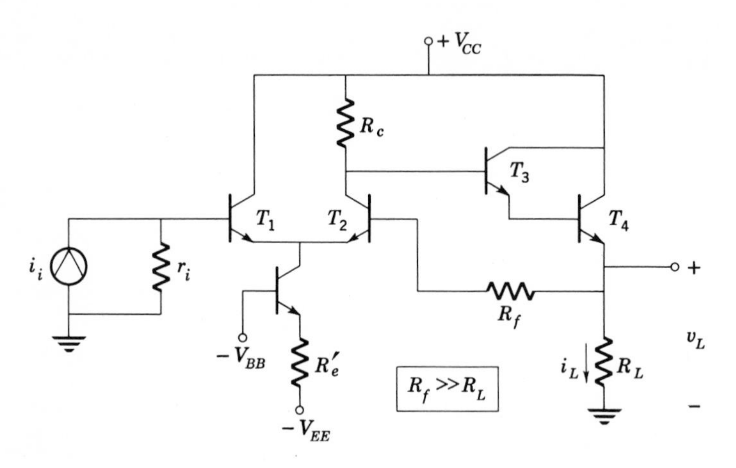

Fig. 8.5-1 A voltage-feedback amplifier using voltage error.

where (see Sec. 7.3)

$$R_A = (h_{fe} + 1)^2 R_L + 2h_{ie3}$$

Assuming $R_A \gg R_c$, and $(h_{fe} + 1)^2 R_L \gg 2h_{ie3}$, the loop gain becomes

$$T \approx \frac{-h_{fe}R_c}{R_f + r_i + 2h_{ie1}}$$

The forward voltage gain without feedback is found by letting $v_L' = 0$, as shown in Fig. 8.5-2c.

$$A_v = \frac{v_L}{v_i}\bigg|_{v'_L=0} = \left(\frac{v_L}{i_{b3}}\right)\left(\frac{i_{b3}}{i_{b2}}\right)\left(\frac{i_{b2}}{v_i}\right) = [(h_{fe} + 1)^2 R_L]$$

$$\left(\frac{-h_{fe}R_c}{R_c + R_A}\right)\left(\frac{-1}{R_f + r_i + 2h_{ie1}}\right) \approx \frac{h_{fe}R_c}{R_f + r_i + 2h_{ie1}} = -T$$

The input impedance without feedback Z_i is easily determined using Fig. 8.5-2c.

$$Z_i = \frac{v_i}{i_{b1}}\bigg|_{v'_L=0} = r_i + R_f + 2h_{ie1}$$

The output impedance without feedback Z_o is obtained by reflecting T_3 and T_4 into the emitter circuit of T_4, as shown in Fig. 8.5-3, rather than into the base of T_3, as was done in Fig. 8.5-2c.

Since $v_L' = v_i = 0$, the current i_{b2} is zero. It follows that

$$Z_o = \frac{v_L}{i_o}\bigg|_{v_i=v'_L=0} = R_L \| \left[2h_{ib4} + \frac{R_c}{(h_{fe} + 1)^2}\right]$$

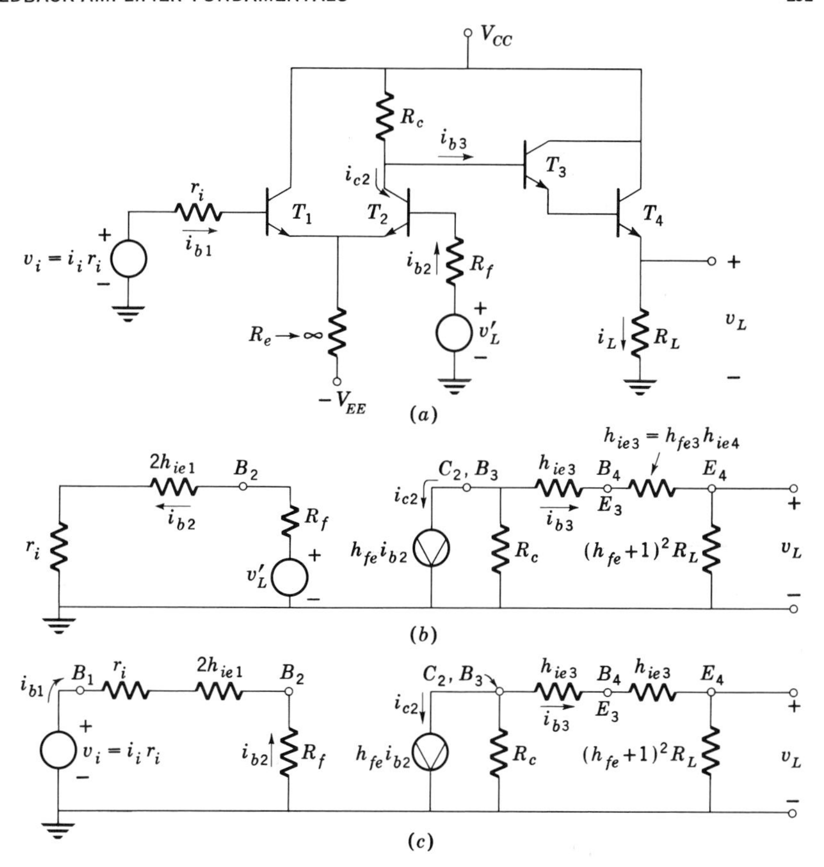

Fig. 8.5-2 The circuit of Fig. 8.5-1 prepared for analysis. (a) Circuit with feedback removed from output; (b) small-signal equivalent circuit for calculating T ($v_i = 0$); (c) small-signal equivalent circuit for calculating A_v ($v_L' = 0$).

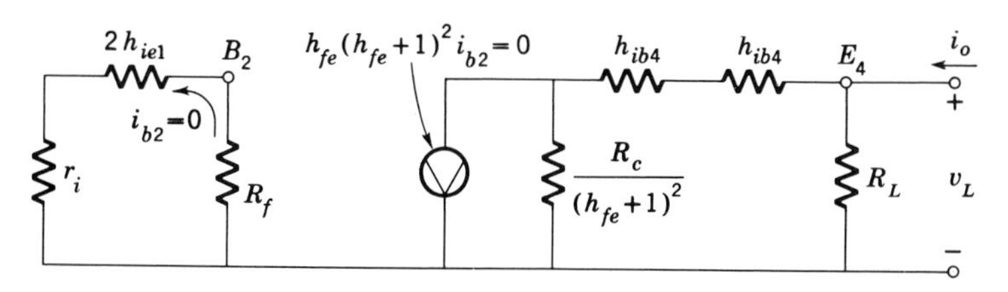

Fig. 8.5-3 Circuit for calculating Z_o ($v_L' = v_i = 0$).

The voltage gain with feedback A_{vf} can now be calculated. Assuming large negative loop gain,

$$A_{vf} = \frac{A_v}{1 - T} \approx \frac{A_v}{-T} = +1$$

The input impedance is

$$Z_{if} = Z_i(1 - T) \approx h_{fe}R_c + r_i + R_f + 2h_{ie1}$$

and the output impedance is

$$Z_{of} = \frac{Z_o}{1 - T} \approx \frac{Z_o}{-T} \approx \frac{R_L \| [2h_{ib4} + (R_c/h_{fe}^2)]}{h_{fe}R_c/(R_f + r_i + 2h_{ie1})}$$

Assume that $h_{fe} = 100$, $R_c = 1$ kΩ, $R_f = 1$ kΩ, $r_i = 1$ kΩ, $h_{ie1} = 1$ kΩ, $h_{ie3} = 10$ kΩ, and $R_L = 10$ Ω. Then

$$-T \approx +\frac{(100)(1000)}{1000 + 1000 + 2000} \approx +25 \ (= 28 \text{ dB})$$

To find the current gain, we make use of the previous result.

$$A_{vf} = \frac{v_L}{v_i} = \frac{i_L R_L}{i_i r_i} = A_{if}\left(\frac{R_L}{r_i}\right) \approx +1$$

Therefore

$$A_{if} = \left(\frac{r_i}{R_L}\right) A_{vf} = \frac{1000}{10} = 100 \ (= 40 \text{ dB})$$

The input impedance

$$Z_{if} \approx h_{fe}R_c + r_i + R_f + 2h_{ie1} \approx h_{fe}R_c = 100 \text{ k}\Omega$$

and the output impedance

$$Z_{of} \approx \frac{10 \| [2 + 1000/(100)^2]}{25} \approx \frac{1.67}{25} \approx 0.067 \ \Omega$$

Thus feedback resulted in a high input impedance, a low output impedance, and a large current gain.

The voltage gain is unity, making this amplifier appear to act like an emitter follower, with $h_{fe} = 10^4$ (to obtain the high input impedance) and $h_{ib} \approx 0.07$ Ω. These characteristics could not be obtained with a single transistor or even with a Darlington emitter follower.

Example 8.5-2 An alternative configuration to the circuit of Fig. 8.5-1 results if the output is taken from the collector rather than from the

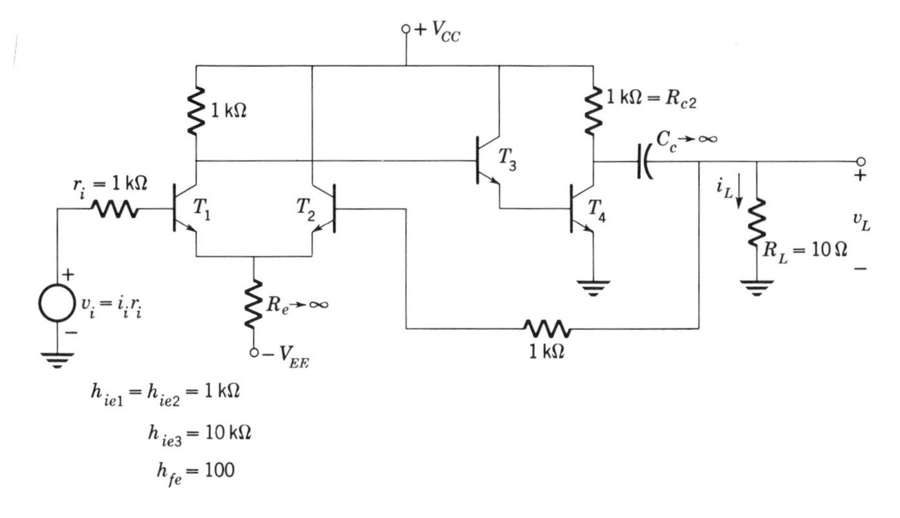

Fig. 8.5-4 A voltage-feedback amplifier using voltage error.

emitter of the last stage, as shown in Fig. 8.5-4. Note that the difference current is taken from the collector of T_1 rather than T_2, as in Fig. 8.5-1. This is necessary to obtain negative feedback.

Calculate T, A_{if}, Z_{if}, and Z_{of}. Assume the same parameter values as in Example 8.5-1, and R_{c2} = 1 kΩ.

Solution Again proceed by first drawing the small-signal equivalent circuits, as shown in Fig. 8.5-5*b* to *d*.

The loop gain is found from Fig. 8.5-5*b*.

$$T = \frac{v_L}{v'_L}\bigg|_{v_i=0} = \left(\frac{v_L}{i_{b3}}\right)\left(\frac{i_{b3}}{i_{b1}}\right)\left(\frac{i_{b1}}{v'_L}\right)$$

$$= \left[-h_{fe}(h_{fe}+1)\left(\frac{R_{c2}R_L}{R_{c2}+R_L}\right)\right]\left[-h_{fe}\left(\frac{R_{c1}}{R_{c1}+2h_{ie3}}\right)\right]\left(-\frac{1}{r_i+R_f+2h_{ie1}}\right)$$

Using the numerical values of Example 8.5-1

$$T \approx [(-10^4)(10)]\left[(-100)\left(\frac{10^3}{20\times 10^3}\right)\right]\left(\frac{-1}{4\times 10^3}\right)$$

and

$$-T \approx 125 \;(= 42 \text{ dB})$$

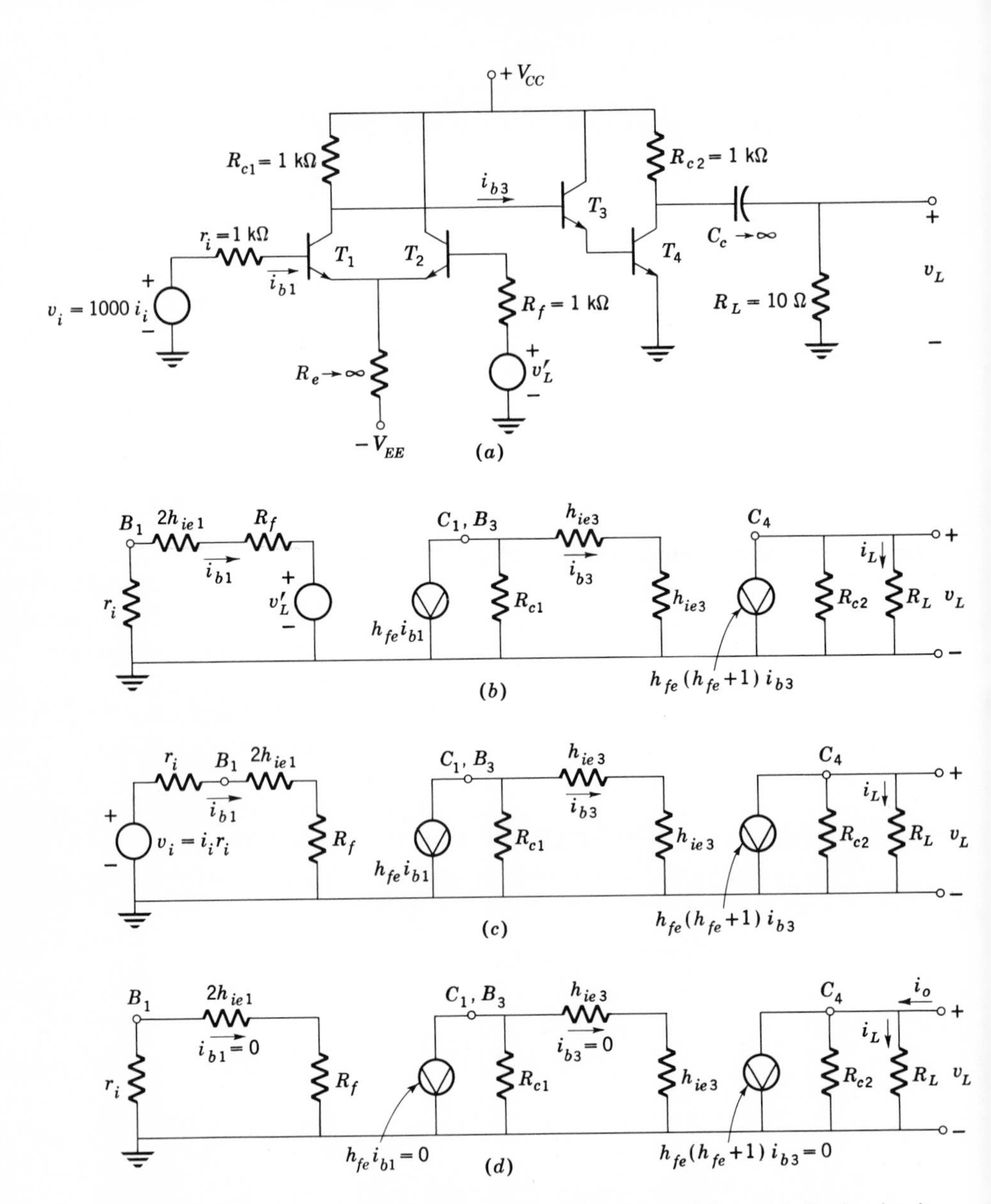

Fig. 8.5-5 Alternative arrangements of the feedback amplifier. (*a*) Basic circuit with the feedback isolated from the output; (*b*) small-signal circuit for calculating T; (*c*) small-signal circuit for calculating A_v and Z_i; (*d*) small-signal circuit for calculating Z_o.

The voltage gain without feedback is calculated using Fig. 8.5-5*c*.

$$A_v = \frac{v_L}{v_i}\bigg|_{v'_L=0} = \left(\frac{v_L}{i_{b3}}\right)\left(\frac{i_{b3}}{i_{b1}}\right)\left(\frac{i_{b1}}{v_i}\right) = -T = 125$$

Thus, as before,

$$A_{vf} \approx +1$$

and

$$A_{if} = \left(\frac{r_i}{R_L}\right)A_{vf} = 100\ (= 40\ \text{dB})$$

The input impedance Z_i can also be calculated using Fig. 8.5-5*c*. This results in

$$Z_{if} = Z_i(1 - T) = (r_i + R_f + 2h_{ie})(1 - T) \approx (4000)(125) = 0.5\ \text{M}\Omega$$

The output impedance can be determined using Fig. 8.5-5*d*. This yields

$$Z_{of} = \frac{Z_o}{1 - T} \approx \frac{R_L}{1 - T} \approx \frac{10}{125} = 0.08\ \Omega$$

Example 8.5-3 Calculate the voltage gains and the input and output impedances of the amplifier shown in Fig. 8.5-6. Consider the outputs to be v_{L1} and v_{L2}. This circuit, if constructed, would probably be unstable. The example is intended to illustrate the principles of feedback, and is not intended as a realistic amplifier configuration.

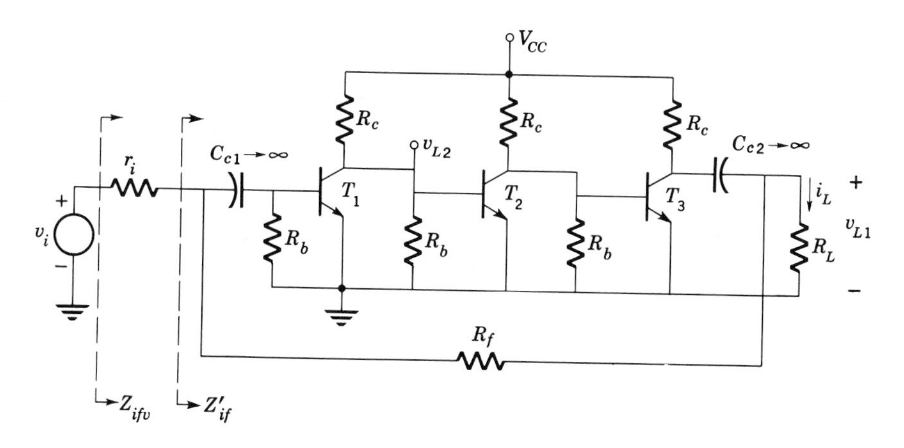

Fig. 8.5-6 Feedback amplifier for Example 8.5-3 (bias networks omitted).

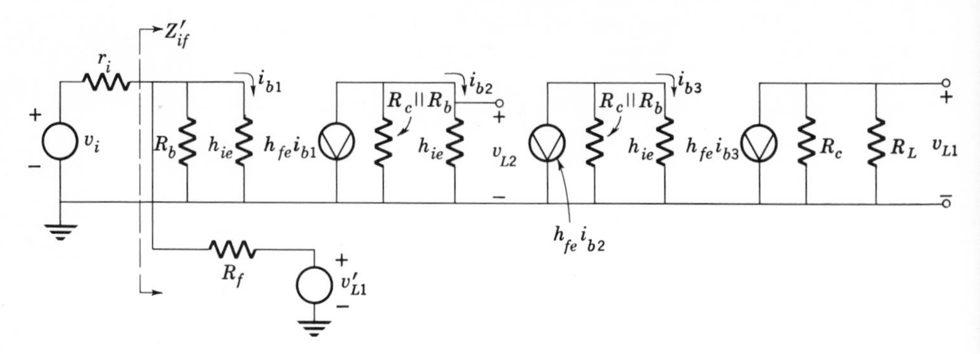

Fig. 8.5-7 Small-signal equivalent circuit for Example 8.5-3.

Assume $R_b = 500\ \Omega$, $R_c = 10\ \text{k}\Omega$, $h_{fe} = 50$, $h_{ie} = 500\ \Omega$, $r_i = R_L = 25\ \Omega$, and $R_f = 1\ \text{k}\Omega$. All transistors are assumed identical.

Solution The loop gain T is invariant; i.e., it does *not* depend on whether the output is v_{L1} or v_{L2}. To calculate T, the gain, and the impedance levels, first redraw Fig. 8.5-6 as shown in Fig. 8.5-7. The loop gain is then

$$T = \left.\frac{v_{L1}}{v'_{L1}}\right|_{v_i=0} = \left(\frac{i_{b1}}{v'_{L1}}\right)\left(\frac{i_{b2}}{i_{b1}}\right)\left(\frac{i_{b3}}{i_{b2}}\right)\left(\frac{v_{L1}}{i_{b3}}\right)$$

$$\approx \left(\frac{r_i/h_{ie}}{R_f}\right)\left(-\frac{h_{fe}}{2}\right)\left(-\frac{h_{fe}}{2}\right)(-h_{fe}R_L)$$

Hence

$$T \approx \left[\frac{25}{(500)(10^3)}\right](-25)(-25)(-50)(25) \approx -39$$

The forward gain without feedback with respect to terminal 1 is

$$A_{v1} = \left.\frac{v_{L1}}{v_i}\right|_{v'_{L1}=0} = \left(\frac{i_{b1}}{v_i}\right)\left(\frac{i_{b2}}{i_{b1}}\right)\left(\frac{i_{b3}}{i_{b2}}\right)\left(\frac{v_{L1}}{i_{b3}}\right)$$

$$\approx \frac{1}{h_{ie}}\left(-\frac{h_{fe}}{2}\right)\left(-\frac{h_{fe}}{2}\right)(-h_{fe}R_L) = \frac{R_fT}{r_i} \approx (40)(-39) = -1560$$

Thus

$$A_{vf1} \approx \frac{-1560}{1-(-39)} = -39\ (\approx 32\ \text{dB})$$

The output impedance without feedback is

$$Z_{o1} \approx R_L = 25\ \Omega$$

Therefore

$$Z_{of1} = \frac{Z_{o1}}{1 - T} \approx \frac{25}{40} = 0.625\ \Omega$$

To calculate the input impedance, refer to Sec. 8.4-1 [see Fig. 8.4-2 and Eq. (8.4-7a)]. The input impedance without feedback is obtained by letting $v'_{L1} = 0$ in Fig. 8.5-7.

$$Z_i = r_i \| R_b \| R_f \| h_{ie} \approx 22\ \Omega$$

Using (8.4-5d),

$$Z_{if} = \frac{Z_i}{1 - T} \approx \frac{22}{40} = 0.55\ \Omega$$

From (8.4-6c),

$$Y'_{if} = Y_{if} - \frac{1}{r_i} = \frac{1}{0.55} - \frac{1}{25}$$

Therefore

$$Z'_{if} = \frac{1}{Y'_{if}} \approx 0.6\ \Omega$$

and from (8.4-7a)

$$Z_{ifv} \approx 25.6\ \Omega$$

To determine the gain and the output impedance when terminal 2 is used as the output, we redraw Fig. 8.5-7 as shown in Fig. 8.5-8.

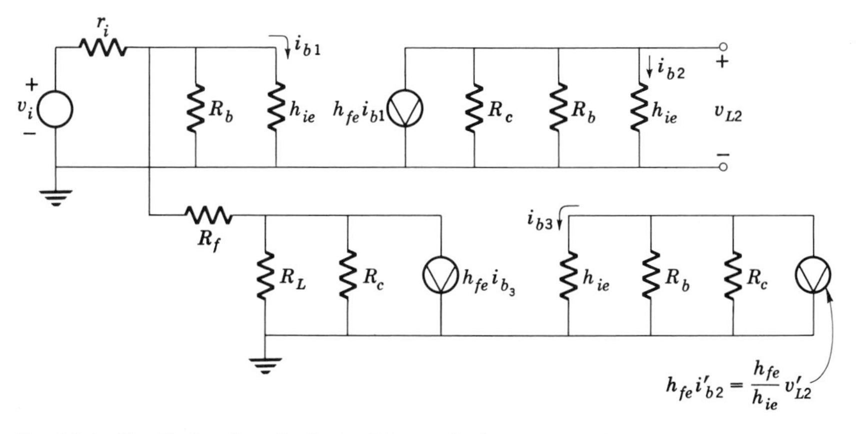

Fig. 8.5-8 Small-signal equivalent with terminal 2 as output.

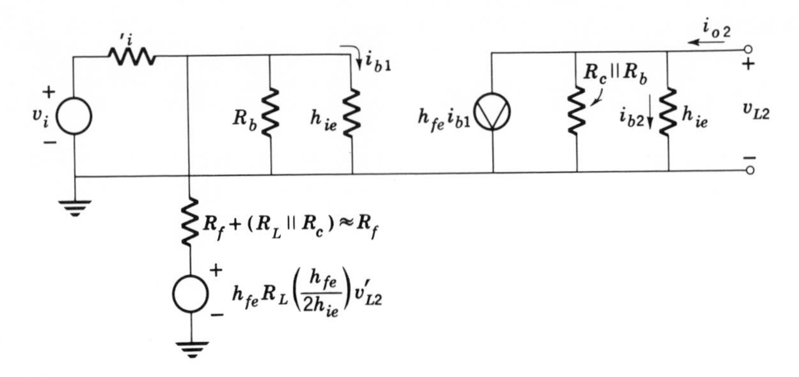

Fig. 8.5-9 Simplification of Fig. 8.5-8.

This figure shows the output and feedback path. It, in turn, can be simplified as shown in Fig. 8.5-9.

Figure 8.5-9 shows the Thévenin equivalent circuit of the feedback path. The forward voltage gain without feedback is now

$$A_{v2} = \left.\frac{v_{L2}}{v_i}\right|_{v'_{L2}=0} = \left(\frac{i_{b1}}{v_i}\right)\left(\frac{v_{L2}}{i_{b1}}\right) \approx \left(\frac{1}{h_{ie}}\right)\left[\left(-\frac{h_{fe}}{2}\right)h_{ie}\right] = -\frac{h_{fe}}{2}$$

Hence

$$A_{vf2} = \frac{A_{v2}}{1-T} \approx -\frac{25}{40} = -0.625$$

Note that $A_{vf1} \gg A_{vf2}$. The ratio is simply the gain provided by transistors T_2 and T_3.

The output impedance, without feedback, is

$$Z_{o2} = \left.\frac{v_{L2}}{i_{o2}}\right|_{v'_{L2}=0} \approx h_{ie}\|R_c\|R_b \approx 250\ \Omega$$

Thus

$$Z_{of2} \approx 250/40 = 6.25\ \Omega$$

The input impedance is, of course, unchanged.

8.6 INTRODUCTION TO THE DESIGN OF FEEDBACK AMPLIFIERS

Two basic feedback-amplifier design problems are presented in this section. The design specifications include the gain, the input and output impedances, and the sensitivity to internal-amplifier gain variations and power-supply ripple. Frequency-response and stability problems are considered in Chap. 15.

Example 8.6-1 Design an amplifier to have a voltage gain of magnitude unity or greater and an output impedance of 0.1 Ω or less. The signal source has 100 Ω internal resistance.

Solution Since the input impedance is not of interest in this problem, use of voltage feedback to obtain low output impedance, with either current or voltage error, is indicated. Let us employ the amplifier shown in Fig. 8.6-1. This configuration represents many standard integrated circuit amplifiers. Note that the β network is connected to obtain negative voltage feedback with a current error. Transistors T_1, T_2, and T_3 form a difference amplifier. T_4 is a CE current amplifier, and T_5 is an emitter follower for low output impedance. The dc biasing of these amplifiers is straightforward, and is not discussed. The equivalent circuit of Fig. 8.6-1 is shown in Fig. 8.6-2.

The loop gain is

$$T = \left.\frac{v_L}{v_L'}\right|_{v_i=0} = \left(\frac{v_L}{i_{b4}}\right)\left(\frac{i_{b4}}{i_{b2}}\right)\left(\frac{i_{b2}}{v_L'}\right)$$

$$\approx (-h_{fe4}R_{c4})(+h_{fe1})\left(\frac{1}{[100\|(R_{b1}+2h_{ie})]+R_f}\right)\left(\frac{100}{100+2h_{ie1}+R_{b1}}\right)$$

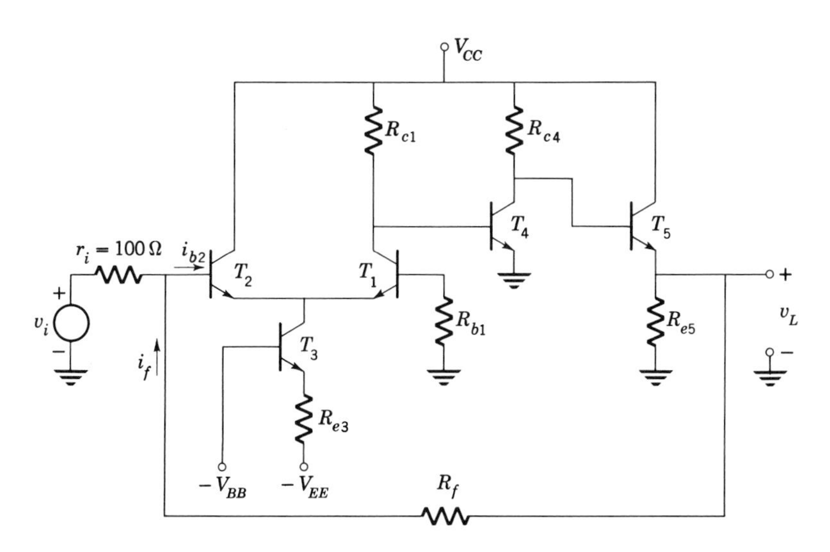

Fig. 8.6-1 Tentative design for Example 8.6-1.

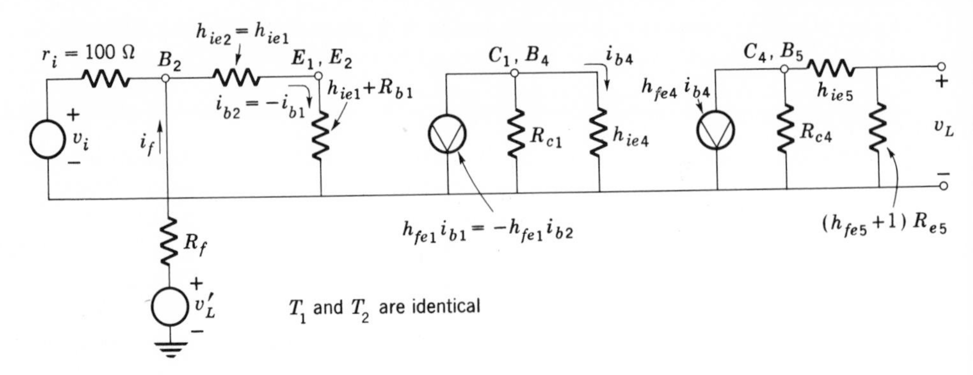

Fig. 8.6-2 Small-signal equivalent for Example 8.6-1.

where it has been assumed that

(a) $(h_{fe5})R_{e5} + h_{ie5} \gg R_{c4}$
(b) $h_{ie5} \ll h_{fe5}R_{e5}$
(c) $h_{ie4} \ll R_{c1}$
(d) $R_f \gg R_{e5} \| [h_{ib5} + R_{c4}/h_{fe5}]$

Let us assume $h_{ie1} \approx 1000\ \Omega$ and $R_{b1} = 100\ \Omega$. Then

$$T \approx -h_{fe1}h_{fe4}\left[\frac{R_{c4}}{22(R_f + 100)}\right]$$

The output impedance without feedback is

$$Z_o \approx R_{e5} \| \left(h_{ib5} + \frac{R_{c4}}{h_{fe5}}\right)$$

and the gain without feedback is

$$A_v = \left.\frac{v_L}{v_i}\right|_{v'_L=0} = \left(\frac{v_L}{i_{b4}}\right)\left(\frac{i_{b4}}{i_{b2}}\right)\left(\frac{i_{b2}}{v_i}\right)$$

$$\approx -h_{fe4}R_{c4}h_{fe1}\left(\frac{1}{100 + [R_f \| 2100]}\right)\left(\frac{R_f}{R_f + 2100}\right)$$

The gain with feedback for the operational amplifier when the loop gain is high was found in Sec. 8.2.

$$A_{vf} \approx \frac{A_v}{-T} = -\frac{R_f}{r_i} = -\frac{R_f}{100}$$

From this result we see that R_f must be greater than $100\ \Omega$ to obtain a minimum gain of magnitude unity.

Next, consider the requirement that the output impedance be less than 0.1 Ω. Thus

$$Z_{of} = \frac{Z_o}{1 - T} \approx \frac{R_{e5}\|(h_{ib5} + R_{c4}/h_{fe5})}{h_{fe1}h_{fe4}\{R_{c4}/[22(R_f + 100)]\}} \leq 0.1\ \Omega$$

It should be noted that, to use these simplified expressions for A_{vf} and Z_{of}, conditions *a* and *b* above must be satisfied. With these conditions the expression for the output impedance reduces to

$$Z_{of} \approx \frac{h_{ib5} + R_{c4}/h_{fe5}}{h_{fe1}h_{fe4}R_{c4}/[22(R_f + 100)]} \leq 0.1\ \Omega$$

Consider selecting transistors so that

$$h_{fe1} = h_{fe2} = h_{fe4} = h_{fe5} = 50$$

and the quiescent current of T_5 is set at 10 mA, so that

$$h_{ib5} = 2.5\ \Omega$$

If, in addition, we choose

$$R_{c4} = 1\ \text{k}\Omega$$

then

$$Z_{of} \approx \frac{2.5 + 20}{(50)(50)\left[\dfrac{1000}{22(R_f + 100)}\right]} \approx \frac{R_f + 100}{5000} \leq 0.1$$

Hence to ensure a minimum gain of unity and an output impedance less than 0.1 Ω,

$$100 \leq R_f \leq 400\ \Omega$$

If we choose $R_f = 200\ \Omega$, then

$$T \approx -380$$

$$A_{vf} \approx -2$$

and

$$Z_{of} \approx 0.06\ \Omega$$

The remaining critical components R_{e5} and R_{c1} can now be selected to satisfy conditions *a* and *b*, and bias networks added where necessary.

Example 8.6-2 Assume that in the circuit of Example 8.6-1, V_{CC} has a ripple component. Calculate the ripple reduction due to the feedback. How could you further reduce this ripple?

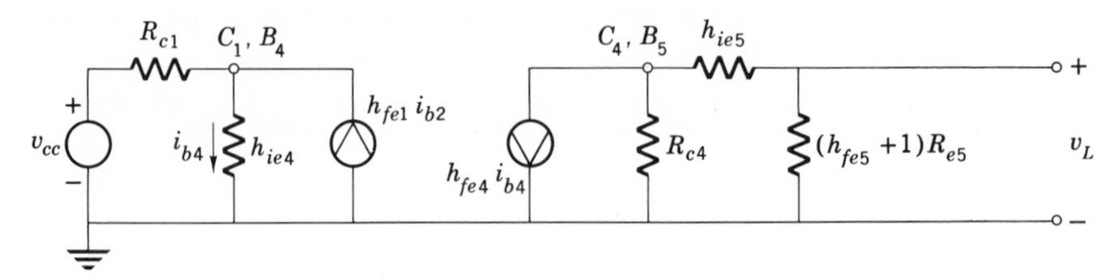

Fig. 8.6-3 Circuit for calculating effect of ripple.

Solution Variations in V_{CC} do not affect i_{c1} since $i_{c1} + i_{c2}$ = constant (T_3 is a constant current source), and variations in V_{CC} affect T_1 and T_2 in the identical way. The major contribution to variations in v_L caused by changes in V_{CC} occurs at the base of T_4, since this signal is amplified most. The equivalent circuit used to compute this gain is shown in Fig. 8.6-3. The ripple voltage is defined as v_{cc}.

$$\left.\frac{v_L}{v_{cc}}\right|_{\substack{v'_L=0\\ v_i=0\\ (i_{b2}=0)}} = \left(\frac{v_L}{i_{b4}}\right)\left(\frac{i_{b4}}{v_{cc}}\right) \approx \frac{-h_{fe4}R_{c4}}{R_{c1}+h_{ie4}} \approx \frac{(-50)(1000)}{R_{c1}+h_{ie4}}$$

If $R_{c1} + h_{ie4} = 2500\ \Omega$, then $v_L/v_{cc} = -20$. When the effect of feedback is included, this gain is reduced by the factor $1 - T$, as shown in Sec. 8.2. Thus

$$\left(\frac{v_L}{v_{cc}}\right)_{\text{with feedback}} = \frac{(v_L/v_{cc})\big|_{v'_L=v_i=0}}{1-T} \approx -\frac{20}{380} \approx 0.05$$

Thus 1 mV of ripple in the power supply will appear as 50 μV across the load. To reduce this further, the loop gain could be increased by a factor of 2 or 3, while still maintaining the overall gain at unity or more.

8.7 OTHER APPLICATIONS OF FEEDBACK

8.7-1 AN AUTOMATIC VOLUME CONTROL (AVC) CIRCUIT*

Automatic volume control (AVC) is used in almost every home radio receiver which utilizes amplitude modulation (AM) and is used extensively in frequency-modulation (FM) receivers to maintain a fixed signal level within the receiver.

In Chap. 1 the basic operation of the AM radio was discussed as an example of the application of transistor circuitry. Let us now use this same example to explain the operation of an AVC system.

Consider the effect of changing stations in an AM radio. As pointed

* Sometimes called automatic gain control (AGC).

out in Chap. 1, this adjusts the tuned input circuit to a different RF frequency. Since each station may transmit with different power, and may be situated at a different distance from the receiver, the received carrier level from different stations will, in general, differ considerably. The audio output being proportional to the carrier level, the volume of sound from the loudspeaker will be different. Thus a volume adjustment may be necessary each time the station is changed.

The AVC system described in this section automatically adjusts the receiver gain by means of a feedback loop, so that the volume of sound from the loudspeaker is approximately the same, regardless of the received carrier level. This is done by converting the received carrier into a "dc" control voltage, which controls the gain of the receiver, and hence the loudspeaker volume. Thus, if the received carrier level increases, the control voltage increases, decreasing the gain of the receiver, thereby maintaining a nearly constant volume of sound over large variations in received signal amplitude.

An AVC amplifier is shown in Fig. 8.7-1. The output of the RF amplifier is rectified, and then filtered, to provide a negative control voltage. This control voltage is proportional to the slowly varying carrier amplitude fluctuation, and is *not* proportional to the *amplitude modulation*.

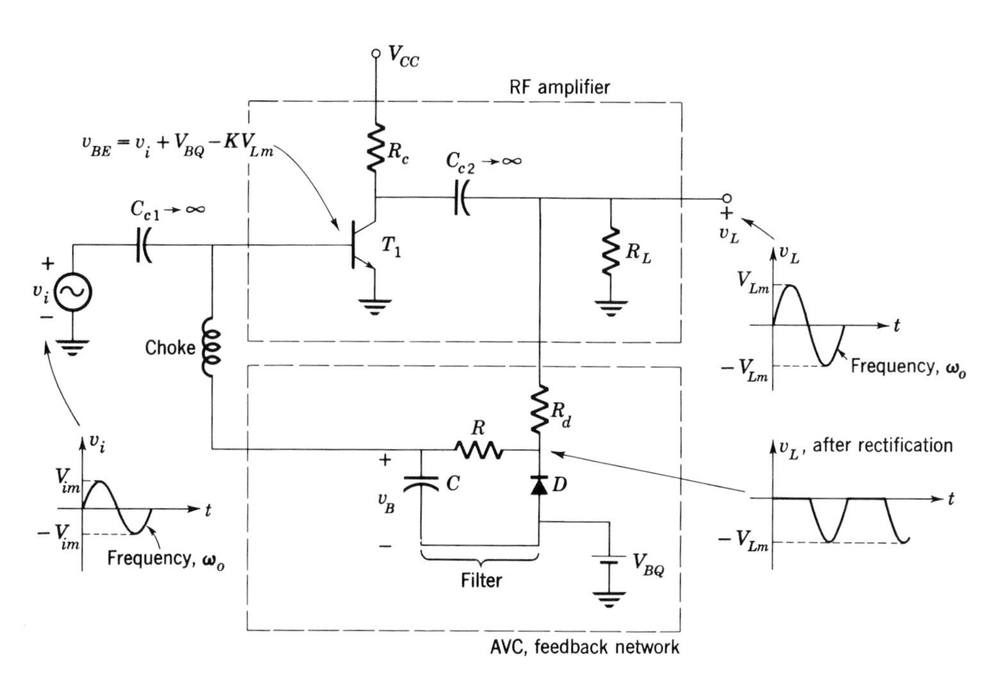

Fig. 8.7-1 AVC circuit.

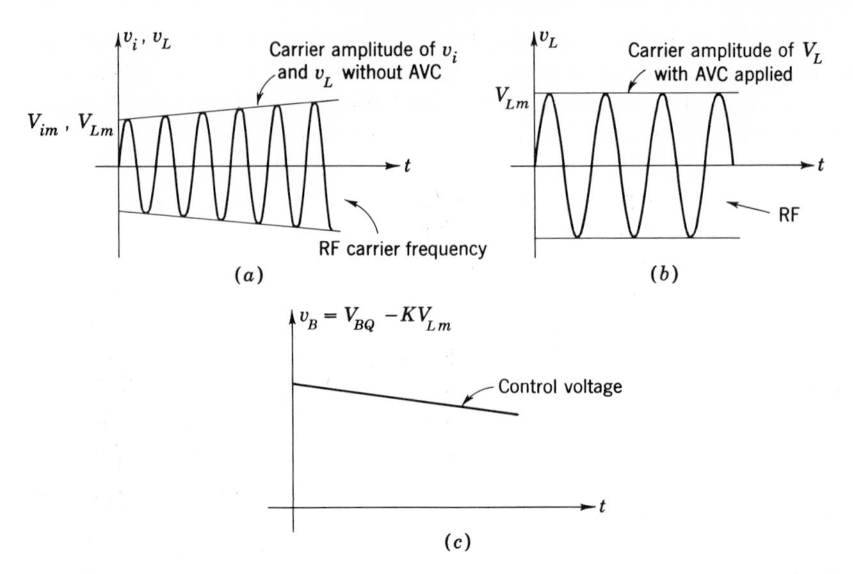

Fig. 8.7-2 Waveforms in the AVC circuit. (*a*) Input signal; (*b*) output signal with AVC; (*c*) control voltage.

Thus, since the AM has frequency components between about 100 Hz and 5 kHz, the RC low-pass filter is designed to pass frequencies much less than 100 Hz.

Consider that an unmodulated carrier is transmitted from a radio station located in a satellite which is slowly approaching the receiver. Then the input carrier level slowly increases as shown in Fig. 8.7-2*a*. If no AVC is applied, v_i and v_L will both increase at the same rate. If, however, AVC is applied, the carrier amplitude of the output V_{Lm} increases at a slower rate than the input amplitude V_{im}. This is illustrated in Fig. 8.7-2*b*. The control voltage v_B (Fig. 8.7-2*c*) decreases proportionally with the amplitude of the output V_{Lm}. This results in decreased base current, and hence an increase in the input impedance of the amplifier, h_{ie}. This, in turn, produces the desired effect of decreased gain.

The use of the *choke* in Fig. 8.7-1 is symbolic, and represents the isolation that must be present between the high-frequency input signal and the "dc" control signal.

The small-signal and dc equivalent circuits are shown in Fig. 8.7-3*a* and *b*.

The carrier amplitude level is, from Fig. 8.7-3*a*,

$$V_{Lm} \approx V_{im}\left(\frac{h_{fe}R_L\|R_c}{h_{ie}}\right) = V_{im}\left(\frac{R_L\|R_c}{h_{ib}}\right) = V_{im}\left(\frac{R_L\|R_c}{V_T}\right)I_{EQ} \tag{8.7-1}$$

Thus the output carrier amplitude V_{Lm} is proportional to the input amplitude V_{im} and the quiescent current I_{EQ}.

We differentiate to determine the change in V_{Lm} due to a change in V_{im}.

$$\Delta V_{Lm} = \left(\frac{R_L \| R_c}{h_{ib}}\right) \Delta V_{im} + V_{im} \left(\frac{R_L \| R_c}{V_T}\right) \Delta I_{EQ} \tag{8.7-2}$$

The quiescent emitter current is related to the output-carrier amplitude level by the diode equation (Sec. 3.1). Also using Fig. 8.7-3*b*,

$$I_{EQ} \approx I_{ES}\epsilon^{v_B/V_T} = I_{ES}\epsilon^{(V_{BQ}-KV_{LM})/V_T} \tag{8.7-3}$$

Then

$$\Delta I_{EQ} = \left(\frac{-K\,\Delta V_{Lm}}{V_T}\right) I_{ES}\epsilon^{(V_{BQ}-KV_{Lm})/V_T} = \left(\frac{-K\,\Delta V_{Lm}}{V_T}\right) I_{EQ}$$
$$= -\left(\frac{K}{h_{ib}}\right) \Delta V_{Lm} \tag{8.7-4}$$

Substituting (8.7-4) into (8.7-2) yields

$$\Delta V_{Lm} = \left(\frac{R_L \| R_c}{h_{ib}}\right) \Delta V_{im} - \left[\frac{(R_L \| R_c) V_{im}}{V_T}\right] \left(\frac{K\,\Delta V_{Lm}}{h_{ib}}\right) \tag{8.7-5}$$

Therefore

$$\Delta V_{Lm} = \frac{[(R_L \| R_c)/h_{ib}]\,\Delta V_{im}}{1 + [(R_L \| R_c)/V_T](KV_{im}/h_{ib})} \tag{8.7-6}$$

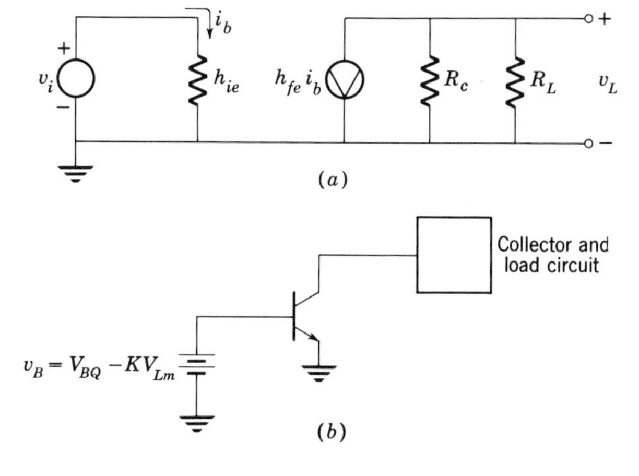

Fig. 8.7-3 Equivalent circuits of AVC amplifier. (*a*) Small-signal equivalent circuit for carrier frequencies; (*b*) equivalent circuit for dc and envelope frequencies.

It is shown in Prob. 8.25 that the voltage gain without feedback

$$A_v = \frac{R_L \| R_c}{h_{ib}} \tag{8.7-7}$$

and the loop gain

$$T = -A_v \left(\frac{K V_{im}}{V_T} \right) \tag{8.7-8}$$

In a well-designed AVC system, $-T \gg 1$, and (8.7-6) reduces to

$$\Delta V_{Lm} \approx \left(\frac{V_T}{K V_{im}} \right) \Delta V_{im} \tag{8.7-9}$$

The sensitivity function $S_{V_{im}}^{V_{Lm}}$ is

$$S_{V_{im}}^{V_{Lm}} = \frac{\Delta V_{Lm}/V_{Lm}}{\Delta V_{im}/V_{im}} = \frac{V_T}{K V_{Lm}} \tag{8.7-10}$$

As an example, let $V_{im} = 1$ V, $I_{EQ} = 1$ mA, $R_L = 100\ \Omega$, $R_c = 5$ kΩ, and $h_{fe} = 100$. Calculate V_{Lm} and $K_{\min}$ so that V_{Lm} changes by less than 0.1 percent when V_{im} changes by 10 percent.

Using (8.7-10), we have

$$\frac{0.1}{10} \geq \frac{V_T}{K V_{Lm}}$$

$$K V_{Lm} \geq 100 V_T = 2.5 \text{ V}$$

Using (8.7-1),

$$V_{Lm} = \frac{100 \| 5000}{25} = 4 \text{ V}$$

Therefore

$$K \geq 0.625$$

Example 8.7-1 An experimental AVC circuit The AVC circuit shown in Fig. 8.7-4 was constructed and tested. Transistors T_1 to T_7 form the forward amplifier, which is represented simply by T_1 in Fig. 8.7-1. Transistors T_1, T_2, and T_3 and T_5, T_6, and T_7 are IC amplifiers. The EF, T_4, is employed to obtain increased voltage gain (Prob. 8.27).

The output voltage v_L is rectified and filtered, and a dc voltage V_{BB} is formed (Fig. 8.7-4). This is the base voltage of T_3, and it is proportional to the output carrier level V_{Lm}. I_{EQ3} is proportional to V_{BB} so that h_{ie1} and h_{ie2} are inversely proportional to V_{BB}, thus providing the AVC action.

The quiescent voltages are given in Fig. 8.7-4 when $V_{Lm} \approx$

1.25 V and $V_{im} = 1.8$ mV. To determine the quiescent operating point, we note that $V_{C2} = 7.4$ V, hence $I_{E1} = I_{E2} \approx 0.125$ mA (neglecting the current in the base of T_4). Thus, $h_{ib1} = h_{ib2} \approx 200\ \Omega$.

The quiescent current in T_4 is found by observing that $V_{E4} \approx 7.4 - 0.7 = 6.7$ V. Hence, $I_{E4} \approx 3.3$ mA. The quiescent point of T_5 and T_6 is found by first calculating I_{E7}. Since $V_{E7} \approx -2.3$ V, $I_{E7} \approx 1.25$ mA. Thus, $h_{ib5} = h_{ib6} = 40\ \Omega$. The voltage gain in T_4 to T_6 can be shown to be

$$A_2 = \frac{v_L}{v_{c2}} \approx 60$$

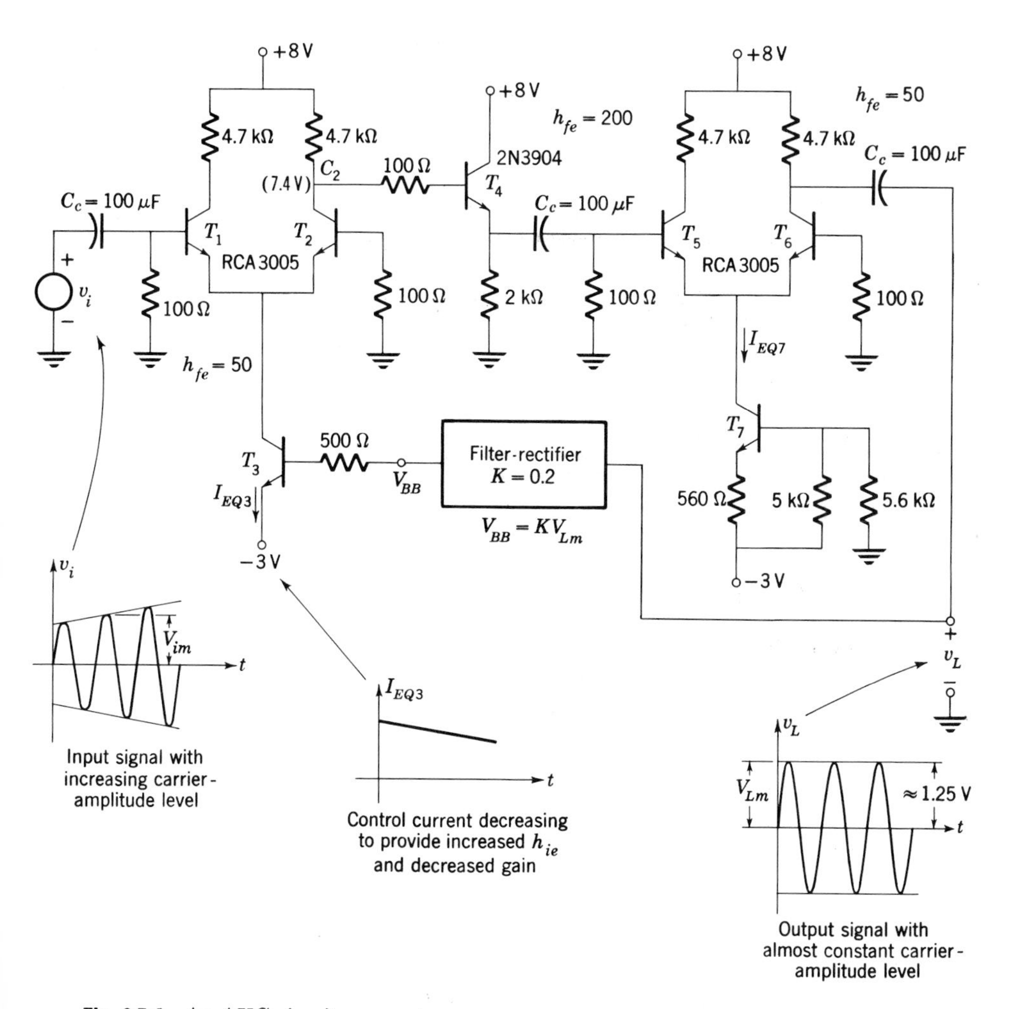

Fig. 8.7-4 An AVC circuit.

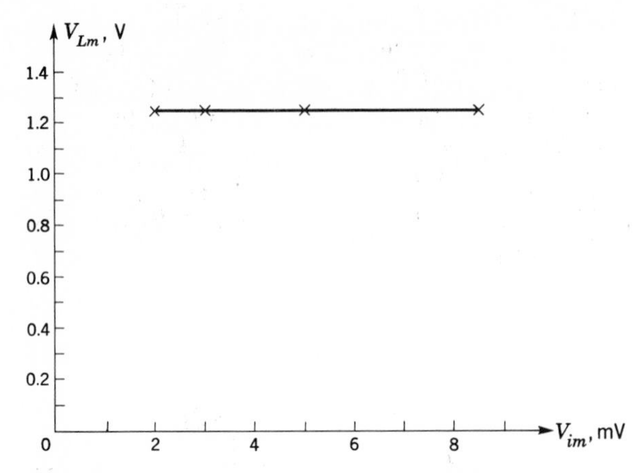

Fig. 8.7-5 Experimental transfer characteristic of the AVC circuit of Fig. 8.7-4.

Let us now calculate the sensitivity of the AVC in this circuit. First refer to (8.7-1). In this amplifier

$$V_{Lm} \approx V_{im} \frac{R_{c2}}{2h_{ib1}} A_2 = \frac{V_{im}}{h_{ib1}} (4700)(30) \approx 14 \times 10^4 \frac{V_{im}}{h_{ib1}}$$

Differentiating as in (8.7-2),

$$\Delta V_{Lm} = \frac{14 \times 10^4}{h_{ib1}} \Delta V_{im} + \left(\frac{14 \times 10^4 V_{im}}{V_T} \right) \Delta I_{EQ1}$$

We find ΔI_{EQ1} from ΔI_{EQ3}

$$\Delta I_{EQ1} = \frac{1}{2} \Delta I_{EQ3} = \frac{1}{2} \left(\frac{\Delta V_{BB}}{500/\beta_3} \right)$$

and since $K \approx 0.2$, $\Delta V_{BB} = -0.2(\Delta V_{Lm})$, and

$$\Delta I_{EQ1} = \frac{1}{2} \left(\frac{-0.2\, \Delta V_{Lm}}{10} \right)$$

This result is analogous to (8.7-4). Thus, $\Delta I_{EQ1} = 10^{-2}\, \Delta V_{Lm}$.

Substituting the result for ΔI_{EQ1} into our equation for ΔV_{Lm} yields

$$\Delta V_{Lm} \approx 700\, \Delta V_{im} - 56{,}000 V_{im}\, \Delta V_{Lm}$$

Solving,

$$\Delta V_{Lm} \approx \frac{700\, \Delta V_{im}}{1 + 56{,}000 V_{im}} \approx 0.0125 \frac{\Delta V_{im}}{V_{im}}$$

Dividing by $V_{Lm} \approx 1.25$ V we have

$$\frac{\Delta V_{Lm}}{V_{Lm}} = 0.01 \frac{\Delta V_{im}}{V_{im}}$$

Thus a 10 percent change in V_{im} results in a 0.1 percent change in V_{Lm}.

The transfer characteristic of the AVC circuit was measured and is shown in Fig. 8.7-5. Note that the variation of V_{Lm} is not noticeable over a 4.5 to 1 variation of V_{im}.

8.7-2 THE REGULATED POWER SUPPLY[2]

A second very important application of feedback is the regulated power supply. The use of a Zener-diode voltage regulator was discussed in Sec. 2.7. The transistor regulators analyzed in this section result in increased stability.

A simple voltage regulator is shown in Fig. 8.7-6. The unregulated supply is represented by a battery V_i in series with a variable voltage v_i, which represents changes in the supply voltage due to ripple or power-line voltage variations, and a series internal resistance r_i. The transistor is seen to be an emitter follower whose base voltage is maintained relatively constant by using a Zener diode. Current is supplied to the diode through R_b. High-stability voltage regulators often use a battery rather than a Zener diode. R_b is then eliminated, and the emitter current remains approximately constant.

$$i_E \approx \frac{V_{\text{battery}} - V_{BE}}{R_L} \tag{8.7-11}$$

independent of changes caused by v_i.

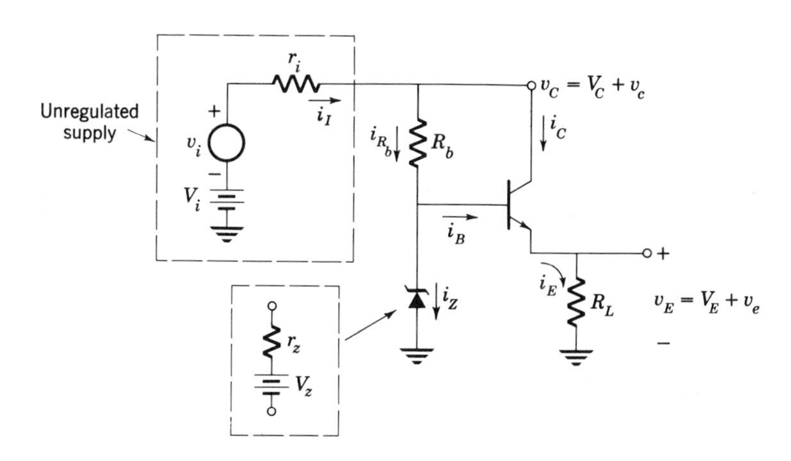

Fig. 8.7-6 Emitter-follower regulator.

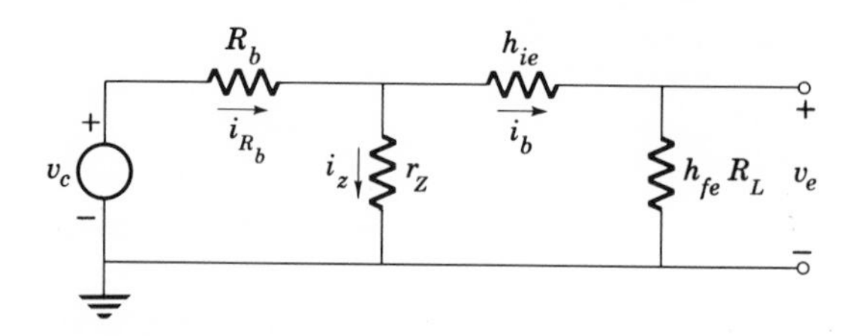

Fig. 8.7-7 Small-signal equivalent of EF regulator of Fig. 8.7-6.

Effect of input-voltage variation The following analyses assume that the Zener diode can be approximated by a battery V_Z in series with a resistance r_z. In addition, we assume that v_e is small (this assumption is to be verified). Let us first calculate the *change* in output voltage v_e, with respect to a change v_c, in the collector voltage. We use Fig. 8.7-6.

Since the transistor is connected as an emitter follower, we begin by reflecting the emitter circuit into the base as shown in Fig. 8.7-7. In most practical cases we can assume that

$$r_z \ll R_b \tag{8.7-12a}$$

$$r_z \ll h_{ie} + h_{fe}R_L \tag{8.7-12b}$$

and

$$h_{ie} \ll h_{fe}R_L \tag{8.7-12c}$$

Then, by inspection,

$$\frac{v_e}{v_c} \approx \frac{r_z}{R_b} \tag{8.7-13}$$

Thus, if the Zener resistance is zero, changes in collector voltage v_c do not affect the emitter voltage. This result is expected since, if $r_z = 0$ (Fig. 8.7-6),

$$V_E = V_Z - V_{BE} \approx \text{constant}$$

To determine the change in output voltage v_e with respect to changes in the unregulated supply voltage v_i, we must calculate the change in the current i_1 (Fig. 8.7-6).

$$i_1 = i_c + i_{R_b}$$

From Fig. 8.7-7

$$i_{R_b} \approx \frac{v_c}{R_b + r_z} \approx \frac{v_c}{R_b} \tag{8.7-14a}$$

and

$$i_c \approx i_e = \left(\frac{v_e}{R_L}\right) \approx \left(\frac{1}{R_L}\right)\left(\frac{r_z}{R_b}\right) v_c \tag{8.7-14b}$$

Referring to Fig. 8.7-6, we see that

$$v_c = v_i - r_i i_1$$
$$= v_i - r_i\left(1 + \frac{r_z}{R_L}\right)\left(\frac{v_c}{R_b}\right) \tag{8.7-15a}$$

and therefore

$$\frac{v_c}{v_i} = \frac{1}{1 + (r_i/R_b)(1 + r_z/R_L)} \tag{8.7-15b}$$

Combining (8.7-13) and (8.7-15*b*) yields

$$\frac{v_e}{v_i} \approx \frac{r_z/R_b}{1 + (r_i/R_b)(1 + r_z/R_L)} \tag{8.7-16}$$

This result indicates that, to achieve good regulation, R_b should be chosen much larger than r_z. In addition, it indicates that R_b should be chosen much smaller than r_i. These are conflicting requirements, since r_z and r_i both usually vary between 1 and 10 Ω. (r_i is often increased by the addition of an external resistance in series with the unregulated supply. However, care must be taken to ensure that the transistor does not saturate.)

Effect of load-resistance variation Let us now determine the change in output voltage v_e due to a change in load resistance R_L. This change in R_L is often very large. (Consider a battery-operated radio. With the radio turned off, the load resistance across the battery is infinite, since the transistors are not drawing current. With the radio turned on, 10 mA may be drawn. If a 6-V battery is used, the transistors "look like" a 600-Ω load.)

Consider that the regulator of Fig. 8.7-6 is very good. Then, even when R_L changes by a large amount, v_e remains small. If r_z is zero,

$$V_Z = V_{BE1} + I_{E1}R_{L1} \tag{8.7-17a}$$

and

$$V_Z = V_{BE2} + I_{E2}R_{L2} \tag{8.7-17b}$$

where R_{L1} and R_{L2} are the initial and final values of the load resistance, and I_{E1}, I_{E2}, V_{BE1}, and V_{BE2} represent the initial and final values of the emitter current and base-emitter voltage. (Note that even though v_e is small, $I_{E2} - I_{E1}$ is large for large changes in load resistance.)

The base-to-emitter voltage and the emitter current are related by the diode equation (Sec. 3.1).

$$I_E \approx I_{ES}\epsilon^{V_{BE}/V_T} \tag{8.7-18a}$$

Thus

$$V_{BE1} - V_{BE2} = V_T\left(\ln\frac{I_{E1}}{I_{E2}}\right) \tag{8.7-18b}$$

Equating (8.7-17*a*) and (8.7-17*b*) and using (8.7-18*b*) yields

$$I_{E2}R_{L2} - I_{E1}R_{L1} = V_T\left(\ln \frac{I_{E1}}{I_{E2}}\right) = v_e \tag{8.7-19}$$

Since v_e is assumed small, let us also assume that

$$V_{E2} = I_{E2}R_{L2} \approx I_{E1}R_{L1} = V_{E1}$$

Then

$$\frac{I_{E1}}{I_{E2}} \approx \frac{R_{L2}}{R_{L1}}$$

and using (8.7-19),

$$v_e \approx V_T\left(\ln \frac{R_{L2}}{R_{L1}}\right) = (25 \times 10^{-3})\left(\ln \frac{R_{L2}}{R_{L1}}\right) \tag{8.7-20}$$

Therefore, if $V_{E1} = 12$ V and $R_{L2}/R_{L1} = 1000$,

$$v_e \approx 0.025 \ln 1000 \approx 173 \text{ mV} \ll V_{E1}$$

If the Zener resistance r_z is not zero, (8.7-19) becomes

$$v_e = I_{E2}R_{L2} - I_{E1}R_{L1} \approx V_T\left(\ln \frac{I_{E1}}{I_{E2}}\right) + \left(\frac{r_z}{h_{fe}+1}\right)(I_{E1} - I_{E2}) \tag{8.7-21}$$

If, again, v_e is assumed small, we have

$$v_e = V_T\left(\ln \frac{R_{L2}}{R_{L1}}\right) + \left(\frac{r_z}{h_{fe}+1}\right)\left(\frac{1}{R_{L1}} - \frac{1}{R_{L2}}\right)V_{E,\text{nominal}} \tag{8.7-22}$$

For example, if $V_{E,\text{nominal}} = 12$ V, $R_{L1} = 100\ \Omega$, $R_{L2} = 100$ kΩ, $r_z = 2\ \Omega$, and $h_{fe} = 100$,

$$v_e \approx 0.173 + (2/101)(1/100)(12) \approx 0.175 \text{ V} \ll V_{E,\text{nominal}}$$

Note that the effect of $r_z/(h_{fe} + 1)$ is negligible.

A two-transistor feedback regulator The simple regulator shown in Fig. 8.7-6 has several disadvantages. Its output voltage is determined solely by the Zener diode. It therefore cannot be varied without changing the Zener diode. In addition, high-voltage Zener diodes have a high temperature coefficient. Therefore, temperature variations result in large changes in diode voltage and hence load voltage.

The two-transistor feedback regulator shown in Fig. 8.7-8 permits the use of a low voltage Zener diode which has a low temperature coefficient. Thus, the load voltage is fairly insensitive to Zener temperature variations. Another advantage of the feedback regulator is that the load

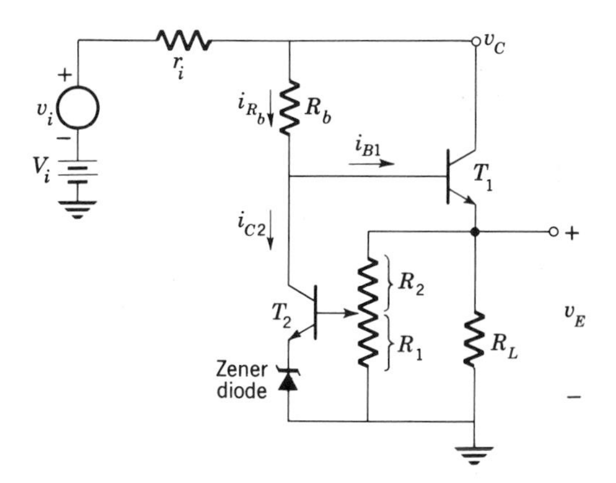

Fig. 8.7-8 Feedback regulator.

voltage is easily adjusted by varying the *arm* of the potentiometer (adjusting R_1 and R_2).

The operation of the feedback regulator is as follows: Transistor T_1 is an emitter follower. The output voltage v_E is fed back to T_2, resulting in a current i_{C2} proportional to v_E. The "error" current i_{B1} is thereby maintained nearly constant with respect to changes in v_C due to v_i.

Effect of input-voltage variation The equivalent circuit of Fig. 8.7-8 is shown in Fig. 8.7-9. Figure 8.7-9*a* illustrates the effect of voltage feedback in the circuit. v_E' is set equal to v_E when directly calculating the voltage gain with feedback. Transistor T_2 is seen to act as a current source which is proportional to v_E'. This is illustrated in Fig. 8.7-9*b*.

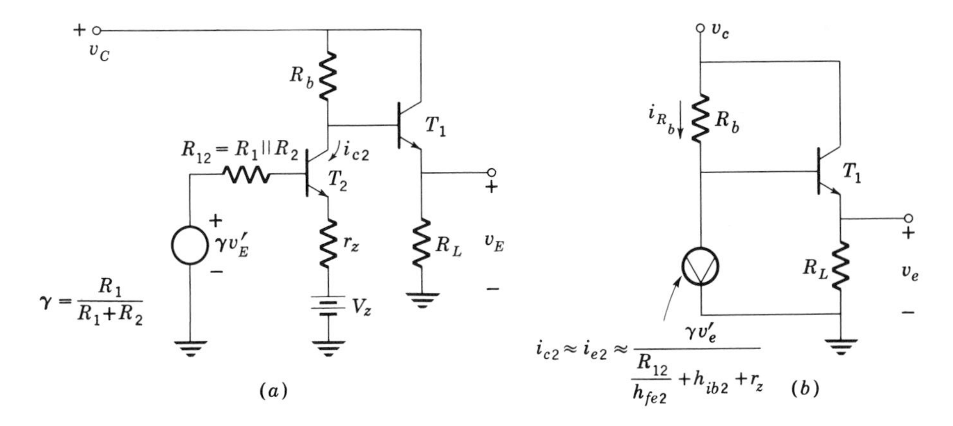

Fig. 8.7-9 Feedback regulator with loop opened. (*a*) Circuit for total currents and voltages; (*b*) small-signal equivalent circuit.

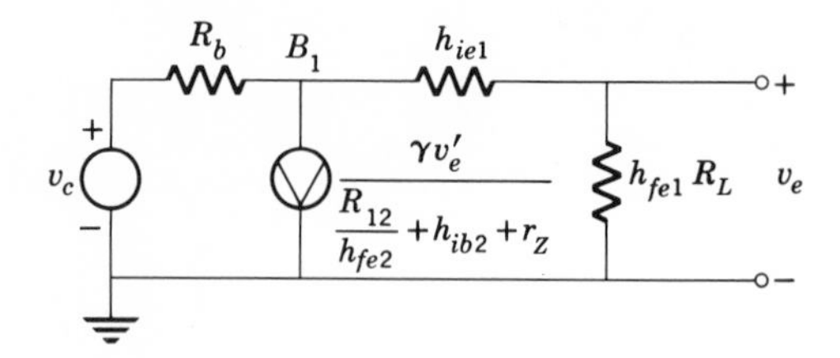

Fig. 8.7-10 Small-signal equivalent circuit of feedback regulator.

The value of the current was obtained using KVL around the base-emitter loop of T_2. The final equivalent circuit shown in Fig. 8.7-10 is obtained by reflecting R_L into the base circuit of T_1.

The voltage gain without feedback A_v is

$$A_v = \left.\frac{v_e}{v_c}\right|_{v'_e=0} = \frac{h_{fe1}R_L}{R_b + h_{ie1} + h_{fe1}R_L} \tag{8.7-23a}$$

and the loop gain T is

$$T = \left.\frac{v_e}{v_e'}\right|_{v_c=0} = -\left[\frac{\gamma}{(R_{12}/h_{fe2}) + h_{ib2} + r_z}\right]\left(\frac{R_b}{R_b + h_{ie1} + h_{fe1}R_L}\right)h_{fe1}R_L \tag{8.7-23b}$$

where

$$\gamma = \frac{R_1}{R_1 + R_2} \tag{8.7-23c}$$

Assuming $-T \gg 1$,

$$A_{vf} = \frac{v_e}{v_c} \approx \left(\frac{1}{\gamma R_b}\right)\left(\frac{R_{12}}{h_{fe2}} + h_{ib2} + r_z\right) \tag{8.7-24}$$

Comparing (8.7-24) with (8.7-13) shows that the simple regulator is more insensitive to changes in input voltage than is the feedback regulator. However, the feedback regulator is adjustable.

Effect of load-resistance variation Let us now determine the variation of v_E caused by a large change in load resistance (when v_C is constant). Since v_e will be small, T_2 can be represented by its small-signal equivalent circuit, and therefore Fig. 8.7-9*b* can be used. Then, using KVL,

$$\Delta i_{R_b} R_b = -(v_{be1} + v_e) \tag{8.7-25a}$$

where

$$\Delta i_{R_b} = \frac{\gamma v_e}{R_{12}/h_{fe2} + h_{ib2} + r_z} + \left(\frac{1}{h_{fe1}}\right)(I_{E2} - I_{E1}) \tag{8.7-25b}$$

Then, combining (8.7-25*a*), (8.7-25*b*), and (8.7-18*b*) yields

$$v_e\left(1+\frac{\gamma R_b}{\dfrac{R_{12}}{h_{fe2}}+h_{ib2}+r_z}\right) \approx -V_T\left(\ln\frac{I_{E2}}{I_{E1}}\right)-\left(\frac{V_{E,\text{nominal}}}{h_{fe1}}\right)\left(\frac{1}{R_{L2}}-\frac{1}{R_{L1}}\right)R_b \qquad (8.7\text{-}26a)$$

This equation simplifies if R_b is made very large [Eq. (8.7-24)]. Then

$$\frac{v_e}{V_E}\approx\left(\frac{R_{L2}-R_{L1}}{\gamma h_{fe1}R_{L1}R_{L2}}\right)\left(\frac{R_{12}}{h_{fe2}}+h_{ib2}+r_z\right) \qquad (8.7\text{-}26b)$$

Consider, for example, that $V_E \approx 12$ V, $R_{L1} = 100\ \Omega$, $R_{L2} = 100$ kΩ, $\gamma = 0.5$, $r_z + h_{ib2} + R_{12}/h_{fe2} \approx 15\ \Omega$, and $h_{fe1} = 50$. Then

$$v_e \approx (12)\left[\frac{2}{(50)(100)}\right](15) = 0.072 \text{ V}$$

Example 8.7-2 The regulated power supply shown in Fig. 8.7-11 was designed to supply between 11 and 12 V to a load which could vary between 250 Ω and 10 kΩ. The input voltage v_C has a total variation between 18 and 20 V.

The following experiments were performed (voltages were read using a digital voltmeter):

	R_L, Ω	V_C, V	V_E, V	V_{BE}, V
1.	246	18	11.3	0.61
2.	246	20	11.45	0.61
3.	10 kΩ	20	11.55	

Calculations show that [Eq. (8.7-13)]

$$\frac{v_e}{v_c}\approx\frac{r_z}{R_b}=0.06$$

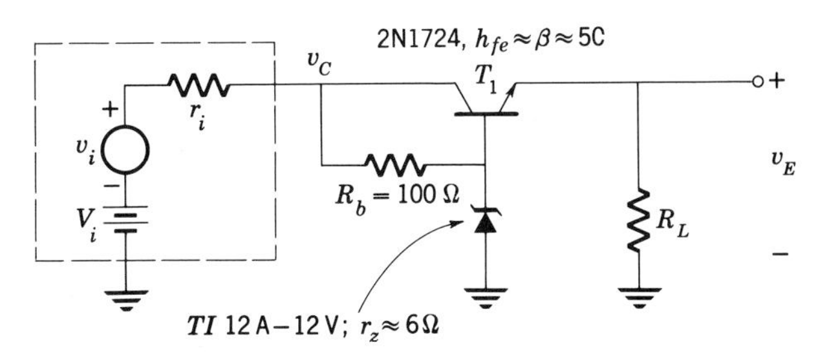

Fig. 8.7-11 Circuit for Example 8.7-2.

Measurements performed when $R_L = 246\ \Omega$ show that

$$\frac{v_e}{v_c} = \frac{v_{E2} - v_{E1}}{v_{C2} - v_{C1}} = \frac{11.45 - 11.3}{20 - 18} = \frac{0.15}{2} = 0.075$$

The expected variation of output voltage with respect to changes in R_L was calculated using (8.7-22).

$$v_e = V_T\left(\ln\frac{R_{L2}}{R_{L1}}\right) + \left(\frac{r_z}{h_{fe}+1}\right)\left(\frac{1}{R_{L1}} - \frac{1}{R_{L2}}\right)V_{E,\text{nominal}}$$

$$= 0.025\left(\ln\frac{10{,}000}{246}\right) + \left(\frac{6}{51}\right)\left(\frac{1}{246} - \frac{1}{10{,}000}\right)(11.5)$$

$$\approx 0.0925 + 0.0055$$

$$\approx 0.098\ \text{V}$$

Notice that v_e varies slightly, depending on the nominal value of V_E chosen for the calculation. Measurements performed with $R_L = 246\ \Omega$ and $R_L = 10\ \text{k}\Omega$ when $v_C = 20$ V yield

$$v_e = V_{E3} - V_{E2} = 11.55 - 11.45 = 0.10\ \text{V}$$

PROBLEMS

SECTION 8.2

8.1. Without making any approximations concerning R_f/R_i and $R_\beta \| R_L \| r_o$, calculate A_{if}. Compare your answer with (8.2-4c).

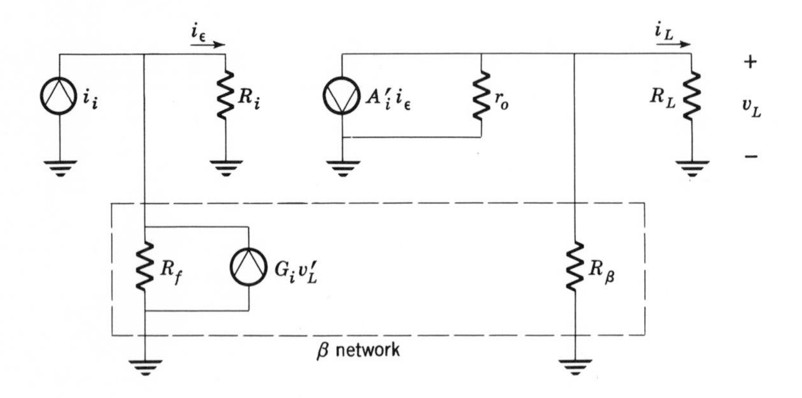

Fig. P8.1

8.2. Calculate the current gain of the amplifier shown in Fig. 8.2-3 when R_f is not much greater than R_L.

8.3. (a) Calculate the loop gain. Assume $h_{ob} = 0$.
(b) Calculate the voltage gain v_L/v_i.

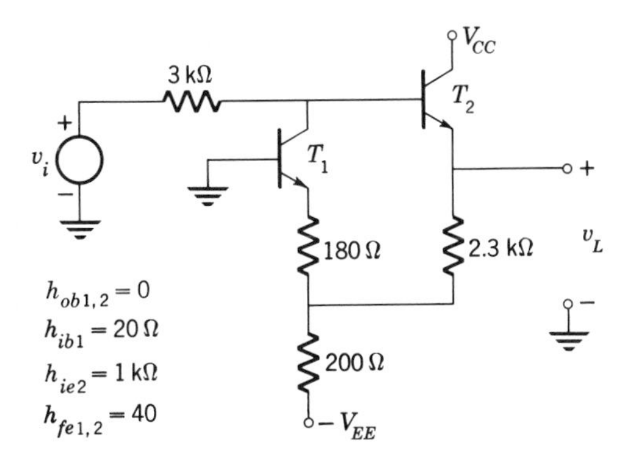

Fig. P8.3

8.4. (a) Calculate the loop gain.
(b) Calculate the current gain i_L/i_i. $\left(Hint{:}\ T = v_1/v_1'\Big|_{i_i=0}.\right)$

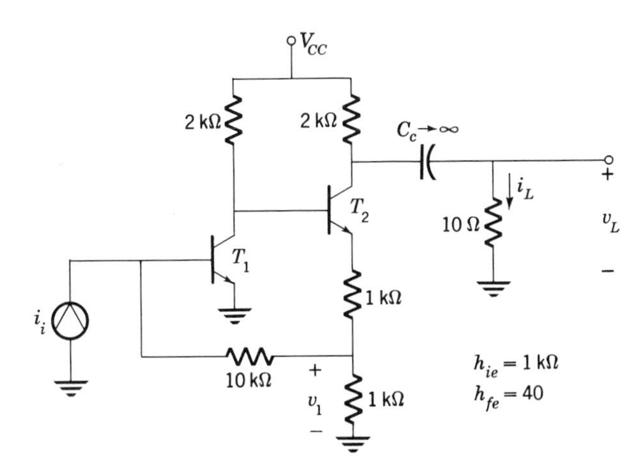

Fig. P8.4

SECTION 8.3.

8.5. (*a*) Calculate v_{c2}/v_{cc}.
(*b*) Calculate $\Delta v_{C2}/\Delta T$, where ΔT is the change in temperature.

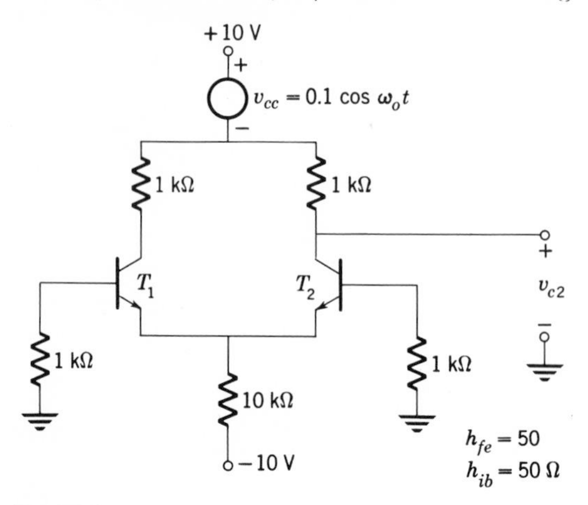

Fig. P8.5

8.6. If the $-V_{EE}$ supply in Fig. P8.3 varies, find v_L/v_{cc}, where v_{cc} is the variable component of the supply.

8.7. If V_{CC} varies in Fig. P8.4, calculate v_L/v_{cc}.

SECTIONS 8.4 AND 8.5

8.8. Calculate the input and output impedances of the amplifier shown in Fig. P8.3.

8.9. Calculate the input and output impedances of the amplifier shown in Fig. P8.4.

8.10. Calculate
(*a*) The loop gain.
(*b*) The input and output impedances.
(*c*) The gain v_L/i_i.

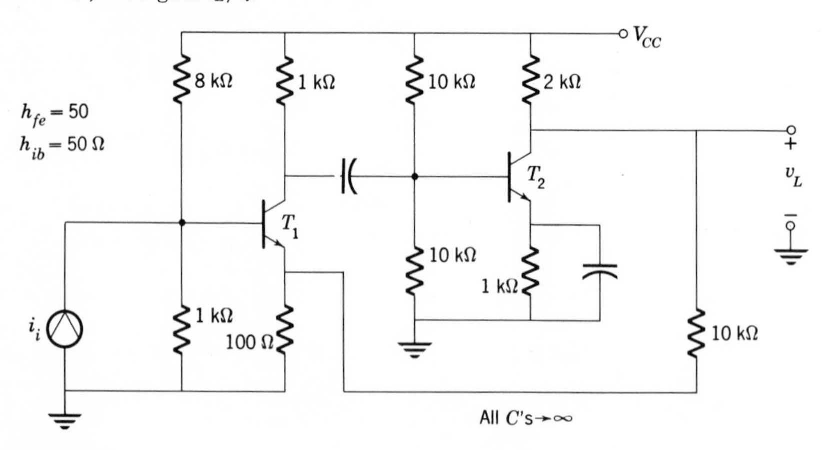

Fig. P8.10

8.11. Calculate

(*a*) The loop gain.

(*b*) The input and output impedances.

(*c*) The current gain i_L/i_i.

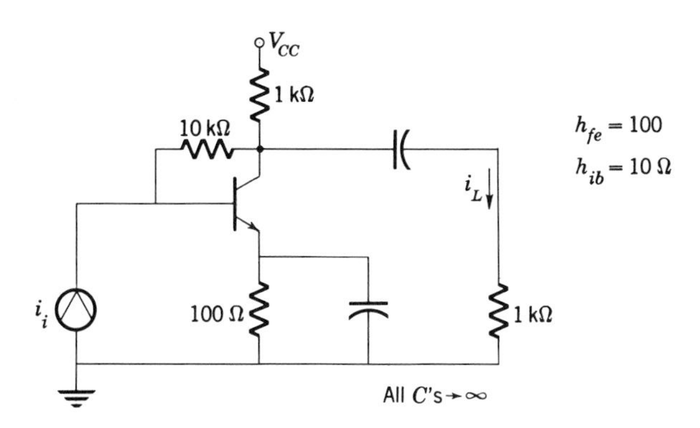

Fig. P8.11

8.12. (*a*) Calculate the loop gain.

(*b*) Calculate Z_{of}.

(*c*) Calculate the voltage gain

(*d*) If a load resistor R_L is placed across the output terminals, calculate v_L as a function of R_L.

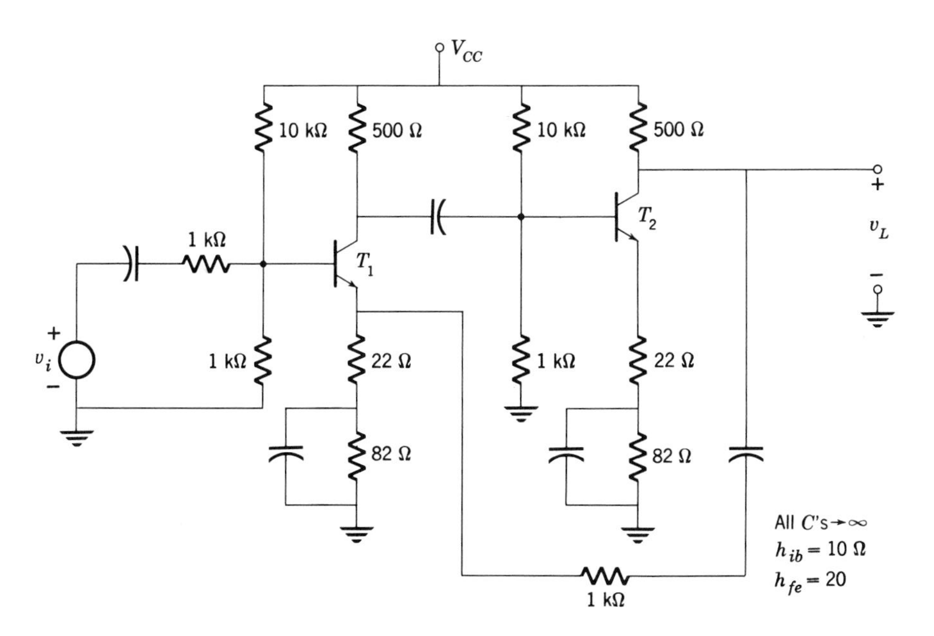

Fig. P8.12

8.13. (*a*) Calculate the loop gain.
(*b*) Calculate the input and output impedances.

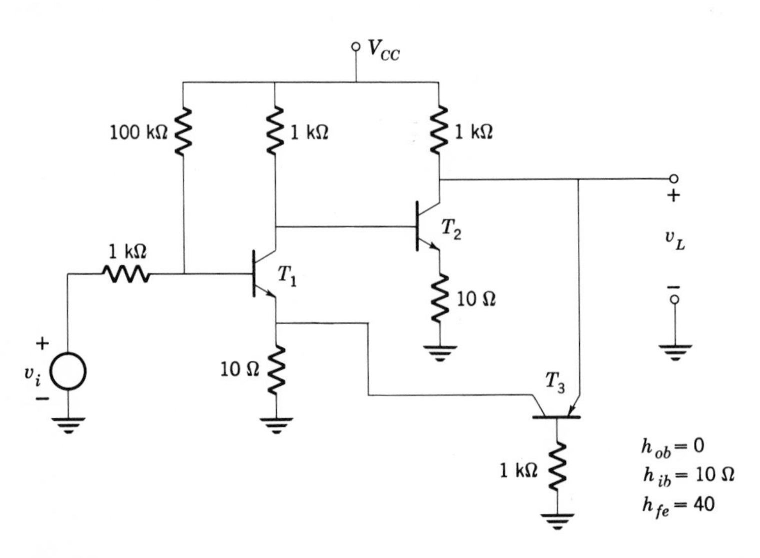

Fig. P8.13

8.14. Using Fig. P8.13, calculate
(*a*) The voltage gain.
(*b*) v_L as a function of R_L, where R_L is placed across the output.

8.15. Calculate
(*a*) The loop gain.
(*b*) Z_{if} and Z_{of}.
(*c*) v_L/v_i.

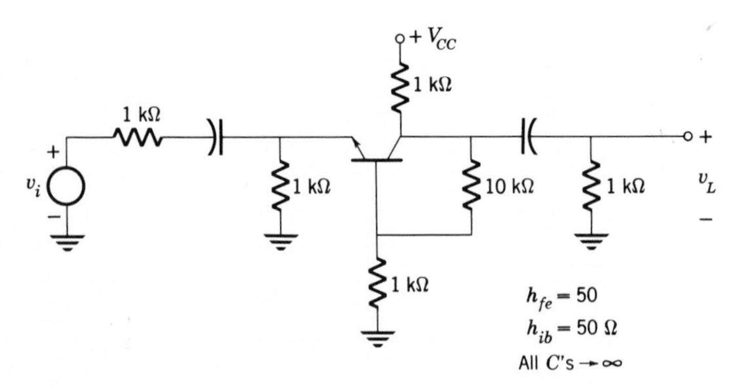

Fig. P8.15

8.16. (*a*) Calculate the loop gain. $\left(Hint: T = v_{L2}/v'_{L2}\Big|_{v_1=0}.\right)$

(*b*) Find Z_{of}, Z_{if}.

(*c*) Find v_{L1}/v_i.

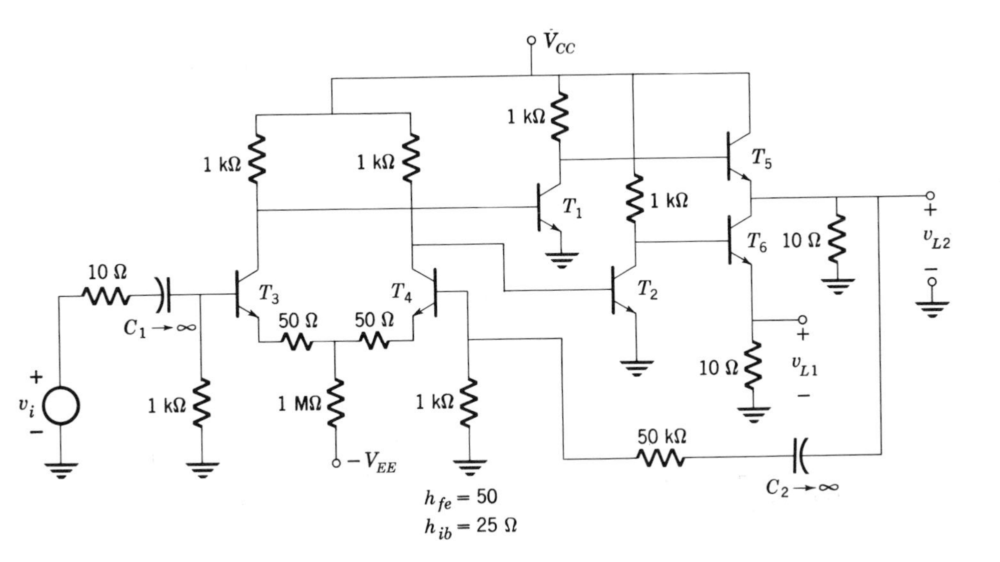

Fig. P8.16

8.17. (*a*) Find v_c/v_i (calculate T) and the impedance seen looking into the collector.

(*b*) Find v_e/v_i (calculate T) and the impedance seen looking into the emitter.

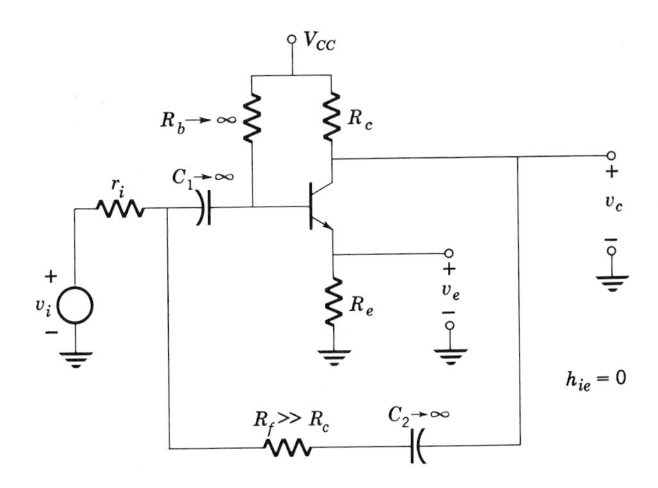

Fig. P8.17

8.18. Repeat Prob. 8.17 using Fig. P8.18.

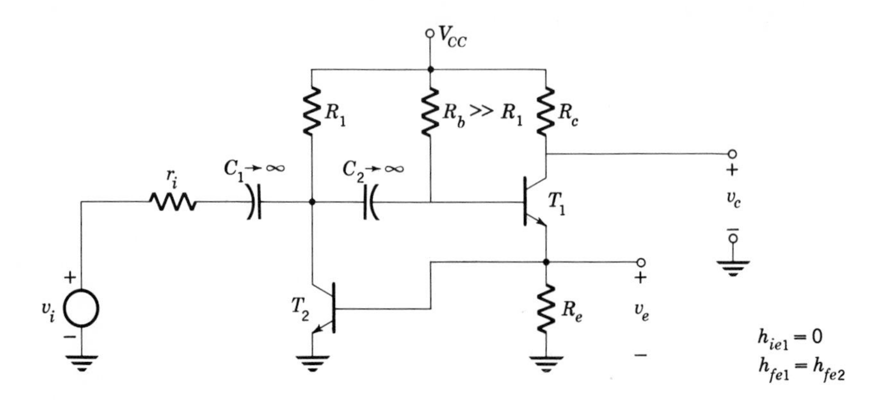

Fig. P8.18

SECTION 8.6

8.19. Design a feedback amplifier to have a voltage gain of 10, a loop gain greater than 10, and an input impedance of 10 kΩ. Use a minimum number of transistors. Each transistor has $h_{fe} = 100$. A 24-V supply is available. (*Hint:* Begin by choosing a configuration. Use the EF where it helps. Some positive feedback is allowed if it helps.)

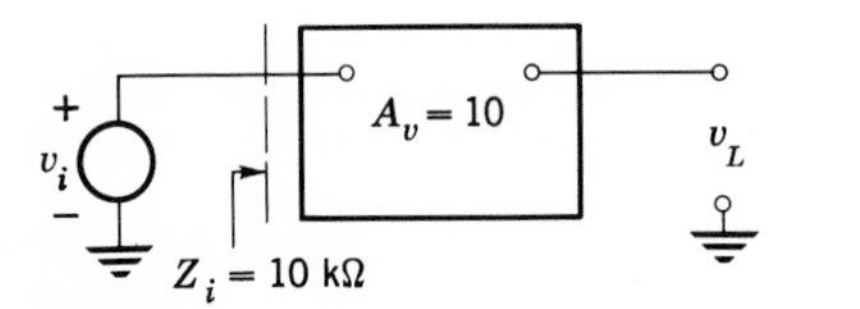

Fig. P8.19

8.20. The amplifier of Prob. 8.19 is to be designed using only IC difference amplifiers of the type shown in Fig. P8.20. How many IC amplifiers are needed? Sketch a possible design.

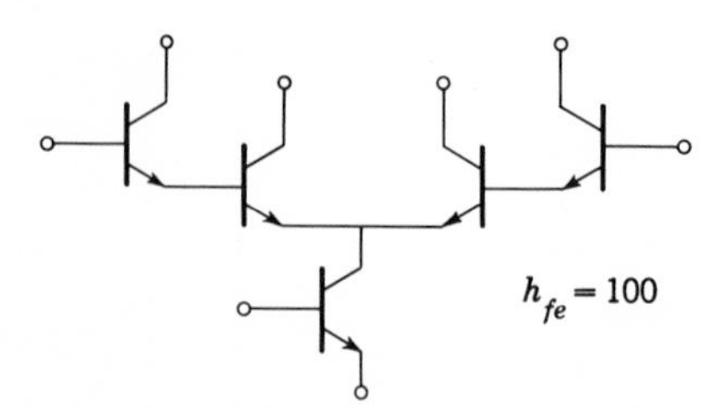

Fig. P8.20

8.21. Show that the emitter follower is an example of voltage feedback and voltage error. Calculate A_v, Z_i, Z_o, T, A_{vf}, Z_{if}, Z_{of}.

SECTION 8.7

8.22. The IC amplifier shown is used as an AVC amplifier by controlling V_B. Calculate and plot v_L/v_i as a function of V_B (positive and negative values).

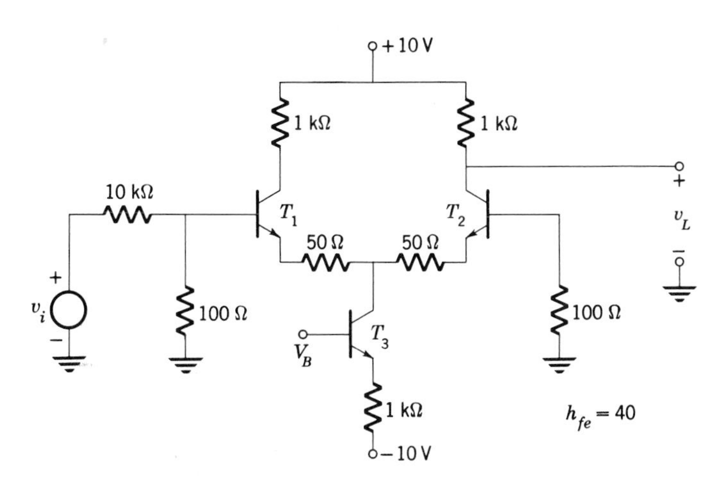

Fig. P8.22

8.23. The amplifier shown can be used as an AVC amplifier by controlling the dc base current to T_1.

Plot v_L/v_i as a function of I_{AVC}.

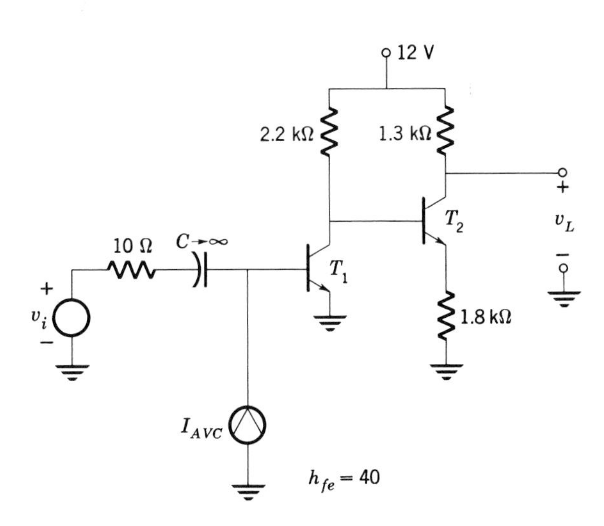

Fig. P8.23

8.24. The circuit shown in Fig. P8.24 is a current-controlled attenuator often used in AVC amplifiers.

Calculate i_L/i_i as a function of I_{AVC}.

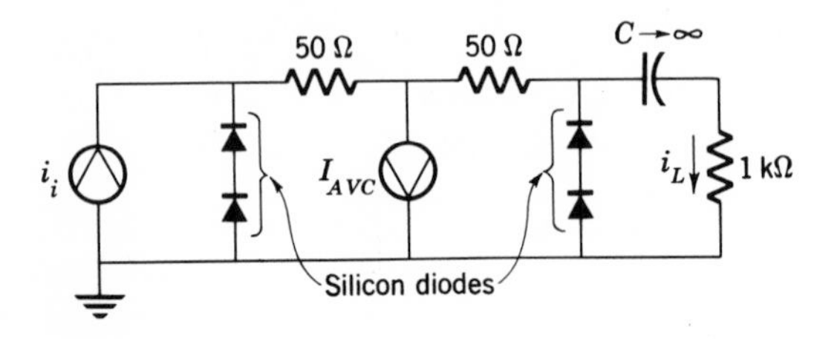

Fig. P8.24

8.25. Verify (8.7-7) and (8.7-8). $\left(Hint: A_v \equiv \Delta V_{Lm}/\Delta V_{im}\Big|_{\Delta V'_{im}=0}; \; T \equiv \Delta V_{Lm}/\Delta V'_{Lm}\Big|_{\Delta V_{im}=0}.\right)$

8.26. For the AVC circuit shown, find

(a) The Q point for T_1, T_2, T_3.

(b) The dynamic impedance of D_1, D_2, D_3, D_4.

(c) The gain I_{AVC}/i_L (peak value).

(d) The small-signal equivalent circuit.

(e) The gain i_L/i_i of the AVC amplifier.

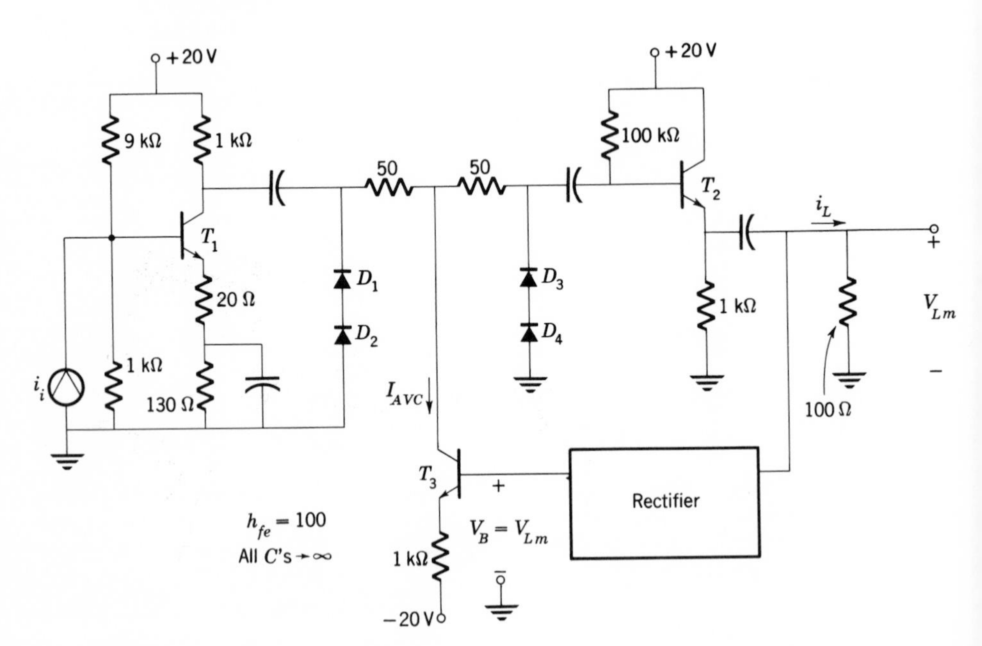

Fig. P8.26

8.27. Show that in Fig. 8.7-4 the EF T_4 is required to obtain increased voltage gain. To do this compare the voltage gain, without AVC, with and without the EF.

8.28. Plot the output voltage v_e in (8.7-22) as a function of the ratio R_{L2}/R_{L1}. Let $r_z/(h_{fe} + 1) = 0.1$ and $V_{E,\text{nominal}}/R_{L1} = 100$ mA.

8.29. Design a voltage regulator using the circuit of Fig. 8.7-6. The load resistance can vary from 100 Ω to 10 kΩ. The output (emitter) voltage is to be nominally 24 V. A 2.5 percent peak output ripple is allowed. V_i is nominally 75 V, and $r_i = 1$ Ω. How much peak input ripple is permitted? *Note:* A resistance can be placed in series with r_i if desired.

8.30. Refer to Example 8.7-2. Verify that when $R_L = 246$ Ω, $v_e/v_c \approx 0.06$. If measurements of r_z and R_b were each incorrect by 10 percent, find the range within which the calculation of v_e/v_c would have to lie.

8.31. The regulation of the EF regulator to changes in load resistance is often measured by calculating a small-signal output resistance and comparing R_L with this resistance. If R_L can vary significantly, this procedure is not valid. Why?

REFERENCES

1. Thornton, R. D., et al.: "Multistage Transistor Circuits," vol. 5, John Wiley & Sons, Inc., New York, 1965.
2. "Silicon Rectifier Handbook" and "Silicon Zener Diode and Rectifier Handbook," Motorola, Inc., Semiconductor Product Division, Phoenix, Ariz., 1961.

9
Integrated Circuits

INTRODUCTION

In this chapter we discuss a new technique being used to construct transistor amplifier circuits and systems.[1] This technique yields *integrated circuits,* which require orders of magnitude of less space than the discrete transistor circuits used previously. Figure 9-1*a* shows a single conventional transistor, while Fig. 9-1*b* is a photograph of a nine-transistor integrated-circuit (IC) amplifier. The transistor and the IC amplifier are each contained in a TO-5 package. This is a standard case used to package transistors and ICs.

Two types of integrated circuits are currently being manufactured, the monolithic and hybrid circuits. The monolithic circuit is "fully integrated"; i.e., the complete amplifier is made at the same time, using diffusion techniques. In the hybrid circuit, separate microminiaturized components, constructed using both diffusion and thin-film techniques, are connected either by deposition techniques, to be described below, or by wire bonds. Thus the hybrid IC has the form of a discrete circuit

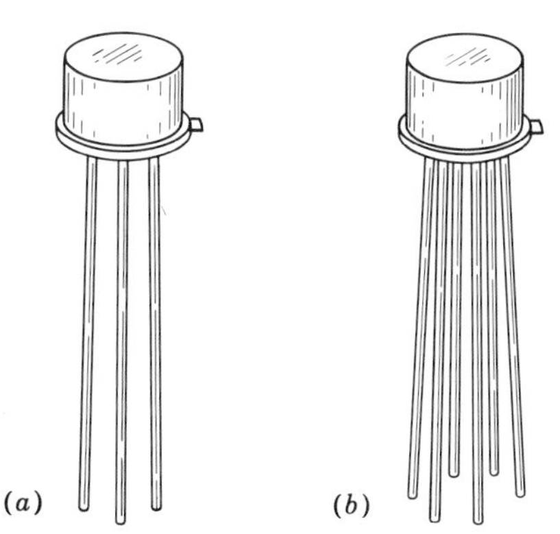

Fig. 9.1 Comparison of a single transistor and a complete integrated-circuit amplifier. (*a*) Single transistor; (*b*) integrated circuit.

which is packaged in a single case. Several fabrication techniques are used to form the monolithic and hybrid IC. In this chapter only the basic fabrication process is discussed. More detail can be found in the References.

A major result of this new technology is seen to be a tremendous saving of space. Consider, for example, a medium-sized digital computer. In 1955, using vacuum tubes, the space which it would have occupied would have been equivalent to several large rooms. In 1965, using discrete transistors, one room might suffice. Today, if the same computer were constructed using integrated circuits, the computer might fit on a desk top.

A second advantage of integrated circuits lies in their increased reliability. This is partially due to the fact that all interconnections, along with the transistors and other elements, are made in the initial manufacturing process. The use of highly refined manufacturing and testing techniques which are economically feasible in the mass production of integrated circuits also contributes to increased reliability.

A third advantage of the integrated-circuit amplifier is its extended frequency response (Chap. 13). Since there are no lengthy wire connections, and since the size of each element is decreased, the frequency response of the integrated circuit is extremely good (very high frequency transistors are currently being constructed by a process similar to that used for integrated circuits).

With each new technology, the engineer is faced with new and different problems. The design of transistor circuits differed greatly from that of vacuum-tube circuits. For example, the first transistor radios did not operate properly in very hot or cold temperatures, as on a beach or on ski slopes, because temperature compensation was not well understood. Because of this fact automobile radios until 1960 used vacuum tubes in the output power amplifier.

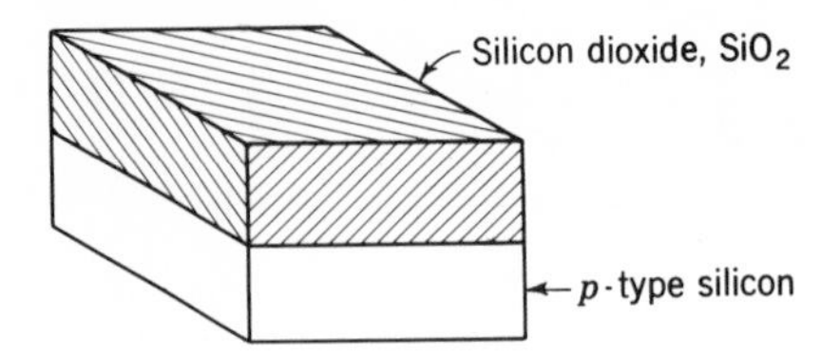

Fig. 9.1-1 Oxidized wafer.

Similar problems are arising now in the use of integrated circuits.[2] They are basically low-power amplifiers. Care must be taken in the design of their biasing networks. Feedback must be used with compensating networks (Chap. 15) in order to avoid oscillation.

The underlying causes of these problems can best be understood from a discussion of the fabrication techniques employed in the manufacture of integrated circuits. This chapter deals primarily with these techniques. The application of integrated circuits, biasing, and related problems have been considered in earlier chapters, and are discussed further in the remaining chapters of this book.

9.1 AN INTRODUCTION TO THE FABRICATION OF AN INTEGRATED-CIRCUIT TRANSISTOR[1]

The fabrication process begins with a *p*-type silicon crystal wafer which is cut and polished until it is 0.005 to 0.007 in. thick. This *p*-type silicon is called the *substrate* material.

The silicon crystal wafer is then placed in an oven having an oxygen atmosphere, and the temperature is raised to 1200°C. The result of this process is oxidation of the silicon, which results in a silicon dioxide (SiO_2) layer on top of the wafer, as shown in Fig. 9.1-1. The wafer is then coated with a photosensitive emulsion, as shown in Fig. 9.1-2. A prescribed mask is then placed over the wafer, as illustrated on the figure. The

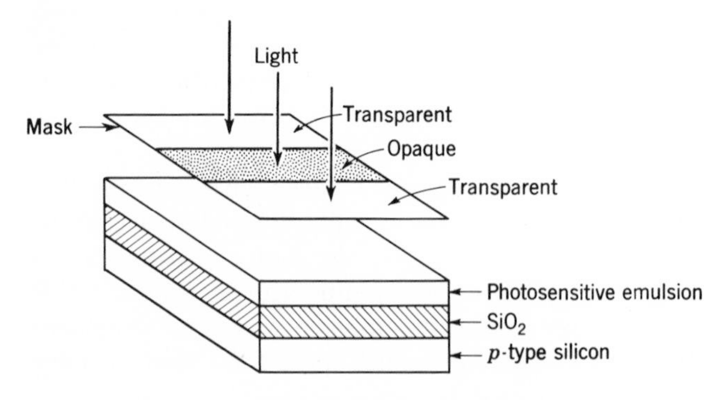

Fig. 9.1-2 Photolithographic isolation masking.

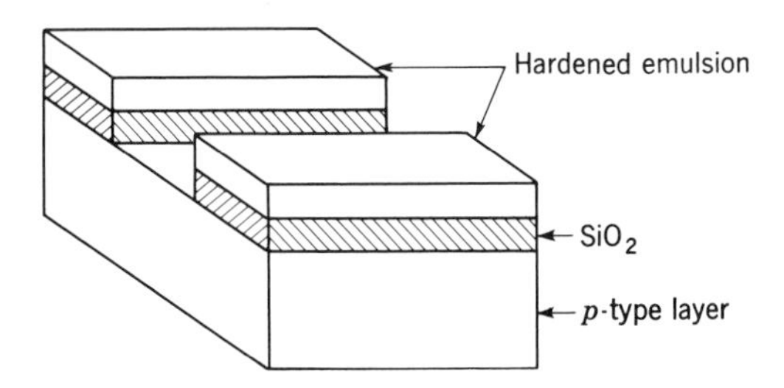

Fig. 9.1-3 The wafer ready for the first diffusion.

masked wafer is then exposed to light. Those portions of the emulsion not exposed to the light are removed, using a solvent, and the corresponding portions of the silicon dioxide coating are also removed. The result is shown in Fig. 9.1-3.

The wafer is then placed into a high-temperature oven and exposed to an n-type dopant such as arsenic, which diffuses into the substrate. The heavy concentration of the doping material results in a heavily doped n region called an $n+$ region. The result of this process is shown in Fig. 9.1-4. An n-type layer is now formed above the $n+$ layer by passing a gas containing an n-type impurity over the $n+$ region. The concentration of the n-type dopant is less than in the $n+$ region. The n-type impurity deposits a *layer* on top of the $n+$ region. This is called an *epitaxial layer*, and will eventually become the collector.

To form the base of the transistor, a p-type diffusion is applied as shown in Fig. 9.1-5. The side layers of SiO_2 have been omitted from this figure. They are essential for proper operation of the integrated circuit, since they provide isolation between transistors, resistors, diodes, and capacitors. An additional n-type diffusion (the emitter) ends the diffusing process. A silicon dioxide protective covering is now applied, except at the terminals of the emitter, base, and collector, where aluminum metallic bonds are made. The final result is shown in Fig. 9.1-6.

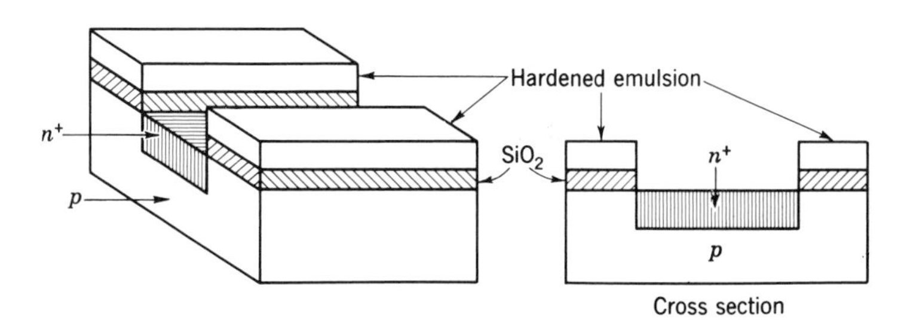

Fig. 9.1-4 The first diffusion.

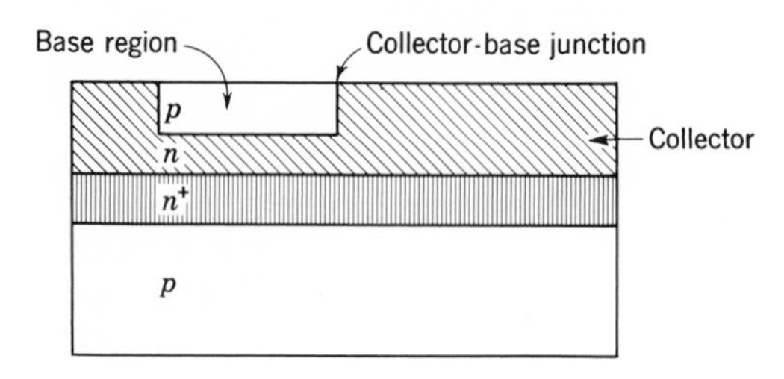

Fig. 9.1-5 Base diffusion.

Each *wafer*, or *slice*, of substrate material is a disk about 1 in. in diameter. This is subdivided into squares or rectangles, called *dice*, or *chips*, of the order of 0.050 to 0.100 in. Each *die*, or *chip*, becomes one complete circuit. Thus each *wafer* may contain as many as three or four hundred complete circuits. In practice, the masks are cut so that all the elements of the desired circuit are formed at the same time.

The interconnection of the emitter, base, and collector terminals determines whether the element becomes a transistor, diode, resistor, or capacitor. These are described individually in succeeding sections. Since the construction of all transistors of a complete amplifier system occurs simultaneously, the transistors can be made almost identical. Thus, matching of integrated transistors on one wafer is accomplished automatically. A problem that does arise, however, is the accuracy of the doping required to achieve a specified h_{fe}. Thus, while all transistors on one wafer will have nearly the same h_{fe}, the value of h_{fe} still may vary by 3:1 from wafer to wafer. Another problem is that resistors cannot be accurately specified (10 percent tolerances are typical), although *ratios* between resistors can be kept to within 3 percent.

In the sections that follow, the connection and interconnection of the

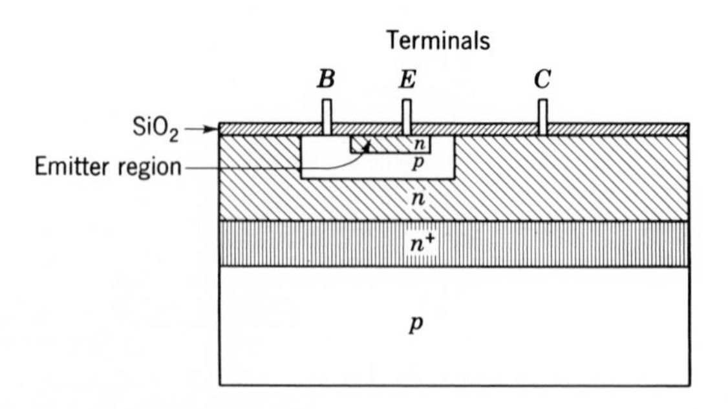

Fig. 9.1-6 Emitter diffusion, protective covering, and location of terminals.

finished elements (as shown in Fig. 9.1-6) are discussed, and some of the resulting problems analyzed.

9.2 THE EQUIVALENT CIRCUIT OF THE INTEGRATED TRANSISTOR

Let us now consider the equivalent circuit of the transistor of Fig. 9.1-6. To do this we represent the circuit as an ideal transistor with the undesirable (parasitic) elements attached externally.

The first "external" element considered is the series collector resistance r_{sc}, shown in Fig. 9.2-1. This resistor can be calculated from the voltage drop v_{ab} and the collector current. The reason for the insertion of the $n+$ material can now be explained. This material shunts the collector n material and has a low resistance. The resulting r_{sc} is then typically less than 1 Ω. The importance of r_{sc} can be seen by considering a transistor in saturation. Let the ideal transistor have a saturation voltage of 0.1 V at a current of 100 mA. Then if r_{sc} is 1 Ω, the apparent saturation voltage is 0.2 V, an increase of 100 percent. Without the $n+$ region, r_{sc} is typically 6 to 15 Ω, which would result in an apparent saturation voltage of 0.7 to 1.6 V.

The second and third external elements are *RC transmission lines* formed by the distributed *RC* circuits along the collector-base and base-emitter junctions. This is shown in Fig. 9.2-2.

A most important external element is the *pn* junction diode, shown in Fig. 9.2-2, which is formed by the *p*-type substrate and *n*-type collector regions. If this diode is forward-biased under any operating conditions, the collector will essentially be short-circuited to the substrate. Thus most integrated circuits have the substrate connected to the most negative part of the circuit in order to ensure that the diode is back-biased.

Since integrated amplifiers are capable of operating at very high frequencies (above 1 GHz), and since most amplifiers operate at fre-

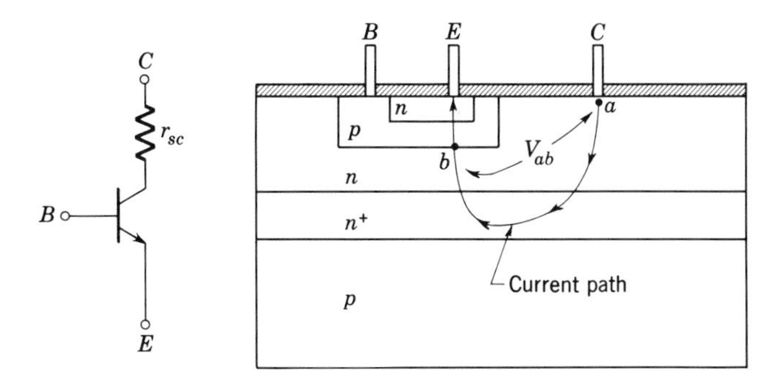

Fig. 9.2-1 The integrated transistor.

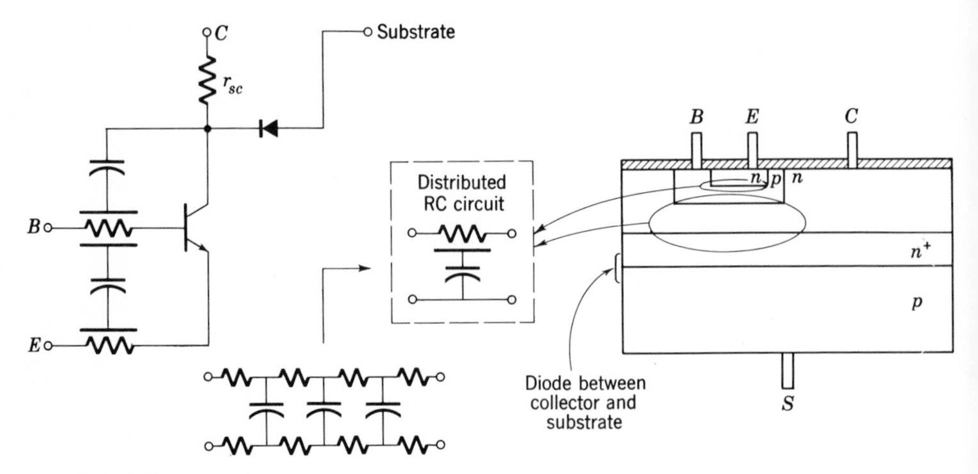

Fig. 9.2-2 Approximate equivalent circuit of the integrated transistor.

quencies well below this value, the midfrequency equivalent circuit of the transistor is generally used. A practical equivalent circuit for Fig. 9.2-2 is shown in Fig. 9.2-3. The transistor shown has the same characteristics as an ordinary transistor.

Using the equivalent circuit of Fig. 9.2-3, let us now determine the equivalent circuit of two adjacent transistors on the same chip, as shown in Fig. 9.2-4. From this figure it is apparent that the two transistors are isolated from each other by the p-type substrate. However, the collectors of T_1 and T_2 and the substrate form two pn junction diodes. These diodes are placed "back to back" and therefore present a high impedance. The resulting equivalent circuit is shown in Fig. 9.2-5. As long as the substrate is connected to the most negative point of the circuit, both diodes are back-biased and will not affect normal circuit operation.

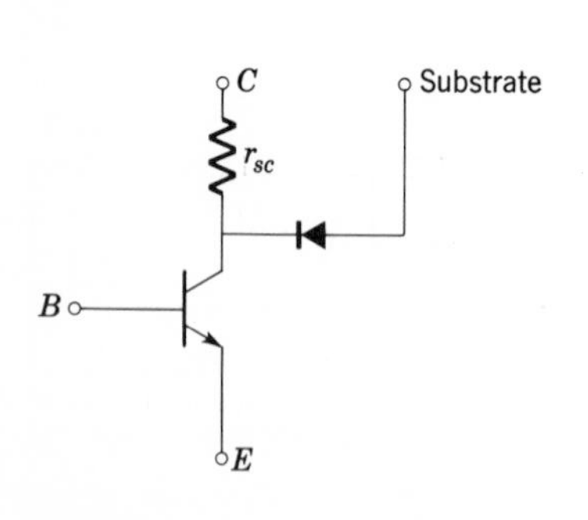

Fig. 9.2-3 Integrated transistor showing the collector-substrate diode.

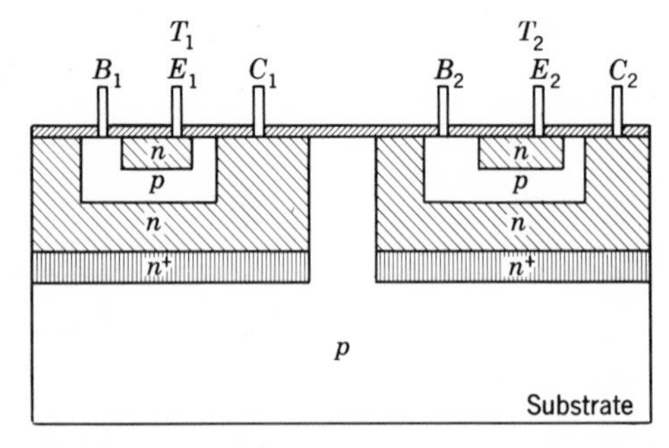

Fig. 9.2-4 Two transistors diffused on one chip.

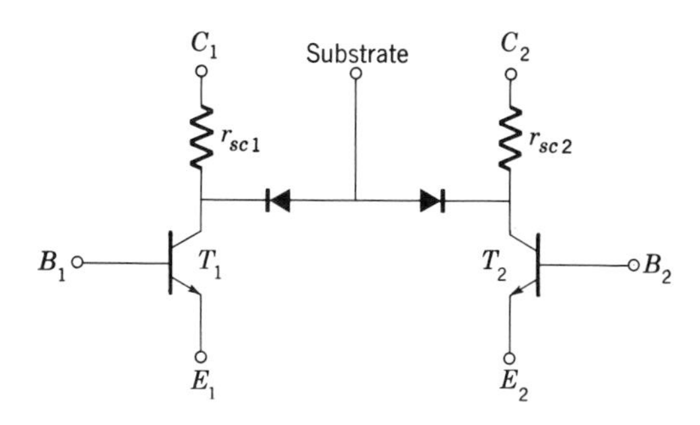

Fig. 9.2-5 Equivalent circuit of the two-transistor chip.

9.3 THE INTEGRATED DIODE

To form a diode, one first constructs a transistor, and then connects the transistor terminals in such a way as to obtain diode action. There are five such possible combinations, as shown in Fig. 9.3-1. Each of the five configurations has its own advantages and disadvantages. They are not discussed here, except for connections *a* and *b*, which are most commonly used. Connection *a* is perhaps the most obvious diode configuration. The problem associated with it is that the reverse breakdown voltage is small (5 to 7 V), because it is associated with the emitter-base junction. Connection *b*, the most common diode configuration, also has a 5- to 7-V reverse breakdown voltage but is able to pass much higher currents.

Let us take a look at the equivalent circuits of these diodes, including the effect of the substrate material, as shown in Fig. 9.3-2.

The diodes D_1 and D_2 formed by the *p* substrate and the collector are reverse-biased. This provides isolation between the various elements (for example, see Fig. 9.2-5 or 9.7-2). Thus, since a back-biased diode is

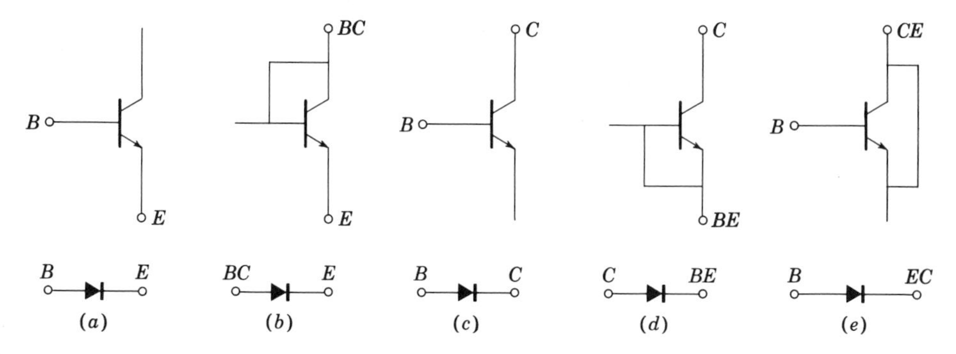

Fig. 9.3-1 The five ways in which a transistor can be connected as a diode.

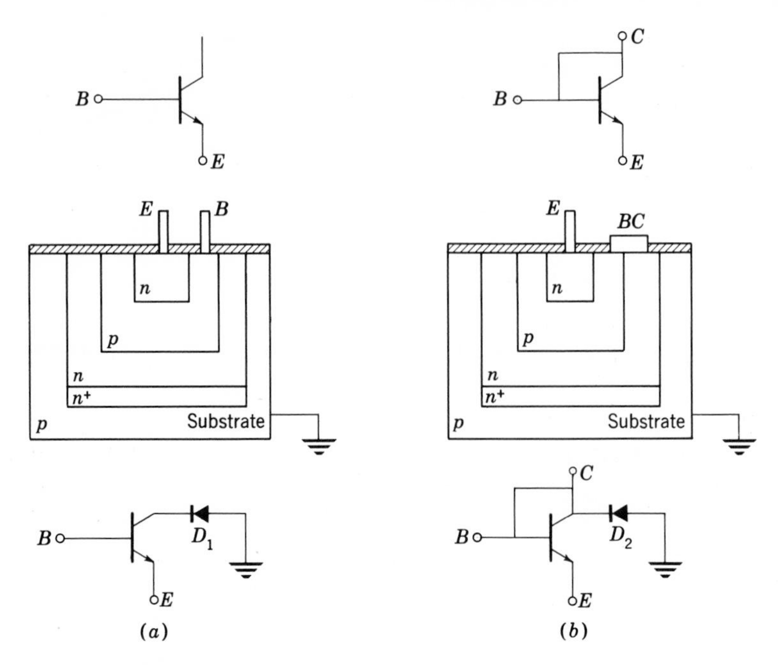

Fig. 9.3-2 The integrated diode.

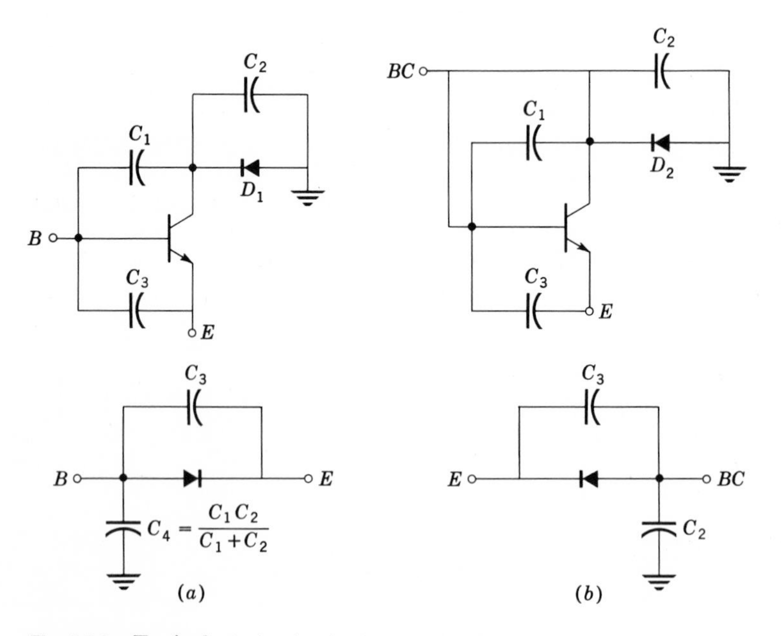

Fig. 9.3-3 Equivalent circuits for integrated diode.

effectively a capacitor, D_1 and D_2 behave as capacitors. Collector-base and base-emitter capacitance being always present, the equivalent circuits of the diodes formed by (*a*) and (*b*) are as shown in Fig. 9.3-3. Values for these capacitances are usually of the order of 0.5 to 3 pF. Discrete diodes always have a capacitance C_3 present. The integrated diode has an extra capacitance (C_4 or C_2).

9.4 THE INTEGRATED CAPACITOR

Two types of capacitors are used in integrated circuits. The first to be discussed is the junction capacitor, which is formed from an integrated transistor (Fig. 9.1-6) by using the collector and base terminals. Two disadvantages of this type of capacitor are that its capacitance varies with collector-to-base voltage and that the maximum capacitance is small.

The second type to be discussed is the thin-film capacitor. This type is capable of yielding larger capacitance values than the junction capacitor, and its capacitance is not a function of the terminal potential. However, its construction is far more expensive, and it uses a large area on the chip.

9.4-1 THE JUNCTION CAPACITOR

The junction capacitor is formed from the collector-base terminals of an ordinary transistor. Thus the circuit of Fig. 9.2-2, after connecting the substrate to ground, becomes the circuit of Fig. 9.4-1. At low frequencies the distributed RC circuit appears capacitive since the impedance of the distributed capacitor, C_1, is considerably higher than that of the distributed resistance, R. The resulting equivalent circuit is shown in Fig. 9.4-2.

Consideration of the physics of the junction yields the result that the effective capacitance of a back-biased diode is inversely proportional either to the square root or to the ⅓ power of the diode voltage. Since C_1 and C_2 represent capacitors formed by back-biased diodes, C_1 is inversely proportional to the voltage V_{CB}, and C_2 is inversely proportional

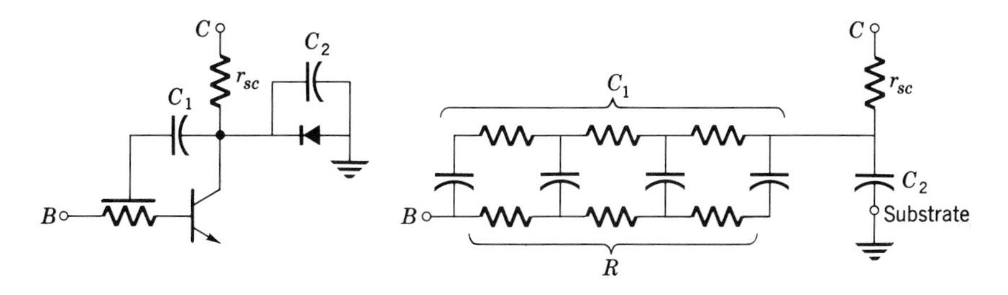

Fig. 9.4-1 The junction capacitor and its equivalent circuit.

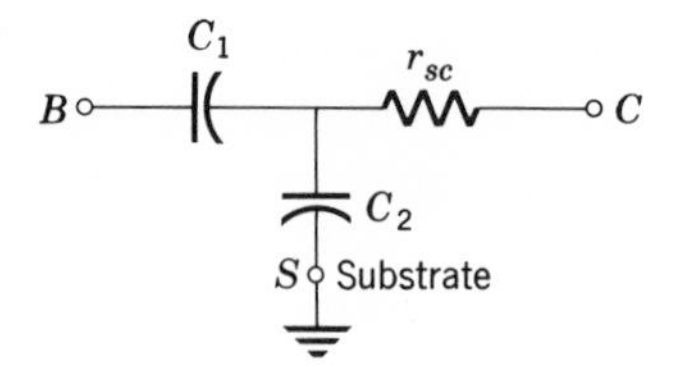

Fig. 9.4-2 Simplified lumped equivalent circuit of the junction capacitor.

to V_{CS}. To make the "capacitor" of Fig. 9.4-2 approach the ideal $C_1 \gg C_2$, the voltage V_{CS} is made as large as possible, and the voltage V_{CB} is made as small as possible. Typical ratios of C_1 to C_2 vary between 1:1 and 7:1.

The junction capacitor of Fig. 9.4-2 is further degraded by the series collector resistor r_{sc}. This resistance is minimized by the use of the $n+$ region shown in Fig. 9.2-2.

9.4-2 THE THIN-FILM CAPACITOR

The thin-film capacitor is constructed by first diffusing an n region and then an $n+$ region. A thin layer of dielectric material (usually silicon dioxide) is applied to cover the $n+$ material. A metallic film (aluminum) which acts as the second terminal of the capacitor is then applied to the dielectric, as shown in Fig. 9.4-3a. The equivalent circuit of this capacitor is shown in Fig. 9.4-3b.

There are two major advantages to using the thin-film capacitor. First, C_1 is fixed and is not a function of collector voltage. This is extremely important, since variations in voltage cause variations in the *junction* capacitance which may result in possible frequency modulation of the signal. Second, the thin-film capacitance is nonpolar, that is, V_{CO} can be positive or negative. Capacitance C_2, due to the substrate material, still varies with variations of the voltage across it.

These capacitors have a capacitance per unit area which varies

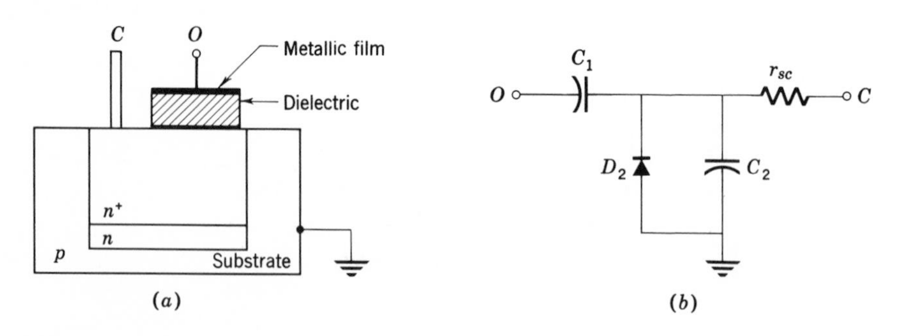

Fig. 9.4-3 Thin-film capacitor and equivalent circuit.

between 0.2 and 0.5 pF/mil^2, depending on the process (and the dielectric used, in the case of thin-film capacitors). Thus, to make a large capacitor requires a significant amount of area, and hence is costly. At the present time maximum capacitance values range from 500 to 5000 pF, again depending on the process employed. This is considerably less than the 20- or 100-μF capacitors used as bypass and coupling capacitors in ordinary transistor circuits. The problems associated with small bypass and coupling capacitors when considering low- and high-frequency circuits are discussed in Chaps. 12 and 13. It is also of interest to note that although adjacent capacitors can be made almost identical on one wafer, their capacitance may vary by ± 20 percent from wafer to wafer.

9.5 THE INTEGRATED RESISTOR

Resistors, as well as capacitors, can be constructed either directly from the basic transistor element or using thin-film techniques. Let us first consider the junction resistor formed from the transistor.

9.5-1 THE JUNCTION RESISTOR

The junction resistor is formed by stopping the diffusion process after depositing the p-type material which forms the base of the transistor, as shown in Fig. 9.1-5. The silicon dioxide and aluminum contacts are then applied as shown in Fig. 9.5-1a. The resistance of the junction resistor is determined by the resistivity ρ, length L, and area A of the p-type base material, as shown in Fig. 9.5-1b. An approximate expression for the resistance, assuming uniform fields, is

$$R_{AB} = \rho \left(\frac{L}{A}\right) \tag{9.5-1}$$

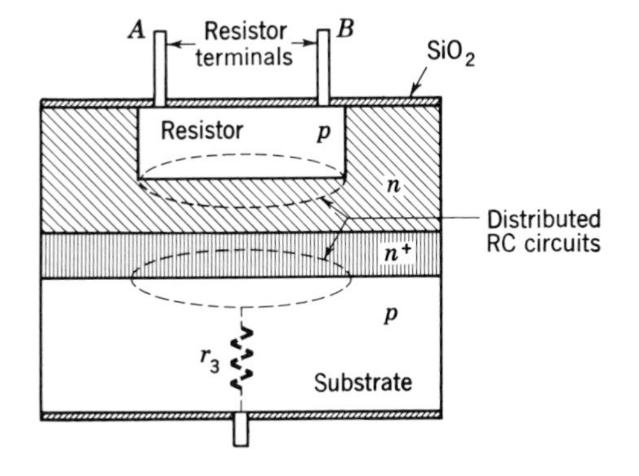

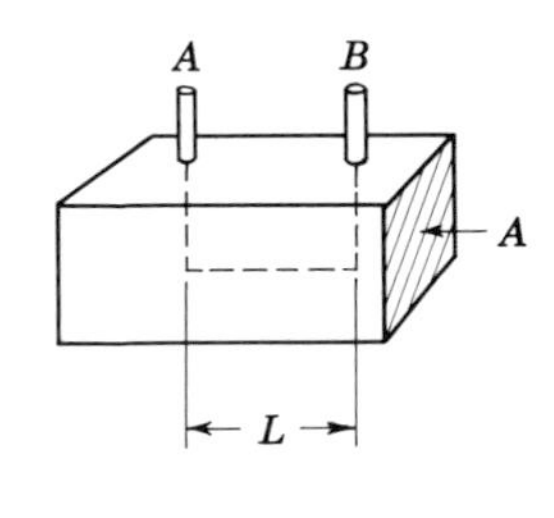

Fig. 9.5-1 The junction resistor.

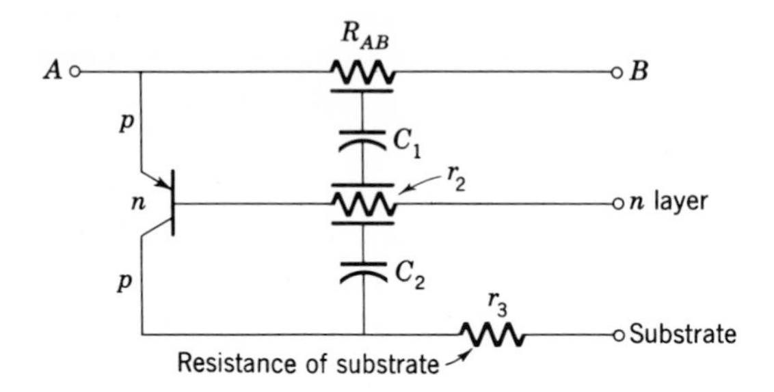

Fig. 9.5-2 Equivalent circuit of junction resistor.

Figure 9.5-2 shows the complete equivalent circuit of the resistance element. We see from Fig. 9.5-1*a* that the *pnp* materials form a transistor. The collector of this transistor is the *p*-type substrate, and the emitter is the *p*-type material to be used as the resistor. (The substrate is always connected to the lowest potential point in the circuit; thus it becomes the collector in the *pnp* configuration.)

Figure 9.5-2 shows that a distributed *RC* circuit exists between the emitter and the base and between the base and the collector. This can be seen from Fig. 9.5-1*a*. Resistor r_3 is the series collector resistance shown diagrammatically in Fig. 9.5-1*a*.

The h_{fe} of the transistor formed is small, since the base *n* region was intended to be a collector and therefore is relatively large. In Sec. 3.1, it was pointed out that the base current of a well-designed transistor is small compared with the collector current because the base region is narrow. In this case the transistor formed in Fig. 9.5-2 has a wide base region, and hence the base current is large, thus resulting in a low h_{fe}.

Typical values for a 4-kΩ resistor are

$$h_{fe} \approx 0.5 \text{ to } 5 \qquad r_3 \approx 1\ \Omega$$
$$R_{AB} = 4\ \text{k}\Omega \qquad C_1 \approx 5\ \text{pF}$$
$$r_2 \approx 50\ \Omega \qquad C_2 \approx 15\ \text{pF}$$

At low frequencies ($f \ll 1/2\pi R_{AB}C_1 = 8$ MHz for this example) capacitors C_1 and C_2 can be neglected. The result is an effective resistance R_{AB}. It is to be noted that the *n* region must be connected to the most positive potential in the circuit to reverse-bias the *pn* emitter-base diode. If this were not done, current would flow in this diode and the resistance R_{AB} would be effectively short-circuited.

Values of resistance typically range from 100 Ω to 30 kΩ. Absolute variations are of the order of ±10 percent, while variations between resistances on the same chip can be less than ±3 percent. Thus resistance *ratios* can be maintained more accurately in production than resistance *values*. For this reason, it is good practice to design integrated circuits so that performance depends on resistance ratios rather than absolute values. Resistance is also a function of temperature, because of the properties of the semiconductor material.

9.5-2 THE THIN-FILM RESISTOR

The thin-film resistor is constructed by depositing a resistive material such as nitrided tantalum, nichrome, or tin oxide over the silicon dioxide covering the p-type substrate, as shown in Fig. 9.5-3a. The resistance obtained using this process is given by the standard resistance equation (9.5-1). Applying this relation to the configuration of Fig. 9.5-3b, the resistance between terminals A and B is

$$R_{AB} = \frac{\rho L}{dW} \tag{9.5-2}$$

If $L = W$, we have a *unit square* of the material, and the resistance per unit square is

$$R = \frac{\rho}{d} \qquad \Omega/\text{square} \tag{9.5-3}$$

Note that the dimension of ρ is in ohm-centimeters, and the dimension of d is in centimeters; thus the "square" is *dimensionless*. R is called the *sheet resistance* of the material, and the overall resistance is

$$\begin{aligned} R_{AB} &= (\Omega/\text{square})(\text{number of squares}) \\ &= R\left(\frac{L}{W}\right) \end{aligned} \tag{9.5-4}$$

Some typical values of R are

Nitrided tantalum = 50 Ω/square
Nichrome = 400 Ω/square
Tin oxide = 1000 Ω/square

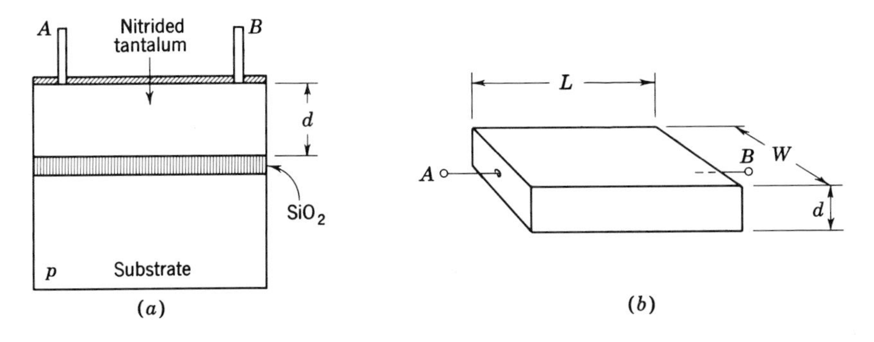

Fig. 9.5-3 Thin-film resistor.

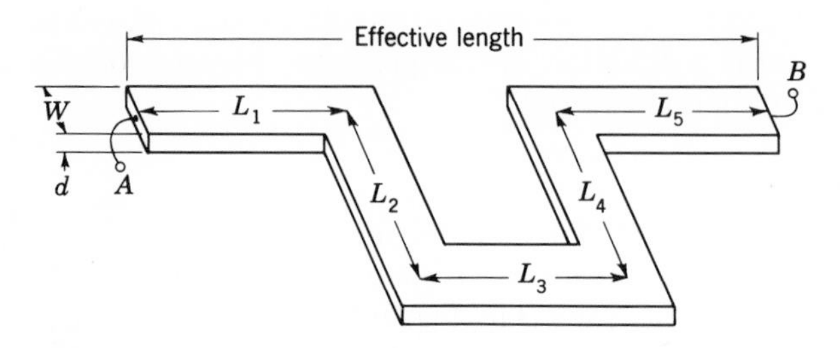

Fig. 9.5-4 Zigzagging a thin-film resistor to reduce the overall length.

If a 5-kΩ resistor is to be designed using nitrided tantalum, the length-to-width ratio L/W is

$$\frac{R_{AB}}{R} = \frac{L}{W} = \frac{5000}{50} = 100 \text{ squares} \tag{9.5-5}$$

If a width of 5 mils is selected, the length will be 500 mils, or ½ in., a very long distance on the integrated-circuit scale. In order to shorten the overall length, the resistor is often constructed in a zigzag pattern, as shown in Fig. 9.5-4.

If

$$L_1 = L_2 = L_3 = L_4 = L_5 \qquad \text{and} \qquad w = 5 \text{ mils}$$

then the effective length required is only about 300 mils. Additional zigzagging will reduce the length further.

Another technique for reducing the length is to reduce the width W. However, power-dissipation limitations preclude decreasing W indefinitely. For example, nitrided aluminum on silicon dioxide has a maximum power rating of 10 W/in.² of film. Therefore, assuming a dc voltage V impressed across the resistor,

$$\frac{P}{A} = 10 = \frac{V^2/R}{LW} \tag{9.5-6}$$

If V is 10 V and $R_{AB} = 5$ kΩ,

$$LW = \tfrac{1}{500} \qquad \text{in.}^2 \tag{9.5-7}$$

Combining (9.5-5) and (9.5-7) yields

$$W \approx 4.5 \text{ mils} \tag{9.5-8a}$$

and

$$L \approx 450 \text{ mils} \tag{9.5-8b}$$

in order not to exceed the power rating of the material.

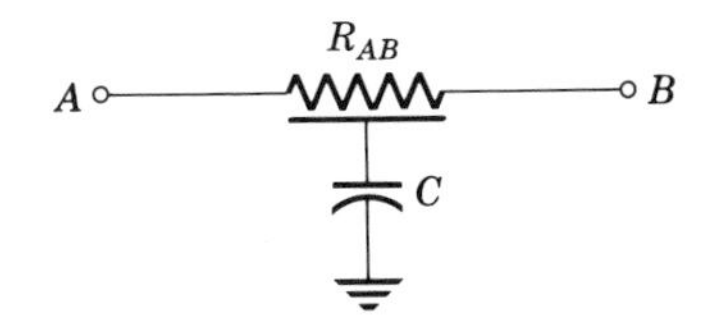

Fig. 9.5-5 Equivalent circuit of the thin-film resistor.

The foregoing analysis assumed that the thin-film resistor was ideal. However, from Fig. 9.5-3*a* we see that a capacitor is formed between the resistive and substrate materials, the dielectric being the silicon dioxide. Thus the thin-film resistor is actually a distributed *RC* circuit, as shown in Fig. 9.5-5. The equivalent circuit reduces to a simple resistance for frequencies less than about 1 GHz.

9.6 THE INTEGRATED INDUCTOR

The integrated inductor has not been perfected at the time of this writing. Inductors as large as only 5 μH can be built, with a Q between 30 and 50. Practical amplifier systems still use external inductors attached to the integrated circuit. These inductors require a great deal of space, so that *RC* or direct-coupled amplifiers are preferred.

9.7 DESIGN OF A SIMPLE INTEGRATED CIRCUIT

Consider designing a simple monolithic integrated-circuit amplifier as shown in Fig. 9.7-1, i.e., an integrated circuit not using thin-film techniques. The first step is to lay out the circuit so that no wires cross, as shown in Fig. 9.7-2*a*. The cross-sectional view is shown in Fig. 9.7-2*b*. After the required masks are designed and constructed, the manufacturing process outlined in Sec. 9.1 follows.

Figure 9.7-3*a* shows a photomicrograph of an integrated differential-amplifier circuit. It is interesting to note that the required diffusion masks can be generated directly by a digital computer. The reader

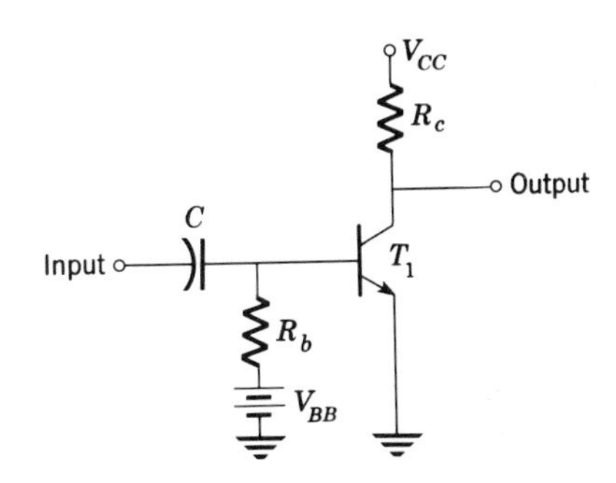

Fig. 9.7-1 An IC amplifier.

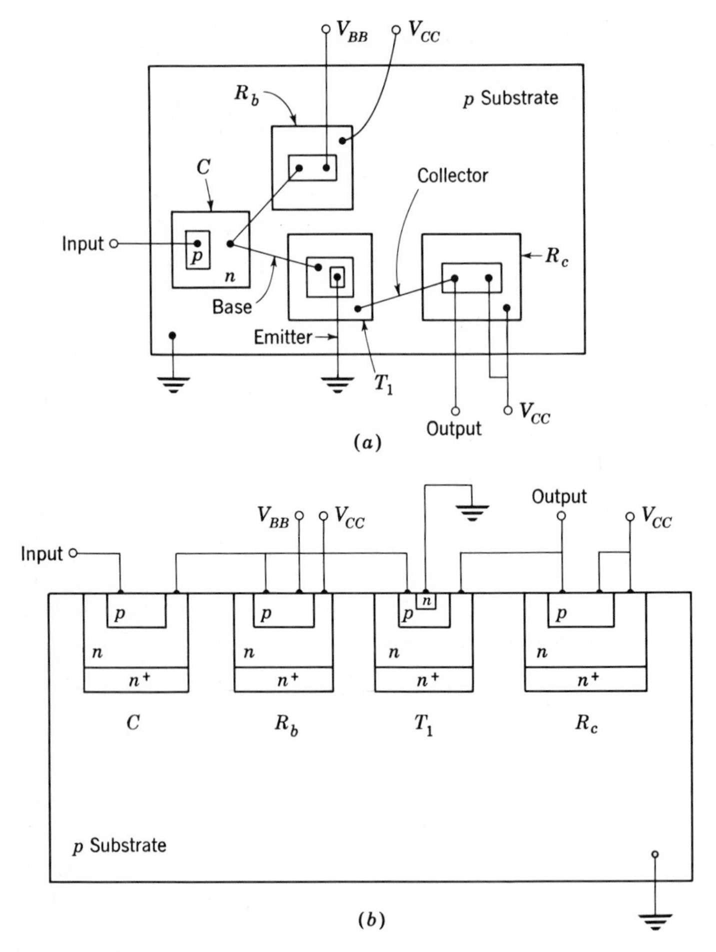

Fig. 9.7-2 Realization of IC amplifier of Fig. 9.7-1 (not to scale). (*a*) Top view; (*b*) distorted cross-sectional view drawn to avoid overlap of R_b and T_1.

should try to identify the individual transistors, diodes, resistors, and capacitors by comparing the photomicrograph with the circuit diagram.

9.8 ANALYSIS OF A TYPICAL INTEGRATED-CIRCUIT AMPLIFIER[3]—THE FAIRCHILD μA702

An integrated-circuit (IC) amplifier which embodies many of the circuits studied in preceding chapters is shown in Fig. 9.8-1. Transistors T_1, T_2, and T_3 form a difference amplifier with a constant-current supply. The

(a)

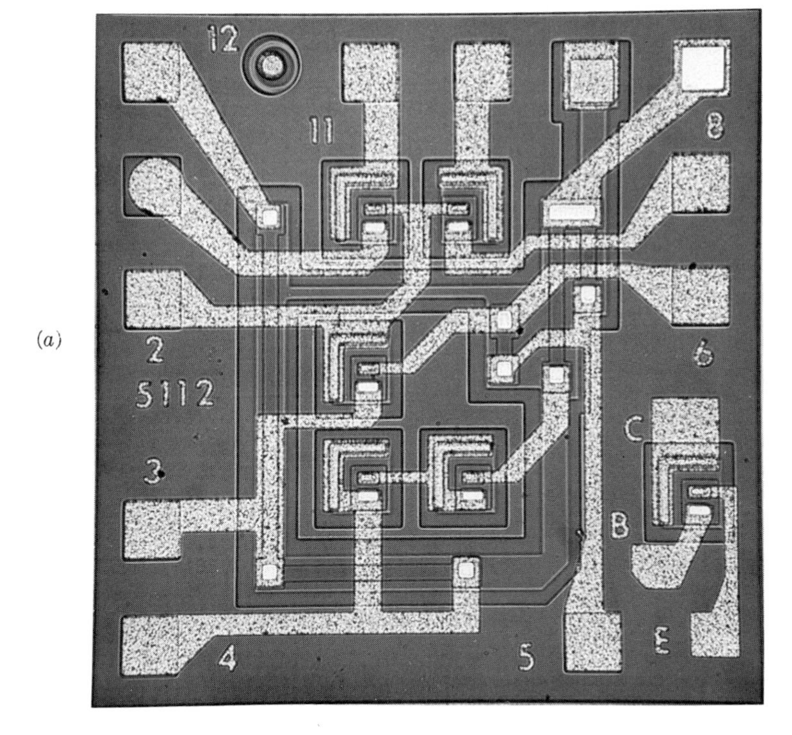

(b)

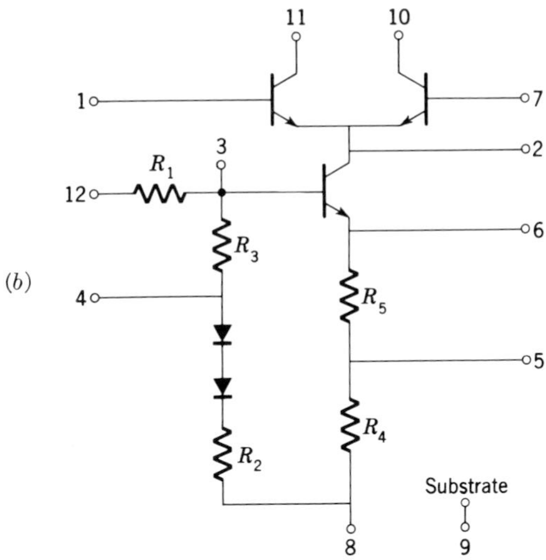

Fig. 9.7-3 IC differential amplifier. (a) Photomicrograph of the RCA 3005 differential amplifier (*Courtesy of RCA*); (b) circuit diagram.

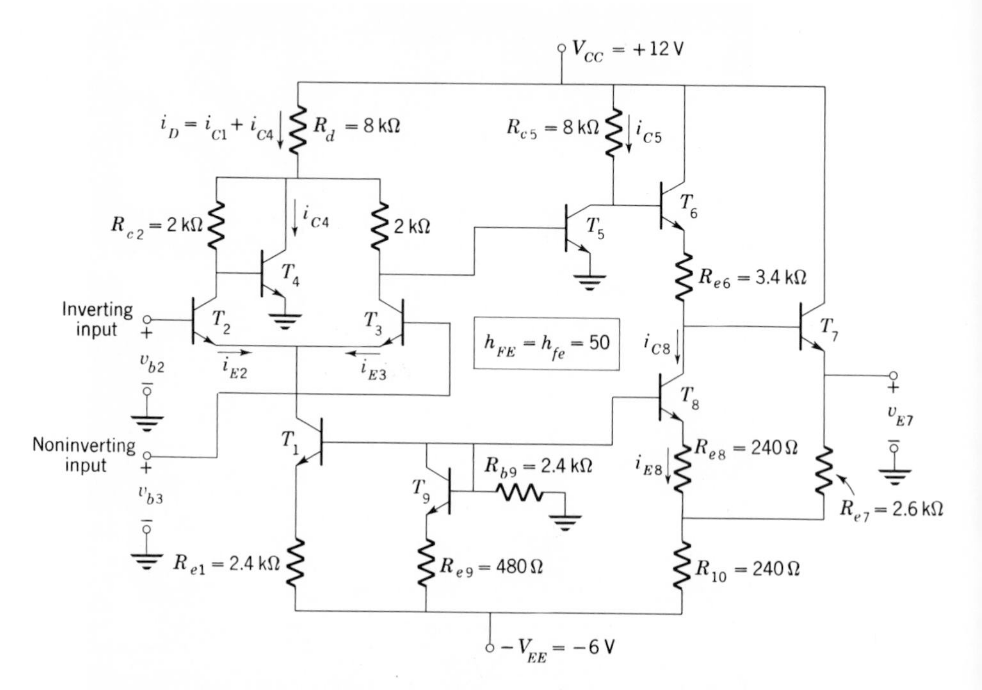

Fig. 9.8-1 μA702 IC operational amplifier. *(Courtesy of Fairchild Semiconductor Corp.)*

base voltage for T_1 and T_8 is supplied by the diode bias circuit (Sec. 4.4), using T_9 connected as a diode.

Transistor T_5 amplifies the output signal from the difference amplifier. Transistors T_6 and T_8 form a cascode amplifier which shifts the signal level from a high dc voltage (the collector of T_5) to approximately ground potential without significant loss of gain (Sec. 7.4).

T_4 serves two functions: The first is to balance approximately the currents in T_2 and T_3. The second is to provide an additional gain of 2 in the difference amplifier (Sec. 9.8-2).

The output transistor T_7 is an emitter follower which provides a low output impedance. Note that the 2.6-kΩ emitter resistor is not returned to ground, but to the emitter circuit of T_8. This provides *positive* feedback, which increases the gain of the amplifier. To see that the feedback is indeed positive, consider that the output voltage increases. This results in an increase in the emitter voltage of T_8. Since the base voltage of T_8 is constant, an increase in the emitter voltage of T_8 causes a *decrease* in the collector current. The voltage drop across the 3.4-kΩ emitter resistor of T_6 then decreases, which results in an *increase* in the base and emitter voltages of T_7. Thus the gain around the T_6, T_7, T_8 loop is positive, and the feedback is positive.

9.8-1 CALCULATION OF THE QUIESCENT OPERATING POINTS

The design of a good integrated circuit is based on the fact that specified resistor *ratios* can be maintained within close limits while specified absolute resistor *values* cannot. In the circuit of Fig. 9.8-1 the output voltage is 0.7 V, independent of V_{CC} and V_{EE} as long as V_{B2} is equal to V_{B3}.

To accomplish this, the IC is designed so that the difference amplifier is completely balanced when $V_{B2} = V_{B3}$. That is,

$$I_{C2} = I_{C3} = \frac{I_{C1}}{2} \tag{9.8-1a}$$

and

$$I_{C4} = I_{C5} \tag{9.8-1b}$$

Then the output voltage V_{E7} is

$$V_{E7} = V_{CC} - R_{c5}I_{C5} - V_{BE6} - R_{e6}I_{C8} - V_{BE7} \tag{9.8-2a}$$

where we have assumed that

$$I_{B6} \ll I_{C5} \tag{9.8-2b}$$

and

$$I_{B7} \ll I_{C8} \tag{9.8-2c}$$

These assumptions will subsequently be verified. In addition, referring to T_4, and assuming balanced conditions,

$$V_{CC} = R_d(I_{C1} + I_{C4}) + R_{c2}\left(\frac{I_{C1}}{2}\right) + V_{BE4} \tag{9.8-3}$$

The collector current I_{C1} is found from the equations

$$V_{B1} = V_{B8} \approx \left(\frac{R_{b9}}{R_{b9} + R_{e9}}\right)(-V_{EE} + V_{BE9}) \tag{9.8-4a}$$

and

$$I_{C1} \approx I_{E1} = \frac{V_{B1} - V_{BE1} + V_{EE}}{R_{e1}} \tag{9.8-4b}$$

Therefore

$$I_{C1} \approx \frac{(V_{BE9} - V_{EE})[R_{b9}/(R_{b9} + R_{e9})] - V_{BE1} + V_{EE}}{R_{e1}} \tag{9.8-4c}$$

The collector current I_{C8} can be calculated from the emitter circuits of T_7 and T_8. Then

$$I_{C8} = \frac{V_{B8} - V_{BE8} - R_{e8}I_{C8} - V_{E7}}{R_{e7}} + \frac{V_{B8} - V_{BE8} - R_{e8}I_{C8} + V_{EE}}{R_{10}} \quad (9.8\text{-}5)$$

Equations (9.8-2a), (9.8-3), (9.8-4c), and (9.8-5) completely describe the quiescent conditions of the amplifier. Equation (9.8-3) relates I_{C1} to I_{C4} ($= I_{C5}$). Substituting these equations into (9.8-2a) yields the following equation, assuming that all V_{BE} are equal:

$$V_{E7} = V_{CC} - \left\{ V_{CC}\left(\frac{R_{c5}}{R_d}\right) - V_{BE} - \left(R_d + \frac{R_{c1}}{2}\right) \left[\frac{(V_{BE} - V_{EE})\left(\dfrac{R_{b4}}{R_{b9} + R_{e9}}\right) - V_{BE} + V_{EE}}{R_{e1}}\right]\right\}$$
$$- V_{BE} - R_{e6}\left\{\frac{\left(\dfrac{R_{b9}}{R_{b9} + R_{e9}}\right)(-V_{EE} + V_{BE})\left(\dfrac{1}{R_{e7}} + \dfrac{1}{R_{10}}\right) - V_{BE}\left(\dfrac{1}{R_{e7}} + \dfrac{1}{R_{10}}\right) - \dfrac{V_{E7}}{R_{e7}} + \dfrac{V_{EE}}{R_{10}}}{1 + R_{e8}/R_{e7} + R_{e8}/R_{10}}\right\}$$
$$- V_{BE} \quad (9.8\text{-}6)$$

Collecting terms yields

$$V_{E7}\left(1 - \frac{R_{e6}/R_{e7}}{1 + \dfrac{R_{e8}}{R_{e7}} + \dfrac{R_{e8}}{R_{10}}}\right) = V_{CC}\left(1 - \frac{R_{c5}}{R_d}\right)$$
$$+ V_{EE}\left\{\left(\frac{R_{c5}}{R_d}\right)\left(\frac{R_d + \dfrac{R_{c2}}{2}}{R_{e1}}\right)\left(1 - \frac{R_{b9}}{R_{b9} + R_{e9}}\right) + \left(\frac{R_{e6}}{1 + \dfrac{R_{e8}}{R_{e7}} + \dfrac{R_{e8}}{R_{10}}}\right)\left[\left(\frac{R_{b9}}{R_{b9} + R_{e9}}\right)\left(\frac{1}{R_{e7}} + \frac{1}{R_{10}}\right) - \frac{1}{R_{10}}\right]\right\}$$
$$+ V_{BE}\left\{\frac{R_{c5}}{R_d}\left[1 - \left(\frac{R_d + \dfrac{R_{c2}}{2}}{R_{e1}}\right)\left(1 - \frac{R_{b9}}{R_{b9} + R_{e9}}\right)\right] - 2 + \frac{R_{e6}\left(\dfrac{1}{R_{e7}} + \dfrac{1}{R_{10}}\right)}{1 + \dfrac{R_{e8}}{R_{e7}} + \dfrac{R_{e8}}{R_{10}}}\left(1 - \frac{R_{b9}}{R_{b9} + R_{e9}}\right)\right\} \quad (9.8\text{-}7)$$

In order to make V_{E7} independent of V_{CC} and V_{EE}, we set the appropriate coefficients equal to zero. Thus, from the coefficient of V_{CC},

$$R_{c5} = R_d \tag{9.8-8}$$

Using (9.8-8) in the coefficient of V_{EE}, we obtain

$$\left(\frac{R_d + \dfrac{R_{c2}}{2}}{R_{e1}}\right)\left(\frac{R_{e9}}{R_{b9} + R_{e9}}\right) + \frac{\left(\dfrac{R_{b9}}{R_{b9} + R_{e9}}\right)\left(\dfrac{R_{e6}}{R_{e7}} + \dfrac{R_{e6}}{R_{10}}\right) - \dfrac{R_{e6}}{R_{10}}}{1 + R_{e8}/R_{e7} + R_{e8}/R_{10}} = 0 \tag{9.8-9}$$

The circuit values are adjusted so that (9.8-9) is satisfied. Note that (9.8-9) depends only on resistance *ratios*, and not on absolute values. Since the manufacturing process maintains resistance ratios within close limits, the output voltage will be independent of the supply voltages from unit to unit, even though the resistance values may be different.

When (9.8-8) and (9.8-9) are substituted into (9.8-7), the result is

$$V_{E7} \approx -V_{BE} \approx -0.7 \text{ V} \tag{9.8-10}$$

This is the output offset voltage, i.e., the output voltage when the differential-mode input is zero.

Using the circuit values indicated with $V_{CC} = +12$ V, $V_{E7} = -0.7$ V, and $-V_{EE} = -6$ V, the following dc values are calculated (Prob. 9.21):

$$I_{E9} \approx 1.84 \text{ mA}$$
$$V_{B1} \approx V_{B8} \approx -4.4 \text{ V}$$
$$I_{C1} \approx 0.38 \text{ mA}$$
$$V_{E8} \approx -5.1 \text{ V}$$
$$I_{E7} \approx 1.8 \text{ mA}$$
$$I_{E8} \approx 1 \text{ mA}$$
$$I_{C4} \approx I_{C5} \approx 1 \text{ mA}$$

9.8-2 THE SMALL-SIGNAL GAIN

The emitter current i_{e2} of the differential input stage is

$$i_{e2} = -i_{e3} = \frac{v_{b2} - v_{b3}}{2h_{ib2}} \tag{9.8-11}$$

where

$$h_{ib1} = h_{ib2} \approx \frac{0.025}{I_{E2}} = \frac{0.025}{0.38/2 \times 10^{-3}} = 130\ \Omega \tag{9.8-12}$$

The base current of T_5 can be found from the equivalent circuit for T_2 to T_5 as shown in Fig. 9.8-2*a*. In the difference amplifier (T_2 and T_3), $i_{e2} = -i_{e3}$, and $h_{ie4} = h_{ie5}$. Thus the circuit is symmetrical about terminal C_4. Because of the symmetry v_{c4} does not depend on i_{e2}. (This

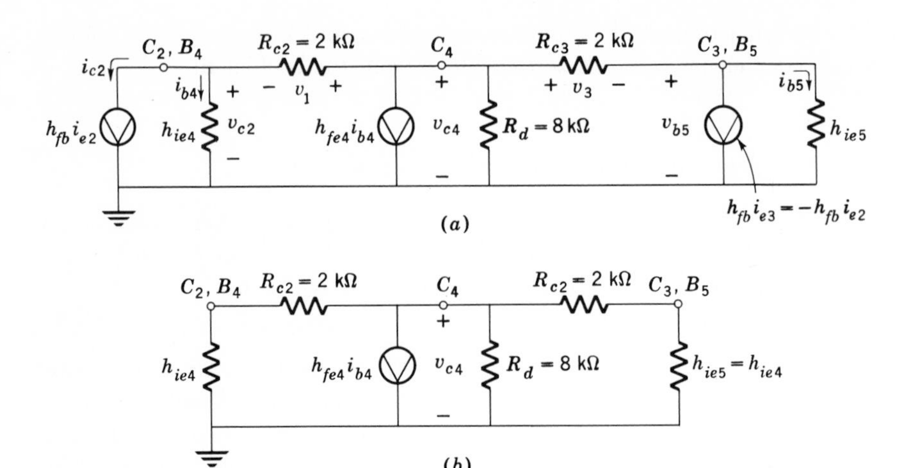

Fig. 9.8-2 Small-signal equivalent of portion of amplifier. (*a*) Complete circuit; (*b*) $i_{e2} = 0$.

can be seen by setting the current source $h_{fe4}i_{b4} = 0$. Then, by symmetry, $v_{c4} = 0$.) We can thus let $i_{e2} = 0$ and reduce the circuit to that shown in Fig. 9.8-2*b*.

From the reduced circuit, with

$$h_{ie4} = h_{ie5} = (0.025)h_{fe}/I_{EQ4} = (0.025)(50)/(10^{-3}) = 1250\ \Omega$$

$$v_{c4} = -h_{fe4}i_{b4}\left(8000 \| \frac{2000 + h_{ie4}}{2}\right) = -67.5 \times 10^3 i_{b4} \qquad (9.8\text{-}13a)$$

To obtain additional relations between v_{c4}, i_{b4}, and i_{b5}, we refer to Fig. 9.8-2*a*:

$$\begin{aligned} v_{c4} = v_1 + v_{c2} &= 2000(h_{fb}i_{e2} + i_{b4}) + h_{ie4}i_{b4} \\ &\approx 2000i_{e2} + 3250i_{b4} \end{aligned} \qquad (9.8\text{-}13b)$$

$$\begin{aligned} v_{c4} = v_3 + v_{b5} &= 2000(i_{b5} - h_{fb}i_{e2}) + h_{ie4}i_{b5} \\ &\approx -2000i_{e2} + 3250i_{b5} \end{aligned} \qquad (9.8\text{-}13c)$$

Solving (9.8-13*a*, *b*, and *c*) for i_{b5} yields

$$i_{b5} \approx 1.23i_{e2} \qquad (9.8\text{-}14a)$$

Using (9.8-11) and (9.8-12), and letting $h_{fe4} = 50$:

$$i_{b5} = 1.23\left[\frac{v_{b2} - v_{b3}}{2(130)}\right] \approx 4.7 \times 10^{-3}(v_{b2} - v_{b3}) \qquad (9.8\text{-}14b)$$

It is interesting to note that if T_4 were removed from the circuit, and a resistor having a value of h_{ie4} inserted between the collector of T_2 and ground, then

$$i_{b5} = \left(\frac{1}{1 + h_{ie5}/2000}\right) i_{e2} \approx 0.615 i_{e2} \tag{9.8-14c}$$

Thus, T_4 provides an additional gain of 2, while improving the balance in T_2 and T_3.

To calculate v_{e7}, it is advantageous to redraw the cascode configuration of T_7 and T_8 as shown in Fig. 9.8-3. To obtain this figure, the base circuit of T_6 was reflected into its emitter circuit.

If we neglect small resistors, the circuit of Fig. 9.8-3 can be reduced to that shown in Fig. 9.8-4. In this figure we have employed (9.8-14*b*) to replace i_{b5} by the difference voltage, $v_{b2} - v_{b3}$.

Figure 9.8-4 places in evidence the positive feedback used in this circuit. If the current source i_{e8} decreases, v_{e7} increases. Then the controlling current i_{e8} in the 240-Ω resistor *decreases*, further decreasing the controlled current source, which is equal to i_{e8}. This circuit can be analyzed using the standard feedback techniques presented in Chap. 8. First we redraw Fig. 9.8-4 as shown in Fig. 9.8-5. We neglect h_{ie7} for simplicity. Then we see that the circuit is an example of voltage feedback with a current error.

The loop gain T is

$$T = \left.\frac{v_{e7}}{v'_{e7}}\right|_{v_i=0} = \left(\frac{v_{e7}}{i_{e8}}\right)\left(\frac{i_{e8}}{v'_{e7}}\right) \approx (3600 \| 2700 h_{fe}) \left(\frac{1}{2700}\right)\left(\frac{240}{513}\right) \approx 0.6 \tag{9.8-15}$$

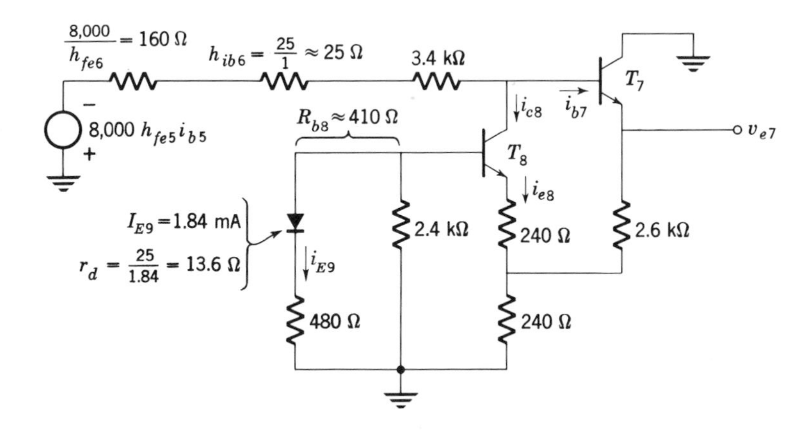

Fig. 9.8-3 Equivalent circuit of output stages.

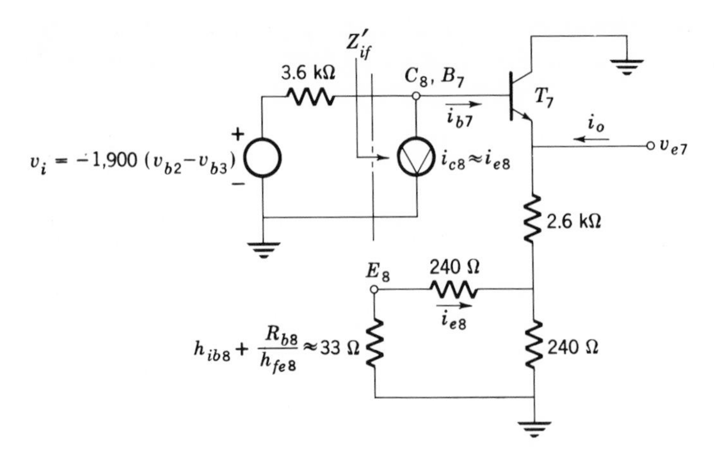

Fig. 9.8-4 Simplification of output circuit: T_8 removed after reflection of impedances.

The forward voltage gain without feedback is

$$A_v = \left.\frac{v_{e7}}{v_i}\right|_{\substack{v'_{e7}=0 \\ i_{e8}=0}} \approx \frac{2700h_{fe}}{3600 + 2700h_{fe}} \approx 1 \tag{9.8-16a}$$

Therefore the gain with feedback is

$$A_{vf} = \frac{A_v}{1 - T} \approx 2.5 \tag{9.8-16b}$$

The total IC amplifier gain is

$$A_{v\text{IC}} = \frac{v_{e7}}{v_{b3} - v_{b2}} = (2.5)(1900) \approx 4750 \; (\simeq 73 \text{ dB}) \tag{9.8-17}$$

The input impedance Z'_{if} seen by the 3.6 kΩ resistor can be found using Fig. 9.8-5. We note that the voltage gain $A_{vf} = v_{e7}/v_i$ can be written in terms of Z'_{if},

$$A_{vf} = \frac{v_{e7}}{v_i} = \frac{Z'_{if}}{Z'_{if} + 3600} \tag{9.8-18}$$

However, we have found from (9.8-16b) that $A_{vf} = 2.5$. Solving for Z'_{if} yields $Z'_{if} \approx -6$ kΩ. Note that Z'_{if} is negative. This is necessary for the emitter follower to provide the gain of 2.5. It must be noted that even though Z'_{if} is negative the amplifier is *stable* and will not oscillate since $T = 0.4 < +1$. Thus, the result that Z'_{if} is negative represents a curiosity, rather than a result of great significance.

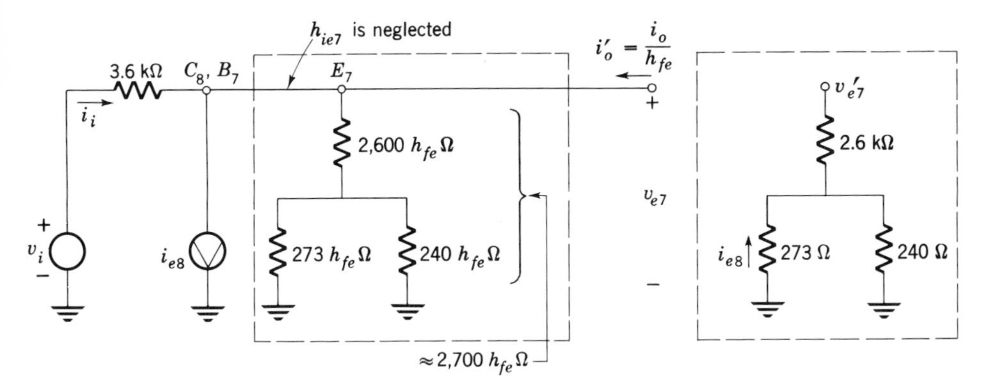

Fig. 9.8-5 Further simplification of output circuit.

The output impedance without feedback (see Fig. 9.8-5) is

$$Z_o = \frac{v_{e7}}{i_o}\bigg|_{\substack{v'_{e7}=0(\therefore i_{e8}=0)\\ v_i=0}} = \frac{v_{e7}}{h_{fe}i_o'} \approx \frac{3600}{h_{fe}} = 72\ \Omega \tag{9.8-19a}$$

and the output impedance with feedback is

$$z_{of} = \frac{Z_o}{1-T} = \frac{72}{0.4} = 180\ \Omega \tag{9.8-19b}$$

We have now determined the small-signal amplifier specifications for the μA 702. They include a voltage gain of approximately 73 dB and an output impedance of 180 Ω. The input impedance of the amplifier with B_3 grounded is 2 $h_{ie2} \approx$ 13 kΩ.

9.9 CASCADING INTEGRATED–CIRCUIT AMPLIFIERS

Often a single integrated-circuit amplifier is not sufficient to satisfy the requirements of a design, and two or more ICs must be interconnected. This is illustrated in the following example.

Example 9.9-1 ICs similar to that shown in Fig. 9.8-1 are available. The pertinent specifications are

Open-loop voltage gain	2600
Input impedance	25 kΩ
Output impedance	200 Ω
Maximum output swing	±5.3 V

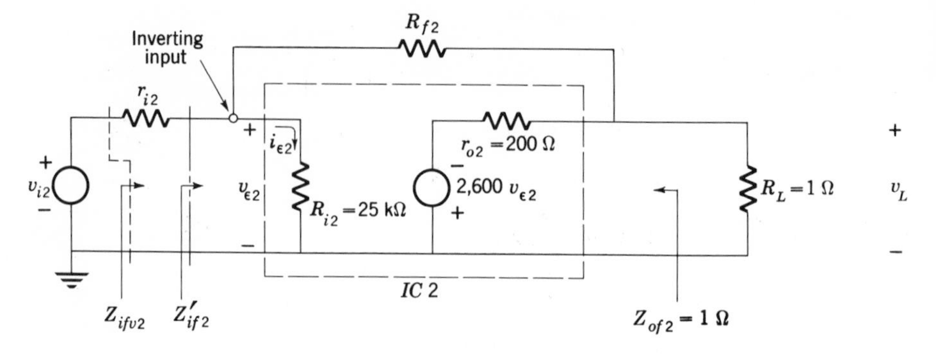

Fig. 9.9-1 Output stage for Example 9.9-1.

These amplifiers are to be used with feedback to design a system with the following specifications:

Overall voltage gain	10
Maximum input voltage	10 mV
Signal-source resistance	500 Ω
Load resistance	1 Ω
Output impedance	1 Ω

Solution It is usually convenient to begin such a design by attempting to satisfy the requirements at the output of the system. Thus we have the equivalent circuit shown in Fig. 9.9-1.

The voltage-feedback–current-error configuration (Sec. 8.2-1) is used. The loop gain will have to be adjusted to achieve an output impedance of 1 Ω with the load disconnected. Thus

$$Z_{of2} = \frac{200}{1 - T_2} = 1\ \Omega \tag{9.9-1}$$

Therefore $T_2 \approx -200$ with the load disconnected. The voltage gain with feedback then becomes (Example 8.2-1)

$$A_{vf2} = \frac{v_L}{v_1} = \frac{A_{v2}}{1 - T_2} \approx \frac{-2600}{200} = -13 = -\frac{R_{f2}}{r_{i2}} \tag{9.9-2}$$

Since this gain is divided by 2 because $R_L = Z_{of2}$, the *net* voltage gain of IC2 is 6.5, which is not enough to meet the specifications. We must now add an additional stage which has a gain of $10/6.5 \approx 1.6$. This stage is shown in Fig. 9.9-2. Note that the

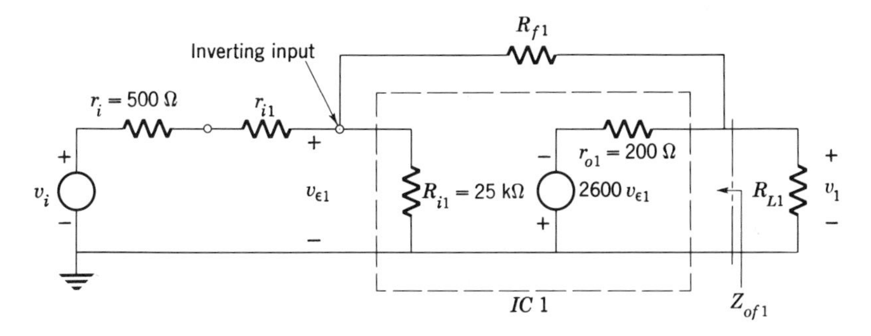

Fig. 9.9-2 Input stage for Example 9.9-1.

load for IC1 is [Eq. (8.4-7a)]

$$R_{L1} = Z_{ifv2} \approx r_{i2}$$

where we have assumed that Z'_{if} will be negligible. If r_{i2} does not load IC1, the voltage gain of IC1 will be

$$A_{vf1} = \frac{v_1}{v_i} \approx -\frac{R_{f1}}{r_{i1} + r_i} \tag{9.9-3}$$

If we choose $R_{f1} = 80\ \text{k}\Omega$ and $r_{i1} = 50\ \text{k}\Omega$, then $A_{vf1} \approx -1.6$. The loop gain of IC1 is

$$T_1 \approx A_v r_{i1}/R_{f1} = (-2600)(50{,}000)/(80{,}000) \approx -1600.$$

Thus the output impedance of IC1 is $Z_{of1} \simeq 200/1600 = 0.125\ \Omega$. To avoid loading IC1, r_{i2} must be chosen much greater than Z_{of1}. For convenience we choose $r_{i2} = 2\ \text{k}\Omega$, and from (9.9-2), $R_{f2} = 26\ \text{k}\Omega$. The complete system is shown in Fig. 9.9-3.

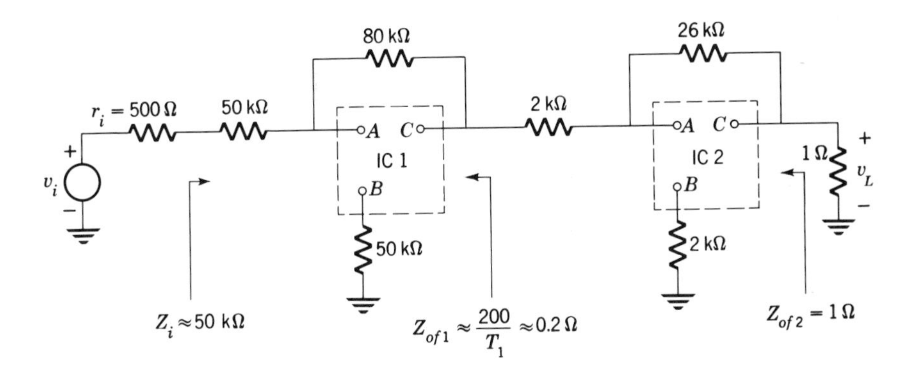

Fig. 9.9-3 Complete system. A—Inverting input terminal; B—noninverting input terminal; C—output terminal.

In an actual realization of this circuit, stabilizing networks would be required to prevent oscillations. This topic is discussed in Chap. 15.

PROBLEMS

SECTION 9.2

9.1. Fig. P9.1 is a model of an IC transistor at low frequencies.

(a) Explain how r_b, r_e, and r_{sc} arise.

(b) Should S be connected to the most positive or most negative voltage?

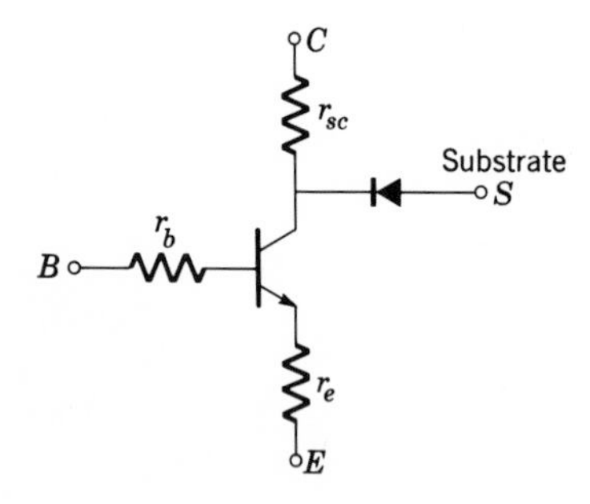

Fig. P9.1

9.2. The IC transistor shown in Fig. P9.1 is connected to the circuit shown in Fig. P9.2.

(a) If $r_b = r_{sc} = r_e = 10\ \Omega$, draw an approximate small-signal equivalent circuit for the amplifier.

(b) Calculate the current gain A_i.

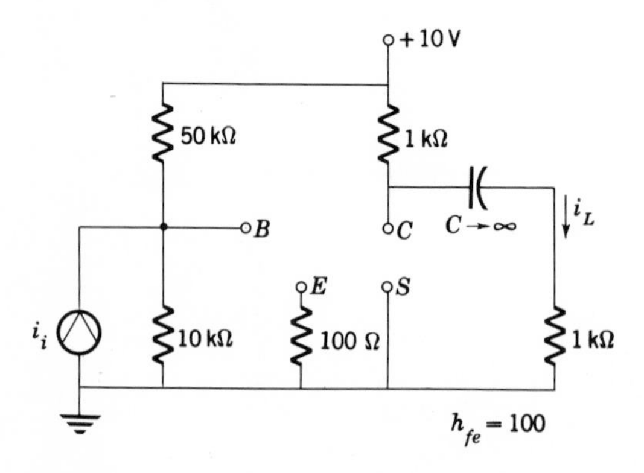

Fig. P9.2

SECTION 9.3

9.3. A 10-V step is applied to the diode circuit of Fig. P9.3. The IC diodes shown in Fig. 9.3-3 are used. Sketch the response.

Compare the results obtained, using each model.

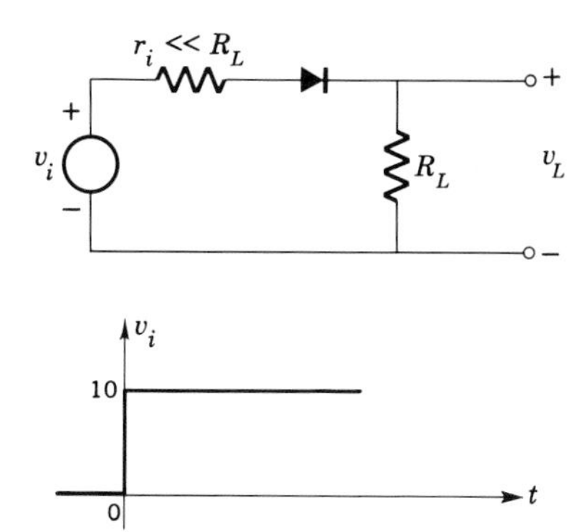

Fig. P9.3

SECTION 9.4

9.4. A junction capacitor having the model shown in Fig. 9.4-2 is used to tune an inductance to 100 Mrad/s in Fig. P9.4. If $C_1 = 500$ pF, $C_2 = 70$ pF, and $r_{sc} = 2\ \Omega$:

(*a*) Determine the voltage v_L when $v_i = \cos \omega t$, where $\omega = \omega_0 = 100$ Mrad/s.

(*b*) Calculate L.

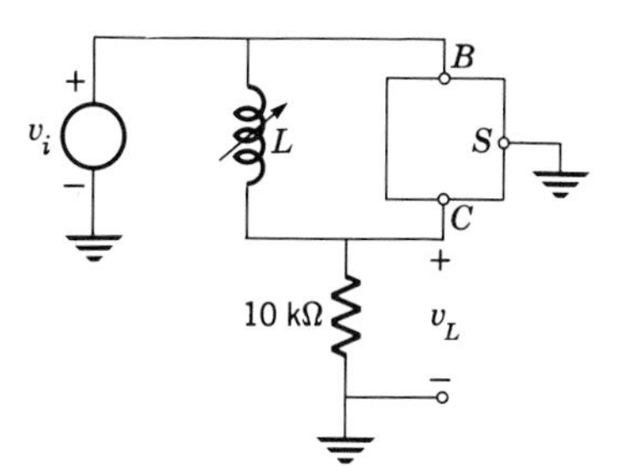

Fig. P9.4

9.5. Plot v_L in Fig. P9.4 as ω varies from zero to infinity.

SECTION 9.5

9.6. Show that at low frequencies the circuit of Fig. 9.5-2 is a resistor. *Hint:* Neglect capacitances, and draw the model for the transistor. Is S connected to a positive or negative voltage? Why?

9.7. Design a 4-kΩ thin-film resistor using nitrided tantalum. Assume a maximum power rating of 10 W/in.^{2} of film. The maximum applied voltage is 10 V. Design for *minimum area.*

SECTION 9.7

9.8. Design the IC amplifier shown in Fig. P9.8. Show a top view. Your layout should avoid overlapping connections.

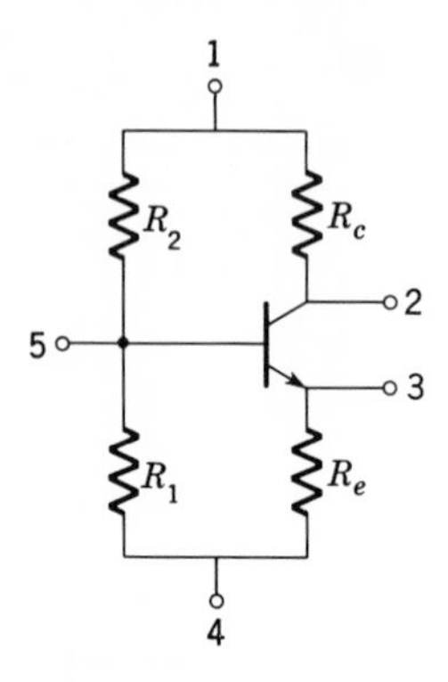

Fig. P9.8

9.9. Design the IC Darlington amplifier.

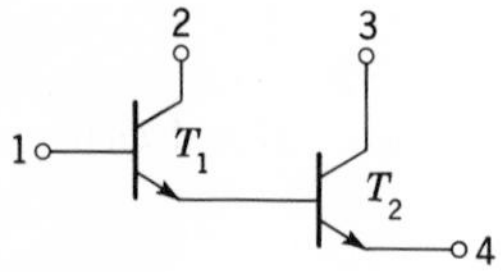

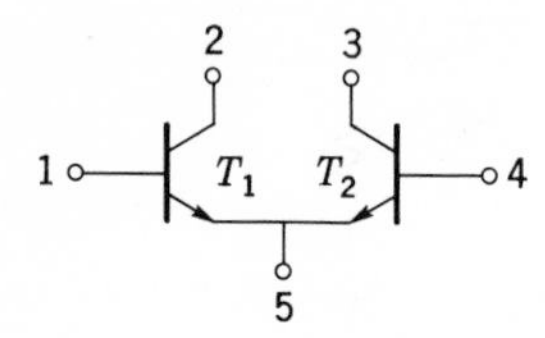

Fig. P9.9

9.10. Design the IC difference amplifier.

Fig. P9.10

9.11. Design the IC difference amplifier with a constant-current source.

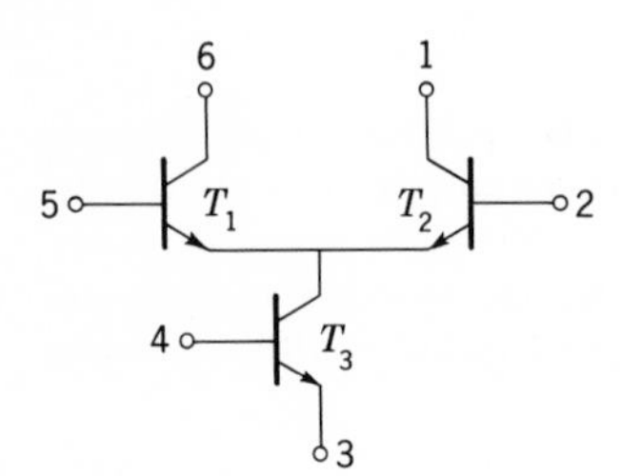

Fig. P9.11

9.12. Design the amplifier shown.

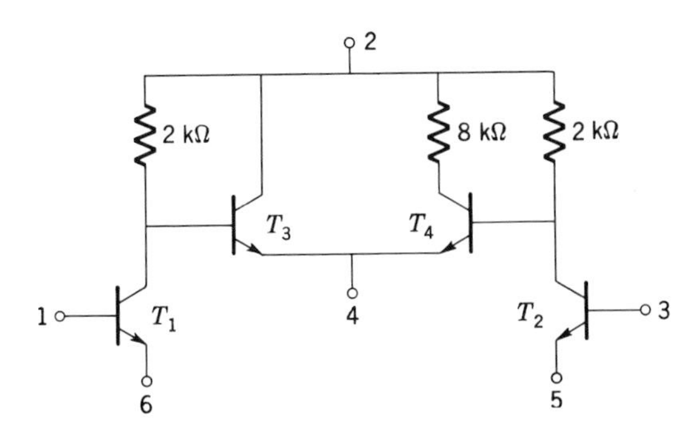

Fig. P9.12

SECTION 9.8

9.13. The IC circuit shown is a wideband amplifier. If $h_{ie} = 1$ kΩ and $h_{fe} = 50$ for each transistor, calculate the voltage gain.

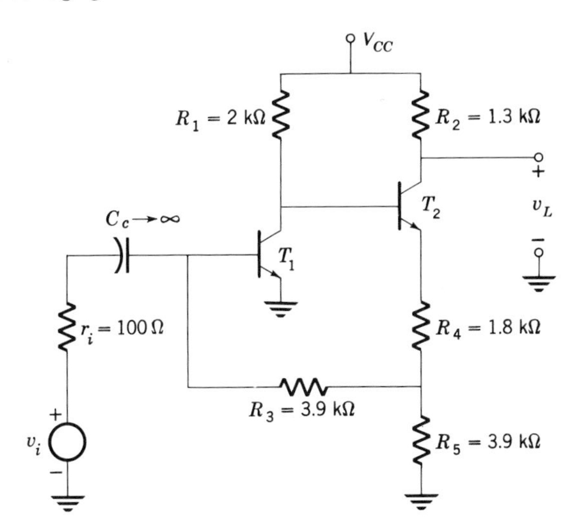

Fig. P9.13

9.14. If in Fig. P9.13 $V_{CC} = +12$ V, find the Q point for each transistor.

9.15. Obtain an expression for V_{LQ} in Fig. P9.13 in terms of V_{BE1}, V_{BE2}, and V_{CC} and resistors $R_1 - R_5$. Is V_{LQ} a function of resistor ratios only? Using the resistance values given in the figure, calculate

$$\frac{\partial V_L}{\partial V_{BE1}}, \frac{\partial V_L}{\partial V_{BE2}}, \text{ and } \frac{\partial V_L}{\partial V_{CC}}$$

Can you choose better resistance values?

9.16. In the circuit of Fig. P9.13 determine the shift in quiescent value of v_L due to a change in temperature ΔT.

9.17. Verify (9.8-5).

9.18. Verify (9.8-7).

9.19. Verify (9.8-9). Substitute the appropriate values into (9.8-9) and determine if the equation is equal to zero.

9.20. Show that in Fig. 9.8-1 $V_{E7} = -0.7$ V and that it cannot be made equal to zero.

9.21. Verify the quiescent-point values presented for Fig. 9.8-1.

SECTION 9.9

9.22. The IC shown in Fig. P9.22*a* is a difference amplifier with a Darlington iuput and an EF output. Two of these circuits are cascaded as shown in Fig. P9.22*b*. Calculate R_i, R_o, and A_i.

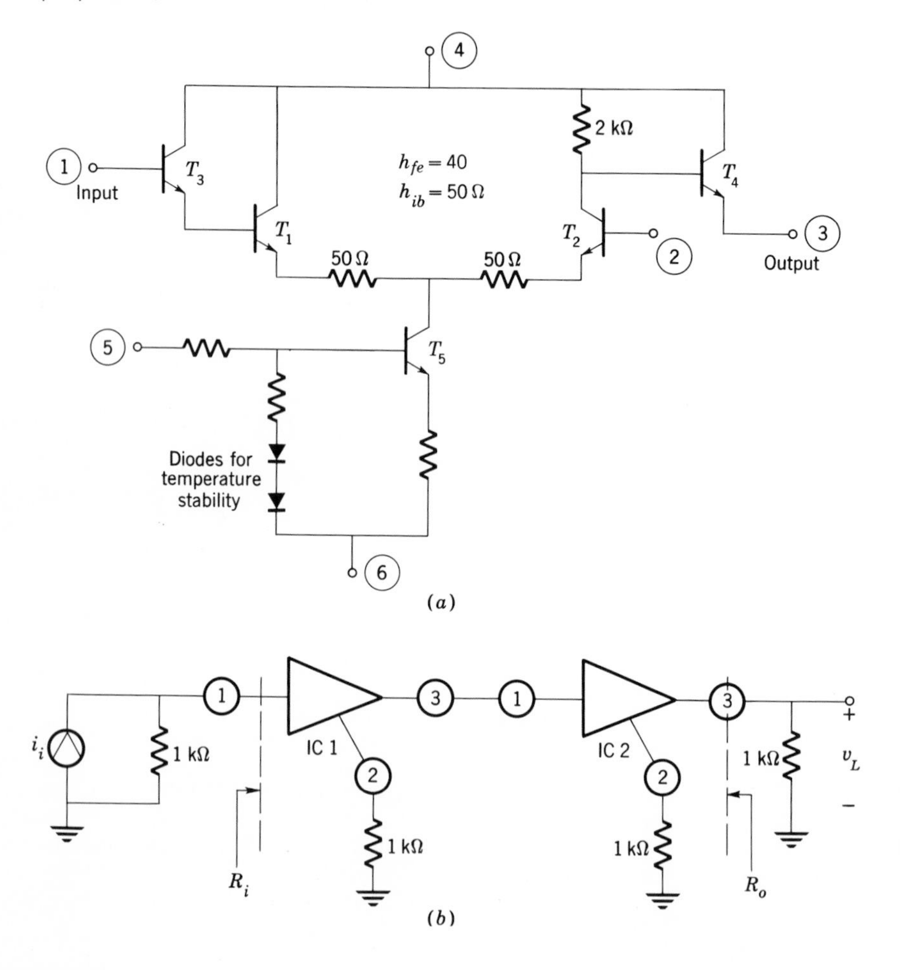

Fig. P9.22

9.23. Calculate for the IC amplifier shown in Fig. P9.23*b*
(*a*) The loop gain T
(*b*) R_{if}
(*c*) R_{of}
(*d*) $A_{vf} = v_L/v_i$

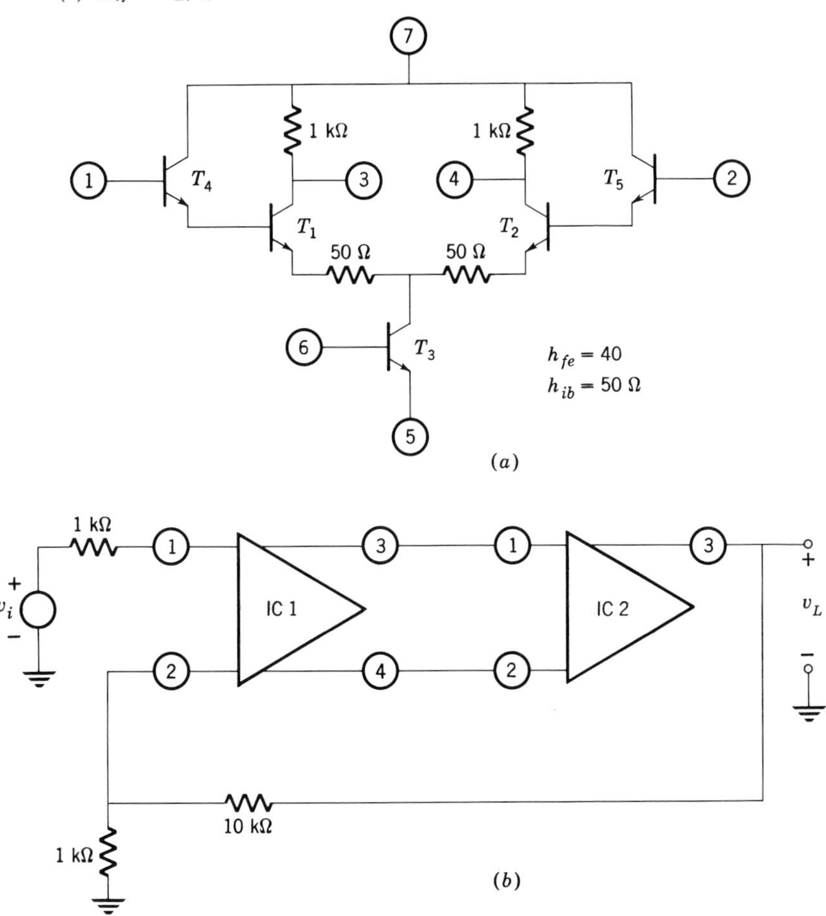

Fig. P9.23

REFERENCES

1. Motorola Inc. Engineering Staff: "Integrated Circuits," McGraw-Hill Book Company, New York, 1965.
2. Motorola Inc. Engineering Staff: "Analysis and Design of Integrated Circuits," McGraw-Hill Book Company, New York, 1967.
3. Giles, J. N.: "Fairchild Semiconductor Linear Integrated Circuits Applications Handbook," Fairchild Semiconductor, Mountain View, Calif., 1967.
   "RCA Linear Integrated Circuit Fundamentals," Radio Corporation of America, Harrison, N.J., 1966.

10
The Field-effect Transistor

INTRODUCTION

The field-effect transistor (FET) is a voltage-sensitive device which has extremely high input impedance (as high as $10^{14}\ \Omega$) as well as high output impedance. It is available in two types, the diffused-junction field-effect transistor (JFET) and the insulated-gate (sometimes called the metal oxide semiconductor) field-effect transistor (IGFET, or MOSFET).

The IGFET is often used in integrated circuits. It has lower input capacitance and higher input impedance than the JFET, and therefore has greater applicability.

10.1 INTRODUCTION TO THE THEORY OF OPERATION OF THE JFET

The JFET is shown schematically and with its circuit symbol in Fig. 10.1-1. The device consists of a thin layer of n-type material with two ohmic contacts, the *source* S and the *drain* D, along with two rectifying contacts, called the *gates* G. The conducting path between the source

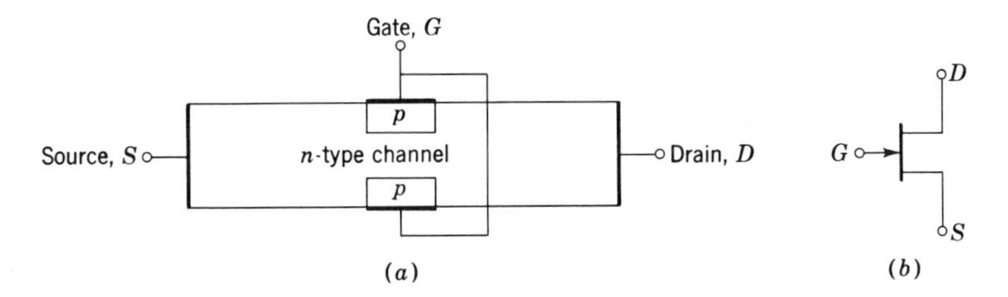

Fig. 10.1-1 The JFET. (*a*) Schematic; (*b*) circuit symbol—*n* channel.

and the drain is called the *channel*. *P*-type JFETs, in which the channel is *p* type and the gate is *n* type, are also available.

To understand the operation of this device, assume the source and the gates to be at ground potential. Now let us study the effect of placing a small positive potential on the drain. Since there is a positive voltage between drain and source, electrons will flow from the source to the drain (current flow is from drain to source). Note that negligible current flows between source (or drain) and gate, since the diode formed by the channel-to-gate junction is reverse-biased. The amount of current flowing from the drain to the source depends initially on the drain-to-source potential difference v_{DS} and the resistance of the *n* material in the channel between the drain and the source. This resistance is a function of the doping of the *n* material and the channel width, length, and thickness.

As the drain potential v_{DS} is increased, the diode formed by the channel-gate junction is further reverse-biased. Let us see what happens to the channel when this happens. Figure 10.1-2 shows a reverse-biased diode. Initially, the holes in the *p* material flow toward the negative terminal of V_r, and the electrons in the *n* material flow toward the positive terminal of V_r. This result is the formation of a central region of length l, which is void of free charges (holes and electrons). Since the region contained within l has been depleted of free charges, it is called the *depletion region*. As the reverse voltage is increased, the free charges (holes and electrons) move farther from the junction and the effective separation length l increases.

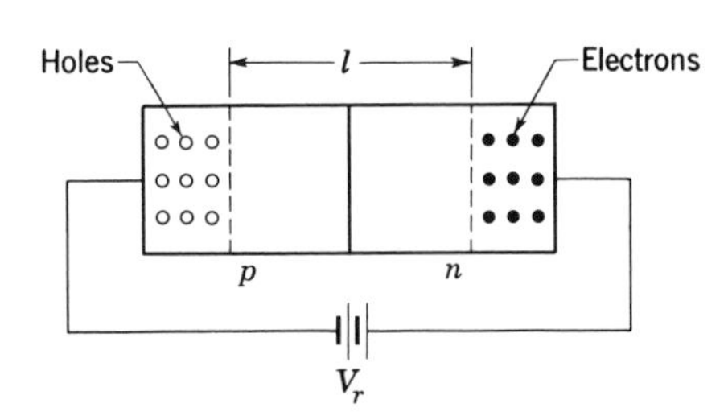

Fig. 10.1-2 A reverse-biased diode.

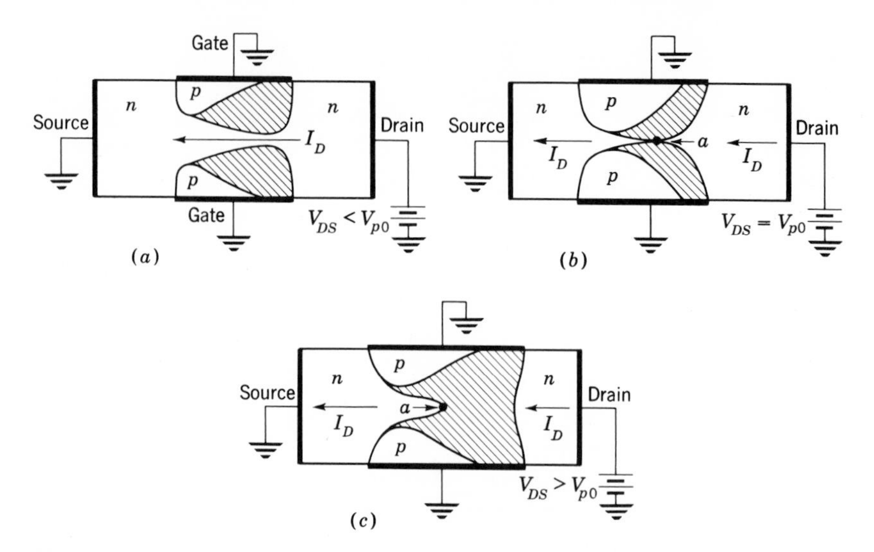

Fig. 10.1-3 The JFET below, at, and above pinch-off. The shaded areas indicate depletion regions. (*a*) $V_{DS} < V_{p0}$; (*b*) $V_{DS} = V_{p0}$; (*c*) $V_{DS} > V_{p0}$.

This result is directly applicable to the JFET under consideration. Sketches of the space-charge regions in the FET for several values of v_{DS} are shown in Fig. 10.1-3.

We see in Fig. 10.1-3*a* that as v_{DS} increases, the depletion region increases, causing narrowing of the channel. The channel area decreases, resulting in an increase in channel resistance; hence the rate of increase of current per unit increase in v_{DS} decreases. This decrease is shown in Fig. 10.1-4.

When $v_{DS} = V_{p0}$ the depletion regions on each side of the channel join together, as shown in Fig. 10.1-3*b*. The voltage V_{p0} is called the

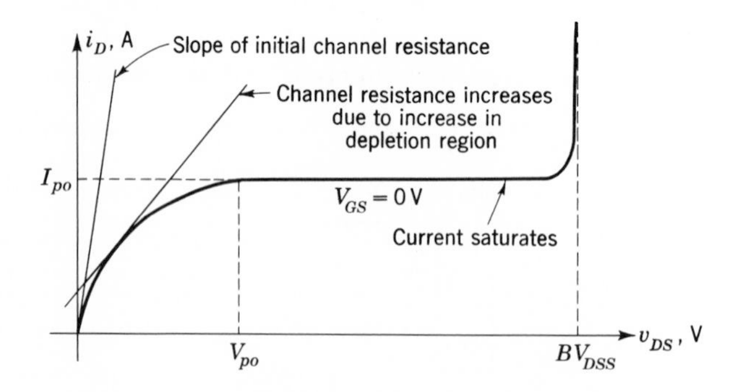

Fig. 10.1-4 Ideal JFET characteristic ($V_{GS} = 0$).

pinch-off voltage* since it "pinches off" the channel connection between drain and source. In Fig. 10.1-3*c* the drain voltage v_{DS} is larger than the pinch-off voltage. In this region the depletion area thickens. However, we find that the potential at point *a* remains essentially at the pinch-off voltage V_{p0}. Thus the current i_D remains almost constant as v_{DS} increases above V_{p0}. This characteristic of the JFET is shown in Fig. 10.1-4.

Note that the drain current i_D increases rapidly as v_{DS} increases toward V_{p0}. Above V_{p0}, the current tends to level off at I_{p0}, and then rises slowly. When v_{DS} equals the breakdown voltage BV_{DSS}, an avalanche breakdown occurs (Sec. 2.7), and the current again rises rapidly.

Let us now consider holding the drain-source voltage fixed and varying the gate-source voltage. As the gate-source voltage is made negative, the *pn* junction is reverse-biased, increasing the depletion region between the gate and the source. This decreases the channel width, increasing the channel resistance. The current i_D therefore decreases. When the gate voltage is made positive, the depletion region decreases until, for large positive gate voltages, the channel opens. Then the *pn* junction between gate and source becomes forward-biased and current flows from the gate to the source. The *n*-type JFET is usually operated so that the gate-to-source potential is either negative or slightly positive to avoid gate-to-source current.

Summarizing, we see that varying the gate voltage varies the channel width, and hence the channel resistance. This in turn varies the current from drain to source, i_D. We note that it is the gate *voltage* variation which varies i_D; thus the FET is a *voltage-sensitive* device as compared with the junction transistor, which is a current-sensitive device.

A typical set of *vi* output characteristics for the JFET is shown in Fig. 10.1-5, with gate-to-source voltage as the parameter. The pinch-off voltage is 5 V when $v_{GS} = 0$, and the drain current at this point is 10 mA. Notice that as the gate potential decreases, the pinch-off voltage also decreases. The drain-source voltage at which pinch-off occurs is approximately given by the equation

$$v_{DS} \text{ (at pinch-off)} = V_p = V_{p0} + v_{GS} \qquad (10.1\text{-}1)$$

Thus, when $v_{GS} = 0$, $V_p = V_{p0}$ as expected. Note that for the *vi* characteristics shown in Fig. 10.1-5, the pinch-off voltage is zero when $v_{GS} = -5$ V. At this negative potential no drain current flows.

The breakdown voltage is also a function of the gate-to-source voltage. This variation is given by

$$BV_{DSX} \approx BV_{DSS} + v_{GS} \qquad (10.1\text{-}2)$$

* The first subscript p represents pinch-off, while the second subscript 0 means that $v_{GS} = 0$.

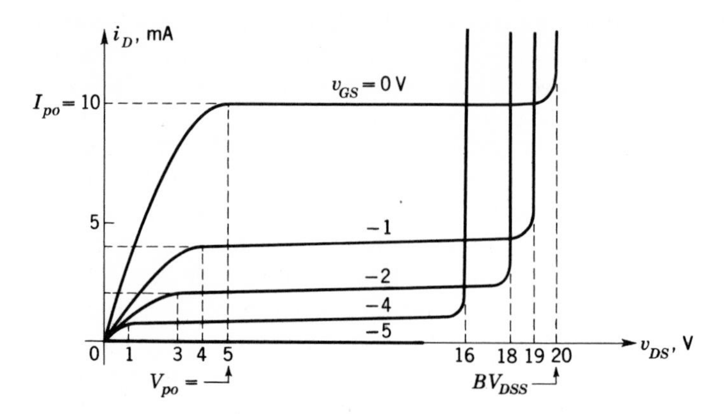

Fig. 10.1-5 JFET *vi* characteristic.

where BV_{DSS} is the breakdown voltage for $v_{GS} = 0$ (for this set of characteristics $BV_{DSS} = 20$ V), and BV_{DSX} is the breakdown voltage for arbitrary v_{GS}.

At drain-to-source potentials between pinch-off and breakdown, called the *saturation* region, the drain current is very nearly constant, and can be approximated as

$$i_D = I_{p0}\left[1 + \frac{3v_{GS}}{V_{p0}} + 2\left(-\frac{v_{GS}}{V_{p0}}\right)^{3/2}\right] \qquad v_{GS} < 0 \tag{10.1-3}$$

We see from (10.1-3) that when $v_{GS} = 0$, $i_D = I_{p0}$, and when $v_{GS} = -V_{p0}$, $i_D = 0$. Note that these results and (10.1-3) are independent of v_{DS}. The current I_{p0} can be shown to be inversely proportional to the $3/2$ power of the temperature ($I_{p0} \sim T^{-3/2}$); hence the current i_D is temperature-dependent. To show this temperature dependence, (10.1-3) can be rewritten

$$i_D = I'_{p0}\left(\frac{T_o}{T}\right)^{3/2}\left[1 + \frac{3v_{GS}}{V_{p0}} + 2\left(-\frac{v_{GS}}{V_{p0}}\right)^{3/2}\right] \tag{10.1-4}$$

where I'_{p0} is the drain current when $v_{GS} = 0$, at an operating temperature T_0.

10.2 INTRODUCTION TO THE THEORY OF OPERATION OF THE IGFET (MOSFET)

The operation of the insulated-gate FET (IGFET) is similar to the operation of the JFET, which was discussed in Sec. 10.1. There are, however, basic differences, which result in the IGFET having lower capacitance and higher input impedance than the JFET.

An *n*-channel IGFET (MOSFET) is shown in Fig. 10.2-1. It

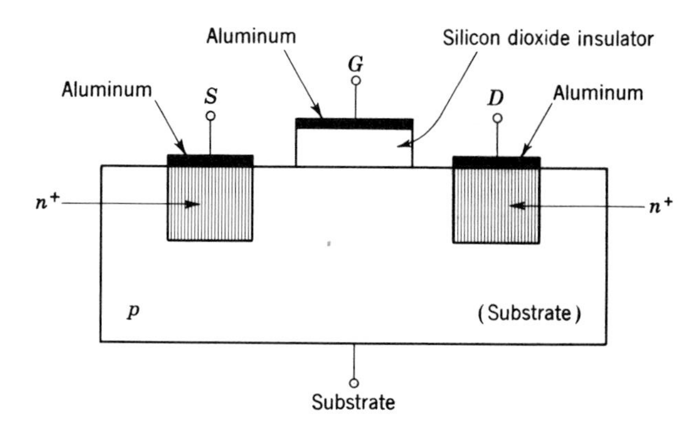

Fig. 10.2-1 Insulated-gate FET (MOSFET).

consists of a p-type substrate into which two $n+$ regions have been diffused. These two regions form the *source* and the *drain*. No channel is actually fabricated in this device, which should be compared with the JFET (Fig. 10.1-1).

The gate is formed by covering the region between the drain and the source with a silicon dioxide layer on top of which is deposited a metal plate. (It is this gate formation of a Metal, Oxide, and Semiconductor that results in the name MOSFET.)

The MOSFET is usually operated with a positive gate-source potential. This is called the *enhancement* mode of operation. When the gate is positive an n-type channel is *induced* between the source and the drain (Fig. 10.2-2). As the drain voltage is increased, a depletion region forms, constricting the channel. The small-signal channel impedance increases until, when $v_{DS} = V_P$, the channel pinches off and the small-signal channel impedance becomes infinite. (Actually, as with the JFET, the

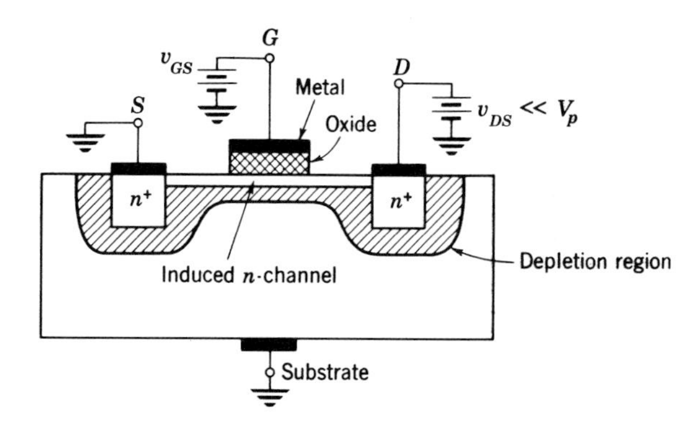

Fig. 10.2-2 The IGFET below pinch-off.

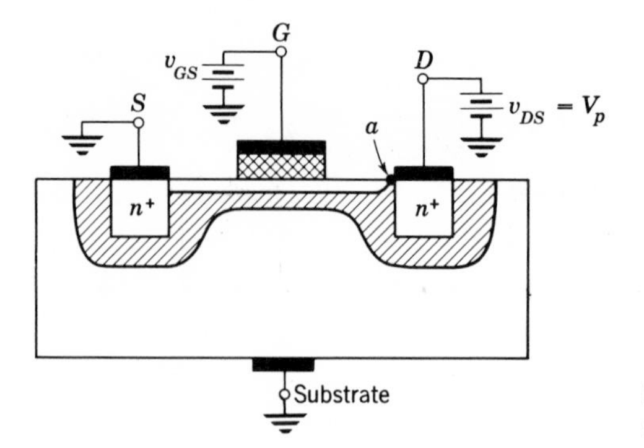

Fig. 10.2-3 The IGFET at pinch-off.

impedance never really becomes infinite; 100 kΩ is a typical value.) This is shown in Fig. 10.2-3. Further increases in drain-source potential result in only a slight increase in the drain-source current.

When the drain-source potential exceeds the pinch-off voltage, a depletion region forms between the drain and the channel as shown in Fig. 10.2-4. Point a, which denotes the pinch-off point, moves only slightly toward the source. Note the similarity to the JFET shown in Fig. 10.1-3.

Now consider varying the gate-source potential, keeping the drain-source potential fixed and above pinch-off. Increasing the gate voltage increases the *conductivity* of the channel and thereby increases the current. Thus, the drain-source *current* is modulated by the gate-source *voltage*.

The basic operation of the MOSFET and the JFET are therefore seen to be similar. Above pinch-off, increases in the drain voltage do not result in a proportional increase in drain current, and the drain *current* is proportional to changes in the gate *voltage*.

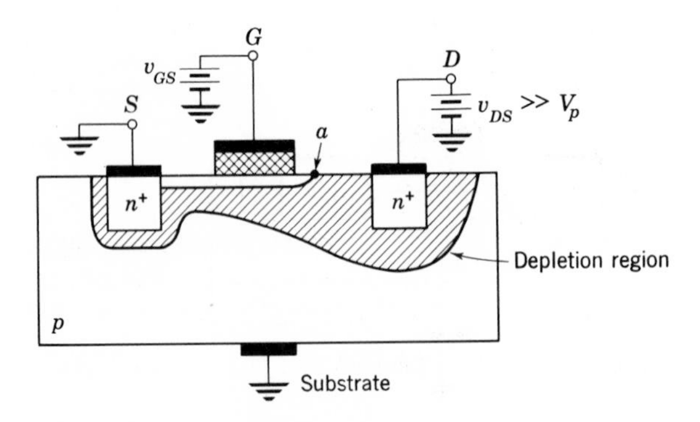

Fig. 10.2-4 The IGFET above pinch-off.

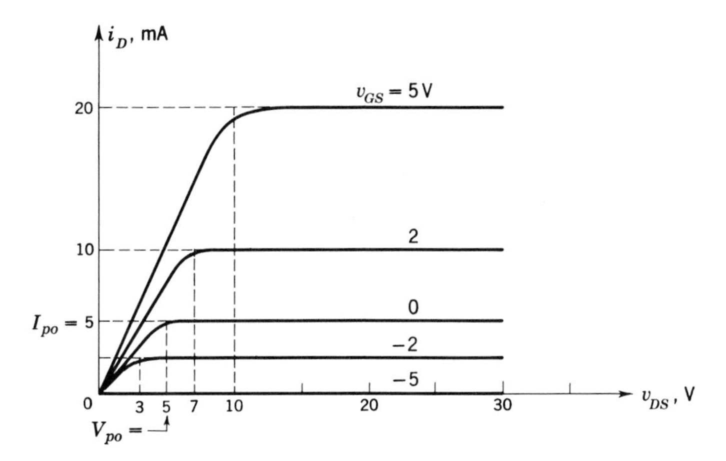

Fig. 10.2-5 IGFET *vi* characteristic.

The operation of the devices differs in that the *n*-channel JFET requires a negative gate voltage, while the MOSFET employs a positive gate voltage. Also, the mechanism which results in a drain current proportional to the gate voltage is different. In the JFET, increasing v_{GS} decreased the depletion region, *widened* the channel, and in this manner increased the channel conductivity. In the MOSFET, increasing v_{GS} increases the mobile charge carrier density in the channel, and in this way increases the channel conductivity.

The *vi* characteristics of a typical IGFET are shown in Fig. 10.2-5. As before, the drain-source voltage at pinch-off is, approximately,

$$v_{DS} \text{ (at pinch-off)} = V_p \approx V_{p0} + v_{GS} \tag{10.2-1}$$

However, in this case, v_{GS} is usually maintained positive.

The pinch-off current i_D is given by the equation

$$i_D \text{ (at and above pinch-off, called the \textit{saturation} region*)}$$
$$\approx I_{p0}\left(1 + \frac{v_{GS}}{V_{p0}}\right)^2 \tag{10.2-2}$$

The current I_{p0} is again proportional to $T^{-3/2}$.

Note that the IGFET is, theoretically, a *square-law* device, while the JFET is a $3/2$-*power-law* device. It is assumed in (10.2-2) that the current above pinch-off is constant at I_p. However, we know that the current does increase slightly as v_{DS} increases above the value given by (10.2-1).

* The *saturation* region for an FET is the region between pinch-off and breakdown, and represents the normal operating region. In contrast, the saturation region of the transistor refers to collector-emitter saturation and is not used for linear amplification.

The breakdown voltage BV_{DSX} is again given approximately by

$$BV_{DSX} \approx BV_{DSS} + v_{GS} \tag{10.2-3}$$

10.3 GRAPHICAL ANALYSIS AND BIASING

The operating ranges of the JFET and IGFET, when used as amplifiers, lie between the drain-to-source pinch-off voltage V_p and the drain-to-source breakdown voltage BV_{DSX}. Between these limits the drain current i_D is approximately independent of changes in v_{DS}, but is in general nonlinear with respect to changes in v_{GS}. The *vi* characteristics for a given FET are usually supplied by the manufacturer, along with specifications of maximum drain current, maximum power dissipation, maximum junction temperature, etc.

BIASING THE JFET

A typical amplifier employing a JFET is shown in Fig. 10.3-1*a*. The *vi* characteristic of the JFET is shown in Fig. 10.3-1*b*. Note that an analogy can be made between the FET and the common-emitter transistor amplifier. The drain acts like the collector, the source like the emitter, and the gate *voltage* like the base *current*. The source resistor R_s is analogous to the emitter resistor R_e, and R_d is analogous to the collector resistor R_c.

The dc-load-line equation for the FET is

$$V_{DD} = v_{DS} + i_D(R_d + R_s) \tag{10.3-1}$$

and the bias equation, assuming $i_G = 0$ (since the gate current in a properly biased JFET is less than 0.01 μA), is

$$v_{GS} = -i_D R_s \tag{10.3-2}$$

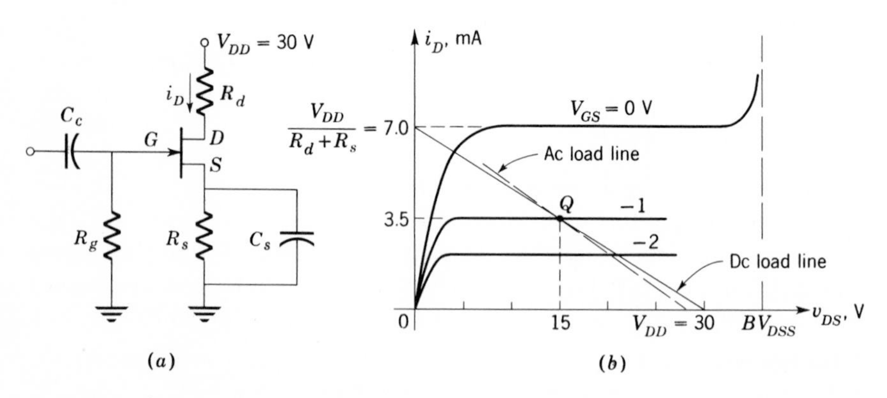

Fig. 10.3-1 Common-source JFET amplifier. (*a*) Schematic; (*b*) *vi* characteristic.

Note that an external bias circuit need not be employed when using a JFET, since the gate-to-source voltage is developed directly across R_s by the drain current. This type of bias is called *self-bias,* and differs considerably from the bias circuit used with a transistor, where V_{BB} is required to ensure that sufficient quiescent base current is flowing.

To design the amplifier of Fig. 10.3-1 for a Q point at

$$V_{DSQ} = 15 \text{ V}$$

and

$$I_{DQ} = 3.5 \text{ mA}$$

we substitute these values (and $V_{DD} = 30$ V) into (10.3-1).

$$R_d + R_s = \frac{V_{DD} - V_{DSQ}}{I_{DQ}} = \frac{30 - 15}{3.5 \times 10^{-3}} = 4.3 \text{ k}\Omega \tag{10.3-3}$$

The vi characteristic of Fig. 10.3-1*b* indicates that the Q point occurs at a gate-to-source voltage

$$V_{GSQ} = -1 \text{ V}$$

Then, using (10.3-2), the source resistor is

$$R_s = -\frac{V_{GSQ}}{I_{DQ}} = \frac{1}{3.5 \times 10^{-3}} = 286 \ \Omega \tag{10.3-4}$$

The drain resistance is

$$R_d = 4300 - R_s \approx 4 \text{ k}\Omega$$

Standard resistors of 270 Ω and 3.9 kΩ would be used.

The dc and ac load lines are calculated, and plotted as shown in Fig. 10.3-1*b*. Note that if v_{GS} varies by ± 1 V about the Q point, significant distortion due to unsymmetrical variation in drain current and drain voltage results. This is due to the unequal spacing of the vi characteristics. While this was also true in the transistor, variations of i_C due to variations of i_B remained fairly linear, as contrasted with the JFET, where i_D increases as $(v_{GS})^{3/2}$. Thus the FET is used primarily as a small-signal linear amplifier.

BIASING THE IGFET

The IGFET is most often operated using a forward-biased gate. The quiescent gate potential is obtained using an external bias circuit (as for the transistor), as shown in Fig. 10.3-2*a*. In this circuit the dc-load-line equation is

$$V_{DD} = v_{DS} + i_D(R_d + R_s) \tag{10.3-5}$$

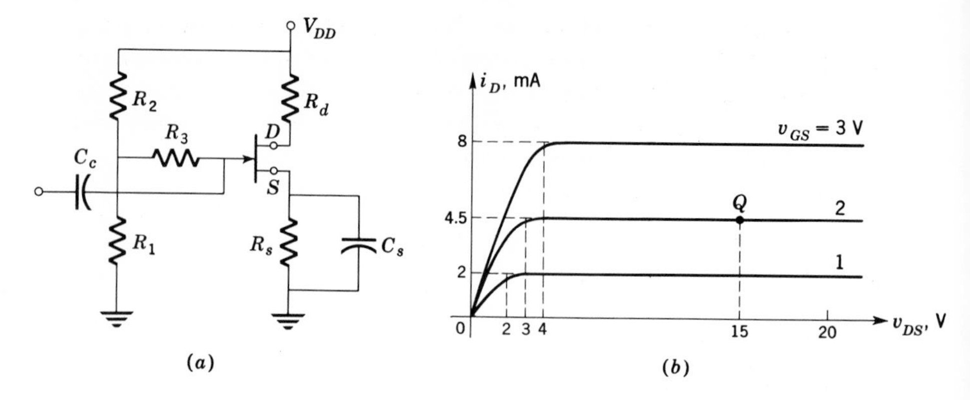

Fig. 10.3-2 Common-source IGFET amplifier. (*a*) Schematic; (*b*) *vi* characteristic.

and the gate voltage is

$$v_{GS} = \left(\frac{R_1}{R_1 + R_2}\right) V_{DD} - i_D R_s \tag{10.3-6a}$$

If we define a gate supply voltage V_{GG} as

$$V_{GG} = \left(\frac{R_1}{R_1 + R_2}\right) V_{DD}$$

then (10.3-6*a*) becomes

$$v_{GS} = V_{GG} - i_D R_s \tag{10.3-6b}$$

Note that we can let the source resistor $R_s = 0$. In practice, however, in addition to the external bias, self-bias is employed to improve *Q*-point stability by virtue of the current feedback it introduces at dc.

The resistor R_3 in the input circuit does not affect the dc bias since negligible gate current flows. Its function is simply to maintain a high ac input impedance. R_3 is usually of the order of several megohms.

Let us determine the bias circuit of the IGFET of Fig. 10.3-2*a*. The supply voltage V_{DD} is given. The gate supply V_{GG}, source resistor R_s, and drain resistor R_d are to be selected to achieve minimum *Q*-point variation due to temperature changes. The equation representing the *vi* characteristic of this IGFET is

$$i_D = I'_{p0} \left(\frac{T_0}{T}\right)^{3/2} \left(1 + \frac{v_{GS}}{V_{p0}}\right)^2 \tag{10.3-7}$$

This equation approximates the characteristic of Fig. 10.3-2*b*.

To determine the minimum variation of i_D with respect to temperature, we differentiate (10.3-7) and (10.3-6b) with respect to T and combine the results. From (10.3-7)

$$di_D = \left(-\frac{3}{2}\right)\left(\frac{dT}{T}\right)\left[I'_{p0}\left(\frac{T_0}{T}\right)^{3/2}\left(1+\frac{v_{GS}}{V_{p0}}\right)^2\right] + 2I'_{p0}\left(\frac{T_0}{T}\right)^{3/2}\left(1+\frac{v_{GS}}{V_{p0}}\right) d\left(\frac{v_{GS}}{V_{p0}}\right) \quad (10.3\text{-}8a)$$

Equation (10.3-6b) yields

$$dv_{GS} = -R_s di_D \quad (10.3\text{-}8b)$$

Combining (10.3-8a) and (10.3-8b), we obtain, after some manipulation,

$$S_T{}^{i_D} = \frac{di_D/i_D}{dT/T} = \frac{-3/2}{1 + 2I'_{p0}\,(T_0/T)^{3/2}\left(1 + \dfrac{V_{GG} - R_s i_D}{V_{p0}}\right)\dfrac{R_s}{V_{p0}}} \quad (10.3\text{-}9a)$$

Note that if no self-bias were employed, $R_s = 0$, and

$$S_T{}^{i_D} = -3/2 \quad (10.3\text{-}9b)$$

Therefore self-bias decreases the sensitivity.

To minimize $S_T{}^{i_D}$, we maximize the function

$$\lambda = 2i'_{p0}\left(\frac{T_0}{T}\right)^{3/2}\left(1 + \frac{V_{GG} - R_s i_D}{V_{p0}}\right)\frac{R_s}{V_{p0}} \quad (10.3\text{-}10)$$

with respect to R_s. Differentiating with respect to R_s and equating to zero results in $\lambda_{\max}$ occurring at

$$R_s = \left.\frac{V_{p0} + V_{GG}}{2i_D}\right|_{i_D = I_{DQ}} \quad (10.3\text{-}11)$$

Thus, from (10.3-6b),

$$V_{GSQ} = V_{GG} - \frac{V_{p0} + V_{GG}}{2} \quad (10.3\text{-}12a)$$

and

$$V_{GG} = 2V_{GSQ} + V_{p0} \quad (10.3\text{-}12b)$$

and from (10.3-11),

$$R_s = \frac{V_{GSQ} + V_{p0}}{I_{DQ}} \quad (10.3\text{-}12c)$$

The values for V_{GG} and R_s given by (10.3-12b) and (10.3-12c) result in a quiescent point with minimal variation due to temperature changes.

For example, if a Q point,

$$V_{DSQ} = 15 \text{ V}$$
$$I_{DQ} = 4.5 \text{ mA}$$

and

$$V_{GSQ} = 2 \text{ V}$$

is required, find V_{GG}, R_s, and R_d.

From Fig. 10.3-2b and (10.2-1), $V_{p0} = 1$ V. Then, using (10.3-12b), we find $V_{GG} = 5$ V. From (10.3-12c)

$$R_s = 667\ \Omega \approx 680\ \Omega \qquad \text{(the nearest standard resistance)}$$

The drain resistance is now selected to reduce the voltage from the supply V_{DD} to the drain voltage V_{DQ}.

$$R_d = \frac{V_{DD} - V_{DQ}}{I_{DQ}} = \frac{V_{DD} - (V_{DSQ} + R_s I_{DQ})}{I_{DQ}}$$

The supply voltage V_{DD} is usually given for the system in which the FET is to be used. If, for example, V_{DD} is 24 V, then

$$R_d = \frac{24 - 15 - 3}{4.5 \times 10^{-3}} \approx 1.3 \text{ k}\Omega$$

10.4 LARGE-SIGNAL ANALYSIS—DISTORTION

Here we investigate the effect of the nonlinear relation between i_D and v_{GS}. Consider an IGFET described above pinch-off by the equation

$$i_D = I_{p0}\left(1 + \frac{v_{GS}}{V_{p0}}\right)^2 \tag{10.4-1}$$

Let the dc gate voltage be V_{GSQ}, and let the dc drain current be I_{DQ}. Now consider an ac signal impressed on the gate as shown in Fig. 10.4-1. Then

$$v_{GS} = V_{GSQ} + v_i = V_{GSQ} + V_{im} \cos \omega_0 t \tag{10.4-2}$$

Substituting (10.4-2) into (10.4-1) yields

$$\begin{aligned} i_D &= I_{p0}\left(1 + \frac{V_{GSQ}}{V_{p0}} + \frac{V_{im}}{V_{p0}} \cos \omega_0 t\right)^2 \\ &= I_{p0}\left[\left(1 + \frac{V_{GSQ}}{V_{p0}}\right)^2 + \frac{1}{2}\left(\frac{V_{im}}{V_{p0}}\right)^2\right] \\ &\quad + 2I_{p0}\left(1 + \frac{V_{GSQ}}{V_{p0}}\right)\left(\frac{V_{im}}{V_{p0}}\right) \cos \omega_0 t \\ &\quad + \left(\frac{I_{p0}}{2}\right)\left(\frac{V_{im}}{V_{p0}}\right)^2 \cos 2\omega_0 t \end{aligned} \tag{10.4-3}$$

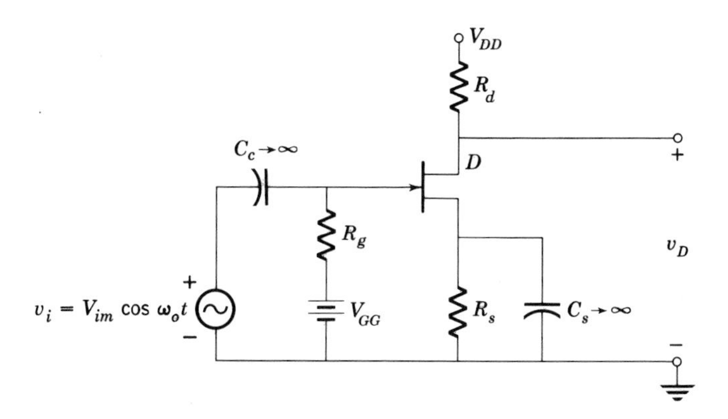

Fig. 10.4-1 An IGFET amplifier.

Note that when $V_{im} = 0$, the quiescent current is

$$I_{DQ} = I_{p0}\left(1 + \frac{V_{GSQ}}{V_{p0}}\right)^2 \tag{10.4-4}$$

With $V_{im} \neq 0$, the nonlinearity of the device causes the dc current to increase to

$$\begin{aligned} I_D &= I_{DQ} + \tfrac{1}{2}I_{p0}\left(\frac{V_{im}}{V_{p0}}\right)^2 \\ &= I_{DQ}\left[1 + \frac{1}{2}\left(\frac{V_{im}}{V_{GSQ} + V_{p0}}\right)^2\right] \end{aligned} \tag{10.4-5}$$

If the signal level is such that

$$V_{im} \ll V_{GSQ} + V_{p0} \tag{10.4-6}$$

then the shift in dc current is negligible.

Equation (10.4-3) also indicates the presence of a *second*-harmonic component of the signal, as expected, because of the square-law relation. The second-harmonic distortion D_2 is often expressed as the ratio in decibels of the amplitude of the second harmonic to the amplitude of the fundamental. Thus

$$D_2 = 20 \log \left[\frac{\left(\dfrac{I_{p0}}{2}\right)\left(\dfrac{V_{im}}{V_{p0}}\right)^2}{2I_{p0}\left(1 + \dfrac{V_{GSQ}}{V_{p0}}\right)\left(\dfrac{V_{im}}{V_{p0}}\right)}\right] = 20 \log \left[\frac{V_{im}/V_{p0}}{4\left(1 + \dfrac{V_{GSQ}}{V_{p0}}\right)}\right] \tag{10.4-7}$$

If

$$V_{p0} = 1 \text{ V}$$

and

$$V_{GSQ} = 4 \text{ V}$$

then

$$D_2 = 20 \log \left(\frac{V_{im}}{20}\right)$$

If the second-harmonic distortion is to be less than -20 dB, the peak signal amplitude must not exceed 2 V.

10.5 SMALL-SIGNAL ANALYSIS

Small-signal analyses of transistor, FET, and vacuum tube circuits are similar. Thévenin or Norton equivalent circuits are obtained at the input and the output of the device. The equivalent circuits are then employed to calculate the input and output impedances and the gain of the amplifier. The difference between the analyses of the transistor and the FET lies in the range of values of the elements of the model and the fact that the FET is a voltage-controlled device, while the transistor is current-controlled.

The low-frequency equivalent circuit of the FET is shown in Fig. 10.5-1. All capacitive effects are neglected in this chapter; they are discussed in Chaps. 12 and 13. Standard symbols for the model parameters have not yet been specified, and the *gate-to-source resistance* r_{gs} is sometimes replaced by the *input admittance* y_{is}; the *mutual transconductance* g_m is often replaced by the symbol y_{fs}; and the *drain-source resistance* r_{ds} is on occasion replaced by the *output admittance* y_{os}. These parameters are defined as follows.

GATE-TO-SOURCE RESISTANCE (OHMS)

$$r_{gs} = \left.\frac{\partial v_{GS}}{\partial i_G}\right|_{Q \text{ point}} \rightarrow \infty \qquad \text{(negligible gate current flows in a FET)} \qquad (10.5\text{-}1)$$

The input circuit of the FET can be represented by an open circuit at low

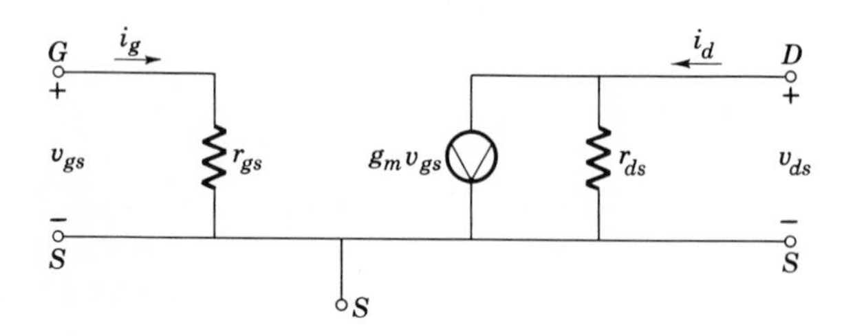

Fig. 10.5-1 Small-signal model of the FET.

frequencies and midfrequencies. This should be compared with the junction transistor for which h_{ie} represents a relatively low input impedance.

TRANSCONDUCTANCE (MHOS)

$$g_m = \left.\frac{\partial i_D}{\partial v_{GS}}\right|_{Q\text{ point}} \tag{10.5-2}$$

The theoretical equation which describes the FET can be used to obtain an idea of the range of values of g_m. Consider the IGFET, for which

$$i_D = I_{p0}\left(1 + \frac{v_{GS}}{V_{p0}}\right)^2 \tag{10.5-3a}$$

Then

$$g_m = \left.\frac{\partial i_D}{\partial v_{GS}}\right|_{Q\text{ point}} = \frac{2I_{p0}}{V_{p0}}\left(1 + \frac{V_{GSQ}}{V_{p0}}\right) \tag{10.5-3b}$$

Let us apply this relation to the IGFET *vi* characteristic shown in Fig. 10.3-2 and given by (10.3-7). For this case $I_{p0} = 0.5$ mA and $V_{p0} = 1$ V. This yields

$$g_m = 10^{-3}(1 + V_{GSQ})$$

If $V_{GSQ} = 2$ V, $g_m = 3 \times 10^{-3}$ mho. Typical values of g_m lie between 10^{-3} and 5×10^{-3} mho. Note that g_m is directly proportional to the quiescent gate-to-source potential.

The g_m of a FET is analogous to $1/h_{ib}$ in a transistor.* This is easily shown from Fig. 10.5-2. The output current is

$$h_{fe}i_b = h_{fe}\left(\frac{v_{be}}{h_{ie}}\right) = \left(\frac{1}{h_{ib}}\right)v_{be}$$

Thus the output-current source in the transistor can be, and often is (Chap. 13), replaced by $g_m v_{be}$, where $g_m = 1/h_{ib}$.

* It is interesting to note that the transconductance of an IGFET is proportional to $\sqrt{I_{DQ}}$ [Eq. (10.6-1)], while the hybrid parameter $1/h_{ib}$ is proportional to I_{EQ}.

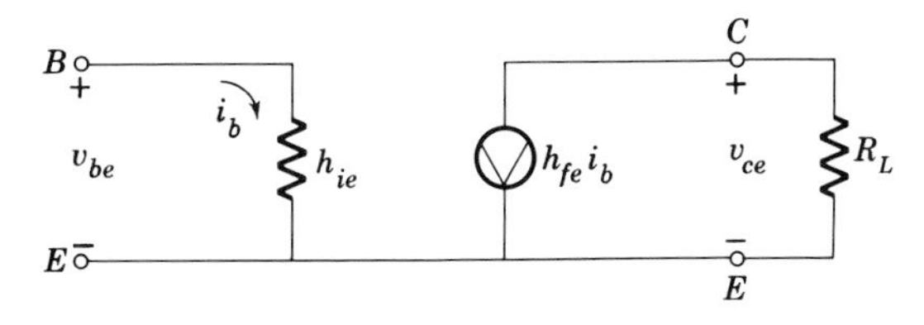

Fig. 10.5-2 Transistor small-signal equivalent circuit.

It is extremely important to note that the voltage gain A_v of a CE transistor amplifier is

$$A_v = \frac{v_{ce}}{v_{be}} = \frac{-R_L}{h_{ib}}$$

while the voltage gain of a common-source FET amplifier is, neglecting r_{ds},

$$A_v = \frac{v_{ds}}{v_{gs}} \approx -g_m R_L$$

The voltage gain of the transistor amplifier is significantly higher than the gain of the FET for the same value of R_L since

$$\left(\frac{1}{h_{ib}}\right)_{\text{transistor}} \gg (g_m)_{\text{FET}}$$

As an example, consider a transistor with $I_{CQ} = 1$ mA. Then

$$\frac{1}{h_{ib}} = 40 \times 10^{-3} \text{ mho}$$

If $I_{CQ} = 10$ mA, $1/h_{ib} = 400 \times 10^{-3}$ mho. These values should be compared with the values of g_m for the FET (10^{-3} to 5×10^{-3} mho).

DRAIN–SOURCE RESISTANCE

$$r_{ds} = \left(\frac{\partial v_{DS}}{\partial i_D}\right)\bigg|_{Q\text{ point}} \tag{10.5-4a}$$

In theory, this resistance should be infinite, since i_D is not a function of drain-source voltage above pinch-off. However, the values of i_D given by (10.1-3) and (10.2-2) represent asymptotic values not actually achieved in practice. Measured output vi characteristics do display a slight slope. This is illustrated in Fig. 10.6-1. The range of values of r_{ds} is similar to that of the transistor output resistance h_{oe}, 20 kΩ to 500 kΩ. The drain-source resistance is found to be, approximately, inversely proportional to the quiescent current.

$$r_{ds} \sim \frac{1}{I_{DQ}} \tag{10.5-4b}$$

This is similar to the transistor, where h_{oe} is directly proportional to I_{CQ}.

AMPLIFICATION FACTOR

An amplification factor μ is often defined as the product $g_m r_{ds}$. It can be calculated directly from the vi characteristics, using the relation

$$\mu = -\frac{\partial v_{DS}}{\partial v_{GS}}\bigg|_{Q\text{ point}}$$

10.5-1 THE COMMON–SOURCE VOLTAGE AMPLIFIER

The common-source voltage amplifier and its small-signal equivalent circuit are shown in Fig. 10.5-3.

The input impedance seen by the source is

$$Z_i = R_3 + (R_1 \| R_2) \tag{10.5-5}$$

The output impedance seen by the load resistance R_i is

$$Z_o = R_d \| r_{ds} \tag{10.5-6}$$

and the voltage gain

$$A_v = \frac{v_L}{v_i} = \left(\frac{v_L}{v_{gs}}\right)\left(\frac{v_{gs}}{v_i}\right) = -g_m(R_L \| Z_o)\left\{\frac{1}{1 + \dfrac{r_i}{[R_3 + (R_1 \| R_2)]}}\right\} \tag{10.5-7a}$$

Usually,

$$r_i \ll R_3 + (R_1 \| R_2)$$

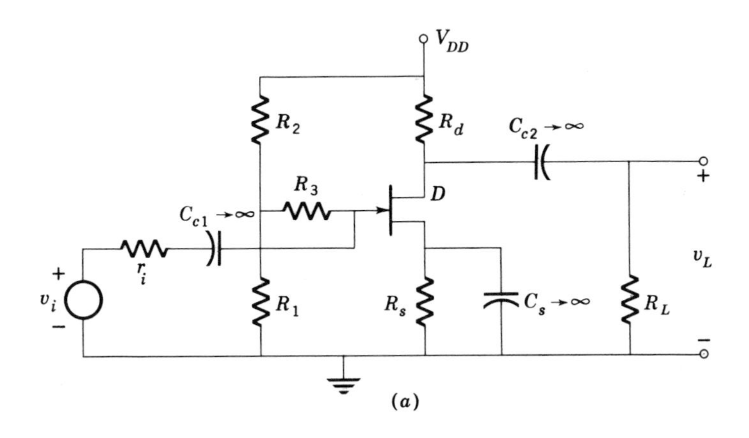

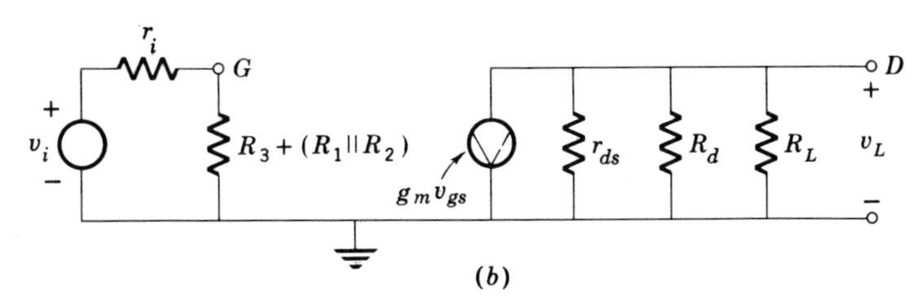

Fig. 10.5-3 The common-source amplifier. (*a*) Schematic; (*b*) small-signal equivalent circuit.

and if

$$R_L \ll Z_o$$

the voltage gain reduces to

$$A_v \approx -g_m R_L \tag{10.5-7b}$$

Example 10.5-1 An IGFET voltage amplifier is shown in Fig. 10.5-4*a*. Calculate the voltage gain, the input impedance, and the output impedance.

Solution The amplifier employs voltage feedback with a current error. A simple technique often employed to analyze this type of circuit is given as follows:

$$i = \frac{v_i - v_{gs}}{r_i} = \frac{v_{gs} - v_d}{R_f} = g_m v_{gs} + \frac{v_d}{r_{ds}\|R_L}$$

$$= \frac{v_i - v_{gs}}{5 \times 10^3} = \frac{v_{gs} - v_d}{10^5} = 2 \times 10^{-3} v_{gs} + \frac{v_d}{6 \times 10^3}$$

Note that

$$\frac{v_d}{v_{gs}} \approx -g_m(r_{ds}\|R_L) = -12$$

Therefore since

$$v_i = v_{gs} + v_{gs}\frac{r_i}{R_f} - v_d\frac{r_i}{R_f}$$

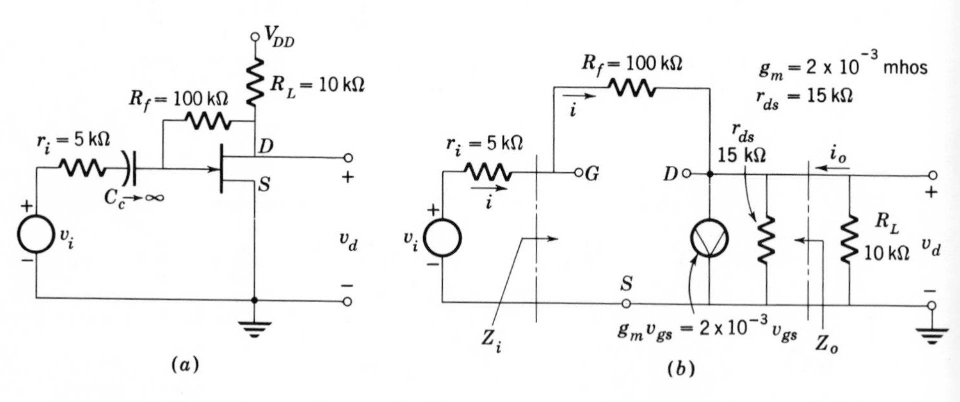

Fig. 10.5-4 IGFET amplifier for Example 10.5-1. (*a*) Schematic; (*b*) small-signal equivalent circuit.

we have

$$v_i = \left[-\frac{1}{12} - \frac{1}{12}\left(\frac{1}{20}\right) - \frac{1}{20} \right] v_d$$

and

$$A_v = \frac{v_d}{v_i} \approx -\frac{240}{33} \approx -7.3$$

The input impedance seen by the source is

$$Z_i = \frac{v_{gs}}{i} = \frac{v_{gs}}{(v_{gs} - v_d)/R_f} = \frac{R_f}{1 - v_d/v_{gs}} = \frac{10^5}{1 + 12} \approx 7.7\ \text{k}\Omega$$

and the output resistance seen by the 10-kΩ load is

$$Z_o = \left. \frac{v_d}{\dfrac{v_d}{r_{ds}} + g_m v_{gs} - i} \right|_{v_i = 0} = \left. \frac{1}{\dfrac{1}{15 \times 10^3} + \dfrac{2 \times 10^{-3}\, v_{gs}}{v_d} - \dfrac{i}{v_d}} \right|_{v_i = 0}$$

When $v_i = 0$,

$$i = -\frac{v_d}{10^5}$$

and

$$v_{gs} \approx \frac{v_d}{20}$$

Therefore

$$Z_o = \frac{1}{1/(15 \times 10^3) + 10^{-4} + 10^{-5}} \approx 5.7\ \text{k}\Omega$$

The same results can be obtained by considering the feedback character of the amplifier, using the techniques of Chap. 8. To do this, we disconnect the 100-kΩ feedback resistor from the output and calculate the loop gain.

$$T = -\left(\frac{r_i}{R_f + r_i}\right) g_m(r_{ds} \| R_L) \approx -0.57$$

The forward gain without feedback is

$$A'_v = -\left(\frac{R_f}{R_f + r_i}\right) g_m(r_{ds} \| R_L) \approx -11.4$$

Hence

$$A_v = \frac{A'_v}{I - T} \approx -7.3$$

The input and output impedances can be calculated in a similar manner.

10.5-2 THE SOURCE FOLLOWER (THE COMMON-DRAIN AMPLIFIER)

A *source-follower* circuit is shown in Fig. 10.5-5. The biasing arrangement of Fig. 10.5-5*a* provides self-bias for negative gate operation. For this type of operation the dc-load-line equation is

$$V_{DD} = v_{DS} + i_D(R_{s1} + R_{s2}) \tag{10.5-8a}$$

and the bias voltage is, assuming zero current in R_1,

$$V_{GSQ} = -I_{DQ}R_{s1} \tag{10.5-8b}$$

Typically, V_{GSQ} will be only a few volts, while V_{SQ} will be, roughly, one-half of V_{DD}, in order to place the operating point near the center of the load line. Therefore $R_{s1} \ll R_{s2}$.

The small-signal characteristics of the source follower are obtained from the small-signal equivalent circuit of Fig. 10.5-5*c*. We determine the Thévenin equivalent circuit for this device and show that the voltage

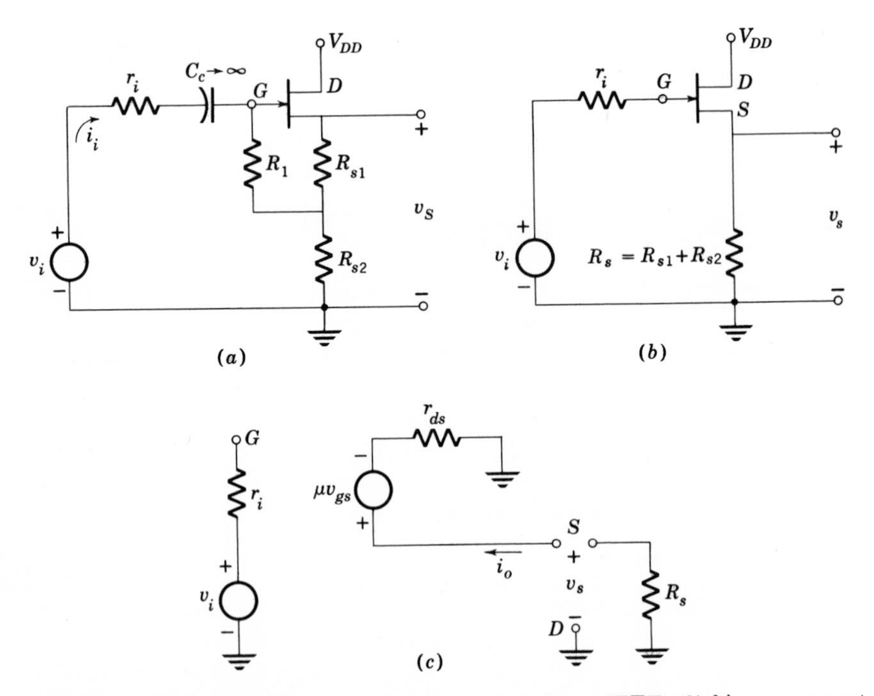

Fig. 10.5-5 The source follower. (*a*) Schematic using a JFET; (*b*) bias components omitted; (*c*) small-signal equivalent circuit.

gain is nearly unity, the input resistance essentially infinite, and the output resistance small.

The *output impedance* Z_o, as seen by the source resistance R_s, is

$$Z_o = \frac{v_s}{i_o}\bigg|_{v_i=0} \tag{10.5-9a}$$

Then, since (Fig. 10.5-5*b*)

$$v_s = -v_{gs} = \mu v_{gs} + i_o r_{ds} \tag{10.5-9b}$$

we find that

$$Z_o = \frac{r_{ds}}{\mu + 1} \tag{10.5-9c}$$

When $\mu = g_m r_{ds} \gg 1$, the output impedance becomes

$$Z_o \approx \frac{1}{g_m} \tag{10.5-9d}$$

The *voltage gain* A_v' with R_s removed is

$$A_v'\bigg|_{R_s \to \infty} = \frac{v_s}{v_g} \tag{10.5-10a}$$

The output voltage v_s is

$$v_s = \mu v_{gs} \tag{10.5-10b}$$

and

$$v_{gs} = v_g - v_s \tag{10.5-10c}$$

Thus

$$A_v' = \frac{\mu}{\mu + 1} \tag{10.5-11a}$$

If $\mu \gg 1$, the open-circuit voltage gain A_v' becomes

$$A_v' \approx 1 \tag{10.5-11b}$$

The *input impedance* Z_i of the source-follower circuit of Fig. 10.5-5*b* is infinite since the bias resistor R_1 was neglected. The input impedance of the source follower shown in Fig. 10.5-5*a*, as seen by r_i, is defined as

$$Z_i = \frac{v_g}{i_i} \tag{10.5-12a}$$

All the input current flows in R_1. Then ($R_1 \gg R_{s2}$)

$$i_i R_1 \approx v_g - \left(\frac{R_{s2}}{R_{s1} + R_{s2}}\right) v_s \tag{10.5-12b}$$

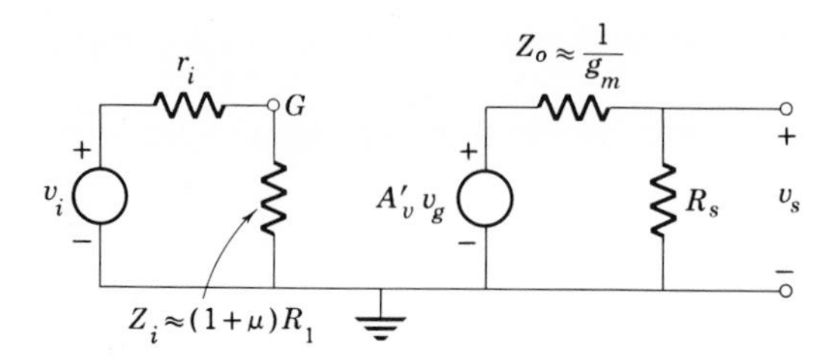

Fig. 10.5-6 Thévenin equivalent circuit of the source follower.

and

$$Z_i \approx \frac{R_1}{1 - (v_s/v_g)[R_{s2}/(R_{s1} + R_{s2})]} \tag{10.5-12c}$$

But from (10.5-11*a*)

$$\frac{v_s}{v_g} = \frac{\mu}{\mu + 1} \tag{10.5-13a}$$

This result tacitly assumes that $Z_o \ll R_s$ [Eq. (10.5-10*a*)]. Then

$$Z_i \approx \frac{R_1}{1 - [\mu/(\mu + 1)][R_{s2}/(R_{s1} + R_{s2})]} \tag{10.5-13b}$$

If $R_{s2} \gg R_{s1}$,

$$Z_i \approx (\mu + 1)R_1 \tag{10.5-13c}$$

The Thévenin equivalent circuit of the source follower is shown in Fig. 10.5-6. Note the similarity between the source follower and the emitter follower. Both devices are characterized by a high input impedance Z_i, a low output impedance Z_o, and a voltage gain almost equal to unity. Note also that returning R_1 to the node formed by R_{s1} and R_{s2} results in a much higher input impedance than would be obtained by returning R_1 to ground.

Example 10.5-2 Design a source follower using the JFET 2N4223 to have a Q point at $I_{DQ} = 3$ mA and $V_{DSQ} = 15$ V. The available supply voltage is 20 V. Calculate the input and output impedances and the voltage gain.

Solution The circuit chosen uses self-bias, as shown in Fig. 10.5-7.

The characteristic shown in Appendix III.3 indicates that $V_{GSQ} \approx -1.2$ V. From (10.5-8*b*)

$$R_{s1} = \frac{V_{GSQ}}{I_{DQ}} = \frac{1.2}{3 \times 10^{-3}} = 400\ \Omega$$

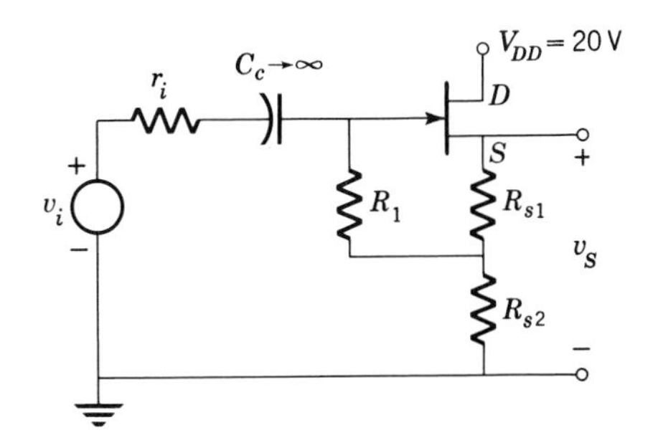

Fig. 10.5-7 Source follower for Example 10.5-2.

A standard 390-Ω resistor is used. R_{s2} is now found using (10.5-8*a*).

$$R_{s2} = \frac{V_{DD} - V_{DSQ}}{I_{DQ}} - R_{s1} = \frac{20 - 15}{3 \times 10^{-3}} - 390 \approx 1280\ \Omega$$

A standard 1.2-kΩ resistor is used.

The Thévenin output impedance Z_o is $1/g_m$. From Appendix III.3 we see that at the specified quiescent drain current (3 mA),

$$g_m \approx 2 \times 10^{-3} \text{ mho}$$

Therefore

$$Z_o \approx \frac{1}{g_m} = 500\ \Omega$$

r_{ds} is the slope of the *vi* characteristic. It is found to be ≈ 83 kΩ. Thus $\mu = g_m r_{ds} \approx 166$. The voltage gain A_v' is $\mu/(\mu + 1)$. Therefore

$$A_v' \approx \frac{166}{167} \approx 1$$

The input impedance is [Eq. (10.5-13*b*)]

$$Z_i \approx \frac{R_1}{1 - \left(\dfrac{\mu}{\mu + 1}\right)\left(\dfrac{R_{s2}}{R_{s1} + R_{s2}}\right)} = \frac{R_1}{1 - \left(\dfrac{166}{167}\right)\left(\dfrac{1280}{390 + 1280}\right)} \approx 4.2R_1$$

Note that if $R_{s2} \gg R_{s1}$, Z_i would be [Eq. (10.5-13*c*)]

$$Z_i \to (\mu + 1)R_1 = 167R_1$$

THE PHASE-SPLITTING CIRCUIT

The FET can be used in a phase-splitting circuit, as shown in Fig. 10.5-8*a*. All biasing resistors have been omitted for simplicity.

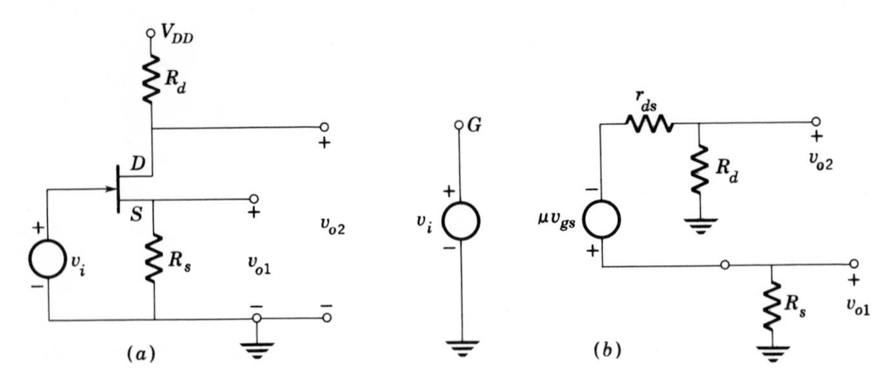

Fig. 10.5-8 FET phase-splitter. (*a*) Schematic; (*b*) small-signal equivalent circuit.

Since the gate current is negligible, the drain and source currents are the same. Thus

$$\frac{v_{o1}}{R_s} = -\frac{v_{o2}}{R_d} \tag{10.5-14}$$

We expect $A_{v1} = v_{o1}/v_i \approx 1$ since the circuit is that of a source follower. However, we must determine the effect of the drain resistor R_d on the gain. Refer to Fig. 10.5-8*b*.

The open-circuit voltage gain is the same as (10.5-11*a*).

$$A'_v = \frac{v_{o1}}{v_g} = \frac{v_{o1}}{v_i} = \frac{\mu}{\mu + 1} \tag{10.5-15a}$$

The output impedance is similar to (10.5-9*c*).

$$Z_o = \frac{r_{ds} + R_d}{\mu + 1} \approx \frac{1}{g_m} + \frac{R_d}{\mu + 1} \tag{10.5-15b}$$

The resulting output equivalent circuit for v_{o1} is shown in Fig. 10.5-9. If

$$R_s \gg \frac{r_{ds} + R_d}{\mu + 1} \approx \frac{1}{g_m} + \frac{R_d}{\mu} \tag{10.5-16a}$$

then

$$v_{o1} \approx v_i \tag{10.5-16b}$$

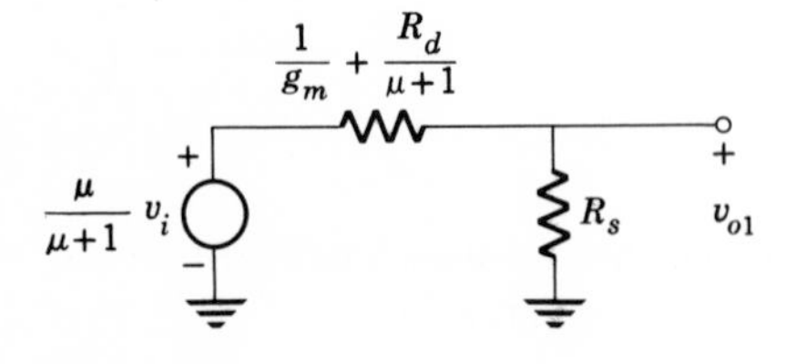

Fig. 10.5-9 Output equivalent circuit looking into the source.

Note that the effect of R_d is to increase the output impedance of the source follower. Referring to Fig. 10.5-9, we see that, if $R_d = R_s$,

$$v_{o1} = -v_{o2} = \left[\frac{\mu R_d}{r_{ds} + (2 + \mu)R_d}\right] v_i \tag{10.5-16c}$$

IMPEDANCE REFLECTION IN THE FET

It is important to note that R_d and r_{ds} reflect into the source as a small resistance $(r_{ds} + R_d)/(\mu + 1)$. In analyzing transistor circuits, impedances and currents were reflected from base to emitter by multiplying base current by $h_{fe} + 1$ and dividing base resistance by $h_{fe} + 1$. In analyzing FET circuits, voltages and resistances in the drain circuit are divided by $1 + \mu$ when reflected into the source circuit (Fig. 10.5-9). This is further illustrated by obtaining the equivalent drain circuit of the phase splitter of Fig. 10.5-8*a*, looking into the drain as shown in Fig. 10.5-10*a*. The open-circuit voltage gain A'_{v2} is

$$A'_{v2} = \left.\frac{v_{o2}}{v_i}\right|_{R_d=\infty} = -\frac{\mu v_{gs}}{v_i} \tag{10.5-17a}$$

Since R_d is removed from the circuit, no current flows; hence $v_{gs} = v_i$ and

$$A'_{v2} = -\mu \tag{10.5-17b}$$

The output impedance is, from Fig. 10.5-10*a*,

$$Z_o = \left.\frac{v_{o2}}{i_o}\right|_{v_i=0} \tag{10.5-18a}$$

where

$$v_{o2} = -\mu v_{gs} + (r_{ds} + R_s)i_o \tag{10.5-18b}$$

and

$$v_{gs} = -R_s i_o \tag{10.5-18c}$$

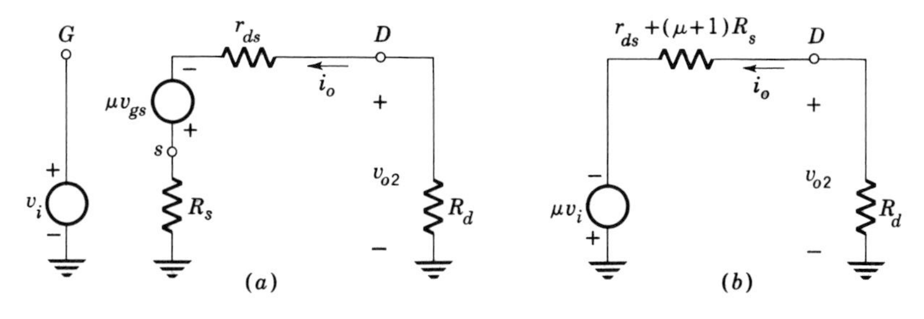

Fig. 10.5-10 FET equivalent circuit for v_{o2} looking into the drain. (*a*) Equivalent circuit of phase splitter; (*b*) equivalent circuit looking into the drain.

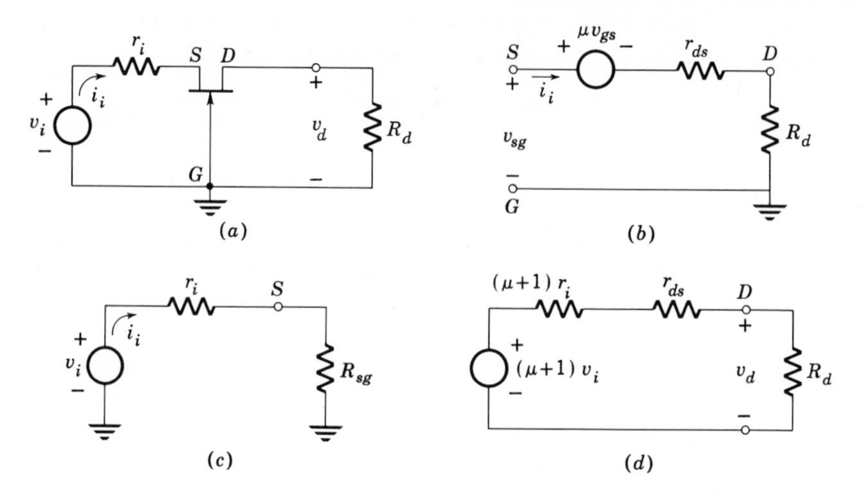

Fig. 10.5-11 The common-gate amplifier. (*a*) Schematic; (*b*) looking into the source; (*c*) equivalent circuit at the input (source); (*d*) equivalent circuit looking into the drain.

Combining (10.5-18*b*) and (10.5-18*c*) yields

$$Z_o = r_{ds} + R_s(\mu + 1) \tag{10.5-18d}$$

The resulting equivalent circuit for v_{o2} is shown in Fig. 10.5-10*b*.

10.5-3 THE COMMON-GATE AMPLIFIER

The common-gate amplifier is analogous to the common-base amplifier, and is used primarily at high frequencies. The circuit is shown in Fig. 10.5-11, where all biasing components have been eliminated for simplicity. The circuit is most easily analyzed by determining the impedance looking into the source at terminals *SG*. This impedance is

$$R_{sg} = \frac{v_{sg}}{i_i} = \frac{r_{ds} + R_d}{\mu + 1} \tag{10.5-19}$$

which can be obtained either from the equivalent circuit of Fig. 10.5-11*b* or by inspection, noting that the resistances in the drain circuit reflect into the source after dividing by $\mu + 1$.

The final equivalent circuit is shown in Fig. 10.5-11*c*, where the current i_i is

$$i_i = \frac{v_i}{r_i + R_{sg}} \tag{10.5-20a}$$

Hence

$$v_d = i_i R_d = \left[\frac{R_d}{r_i + (r_{ds} + R_d)/(\mu + 1)}\right] v_i \tag{10.5-20b}$$

The complete circuit, reflected into the drain circuit, is shown in Fig. 10.5-11*d*. The derivation of this circuit comes about directly by reflecting the source circuit into the drain circuit. The source voltage is multiplied by $\mu + 1$.

10.6 TYPICAL MANUFACTURERS' SPECIFICATIONS

The specifications given below are for the *n*-channel IGFET 2N3796. This is a low-power audio-frequency device.

MAXIMUM RATINGS (T_a = 25°C)

$$V_{DS} = 25 \text{ V}$$
$$V_{GS} = \pm 10 \text{ V}$$
$$I_D = 20 \text{ mA}$$
$$P_D = 200 \text{ mW}$$
$$\theta_{jc} = 1.14°\text{C/W}$$
$$T_j = 200°\text{C}$$

EXPLANATION OF SPECIFICATIONS

The maximum-rating specifications define the breakdown voltages, the maximum current, and the maximum power and derating characteristics. These are similar to the transistor specifications presented in Secs. 4.6 and 6.6, and are used in the same way.

Figure 10.6-1 shows the *vi* characteristic for this IGFET. This characteristic is "typical" (i.e., the actual 2N3796 FET used may differ by up to 15 percent from the "typical" characteristic). Thus, unlike the transistor, the *vi* characteristics can be used to set the Q point.

The transfer characteristic of Fig. 10.6-2 is sometimes provided by manufacturers. This characteristic can be considered an input-output voltage curve from which we see that distortion results when v_{GS} is large.

Figure 10.6-3 is an extremely useful, although not a necessary, characteristic. It shows that at $V_{DS} = 10$ V, $1/r_{ds}$ increases linearly with current [Eq. (10.5-4*b*)]. In addition, it gives the variation of g_m with drain current. Referring to (10.5-3*b*), we see that g_m increases with v_{GS}, as

$$g_m = \left(\frac{2I_{p0}}{V_{p0}}\right)\left(1 + \frac{V_{GSQ}}{V_{p0}}\right)$$

Substituting (10.5-3*a*) into (10.5-3*b*) yields

$$g_m = \left(\frac{2\sqrt{I_{p0}}}{V_{p0}}\right)\sqrt{I_{DQ}} \tag{10.6-1}$$

Note that the product $g_m r_{ds} = \mu$ decreases with increasing drain current. This can be shown by combining (10.6-1) and (10.5-4*b*) to get

$$g_m r_{ds} \sim \frac{1}{\sqrt{I_{DQ}}} \tag{10.6-2}$$

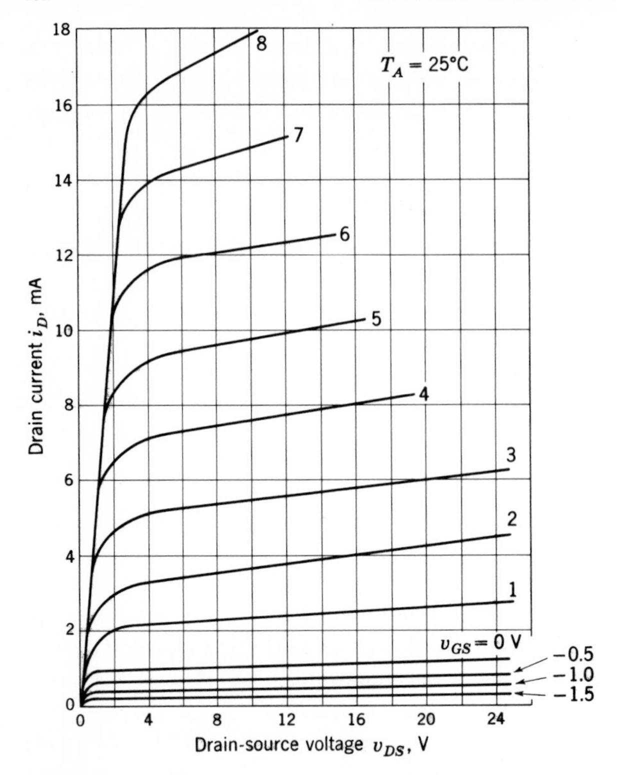

Fig. 10.6-1 IGFET output *vi* characteristic.

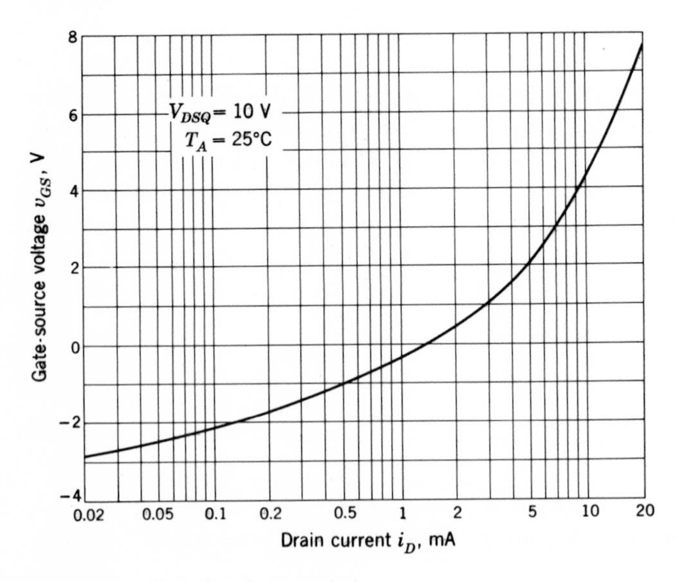

Fig. 10.6-2 Transfer characteristic.

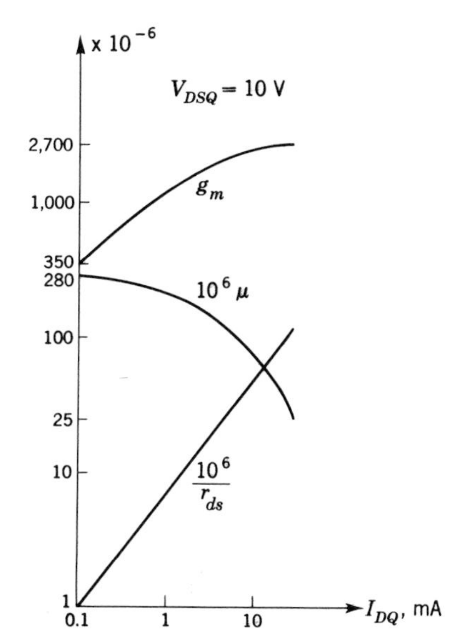

Fig. 10.6-3 FET parameter variation with drain current.

PROBLEMS

SECTION 10.1

10.1. Using (10.1-3), find i_D when

$$v_{GS} = V_{GSQ} + \epsilon \cos \omega t$$

where $\epsilon \ll V_{p0}$.

Show that i_D consists of an ac term and a quiescent term. Comment on the ratio ϵ/V_{p0} with regard to a linear model representing (10.1-3).

10.2. (*a*) Sketch i_D versus T with v_{GS} = constant [Eq. (10.1-4)].

(*b*) Sketch v_{GS} versus T with i_D = constant.

(*c*) Sketch i_D versus v_{GS} with T = constant. Find the quiescent point at which the approximation $i_d = g_m v_{gs}$ is best.

10.3. At small values of drain-source voltage (i.e., below pinch-off), the FET acts as a resistance. The *vi* characteristic is given approximately by the equation

$$i_D = \left(\frac{1}{r_{ds}}\right)\left[1 - \left(-\frac{v_{GS}}{V_{p0}}\right)^{1/2}\right] v_{DS}$$

where

$$\frac{1}{r_{ds}} = \left.\frac{\partial i_D}{\partial v_{DS}}\right|_{v_{GS}=0}$$

(*a*) Plot the *vi* characteristics, using this equation and (10.1-3).

(*b*) Determine the drain-to-source pinch-off voltage, and plot as a function of v_{GS} by noting the intersection of the two equations. Evaluate r_{ds} in terms of V_{p0}.

(*c*) Compare with (10.1-1). Explain why the results differ.

SECTION 10.2

10.4. Assuming that

$$v_{GS} = V_{GSQ} + \epsilon \cos \omega_0 t$$

and using (10.2-2), show that the IGFET can "amplify" signals. Explain your answer.

SECTION 10.3

10.5. In Sec. 10.3 the JFET of Fig. 10.3-1 was designed to operate at the Q point

$$V_{DSQ} = 15 \text{ V}$$

and

$$I_{DQ} = 3.5 \text{ mA}$$

The drain and source resistors were found to be 3.9 kΩ and 270 Ω.

(*a*) Redraw the vi characteristics of Fig. 10.3-1*b* if $T = 2T_0$.

(*b*) Find the new Q point.

(*c*) If the original Q point is to be maintained, find the new quiescent grid-to-source voltage and R_d and R_s.

10.6. The IGFET shown in Fig. 10.6-1 is designed to have a quiescent point at $V_{GSQ} =$ 4 V and $V_{DSQ} = 16$ V.

(*a*) Find I_{DQ}.

(*b*) Calculate R_s to result in minimal Q-point variation due to temperature changes.

(*c*) Let $V_{DD} = 24$ V. Calculate R_d.

(*d*) Complete the design by finding R_1 and R_2.

10.7. If $T = 4T_0$, the quiescent point specified in Prob. 10.6 changes. Find the new Q point using the values of R_1, R_2, R_d, R_s, and V_{DD} obtained in Prob. 10.6.

10.8. If $T = 4T_0$ and the Q point in Prob. 10.6 is to be maintained, find

(*a*) The new I_{DQ}.

(*b*) The new values of R_1, R_2, R_d, and R_s ($V_{DD} = 24$ V).

SECTION 10.4

10.9. The vi characteristic of the IGFET amplifier shown in Fig. P10.9 is given in Fig. 10.6-1. The Q point is at $V_{GSQ} = 4$ V, $V_{DSQ} = 10$ V.

Find the maximum value of V_i for which the relative distortion is less than 20 dB.

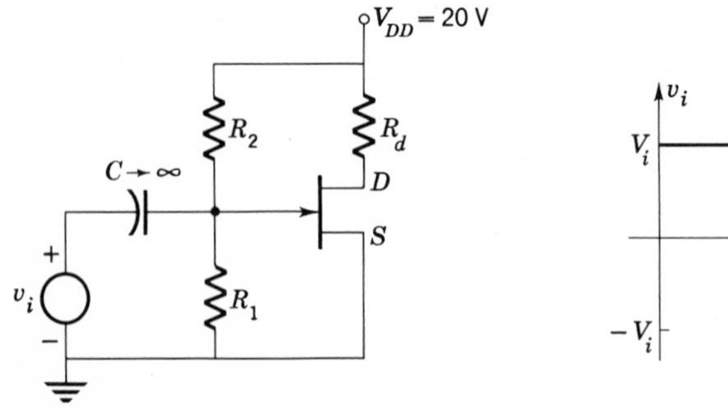

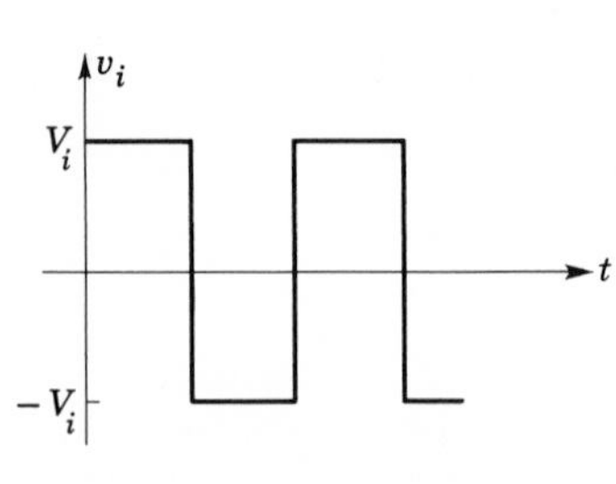

Fig. P10.9

SECTION 10.5

10.10. Expand (10.5-3a) in a Taylor series expansion and show that, for (10.5-1) to be valid,

$$\Delta v_{GS} = v_{gs} \ll V_{p0}\left(1 + \frac{V_{GSQ}}{V_{p0}}\right)$$

10.11. Using (10.1-3), find an expression for the g_m of a JFET in terms of I_{p0}, V_{p0}, and V_{GSQ}.

10.12. Plot g_m for the IGFET and JFET as a function of quiescent drain current I_{DQ}.

10.13. Two IGFETs are connected in parallel as shown in Fig. P10.13. If the resulting three-terminal device is considered to be an IGFET, find the resulting g_m. For each IGFET,

$$i_D = (5 \times 10^{-3})\left(1 + \frac{v_{GS}}{5}\right)^2$$

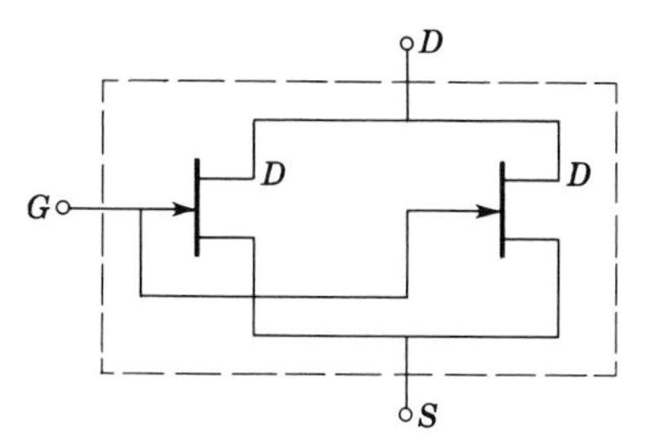

Fig. P10.13

10.14. The JFET shown in Fig. P10.14 is connected as a two-terminal element. At drain-to-source voltages less than V_{p0}, it behaves like a resistance to ac signals. At drain-to-source voltages greater than V_{p0}, it behaves like a "current limiter." (*Note:* A Zener diode is a "voltage limiter.")

(a) If the JFET is described at voltages below pinch-off by the equation

$$i_D = G_o(v_{DS} - v_{GS})\left[1 - \left(\frac{2}{3}\right)\left(\frac{v_{DS} - v_{GS}}{V_{p0}}\right)^{1/2}\right] + G_o(v_{GS})\left[1 - \left(\frac{2}{3}\right)\left(-\frac{v_{GS}}{V_{p0}}\right)^{1/2}\right]$$

find the ac (small-signal) resistance.

(b) The JFET is described above pinch-off by the equation

$$i_D = I_{p0}\left[1 + \frac{3v_{GS}}{V_{p0}} + 2\left(-\frac{v_{GS}}{V_{p0}}\right)^{3/2}\right]$$

Plot v_L versus V_{DD}.

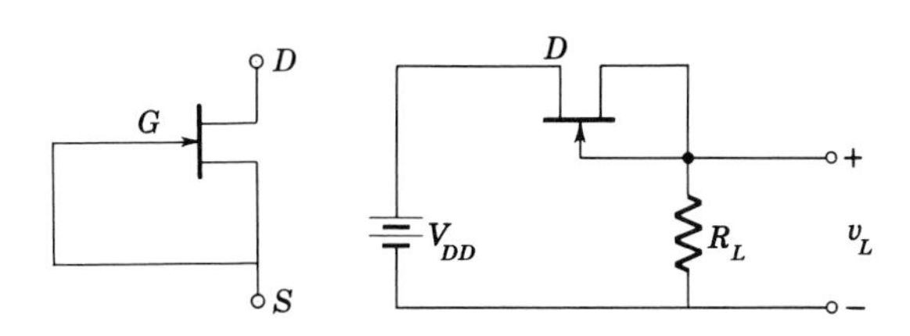

Fig. P10.14

10.15. A 2N3796 IGFET (Fig. 10.6-1) is to be used as an amplifier. The quiescent point is to be set at $V_{GSQ} = 3$ V and $V_{DSQ} = 10$ V. Calculate μ, r_{ds}, and g_m.

10.16. (*a*) Using the 2N3796 find the Q point.
(*b*) Calculate μ, r_{ds}, and g_m.
(*c*) Calculate the gain v_L/v_i.

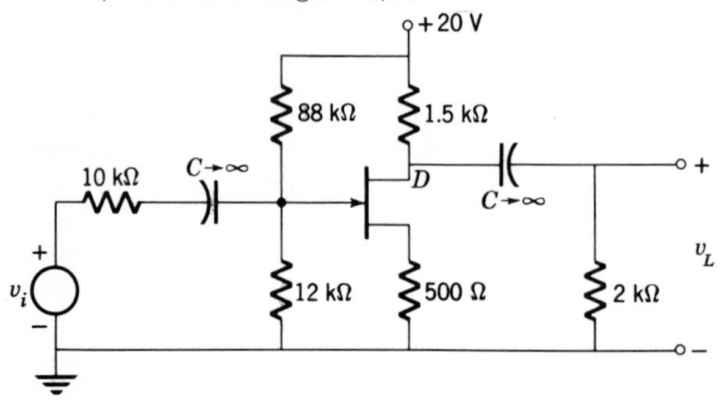

Fig. P10.16

SECTION 10.5-1

10.17. Design an amplifier having a gain of 10. The supply voltage is 24 V. Use the 2N3796.

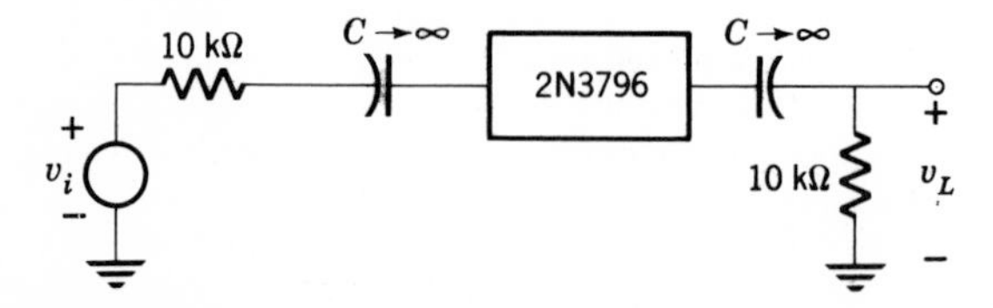

Fig. P10.17

SECTION 10.5-2

10.18. Plot v_S versus v_G. Find v_G when $v_{DS} = 20$ V and
(*a*) The FET is described by (10.1-3).
(*b*) The FET is described by (10.2-2).
(*c*) The FET is a 2N3796.

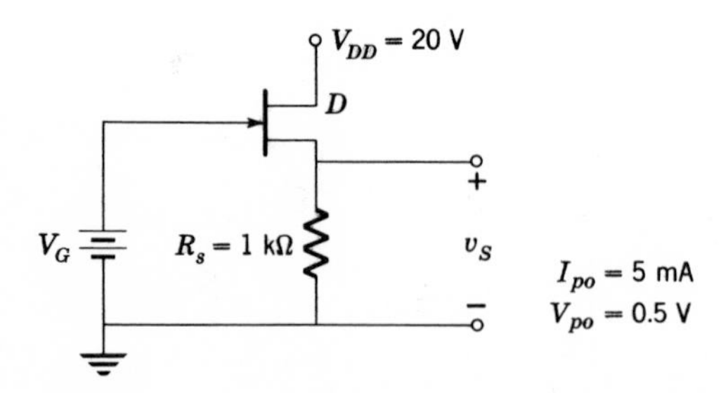

Fig. P10.18

10.19. Derive an expression for the input impedance Z_{if} of the source follower shown in Fig. P10.19. Use the principles of feedback.

Show that when $R_1 \gg R_A$, the result is (10.5-13*b*). Find Z_{if} when this assumption is not valid.

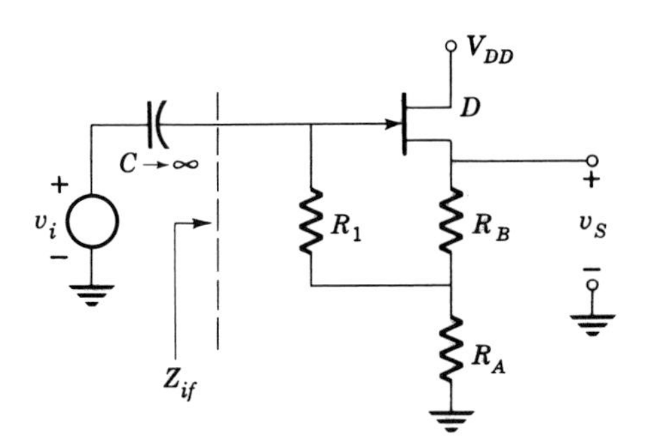

Fig. P10.19

10.20. Find the output impedance and voltage gain of the source follower shown in Fig. P10.19 using the principles of feedback. Compare your results with (10.5-9*c*) and (10.5-11*a*).

10.21. A source follower is to be designed using the circuit of P10.21. A 2N4223 JFET is to be used. The gain is to be greater than 0.8. Find R_1, R_{s1}, and R_{s2}.

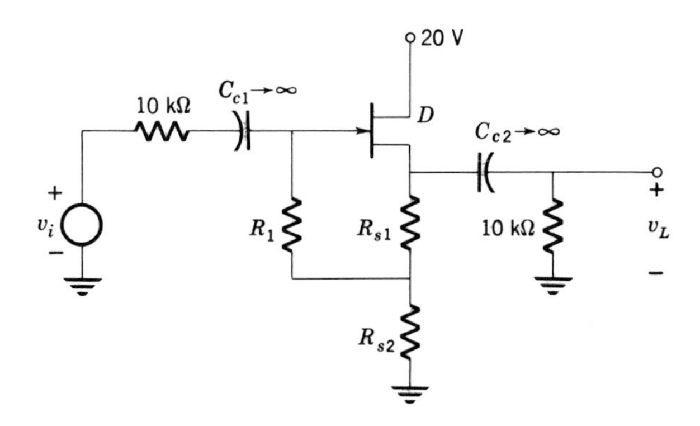

Fig. P10.21

SECTION 10.5-3

10.22. The *vi* characteristic of the CG amplifier shown in Fig. P10.22 is given approximately by the equation

$$i_D = (1 + v_{GS})^2 \times 10^{-4}$$

(*a*) Plot the characteristics.
(*b*) Find the Q point graphically.

(c) Calculate g_m.
(d) Let $r_{ds} = 10\ \text{k}\Omega$; calculate μ.
(e) Determine Z_i, Z_o, and the gain, v_d/v_i.

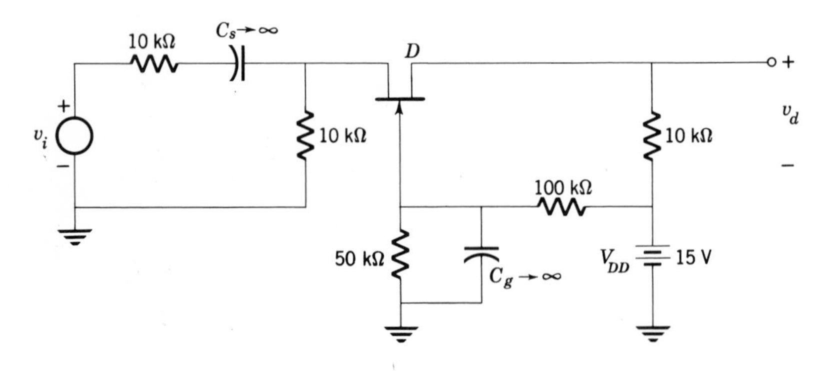

Fig. P10.22

10.23. Both FETs have identical g_m, μ, and r_{ds}. Calculate
(a) i_L as a function of v_1 and v_2.
(b) v_{o1}, v_{o2}, and v_{o3}.
(c) The output resistance looking into terminals S_2S_2'.

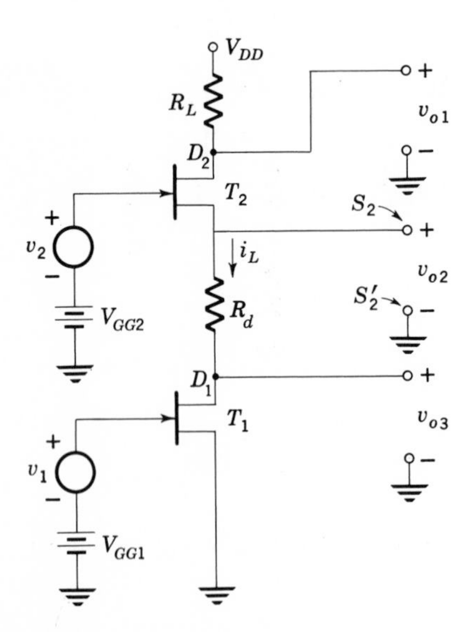

Fig. P10.23

10.24. T_1 and T_2 are identical.
(a) Find v_{d2} as a function of v_1 and v_2.

(*b*) Do you expect the CMRR to be lower for the FET difference amplifier than for the transistor difference amplifier? Why? *Hint:* Compare $(1/g_m)_{\text{FET}}$ with h_{ib}.

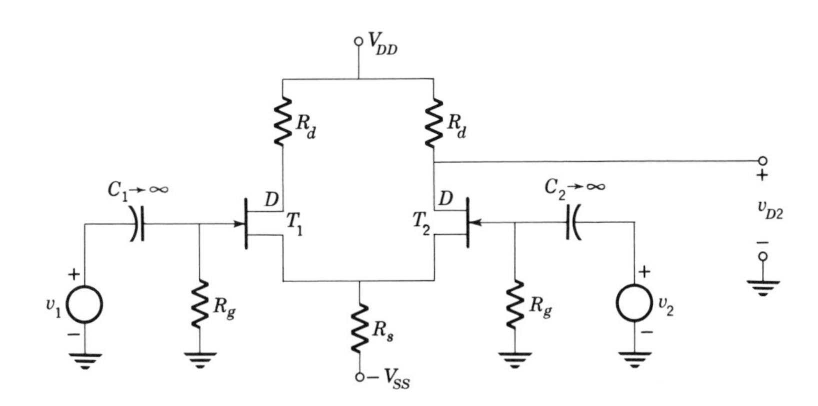

Fig. P10.24

10.25. Derive an expression for the output impedance at the drain of T_2, for the circuit shown in Fig. P10.24.

10.26. If in Fig. P10.24, $R_s = 10\ \text{k}\Omega$, $R_d = 10\ \text{k}\Omega$, $g_m = 5 \times 10^{-3}$ mhos, and $r_{ds} = 10\ \text{k}\Omega$, calculate

(*a*) CMRR.

(*b*) Z_o at the drain of T_2.

10.27. A Darlington configuration can be constructed using the JFET and transistor shown.

(*a*) Determine Z_{if} and Z_{of}.

(*b*) Obtain an expression for the gain v_e/v_i.

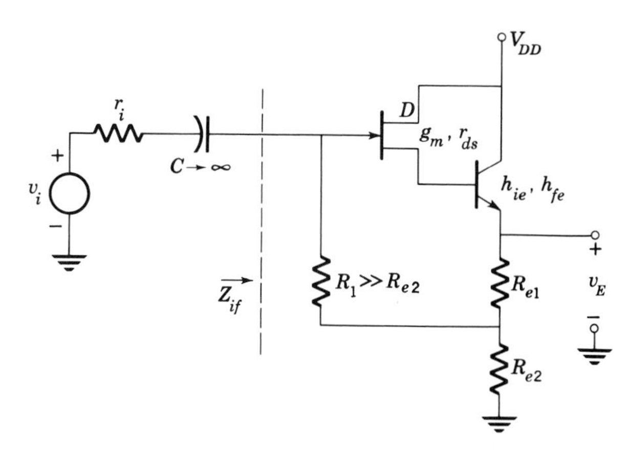

Fig. P10.27

10.28. Transistors T_1, T_2, and T_3 form a difference amplifier. The JFET T_4 is a Darlington amplifier used to provide a high input impedance.

(*a*) If the resistance looking into the collector of T_3 is infinite, calculate Z_i.

(*b*) Calculate v_L/v_i.

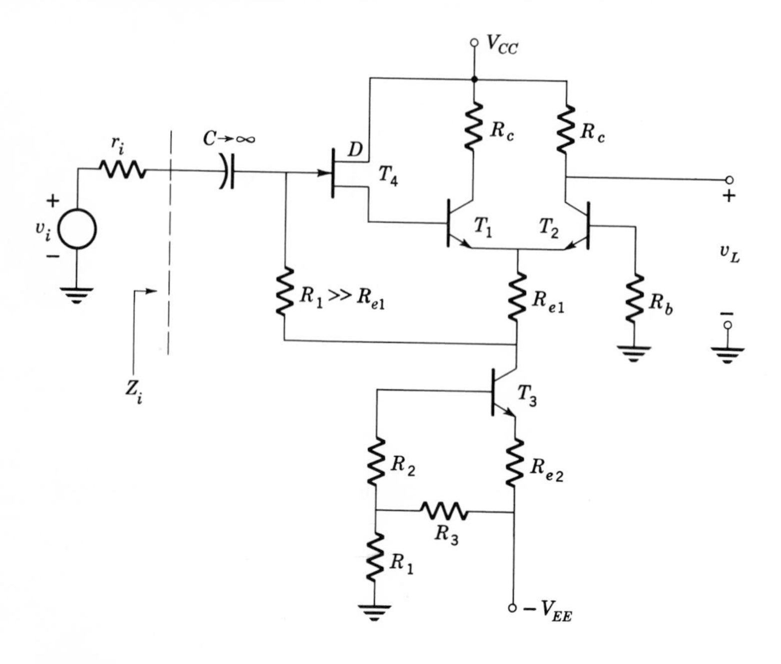

Fig. P10.28

REFERENCES

1. Grove, A. S.: "Physics and Technology of Semiconductor Devices," pp. 243–259, 317–333, John Wiley & Sons, Inc., New York, 1967.
2. Sevin, L. J.: "Field-effect Transistors," McGraw-Hill Book Company, New York, 1965.
3. Crawford, R. H.: "Mosfet in Circuit Design," McGraw-Hill Book Company, New York, 1967.

11
The Vacuum Tube

INTRODUCTION

The first active device used for amplification was the triode vacuum tube invented in 1907. From that time until the early 1950s the vacuum tube underwent continuous development and refinement, forming the cornerstone of many vast industries. The transistor, invented in 1948, has at the present time supplanted the vacuum tube in so many areas that the vacuum tube can now be considered a "special device" which is used primarily at very high frequencies and very high power levels.[1]

The greatest disadvantage of the vacuum tube as compared with the transistor is the fact that power is required to activate it; i.e., the vacuum tube depends on thermal emission of electrons in vacuum for its action. Several watts is required to heat the cathode which emits the electrons. This power is not required when using semiconductor devices. Another disadvantage of the vacuum tube is its relatively short life, due to the effects of constant heating of the cathode. The semiconductor device, on the other hand, has almost infinite life.

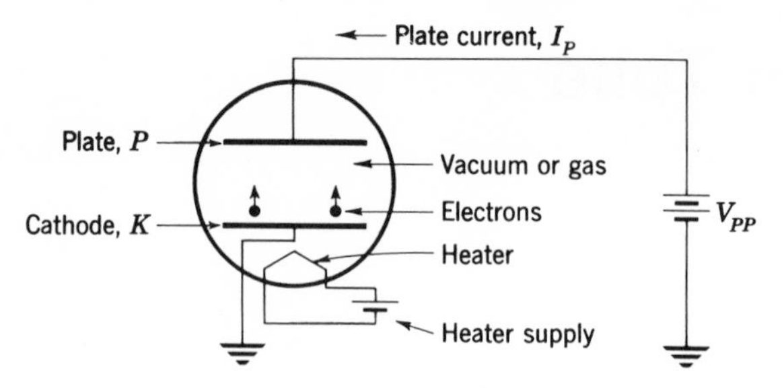

Fig. 11.1-1 The vacuum diode.

From the point of view of the circuit designer, the vacuum tube is similar to the FET, being a voltage-sensitive device with a high input impedance ($\approx$ 100 MΩ). However, the supply voltages used lie typically between 100 and 400 V, and typical current levels vary from 1 to 100 mA for medium-power vacuum tubes.

In this chapter we discuss the theory and operation of the most commonly used vacuum tubes, the diode, the triode, and the pentode. Tubes such as the klystron and the traveling-wave tube, which are used above 1-kW dissipation and 1 GHz, are not discussed here, because of their specialized nature.

11.1 INTRODUCTION TO THE VACUUM TUBE

11.1-1 THE DIODE

The vacuum diode consists of two electrodes, called the plate (or anode) and the cathode, as shown schematically in Fig. 11.1-1. The cathode is usually constructed from tungsten, thoriated tungsten, or an oxide-coated layer of Kovar or other similar alloy. When heated, these materials emit large quantities of electrons into the surrounding space. The plate, if kept at a positive potential as shown, will attract and capture these electrons, thus forming a current flow. This type of conduction takes place most efficiently in a vacuum or in certain gases. Thus the envelope containing the electrodes is highly evacuated.*

The cathode is usually heated indirectly by separately powered heater wires called filaments. A small receiving-type diode might require 300 mA at 6.3 V, or 1.9 W, to power the filament, while an equivalent junction diode requires no power at all for this purpose and is ready to operate instantaneously. The vacuum diode must wait a few seconds

* Some tubes are filled with a gas or mixture of gases such as argon and mercury at low pressures. These tubes have many uses, some of which are rectifiers, voltage regulators, and light sources.[2]

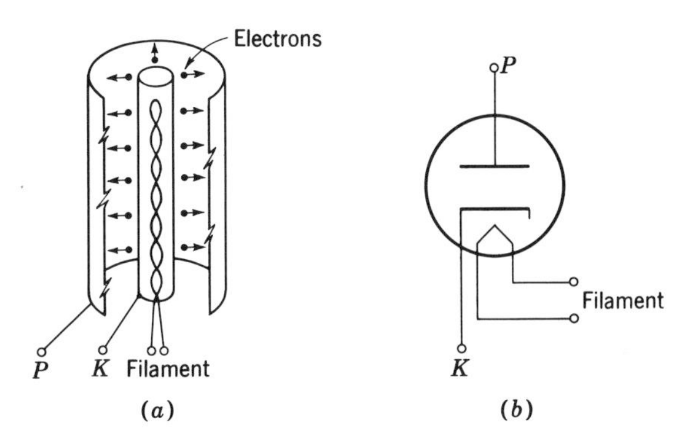

Fig. 11.1-2 Diode construction and circuit symbol. (*a*) Pictorial representation; (*b*) vacuum-diode symbol.

for the filament to reach operating temperature so that the electrons have sufficient energy to leave the cathode for the plate.

Vacuum tubes are often constructed in cylindrical form, as shown in Fig. 11.1-2*a*. The filament wire is within the cathode, which is centrally located inside the plate. The vacuum-diode symbol is shown in Fig. 11.1-2*b*. Note that the symbol for cathode is K. The filament is usually *not* shown, but is included here for completeness.

The circuit operation of the vacuum diode can be explained using Fig. 11.1-3*a*. When V_{PP} is positive, electrons flow from the cathode to the plate, resulting in a current I_P as shown. The variation of I_P with V_{PK} is shown in Fig. 11.1-3*b*. Note that the break voltage occurs at approximately 50 V as compared with 0.7 V for a silicon diode.

When V_{PP} is negative, the plate is at a negative potential with

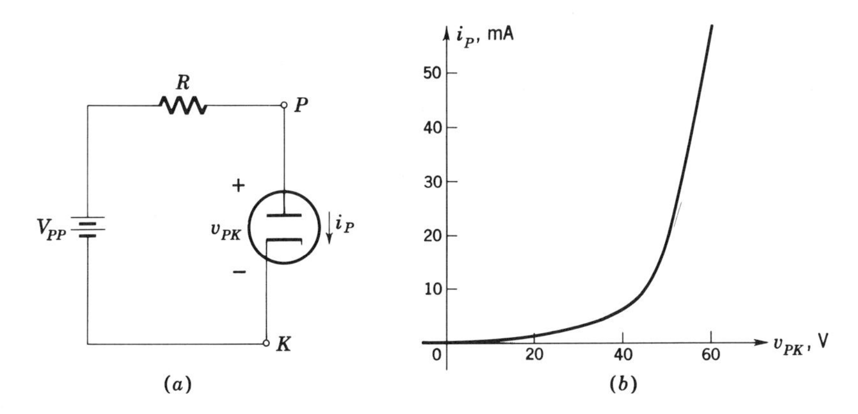

Fig. 11.1-3 (*a*) Basic diode circuit; (*b*) *vi* characteristic.

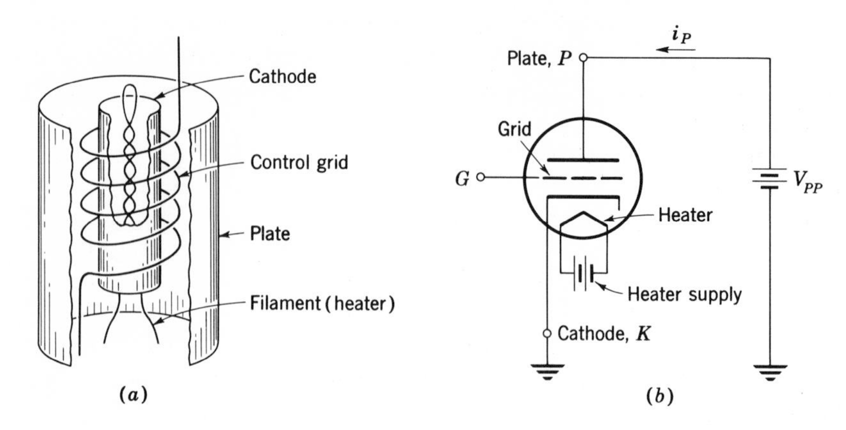

Fig. 11.1-4 The vacuum triode. (*a*) Pictorial representation; (*b*) circuit symbol.

respect to the cathode. Thus an electron leaving the cathode will be repelled and returned toward the cathode, resulting in a space charge, unless it has an extremely high initial velocity (very high energy). The probability of this occurrence is extremely small; hence the reverse current is negligible, usually less than 0.01 μA.

Consideration of the physics of the conduction process leads to the relation governing current flow in a vacuum diode.[2] This is known as the *Langmuir-Childs Law,*

$$i_P \approx G v_{PK}^{3/2} \tag{11.1-1}$$

where i_P = instantaneous value of plate current
v_{PK} = instantaneous value of plate-cathode voltage
G = constant, depending primarily on geometry and cathode material, called *perveance*

This expression should be compared with (2.2-1), which describes current flow in a semiconductor diode. The semiconductor-diode current is an exponential function of diode voltage v_D, while the vacuum diode is approximately a $3/2$-power-law device.

11.1-2 THE TRIODE

A typical vacuum *triode* is a diode to which a third electrode, called the *control grid,* has been added. Physically, this grid is usually a loosely wound wire helix placed very close to the cathode, as shown in Fig. 11.1-4.

To explain the action of this control grid, we note that if the grid-to-cathode potential is negative, some of the electrons leaving the cathode will be repelled by the grid and will return toward the cathode. Other electrons, having more energy, will pass through the grid structure and reach the plate. As the grid-to-cathode potential is made more negative, a smaller number of electrons possess sufficient energy to pass through the grid and reach the plate. Variations in the grid-to-cathode voltage therefore result in variations in the plate current, i_P.

Amplification occurs because the grid is much closer to the cathode than to the plate. Hence the plate voltage exerts a much smaller effect on the electron flow than the grid voltage. For example, a 1-V reduction in grid voltage will reduce the plate current to a certain level. In order to bring the plate current back to its original level with the new lower value of grid voltage, the plate voltage must be raised by a voltage of from 20 to 200 V. This increase in plate-to-cathode voltage for a unit decrease in grid-to-cathode voltage is called the amplification factor μ [this parameter is defined by (11.4-3)].

The output characteristics of a typical triode are sketched in Fig. 11.1-5. We see that the plate current i_P is a function of the plate-to-cathode voltage v_{PK}, as well as the grid-to-cathode voltage v_{GK}. This can be explained from the previous description of the diode and triode action. With v_{GK} held constant, the triode acts as a diode, and an increase in plate-to-cathode voltage results in an increase in plate current, as shown in Fig. 11.1-5*b* (see also Fig. 11.1-3*b*). With the plate-to-cathode voltage maintained constant, the plate current is a function of the grid-to-cathode voltage, due to the triode action. Thus the triode is a voltage-sensitive device. The relationship between the plate current, plate-to-cathode voltage, and grid-to-cathode voltage is

$$i_P \approx G(v_{PK} + \mu v_{GK})^{3/2} \qquad v_{GK} \leq 0,\ v_{PK} \geq 0,\ i_P \geq 0 \tag{11.1-2}$$

Note that with v_{GK} equal to a constant, (11.1-2) is similar to (11.1-1).

When $v_{GK} > 0$, current flows from grid to cathode, as well as from

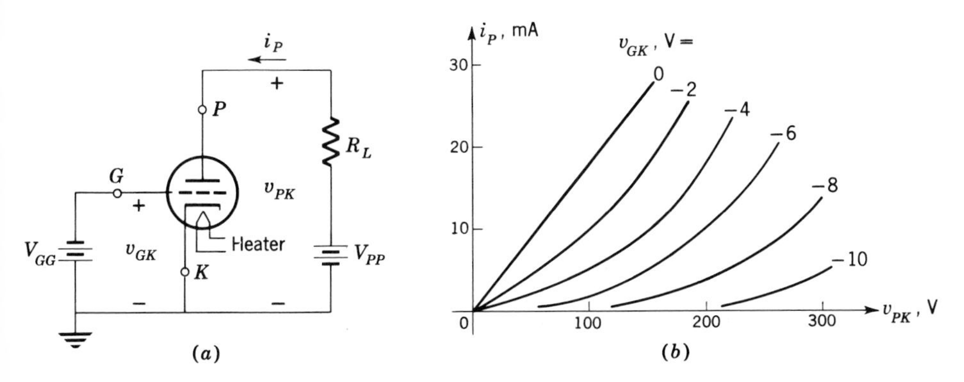

Fig. 11.1-5 (*a*) Basic triode circuit; (*b*) *vi* characteristic.

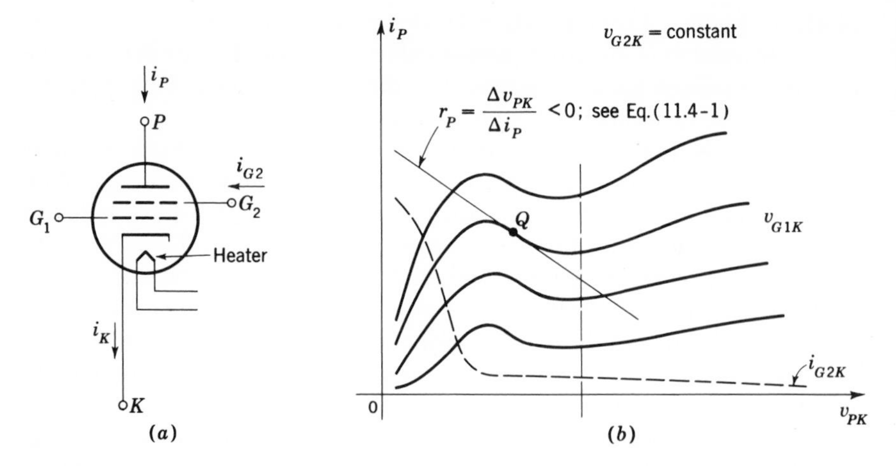

Fig. 11.1-6 Tetrode output characteristic. (*a*) Circuit symbol; (*b*) output characteristic.

plate to cathode. The grid-to-cathode circuit now appears as a relatively low impedance (100 Ω to 1 kΩ). Thus the grid-to-cathode circuit can be represented as a diode, the grid *acting* like the plate.

In the following discussion the grid-to-cathode voltage is maintained negative, and the current flow from grid to cathode is assumed to be zero.

11.1-3 THE TETRODE

The high-frequency response of the triode is limited by the capacitance between the grid and the plate (an explanation of how this capacitance affects the high-frequency behavior of the triode is postponed until Chap. 13). In order to improve this high-frequency response, a second grid, commonly called the *screen grid* (G_2), is inserted between the control grid and the plate to reduce the grid-to-plate capacitance. The screen grid is a relatively coarse wire helix, to permit easy passage of electrons. It is maintained at ac ground potential, and thus provides electrostatic shielding between the plate and the control grid, thereby reducing the grid-to-plate capacitance by a factor of 100 or more. Typical values of grid-to-plate capacitance in a triode are 1 to 5 pF, while in this new device, called a *tetrode*, the control grid-to-plate capacitance is 0.01 pF or less.

The tetrode (four-electrode tube) is shown schematically in Fig. 11.1-6*a*. The control grid G_1 is operated at a negative voltage with respect to the cathode to prevent grid-to-cathode current. The screen grid G_2 is usually connected to a positive voltage V_{GG2} and is bypassed to the cathode by a large capacitor to maintain its voltage nearly constant.

To see how the tetrode operates, consider that v_{G1K} and v_{G2K} are held constant. Now increase v_{PK}. When v_{PK} is very small, the total cathode current consists primarily of screen current, since electrons leaving the cathode reach the screen and find it the most positive terminal. As v_{PK} increases, the plate current increases because electrons from the cathode begin to pass through the screen and reach the plate. As v_{PK} increases further, electrons begin to reach the plate with enough energy to free other electrons (called *secondary* electrons),[1,2] which are attracted to and collected by the screen, which is still at a higher potential than the plate. This results in a decrease in plate current, as shown in Fig. 11.1-6*b*. As v_{PK} increases above v_{G2K} the secondary electrons are again collected by the plate. In addition, since the plate is now at a higher potential than the screen, most of the electrons leaving the cathode pass through the screen to the plate. Further increases in plate potential are not reflected in increased cathode current because of the constant screen potential. An electron leaving the cathode "sees" the positive screen potential and "thinks" that it is the plate potential. Hence, since the screen potential is constant, the plate current remains nearly constant. The electron passes through the coarse screen to the more positive plate. Since the electrons leaving the cathode are not influenced by the plate potential, the cathode current is constant. Hence an increase in plate current due to an increase in plate voltage will produce a corresponding decrease in screen current.

$$i_K = i_P + i_{G2} \approx \text{constant} \qquad v_{GK} \leq 0 \tag{11.1-3}$$

as shown in Fig. 11.1-6*b*.

Referring to Fig. 11.1-6*b*, one sees the basic drawback of the tetrode, the incremental *decrease* in plate current with v_{PK} for small values of v_{PK}. This represents a negative output resistance, and linear operation requires that this region be avoided.

11.1-4 THE PENTODE

In order to suppress the dip in plate current which results from secondary electrons leaving the plate and being collected by the screen, a third grid, G_3, is inserted between screen and plate, resulting in a *five*-electrode tube called a *pentode*. This grid, called the *suppressor grid,* is often connected directly to the cathode, as shown in Fig. 11.1-7*a*.

The suppressor grid is a widely spaced mesh capable of repelling the low-velocity secondary electrons leaving the plate, but not capable of deterring electrons which leave the cathode, and pass the screen, from reaching the plate. Thus the pentode not only has a very small grid-to-plate capacitance, but in addition, the plate current increases monotonically with increasing v_{PK}. Note that at large values of v_{PK} the current increases slowly. This is similar to the *vi* characteristic of the transistor and FET. The pentode is very similar in operating characteristics to the

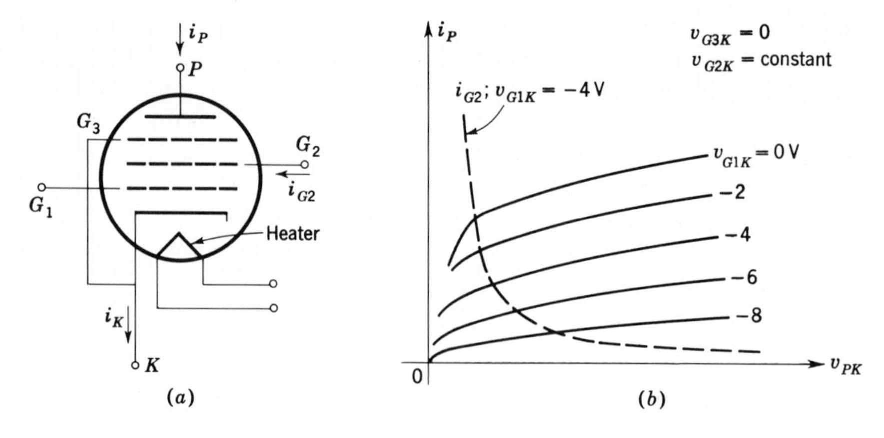

Fig. 11.1-7 The pentode. (*a*) Circuit symbol; (*b*) output characteristic.

FET, being a voltage-sensitive device with a very high input impedance. The *vi* characteristic of the pentode can be approximated by a generalized 3⁄2 power law:

$$i_P = i_P(v_{G1K}, v_{G2K}, v_{G3K}, v_{PK})^{3/2} \tag{11.1-4}$$

This generalized equation is employed in Sec. 11.4 to obtain information about the small-signal parameters of the triode.

11.2 GRAPHICAL ANALYSIS AND BIASING

The vacuum tube, like the FET, is a highly nonlinear device. Graphical analysis and biasing are accomplished with the aid of the *vi* characteristics supplied by the manufacturer. The actual characteristics usually differ by as much as 10 to 20 percent from the published characteristics.

BIASING THE TRIODE

A typical triode amplifier is shown in Fig. 11.2-1. Notice that self-bias is employed to keep the grid-to-cathode voltage negative (this is identical with the procedure employed to bias the JFET). There are, as expected, two equations which specify the Q point of the triode amplifier:

Load-line equation: $$V_{PP} = v_{PK} + i_P R_p + i_P R_k \tag{11.2-1}$$

Grid-bias equation: $$v_{GK} = -i_P R_k \tag{11.2-2}$$

A glance at the *vi* characteristic shows that, for this tube, approximately,

$$-16 < v_{GK} < 0$$

for class A operation (Chap. 5). Since $-v_{GK}$ is much smaller than V_{PP}, it is normal operating procedure either to neglect $i_P R_k$ in the load-line equation (11.2-1) or to substitute a constant, equal to the average value of v_{GK} (which in this case is 8 V), for this term. The percent error when compared with V_{PP} is certainly tolerable (<10 percent). For example, the amplifier of Fig. 11.2-1 has $V_{PP} = 300$ V. If we assumed $i_P R_k$ was zero and it turned out to be -16 V, the error would be about 5 percent.

Assuming $i_P R_k$ is negligible ($R_k \ll R_p$), we draw the dc load line as shown in the figure. Consider that $V_{GKQ} = -4$ V. The point where this curve crosses the dc load line represents the Q point. In this case

$$I_{PQ} = 17 \text{ mA}$$
$$V_{PKQ} = 130 \text{ V}$$

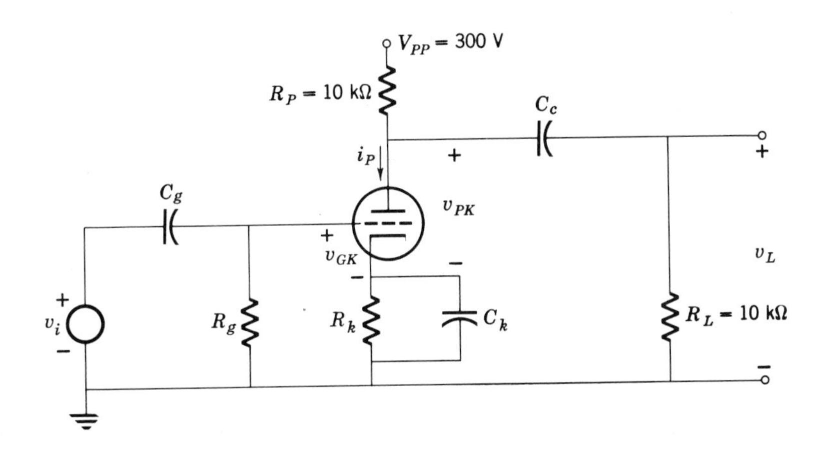

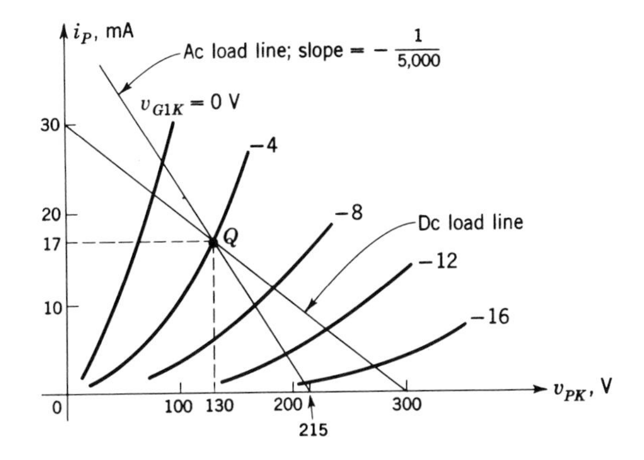

Fig. 11.2-1 Triode amplifier and load lines.

Hence

$$R_k = \frac{4}{17 \times 10^{-3}} \approx 235\ \Omega \qquad \text{use } 220\ \Omega$$

(Note that $R_k \ll R_p$, thereby verifying our assumption.)

Let us digress for a moment and consider the function of R_g. This resistor provides a current path to ground for the signal. In addition, as in the FET, it provides a dc path for the few electrons which leave the cathode and are collected by the control grid. Since the dc grid current is much less than 1 μA and R_g is usually 0.5 to 1 MΩ, the dc drop in R_g is then much less than 1 V and has negligible effect on the quiescent grid-to-cathode voltage.

THE CATHODE FOLLOWER

The cathode follower, as the name implies, is similar to the source follower and the emitter follower. A self-biased cathode follower is shown in Fig. 11.2-2*a*.

The dc-load-line equation for this circuit is

$$V_{PP} = v_{PK} + i_P(R_1 + R_2) \tag{11.2-3}$$

and the grid-bias equation when self-bias is employed is

$$v_{GK} = -i_P R_1 \tag{11.2-4}$$

The dc-load-line and the grid-bias equations are plotted in Fig. 11.2-2*b* [the simplest way to plot (11.2-4) is to choose values of v_{GK} and calculate the corresponding value of i_P]. The Q point is the intersection of the two curves as shown in the figure. Scales for v_K and v_k have been added for convenience.

The variation of v_k with respect to variations in v_i can be found using the *vi* characteristic. Using KVL around the grid circuit, we have for ac

$$v_i = v_{gk} + v_k \tag{11.2-5a}$$

where

$$v_k = -v_{pk} \tag{11.2-5b}$$

Since v_{gk} is very small compared with v_k, we see that $v_k \approx v_i$. The following calculation will serve to demonstrate this.

When $v_{GK} = 0$ V ($v_{gk} = 2$ V), $v_{PK} = 75$ V ($v_{pk} = 75 - 200 = -125$ V) and

$$v_k = +125\text{ V}$$

From (11.2-5*a*)

$$v_i = +2 + 125 = 127\text{ V}$$

When the tube is cut off,

$$v_{GK} = -5 \text{ V } (v_{gk} = -3\text{V})$$

and

$$v_k = -100 \text{ V}$$

Using (11.2-5*a*),

$$v_i = -3 - 100 = -103 \text{ V}$$

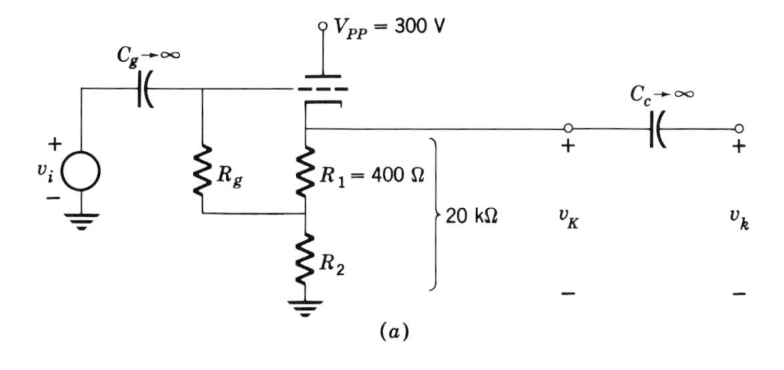

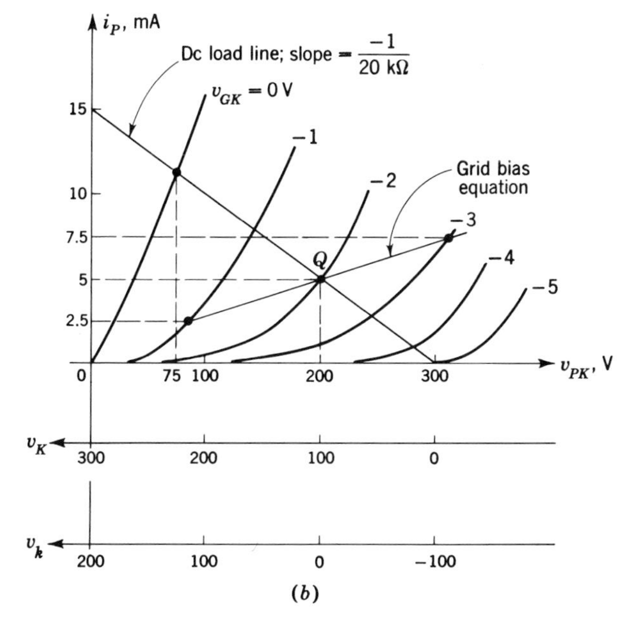

Fig. 11.2-2 The cathode follower and load lines. (*a*) Circuit; (*b*) load line and auxiliary voltage scales.

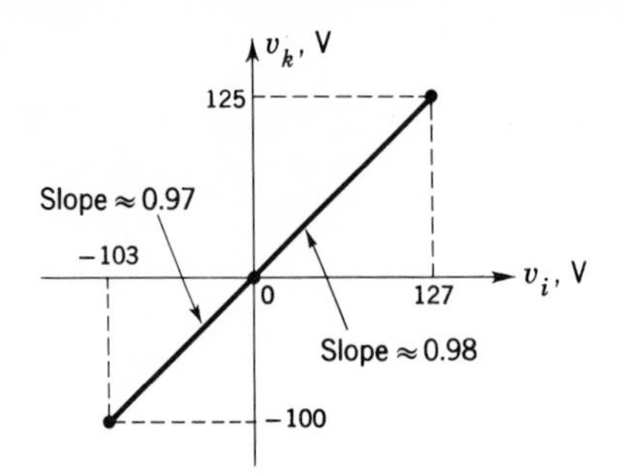

Fig. 11.2-3 Transfer characteristic of cathode follower.

Continuing in this way, a plot of the variation of v_k as a function of v_i can be obtained, as shown in Fig. 11.2-3. It should be noted that the resulting curve is not exactly linear, but is quite close to a straight line.

BIASING THE PENTODE

A typical pentode amplifier is shown in Fig. 11.2-4*a*. The variations of the plate current i_P and the screen current i_{G2} with plate voltage v_{PK} and control-grid voltage v_{G1K} are shown in Fig. 11.2-4*b*. These *vi* characteristics are plotted for $V_{G2KQ} = 100$ V and $V_{G3KQ} = 0$ V.

The screen-current variation is often omitted from the characteristics. When this occurs, a rule of thumb is to let

$$i_{G2} \approx 0.4 i_P \tag{11.2-6}$$

This approximation is valid only when $v_{PK} > v_{G2K}$.

Notice that the characteristic is drawn for $v_{G2K} = 100$ V. If, for a particular application, $v_{G2K} = 200$ V is required, each axis must be rescaled using the generalized $\frac{3}{2}$ power law given in (11.1-4). Thus we should multiply all voltage scales by 2, and all current scales by $(2)^{3/2} = 2.82$.

Example 11.2-1 Determine R_p, R_2, and R_k to set the Q point of the pentode amplifier shown in Fig. 11.2-4 at $V_{G1KQ} = -2$ V,

$$V_{PKQ} = 150 \text{ V}$$

and $V_{G2KQ} = 100$ V, when $V_{PP} = 300$ V.

Solution We see that when $V_{PK} = 150$ V and $V_{G1KQ} = -2$ V, then

$$I_P = 11 \text{ mA}$$

and

$$I_{G2} = 5 \text{ mA}$$

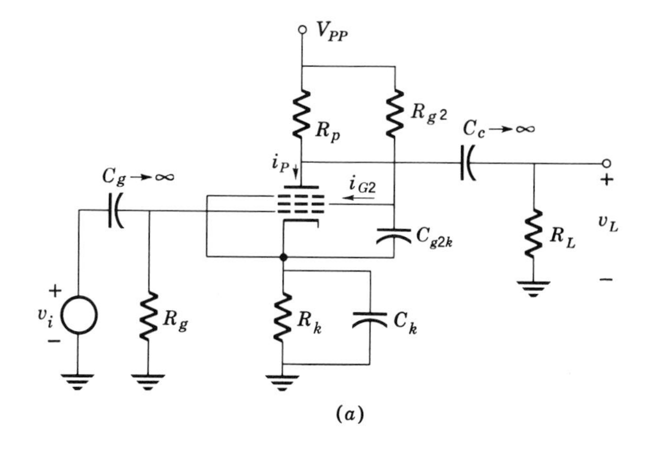

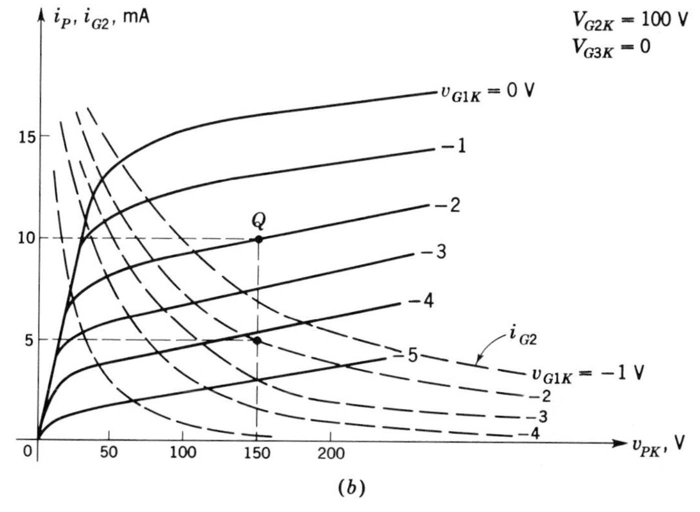

Fig. 11.2-4 (*a*) Pentode amplifier circuit; (*b*) output characteristic.

Note that (11.2-6) predicts an approximate screen current of 4.4 mA, which is reasonably close to the value obtained from the curves.

The cathode current I_K is

$$I_K = I_P + I_{G2} = 11 + 5 = 16 \text{ mA}$$

Thus, to obtain a 2-V cathode-bias voltage requires

$$R_k = \frac{2}{16 \times 10^{-3}} = 125 \ \Omega \qquad \text{use } 120 \ \Omega$$

The voltage drop across R_p is easily shown to be

$$V_{Rp} = V_{PP} - V_{PKQ} - V_{KQ} = 300 - 150 - 2 = 148 \text{ V}$$

Thus the required plate-load resistance is

$$R_p = \frac{148}{11 \times 10^{-3}} \approx 13.5 \text{ k}\Omega \qquad \text{use } 12 \text{ k}\Omega$$

The screen-grid resistance R_{G2} must produce a voltage drop

$$I_{G2}R_{G2} = V_{PP} - V_{G2KQ} - V_{KQ} = 300 - 100 - 2 = 198 \text{ V}$$

Thus

$$R_{G2} = \frac{198}{5 \times 10^{-3}} = 39.6 \text{ k}\Omega \qquad \text{use } 39 \text{ k}\Omega$$

11.3 DISTORTION

The vacuum tube, the FET, and the transistor, although considered linear for small changes in current, are basically nonlinear devices. The distortion resulting from the nonlinearities of the transistor is negligible when compared with the nonlinearities in the FET and the vacuum tube. The distortion present in an IGFET amplifier was calculated in Sec. 10.4, using the approximate analytical equation for the device.

In this section, harmonic distortion in a triode amplifier is calculated using the *vi* characteristics. The procedure employed is directly applicable to the FET amplifier and the transistor amplifier. Consider the triode amplifier shown in Fig. 11.3-1*a*. The dc and ac load lines are sketched on the *vi* characteristic in Fig. 11.3-1*b*. Let us assume that v_i is sinusoidal, with a peak value of 2 V.

$$v_i = 2 \cos \omega_0 t \tag{11.3-1}$$

Then, since

$$v_{GK} = -V_{GG} + v_i = -2 + 2 \cos \omega_0 t \tag{11.3-2}$$

the grid-to-cathode voltage varies from $v_{GK} = 0$ V to $v_{GK} = -4$ V.

The output voltage v_L ($= v_{pk}$) is obtained directly from the characteristic, as shown in Fig. 11.3-1*b*. The resulting waveform is redrawn in Fig. 11.3-2. This waveform is clearly different from the sinusoidal input signal. The distortion which produced this difference is called *harmonic* distortion, since v_L can be represented by a Fourier series,

$$v_L(t) = A_0 + A_1 \cos \omega_0 t + A_2 \cos 2\omega_0 t + A_3 \cos 3\omega_0 t + \cdots + A_N \cos N\omega_0 t \tag{11.3-3}$$

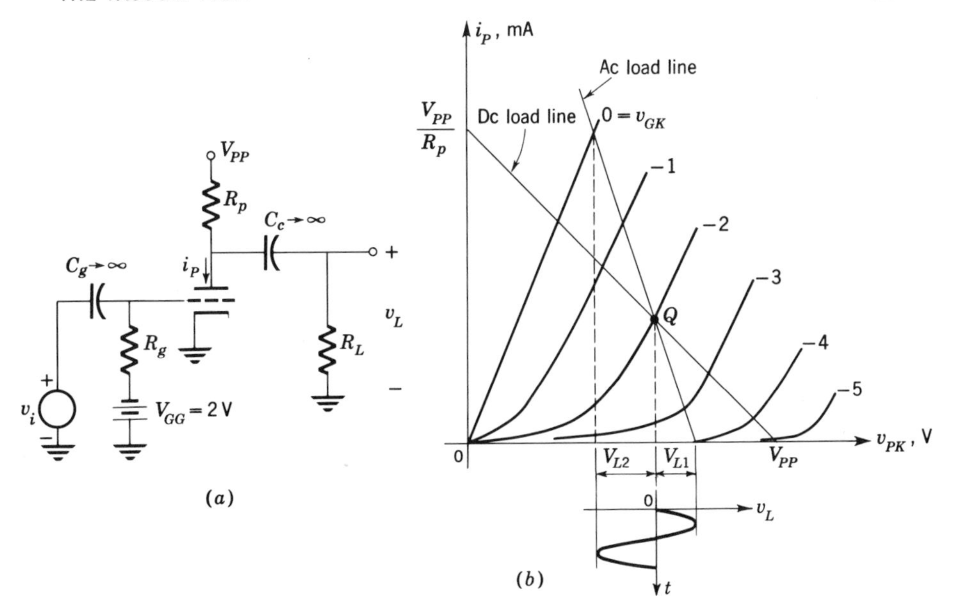

Fig. 11.3-1 Distortion in a triode amplifier. (*a*) Circuit; (*b*) load lines.

Each frequency present in the Fourier series is a harmonic of the input-signal frequency ω_0. In theory, there are an infinite number of harmonics present. However, the distortion caused by the second and third harmonics is usually the most significant in a triode.

In order to determine A_0, A_1, A_2, A_3, we assume

$$A_N = 0 \qquad N = 4, 5, \ldots \tag{11.3-4}$$

Then four equations are required to determine the required coefficients.

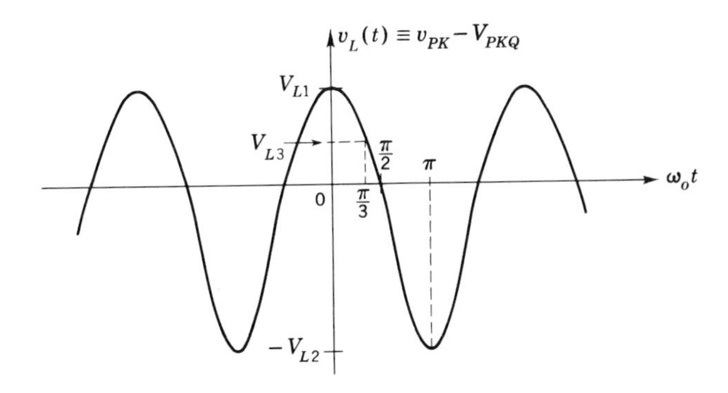

Fig. 11.3-2 Distorted waveform in the triode amplifier of Fig. 11.3-1.

For convenience, we choose four values of $\omega_0 t$ between 0 and π, such as

$$\omega_0 t = 0, \frac{\pi}{3}, \frac{\pi}{2}, \pi$$

Then we determine v_L, using Fig. 11.3-2, at each of these times.

$$v_L(0) = V_{L1} \tag{11.3-5a}$$

$$v_L\left(\frac{\pi}{3}\right) = V_{L3} \tag{11.3-5b}$$

$$v_L\left(\frac{\pi}{2}\right) = 0 \tag{11.3-5c}$$

$$v_L(\pi) = -V_{L2} \tag{11.3-5d}$$

Substituting these values into (11.3-3), we have

$$V_{L1} = A_0 + A_1 + A_2 + A_3 \tag{11.3-6a}$$

$$V_{L3} = A_0 + \tfrac{1}{2}A_1 - \tfrac{1}{2}A_2 - A_3 \tag{11.3-6b}$$

$$0 = A_0 + 0A_1 - A_2 + 0A_3 \tag{11.3-6c}$$

$$-V_{L2} = A_0 - A_1 + A_2 - A_3 \tag{11.3-6d}$$

Solving yields

$$A_0 = \frac{V_{L1} - V_{L2}}{4} \tag{11.3-7a}$$

$$A_1 = \frac{V_{L1}}{4} + \tfrac{5}{12}V_{L2} + \tfrac{2}{3}V_{L3} \tag{11.3-7b}$$

$$A_2 = \frac{V_{L1} - V_{L2}}{4} \tag{11.3-7c}$$

$$A_3 = \frac{V_{L1}}{4} + \frac{V_{L2}}{12} - \tfrac{2}{3}V_{L3} \tag{11.3-7d}$$

Note that if there were no distortion,

$$V_{L1} = V_{L2} = 2V_{L3} \tag{11.3-8a}$$

Then

$$A_0 = A_2 = A_3 = 0 \tag{11.3-8b}$$

and

$$A_1 = V_{L1} \tag{11.3-8c}$$

The total distortion caused by the second and third harmonics is defined as the ratio of the rms value of the harmonics to the fundamental.

$$D = \frac{\sqrt{(A_2/\sqrt{2})^2 + (A_3/\sqrt{2})^2}}{A_1/\sqrt{2}} = \frac{\sqrt{A_2^2 + A_3^2}}{A_1} \tag{11.3-9}$$

Distortion is often specified either as a percent or in decibels.

Example 11.3-1 The output of the amplifier of Fig. 11.3-1 was measured, and it was found that

$$V_{L1} = 40 \text{ V}$$

$$V_{L2} = 60 \text{ V}$$

$$V_{L3} = 30 \text{ V}$$

Calculate the second- and third-harmonic distortion.

Solution Using (11.3-7),

$$A_1 = 55 \text{ V}$$

$$A_0 = A_2 = -5 \text{ V}$$

and

$$A_3 = -5 \text{ V}$$

Then, using (11.3-9),

$$D = \frac{\sqrt{50}}{55} \approx 13\%$$

The fact that A_0, the average value, is -5 V indicates that the Q point will shift when this signal is applied. Since V_{PKQ} is usually 100 V or more, we should be inclined to neglect the shift in this case.

11.4 SMALL-SIGNAL ANALYSIS

The small-signal model for a vacuum tube is similar to that of the FET. At low frequencies it has an almost infinite input impedance. Its output equivalent circuit consists of a current source proportional to the input voltage, in parallel with a *plate resistance* r_p, as shown in Fig. 11.4-1. The model of Fig. 11.4-1 is valid for either triode or pentode amplifiers, if the screen grid and cathode are connected together capacitively (see

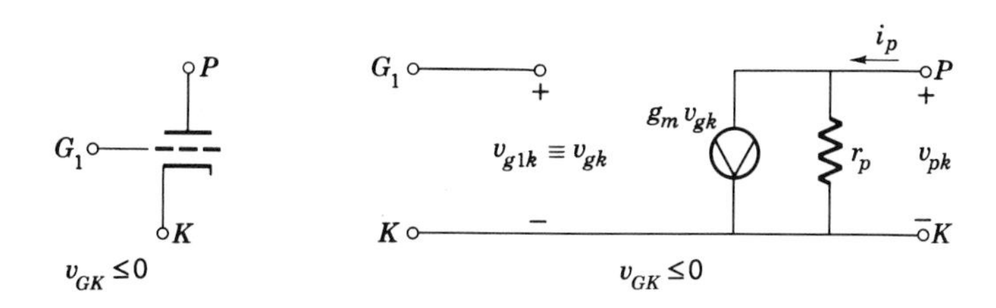

Fig. 11.4-1 Small-signal equivalent-circuit model.

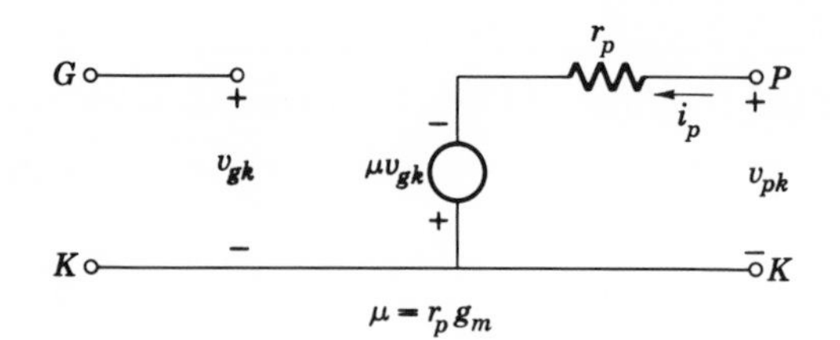

Fig. 11.4-2 Triode small-signal model.

C_{g2k} in Fig. 11.2-4) and if the suppressor grid is tied directly to the cathode (a capacitive short circuit is also sufficient). When the screen and suppressor grids are so bypassed, no ac current flows in these circuits. If they were not bypassed, ac current would flow, varying the screen (and/or suppressor)-to-cathode potential. This results in a "modulation" of the *vi* characteristics.

In the following discussion, we assume that the screen grid and the suppressor grid are adequately bypassed, so that the equivalent circuit shown in Fig. 11.4-1 is valid for both the triode and the pentode. The difference between the triode and the pentode is the relative value of r_p. In a triode, r_p typically lies between 2 and 100 kΩ, while in a pentode r_p lies between 100 kΩ and 1 MΩ.

The load resistances considered usually vary from 1 to 500 kΩ. Thus the output-current-source representation shown in Fig. 11.4-1 is useful when considering a pentode. However, the circuit of Fig. 11.4-2 is generally used when analyzing a circuit employing a triode.

SMALL-SIGNAL PARAMETERS

The *plate resistance* r_p is defined as the change in plate-to-cathode voltage for a unit change in plate current, with the grid-to-cathode voltage held constant.

$$r_p = \left.\frac{\Delta v_{PK}}{\Delta i_P}\right|_{v_{GK}=\text{constant}} \tag{11.4-1}$$

The plate resistance is thus the slope of the *vi* characteristic measured at the Q point. This is shown for a typical triode in Fig. 11.4-3*a*.

The *mutual conductance* g_m is defined (Fig. 11.4-1) as the change in the plate current for a unit change in the grid-to-cathode voltage, with the plate-to-cathode potential maintained constant.

$$g_m = \left.\frac{\Delta i_P}{\Delta v_{GK}}\right|_{v_{PK}=\text{constant}} \tag{11.4-2}$$

g_m can be calculated using the *vi* characteristics as shown in Fig. 11.4-3*b*. Typical values of g_m vary from 10^{-3} to 20×10^{-3} mho.

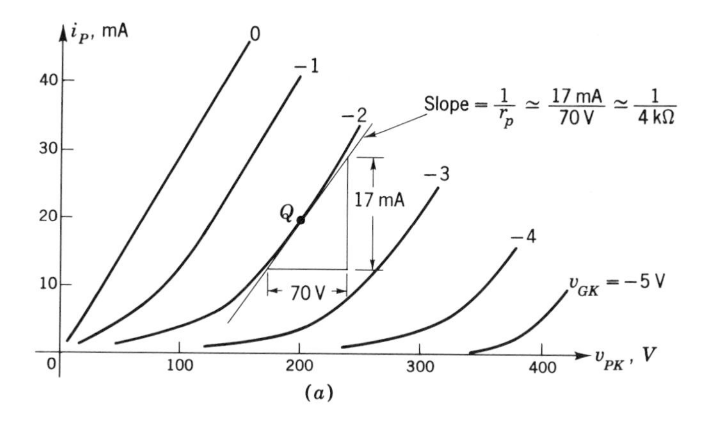

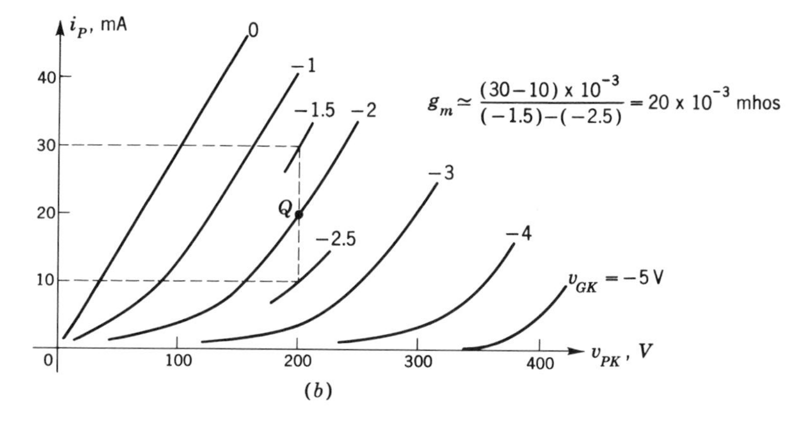

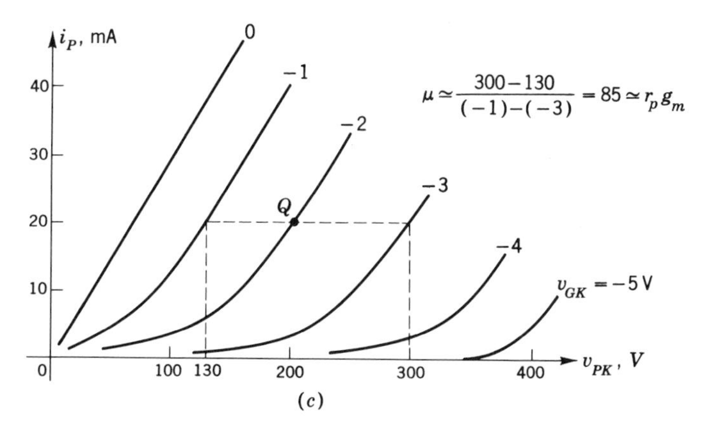

Fig. 11.4-3 Evaluation of triode small-signal parameters. (*a*) Plate resistance r_p; (*b*) mutual conductance g_m; (*c*) amplification factor μ.

The *amplification factor* μ is used as a parameter in the analysis of triode circuits. It is seldom used when dealing with pentodes. By definition,

$$\mu = g_m r_p = -\left.\frac{\Delta v_{PK}}{\Delta v_{GK}}\right|_{i_P=\text{constant}} \tag{11.4-3}$$

This parameter can be calculated directly from the characteristics, as shown in Fig. 11.4-3*c*. Values of μ lie between 2 for very high power triodes and 100 for typical "receiving-type" triodes.

VARIATION OF r_p, g_m, AND μ WITH Q POINT

The plate current varies according to the $\frac{3}{2}$ power of the electrode voltages, as indicated in (11.1-4). Thus the triode and pentode *vi* characteristics can be approximated by (11.1-2).

$$i_P \approx G(v_{PK} + \mu v_{GK})^{3/2} \tag{11.1-2}$$

Then, from (11.4-1),

$$\frac{1}{r_p} = \frac{\partial i_P}{\partial v_{PK}} = \tfrac{3}{2}G(v_{PK} + \mu v_{GK})^{1/2} \tag{11.4-4}$$

The plate resistance is therefore seen to vary inversely with changes in v_{PK} and v_{GK}. For example, if v_{PK} and v_{GK} were both decreased by a factor of 4, r_p would double.

The transconductance g_m is [Eq. (11.4-2)]

$$g_m = \frac{\partial i_P}{\partial v_{GK}} = \tfrac{3}{2}G\mu(v_{PK} + \mu v_{GK})^{1/2} \tag{11.4-5}$$

Thus if v_{PK} and v_{GK} were reduced by a factor of 4, g_m would be reduced by a factor of 2.

Since the amplification factor μ is

$$\mu = g_m r_p$$

it is unaffected, to a large degree, by changes in Q point.

Typical values and curves of the variation of μ, g_m, and r_p with the grid-to-cathode voltage (Fig. 11.5-1*b*) are often provided by tube manufacturers.

11.4-1 THE GROUNDED-CATHODE AMPLIFIER

A grounded-cathode amplifier is shown in Fig. 11.4-4*a*. The amplifier can use either a triode or a pentode, and the equivalent circuit is shown in Fig. 11.4-4*b*.

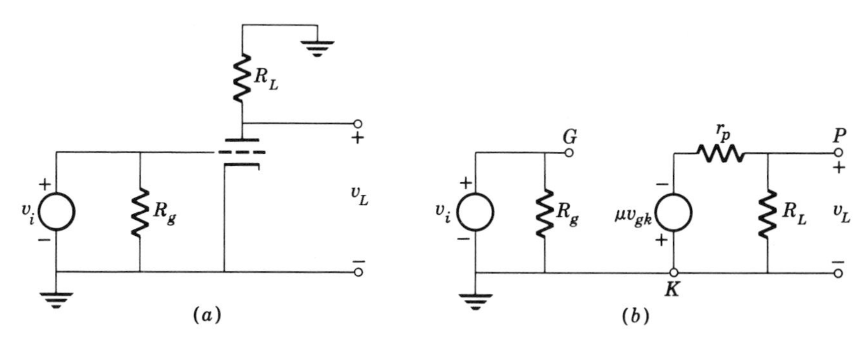

Fig. 11.4-4 Grounded-cathode amplifier. (*a*) Triode or pentode circuit (bias networks omitted); (*b*) small-signal equivalent circuit.

The input impedance of this amplifier is simply R_g. The output impedance, exclusive of R_L, is r_p. The voltage gain A_v is

$$A_v = \frac{v_L}{v_i} = -\frac{\mu R_L}{r_p + R_L} \tag{11.4-6}$$

When analyzing a triode circuit, r_p and R_L are comparable and (11.4-6) must be employed. If R_L is increased so that it is much larger than r_p, the maximum voltage gain is nearly μ. It should be noted that increasing R_L (with V_{PP} = constant) results in a reduction of the quiescent current. When considering a pentode amplifier,

$$r_p \gg R_L$$

and a slightly different form of (11.4-6) is used to calculate the gain,

$$A_v = -\frac{\mu R_L}{r_p + R_L} = -\left(\frac{\mu}{r_p}\right)\left(\frac{r_p R_L}{r_p + R_L}\right) = -g_m(r_p \| R_L) \approx -g_m R_L \tag{11.4-7}$$

Example 11.4-1 A triode amplifier has the following parameters: $\mu = 100$ and $r_p = 20$ kΩ. If $R_L = 10$ kΩ, calculate A_v.

Solution Using (11.4-6),

$$A_v = \frac{(-100)(10)}{10 + 20} = -33$$

Example 11.4-2 A pentode amplifier has the parameters $g_m = 3 \times 10^{-3}$ mho and $r_p = 200$ kΩ. If $R_L = 10$ kΩ, calculate A_v.

Solution Using (11.4-7) and noting that $r_p \| R_L \approx R_L$,

$$A_v \approx -g_m R_L = -30$$

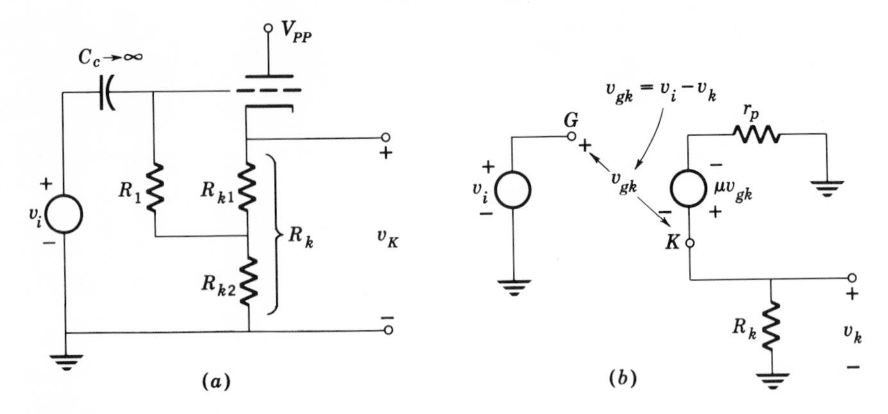

Fig. 11.4-5 Cathode follower. (*a*) Triode or pentode circuit; (*b*) approximate small-signal equivalent, neglecting effect of R_1.

Note that the voltage gains of the triode and pentode of the examples are comparable. The pentode has a higher amplification factor ($\mu = r_p g_m = 600$) and a larger plate resistance.

11.4-2 THE CATHODE FOLLOWER

A cathode follower is shown in Fig. 11.4-5*a*. The equivalent circuit is redrawn in Fig. 11.4-5*b*. This equivalent circuit is identical with the circuit of the source follower shown in Fig. 10.5-5, with r_{ds} replaced by r_p and R_s by R_k.

The Thévenin equivalent circuits of the cathode follower and the source follower are identical. The cathode-follower Thévenin equivalent circuit is shown in Fig. 11.4-6. The output impedance, excluding R_k, is

$$Z_o = \frac{r_p}{\mu + 1} \approx \frac{1}{g_m} \tag{11.4-8}$$

when $\mu \gg 1$. Typically, g_m is about 5×10^{-3} mho; therefore the output resistance of the cathode follower is approximately 200 Ω, which is com-

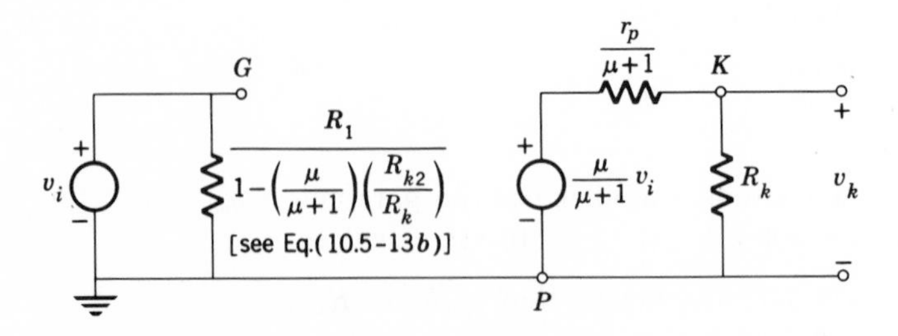

Fig. 11.4-6 Cathode follower with reflected elements (see Fig. 10.5-6).

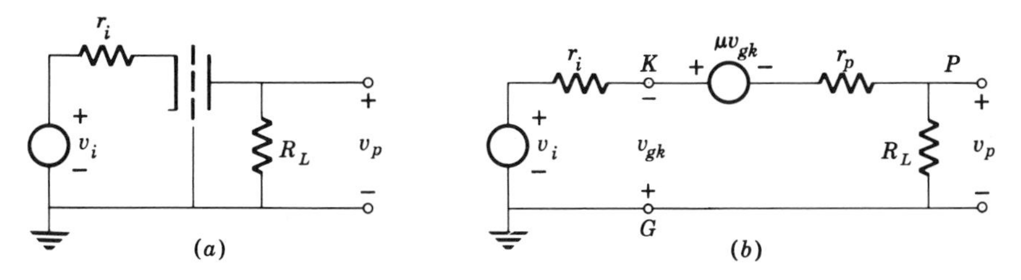

Fig. 11.4-7 The grounded-grid amplifier. (a) Triode or pentode circuit (bias supplies omitted); (b) small-signal equivalent circuit.

parable with the value obtained for the FET source follower. In contrast to this, the emitter follower has an output impedance of

$$Z_o = h_{ib} + \frac{R_b}{h_{fe} + 1}$$

which is often less than 10 Ω.

If $R_k \gg 1/g_m$, the voltage gain of the cathode follower is approximately unity.

11.4-3 THE GROUNDED-GRID AMPLIFIER

The grounded-grid amplifier, which is identical in operation with the grounded-gate amplifier and similar to the common-base amplifier, is used predominantly at high frequencies. The amplifier is shown in Fig. 11.4-7*a*, and its equivalent circuit is drawn in Fig. 11.4-7*b*.

The equivalent circuit of Fig. 11.4-7*b* is identical with the common-gate FET amplifier shown in Fig. 10.5-11. The Thévenin equivalent circuit is therefore the same, and is shown in Fig. 11.4-8. The input impedance, excluding r_i, is

$$Z_i = \frac{r_p + R_L}{\mu + 1} \tag{11.4-9}$$

The output impedance, excluding R_L, is

$$Z_o = r_p + (\mu + 1)r_i \tag{11.4-10}$$

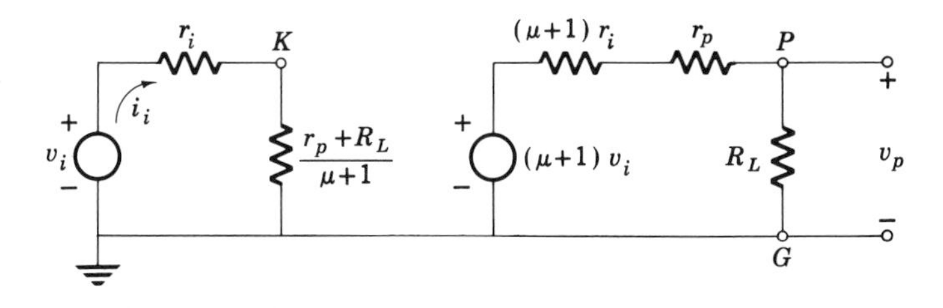

Fig. 11.4-8 Grounded-grid amplifier with reflected elements.

The voltage gain A_v is

$$A_v = \frac{(\mu + 1)R_L}{(\mu + 1)r_i + r_p + R_L} \tag{11.4-11}$$

Note that if

$$(\mu + 1)r_i \gg r_p + R_L \tag{11.4-12a}$$

then

$$A_v \approx \frac{R_L}{r_i} \tag{11.4-12b}$$

11.5 MANUFACTURERS' SPECIFICATIONS

The characteristics below refer to a type 7025 nine-pin miniature twin triode (i.e., two triode units enclosed in a single envelope) intended for low-power audio applications.

1. Heater arrangement

	Series	*Parallel*
Voltage, V, ac or dc	12.6	6.3
Current, A	0.15	0.3

2. Direct interelectrode capacitance (approximate)

	Unit 1	*Unit 2*
Grid-to-plate C_{gp}, pF	1.7	1.7
Grid-to-cathode and heater C_{gk}, pF	1.6	1.6
Plate-to-cathode and heater C_{pk}, pF	0.46	0.34

3. Maximum ratings

Plate voltage v_{PK}	300 V
Plate dissipation P_P	1.2 W
Grid voltage v_{GK}	
Negative-bias value	−55 V
Positive-bias value	0 V

4. Characteristics, class A amplifier (each unit)

	Unit 1	*Unit 2*
Plate voltage V_{PK}, V	100	250
Grid voltage V_{GK}, V	−1	−2
Amplification factor μ	100	100
Plate resistance, approximate, r_p, kΩ	80	62.5
Transconductance g_m, μmhos	1,250	1,600
Plate current I_p, mA	0.5	1.2

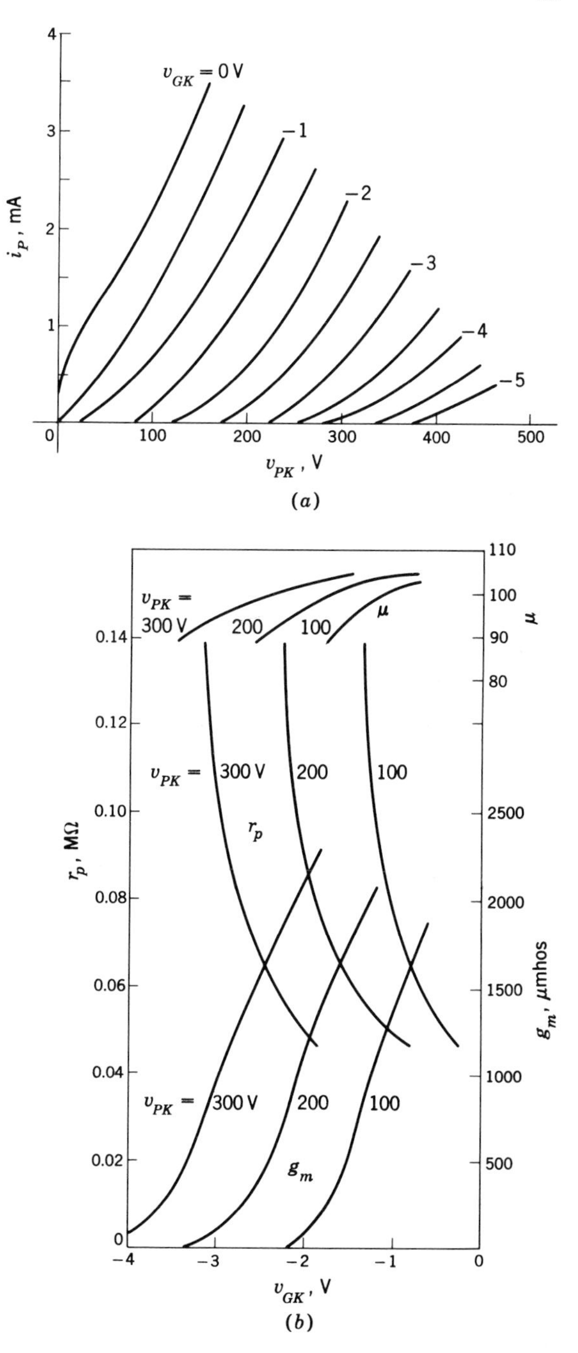

Fig. 11.5-1 Small-signal-parameter variations. (*a*) Plate characteristics; (*b*) parameter variations.

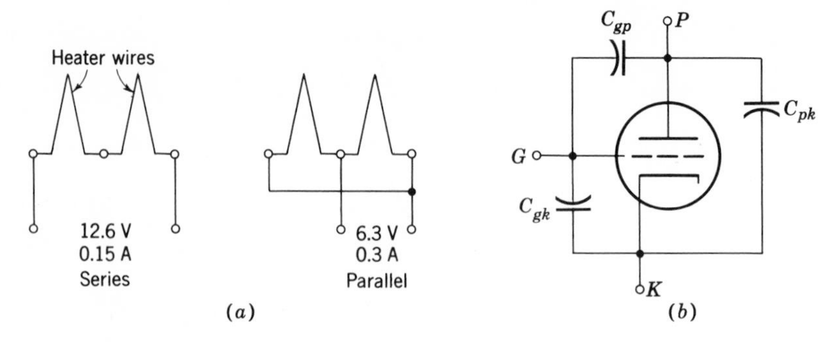

Fig. 11.5-2 (*a*) Heater arrangement; (*b*) interelectrode capacitances.

EXPLANATION OF SPECIFICATIONS

1. The heater (filament) can be wired either in series or in parallel, as shown in Fig. 11.5-2*a*. In either case the power required is 1.9 W.
2. The interelectrode capacitances are shown schematically in Fig. 11.5-2*b*. Their effect on amplifier response is discussed in Chap. 13.
3. The maximum ratings should not be exceeded under normal operating conditions, and it is good practice to add a safety factor of about 20 percent to these figures to take account of variations between units.
4. The typical characteristics give the small-signal parameters at selected operating points. They are useful for initial placement of the Q point and for obtaining quick estimates of amplifier performance.
5. The plate characteristics can be used to calculate input and output swings and distortion for selected operating conditions. These characteristics are almost always used in design, as contrasted to the transistor, where the typical collector characteristics (only occasionally furnished by the manufacturer) are seldom used.

PROBLEMS

SECTION 11.1

11.1. A vacuum diode has a plate characteristic

$$i_P = \begin{cases} 3 \times 10^{-4}\, v_{PK}^{3/2} & v_{PK} \geq 0 \\ 0 & v_{PK} < 0 \end{cases}$$

(*a*) Sketch the characteristic.

(*b*) Plot the dynamic diode resistance $\partial v_{PK}/\partial i_P$ as a function of plate current.

11.2. An analytic expression for the *vi* characteristic of a triode is given by (11.1-2). Consider the *vi* characteristic of a 6SN7 triode.

(*a*) Take two points on this characteristic and determine G and μ.

(*b*) Plot the analytic *vi* characteristic on the same axes as the graphical *vi* characteristic for $v_{GK} = 0, -2, -4$, and compare.

11.3. The 6CY5 is a tetrode. Using the *vi* characteristics provided in Appendix III, measure the maximum negative resistance occurring when $v_{GK} = 0, -0.5, -1.5, -2, -2.5$, and -3 V. Plot your results.

11.4. The 6CL6 is a power pentode.

(*a*) Calculate the cathode current i_K, where

$$i_K = i_P + i_{G2}$$

when $v_{G1K} = 0$ V, as a function of plate-to-cathode voltage v_{PK}. Assume $v_{G2K} = 150$ V and $v_{G3K} = 0$ V.

(*b*) Sketch the maximum-power-dissipation hyperbola on the *vi* characteristic. When $V_{G1K} = -2$ V, find the maximum quiescent plate-to-cathode voltage.

SECTION 11.2

11.5. One section of a 12AT7 triode is to be used as an amplifier in Fig. P11.5. If $V_{GKQ} = -2$ V and if maximum power dissipation is not to be exceeded:

(*a*) Find $I_{PQ,\max}$.

(*b*) Let $V_{PP} = 300$ V; calculate R_p and R_k.

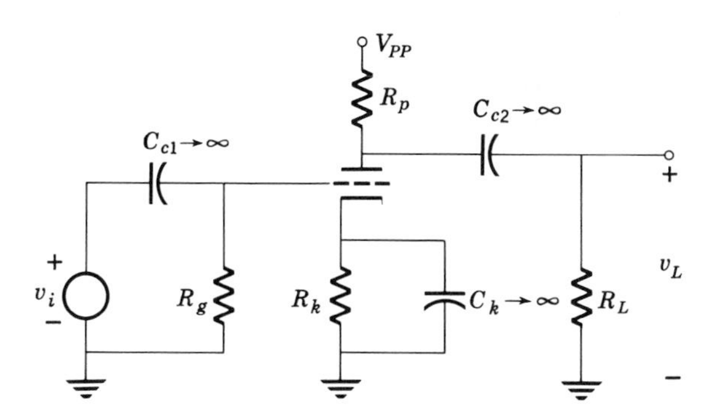

Fig. P11.5

11.6. If in Fig. P11.5 $V_{PP} = 250$ V, $R_p = 15$ kΩ, and $R_k = 100$ Ω, determine the quiescent point.

11.7. Plot v_K versus v_G as v_G varies from -5 to $+150$ V. Discuss your results.
(*a*) Find Q point.
(*b*) When $v_{GK} = 0$ V, find i_P. Calculate v_K and v_G.
(*c*) When $i_P = 0$, find v_{GK} and v_G.
(*d*) Find maximum peak-to-peak symmetrical output swing.
(*e*) Find voltage gain.

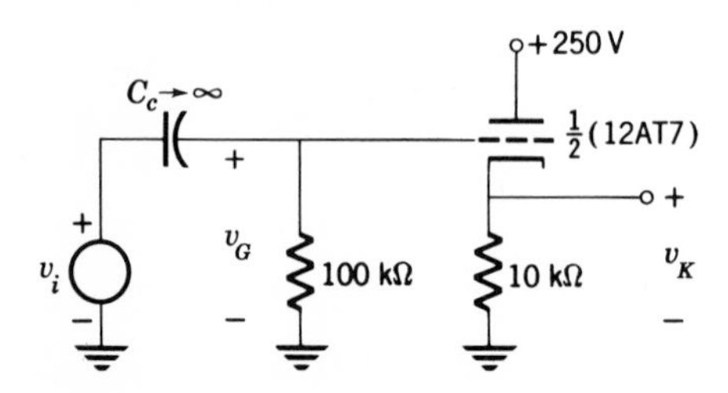

Fig. P11.7

11.8. The Q point of this cathode follower is to be set at $I_{PQ} = 15$ mA and $V_{GK} = -6$ V.
(*a*) Is the maximum dissipation rating exceeded?
(*b*) Find R_1 and R_2.

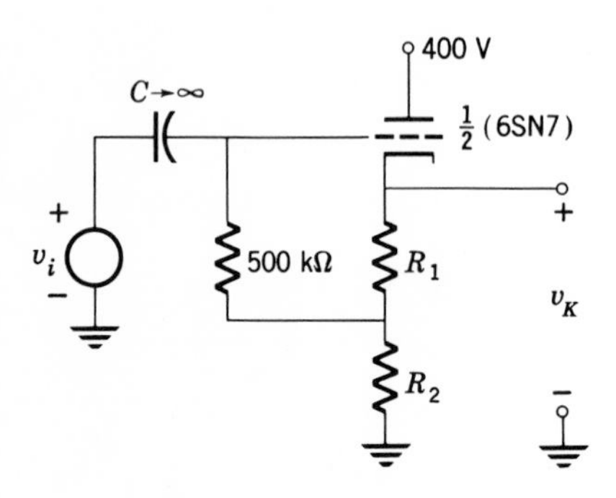

Fig. P11.8

11.9. In Prob. 11.8 let $R_1 = 1$ kΩ and $R_2 = 9$ kΩ.
(*a*) Find the Q point.
(*b*) Plot v_K versus v_G as v_G varies from -26 V to $+230$ V. Comment on the maximum peak-to-peak symmetrical output swing and the voltage gain.

11.10. The pentode amplifier is to have the Q point $V_{PKQ} = 200$ V, $V_{G1KQ} = -2$ V, and $V_{G2KQ} = 150$ V.
(*a*) Find the ratio

$$\left.\frac{I_{G2}}{I_P}\right|_{V_{PK}=200\text{ V}}$$

(*b*) Determine I_{PQ}.

(c) Calculate I_{G2}.
(d) Calculate R_{g2}, R_k, and R_p.

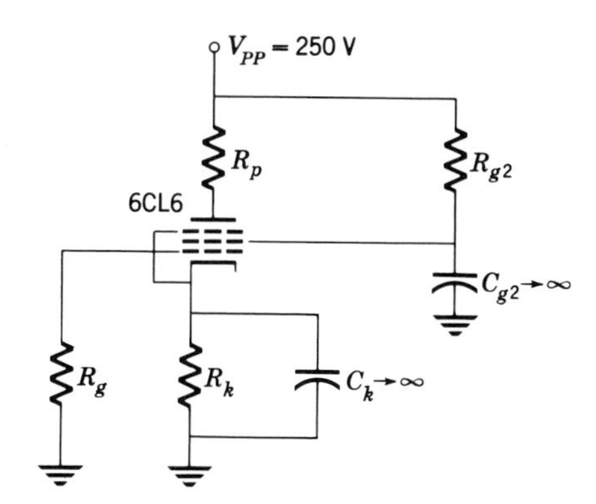

Fig. P11.10

11.11. Repeat Prob. 11.10 if $V_{G2KQ} = 75$ V. *Hint:* Redraw the characteristic.

11.12. In the amplifier of Fig. P11.10, $R_p = 1.5$ kΩ and $R_k = 100$ Ω, $V_{G2KQ} = 150$ V, and $V_{PP} = 250$ V.
Find the Q point.

SECTION 11.3

11.13. If v_i is a square wave with a peak-to-peak voltage of 12 V:
(a) Find the Q point.
(b) Let R_L be infinite. Find the output voltage.
(c) Is there distortion? Explain.

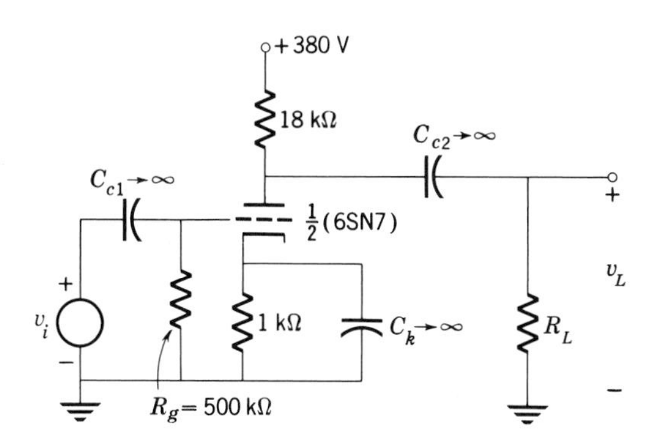

Fig. P11.13

11.14. The input voltage in Fig. P11.13 is sinusoidal, with a peak voltage equal to 12 V.
(a) With R_L infinite, calculate the second- and third-harmonic distortion.
(b) Calculate the total distortion.

11.15. Repeat Prob. 11.14 with $R_L = 18$ kΩ.

11.16. Calculate the distortion obtained from the cathode follower shown in Fig. P11.7 when v_i is sinusoidal with a peak value of 4 V.

11.17. The cathode follower in Fig. P11.8 has $R_1 = 1$ kΩ and $R_2 = 9$ kΩ (see Prob. 11.9). If v_i is sinusoidal with a peak voltage of 100 V, calculate the distortion present in the output.

SECTION 11.4

11.18. The 6SN7 is biased at $V_{PK} = 200$ V and $V_{GK} = -4$ V. Calculate r_p, μ, and g_m.

11.19. Plot μ as a function of V_{GKQ} with $V_{PK} = 250$ V for the 6SN7.

11.20. For the 6CL6, plot r_p as a function of V_{PKQ} when $V_{G1K} = -2$ V and $V_{G2K} = 80$ V.

11.21. Plot r_p as a function of V_{GKQ} with $V_{PK} = 250$ V for the 6SN7.

11.22. Plot r_p as a function of V_{GKQ} with $V_{PK} = 125$ V for the 6SN7.

11.23. Plot g_m as a function of V_{GK} with $V_{PK} = 250$ V for the 6SN7.

11.24. Plot r_p as a function of V_{G1KQ} when $V_{PKQ} = 250$ V and $V_{G2K} = 150$ V for the 6CL6.

11.25. Plot μ as a function of V_{G1KQ} when $V_{PKQ} = 250$ V and $V_{G2K} = 150$ V for the 6CL6.

11.26. (*a*) Find the Q point.
(*b*) Calculate μ and r_p.
(*c*) Calculate the small-signal voltage gain.

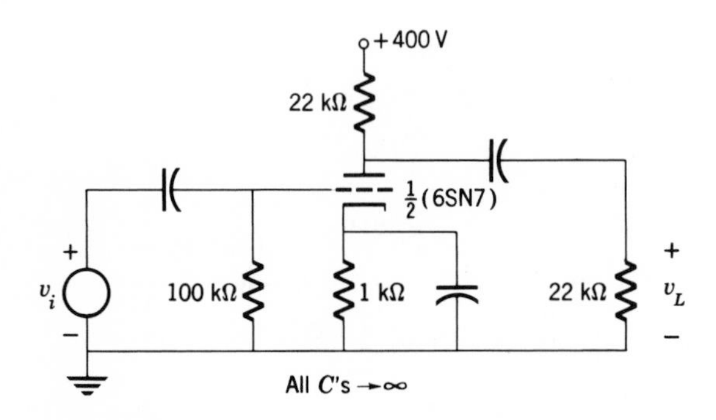

Fig. P11.26

11.27. (*a*) Find the Q point.
(*b*) Determine μ, r_p, and g_m.
(*c*) Calculate Z_i and Z_o.
(*d*) Determine the voltage gain.

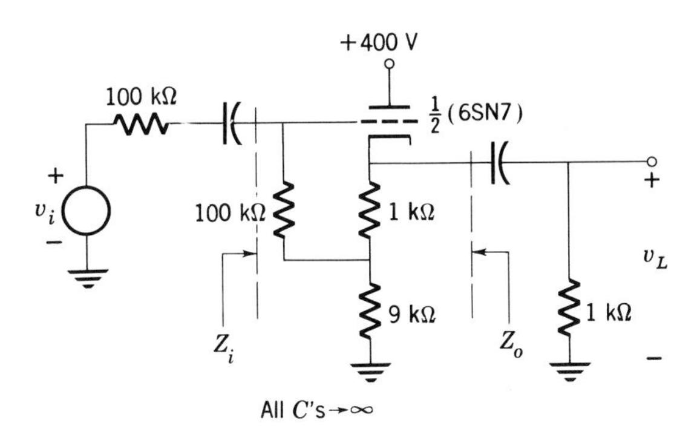

Fig. P11.27

11.28. (*a*) Determine the Q points for T_1 and T_2.
(*b*) Calculate μ, r_p, and g_m for T_1 and T_2.
(*c*) Calculate the gain v_L/v_i.

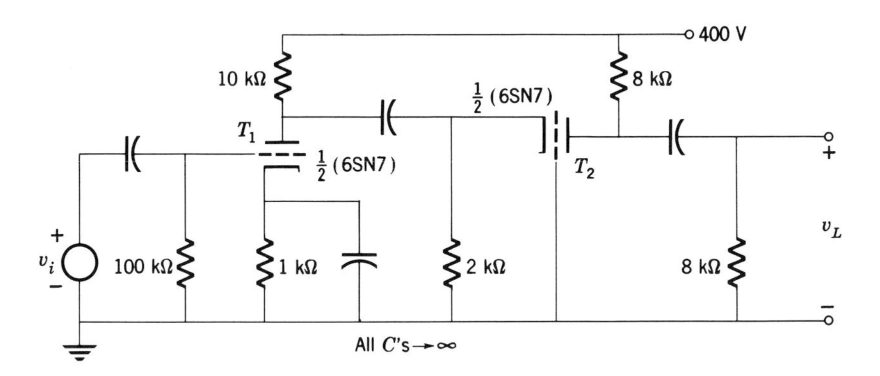

Fig. P11.28

11.29. (*a*) Calculate v_{o1}, v_{o2}, v_{o3}, and v_{o4}.
(*b*) Determine Z_o between each output terminal and ground.

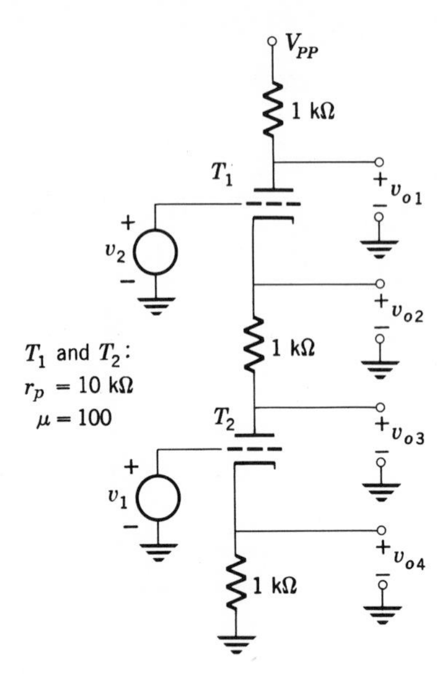

Fig. P11.29

11.30. (*a*) Find the Q point.
(*b*) Calculate v_L/v_i. (Use typical values for r_p and μ.)

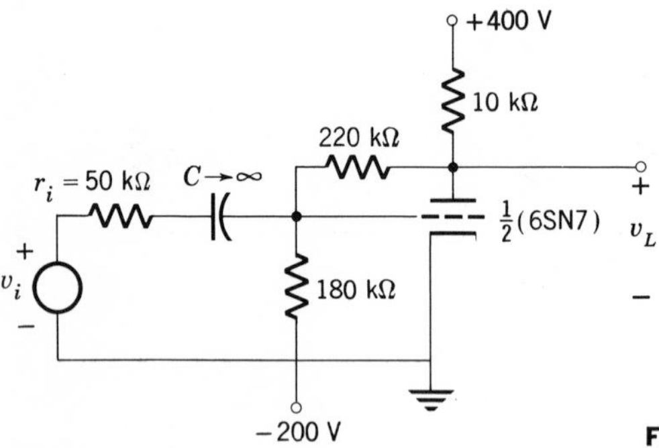

Fig. P11.30

REFERENCES

1. Spangenberg, K. R.: "Fundamentals of Electron Devices," McGraw-Hill Book Company, New York, 1957.
2. Millman, J., and S. Seeley: "Electronics," 2d ed., McGraw-Hill Book Company, New York, 1951.

12
Low-frequency Response of *RC*-coupled Amplifiers

INTRODUCTION

As noted in Chap. 3, the usual common-emitter *RC*-coupled amplifier employs coupling ("blocking") capacitors between stages, and an emitter bypass capacitor to provide an ac short circuit from the emitter to ground. In this chapter we take account of the fact that at low frequencies the impedance of each of these capacitors increases, reducing the overall gain of the amplifier.

A typical variation of amplifier current gain or voltage gain with frequency is illustrated in Fig. 12-1. A_m is the maximum gain of the amplifier, which occurs in that frequency range where all capacitors can be neglected.

The fall-off at low frequencies is a result of using coupling and bypass capacitors. Integrated circuits and other direct-coupled amplifiers do not exhibit fall-off. These amplifiers, as their name implies, are capable of amplifying uniformly all the way down to dc.

The high-frequency fall-off is caused by the presence of internal

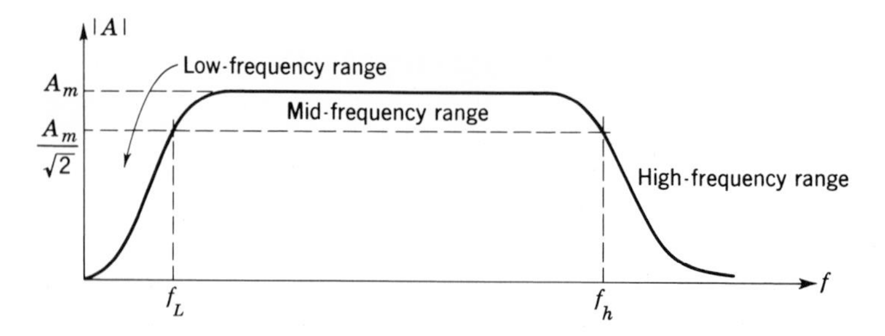

Fig. 12.1 Gain vs. frequency for a typical RC-coupled amplifier.

capacitance and external stray capacitance. For example, capacitance exists across the collector-base junction of a transistor. This capacitance is internal to the transistor, and is not intentionally placed across the junction. The emitter bypass capacitor, on the other hand, is an external capacitor. The effect of the internal capacitances on the circuit response is discussed in Chap. 13.

When the midfrequency range is a few octaves or more, the problem of determining the frequency response of a typical RC-coupled amplifier breaks down into essentially three relatively simple problems. The coupling and bypass capacitors affect only the low-frequency response, while the transistor capacitances affect only the high-frequency response. Thus, for convenience, we divide the frequency spectrum into three regions: low, middle, and high. In each of these frequency bands, a different simplified equivalent circuit is used to calculate the response. In the low-frequency band, the large coupling and bypass capacitors are important, while the small transistor and stray capacitances are effectively open circuits. In the midfrequency range, the large capacitors are effectively short circuits, and the small capacitances are open circuits, so that no capacitances appear in the midband equivalent circuit. At high frequencies, the large capacitors are replaced by short circuits, and the transistor and stray capacitances help determine the response.

Throughout the chapter, transfer functions are written in terms of the Laplace transform frequency variable s (complex frequency). Thus the response to any excitation can be found using Laplace transform techniques. The steady-state sinusoidal response, which is our main concern, is obtained simply by replacing s by $j\omega$ in the transfer function.

The boundaries separating the midfrequency range from the low- and high-frequency ranges are not well defined, and the useful operating range of an amplifier is specified in various ways, depending on the application. In some applications, the exact shape of the gain-frequency characteristic is important. For example, consider an amplifier designed for video signals. Since the *eye* is sensitive to light-intensity variations of the order of 1 dB, the response in the midfrequency range should not depart from A_m by more than about ½ dB if high-quality pictures are to be obtained. The edges of the midfrequency region are determined by the useful frequency content of the signal. If the video signal contains important frequency components ranging from dc to 4 MHz, then no more than ½-dB maximum variation will have to be maintained to 4 MHz. The rate at which the response can fall off above 4 MHz will depend on other factors, such as the required transient response.

For high-quality reproduction of audio signals, a frequency band from about 20 Hz to 20 kHz is required. In the usual application, the boundaries of the midfrequency range are defined as those frequencies at which the response has fallen to 3 dB below A_m. These are shown as f_L and f_h in Fig. 12-1, and are referred to as the 3-dB frequencies, or simply the *cutoff frequencies*. The total midfrequency range is called the *bandwidth* B (it is actually the *half-power* bandwidth), and from the figure, $B = f_h - f_L$ (when $f_h \gg f_L$, the bandwidth $B \approx f_h$). These definitions are used throughout this text.

12.1 THE LOW-FREQUENCY RESPONSE OF THE TRANSISTOR AMPLIFIER

The low-frequency response of the CE amplifier is determined by the emitter bypass capacitor and the coupling capacitors. In practical amplifier circuits, the emitter bypass capacitor usually limits the low-frequency response. In the following sections we consider the effect of each capacitor separately, and then the combined effect.

12.1-1 THE EMITTER BYPASS CAPACITOR

Consider the single-stage amplifier of Fig. 12.1-1a. Coupling capacitors in the input and collector circuits are omitted in order to focus attention on the emitter bypass capacitor. The small-signal equivalent circuit is shown in Fig. 12.1-1b, where the base circuit has been reflected into the emitter circuit. This circuit is valid for the low-frequency and midfrequency regions.

The current-gain transfer function is obtained from the simplified equivalent circuit of Fig. 12.1-1c by routine analysis.

$$A_i = \frac{i_c}{i_i} \approx \frac{i_e}{i_i} = \left(\frac{R_b}{\dfrac{R_b}{h_{fe}+1} + h_{ib}}\right)\left(\frac{s + 1/R_eC_e}{s + \dfrac{1}{\left[R_e \| \left(\dfrac{R_b}{h_{fe}+1} + h_{ib}\right)\right] C_e}}\right) \tag{12.1-1a}$$

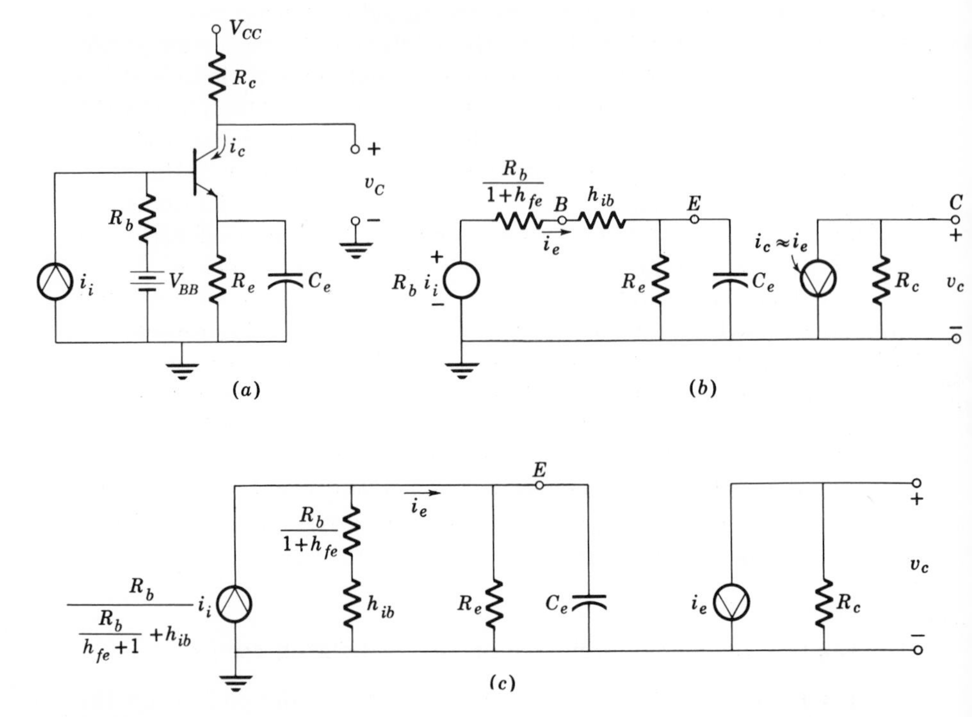

Fig. 12.1-1 CE amplifier with emitter bypass capacitor. (*a*) A single-stage amplifier; (*b*) small-signal equivalent circuit valid for low frequencies; (*c*) reduced equivalent circuit.

When ω is very large, the gain is at its midfrequency value A_{im}.

$$A_{im} = \left|\frac{i_c}{i_i}\right|_{\omega\to\infty} \approx \frac{R_b}{R_b/(h_{fe}+1) + h_{ib}} \approx \frac{h_{fe}}{1 + h_{ie}/R_b} \tag{12.1-1b}$$

The emitter resistor R_e is usually specified so that

$$R_e \gg \frac{R_b}{h_{fe}+1} + h_{ib} \tag{12.1-1c}$$

in order that the zero of (12.1-1) occur at a much lower frequency than the pole.

The zero of (12.1-1*a*) occurs at

$$\omega_1 = \frac{1}{R_eC_e} \tag{12.1-2a}$$

and the pole at

$$\omega_2 = \frac{1}{\{R_e \| [R_b/(h_{fe}+1) + h_{ib}]\}C_e} \tag{12.1-2b}$$

When the condition of (12.1-1*c*) is valid,

$$\omega_1 \ll \omega_2 \tag{12.1-2c}$$

and (12.1-1*a*) becomes [using (12.1-1*b*)], when $\omega > \omega_1$,

$$|A_i| = A_{im}\left|\frac{j\omega + \omega_1}{j\omega + \omega_2}\right| \approx \frac{A_{im}\omega}{\sqrt{\omega^2 + \omega_2^2}} \tag{12.1-3a}$$

The frequency $f = f_L$, at which the current gain is down 3 dB from its value in the midfrequency range, is found by setting

$$|A_i| = A_{im}\left(\frac{\omega_L}{\sqrt{\omega_L^2 + \omega_2^2}}\right) = \frac{A_{im}}{\sqrt{2}} \tag{12.1-3b}$$

Solving, we have

$$f_L \approx f_2 \tag{12.1-3c}$$

If (12.1-1*c*) is not satisfied, the response may *never* be 3 dB down at low frequencies. This can be seen from (12.1-1*a*). For a 3-dB frequency to exist, we must have

$$|A_i(\omega = \omega_L)|^2 = A_{im}^2\left(\frac{\omega_L^2 + \omega_1^2}{\omega_L^2 + \omega_2^2}\right) = \frac{A_{im}^2}{2}$$

Then

$$\omega_L^2 + \omega_2^2 = 2\omega_L^2 + 2\omega_1^2$$

and

$$\omega_L^2 = \omega_2^2 - 2\omega_1^2 \tag{12.1-3d}$$

Thus, if

$$2\omega_1^2 > \omega_2^2$$

the 3-dB frequency f_L *does not exist.*

12.1-2 ASYMPTOTIC (BODE) PLOTS OF AMPLIFIER TRANSFER FUNCTIONS

It is extremely helpful to be able to exhibit the frequency dependence of equations such as (12.1-1*a*) graphically. This is easily done using asymptotic logarithmic characteristics and logarithmic scales. To illustrate the method, we rewrite (12.1-1) in the form

$$A_i(j\omega) = \frac{i_c}{i_i} = A_{im}\left(\frac{j\omega + \omega_1}{j\omega + \omega_2}\right) = A_{io}\left(\frac{1 + j\omega/\omega_1}{1 + j\omega/\omega_2}\right) = |A_i|\epsilon^{j\theta} \tag{12.1-4}$$

where

$$A_{io} = A_{im}\left(\frac{\omega_1}{\omega_2}\right)$$

In most practical problems, interest centers on the magnitude and phase angle of $A_i(j\omega)$. These are

$$|A_i| = A_{io}\frac{\sqrt{1 + (\omega/\omega_1)^2}}{\sqrt{1 + (\omega/\omega_2)^2}} \tag{12.1-5}$$

$$\theta = \left(\tan^{-1}\frac{\omega}{\omega_1}\right) - \left(\tan^{-1}\frac{\omega}{\omega_2}\right) \tag{12.1-6}$$

Let us first find A_i, in decibels.

$$|A_i|_{\mathrm{dB}} = 20(\log A_{io}) + 20\left[\log\sqrt{1 + \left(\frac{\omega}{\omega_1}\right)^2}\right] - 20\left[\log\sqrt{1 + \left(\frac{\omega}{\omega_2}\right)^2}\right] \tag{12.1-7}$$

The problem is now considerably simplified. We plot each of the factors of (12.1-7), and then add the individual curves graphically.

To illustrate this, consider (12.1-7). The first factor, $20 \log A_{io}$, is a constant. To plot the second factor, consider the asymptotic behavior at very low and very high frequencies.

At low frequencies ($\omega \to 0$):

$$20\log\sqrt{1 + \left(\frac{\omega}{\omega_1}\right)^2} \to 20 \log 1 = 0 \text{ dB} \tag{12.1-8a}$$

At high frequencies ($\omega \gg \omega_1$):

$$20\log\sqrt{1 + \left(\frac{\omega}{\omega_1}\right)^2} \to 20\log\sqrt{\left(\frac{\omega}{\omega_1}\right)^2} = 20\log\frac{\omega}{\omega_1} \quad \text{dB} \tag{12.1-8b}$$

The high-frequency asymptote is a straight line when plotted against ω on a logarithmic scale, and the low-frequency asymptote is simply a constant. They intersect at $\omega = \omega_1$ because the high-frequency asymptote is 0 dB at this point. ω_1 is called the *break*, or *corner*, frequency. The asymptotes are sketched in Fig. 12.1-2.

The slope of the high-frequency asymptote is usually expressed in terms of octave (2:1) or decade (10:1) frequency ratios. Thus a frequency increase of an octave results in an increase of gain [Eq. (12.1-8*b*)] of

$$\Delta|A_i|_{\mathrm{dB}} = 20 \log 2 = +6 \text{ dB} \tag{12.1-9a}$$

A frequency increase of a decade results in a change in gain of

$$\Delta|A_i|_{\mathrm{dB}} = 20 \log 10 = +20 \text{ dB} \tag{12.1-9b}$$

Knowing the slope and the frequency at which the asymptote goes through 0 dB, equations such as (12.1-8*b*) can be easily sketched on semilog graph paper.

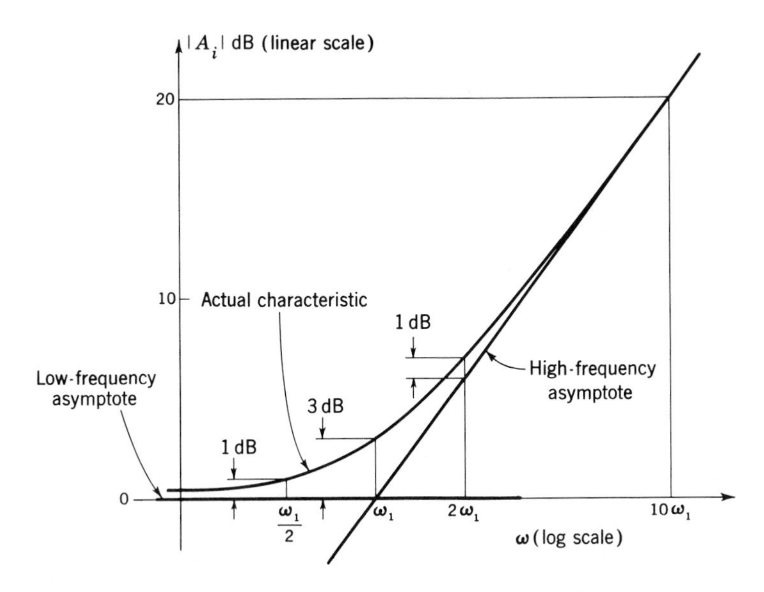

Fig. 12.1-2 Asymptotic and actual frequency characteristics.

Consider the third factor in (12.1-11). We see that the previous discussion holds, with the exception that the break frequency is at ω_2 and the slope above ω_2 is -6 dB/octave.

Example 12.1-1 Plot the asymptotic magnitude for the gain function

$$A_i = 40\left(\frac{1 + j\omega/10}{1 + j\omega/50}\right)$$

Solution The constant is 20 log 40 = 32 dB.

The first break occurs at the numerator corner frequency, $\omega_1 = 10$ rad/s, and the second at the denominator corner frequency, $\omega_2 = 50$ rad/s. The individual factors are shown in Fig. 12.1-3*a*, and their sum in Fig. 12.1-3*b*.

The high-frequency asymptote for $|A_i|$ is a constant, which can be found graphically, or from the asymptotic form,

$$A_i(\omega \gg \omega_2) \approx 40\left(\frac{j\omega/10}{j\omega/50}\right) = (40)(5) = 200$$

or in decibels,

$$|A_i(\omega \gg \omega_2)| = 20 \log 200 = 46 \text{ dB}$$

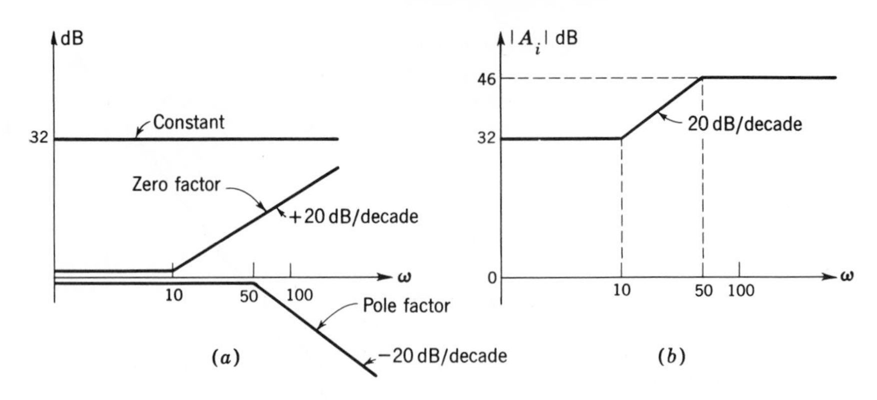

Fig. 12.1-3 Asymptotic amplitude plots for Example 12.1-1. (*a*) Sketch of individual factors; (*b*) asymptotic plot.

Often the asymptotic curve is sufficient to supply the required information. If a more accurate curve is required, simple corrections can be applied to the asymptotic curve. The difference between the actual and asymptotic curves for a single-frequency factor of the form $1 + j\omega/\omega_1$ is usually negligible at frequencies beyond an octave on either side of the break frequency. The corrections to be applied can be found as follows: At ω_1, the asymptotic curve is at 0 dB, while the actual value is

$$20 \log \sqrt{1 + \left(\frac{\omega}{\omega_1}\right)^2} = 10 \log 2 \approx 3 \text{ dB}$$

Thus the actual curve is 3 dB above the asymptote at the break. Similar calculations at $2\omega_1$ and $\omega_1/2$ indicate that the actual curve lies about 1 dB above the asymptotic curve (note that 1 dB represents a 10 percent error). These results are illustrated in Fig. 12.1-2.

PHASE-ANGLE VARIATION

The phase angle of the gain function can also be approximated by straight-line asymptotes. For a single-frequency factor the phase angle is $\theta = \tan^{-1}(\omega/\omega_1)$. The asymptotes often used for this are

$$\begin{aligned} \theta &= 0^\circ && \omega < \omega_1/10 \\ \theta &= 45\left(1 + \log\frac{\omega}{\omega_1}\right) && \frac{\omega_1}{10} < \omega < 10\omega_1 \\ \theta &= 90^\circ && \omega > 10\omega_1 \end{aligned}$$

(when $\omega = \omega_1/10$, $\theta = 5.7^\circ$, and when $\omega = 10\omega_1$, $\theta = 84.3^\circ$). The actual and asymptotic phase curves are shown in Fig. 12.1-4. Through the use

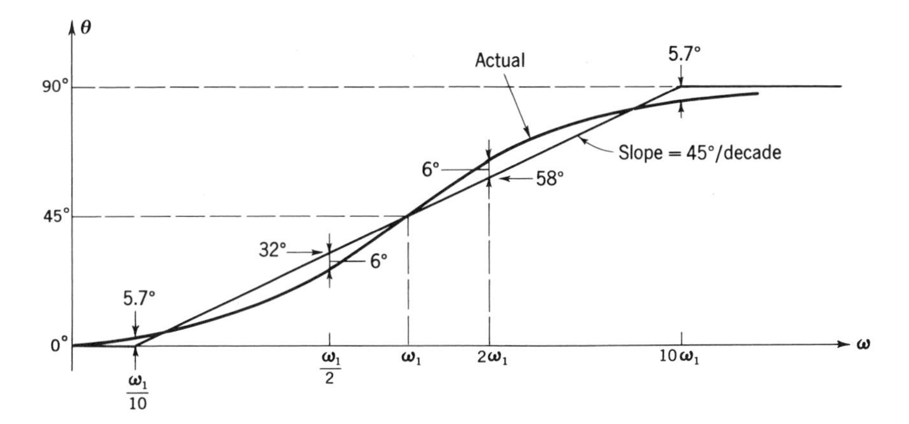

Fig. 12.1-4 Phase of $1 + j\omega/\omega_1$.

of these asymptotic characteristics, both amplitude and phase curves can be sketched quickly and accurately when the transfer function has been reduced to the factored form of (12.1-4).

Example 12.1-2 Sketch the asymptotic phase curve for the gain function of Example 12.1-1

Solution The phase is

$$\theta(\omega) = \left(\tan^{-1}\frac{\omega}{10}\right) - \left(\tan^{-1}\frac{\omega}{50}\right)$$

The curves for the individual factors and their sum are shown in Fig. 12.1-5 (page 438).

Example 12.1-3 Plot the magnitude and phase of A_i as a function of frequency for the amplifier shown in Fig. 12.1-6 (page 438).

Solution The equivalent circuit of Fig. 12.1-7 (page 439) yields the gain A_i. Thus, from (12.1-1a),

$$A_i = \frac{\dot{i}_c}{\dot{i}_i} = \left(\frac{\dot{i}_c}{\dot{i}_e}\right)\left(\frac{\dot{i}_e}{\dot{i}_i}\right) \approx \frac{\dot{i}_e}{\dot{i}_i}$$

$$\approx \left(\frac{600}{8}\right)\left[\frac{j\omega + 1/10^{-3}}{j\omega + 1/(80 \times 10^{-6})}\right]$$

$$\approx 6\left(\frac{1 + j10^{-3}\omega}{1 + j80 \times 10^{-6}\omega}\right)$$

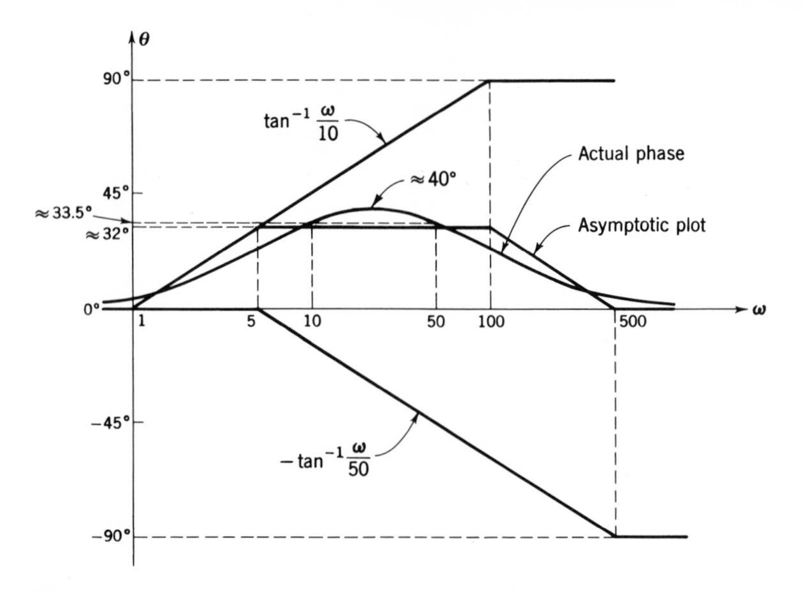

Fig. 12.1-5 Asymptotic and actual phase curves for Example 12.1-2.

Hence

$$|A_i| = 6 \frac{\sqrt{1 + 10^{-6}\omega^2}}{\sqrt{1 + (80^2)(10^{-12}\omega^2)}}$$

and

$$\theta_{A_i} = (\tan^{-1} 10^{-3}\omega) - [\tan^{-1} (80 \times 10^{-6}\omega)]$$

These results are sketched in Fig. 12.1-7*b* and *c*. A simple way of sketching $|A_i|$ is to first plot the low-frequency and high-frequency ($\omega = 0$ and ∞) asymptotes. Mark the zero break

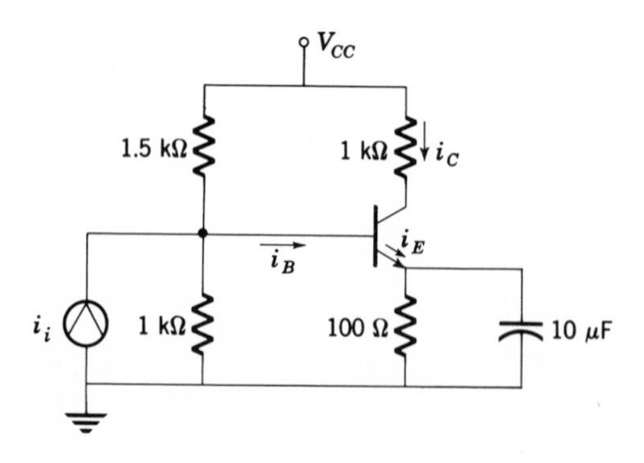

Fig. 12.1-6 Circuit for Example 12.1-3. $A_i = i_c/i_i$; $h_{fe} = 200$; $h_{ie} = 1$ kΩ.

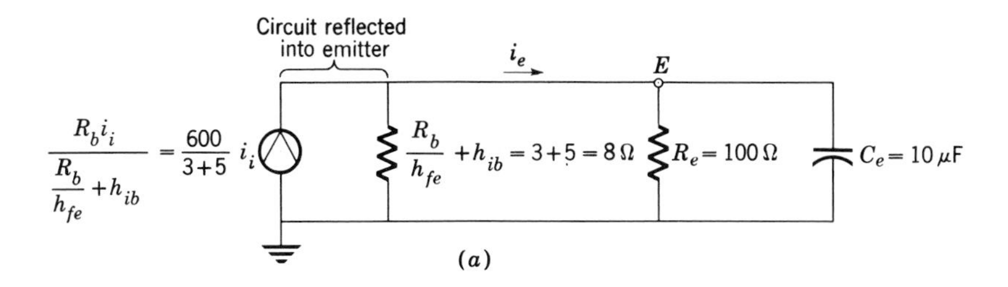

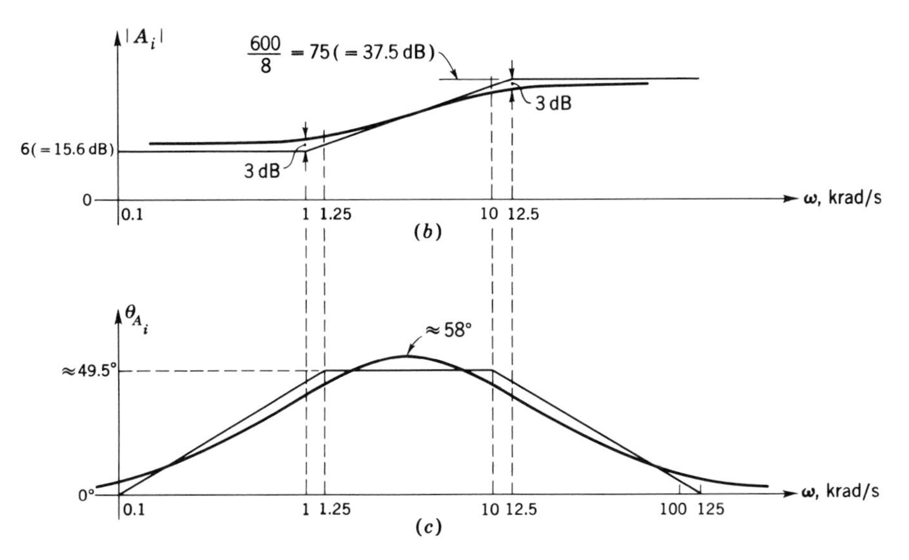

Fig. 12.1-7 (a) Equivalent circuit for Example 12.1-3; (b) gain versus ω; (c) phase versus ω.

($\omega_1 = 1/R_eC_e$) and the pole break ($\omega_2 \approx 1/[(R_b/h_{fe}) + h_{ib}]C_e$). Then connect these two points with a straight line. The phase can be sketched using a similar technique. Mark the phase at 0.1 and 10 times the zero and the pole frequencies. Connect the resulting four points with straight lines.

Figure 12.1-7b shows that the lower 3-dB frequency occurs at $\omega_L = 12.5 \times 10^3$ rad/s, or

$$f_L \approx 2\text{ kHz}$$

For audio applications this is an extremely high frequency at which to end the low-frequency region. If C_e were increased to 1000 μF, f_L would be reduced to 20 Hz, which is a more reasonable value.

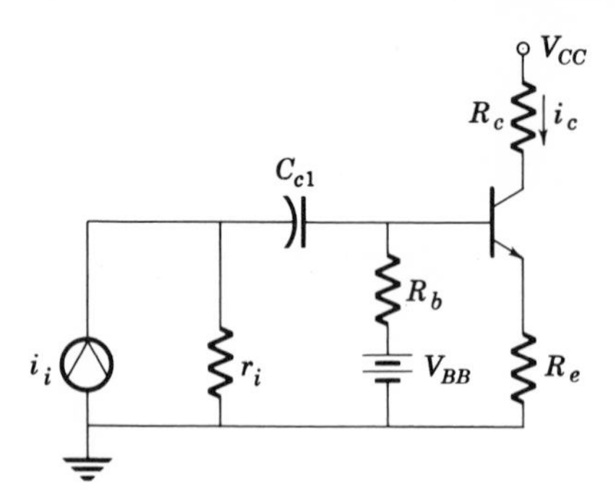

Fig. 12.1-8 CE amplifier with input coupling capacitor.

12.1-3 THE COUPLING CAPACITOR

The capacitor C_{c1} in Fig. 12.1-8 is often needed to couple the ac signal from the source to the base of the transistor, and also to block any dc so that bias conditions are not upset. Let us study the effect of C_{c1} on the circuit response, assuming that the emitter is unbypassed.*

The effect of C_{c1} on the low-frequency response of the amplifier of Fig. 12.1-8 is obtained by reflecting R_e into the base circuit, as shown in Fig. 12.1-9*a*, which is then transformed, using Thévenin's theorem, to the form shown in Fig. 12.1-9*b*.

* The circuit shown in Fig. 12.1-8 also represents the response of an amplifier circuit having a bypassed emitter resistor, in the frequency region where $\omega < 1/R_eC_e$. To investigate the response of this circuit, assuming that C_e represents a perfect bypass capacitor ($C_e \to \infty$), simply let $R_e = 0$.

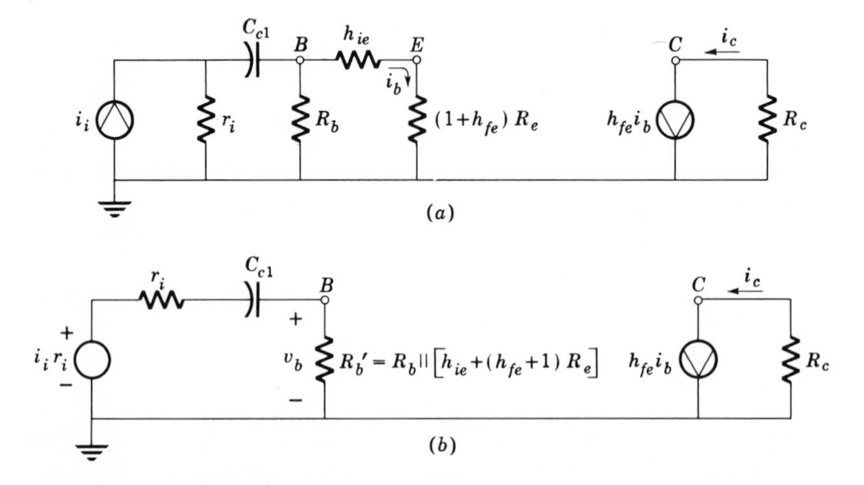

Fig. 12.1-9 (*a*) Small-signal low-frequency equivalent circuit for the amplifier of Fig. 12.1-8; (*b*) reduced equivalent circuit.

The current gain of the amplifier can now be obtained by inspection.

$$A_i = \frac{i_c}{i_i} = \left(\frac{i_c}{i_b}\right)\left(\frac{i_b}{v_b}\right)\left(\frac{v_b}{i_i}\right)$$

$$= h_{fe}\left[\frac{1}{h_{ie} + (1 + h_{fe})R_e}\right]\left(\frac{r_i R_b'}{r_i + R_b' + 1/sC_{c1}}\right) \qquad (12.1\text{-}10a)$$

where

$$R_b' = R_b \| [h_{ie} + (1 + h_{fe})R_e] \qquad (12.1\text{-}10b)$$

This equation can now be simplified to

$$A_i \approx \left(\frac{r_i \| R_b'}{h_{ib} + R_e}\right)\left\{\frac{s}{s + 1/[(r_i + R_b')C_{c1}]}\right\} \qquad (12.1\text{-}11)$$

The 3-dB frequency of this circuit is

$$f_L = \frac{1}{2\pi(r_i + R_b')C_{c1}} \qquad (12.1\text{-}12)$$

A Bode plot of the current-gain magnitude is shown in Fig. 12.1-10.

Example 12.1-4 In the amplifier of Fig. 12.1-8, $h_{ie} = 1\ \text{k}\Omega$, $r_i = 10\ \text{k}\Omega$, $R_b = 1\ \text{k}\Omega$, $R_e = 100\ \Omega$, $C_{c1} = 10\ \mu\text{F}$, and $h_{fe} = 100$.

(*a*) Find the 3-dB frequency f_L of the amplifier.

(*b*) Assume that R_e is perfectly bypassed ($C_e \to \infty$). What is the lower 3-dB frequency f_L now?

Solution (*a*) Using (12.1-12) and (12.1-10*b*),

$$f_L = \frac{1}{2\pi\{10 \times 10^3 + (10^3 \| [10^3 + (100)(100)])\}10^{-5}} \approx 1.6\ \text{Hz}$$

(*b*) Again, using (12.1-12) and (12.1-10*b*) with $R_e = 0$,

$$f_L = \frac{1}{2\pi\{(10 \times 10^3) + (10^3 \| 10^3)\}10^{-5}} \approx 1.6\ \text{Hz}$$

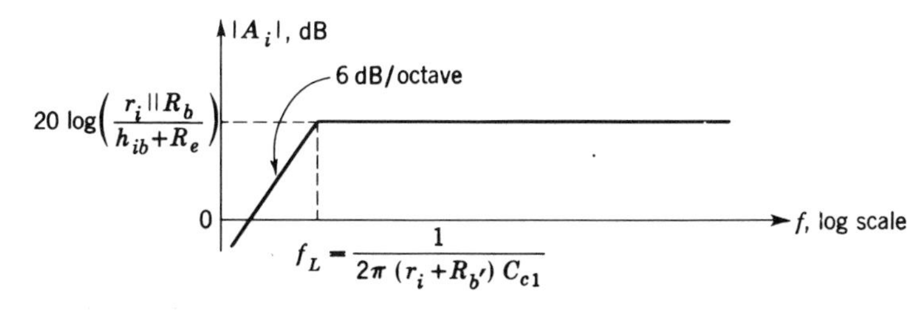

Fig. 12.1-10 Bode plot for CE circuit.

The results are the same since they are both given approximately by the equation

$$f_L \approx \frac{1}{2\pi r_i C_{c1}}$$

when $r_i \gg R'_b$.

12.1-4 THE BASE AND COLLECTOR COUPLING CAPACITORS

If the collector signal in Fig. 12.1-8 is capacitively coupled to a resistive load, the frequency response will depend on both coupling capacitors. Figure 12.1-11*a* shows this configuration, and Fig. 12.1-11*b* shows the resulting equivalent circuit. The gain is

$$A_i = \frac{i_L}{i_i} = \left(\frac{i_L}{i_b}\right)\left(\frac{i_b}{v_b}\right)\left(\frac{v_b}{i_i}\right)$$

$$= -h_{fe}\left(\frac{R_c}{R_c + R_L + \dfrac{1}{sC_{c2}}}\right)\left[\frac{1}{h_{ie} + (1 + h_{fe})R_e}\right]\left(\frac{r_i R'_b}{r_i + R'_b + \dfrac{1}{sC_{c1}}}\right) \qquad (12.1\text{-}13a)$$

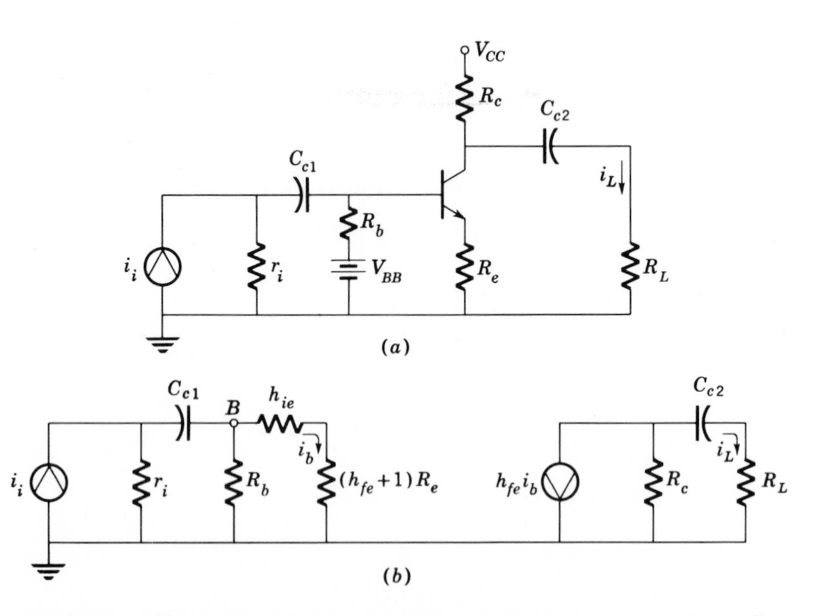

Fig. 12.1-11 CE amplifier with base and collector (input and output) coupling capacitors. (*a*) Circuit; (*b*) equivalent circuit $R'_b = R_b||[h_{ie} + (h_{fe} + 1)R_e]$.

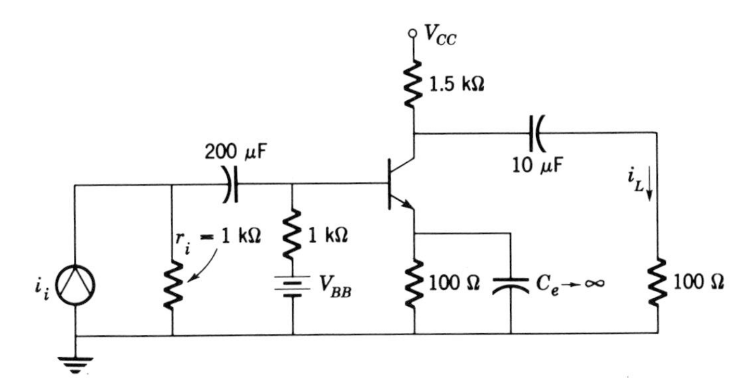

Fig. 12.1-12 Amplifier for Example 12.1-5. $h_{fe} = 100$; $h_{ie} = 1$ kΩ.

which can be simplified to

$$A_i \approx -\left(\frac{R_c}{R_c + R_L}\right)\left(\frac{r_i \| R_b'}{h_{ib} + R_e}\right)\left[\frac{s}{s + \dfrac{1}{(r_i + R_b')C_{c1}}}\right]\left[\frac{s}{s + \dfrac{1}{(R_c + R_L)C_{c2}}}\right] \tag{12.1-13b}$$

Equation (12.1-13*b*) and Fig. 12.1-11*b* indicate that the two coupling circuits do not interact.

Example 12.1-5 Plot the magnitude and phase of the current gain as a function of ω for the amplifier shown in Fig. 12.1-12. The equivalent circuit of this amplifier is shown in Fig. 12.1-13.

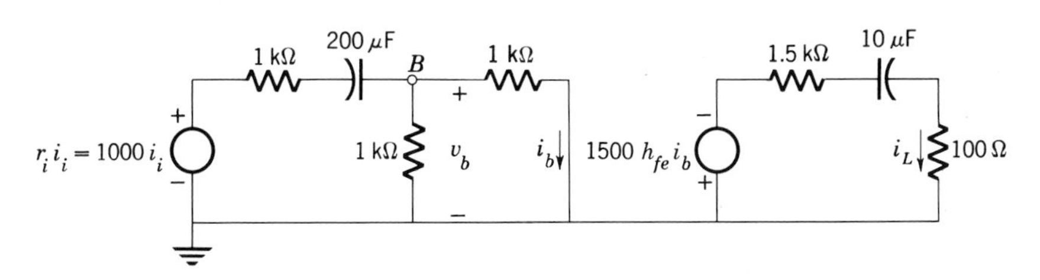

Fig. 12.1-13 Equivalent circuit of amplifier shown in Fig. 12.1-12.

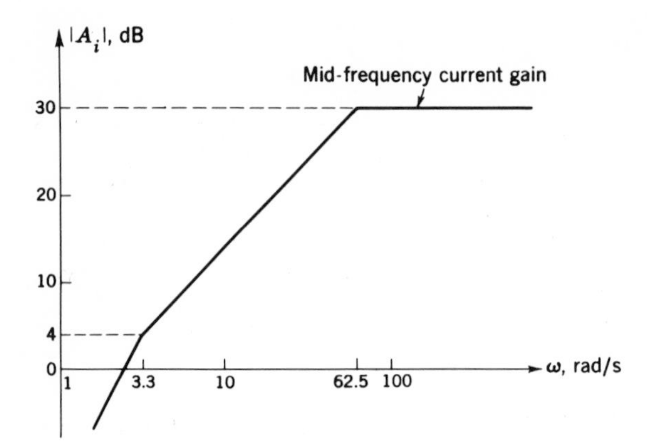

Fig. 12.1-14 Gain vs. frequency for Example 12.1-5.

Solution The gain

$$A_i = \frac{i_L}{i_i} = \left(\frac{i_L}{i_b}\right)\left(\frac{i_b}{i_i}\right)$$

$$= \left(-\frac{1500\, h_{fe}}{1600 + 1/j\omega 10^{-5}}\right)\left[\frac{1000}{1500 + 1/(j\omega 2 \times 10^{-4})}\right]\left(\frac{1}{2}\right)$$

$$= (-75 \times 10^{6})\,\frac{(j\omega 10^{-5})(j\omega 2 \times 10^{-4})}{(1 + j\omega 16 \times 10^{-3})(1 + j\omega 30 \times 10^{-2})}$$

The magnitude of the current gain is plotted in Fig. 12.1-14.

The phase angle can be roughly sketched as a function of ω, by inspection, using the technique outlined in Sec. 12.1-2. We note, for example, that

When $\omega \to \infty$: $\theta \approx -\pi$

When $\omega \to 0$: $\theta \approx 0$

When $\omega = 3.3$ rad/s: $\theta \approx -\dfrac{\pi}{4}$

When $\omega = 62.5$ rad/s: $\theta = -\dfrac{3\pi}{4}$

The plot of θ versus ω is shown in Fig. 12.1-15.

Example 12.1-5 considered the frequency response of an amplifier with two poles widely separated from each other. To achieve this separation, C_{c1} was a very large capacitor, 200 μF. Most of the time we are interested only in the 3-dB frequency. Then we should not use a 200-μF capacitor for C_{c1}, but should use a 20-μF capacitor instead, since it is smaller and less expensive. The resulting poles are now not widely separated, and the 3-dB point is *not* located at the first pole, but actually occurs at a *higher* frequency. For example, consider that the two poles in (12.1-13*b*) coincide. Then

$$A_i = A_{im}\left(\frac{s}{s + \omega_0}\right)^2 \tag{12.1-14a}$$

where

$$A_{im} = \left(\frac{-R_c}{R_c + R_L}\right)\left(\frac{r_i \| R_b'}{h_{ib} + R_e}\right) \tag{12.1-14b}$$

and

$$\omega_0 = \frac{1}{(r_i + R_b')C_{c1}} = \frac{1}{(R_c + R_L)C_{c2}} \tag{12.1-14c}$$

Hence

$$\left|\frac{A_i}{A_{im}}\right|^2 = \frac{\omega^4}{(\omega^2 + \omega_0{}^2)^2} = \frac{1}{\left[1 + \left(\dfrac{\omega_0}{\omega}\right)^2\right]^2} \tag{12.1-15}$$

The 3-dB frequency ω_L is defined in (12.1-3*b*) as the frequency at which $|A_i/A_{im}|^2 = \frac{1}{2}$. Then

$$\left[1 + \left(\frac{\omega_0}{\omega_L}\right)^2\right]^2 = 2 \tag{12.1-16a}$$

and

$$\omega_L = \frac{\omega_0}{\sqrt{0.414}} \approx 1.55\omega_0 \tag{12.1-16b}$$

If the poles do not coincide, the quadratic equation which results when combining the two factors in (12.1-16*a*) yields the 3-dB point.

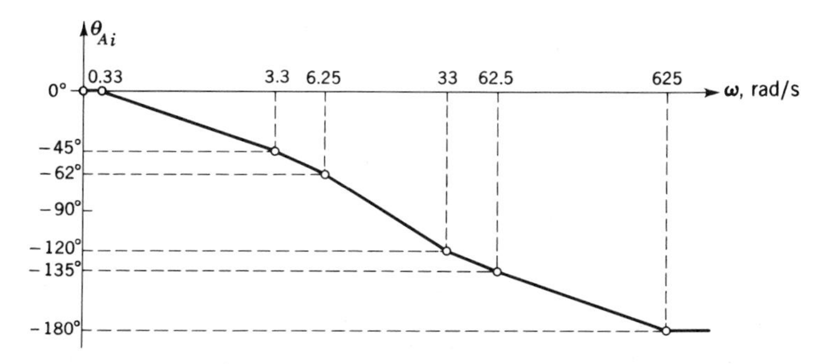

Fig. 12.1-15 Phase angle versus ω for Example 12.1-5.

Thus, referring to (12.1-13*b*),

$$\left|\frac{A_i}{A_{im}}\right|^2 = \frac{\omega_L^4}{(\omega_L^2 + \omega_1^2)(\omega_L^2 + \omega_2^2)} = \frac{1}{2} \tag{12.1-17a}$$

where

$$\omega_1 = \frac{1}{(r_i + R_b')C_{c1}} \tag{12.1-17b}$$

and

$$\omega_2 = \frac{1}{(R_c + R_L)C_{c2}} \tag{12.1-17c}$$

The 3-dB frequency is then found by solving the equation

$$\omega_L^4 + (\omega_1^2 + \omega_2^2)\omega_L^2 + \omega_1^2\omega_2^2 = 2\omega_L^4 \tag{12.1-18a}$$

which yields

$$\omega_L^2 = \frac{\omega_1^2 + \omega_2^2}{2} + \frac{\sqrt{\omega_1^4 + 6\omega_1^2\omega_2^2 + \omega_2^4}}{2} \tag{12.1-18b}$$

12.1-5 COMBINED EFFECT OF BYPASS AND COUPLING CAPACITORS

In most applications, C_{c1}, C_{c2}, and C_e are all present, and the analysis becomes more complicated. Design information is then difficult to obtain unless simplifications are made. Our objective is to find a way to select bypass and coupling capacitors so as to achieve a specified low-frequency response. Often the design specifications will require only that the response be maintained at the midband value down to a certain frequency, ω_L, at which point it may be down 3 dB. Below this break frequency the shape of the response curve is often unimportant, provided the gain continues to decrease with decreasing frequency.

When this is the case, the circuit is designed so that the emitter bypass capacitor C_e determines the specific break frequency. This is done to minimize component size. Equation (12.1-2*b*) shows that the effective resistance, which "works" with C_e to yield the 3-dB frequency, is small ($\approx h_{ib} + R_b/h_{fe}$). Thus, to obtain a low 3-dB frequency, C_e will be large. If the 3-dB frequency is obtained using a different RC combination, then C_e must be *still larger*, which results in increased size and cost.

Coupling capacitors C_{c1} and C_{c2} are then chosen to yield break frequencies well below this point. When the circuit is designed in this manner, the capacitors C_{c1} and C_{c2} are usually much less than C_e.

Example 12.1-6 The amplifier shown in Fig. 12.1-16 is to have a lower 3-dB frequency at 20 Hz. Select C_{c1}, C_{c2}, and C_e to meet this specification.

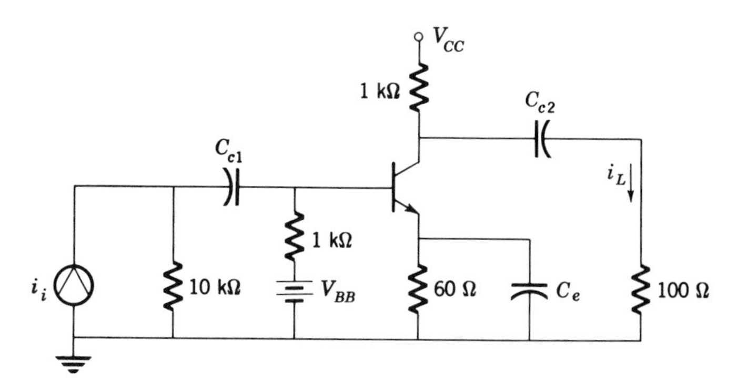

Fig. 12.1-16 Circuit for Example 12.1-6. $h_{fe} = 100$; $h_{ie} = 1$ kΩ.

Solution We begin by selecting C_e to achieve the required 20-Hz 3-dB frequency. From (12.1-2*b*)

$$f_L = 20 = \left(\frac{1}{2\pi}\right)\left\{\frac{1}{C_e[60\|(10+10)]}\right\}$$

$C_e \approx 530\ \mu\text{F}$ (we use a 500- or 1000-μF standard capacitor)

The coupling capacitors C_{c1} and C_{c2} are selected so that the break frequencies occur well below 20 Hz. We see from Fig. 12.1-2 that to be well below the 3-dB frequency means that the break frequencies should occur more than an octave lower. As a practical matter, a decade below the 3-dB frequency represents more than adequate separation. Thus, in this example, the break frequencies due to C_{c1} and C_{c2} are each chosen to be 2 Hz.

Then, from (12.1-17*b*),

$$C_{c1} \approx \frac{1}{2\pi(2)(10^4)} = 8\ \mu\text{F} \qquad \text{(we use a 10-}\mu\text{F capacitor)}$$

and from (12.1-17*c*),

$$C_{c2} \approx \frac{1}{2\pi(2)(10^3)} = 80\ \mu\text{F} \qquad \text{(we use a 100-}\mu\text{F capacitor)}$$

12.2 THE LOW-FREQUENCY RESPONSE OF THE FET AMPLIFIER

In this section we show that the low-frequency response of the FET amplifier is determined primarily by the source bypass capacitor because of the relative magnitudes of the circuit resistances.

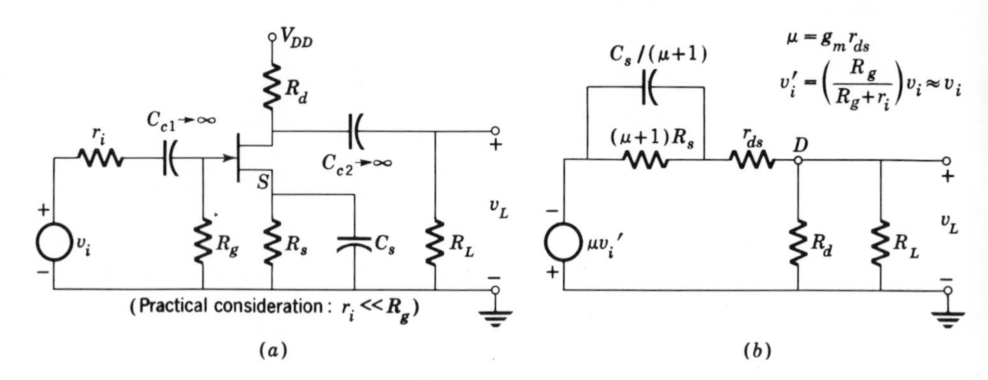

Fig. 12.2-1 (a) FET amplifier circuit; (b) small-signal equivalent circuit.

Here, as in Sec. 12.1, rather than present a general analysis, we consider the effect of each of these capacitors acting separately, and then outline a design procedure.

12.2-1 THE SOURCE BYPASS CAPACITOR

Consider the FET amplifier shown in Fig. 12.2-1a. The coupling capacitors are assumed infinite, so that the 3-dB frequency is caused by the source bypass capacitor. It is shown below that even when the coupling capacitors are considerably smaller than the bypass capacitor, it is the source bypass capacitor that determines the break frequency f_L.

The equivalent circuit of the amplifier is shown in Fig. 12.2-1b. In this figure, the source circuit has been reflected into the drain circuit. The voltage gain is found to be

$$A_v' = \frac{v_L}{v_i'} = -\frac{\mu(R_d \| R_L)}{(R_d \| R_L) + r_{ds} + (\mu + 1)R_s/(1 + sR_sC_s)} \qquad (12.2\text{-}1)$$

This can be put in the form

$$A_v' = \frac{v_L}{v_i'} = -\left[\frac{\mu(R_d \| R_L)}{(R_d \| R_L) + r_{ds}}\right] \left[\frac{s + 1/R_sC_s}{s + \dfrac{1}{C_s\left\{\dfrac{R_s[(R_d \| R_L) + r_{ds}]}{(\mu + 1)R_s + r_{ds} + (R_d \| R_L)}\right\}}}\right] \qquad (12.2\text{-}2)$$

We see that the form of (12.2-2) is similar to the form of (12.1-1). Substituting $\mu = g_m r_{ds}$ into (12.2-2) yields the final expression for the voltage gain:

$$A_v = \frac{v_L}{v_i} \approx \frac{v_L}{v_i'} = -g_m R_{\|} \frac{s + 1/R_s C_s}{s + \dfrac{1}{C_s \left\{ \dfrac{R_s[(R_d \| R_L) + r_{ds}]}{(\mu + 1)R_s + r_{ds} + (R_d \| R_L)} \right\}}} \qquad (12.2\text{-}3)$$

where

$$R_{\|} = r_{ds} \| R_d \| R_L$$

To find the midband gain, let $s = j\omega \to j\infty$. Then

$$A_{vm} = \left. \frac{v_L}{v_i} \right|_{\omega \to \infty} = -g_m R_{\|} \frac{R_g}{r_i + R_g} \approx -g_m R_{\|} \qquad r_i \ll R_g \qquad (12.2\text{-}4)$$

This result can be obtained by inspection of Fig. 12.2-1, and it is the midfrequency voltage gain obtained in Sec. 10.5-1.

The angular frequency at which the voltage gain is 3 dB below the midfrequency value is found as in (12.1-3). Assuming $\omega_L \gg 1/R_s C_s$,

$$\omega_L = \frac{1}{C_s R_s \left[\dfrac{r_{ds} + (R_d \| R_L)}{(\mu + 1)R_s + r_{ds} + (R_d \| R_L)} \right]} \qquad (12.2\text{-}5)$$

Example 12.2-1 Find C_s to obtain a 10-Hz break frequency, when

$$R_s = 1 \text{ k}\Omega \quad r_{ds} = 10 \text{ k}\Omega \quad R_d = 5 \text{ k}\Omega \quad R_L = 100 \text{ k}\Omega$$

and

$$g_m = 3 \times 10^{-3} \text{ mho}$$

Also calculate the midfrequency gain.

Solution From (12.2-5)

$$C_s = \frac{1}{2\pi(10)(10^3)\left(\dfrac{10 \times 10^3 + 5 \times 10^3}{30 \times 10^3 + 10 \times 10^3 + 5 \times 10^3}\right)} = 47.8 \ \mu\text{F}$$

We should use a standard 56-μF capacitor. Note that R_L does not influence the calculation since it is large relative to R_d. Also note that $(\frac{1}{2}\pi)R_s C_s \approx 3.3$ Hz; hence the true 3-dB frequency is 8.8 Hz [Eq. (12.1-3*d*)].

The midfrequency voltage gain is, from (12.2-4),

$$A_v \approx (-3 \times 10^{-3})[(10 \times 10^3) \| (5 \times 10^3)] = -10$$

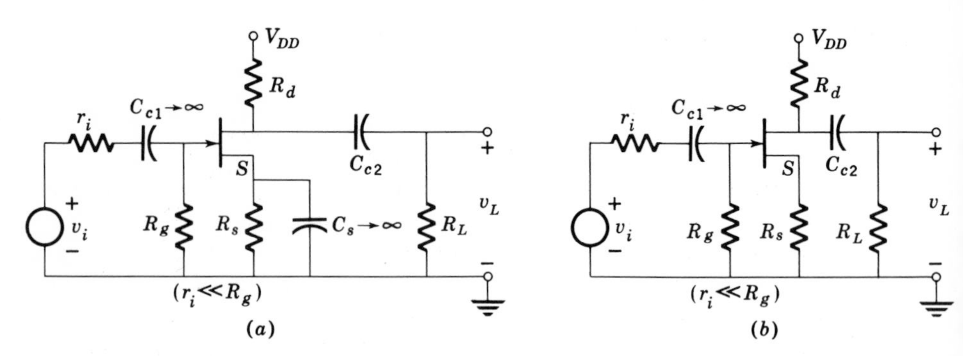

Fig. 12.2-2 The FET amplifier. (*a*) Effect of drain coupling capacitor; (*b*) circuit at very low frequencies.

12.2-2 THE DRAIN COUPLING CAPACITOR

If C_{c2} is sufficiently small, it can influence the 3-dB frequency. Consider the FET amplifier with an infinite bypass capacitor, shown in Fig. 12.2-2*a*. The voltage gain of this amplifier is $A_{v(a)}$.

$$A_{v(a)} \approx -\frac{g_m R_L (r_{ds}\|R_d)}{R_L + (r_{ds}\|R_d) + 1/sC_{c2}} \tag{12.2-6a}$$

Simplifying, we have

$$A_{v(a)} \approx -\frac{(g_m R_\|)s}{s + 1/\{C_{c2}[R_L + (r_{ds}\|R_d)]\}} \tag{12.2-6b}$$

or approximately,

$$A_{v(a)} \approx -\frac{s[g_m(R_d\|r_{ds})]}{s + 1/C_{c2}R_L} = -\frac{s[\mu R_d/(R_d + r_{ds})]}{s + 1/C_{c2}R_L} \tag{12.2-6c}$$

Referring to Example 12.2-1, we note that R_L is much larger than R_d and r_{ds}. This is usually the case in practice, since large voltage gain is usually desired. In addition, R_L is often the gate resistance of the following FET. R_L can then be as large as several megohms, while R_d and r_{ds} are of the order of several kilohms.

Example 12.2-2 Calculate the 3-dB frequency of the amplifier of Fig. 12.2-2*a* when $r_{ds} = 10$ kΩ, $R_d = 5$ kΩ, $R_L = 100$ kΩ, $C_{c2} = 10$ μF, and $g_m = 3 \times 10^{-3}$ mho.

Solution Using (12.2-6*b*), the 3-dB frequency is

$$f_L = \frac{1}{2\pi(10^{-5})(10^5 + 3.3 \times 10^3)} \approx 0.16 \text{ Hz}$$

It is apparent from this example that even relatively small values of C_{c2} result in a very low break frequency. To achieve this same result in the source circuit requires an extremely large value of C_s. Thus the value of C_s usually determines the break frequency.

If the source resistor is unbypassed, or if the amplifier response in the very low frequency region is desired, the FET configuration is that shown in Fig. 12.2-2*b*. If the source resistance is reflected into the drain circuit, the drain-source impedance is effectively increased by $(\mu + 1)R_s$. Let us define the gain under these conditions as $A_{v(b)}$, where

$$A_{v(b)} \approx -\left[\frac{-\mu R_d R_L}{R_d + r_{ds} + (\mu + 1)R_s}\right] \left\{\frac{1}{R_L + [R_d \| (r_{ds} + (\mu + 1)R_s)] + \dfrac{1}{sC_{c2}}}\right\} \qquad (12.2\text{-}7a)$$

Assuming $R_L \gg R_d$, (12.2-7*a*) reduces to

$$A_{v(b)} \approx -\left[\frac{\mu R_d}{R_d + r_{ds} + (\mu + 1)R_s}\right]\left(\frac{s}{s + 1/C_{c2}R_L}\right) \qquad (12.2\text{-}7b)$$

Comparing this expression with (12.2-6*c*), we see that the break frequency is unchanged. The midfrequency gain, as expected, is reduced.

12.2-3 THE GATE COUPLING CAPACITOR

Figure 12.2-3 shows a FET amplifier with both coupling capacitors and the bypass capacitors included. The approximate overall gain of this

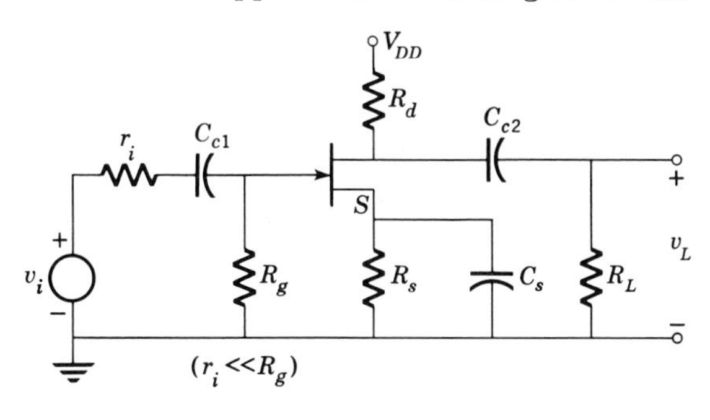

Fig. 12.2-3 The FET amplifier.

amplifier is simply the product of the gain calculated in (12.2-3) (which assumes that the break frequency due to C_{c2} is well below the break point) and the factor

$$\frac{s}{s + 1/[C_{c1}(r_i + R_g)]} \approx \frac{s}{s + 1/C_{c1}R_g} \tag{12.2-8}$$

Since R_g is comparable with R_L (in multiple-stage FET amplifiers the R_g of the second stage is the R_L of the first stage), the break frequency due to C_{c1} is approximately the same as the break frequency caused by C_{c2}. Hence C_{c1} and C_{c2} do not usually influence the 3-dB frequency, C_s being the determining factor.

Example 12.2-3 Determine C_s, C_{c1}, and C_{c2} for the FET amplifier shown in Fig. 12.2-4. The break frequency is to be set at 10 Hz.

Solution C_s is chosen to fix the 10-Hz break frequency. From (12.2-5) we find that

$$C_s = \frac{1}{2\pi(10)(500)\left[\dfrac{5 \times 10^4 + 5 \times 10^3}{(250 + 1)(500) + 5 \times 10^4 + 5 \times 10^3}\right]} = 110\ \mu\text{F}$$

We choose $C_s = 120\ \mu$F. The break frequencies due to C_{c1} and C_{c2} are approximately the same.

$$\omega_{C_{c1}} \approx \omega_{C_{c2}} \approx \frac{1}{C_{c1}(10^5)}$$

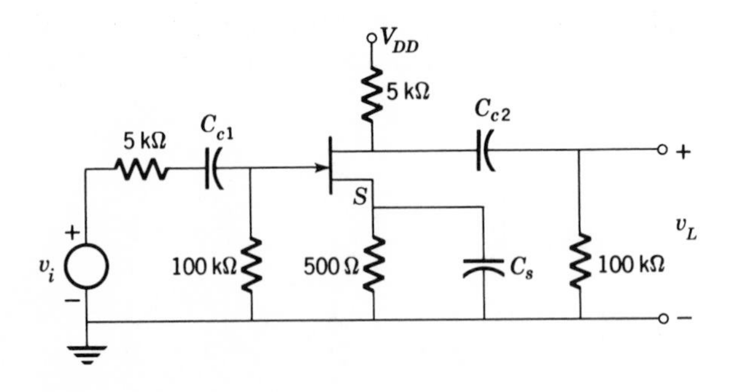

Fig. 12.2-4 Amplifier for Example 12.2-3. $R_{ds} = 50\ \text{k}\Omega$; $g_m = 5 \times 10^{-3}$ mho.

Choosing $C_{c1} = C_{c2} = 2\ \mu\text{F}$ yields break frequencies at 0.8 Hz for each of the coupling capacitors.

It should be noted that C_{c1} and C_{c2} could have been reduced further until they influenced the 3-dB frequency determined by C_s. C_s would then have to be increased above 120 μF to maintain the 10-Hz 3-dB frequency. The resulting increase in size and cost would more than offset the size and cost resulting from the decrease in C_{c1} and C_{c2}.

12.3 THE LOW-FREQUENCY RESPONSE OF THE VACUUM TUBE

The vacuum tube and the FET have identical equivalent circuits (Chaps. 10 and 11). Therefore we expect the equations for the break frequencies to be similar. This is easily verified. The difference between the two devices lies in the relative magnitudes of the load and drain resistors, and in the case of a pentode, between the plate and drain-source resistor. For example, both the FET and triode might have a quiescent operating current of 2 mA. The FET, however, might use a supply voltage of 20 V and have a drain-source voltage of 10 V, resulting in a drain resistance of 5 kΩ. The vacuum tube, on the other hand, may have a 150-V drop across R_p, which results in $R_p = 75$ kΩ. Thus the value of R_p often lies between 10 and 100 times greater than the R_d of a comparable FET amplifier. The other resistors are usually comparable in size.

Example 12.3-1 Consider the triode amplifier shown in Fig. 12.3-1. If a 3-dB frequency of $f_L = 10$ Hz is required, we set the break frequency of C_k at 10 Hz and the break frequency of C_{c1} and C_{c2} at

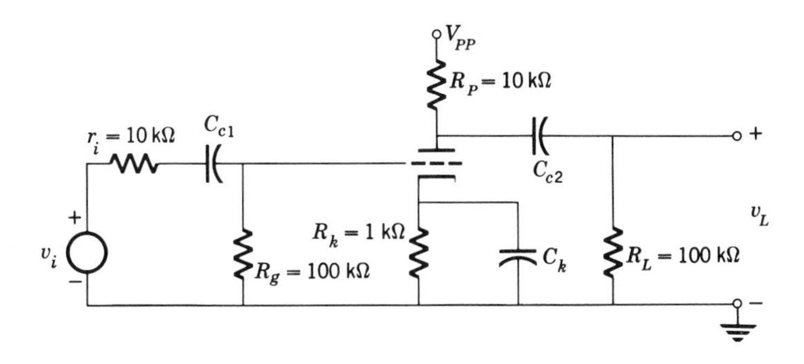

Fig. 12.3-1 Triode amplifier at low frequencies. $r_p = 5$ kΩ; $\mu = 50$.

1 Hz. Then, using (12.2-5),

$$\omega_L = \frac{1}{C_k R_k \left[\dfrac{r_p + (R_p \| R_L)}{(\mu + 1)R_k + r_p + (R_p \| R_L)} \right]} \tag{12.3-1}$$

Therefore

$$C_k = \frac{1}{2\pi(10)(10^3)\left(\dfrac{5 \times 10^3 + 10 \times 10^3}{50 \times 10^3 + 5 \times 10^3 + 10 \times 10^3}\right)} \approx 70\ \mu\text{F}$$

Note that this is comparable with the value of C_s found in Example 12.2-3.

The coupling capacitors are found using (12.2-7*b*) and (12.2-8).

$$\omega_{C_{c2}} \approx \frac{1}{C_{c2} R_L} \tag{12.3-2}$$

$$\omega_{C_{c1}} \approx \frac{1}{C_{c1}(r_i + R_g)} \tag{12.3-3}$$

In this example

$$C_{c1} \approx C_{c2} \approx \frac{1}{2\pi(1)(10^5)} = 1.6\ \mu\text{F}$$

We should use 2-μF capacitors for C_{c1} and C_{c2}. Note the similarity between these results and those that were obtained in Example 12.2-3.

Example 12.3-2 A pentode with $r_p = 500$ kΩ and $g_m = 5 \times 10^{-3}$ mho is used instead of the triode in Fig. 12.3-1. Determine C_k, C_{c1}, and C_{c2} to achieve a 3-dB frequency of 10 Hz.

Solution From (12.3-1)

$$\omega_L = \frac{1}{C_k R_k \left[\dfrac{r_p + (R_p \| R_L)}{(\mu + 1)R_k + r_p + (R_p \| R_L)} \right]}$$

Since $r_p \gg R_p$, (12.3-1) can be simplified.

$$\omega_L \approx \frac{1}{C_k[R_k/(1 + g_m R_k)]}$$

Thus, for this example,

$$C_k = \frac{1}{2\pi(10)(10^3/6)} = 95\ \mu\text{F}$$

Using (12.3-2) and (12.3-3), with $\omega = 2\pi(1)$,

$$C_{c1} \approx \frac{1}{\omega R_g} = \frac{1}{2\pi(1)(10^5)} \approx 1.6\ \mu\text{F}$$

$$C_{c2} \approx \frac{1}{\omega R_L} \approx 1.6\ \mu\text{F}$$

This example shows that the pentode, FET, and triode, all require comparable coupling and bypass capacitors.

PROBLEMS

SECTION 12.1

12.1. (*a*) Draw the small-signal equivalent circuit.

(*b*) Find the transfer function $A_i = i_L(s)/i_i(s)$. *Hint:* Reflect the emitter circuit into the base.

(*c*) Sketch the asymptotic magnitude plot for A_i.

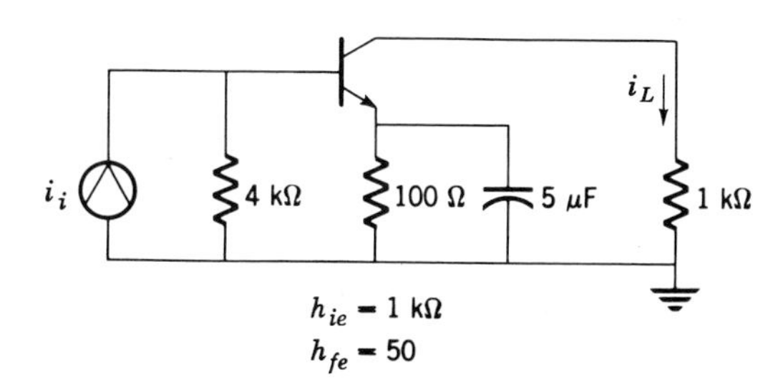

Fig. P12.1

12.2. (*a*) Find R_1 and R_2 for maximum symmetrical swing (recall that, for good dc stability, $h_{ie} < R_b < h_{fe}R_e$). Assume $h_{fe} = 50$.

(*b*) Determine C_e so that the lower 3-dB frequency is at $\omega = 10$ rad/s.

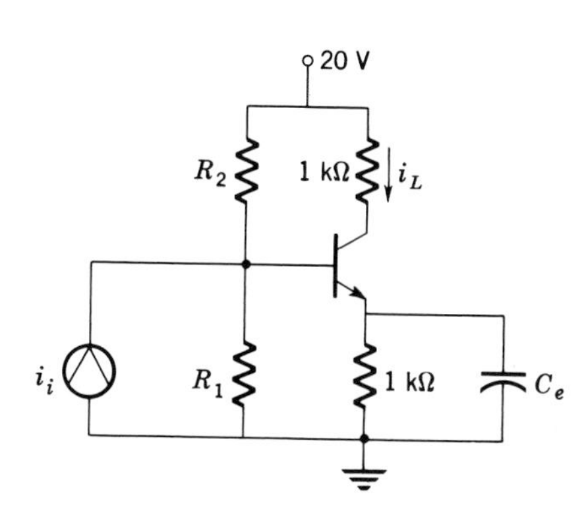

Fig. P12.2

12.3. Plot the asymptotic magnitude and phase for the transfer function

$$A = 10^4 \left[\frac{(s + 10)(s + 300)(s + 4000)}{(s + 2)(s + 12)(s + 2000)} \right]$$

Use semilog paper and plot asymptotes of the gain in decibels versus ω in radians per second and the phase θ in degrees versus ω in radians per second. On the same sheet plot the actual characteristics.

12.4. Find C_{c1} so that the lower 3-dB frequency is at $f = 10$ Hz. Plot v_L/i_i (asymptotic plot).

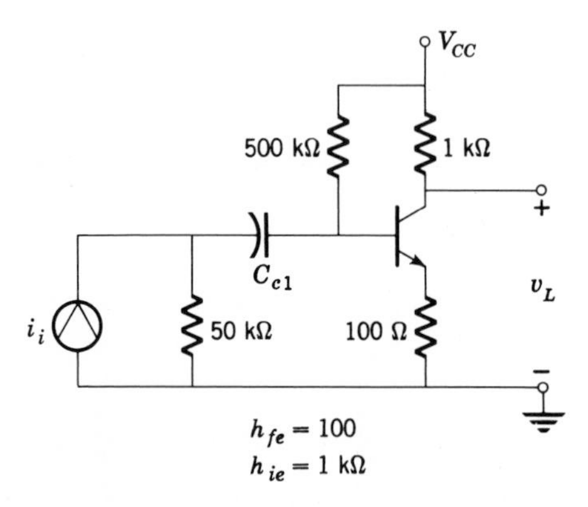

Fig. P12.4

12.5. Find the transfer function v_L/i_i. Plot the asymptotic magnitude and phase.

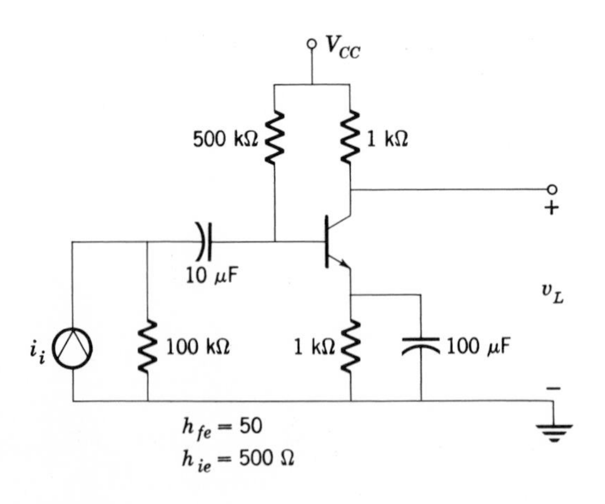

Fig. P12.5

12.6. (*a*) Find $A_i = i_L/i_i$.

(*b*) Plot the asymptotic and actual magnitude of the gain and the phase of A_i.

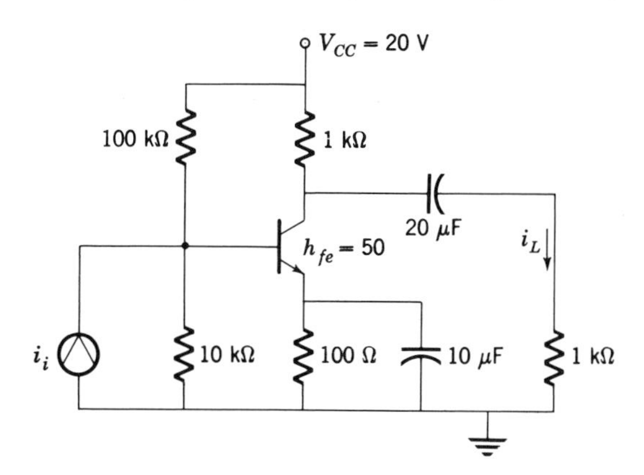

Fig. P12.6

12.7. Find C_{c1} so that the lower 3-dB frequency is at $\omega = 5$ rad/s.

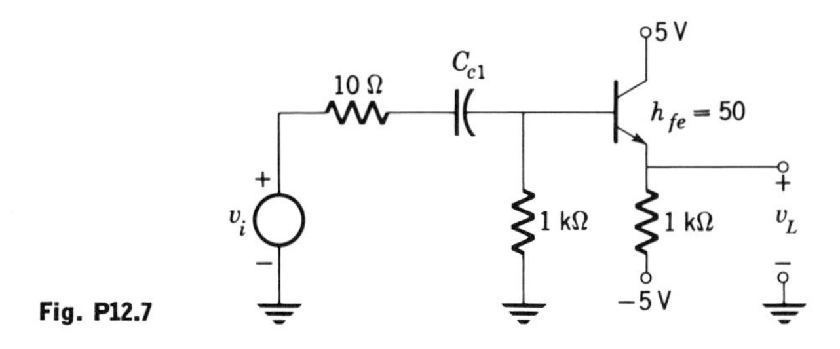

Fig. P12.7

12.8. (*a*) Find C_{c1} and C_{c2} to obtain a double pole, in $A_v = v_L/v_i$, at 5 rad/s. What is the lower 3-dB frequency?

(*b*) Find new values of $C_{c1} = C_{c2}$ so that the lower 3-dB frequency is at 5 rad/s.

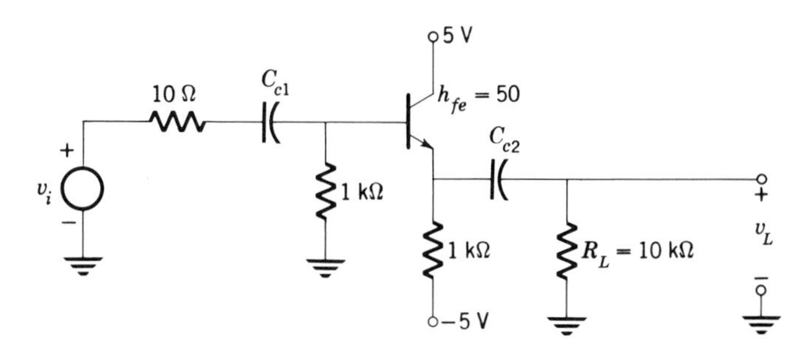

Fig. P12.8

12.9. Repeat Prob. 12.8*a*, if $R_L = 1$ kΩ instead of 10 kΩ.

12.10. In Fig. P12.8, $R_L = 1$ kΩ, $C_{c1} = 100\ \mu$F, and $C_{c2} = 10\ \mu$F.
(*a*) Find the lower 3-dB frequency.
(*b*) Find and plot $A_v = v_L/v_i$.

12.11. (*a*) Find C_e and C_{c2} so that A_i has a double pole at 10 rad/s.
(*b*) Find C_e and C_{c2} so that the lower 3-dB frequency is at 10 rad/s.

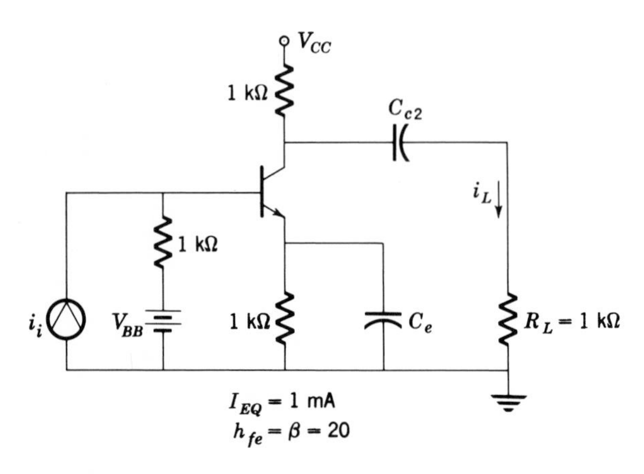

Fig. P12.11

12.12. Plot A_i (magnitude and phase), and find the lower 3-dB frequency.

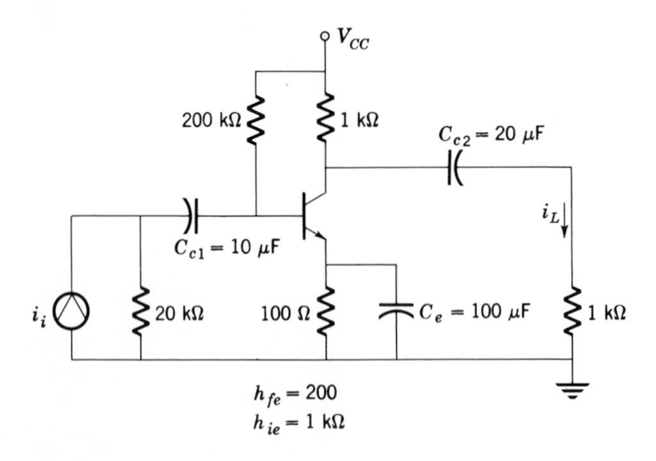

Fig. P12.12

12.13. Find and plot $A_i = i_L/i_i$. Find the lower 3-dB frequency.

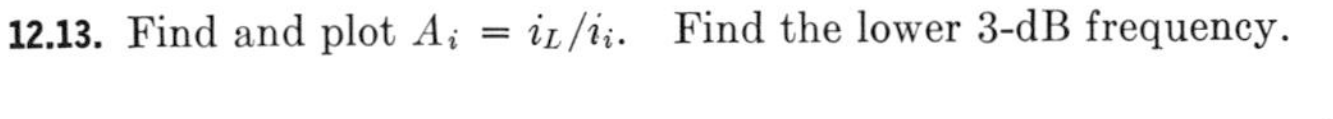

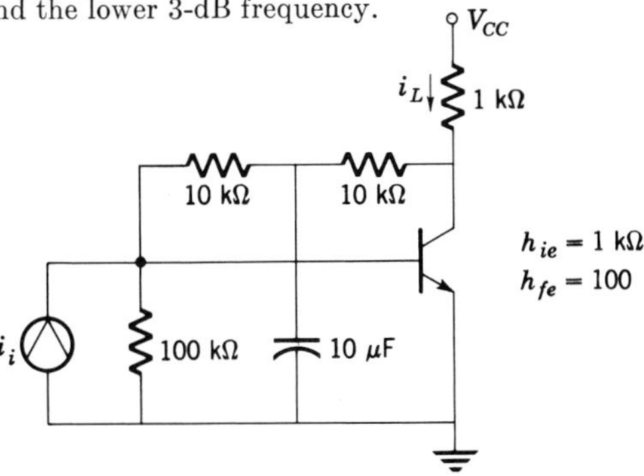

Fig. P12.13

12.14. For the circuit of Fig. P12.12 use the technique of Sec. 12.1-5 to find new values of C_{c1}, C_{c2}, and C_e so that $f_L \leq 10$ Hz.

12.15. Find the steady-state response of the circuit of Fig. P12.4 to the waveform of Fig. P12.15. Comment on the effect of the low-frequency behavior of the amplifier on the response of the amplifier to the square-wave signal.

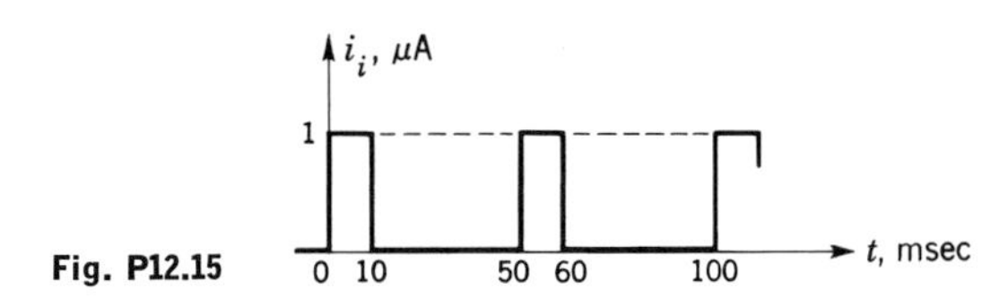

Fig. P12.15

12.16. Find $A_i = i_L/i_i$. Comment on the effect of the split emitter resistance.

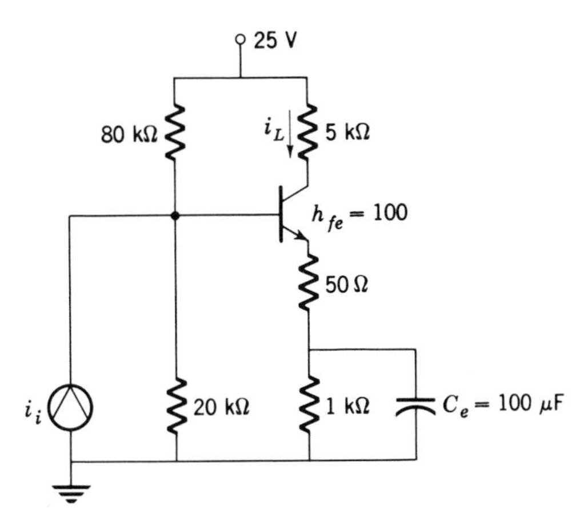

Fig. P12.16

12.17. (*a*) Find $A_v = v_e/v_i$.
(*b*) Select C_c so that the lower 3-dB frequency is 5 Hz.

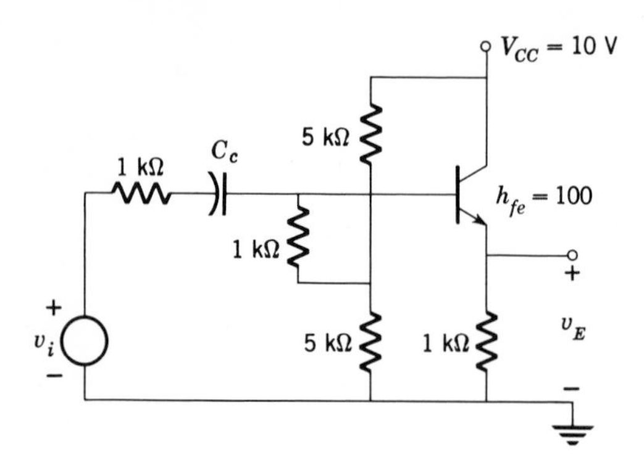

Fig. P12.17

SECTION 12.2

12.18. Draw the small-signal equivalent circuit for Fig. 12.2-1*a*, and verify (12.2-1) and (12.2-2).

12.19. Draw the small-signal equivalent circuit for Fig. 12.2-2*b*, and verify equation (12.2-7*a*).

12.20. Using the 2N3796 IGFET (Sec. 10.6), design a single-stage common-source amplifier to have a midband voltage gain of at least 14 dB. The signal-source impedance is 1 kΩ, and it is capacitively coupled to the gate. The load resistance is 20 kΩ, and is ac-coupled. The lower 3-dB frequency is to be less than 5 Hz.

12.21. (*a*) Find $A_v = v_L/v_i$.
(*b*) Find A_v if the source bypass capacitor is connected across both 250-Ω resistors. Compare with (*a*).

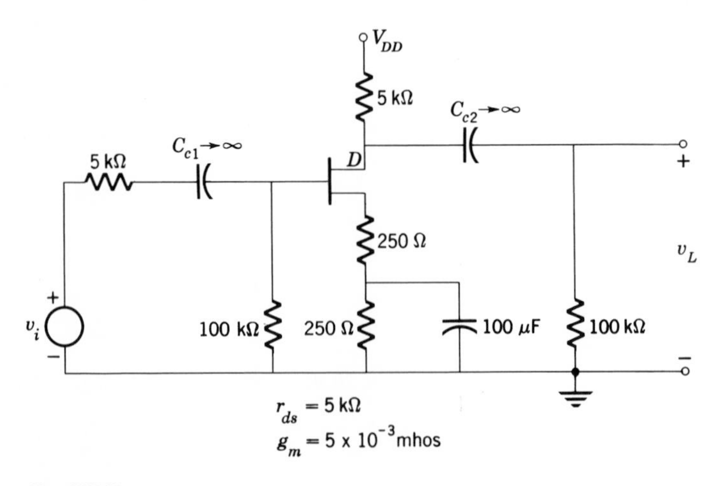

Fig. P12.21

SECTION 12.3

12.22. Draw the small-signal equivalent circuit of the triode amplifier of Fig. 12.3-1, reflecting the cathode circuit into the plate circuit. Verify Eqs. (12.3-1) to (12.3-3).

12.23. Design a two-stage triode amplifier as shown in Fig. P12.23. The 12AT7 (Appendix III) is to be used. The midband voltage gain is to be 60 dB, and $f_L <$ 5 Hz. Specify V_{PP}, all resistors, and all capacitors.

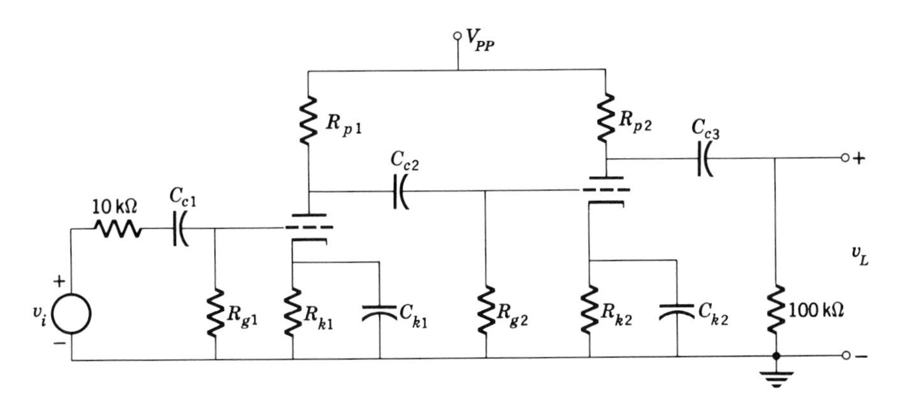

Fig. P12.23

12.24. Repeat Prob. 12.23, except that the tube to be used is the 6CL6 pentode.

12.25. In the pentode amplifier of Fig. 11.2-4, evaluate the effect of C_{g2k} on the low-frequency response, assuming that C_k, C_{c1}, and C_{c2} are infinite.

13 High-frequency Response of *RC*-coupled Amplifiers[1–3]

13.1 THE TRANSISTOR AMPLIFIER AT HIGH FREQUENCIES

We have seen that the low-frequency behavior of the transistor circuit is determined by the *external* capacitors used for coupling and for emitter bypass. The upper limit on the high-frequency response of the device is limited by *internal* capacitance.

Consider the *pnp* transistor shown in Fig. 13.1-1*a*. The collector-base junction appears as a reverse-biased diode, as shown in Fig. 13.1-1*b*. When the collector is biased negatively, with respect to the base, the holes in the base region move into the collector region, and the electrons in the collector region move into the base region. The electrons in the base move away from the base-collector junction, and the holes in the collector move away from base-collector junction, thereby forming a depletion region. The effective length ℓ of the depletion region becomes larger as the reverse voltage increases. Since the electrons and holes have moved away from the junction, the base depletion region becomes positively charged, and the collector depletion region becomes negatively charged

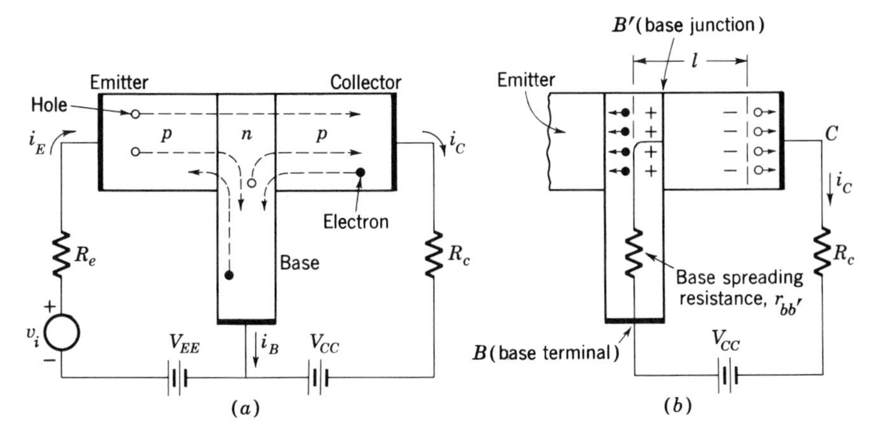

Fig. 13.1-1 Collector-base capacitance. (*a*) *pnp* transistor; (*b*) collector-base region.

(Fig. 13.1-1*b*). The junction therefore behaves like a capacitor, whose capacitance theoretically varies inversely with V_{CB}. Actually, the collector-base junction capacitance $C_{b'c}$ is inversely proportional to either the $\frac{1}{2}$ or $\frac{1}{3}$ power of V_{CB}, depending on whether the transistor is an alloy transistor or a grown-junction transistor. This capacitance is rather small, and varies from about 30 pF in low-frequency transistors to less than 1 pF in high-frequency transistors.

In addition, a diffusion capacitance exists across the base-emitter junction. The diffusion capacitance is a result of the time delay that occurs when a hole moves from emitter to collector by "diffusing" across the base. Consider a hole moving across the base because of a command from the signal voltage v_i. If, before the hole crosses the base region, the voltage reverses polarity, the hole will try to return to the emitter, and the collector will not record this change in signal current. Thus the time required to cross the base must be small compared with the period of the signal. As the signal frequency increases, the collector current decreases, because some charges are "trapped" in the base. Since the current decreases with increasing frequency, the effect is accounted for by a lumped capacitor $C_{b'e}$. Since this capacitance depends on the number of charges in the region, it increases almost linearly with the quiescent emitter current I_{EQ}. $C_{b'e}$ is usually much larger than $C_{b'c}$. Typical values lie in the range of 100 to 5000 pF.

13.1-1 THE HYBRID–PI EQUIVALENT CIRCUIT

The most useful high-frequency model of the transistor is called the *hybrid pi*, and is shown in Fig. 13.1-2. This circuit represents a refinement

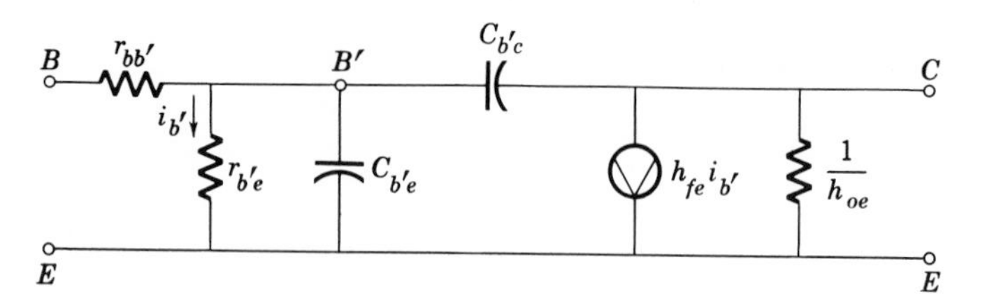

Fig. 13.1-2 Hybrid-pi model of a transistor.

of the common-emitter hybrid equivalent circuit of Fig. 6.2-3*a*. In this circuit, the symbol B' represents the base *junction*, and B represents the base *terminal*. Between these two we have the base ohmic resistance $r_{bb'}$, which is usually considered as a constant in the range 10 to 50 Ω. This resistance is directly proportional to the base width. High-frequency transistors have smaller base widths, and thus a smaller $r_{bb'}$, than low-frequency transistors. The resistance $r_{b'e}$ is the base-emitter junction resistance ($\approx 0.025h_{fe}/I_{EQ}$ at room temperature) and is usually much larger than $r_{bb'}$. The reader will recognize that $r_{b'e}$ is equivalent to the h_{ie} that we have been using as the total base-emitter resistance up to now. Thus a more accurate expression for h_{ie} is

$$h_{ie} = r_{bb'} + r_{b'e} \approx r_{bb'} + \frac{0.025h_{fe}}{I_{EQ}} \qquad T = 300°\text{K} \tag{13.1-1}$$

The approximation $h_{ie} \approx r_{b'e}$ is usually valid.

The output impedance $1/h_{oe}$ can often be neglected at high frequencies, since it is usually much larger than the impedance of the external load, R_L.

Cutoff frequency The capacitors in the equivalent circuit were explained qualitatively at the beginning of this section. Their effect at high frequencies is often given in terms of the β cutoff frequency f_β, which is defined as follows: Let $v_{ce} = 0$; then the hybrid-pi model shown in Fig. 13.1-2 reduces to the form shown in Fig. 13.1-3. From this circuit, we see that the short-circuit current gain $i_c/i_i\big|_{v_{ce}=0}$ will be down 3 dB at a frequency

$$f_\beta = \frac{1}{2\pi r_{b'e}(C_{b'e} + C_{b'c})} \approx \frac{1}{2\pi r_{b'e}C_{b'e}} \tag{13.1-2}$$

Thus f_β is the common-emitter short-circuit 3-dB frequency.

The upper-frequency limit of a transistor is sometimes defined in terms of the frequency f_T, at which the common-emitter current gain is

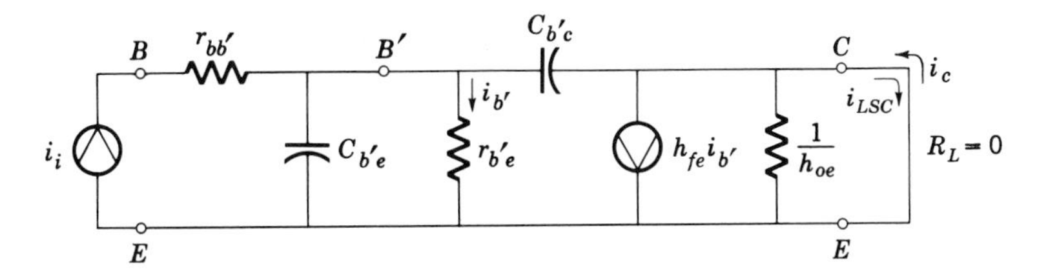

Fig. 13.1-3 The hybrid-pi model used to calculate f_β.

unity. The short-circuit current gain of the ideal amplifier shown in Fig. 13.1-3 is

$$\frac{i_c}{i_i} = -\frac{h_{fe}}{1 + j\omega/\omega_\beta} \tag{13.1-3}$$

This gain becomes unity when

$$f = f_T = f_\beta\sqrt{h_{fe}^2 - 1} \approx f_\beta h_{fe} \tag{13.1-4}$$

The frequency f_T is often called the gain-bandwidth product of the amplifier, although, strictly speaking, this is not correct. This is considered in Sec. 13.5.

If we consider the common-base configuration at midfrequency and then include the base-emitter capacitance and the collector-base capacitance, a high-frequency common-base model similar to that shown in Fig. 13.1-3 is obtained. This model is shown in Fig. 13.1-4. The short-circuit current gain is

$$A_i = \left.\frac{i_{sc}}{i_i}\right|_{v_{cb}=0} \approx \frac{h_{fb}}{1 + j\omega/h_{fe}\omega_\beta} \tag{13.1-5}$$

and the 3-dB bandwidth of this amplifier is called f_α. From (13.1-5)

$$f_\alpha = h_{fe}f_\beta \tag{13.1-6}$$

Comparing (13.1-4) and (13.1-6), we note that $f_\alpha \approx f_T$. This result is not really correct because the circuit of Fig. 13.1-4 is not valid at f_T.

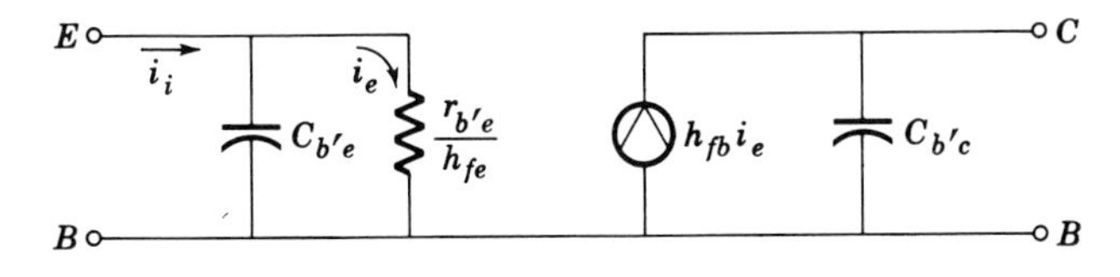

Fig. 13.1-4 High-frequency common-base model.

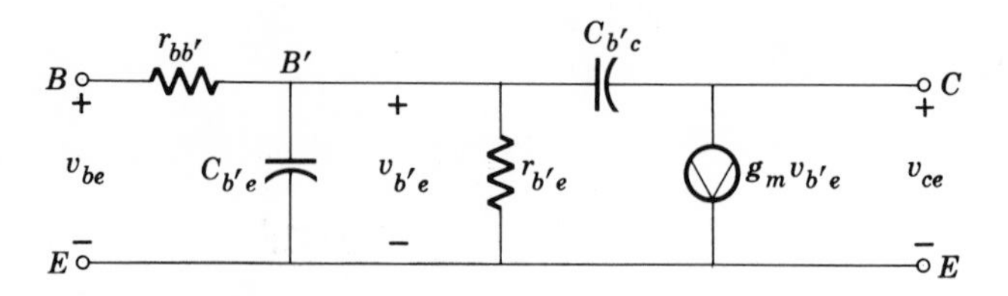

Fig. 13.1-5 Hybrid-pi equivalent with voltage-controlled current source ($h_{oe} = 0$).

The α cutoff frequency can be shown to be[2,3]

$$f_\alpha = (1 + \lambda)f_T \approx (1 + \lambda)h_{fe}f_\beta \tag{13.1-7}$$

where λ is found, empirically, to lie between 0.2 and 1. A typical value is 0.4.

It is often easier, when using the equivalent circuit of Fig. 13.1-2, to calculate the voltage $v_{b'e}$, rather than the current $i_{b'}$ which flows in $r_{b'e}$. The output-current source $h_{fe}i_{b'}$ can be transformed to a voltage-controlled current source $g_m v_{b'e}$, where

$$h_{fe}i_{b'} = h_{fe}\left(\frac{v_{b'e}}{r_{b'e}}\right) = g_m v_{b'e} \tag{13.1-8}$$

and

$$g_m = \frac{h_{fe}}{r_{b'e}} \approx \frac{I_{EQ}}{0.025} = 40I_{EQ} \qquad \text{at } T = 300°\text{K} \tag{13.1-9}$$

Note that g_m is approximately equal to $1/h_{ib}$. In order to utilize similar notation for the transistor, FET, and vacuum tube, we use the symbol g_m.

The resulting common-emitter equivalent circuit is shown in Fig. 13.1-5.

SUMMARY OF ELEMENTS OF HYBRID–PI EQUIVALENT CIRCUIT

$$r_{bb'} \approx 10 \text{ to } 50\ \Omega \qquad \text{(smaller values for high-frequency transistors)}$$

$$r_{b'e} = \frac{h_{fe}}{40I_{EQ}}$$

$$g_m = \frac{I_{EQ}}{0.025} = 40I_{EQ}$$

$$h_{oe} \sim I_{EQ}$$

$$C_{b'e} \approx \frac{h_{fe}}{\omega_T r_{b'e}} = \frac{40I_{EQ}}{\omega_T} = \frac{g_m}{\omega_T} \qquad \text{Eqs. (13.1-2) and (13.1-4)}$$

$$C_{b'c} \sim v_{cb'}^{-p} \qquad \text{where } p \text{ lies between } \tfrac{1}{2} \text{ to } \tfrac{1}{3}$$

($C_{b'c}$ is usually specified by the manufacturer as C_{ob}, the output capacitance of the common-base configuration.)

Example 13.1-1 Manufacturers' specifications for a 2N3647 silicon transistor include, at $I_{CQ} = 150$ mA, $V_{CEQ} = 1$ V:

$$f_T = 350 \text{ MHz}$$
$$h_{fe} \approx 150$$
$$h_{oe} \approx 10^{-4} \text{ mho}$$
$$C_{ob} = 4 \text{ pF}$$

From these data, deduce the values of the components of the high-frequency equivalent circuit of Fig. 13.1-5 if the transistor is to be operated at $I_{CQ} = 300$ mA.

Solution

$$r_{b'e} \approx \left(\frac{h_{fe}}{40I_{CQ}}\right) = \left[\frac{150}{(40)(300)(10^{-3})}\right] \approx 12.5\ \Omega$$

$r_{bb'}$ is not given; so we assume a value of 10 Ω.

$$h_{fe} \approx 150$$
$$g_m = (40)(0.3) = 12 \text{ mhos}$$
$$h_{oe} \approx 2 \times 10^{-4} \text{ mho}$$
$$C_{b'e} = \frac{g_m}{\omega_T} = \frac{12}{2\pi(350)(10^6)} = 5450 \text{ pF}$$
$$C_{b'c} = C_{ob} = 4 \text{ pF}$$

13.1-2 HIGH–FREQUENCY BEHAVIOR OF THE COMMON–EMITTER AMPLIFIER—MILLER CAPACITANCE

The common-emitter amplifier represents the type of high-frequency amplifier most often used. Its high-frequency response is investigated in this section. As will be seen, this response is dominated by a single pole due to the input circuit.

A complete CE stage is shown in Fig. 13.1-6*a*, and its high-frequency equivalent circuit is shown in Fig. 13.1-6*b*. In this equivalent circuit all coupling and bypass capacitors have been assumed to be short circuits at the frequencies of interest. Also, wiring and other stray capacitances have been ignored, even though they sometimes inadvertently prove to be important.

For simplicity, we let R_b represent the parallel combination of r_i and R_b, and also let R_L represent the parallel combination of R_c and R_L. The simplified equivalent circuit is shown in Fig. 13.1-6*c*. This circuit

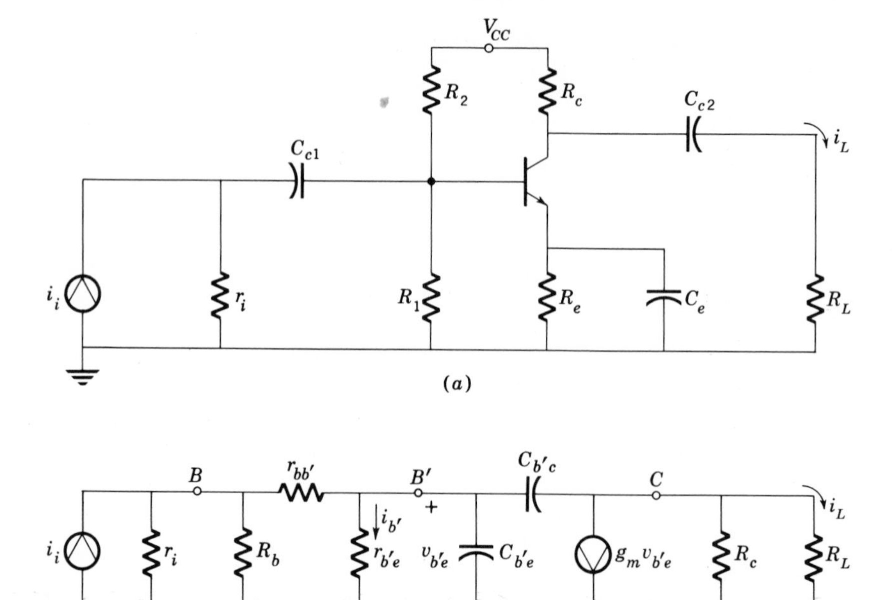

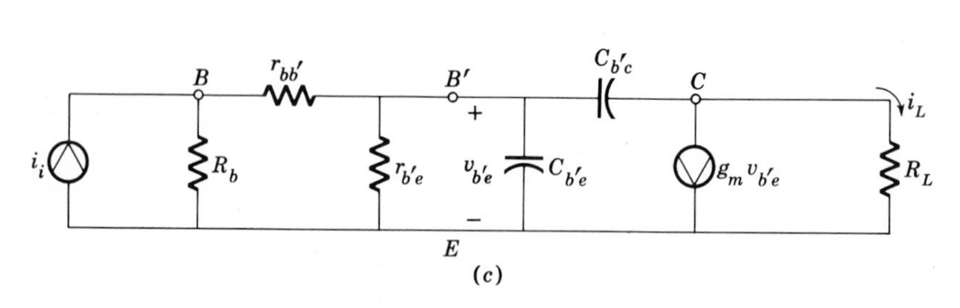

Fig. 13.1-6 The CE amplifier at high frequencies ($h_{oe} = 0$). (a) Circuit; (b) high-frequency equivalent circuit; (c) simplified equivalent circuit.

is identical to the voltage-feedback current-error amplifier shown in Fig. 8.2-3, with $C_{b'c}$ replaced by R_f.

The equivalent circuit prepared for exact analysis using feedback techniques is shown in Fig. 13.1-7a. (This circuit is identical to the circuit shown in Fig. 8.2-4b.) In this figure we have replaced i_i, R_b, $r_{bb'}$ and $r_{b'e}$ by

$$i_i' = i_i\left(\frac{R_b}{R_b + r_{bb}'}\right) \approx i_i \qquad \text{since } R_b \gg r_{bb'} \tag{13.1-10a}$$

and

$$R_{b'e} = r_{b'e}\|(R_b + r_{bb'}) \tag{13.1-10b}$$

Proceeding as in Example 8.2-1, we assume that

$$\frac{1}{\omega C_{b'c}} \gg R_L \tag{13.1-11a}$$

and

$$\omega C_{b'c} \ll g_m \tag{13.1-11b}$$

for all frequencies of interest. With these assumptions, the circuit reduces to that shown in Fig. 13.1-7*b*.

Before continuing the analysis, let us determine the frequency range over which the inequalities of (13.1-11) are valid. If, for example, $g_m = 0.01$ mho, $R_L = 1\ \text{k}\Omega$, and $C_{b'c} = 5$ pF, then $f \ll 32$ MHz defines the frequency range over which both inequalities are valid. Note that (13.1-11*a*) is almost always more restrictive than (13.1-11*b*) since $R_L > 1/g_m$. We will restrict the discussion to those cases where (13.1-11) applies.

Loop gain The loop gain of this amplifier is

$$T = \left.\frac{v_L}{v'_L}\right|_{i'_i=0} = \left(\frac{v_L}{v_{b'e}}\right)\left(\frac{v_{b'e}}{v'_L}\right)$$
$$= -g_m R_L \left[\frac{j\omega R_{b'e} C_{b'c}}{1 + j\omega R_{b'e}(C_{b'e} + C_{b'c})}\right] \tag{13.1-12}$$

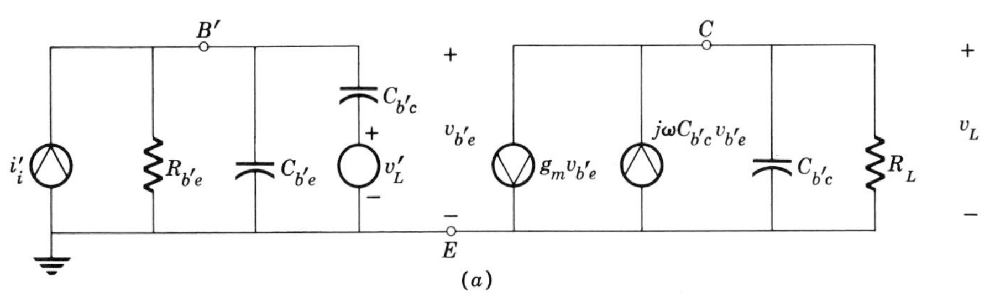

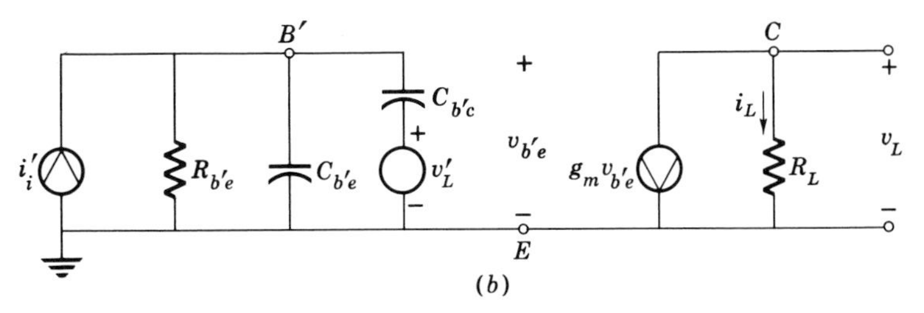

Fig. 13.1-7 Feedback formulation for CE stage at high frequencies.

Input admittance The input admittance *without* feedback Y_i is

$$Y_i = \frac{1}{Z_i} = \frac{i'_i}{v_{b'e}}\bigg|_{v'_L=0} = \frac{1}{R_{b'e}} + j\omega(C_{b'e} + C_{b'c}) \tag{13.1-13a}$$

Therefore the input admittance *with* feedback is

$$Y_{if} = Y_i(1 - T) = \left[\frac{1}{R_{b'e}} + j\omega(C_{b'e} + C_{b'c})\right] \left[1 + \frac{j\omega g_m R_L R_{b'e} C_{b'c}}{1 + j\omega R_{b'e}(C_{b'e} + C_{b'c})}\right]$$

$$= \frac{1}{R_{b'e}} + j\omega[C_{b'e} + (1 + g_m R_L)C_{b'c}] \tag{13.1-13b}$$

We see that the Thévenin input circuit, including feedback, consists of

$$R_{b'e} \| C_{b'e} \| (1 + g_m R_L)C_{b'c} \tag{13.1-13c}$$

Hence the input capacitance is increased by the *Miller capacitance*

$$C_M = (1 + g_m R_L)C_{b'c} \tag{13.1-13d}$$

Thus the collector-base feedback capacitance is multiplied by $1 + g_m R_L$ when reflected into the input circuit. This is called the *Miller effect,* and, as will be seen subsequently, it reduces the high-frequency 3-dB bandwidth. It should be noted that, if the load is an impedance Z_L, then

$$Y_M = j\omega(1 + g_m Z_L)C_{b'c}$$

Output admittance The Thévenin output admittance with feedback is

$$Y_{of} = \left(\frac{1}{R_L}\right)(1 - T)$$

$$= \frac{1}{R_L} + \frac{j\omega g_m R_{b'e} C_{b'c}}{1 + j\omega R_{b'e}(C_{b'e} + C_{b'c})} \tag{13.1-14a}$$

This expression can be simplified so that the output circuit is recognizable.

$$Y_{of} = \frac{1}{R_L} + \frac{1}{(C_{b'e} + C_{b'c})/g_m C_{b'c} + 1/j\omega g_m R_{b'e} C_{b'c}}$$

$$= \frac{1}{R_L} + \frac{1}{R + 1/j\omega C} \tag{13.1-14b}$$

The output admittance consists of the load resistance R_L, in parallel with a series-connected RC circuit, where

$$R = \left(\frac{1}{g_m}\right)\left(1 + \frac{C_{b'e}}{C_{b'c}}\right) \approx \frac{C_{b'e}}{g_m C_{b'c}} \tag{13.1-14c}$$

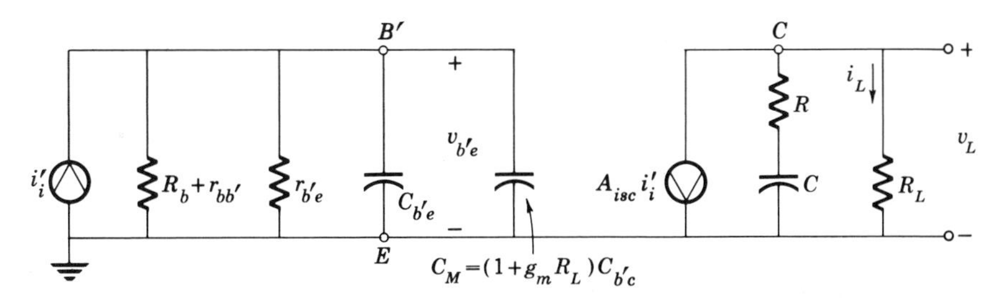

Fig. 13.1-8 Complete Norton's equivalent circuit. (*a*) Complete circuit; (*b*) simplified circuit.

and

$$C = g_m R_{b'e} C_{b'c} \tag{13.1-14d}$$

Short-circuit current gain To obtain a Norton equivalent circuit we next determine the short-circuit current gain A_{isc}. From Fig. 13.1-7*b*, with $v_L = v'_L = 0$,

$$A_{isc} = \frac{i_{sc}}{i'_i} = \frac{i_{sc}}{v_{b'e}} \frac{v_{b'e}}{i'_i} = \frac{-g_m R_{b'e}}{1 + j\omega R_{b'e}(C_{b'e} + C_{b'c})} \tag{13.1-15}$$

A Thévenin equivalent circuit can be drawn using Fig. 13.1-7 and Eqs. (13.1-13*b*), (13.1-14*b*), and (13.1-15). This is shown in Fig. 13.1-8.

Figure 13.1-8 is unnecessarily complicated for most engineering calculations. A simpler equivalent circuit can be obtained from Fig. 13.1-7*b* by using the *Miller effect* (13.1-13*d*), as shown in Fig. 13.1-9. This circuit can only be used to calculate gain and input impedance. This is the high-frequency circuit to be employed in the rest of this text.

Current gain and bandwidth The current gain of the amplifier shown in Fig. 13.1-6 can be calculated either directly from Fig. 13.1-9 or using the expression

$$A_{if} = \frac{A_i}{1 - T} \tag{13.1-16}$$

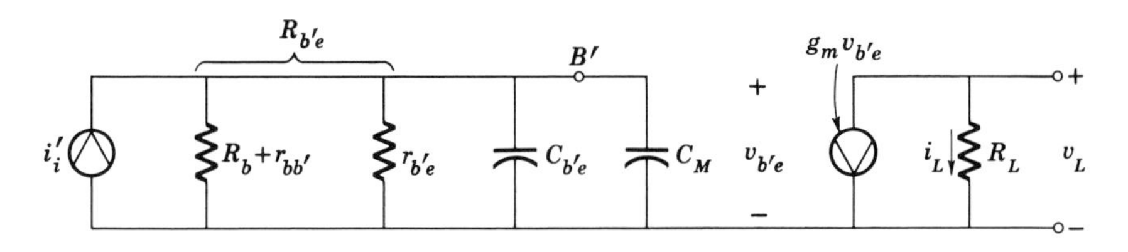

Fig. 13.1-9 Alternate CE high-frequency equivalent circuit.

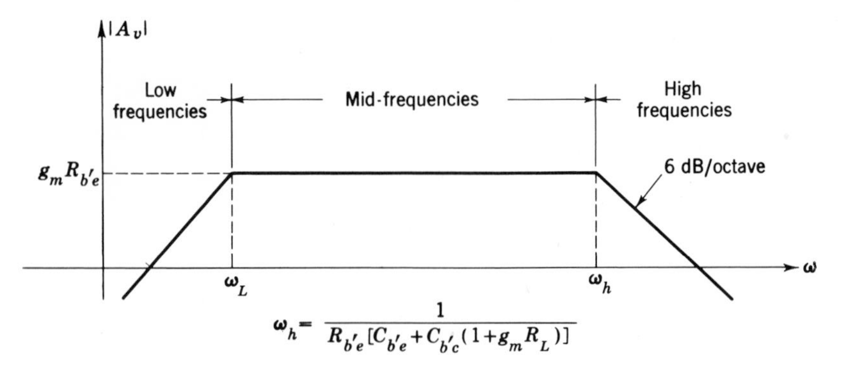

Fig. 13.1-10 Frequency response of CE stage.

The result is

$$A_{if} = \frac{i_L}{i'_i} = -\frac{g_m R_{b'e}}{1 + j\omega R_{b'e}(C_{b'e} + C_M)} \qquad (13.1\text{-}17)$$

This equation indicates that the midfrequency current gain is

$$A_{im} = -g_m R_{b'e} \qquad (13.1\text{-}18a)$$

The current gain is reduced by 3 dB at the frequency

$$f_h = \frac{1}{2\pi R_{b'e}(C_{b'e} + C_M)} \qquad (13.1\text{-}18b)$$

f_h is called the upper 3-dB frequency. When discussing *RC* amplifiers, f_h is usually much greater than f_L (Chap. 12). It is therefore common practice to define the 3-dB *bandwidth* (BW) of the amplifier, which is $f_h - f_L$, by f_h.

An asymptotic plot of this gain function is shown in Fig. 13.1-10. It is interesting to note that the high-frequency response exhibits only one pole. A glance at the original circuit shown in Fig. 13.1-6 indicates that a second pole and a zero should also be present. However, this pole and zero have been eliminated as a result of our initial assumption (13.1-11).

Example 13.1-2 The amplifier shown in Fig. 13.1-6*a* has the following component values:

$r_i = 10\ \text{k}\Omega$ $\qquad C_{b'c} = 2\ \text{pF}$
$R_b = 2\ \text{k}\Omega$ $\qquad C_{b'e} = 200\ \text{pF}$
$r_{bb'} = 25\ \Omega$ $\qquad g_m = 0.5\ \text{mho}$
$r_{b'e} = 150\ \Omega$ $\qquad R_L = 200\ \Omega$

Find the midband current gain and the 3-dB frequency f_h.

Solution MIDBAND CURRENT GAIN From (13.1-18*a*)

$$A_{im} = -g_m R_{b'e} = -g_m(R_b + r_{bb'})\|r_i\|r_{b'e} \approx -(0.5)(150) = -75$$

3-dB FREQUENCY f_h (13.1-18*b*)

$$C_{b'e} + C_M = [200 + 2\ (1 + (\tfrac{1}{2})(200))] = 400 \text{ pF}$$

$$f_h = \frac{1}{2\pi R_{b'e}(C_{b'e} + C_M)} \approx 2.6 \text{ MHz}$$

VALIDITY OF THE CIRCUIT OF FIG. 13.1-9 The 3-dB frequency f_h, calculated using Fig. 13.1-9, is valid when (13.1-11) applies; i.e.,

$$f_h \ll \frac{1}{2\pi R_L C_{b'c}} \approx 400 \text{ MHz}$$

Since $f_h = 2.6$ MHz $\ll 400$ MHz, (13.1-18*b*) is valid and should accurately predict the upper 3-dB frequency.

APPROXIMATING f_h BY f_β Sometimes f_β is used as a rough approximation to f_h. In this example,

$$f_\beta = \frac{1}{2\pi r_{b'e} C_{b'e}} = 5.3 \text{ MHz}$$

and the approximation is not very good. As a matter of fact, the approximation is good only when [Eq. (13.1-18*b*)]

$$C_{b'e} \gg C_{b'c}(1 + g_m R_L)$$

and

$$R_{b'e} \approx r_{b'e}$$

13.1-3 THE EMITTER FOLLOWER AT HIGH FREQUENCIES

In this section, we calculate the frequency dependence of the input impedance, the output impedance, and the gain of the emitter follower.

The circuit and its small-signal high-frequency equivalent are shown in Fig. 13.1-11.

In most practical emitter-follower circuits, a voltage gain near unity is desired; so r_i is chosen to be as small as possible. In order for r_i to see a high input impedance, R_b is chosen to be large. To simplify calculations, we assume that R_b can be considered infinite.

The input and output impedance at midfrequency and the midfrequency voltage gain are, from Sec. 6.4,

$$Z_{im} = h_{ie} + (h_{fe} + 1)R'_e \approx h_{fe}R'_e \qquad R'_e = R_e\|R_L \tag{13.1-19a}$$

$$Z_{om} \approx \frac{r_i + h_{ie}}{h_{fe} + 1} = \frac{r_i + r_{bb'} + r_{b'e}}{h_{fe} + 1} \tag{13.1-19b}$$

$$A_{vm} = \frac{v_e}{v_i} \approx 1 \tag{13.1-19c}$$

Our task now is to modify these expressions to include the effects of $C_{b'c}$ and $C_{b'e}$.

Input impedance In Prob. 13-9 the impedance Z_i' (Fig. 13.1-11) is found to be

$$Z_i' = R_e' + \frac{r_{b'e} + h_{fe}R_e'}{1 + s/\omega_\beta} \tag{13.1-20}$$

This is combined with $r_{bb'}$ and $C_{b'c}$ to yield the actual input impedance Z_i, shown in Fig. 13.1-12. If $R' \gg R_e'$, we can neglect R_e'. Then $|Z_i|$ is 0.707 times its midband value at the frequency

$$\omega_1 \approx \frac{1}{(r_{b'e} + h_{fe}R_e')(C_{b'c} + C')} \tag{13.1-21a}$$

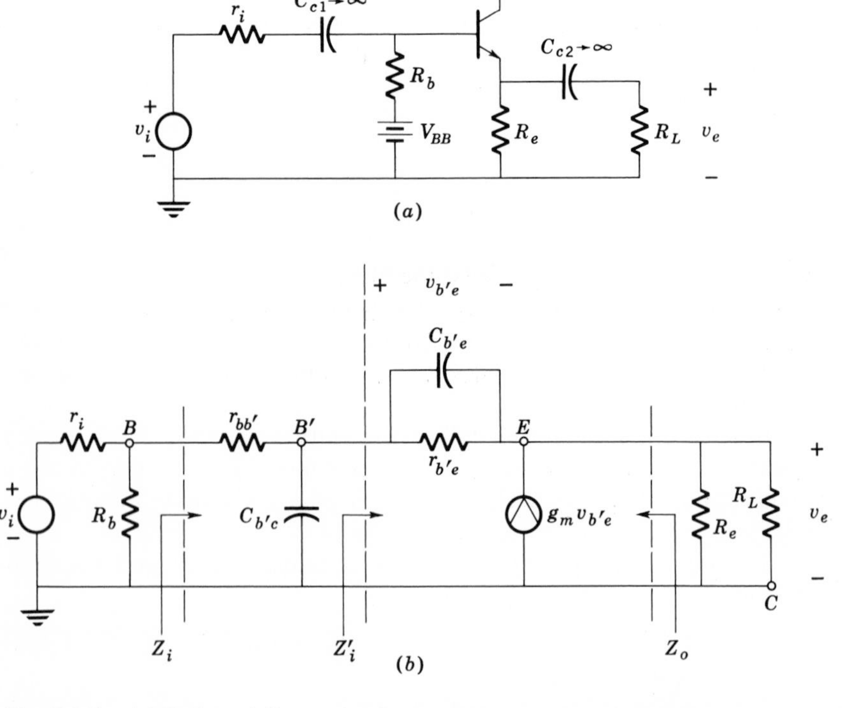

Fig. 13.1-11 (a) Emitter-follower circuit; (b) high-frequency equivalent circuit.

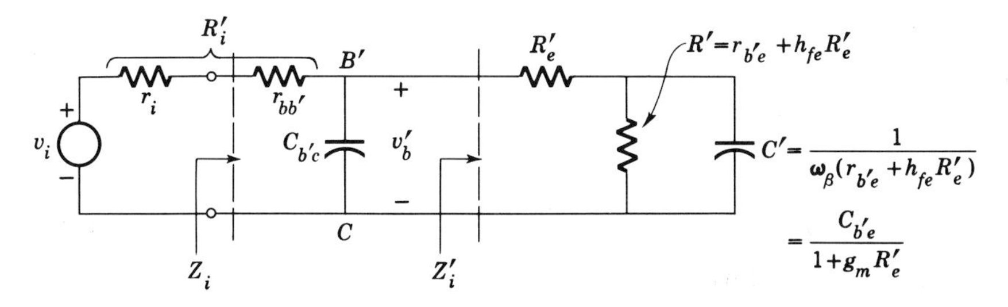

Fig. 13.1-12 High-frequency input impedance of emitter follower.

where

$$C' = \frac{1}{\omega_\beta(r_{b'e} + h_{fe}R_e')} = \frac{C_{b'e}}{1 + g_m R_e'} \tag{13.1-21b}$$

From the circuit of Fig. 13.1-12 it is seen that Z_i has one pole, in addition to the pole at ω_1, and two zeros. These can be found by routine analysis, if required. In many cases it is sufficient to observe that, at very high frequencies (i.e., at frequencies above $1/r_{bb'}C_{b'c}$), $Z_i \approx r_{bb'}$, which will be much less than the midband value (see Prob. 13.12).

Output impedance In Prob. 13.11 the output impedance is found to be

$$Z_o \approx \left(\frac{r_{b'e} + R_i'}{h_{fe} + 1}\right)\left[\frac{1 + sC_{b'e}(r_{b'e}\|R_i')}{(1 + s/\omega_T)(1 + s/\omega_i)}\right] \tag{13.1-22}$$

where

$$R_i' = r_i + r_{bb'} \tag{13.1-23a}$$

$$\omega_i = \frac{1}{R_i'C_{b'c}} \tag{13.1-23b}$$

If R_i' is much less than $r_{b'e}$, the zero in (13.1-22) occurs at a frequency which is higher than ω_β but lower than ω_i. At this frequency the output impedance begins to increase above its midband value. At much higher frequencies the poles dominate (13.1-22), and the impedance decreases.

Voltage gain The voltage gain of the emitter follower in the high-frequency region can be found from Fig. 13.1-11*b*. The gain is calculated by noting that

$$A_v = \left(\frac{v_e}{v_{b'}}\right)\left(\frac{v_{b'}}{v_i}\right)$$

We first calculate $v_e/v_{b'}$ by writing a node equation at the emitter node.

$$\frac{v_{b'e}}{Z_{b'e}} + g_m v_{b'e} = \frac{v_e}{R_e'} \tag{13.1-24a}$$

where

$$Z_{b'e} = \frac{r_{b'e}}{1 + s/\omega_\beta}$$

Since

$$v_{b'e} = v_{b'} - v_e \tag{13.1-24b}$$

(13.1-24*a*) can be written

$$(v_{b'} - v_e)\left(g_m + \frac{1}{Z_{b'e}}\right) = \frac{v_e}{R'_e} \tag{13.1-24c}$$

and

$$v_e = v_{b'} \frac{g_m + 1/Z_{b'e}}{g_m + 1/R'_e + 1/Z_{b'e}} \approx v_{b'} \tag{13.1-25a}$$

since

$$R'_e \gg \frac{1}{g_m} \tag{13.1-25b}$$

Note that the voltage gain from B' to E is approximately equal to unity, independent of frequency.

The gain $v_{b'}/v_i$ can be found by making use of the equivalent input circuit of Fig. 13.1-12. At frequencies less than ω_β, the impedance of C' is greater than the impedance of R'. Thus C' can be neglected. The gain $v_{b'}/v_i$ is then

$$\frac{v_{b'}}{v_i} \approx \frac{1}{1 + sC_{b'c}R'_i} = \frac{1}{1 + s/\omega_i} \qquad s = j\omega < j\omega_\beta \tag{13.1-26a}$$

where we have assumed that

$$R'_i \ll r_{b'e} + h_{fe}R_{e'}$$

Letting $s = j\omega = j\omega_\beta$ yields

$$\frac{v_{b'}}{v_i} = \frac{1}{\sqrt{1 + (\omega_\beta/\omega_i)^2}} \approx 1 \tag{13.1-26b}$$

since $\omega_\beta \ll \omega_i$. (This follows from the fact that $C_{b'c} \ll C_{b'e}$ and the assumption that $R'_i \ll r_{b'e}$.)

At frequencies greater than ω_β, the impedance of C' is less than that of $r_{b'e} + h_{fe}R'_e$. Figure 13.1-12 can therefore be simplified as shown in Fig. 13.1-13. The gain $v_{b'}/v_i$ is then

$$\frac{v_{b'}}{v_i} = \frac{1 + sR'_eC'}{1 + s[(R'_e + R'_i)C' + R'_iC_{b'c}] + s^2(R'_eC')(R'_iC_{b'c})} \tag{13.1-26c}$$

Since $v_e/v_{b'} \approx 1$ for all frequencies [Eq. (13.1-25*a*)], the 3-dB frequency of the emitter follower is found from (13.1-26*c*).

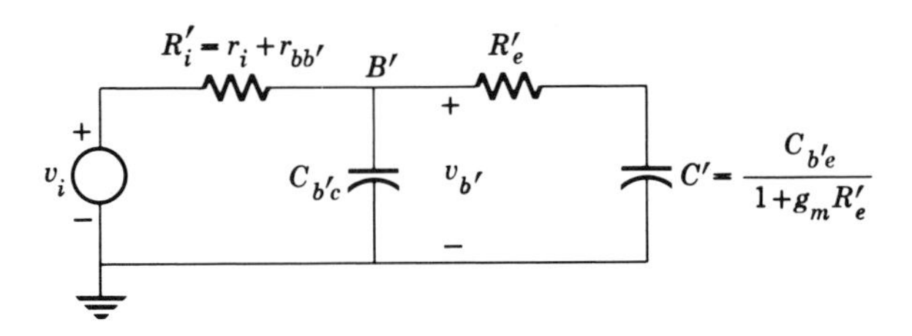

Fig. 13.1-13 Simplified EF input circuit at high frequencies.

Typical orders of magnitude will be obtained from the following example.

Example 13.1-3 The emitter follower of Fig. 13.1-11 has the component values

$$C_{b'e} = 1000 \text{ pF} \qquad h_{fe} = 100$$
$$C_{b'c} = 10 \text{ pF} \qquad R_e' = 100\ \Omega$$
$$r_{b'e} = 100\ \Omega \qquad R_i' = 100\ \Omega$$
$$r_{bb'} = 30\ \Omega$$

Find Z_i, Z_o, and A_v.

Solution INPUT IMPEDANCE The midband input impedance is

$$\begin{aligned} Z_{im} &= r_{bb'} + r_{b'e} + (1 + h_{fe})R_e' \\ &= 30 + 100 + (101)(100) \\ &\approx 10 \text{ k}\Omega \end{aligned}$$

Since $r_{b'e} + h_{fe}R_e' \gg R_e'$, we use the approximate expression (13.1-21a) to find the lowest frequency pole of Z_i. Thus, using (13.1-21b),

$$C' = \frac{C_{b'e}}{1 + g_m R_e'} = \frac{C_{b'e}}{1 + (h_{fe}/r_{b'e})R_e'} = \frac{1000 \times 10^{-12}}{1 + 100} \approx 10 \text{ pF}$$

and

$$\omega_1 = \frac{1}{(r_{b'e} + h_{fe}R_e')(C_{b'c} + C')} \approx \frac{1}{(100 + 10^4)(10 + 10)(10^{-12})} \approx 5 \times 10^6 \text{ rad/s}$$

An asymptotic plot of the magnitude of the input impedance is shown in Fig. 13.1-14. The curve is shown in dashed lines at frequencies above ω_β, because of the additional zero and pole which have been neglected. The reader will observe that an exact calcula-

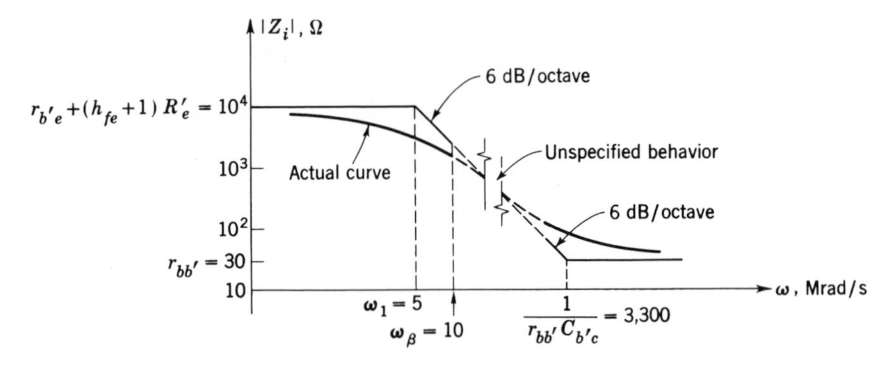

Fig. 13.1-14 Input impedance of EF.

tion of the impedance of the circuit of Fig. 13.1-12, although straightforward, will be tedious. The approximate calculation does provide useful results which are accurate up to at least ω_β.

OUTPUT IMPEDANCE From (13.1-22), with $\omega_T = g_m/C_{b'e} = 10^9$ rad/s and $\omega_i = 1/R_i'C_{b'c} = 10^9$ rad/s,

$$Z_o = \left(\frac{r_{b'e} + R_i'}{h_{fe} + 1}\right)\left[\frac{1 + sC_{b'e}(r_{b'e}\|R_i')}{(1 + s/\omega_T)(1 + s/\omega_i)}\right]$$

$$= \left(\frac{100 + 100}{101}\right)\left[\frac{1 + s(10^{-9})(50)}{(1 + s/10^9)(1 + s/10^9)}\right] \approx 2\,\frac{1 + s/(2 \times 10^7)}{(1 + s/10^9)^2}$$

The magnitude of this impedance ($s = j\omega$) is plotted in Fig. 13.1-15. Note that the output impedance maintains its midfrequency value to 20 Mrad/s, while the input impedance maintains its midfrequency value only to 5 Mrad/s.

VOLTAGE GAIN The voltage gain is approximately equal to unity for $\omega \leq \omega_\beta$ [Eqs. (13.1-25*a*) and (13.1-26*b*)]. To determine

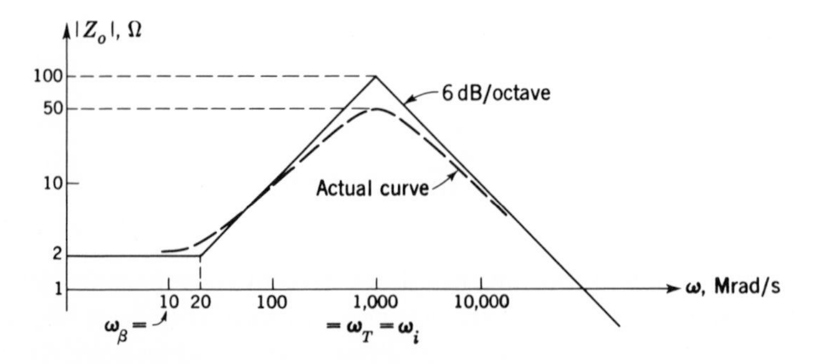

Fig. 13.1-15 Output impedance of EF.

the 3-dB frequency we use (13.1-26*c*).

$$A_v \approx \frac{1 + s(100)(10 \times 10^{-12})}{1 + s[(200)(10 \times 10^{-12}) + (100)(10 \times 10^{-12})] + s^2(100)(10 \times 10^{-12})(100)(10 \times 10^{-12})}$$

$$= \frac{1 + 10^{-9}s}{1 + 3 \times 10^{-9}s + 10^{-18}s^2}$$

$$= \frac{1 + s/10^9}{\left(1 + \dfrac{s}{0.38 \times 10^9}\right)\left(1 + \dfrac{s}{2.62 \times 10^9}\right)}$$

An asymptotic plot of $|A_v|$ is shown in Fig. 13.1-16.

The 3-dB frequency ω_h can be shown to be 440 Mrad/s. The second pole occurs at a higher frequency than ω_T, at which point our high frequency model is no longer valid.[4]

High-frequency Oscillations Very often, in practice, the emitter follower becomes unstable and produces oscillations at very high frequencies. The frequencies of oscillation are so high that the oscillations often go undetected when using an ordinary laboratory oscilloscope, and the resulting nonlinear behavior of the emitter follower is blamed on other causes. The oscillations are due to the fact that the stray wiring capacitance from emitter to ground along with power-supply internal impedance (which results in imperfect ac grounding of the collector) results in a loop gain greater than unity.

The instability can be eliminated by increasing r_i until the oscillations cease. This has the disadvantage of reducing the gain and increasing the output impedance, but these changes may be tolerable, depending on the application. Another solution is to insert a low-Q inductance in series with r_i. This inductance is chosen so that its impedance in the frequency band of interest is negligible, while at the frequency of oscillation it presents a high

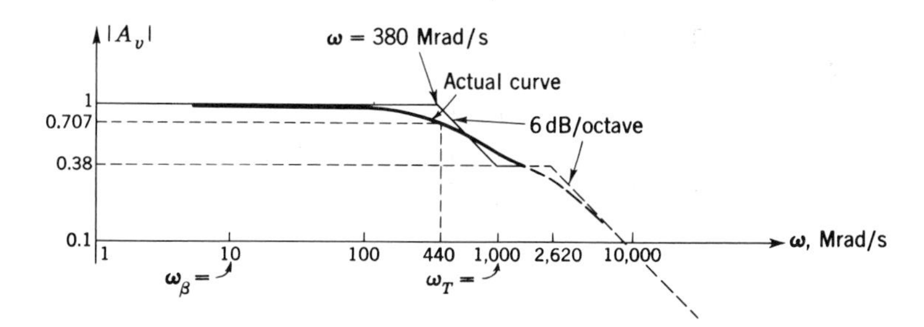

Fig. 13.1-16 Voltage gain of EF.

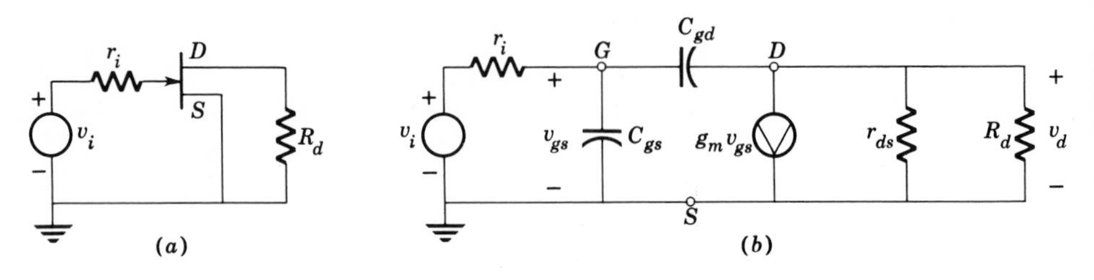

Fig. 13.2-1 The FET at high frequencies. (a) Circuit (bias components omitted); (b) high-frequency equivalent circuit.

impedance. This effectively increases the value of r_i at high frequencies.

13.2 THE FIELD-EFFECT TRANSISTOR AT HIGH FREQUENCIES

The FET at high frequencies can be described in terms of a hybrid-pi equivalent circuit as shown in Fig. 13.2-1. The capacitors C_{gs} and C_{gd} in a junction FET are a consequence of the back-biased gate. Figure 13.2-2 shows a JFET operating above pinch-off. The capacitance between the gate and the source and between the gate and the drain are similar to the collector-base capacitance $C_{b'c}$ since they all result from a reverse-biased *pn* junction (Fig. 13.1-1b). Therefore these capacitors vary, as does $C_{b'c}$.

$$C_{gs} \sim (-V_{GS})^{-1/2} \qquad V_{GS} \leq 0 \tag{13.2-1a}$$

and

$$C_{gd} \sim (-V_{GD})^{-1/2} \qquad V_{GD} \leq 0 \tag{13.2-1b}$$

Since $|V_{GD}| \gg |V_{GS}|$, we have

$$C_{gd} \ll C_{gs} \tag{13.2-1c}$$

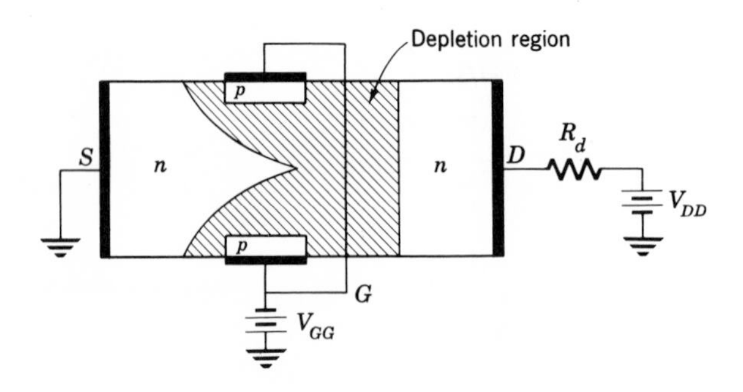

Fig. 13.2-2 The JFET above pinch-off.

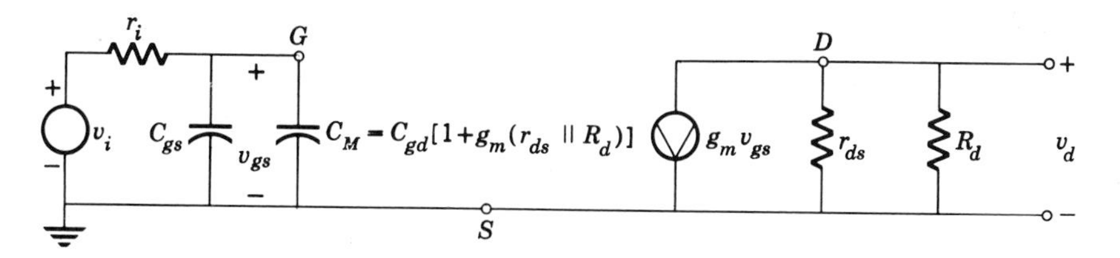

Fig. 13.2-3 Common-source circuit with feedback removed.

Typical values of C_{gs} vary from 50 pF for a low-frequency FET to less than 5 pF for a high-frequency FET. The feedback capacitor C_{gd} is usually less than 5 pF, and is often less than 0.5 pF, for a high-frequency IGFET.

13.2-1 HIGH–FREQUENCY BEHAVIOR OF THE COMMON–SOURCE AMPLIFIER—MILLER CAPACITANCE

The high-frequency equivalent circuit of the FET shown in Fig. 13.2-1*b* is similar to that shown in Fig. 13.1-6 for the transistor. Proceeding with a similar analysis, we find that the Miller capacitance is

$$C_M = C_{gd}[1 + g_m(r_{ds}\|R_d)] \tag{13.2-2a}$$

This result is valid [Eq. (13.1-11)] for frequencies such that

$$\omega \ll \frac{1}{C_{gd}(r_{ds}\|R_d)} \qquad \omega \ll \frac{g_m}{C_{gd}} \tag{13.2-2b}$$

Using the above approximations, we obtain the high-frequency equivalent circuit shown in Fig. 13.2-3 (compare with Fig. 13.1-9).

The voltage gain of the FET amplifier is

$$A_v = \frac{v_d}{v_i} = -g_m(r_{ds}\|R_d)\left[\frac{1}{1 + j\omega r_i(C_{gs} + C_M)}\right] \tag{13.2-3a}$$

The 3-dB frequency is

$$f_h = \frac{1}{2\pi r_i(C_{gs} + C_M)} \tag{13.2-3b}$$

The gain-frequency plot for the FET amplifier is shown in Fig. 13.2-4.

Example 13.2-1 The FET amplifier shown in Fig. 13.2-1*a* has the following component values: $R_d = 10$ kΩ, $r_{ds} = 15$ kΩ, $g_m = 3 \times 10^{-3}$ mho, $C_{gs} = 50$ pF, and $C_{gd} = 5$ pF.

Find r_i to ensure a 3-dB bandwidth of at least 100 kHz.

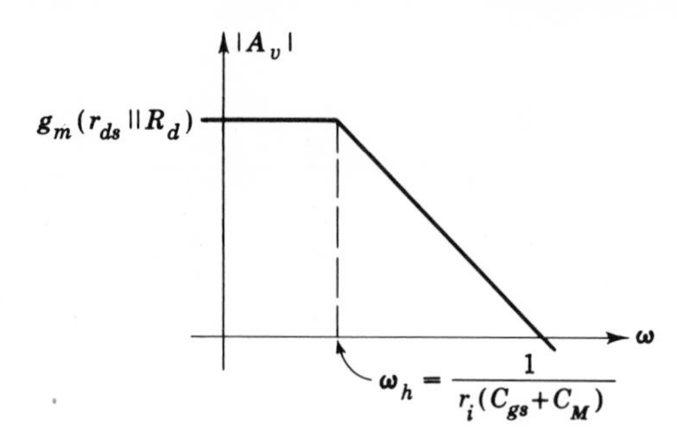

Fig. 13.2-4 Gain vs. frequency for the common-source amplifier.

Solution From (13.2-3*b*) we see that

$$r_i = \frac{1}{2\pi f_h(C_{gs} + C_M)}$$

$$\leq \frac{1}{2\pi(10^5)(50 \times 10^{-12} + 5 \times 10^{-12}[1 + 3 \times 10^{-3}(6 \times 10^3)])} = 11\ \text{k}\Omega$$

We can therefore let $r_i = 10\ \text{k}\Omega$.

Voltage gain when $r_i = 0$ Consider the FET amplifier shown in Fig. 13.2-1*b*, with $r_i = 0$. The equivalent circuit is shown in Fig. 13.2-5*a*. The voltage gain A_v is found from the simplified equivalent circuit of Fig. 13.2-5*b*, where we have used the fact that $v_{gs} = v_i$. Then

$$A_v = \frac{v_d}{v_i} = (-g_m + j\omega C_{gd})\left[\frac{r_{ds}\|R_d}{1 + j\omega C_{gd}(r_{ds}\|R_d)}\right]$$

$$= g_m(r_{ds}\|R_d)\,\frac{-1 + j\omega C_{gd}/g_m}{1 + j\omega C_{gd}(r_{ds}\|R_d)} \tag{13.2-4}$$

where

$$g_m(r_{ds}\|R_d) \gg 1 \tag{13.2-5}$$

The voltage gain vs. frequency is plotted in Fig. 13.2-6.

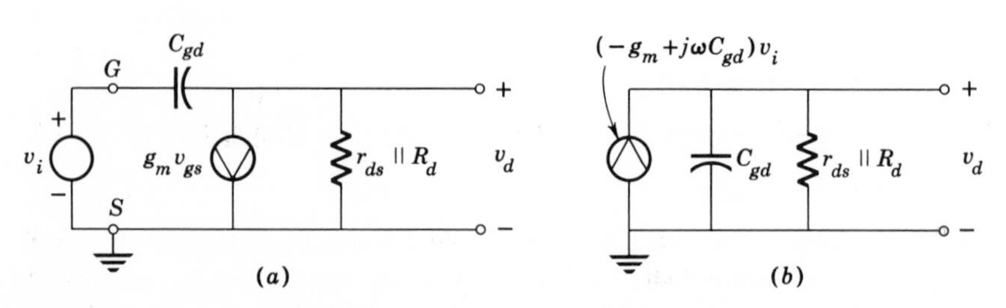

Fig. 13.2-5 FET amplifier with zero input resistance ($r_i = 0$). (*a*) Equivalent circuit; (*b*) simplified equivalent circuit.

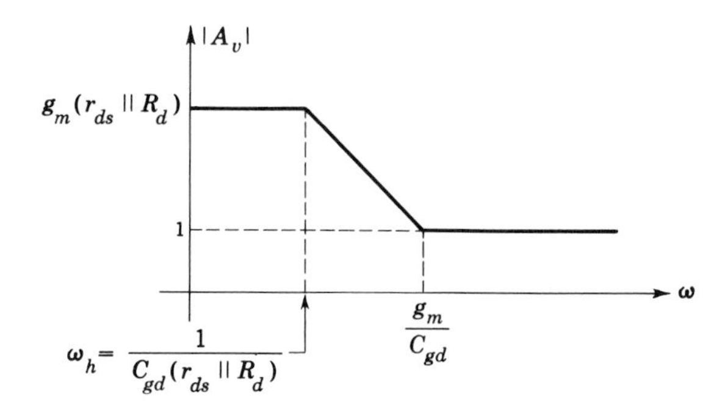

Fig. 13.2-6 Voltage gain of FET amplifier of Fig. 13.2-5.

Note that if R_d is infinite, the 3-dB bandwidth of the common-source amplifier is limited by the channel resistance r_{ds} and the gate drain capacitance C_{gd}.

13.2-2 HIGH-FREQUENCY BEHAVIOR OF THE SOURCE FOLLOWER

The equivalent circuit of the source follower is shown in Fig. 13.2-7. This circuit differs from that of the emitter follower (Fig. 13.1-11*b*) since there is an *infinite* resistance across C_{gs} rather than the relatively small resistance $r_{b'e}$.

Input impedance The input impedance Z_i' is found as in Sec. 13.1-3. The result is

$$Z_i' = \frac{v_{gd}}{i} = (r_{ds}\|R_s) + \left(\frac{1}{j\omega C_{gs}}\right)[1 + g_m(r_{ds}\|R_s)] \tag{13.2-6}$$

The total input equivalent circuit is shown in Fig. 13.2-8.

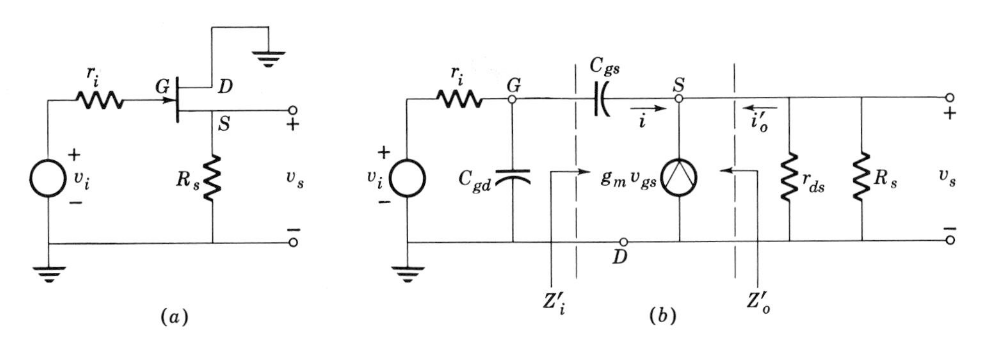

Fig. 13.2-7 Source follower at high frequencies. (*a*) Circuit (bias components omitted); (*b*) high-frequency equivalent circuit.

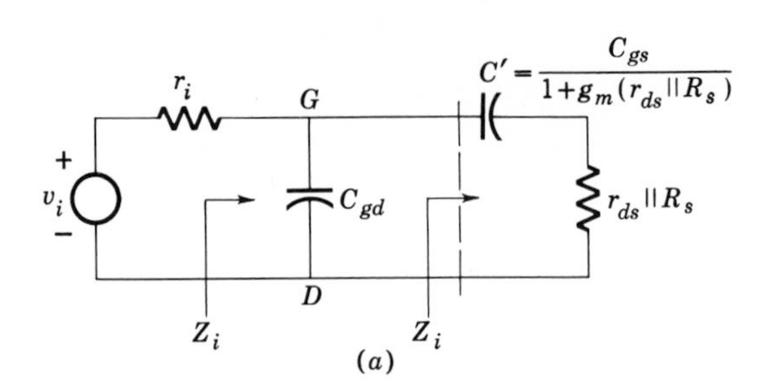

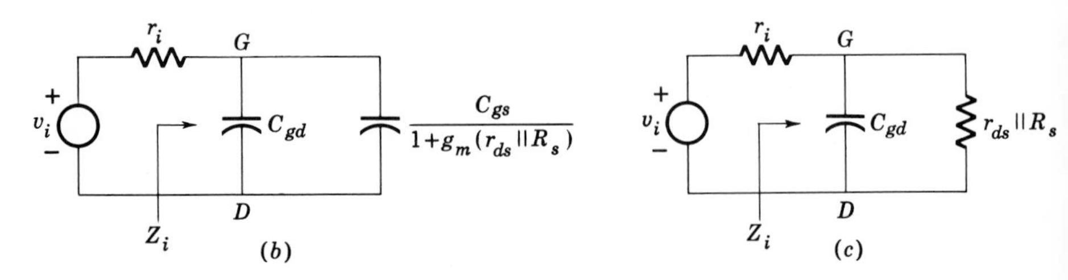

Fig. 13.2-8 Source-follower input impedance. (*a*) Complete circuit; (*b*) circuit for $\omega \ll g_m/C_{gs}$; (*c*) circuit for $\omega \gg g_m/C_{gs}$.

We see from Fig. 13.2-8*a* that the input impedance consists of C_{gd} in parallel with a series RC circuit. Let us determine the relative importance of $r_{ds}||R_s$ and C'. Assuming

$$g_m(r_{ds}||R_s) \gg 1 \tag{13.2-7a}$$

we have

$$|Z_i'|^2 \approx (r_{ds}||R_s)^2 + \left(\frac{g_m}{\omega C_{gs}}\right)^2 (r_{ds}||R_s)^2 \tag{13.2-7b}$$

The capacitance C' is seen to have a higher impedance than the resistance $r_{ds}||R_s$, as long as

$$\omega \ll \frac{g_m}{C_{gs}}$$

In this frequency range the equivalent input circuit can be reduced to that shown in Fig. 13.2-8*b*. For frequencies at which

$$\omega \gg \frac{g_m}{C_{gs}}$$

the equivalent input circuit becomes that shown in Fig. 13.2-8*c*.

Using Fig. 13.2-8*a*, the input impedance Z_i is

$$Z_i = \frac{1}{j\omega C_{gd}}\left\{\frac{1 + 1/[j\omega C'(r_{ds}\|R_s)]}{1 + \dfrac{1}{j\omega(r_{ds}\|R_s)}\left(\dfrac{1}{C'} + \dfrac{1}{C_{gd}}\right)}\right\} \qquad (13.2\text{-}8)$$

The magnitude of the input impedance is shown in Fig. 13.2-9*a*. Note that for frequencies less than $1/C'(r_{ds}\|R_s)$ the input circuit can be approximated by the capacitor C_{gd}.

Output impedance The output impedance is found from Fig. 13.2-7*b*.

$$Z_0' = \left.\frac{v_s}{i_0'}\right|_{v_i=0} = \left(\frac{1}{g_m}\right)\left[\frac{1 + j\omega(C_{gd} + C_{gs})r_i}{(1 + j\omega C_{gs}/g_m)(1 + j\omega r_i C_{gd})}\right] \qquad (13.2\text{-}9a)$$

Note that if $r_i \to 0$, the asymptote of the output impedance Z_o' equals

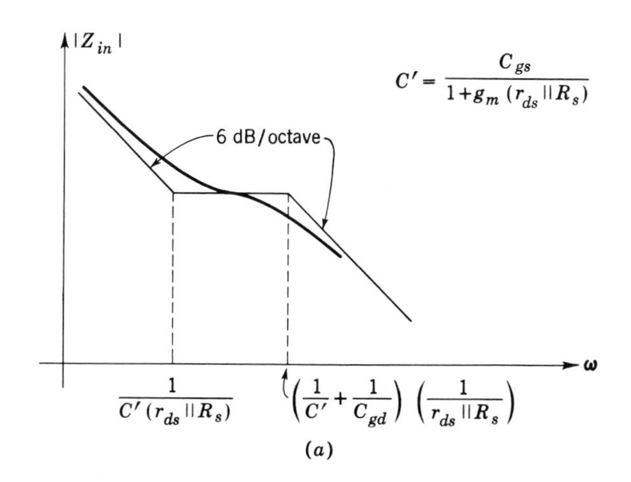

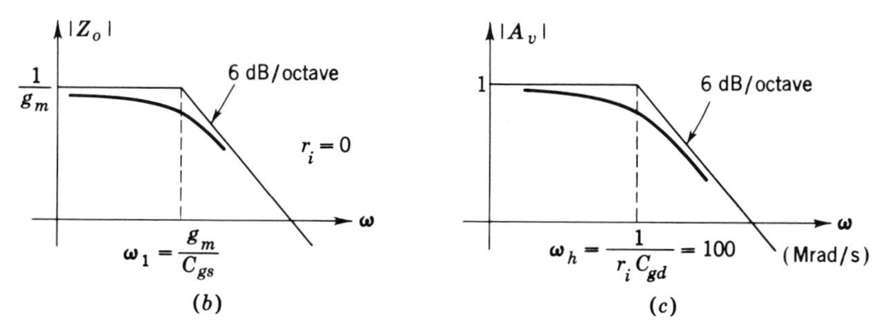

Fig. 13.2-9 High-frequency response of the source follower. (*a*) Input impedance; (*b*) output impedance; (*c*) voltage gain.

$1/g_m$, up to the frequency

$$\omega_1 = \frac{g_m}{C_{gs}} \tag{13.2-9b}$$

and then decreases at the rate of 6 dB/octave. The magnitude of the output impedance is shown in Fig. 13.2-9b.

Voltage gain The voltage gain A_v can be obtained from Fig. 13.2-7b. We note that

$$A_v = \frac{v_s}{v_i} = \left(\frac{v_s}{v_g}\right)\left(\frac{v_g}{v_i}\right) \tag{13.2-10}$$

The gain v_s/v_g is calculated directly from Fig. 13.2-7b using KCL.

$$(g_m v_{gs} + i)(r_{ds}\|R_s) = v_s \tag{13.2-11a}$$

where

$$v_{gs} = v_g - v_s = \frac{i}{j\omega C_{gs}} \tag{13.2-11b}$$

Substituting (13.2-11b) into (13.2-11a) yields

$$\frac{v_s}{v_g} = \left[\frac{g_m(r_{ds}\|R_s)}{1 + g_m(r_{ds}\|R_s)}\right]\left\{\frac{1 + j\omega C_{gs}/g_m}{1 + j\omega C_{gs}\left[\frac{r_{ds}\|R_s}{1 + g_m(r_{ds}\|R_s)}\right]}\right\} \approx 1 \tag{13.2-11c}$$

since

$$g_m(r_{ds}\|R_s) \gg 1$$

The gain v_g/v_i is found using Fig. 13.2-8a to c. At radian frequencies less than g_m/C_{gs}, Fig. 13.2-8b applies, and

$$\frac{v_g}{v_i} \approx \frac{1}{1 + j\omega r_i\{C_{gd} + C_{gs}/[1 + g_m(r_{ds}\|R_s)]\}} \tag{13.2-12a}$$

while at radian frequencies much higher than g_m/C_{gs}, Fig. 13.2-8c can be used, and

$$\frac{v_g}{v_i} \approx \left(\frac{1}{r_i}\right)\left[\frac{r_i\|r_{ds}\|R_s}{1 + j\omega C_{gd}(r_i\|r_{ds}\|R_s)}\right] \tag{13.2-12b}$$

Note that if $r_i \to 0$, v_g/v_i approaches unity, independent of frequency.

The use of (13.2-11c), (13.2-12a), and (13.2-12b) to determine the voltage gain is illustrated in the following example.

Example 13.2-2 A JFET is used as a source follower. It has the following parameters: $C_{gs} = 6$ pF, $C_{gd} = 2$ pF, $r_{ds} = 70$ kΩ, and

$g_m = 3 \times 10^{-3}$ mho. If $R_s = 10$ kΩ and $r_i = 5$ kΩ, plot the voltage gain as a function of frequency and determine the 3-dB frequency.

Solution The voltage gain A_v is

$$A_v = \frac{v_s}{v_i} = \left(\frac{v_s}{v_g}\right)\left(\frac{v_g}{v_i}\right)$$

From (13.2-11*c*) we note that

$$\frac{v_s}{v_g} \approx 1$$

To determine v_g/v_i, we first consider radian frequencies less than

$$\frac{g_m}{C_{gs}} = \frac{3 \times 10^{-3}}{6 \times 10^{-12}} = 500 \text{ Mrad/s}$$

Then, using (13.2-12*a*) and noting that $C_{gd} \gg C_{gs}/[1 + g_m(r_{ds}\|R_s)]$,

$$A_v \approx \frac{v_g}{v_i} = \frac{1}{1 + j\omega r_i\{C_{gd} + C_{gs}/[1 + g_m(r_{ds}\|R_s)]\}}$$

$$\approx \frac{1}{1 + j\omega r_i C_{gd}} = \frac{1}{1 + j\omega 10^{-8}}$$

Thus the 3-dB frequency is

$$\omega_h = 100 \text{ Mrad/s} < \frac{g_m}{C_{gs}}$$

The voltage gain is plotted in Fig. 13.2-9*c*.

13.3 THE VACUUM TUBE AT HIGH FREQUENCIES

The vacuum tube can also be represented in terms of a hybrid-pi equivalent circuit at high frequencies, as shown in Fig. 13.3-1. Note that the equivalent circuit differs slightly from that obtained for the FET (Fig. 13.2-1*b*)

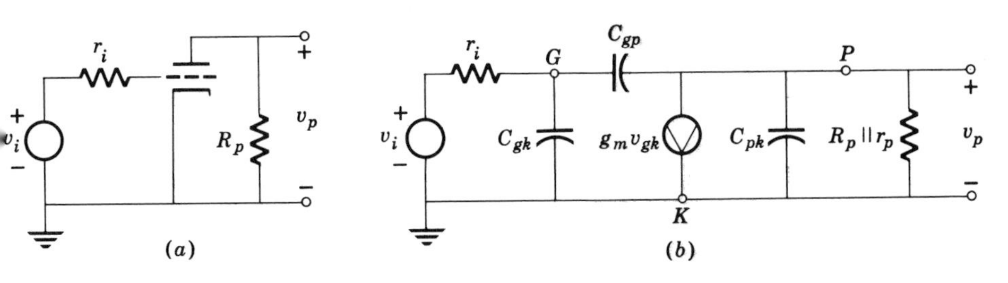

Fig. 13.3-1 Vacuum-tube amplifier at high frequencies. (*a*) Circuit: triode, or pentode (bias components omitted); (*b*) high-frequency equivalent circuit.

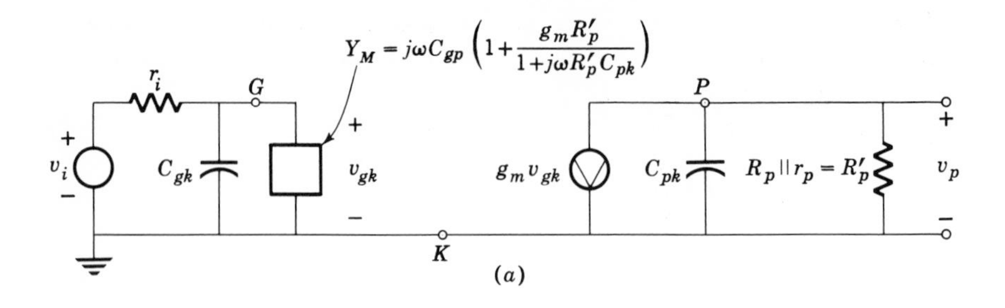

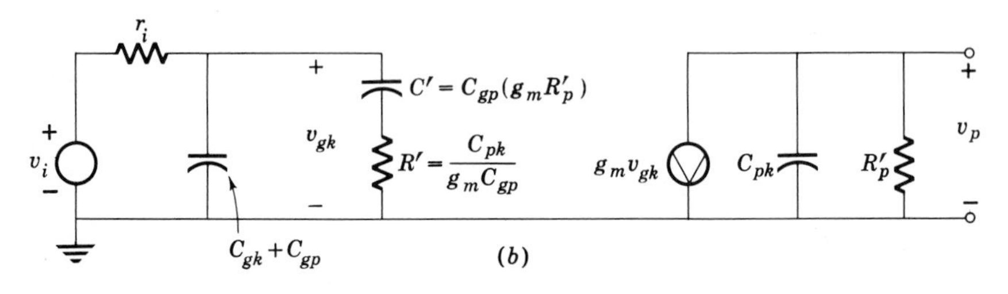

Fig. 13.3-2 Equivalent circuits for the vacuum tube. (*a*) Circuit with Miller impedance; (*b*) final equivalent circuit.

because of the inclusion of the comparatively large output capacitance C_{pk}. The vacuum-tube capacitances are C_{gk}, C_{pk}, and C_{gp}. They are interelectrode capacitances and are *internal* to the tube. Typical values for a triode are $C_{gk} \approx 5$ pF, $C_{gp} \approx 2$ pF, and $C_{pk} \approx 1$ pF. Typical values for a pentode are similar to those of the triode except with regard to C_{gp}. In a pentode, C_{gp} can be as small as 0.01 pF (Chap. 11). Thus the Miller effect is negligible in a pentode.

In addition to these capacitors, stray wiring capacitance and tube-socket capacitance increase the effective total capacitance. To simplify calculations we ordinarily neglect these capacitors. Their effect can be included by adding the value of the stray capacitance to C_{gk}, C_{gp}, and/or C_{pk}.

The Thévenin equivalent circuit of the vacuum tube is shown in Fig. 13.3-2. Note that the Miller impedance is not simply a capacitance. It is easily shown that, for the case of the triode,

$$Y_M = j\omega C_M = j\omega C_{gp} + \frac{1}{1/j\omega C_{gp} g_m R_p' + C_{pk}/g_m C_{gp}}$$

$$= j\omega C_{gp} + \frac{1}{1/j\omega C' + R'} \qquad (13.3\text{-}1)$$

where $R'_p = R_p \| r_p$. The Miller admittance in a pentode is negligible since C_{gp} is small.

The final equivalent circuit for the VT is shown in Fig. 13.3-2*b*. The equivalent circuits are valid for frequencies at which

$$\left|\frac{1}{\omega C_{gp}}\right| \gg \left|\frac{R'_p}{1 + j\omega R'_p C_{pk}}\right| \qquad \omega C_{gp} \ll g_m \tag{13.3-2}$$

VOLTAGE GAIN OF A VACUUM-TUBE AMPLIFIER

Before calculating the gain A_v, let us determine the relative impedances of R' and $1/\omega C'$. Note that the output circuit has a pole at

$$\omega_0 = \frac{1}{R'_p C_{pk}} \tag{13.3-3a}$$

We shall demonstrate that the 3-dB frequency is less than ω_0, and is due to a pole in the input circuit. At frequencies below ω_0,

$$R' < \frac{1}{\omega C'} \qquad \omega < \omega_0 \tag{13.3-3b}$$

Hence R' can be neglected. The Miller impedance is then simply a capacitance,

$$C_M \approx C_{gp}(1 + g_m R'_p) \qquad \omega < \omega_0 \tag{13.3-3c}$$

The gain of the amplifier for frequencies less than ω_0 is then

$$A_v = \frac{v_p}{v_i} = -g_m R'_p \left[\frac{1}{1 + j\omega r_i(C_{gk} + C_M)}\right] \tag{13.3-4a}$$

The 3-dB frequency is

$$\omega_h = \frac{1}{r_i(C_{gk} + C_M)} < \frac{1}{R'_p C_{pk}} = \omega_0 \tag{13.3-4b}$$

Typical orders of magnitude of ω_h and ω_0 are found in the following example.

Example 13.3-1 Find the 3-dB frequency of the VT amplifier shown in Fig. 13.3-1. Let $r_i = 5\ \text{k}\Omega$, $C_{gk} = 5$ pF, $C_{pk} = 2$ pF, $C_{gp} = 1$ pF, $R_p = r_p = 10\ \text{k}\Omega$, and $\mu = 50$.

Solution From (13.3-3*a*)

$$\omega_0 = \frac{1}{R'_p C_{pk}} = \frac{1}{(5 \times 10^3)(2 \times 10^{-12})} = 10^8 \text{ rad/s}$$

and from (13.3-4b)

$$\omega_h = \frac{1}{r_i(C_{gk} + C_M)}$$

$$= \frac{1}{(5 \times 10^3)\{5 \times 10^{-12} + (10^{-12})[1 + (50/10^4)(5 \times 10^3)]\}}$$

$$= 0.64 \times 10^7 \text{ rad/s} \ll \omega_0$$

13.3-1 THE CATHODE FOLLOWER AT HIGH FREQUENCIES

The equivalent circuit of the cathode follower at high frequencies is shown in Fig. 13.3-3. The circuit of the cathode follower is similar to that of the source follower (Fig. 13.2-7). The input impedance, output impedance, and voltage gain can be calculated using the same techniques as before (Prob. 13.27). The cathode follower can be unstable because of the presence of the plate-to-cathode capacitance C_{pk}, stray wiring capacitance, and power-supply impedance.

The oscillations can be eliminated using techniques similar to those described in Example 13.1-3.

13.4 CASCADED RC AMPLIFIERS

In Sec. 7.1 we discussed the analysis and design of cascaded transistor amplifiers. There consideration was given to the midband aspect of the problem, where capacitance effects could be neglected. Now we are faced with the problem of accounting for the effects of emitter bypass and coupling capacitors on the low-frequency response and of internal transistor capacitances on the high-frequency response. The interaction between stages causes the problem to be much more complicated than the single-stage calculations presented in the earlier sections of this chapter. Calculations must start with the last stage and proceed toward the input. In this way we can account for the Miller impedances.

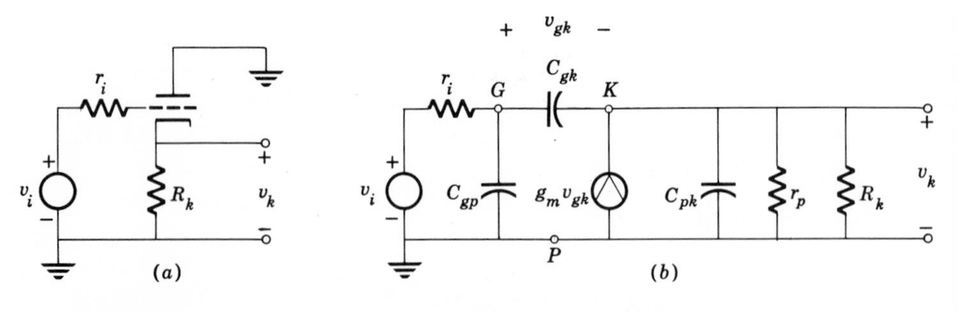

Fig. 13.3-3 The cathode follower at high frequencies. (a) Circuit (bias components omitted); (b) equivalent circuit.

Cascaded amplifiers may be classified as either wideband or narrowband. In this section we consider wideband RC-coupled common- emitter stages. The fact that they are wideband implies that there is a well-defined midband region over which the amplification is approximately constant. As a consequence of this fact, the low- and high-frequency portions of the response can be calculated separately. In this section we are concerned only with the high-frequency response; the low-frequency response can easily be found using the methods outlined in Chap. 12. Our calculations are aimed primarily at determining the 3-dB bandwidth of the amplifier. It will be seen that as we add amplifier stages to our cascade, the gain increases, as expected, but the bandwidth decreases. We first investigate a two-stage cascade, and then consider some of the properties of the gain-bandwidth product for multistage amplifiers. The amplifier and its high-frequency equivalent circuit using identical transistors are shown in Fig. 13.4-1.

To make the calculation of current gain $A_i = i_L/i_i$ tractable, we assume at the outset that $r_{bb'}$ is negligible. Then we use the results of Sec. 13.1-2 to determine the Miller capacitance of the second stage. The result is shown in Fig. 13.4-2. The circuit is still quite complicated. Further simplification is possible if we make the following substitutions:

$$R_2 = R_{c1} \| R_{b2} \| r_{b'e} \tag{13.4-1}$$

$$C_2 = C_{b'e} + C_M = C_{b'e} + C_{b'c}[1 + g_m(R_L \| R_{c2})] \tag{13.4-2}$$

and

$$R_1 = r_i \| R_{b1} \| r_{b'e} \tag{13.4-3}$$

The circuit now takes the form shown in Fig. 13.4-3. We can simplify still further by determining the Miller impedance looking into terminals AA'. The result does not lead to a single Miller capacitance, as in Sec. 13.1-2, because the load is not a pure resistance (Sec. 13.3). The impedance looking to the right from AA' is [Eq. (13.3-1)]

$$\frac{1}{Z_{AA'}} = Y_{AA'} = sC_{b'c} + \frac{1}{1/sC_{b'c}g_mR_2 + C_2/g_mC_{b'c}} \tag{13.4-4}$$

Equation (13.4-4) indicates that $Z_{AA'}$ consists of a capacitor $C_{b'c}$ in parallel with a series RC circuit having the component values indicated in Fig. 13.4-4. It should be noted that the equivalent circuits chosen in Figs. 13.4-4, 13.3-2*b*, and elsewhere were selected so that the resistors and capacitors would not vary with frequency. It is for this reason that $Y_{AA'}$ is put in the form shown in (13.4-4) rather than in some algebraically simpler form.

The complete circuit is shown in Fig. 13.4-5.

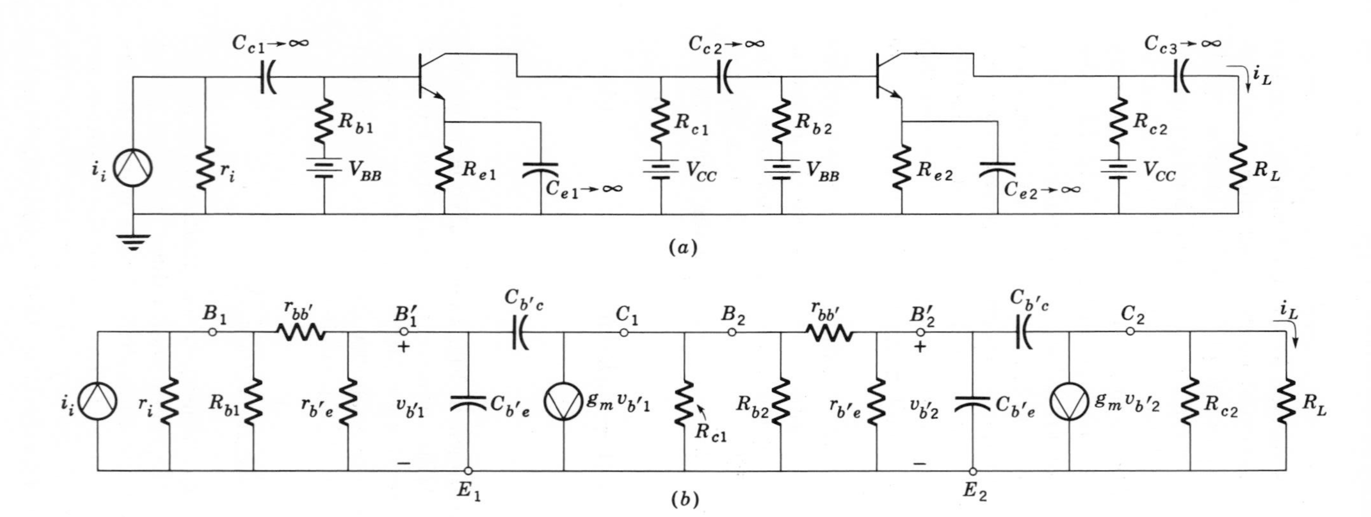

Fig. 13.4-1 (*a*) Two-stage CE amplifier circuit; (*b*) high-frequency equivalent circuit.

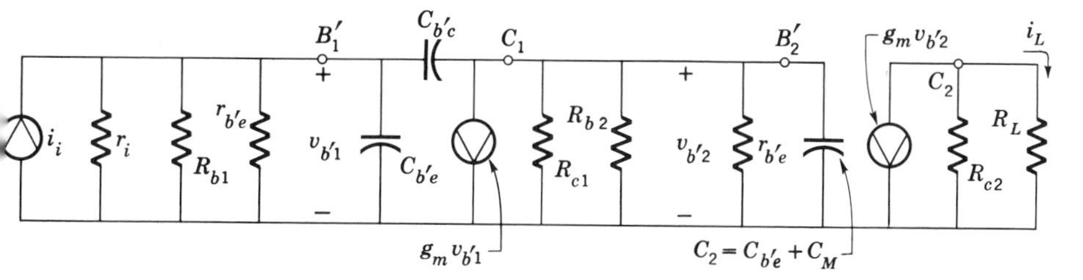

Fig. 13.4-2 First simplification of CE amplifier.

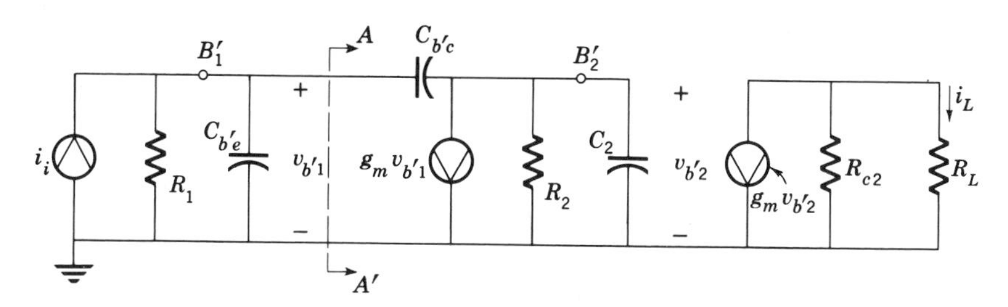

Fig. 13.4-3 Further simplification of CE amplifier.

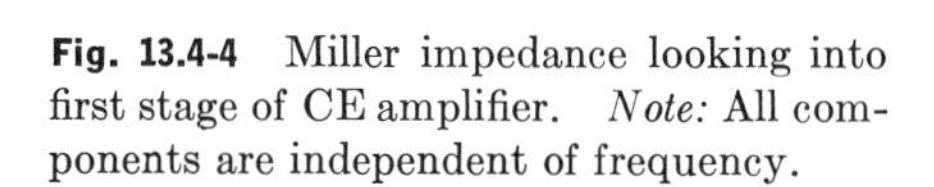
Fig. 13.4-4 Miller impedance looking into first stage of CE amplifier. *Note:* All components are independent of frequency.

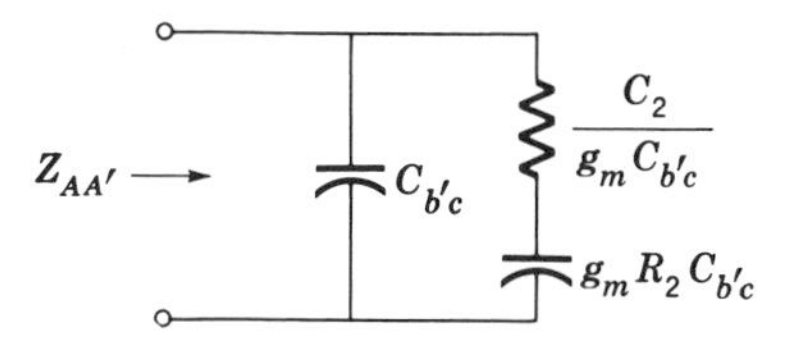

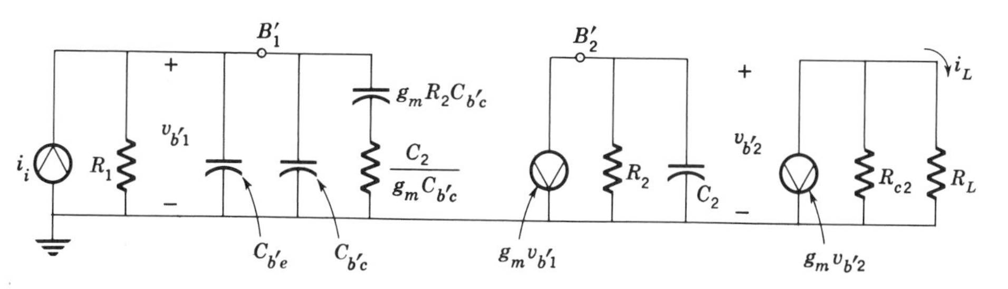

Fig. 13.4-5 Complete equivalent circuit for two-stage CE amplifier.

The current gain A_i is

$$A_i = \frac{i_L}{i_i} = \left(\frac{i_L}{v_{b'2}}\right)\left(\frac{v_{b'2}}{v_{b'1}}\right)\left(\frac{v_{b'1}}{i_i}\right) \approx \left(\frac{-g_m R_{c2}}{R_{c2} + R_L}\right)\left(\frac{-g_m R_2}{\cancel{1 + \frac{s}{\omega_2}}}\right)$$

$$\times \left[\frac{R_1 \cancel{\left(1 + \frac{s}{\omega_2}\right)}}{1 + s\left(\frac{1}{\omega_1} + \frac{1}{\omega_2}\right) + \left(\frac{s^2}{\omega_1 \omega_2}\right)\left(\frac{C_{b'e} + C_{b'c}}{C_1}\right)}\right] \tag{13.4-5a}$$

where

$$\omega_1 = \frac{1}{R_1 C_1} \tag{13.4-5b}$$

$$\omega_2 = \frac{1}{R_2 C_2} \tag{13.4-5c}$$

$$C_1 = C_{b'e} + (1 + g_m R_2) C_{b'c} \tag{13.4-5d}$$

The 3-dB bandwidth of the amplifier of Fig. 13.4-1 can be obtained from (13.4-5*a*) by letting

$$\left[1 - \left(\frac{\omega_h^2}{\omega_1 \omega_2}\right)\left(\frac{C_{b'e} + C_{b'c}}{C_1}\right)\right]^2 + \omega_h^2 \left(\frac{1}{\omega_1} + \frac{1}{\omega_2}\right)^2 = 2 \tag{13.4-6}$$

Solving, we have

$$\omega_h^2 = \left(\frac{\omega_1 \omega_2}{2\gamma^2}\right)\left\{-\left[\frac{\omega_2}{\omega_1} + \frac{\omega_1}{\omega_2} + 2(1 - \gamma)\right] + \sqrt{\left[\frac{\omega_2}{\omega_1} + \frac{\omega_1}{\omega_2} + 2(1 - \gamma)\right]^2 + 4\gamma^2}\right\} \tag{13.4-7a}$$

where

$$\gamma = \frac{C_{b'e} + C_{b'c}}{C_1} \tag{13.4-7b}$$

Example 13.4-1 Calculate the gain and bandwidth of the two-stage common-emitter amplifier shown in Fig. 13.4-1*a*. The component values are $r_{b'e} = 100\ \Omega$, $r_{bb'} = 0$, $C_{b'e} = 100$ pF, $C_{b'c} = 1$ pF, $R_{c1} = R_{c2} = R_{b1} = R_{b2} = r_i = 2$ kΩ, $R_L = 100\ \Omega$, and $h_{fe} = 100$.

Solution The midband gain can be calculated from (13.4-5*a*) by letting $s = j\omega = 0$. This is a valid procedure since all coupling and bypass capacitors are neglected when using the high-frequency equivalent circuit. Then

$$A_{im} = \left(\frac{g_m^2 R_{c2}}{R_{c2} + R_L}\right) R_1 R_2 \approx (g_m r_{b'e})^2 = h_{fe}^2 = 10^4$$

This is the maximum midband current gain possible for this two-stage CE amplifier. To determine the 3-dB frequency, we first calculate ω_1, ω_2, and γ. In this example

$$C_M \approx C_{b'c}(1 + g_m r_{b'e}) \approx h_{fe}C_{b'c} = 100 \text{ pF}$$

Therefore

$$\omega_1 = \omega_2 = \frac{1}{R_2C_2} \approx \frac{1}{r_{b'e}(C_{b'e} + C_M)} = \frac{1}{(100)(10^{-10} + 10^{-10})} = \frac{1}{2 \times 10^{-8}}$$

Using (13.4-7*b*),

$$\gamma = \frac{10^{-10} + 10^{-12}}{10^{-10} + (100)(10^{-12})} \approx \frac{1}{2}$$

The break frequency is then found from (13.4-7*a*).

$$\omega_h^2 = \frac{10^{16}}{(8)(\frac{1}{2})^2}\{-[1 + 1 + 2(\tfrac{1}{2})] + \sqrt{[1 + 1 + 2(\tfrac{1}{2})]^2 + (4)(\tfrac{1}{2})^2}\}$$
$$= 0.08 \times 10^{16}$$

and

$$\omega_h \approx 28 \text{ Mrad/s } (f_h = 4.5 \text{ MHz})$$

13.4-1 CASCADING THE FET

Consider cascading two field-effect transistors. The equivalent circuit is shown in Fig. 13.4-6. Note that Fig. 13.4-1*b* is identical with Fig. 13.4-6*b* if $r_{bb'} = 0$. Hence the analysis for the transistor is valid for the FET. The 3-dB frequency is therefore given by (13.4-7), where

$$R_1 = r_i \| R_g \tag{13.4-8a}$$

$$R_2 = R_d \| r_{ds} \| R_g \tag{13.4-8b}$$

$$C_1 = C_{gs} + C_{gd}[1 + g_m(r_{ds} \| R_d \| R_g)] \tag{13.4-8c}$$

$$C_2 = C_{gs} + C_{gd}[1 + g_m(r_{ds} \| R_d \| R_L)] \tag{13.4-8d}$$

13.4-2 CASCADING THE VACUUM TUBE

Cascading two triode amplifiers is similar to cascading two FETs. The analysis and design are considered in Prob. 13.32. It is instructive to consider cascading two pentode amplifiers. The grid-to-plate capacitance C_{gp} being negligible, the Miller impedance can be neglected. Since there is no feedback, the design can proceed from the input to the output as well as from the output to the input. Two cascaded pentodes are shown in Fig. 13.4-7.

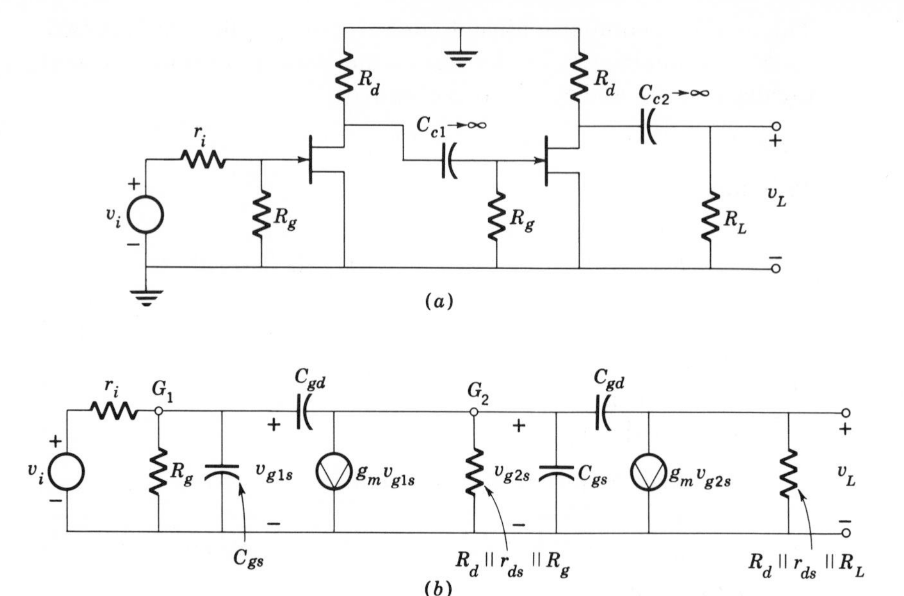

Fig. 13.4-6 (a) Cascaded FET amplifier circuit (bias components omitted); (b) high-frequency equivalent circuit.

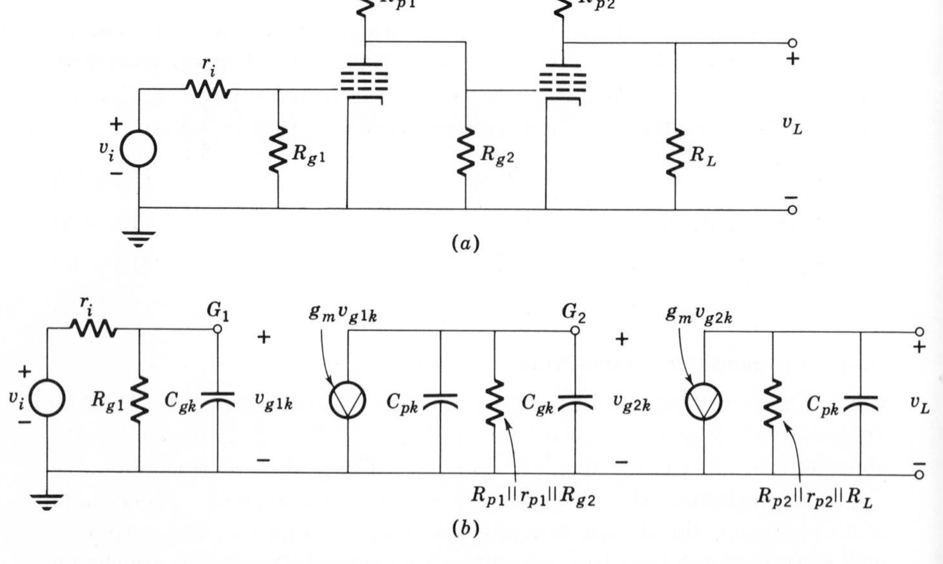

Fig. 13.4-7 (a) Cascaded pentode amplifier circuit (bias components omitted); (b) high-frequency equivalent circuit.

The following simplifications are used:

$$R'_{p1} = R_{p1}\|r_{p1}\|R_{g2}$$
$$R'_{p2} = R_{p2}\|r_{p2}\|R_L$$

The voltage gain of the pentode amplifier is

$$A_v = \left(\frac{v_L}{v_{g2}}\right)\left(\frac{v_{g2}}{v_{g1}}\right)\left(\frac{v_{g1}}{v_i}\right)$$
$$= \left[\frac{-g_m R'_{p2}}{1 + sC_{pk}(R'_{p2})}\right]\left[\frac{-g_m(R'_{p1})}{1 + s(C_{pk} + C_{gk})(R'_{p1})}\right]$$
$$\left[\left(\frac{1}{r_i}\right)\left(\frac{r_i\|R_{g1}}{1 + sC_{gk}(r_i\|R_{g1})}\right)\right] \quad (13.4\text{-}9)$$

The midband gain is

$$A_{vm} = g_m{}^2(R'_{p1}R'_{p2})\left(\frac{R_{g1}}{r_i + R_{g1}}\right) \quad (13.4\text{-}10)$$

The 3-dB frequency is found by solving the cubic equation

$$[1 + \omega^2 C_{pk}{}^2(R'^2_{p2})][1 + \omega^2(C_{pk} + C_{gk})^2(R'^2_{p1})][1 + \omega^2 C_{gk}{}^2(r_i\|R_{g1})^2] = 2 \quad (13.4\text{-}11)$$

If each pole is the same, we have

$$\left(1 + \frac{\omega_h{}^2}{\omega_0{}^2}\right)^3 = 2$$

and

$$\omega_h = \omega_0\sqrt{2^{1/3} - 1} \quad (13.4\text{-}12a)$$

where

$$\omega_0 = \frac{1}{C_{pk}R'_{p2}} = \frac{1}{(C_{pk} + C_{gk})R'_{p1}} = \frac{1}{C_{gk}(r_i\|R_{g1})} \quad (13.4\text{-}12b)$$

The voltage gain of the two-stage pentode amplifier is shown in Fig. 13.4-8. The midband gain is given by (13.4-10), and the 3-dB frequency ω_h is given by (13.4-12a). Note that at frequencies above ω_h, the gain decreases by 18 dB/octave.

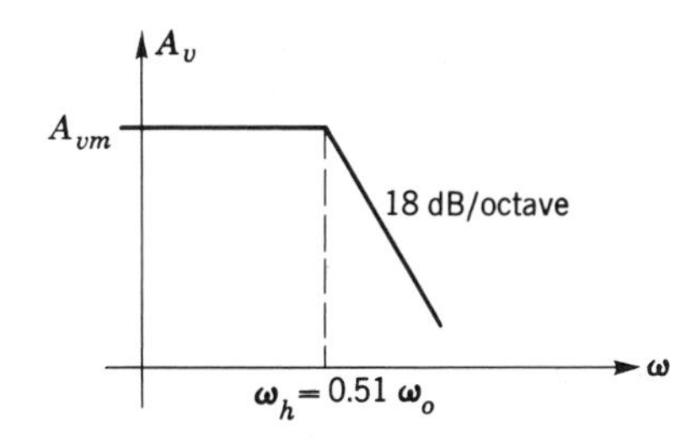

Fig. 13.4-8 Frequency response of three-stage pentode amplifier.

13.5 THE GAIN-BANDWIDTH PRODUCT

13.5-1 GAIN-BANDWIDTH PRODUCT FOR A SINGLE-STAGE AMPLIFIER

In the preliminary design of a multistage wideband amplifier it is very helpful to have some guideposts, or rules of thumb, which can be used to set up a tentative circuit. The gain-bandwidth product is a figure of merit which is often used for this purpose. It is defined in terms of midband gain and upper 3-dB frequency f_h as

$$\mathrm{GBW} = |A_{im}f_h| \tag{13.5-1}$$

For an "ideal" single-stage CE amplifier ($R_L \to 0$), the midband gain is, approximately, h_{fe}, and the upper 3-dB frequency is f_β. Thus

$$\mathrm{GBW} = h_{fe}f_\beta = f_T \approx \frac{h_{fe}}{2\pi r_{b'e}C_{b'e}} = \frac{g_m}{2\pi C_{b'e}} \tag{13.5-2}$$

Manufacturers generally specify f_T, and it is used as a rough estimate of the gain-bandwidth product for a given transistor. The estimate is an upper bound, the actual value being reduced because of the Miller capacitance, which was neglected in arriving at (13.5-2). To refine the estimate, we refer to (13.1-18). Then

$$\mathrm{GBW} = g_mR_{b'e}\left[\frac{1}{2\pi R_{b'e}(C_{b'e}+C_M)}\right] = \frac{g_m}{2\pi(C_{b'e}+C_M)} \tag{13.5-3}$$

Comparing (13.5-3) and (13.5-2), we see that the Miller capacitance reduces the gain-bandwidth product. Note that the GBW product is a function of g_m, $C_{b'e}$, $C_{b'c}$, and R_L (since C_M depends on R_L). Varying $R_{b'e}$ results in a trade-off between the gain and the bandwidth of the amplifier. Let us consider a numerical example.

Example 13.5-1 Find the gain-bandwidth product of the transistor amplifier shown in Fig. 13.5-1. All bias components have been removed for simplicity. The component values are $r_i = 1\ \mathrm{k}\Omega$, $R_c = r_{b'e} = 100\ \Omega$, $C_{b'e} = 100$ pF, $C_{b'c} = 1$ pF, and $h_{fe} = 100$.

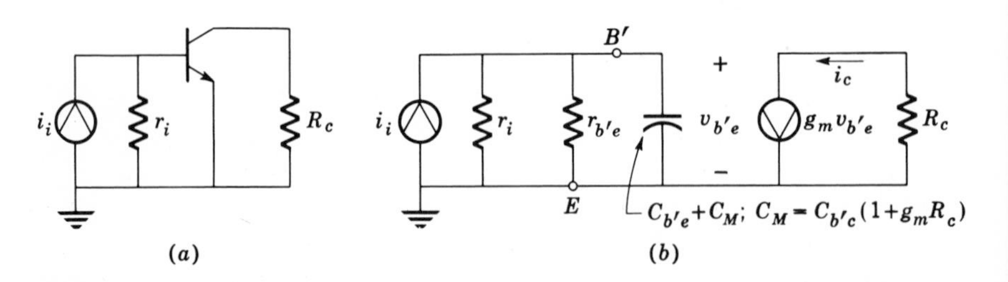

Fig. 13.5-1 (*a*) Amplifier circuit for Example 13.5-1; (*b*) equivalent circuit.

Solution The gain-bandwidth product (13.5-3) is

$$\text{GBW} = \frac{g_m}{2\pi(C_{b'e} + C_M)} = \frac{1}{2\pi(10^{-10} + 10^{-10})} = \frac{10^{10}}{4\pi} = 0.8 \text{ GHz}$$

Note that

$$f_T = \frac{g_m}{2\pi C_{b'e}} = \frac{10^{10}}{2\pi} = 1.6 \text{ GHz}$$

GBW OF A FET AMPLIFIER

The GBW product of a FET amplifier is found from (13.2-3).

$$\text{GBW}_{\text{FET}} = g_m(r_{ds}\|R_d)\left[\frac{1}{2\pi r_i(C_{gs} + C_M)}\right] \tag{13.5-4}$$

Equation (13.5-4) is usually normalized by assuming that

$$r_i = r_{ds}\|R_d$$

(This condition could result when the FET is preceded by another FET.) The GBW then becomes

$$\text{GBW}_{\text{FET}} = \frac{g_m}{2\pi(C_{gs} + C_M)} \tag{13.5-5}$$

A similar expression is obtained for the vacuum tube.

Example 13.5-2 Find the GBW product of a JFET amplifier having the parameters $g_m = 3 \times 10^{-3}$ mho, $C_{gs} = 6$ pF, $C_{gd} = 2$ pF, $r_{ds} = 70$ kΩ, and $R_d = 10$ kΩ.

Solution We first determine the Miller capacitance C_M.

$$C_M = C_{gd}[1 + g_m(r_{ds}\|R_d)] = (2 \times 10^{-12})[1 + (3)(70/8)] \approx 54 \text{ pF}$$

Note that the Miller capacitor is not at all insignificant when using a JFET. The GBW product is, from (13.5-5),

$$\text{GBW} = \frac{g_m}{2\pi(C_{gs} + C_M)} = \frac{3 \times 10^{-3}}{2\pi(60 \times 10^{-12})} \approx 8 \text{ MHz}$$

By comparing this example with Example 13.5-1, the transistor is seen to have a much higher GBW product. The difference is due to the much larger g_m in the transistor.

Example 13.5-3 An IGFET is used instead of the JFET in the preceding example. The FET parameters are $g_m = 2.5 \times 10^{-3}$ mho, $C_{gs} = 6$ pF, $C_{gd} = 0.6$ pF, and $r_{ds} = 60$ kΩ. $R_d = 10$ kΩ, as in Example 13.5-2. Find the GBW product.

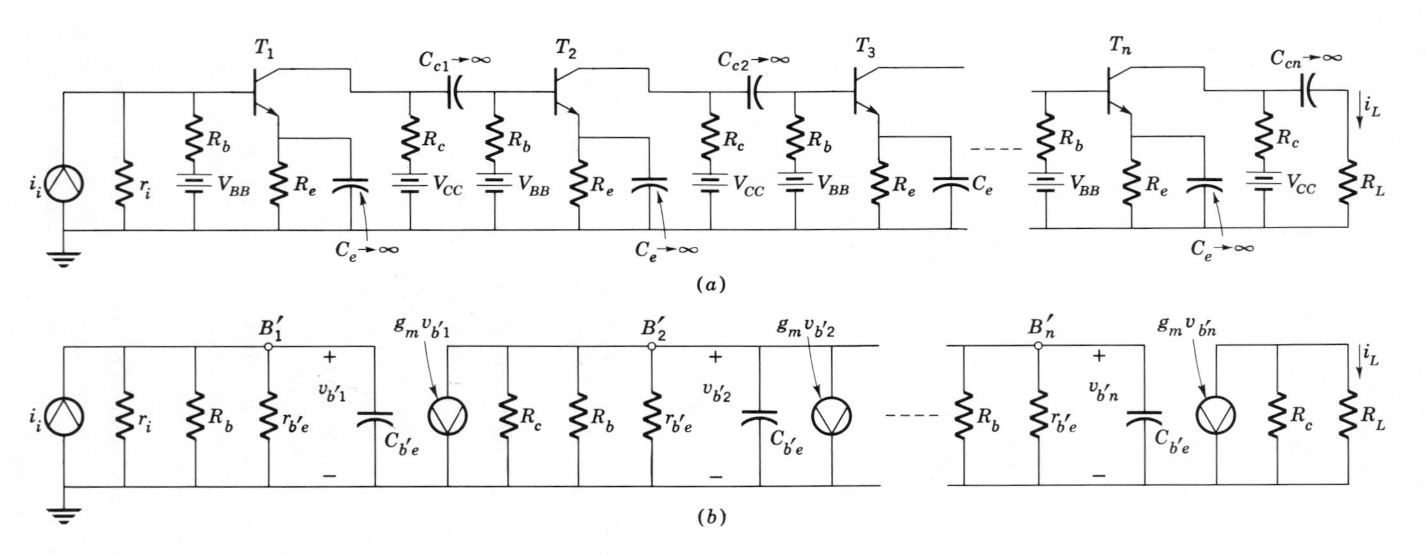

Fig. 13.5-2 (*a*) Multistage CE amplifier circuit; (*b*) high-frequency equivalent circuit.

Solution The Miller capacitance is now

$$C_M = C_{gd}[1 + g_m(r_{ds}\|R_d)] = (0.6 \times 10^{-12})[1 + (2.5)(60/7)] \approx 14 \text{ pF}$$

The GBW product (13.5-5) is then

$$\text{GBW} = \frac{g_m}{2\pi(C_{gs} + C_M)} = \frac{2.5 \times 10^{-3}}{2\pi(20 \times 10^{-12})} \approx 20 \text{ MHz}$$

Note that, while the GBW of the IGFET is significantly higher than that of the JFET, it is still much less than the GBW of the transistor.

13.5-2 GAIN-BANDWIDTH PRODUCT IN A CASCADED AMPLIFIER

Consider the cascade of n identical ideal common-emitter stages shown in Fig. 13.5-2. In this circuit we have assumed $r_{bb'} = C_{b'c} = 0$; so we expect an extremely optimistic estimate of the gain-bandwidth product. To keep the gain expression reasonable in size, we make the following assumptions and simplifications:

$$R_L \ll R_c \tag{13.5-6a}$$

$$R_{b'e} = R_c\|R_b\|r_{b'e} \approx r_i\|R_b\|r_{b'e} \tag{13.5-6b}$$

$$\omega_1 = \frac{1}{R_{b'e}C_{b'e}} \tag{13.5-6c}$$

Then

$$A_i = \frac{i_L}{i_i} = \left(\frac{i_L}{v_{bn}}\right)\left(\frac{v_{bn}}{v_{bn-1}}\right) \cdots \left(\frac{v_{b2}}{v_{b1}}\right)\left(\frac{v_{b1}}{i_i}\right)$$

$$\approx -g_m\left(\frac{-g_mR_{b'e}}{1 + s/\omega_1}\right) \cdots \left(\frac{-g_mR_{b'e}}{1 + s/\omega_1}\right)\left(\frac{R_{b'e}}{1 + s/\omega_1}\right)$$

$$= \frac{(-g_mR_{b'e})^n}{(1 + s/\omega_1)^n} \tag{13.5-7}$$

The midband gain is

$$A_{im} = (-g_mR_{b'e})^n \tag{13.5-8}$$

and the 3-dB frequency f_h is found by setting $|A_i| = A_{im}/\sqrt{2}$. This yields

$$\left[1 + \left(\frac{\omega_h}{\omega_1}\right)^2\right]^{n/2} = 2^{1/2} \tag{13.5-9}$$

Solving,

$$\frac{\omega_h}{\omega_1} = \frac{f_h}{f_1} = \sqrt{2^{1/n} - 1} \tag{13.5-10}$$

Table 13.5-1

n	1	2	3	4	5
$\frac{f_h}{f_1}$	1.0	0.64	0.51	0.44	0.39

This expression is tabulated in Table 13.5-1 for values of n from 1 to 5. The table indicates that the decrease in the 3-dB bandwidth is slow after the second stage is added. Note that we must expect this estimate to be optimistic, representing an upper bound, and that when the Miller capacitance is taken into account, the bandwidth will decrease more rapidly.

Example 13.5-4 Find the midband gain and bandwidth of the two-stage amplifier of Example 13.4-1, using the above approach, and compare with the value obtained when the Miller effect is taken into account.

Solution The midband gain as found in Example 13.4-1 is

$$A_{im} \approx 10^4$$

Using (13.5-6*b*),

$$R_{b'e} = r_{b'e} \| R_c \| R_b \approx r_{b'e} = 100\ \Omega$$

Therefore

$$f_1 = \frac{1}{2\pi(100)(100)(10^{-12})} = 15.9\text{ MHz}$$

From Table 13.5-1, for $n = 2$,

$$f_h = (0.64)(15.9\text{ MHz}) \approx 10\text{ MHz}$$

This is to be compared with $f_h = 4.5$ MHz found in Example 13.4-1, where the Miller effect was taken into account. Note that the overall GBW product, including the Miller capacitance, is $(10^4)(4.5 \times 10^6) = 45$ GHz.

In a given design problem, specifications may include midband gain, 3-dB frequency f_h, and transistor type. We then have to decide whether the particular transistor will do the job and, if so, how many stages will be required.

An alternative problem might specify only midband gain and bandwidth; so the designer is free to choose both the transistor and the number of stages.

In any case, the gain-bandwidth expressions of (13.5-3) and (13.5-5), along with Table 13.5-1, which gives the decrease in bandwidth with n, can be used to establish preliminary design figures. This is illustrated in the following example.

Example 13.5-5 An amplifier is required to have a midband gain of 5000 and a 3-dB frequency f_h of 200 kHz. Transistors having $h_{fe} = 100$, $f_T = 10$ MHz, and $C_{b'c} = 10$ pF are to be used. Design the amplifier.

Solution We first establish a preliminary design, using the crude approximations that GBW $= f_T$ and $A_{im} \approx h_{fe}$ for each stage. Then, to achieve $A_{im} = 5000$, we need at least two stages. Assuming two identical stages, a midband gain per stage of $\sqrt{5000} = 71$ will be required. Since we are assuming a GBW $= f_T$, the 3-dB frequency for each stage will be

$$f_h < \frac{f_T}{A_{im}} = \frac{10^7}{71} = 140 \text{ kHz}$$

For two stages, $n = 2$, and from Table 13.5-1,

$$f_h < (0.64)(140) \times 10^3 = 90 \text{ kHz} \qquad \text{(for the cascade)}$$

We see that two identical stages will not furnish the required gain and bandwidth simultaneously.

Let us investigate the possibility of three identical stages. The midband gain per stage is $\sqrt[3]{5000} \approx 17$. Now the bandwidth per stage is

$$f_h < \frac{10^7}{17} \approx 590 \text{ kHz}$$

According to the table, with $n = 3$,

$$f_h < (0.51)(590) \times 10^3 = 300 \text{ kHz} \qquad \text{(for the cascade)}$$

Thus, if the Miller capacitance does not decrease this figure below the required 200 kHz, three amplifier stages will suffice.

The next step is to estimate the effect of the Miller capacitance. As a first approximation, assume that this effect is the same for all stages, and use (13.5-3) to estimate the GBW product of each stage. To proceed, we assume a quiescent current so that $r_{b'e}$ and $C_{b'e}$ can be estimated. We arbitrarily assume that $I_{EQ} = 2$ mA. Then $r_{b'e} \approx 1200\ \Omega$, $g_m \approx 0.08$ mho, and

$$C_{b'e} = \frac{h_{fe}}{2\pi r_{b'e} f_T} = \frac{100}{2\pi(1200)(10^7)} \approx 1300 \text{ pF}$$

From (13.5-3) the GBW per stage is

$$A_{im}f_h = \frac{g_m}{2\pi(C_{b'e} + C_M)} = \frac{0.08}{2\pi[(13 \times 10^{-10}) + (1 + 0.08\,R_{b'e})(10^{-11})]}$$

where $R_{b'e}$ is the interstage resistance, $R_b \| R_c \| r_{b'e}$. Thus $g_m R_{b'e}$ is the midband gain per stage, which we have already calculated to be 17. Hence $R_{b'e} = 17/0.08 \approx 210\ \Omega$. The single-stage GBW product becomes

$$\text{GBW} = 17f_h = \frac{0.08}{2\pi(13 \times 10^{-10} + 18 \times 10^{-11})} \approx 8.6\text{ MHz}$$

and

$$f_h = 507\text{ kHz} \qquad \text{for each amplifier stage}$$

instead of 590 kHz. From Table 13.5-1, with $n = 3$, the 3-dB frequency for the three stages is

$$f_h = (0.51)(507) \times 10^3 = 258\text{ kHz} \qquad \text{(for the cascade)}$$

which satisfies the bandwidth requirement.

We can also estimate the resistance required between stages and at the input and output by noting that $R_{b'e} = 210\ \Omega$ and $r_{b'e} = 1200\ \Omega$. The resistive load on each collector should be

$$R_{b'e} = R_b \| R_c \| 1200 = 210\ \Omega \qquad \text{or} \qquad R_c \| R_b \approx 260\ \Omega$$

The 3-dB frequency determined above is undoubtedly somewhat high because of the *cumulative* Miller effect, which was neglected. However, there is sufficient margin above the specification to cover this.

13.6 THE TRANSISTOR SWITCH[4–7]

Until now, we have considered only the linear operation of the transistor. In many electronic systems transistors are used as nonlinear elements, i.e., as *controlled switches*. A digital computer, for example, will use, literally, several thousand transistor switches. The speed with which the switches operate is of paramount importance. In this section we consider the time response of a simple transistor switch.

The transistor can be made to operate as a switch by designing the associated circuit so that the transistor is either in the cutoff region or in the saturation region. When the transistor is cut off, no collector current flows and the switch is "open." When the transistor is in *saturation*, maximum collector current flows and the switch is "closed." The switch is "controlled" by the current applied to the base.

Consider the transistor switch shown in Fig. 13.6-1. When the input signal v_i is large and positive, the transistor saturates (Example 3.2-1). The collector current is

$$I_{C,\text{sat}} = \frac{V_{CC} - V_{CE,\text{sat}}}{R_c} \tag{13.6-1a}$$

the base current is

$$I_B = \frac{v_i - V_{BE,\text{sat}}}{r_i} \tag{13.6-1b}$$

and the emitter current is

$$I_E = I_B + I_{C,\text{sat}} \tag{13.6-1c}$$

We note that, because of the saturation condition,

$$I_B \geq \frac{I_{C,\text{sat}}}{h_{FE}} \tag{13.6-1d}$$

When the input signal v_i is negative, the transistor is cut off and i_C goes to zero. Note that if v_i is approximately $+0.5$ V, the collector current in a silicon transistor is approximately zero; hence it is not necessary for v_i actually to be negative in order to obtain the cutoff condition.

The response of the transistor switch to the input voltage v_i is shown in Fig. 13.6-2. In the analysis to follow, we consider each section of this waveform separately. Notice that there is a delay between the leading edge of the input voltage pulse and the time that the collector current takes to reach 90 percent of its maximum value. This time is called the *on time* t_{on}. It is divided into two time intervals, the first called the *delay time* t_d, and the second called the *rise time* t_r.

DELAY TIME

The *delay time* is the time required for the collector current to increase to $0.1I_{C,\text{sat}}$. Another way of describing t_d is to note that it is approxi-

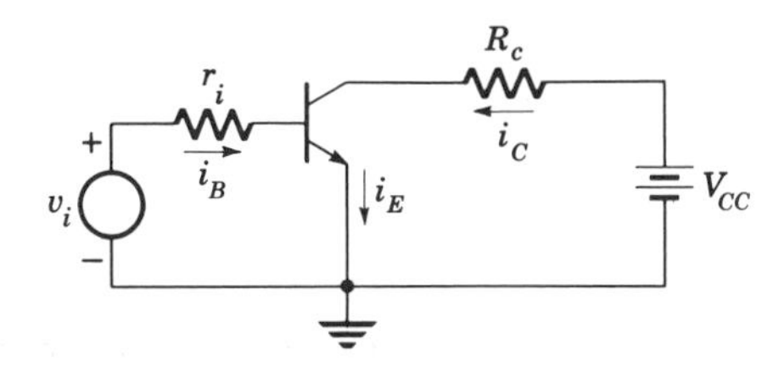

Fig. 13.6-1 The transistor switch.

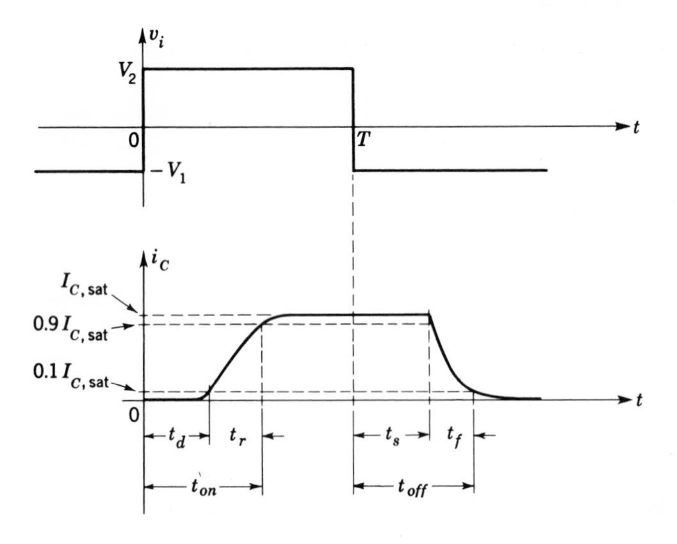

Fig. 13.6-2 Response of a transistor switch to a voltage pulse.

mately equal to the time required for the base-to-emitter (diode) voltage to increase from $-V_1$ to approximately $+0.7$ V. Actually, we should determine the base-to-emitter voltage required to yield a collector current equal to $0.1I_{C,\text{sat}}$. However, this complication is not usually considered necessary.

When negligible collector current flows, the base-to-emitter diode can be approximated by a capacitor, C_{ibo}, which is often specified by the manufacturer. The equivalent circuit used to calculate delay time is shown in Fig. 13.6-3. The voltage $v_{B'E}$ rises exponentially from $-V_1$ to $+V_2$ with a time constant $\tau_d = r_iC_{ibo}$. Thus

$$v_{B'E} \approx V_2 - (V_1 + V_2)\epsilon^{-t/\tau_d} \tag{13.6-2a}$$

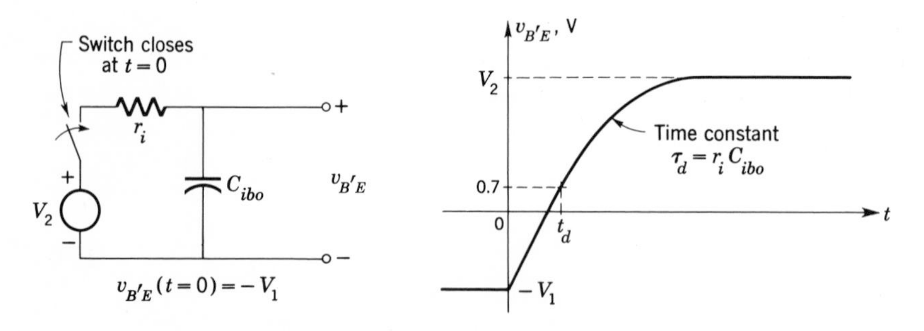

Fig. 13.6-3 Circuit used to calculate the delay time t_d.

and t_d is found by setting $v_{B'E} = 0.7$ V when $t = t_d$. This yields

$$t_d \approx \tau_d \left(\ln \frac{V_1 + V_2}{V_2 - 0.7} \right) \tag{13.6-2b}$$

Example 13.6-1 The 2N3903 (Appendix III) has a maximum value of $C_{ibo} = 8$ pF. If $V_1 = 0.5$ V, $V_2 = 10.6$ V, and $r_i = 10$ kΩ, calculate t_d.

Solution Using (13.6-2*b*), we have

$$t_d \approx (80 \times 10^{-9}) \left(\ln \frac{11.1}{9.9} \right) \approx 9 \text{ nsec}$$

This value is well below the maximum (worst-case) delay time of 35 nsec specified by the manufacturer for the same conditions. However, we have made several assumptions in the derivation of (13.6-2*a*). One assumption is that C_{ibo} is a constant independent of $v_{B'E}$, which is not strictly correct. Actually, C_{ibo} varies inversely with the reverse voltage. Another assumption is that $v_{B'E} = 0.7$ V corresponds to a collector current of $0.1I_{C,\text{sat}}$. Still another approximation is the equivalent circuit employed to calculate t_d. This circuit neglects $r_{b'e}$ and $C_{b'e}$, as well as $C_{b'c}$ and R_c. While these assumptions are valid when $v_{B'E}$ is negative, they are not valid during the time required for $v_{B'E}$ to rise from 0 to 0.7 V. As a result of these approximations and the large spread in transistor parameters, the *typical* delay time calculated, using (13.6-2*b*), will in general differ from the *maximum* (worst-case) expected delay time by a factor of 3 to 4.[6] This is also true for the rise, storage, and fall times.

RISE TIME

The *rise time* is the time required for the collector current to increase from $0.1I_{C,\text{sat}}$ to $0.9I_{C,\text{sat}}$. During this interval the transistor is in the normal active region, and the collector current can be calculated using the small-signal hybrid-pi equivalent circuit shown in Fig. 13.6-4.

To simplify calculations we assume that

$$r_i \| r_{b'e} \approx r_{b'e} \tag{13.6-3a}$$

and

$$C_M = C_{b'c}(1 + g_m R_c) \approx g_m R_c C_{b'c} \tag{13.6-3b}$$

Then

$$\tau_r \approx r_{b'e} C_{b'e} + g_m r_{b'e} R_c C_{b'c} = h_{fe} \left(\frac{1}{\omega_T} + R_c C_{b'c} \right) \tag{13.6-3c}$$

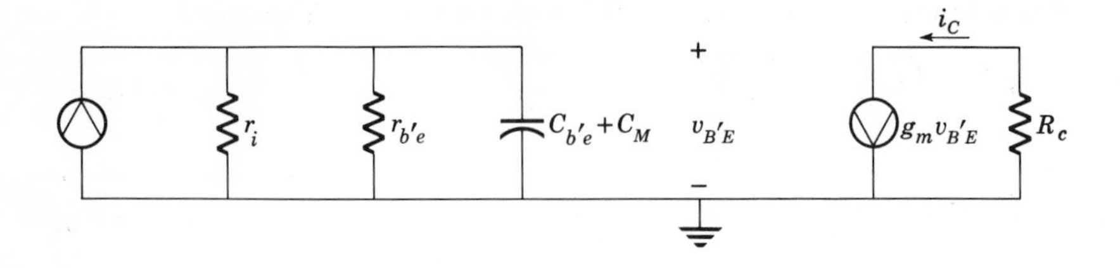

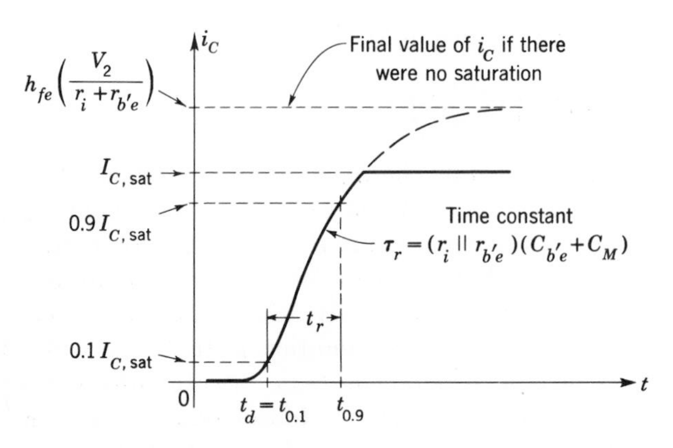

Fig. 13.6-4 Approximate circuit used to calculate $t_r(r_{bb'} = 0)$.

It is now easily shown that the collector current is

$$i_C \approx h_{fe}\left(\frac{V_2}{r_i}\right)(1 - \epsilon^{-t/\tau_r}) \qquad i_C < I_{C,\text{sat}} \tag{13.6-4a}$$

To determine the rise time we note from Fig. 13.6-4 that

$$t_r = t_{0.9} - t_{0.1}$$

where $t_{0.9}$ is the time at which $i_C = 0.9I_{C,\text{sat}}$ and $t_{0.1}$ is the time at which $i_C = 0.1I_{C,\text{sat}}$.

Then

$$t_r = \tau_r\left[\ln\frac{1 - 0.1I_{C,\text{sat}}/(h_{fe}V_2/r_i)}{1 - 0.9I_{C,\text{sat}}/(h_{fe}V_2/r_i)}\right] \tag{13.6-4b}$$

where $I_{C,\text{sat}}$ is given by (13.6-1a).

Example 13.6-2 The 2N3903 has a minimum $\omega_T \approx 1.6$ Grad/s. Calculate the rise time when $V_{CE,\text{sat}} = 0.3$ V, $V_{CC} = 3$ V, $R_c = 270\ \Omega$, $C_{b'c} = 4$ pF, $h_{fe} = 100$, $V_2 = 10.6$ V, and $r_i = 10$ kΩ.

Solution Using (13.6-4*b*), with (13.6-1*a*) and (13.6-3*c*), yields

$$I_{C,\text{sat}} = \frac{3 - 0.3}{270} = 10 \text{ mA}$$

$$\tau_r = 100\left[\frac{10^{-9}}{1.6} + (270)(4 \times 10^{-12})\right] = 170 \text{ nsec}$$

and

$$t_r \approx (170 \times 10^{-9})\left[\ln \frac{1 - \dfrac{10^{-3}}{0.106}}{1 - \left(\dfrac{9}{0.106}\right)(10^{-3})}\right] \approx 14 \text{ nsec}$$

The maximum (worst-case) t_r specified for this transistor is 35 nsec. The major assumption in this calculation is that the small-signal circuit shown in Fig. 13.6-4 can be used when calculating large-signal currents.

It should be noted that (13.6-4*b*) can be simplified by using the expansion $\ln(1 + x) = x - x^2/2 + \cdots$. Thus

$$t_r \approx \left(\frac{1}{\omega_T} + R_c C_{b'c}\right)\left(\frac{0.8 I_{C,\text{sat}}}{V_2/r_i}\right) \tag{13.6-4c}$$

where ω_T is a function of collector current, and $C_{b'c}$ is a function of the collector-base voltage. We have assumed that these parameters are constant.

TURN-ON TIME

The turn-on time t_{on} is

$$t_{\text{on}} = t_d + t_r \tag{13.6-5}$$

Using the results of Examples 13.6-1 and 13.6-2, we have for the 2N3903 transistor

$$t_{\text{on}} = (9 + 14) \text{ nsec} = 23 \text{ nsec}$$

TURN-OFF TIME

The time required for the collector current to decrease from $I_{C,\text{sat}}$ to $0.1 I_{C,\text{sat}}$ when v_i goes negative (Fig. 13.6-2) is called the *turn-off time* t_{off}. The turn-off time is the sum of the *storage time* t_s and the *fall time* t_f, as shown in Fig. 13.6-2.

STORAGE TIME[7]

The storage time is the elapsed time from the trailing edge of the input pulse ($t = T$) to the point where i_C just starts to decrease toward zero.

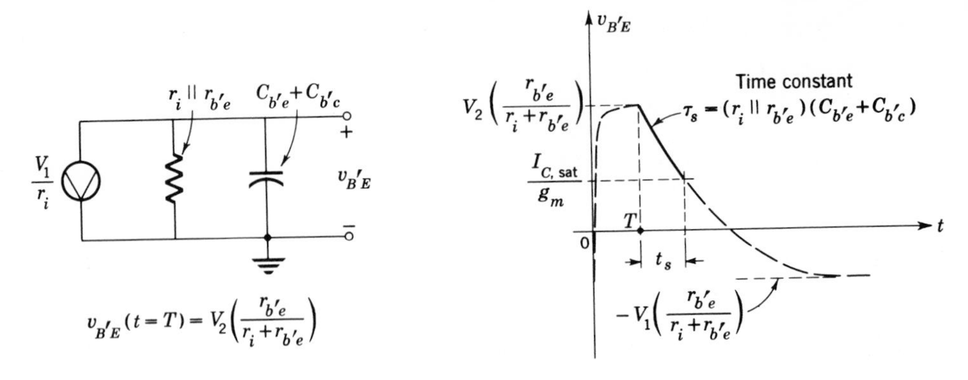

Fig. 13.6-5 Approximate circuit used to calculate $t_s(r_{bb'} = 0)$.

Referring to Fig. 13.6-4, we see that the voltage $v_{B'E}$ increases to a maximum value

$$\max v_{B'E} = \left(\frac{V_2}{r_i}\right)(r_{b'e} \| r_i) = \frac{V_2 r_{b'e}}{r_i + r_{b'e}} \tag{13.6-6}$$

If the transistor did not saturate, i_C would reach the maximum value g_m $(\max v_{B'E})$. However, saturation limits the collector current to the value $I_{C,\text{sat}}$. Now, when v_i reverses, $v_{B'E}$ starts to decrease exponentially. Only when $v_{B'E}$ decreases below the value $I_{C,\text{sat}}/g_m$ does the collector current begin to decrease exponentially from $I_{C,\text{sat}}$. Thus the storage time is the time required for $v_{B'E}$ to decrease from $(V_2/r_i)(r_{b'e}\|r_i)$ to $I_{C,\text{sat}}/g_m$. An equivalent circuit which can be used to calculate the storage time is shown in Fig. 13.6-5. Note that this circuit is similar to the circuit of Fig. 13.6-4, except that the parallel capacitance is $C_{b'e} + C_{b'c}$ rather than $C_{b'e} + C_M$. The reason for this is that, in saturation, i_C does not change and $v_C \approx 0$ V. Thus $C_{b'c}$ is, in effect, placed in parallel with $C_{b'e}$. If we simplify the circuit by assuming that

$$C_{b'c} \ll C_{b'e} \tag{13.6-7a}$$

and

$$r_i \| r_{b'e} \approx r_{b'e} \tag{13.6-7b}$$

then $v_{B'E}$ becomes

$$v_{B'E} \approx -\frac{V_1 r_{b'e}}{r_i} + \left[\frac{(V_1 + V_2) r_{b'e}}{r_i}\right] \epsilon^{-(t-T)/\tau_s} \qquad t \geq T \tag{13.6-8a}$$

where

$$\tau_s \approx r_{b'e} C_{b'e} = \frac{h_{fe}}{\omega_T} \tag{13.6-8b}$$

The storage time is found by setting $v_{B'E}(t = T + t_s) = I_{C,\text{sat}}/g_m$.

$$t_s \approx \tau_s \left\{ \ln \frac{[(V_1 + V_2)r_{b'e}]/r_i}{I_{C,\text{sat}}/g_m + V_1 r_{b'e}/r_i} \right\}$$

$$= \tau_s \left[\ln \frac{h_{fe}(V_1 + V_2)}{I_{C,\text{sat}} r_i + h_{fe} V_1} \right] \tag{13.6-8c}$$

Example 13.6-3 Calculate the storage time of the 2N3903 transistor if $I_{C,\text{sat}} = 10$ mA, $V_2 = 11$ V, $V_1 = 9$ V, $r_i = 10$ kΩ, and $h_{fe} = 100$.

Solution Using (13.6-8*b*) and (13.6-8*c*), we have

$$t_s = \left(\frac{100}{1.6 \times 10^9} \right) \left\{ \ln \frac{(100)(20)}{(10 \times 10^{-3})[10^4 + (100)(9)]} \right\}$$

$$= (62 \times 10^{-9})(\ln 2) \approx 43 \text{ nsec}$$

The maximum (worst-case) storage time is specified by the manufacturer as 175 nsec. The major assumptions in the above analysis are that $C_{b'e}$, $r_{b'e}$, h_{fe}, and ω_T are constants. They do vary with current, and when this variation is included, a somewhat larger value for t_s is found.

FALL TIME

The fall time t_f is the time it takes for the collector current to decrease from $I_{C,\text{sat}}$ to $0.1 I_{C,\text{sat}}$. Calculations are performed using the circuit shown in Fig. 13.6-6. As expected, this is the same circuit used to calculate the rise time, because the transistor is operating in the normal active region during both of these time intervals.

Using this circuit and assuming

$$r_i \gg r_{b'e} \tag{13.6-9a}$$

and

$$C_M \approx g_m R_c C_{b'c} \tag{13.6-9b}$$

we have

$$i_C = -g_m \left(\frac{V_1 r_{b'e}}{r_i} \right) + \left(I_{C,\text{sat}} + \frac{g_m V_1 r_{b'e}}{r_i} \right) \epsilon^{-(t-T-t_s)/\tau_f} \tag{13.6-10a}$$

Letting $i_C\ (t = T + t_s + t_f) = 0.1 I_{C,\text{sat}}$ yields

$$t_f \approx \tau_f \left[\ln \frac{I_{C,\text{sat}} + h_{fe} V_1/r_i}{0.1 I_{C,\text{sat}} + (h_{fe} V_1/r_i)} \right] \tag{13.6-10b}$$

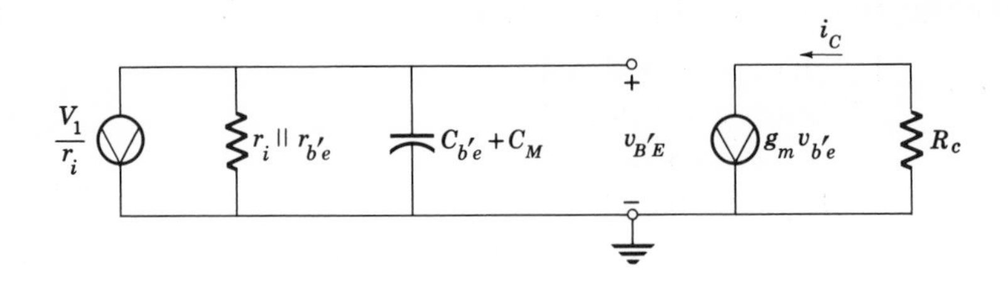

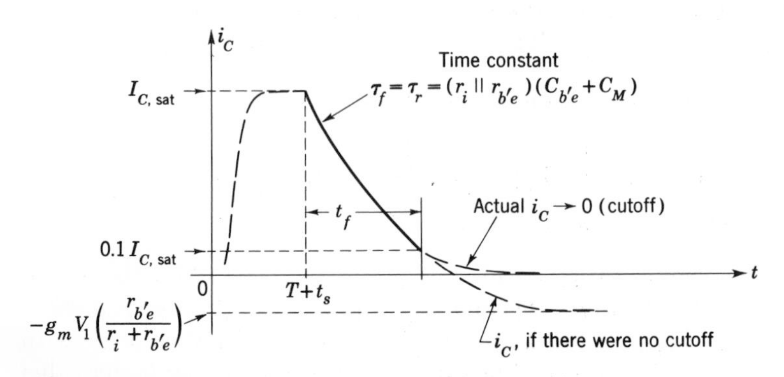

Fig. 13.6-6 Circuit used to calculate the fall time t_f.

where

$$\tau_f \approx h_{fe}\left(\frac{1}{\omega_T} + R_c C_{b'c}\right) \tag{13.6-10c}$$

Example 13.6-4 Calculate the fall time of the 2N3903 transistor. $I_{C,\text{sat}} = 10$ mA, $h_{fe} = 100$, $\omega_T = 1.6$ Grad/s, $R_c = 270\ \Omega$, $C_{b'c} = 4$ pF, $V_1 = 9$ V, and $r_i = 10$ kΩ.

Solution Using (13.6-10*b*) and (13.6-10*c*) yields

$$\tau_f = 170 \text{ nsec}$$

and

$$t_f = (170 \times 10^{-9})\left\{\ln \frac{10^{-2} + [(100)(9)]/(10^4)}{10^{-3} + [(100)(9)]/(10^4)}\right\}$$

$$= (170 \times 10^{-9})\left(\ln \frac{0.1}{0.091}\right) \approx 17 \text{ nsec}$$

Note that this result is approximately one-third of the maximum worst-case value of 50 nsec. The total off time is

$$t_{\text{off}} = t_s + t_f = (120 + 17) \text{ nsec} = 137 \text{ nsec}$$

PROBLEMS

SECTION 13.1

13.1. Measurements indicate that the amplifier of Fig. P13.1 has a midband gain i_L/i_i of 32 dB, an upper 3-dB frequency of 800 kHz, and a quiescent emitter current of 2 mA. Assuming $r_{bb'} = C_{b'c} = 0$, find h_{fe}, $r_{b'e}$, and $C_{b'e}$.

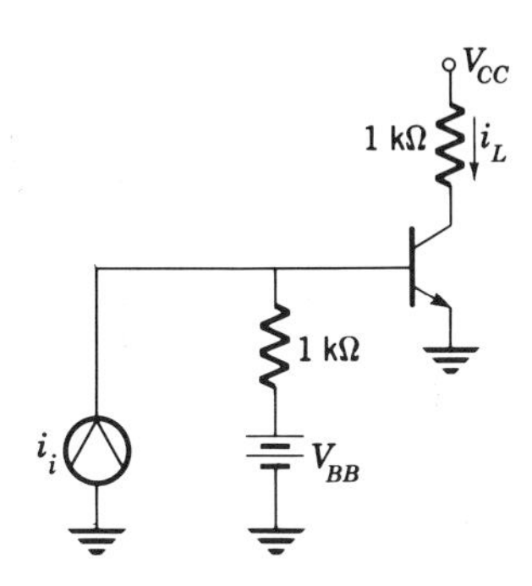

Fig. P13.1

13.2. For the transistor shown, $\omega_T = 10^9$ rad/s, $h_{fe} = 100$, $C_{b'c} = 5$ pF, $r_{bb'} = 0$, and $I_{EQ} = 10$ mA. Find

(*a*) $A_{im} = i_L/i_i$ at midband.
(*b*) The upper 3-dB frequency f_h.

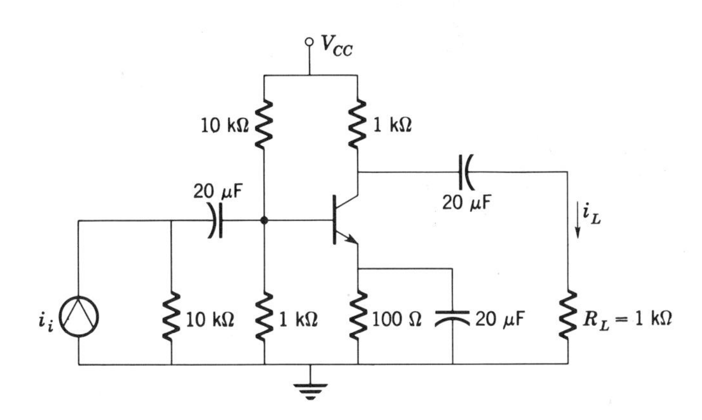

Fig. P13.2

13.3. Repeat Prob. 13.2 if $C_{b'c} = 2$ pF, $h_{fe} = 20$, and $I_{EQ} = 1$ mA.

13.4. Design a single-stage CE amplifier using the MPS6507 (Appendix III). The midband gain is to be 14 dB, and the upper 3-dB frequency is to be as high as possible. The signal source has an internal resistance of 1 kΩ, and the load resistance is 50 Ω.

Maximum required load current swing is ± 1 mA. Specify all resistors, and find the upper 3-dB frequency.

13.5. For the transistor shown, $\omega_T = 10^9$ rad/s, $C_{ob} = 6$ pF, $r_{bb'} = 0$, $I_{EQ} = 1$ mA, and $h_{fe} = 20$.

Find the midband voltage gain and the upper 3-dB frequency.

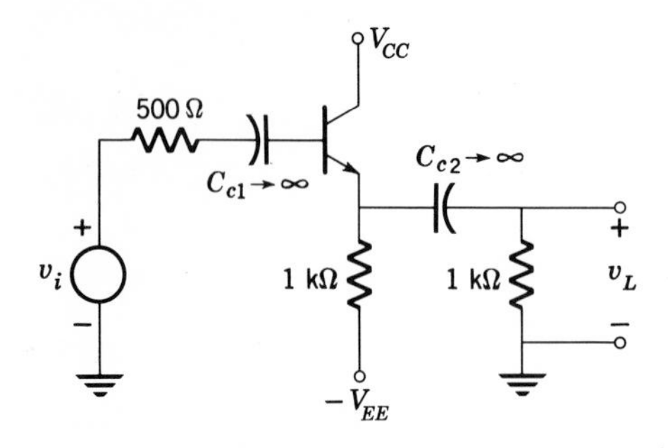

Fig. P13.5

13.6. Find the midband current gain and the upper 3-dB frequency for the cascade amplifier of Fig. P13.6. Assume transistors with the characteristics as given in Prob. 13.5.

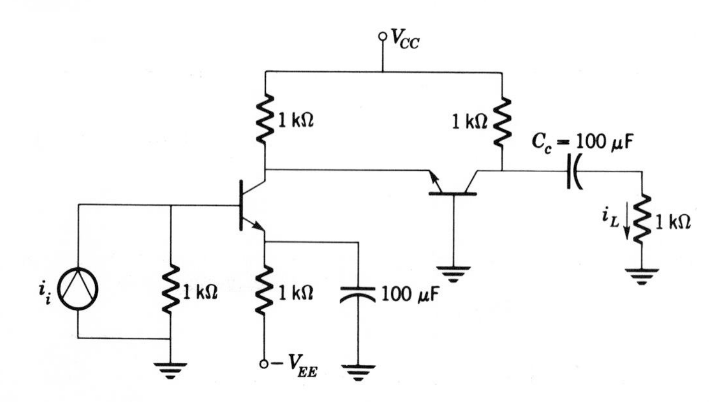

Fig. P13.6

13.7. For the transistor shown, $r_{bb'} = 20\ \Omega$, $r_{b'e} = 1$ kΩ, $C_{b'e} = 1000$ pF, $C_{b'c} = 10$ pF, and $g_m = 0.05$ mho.

Find and sketch the asymptotic voltage-gain characteristic.

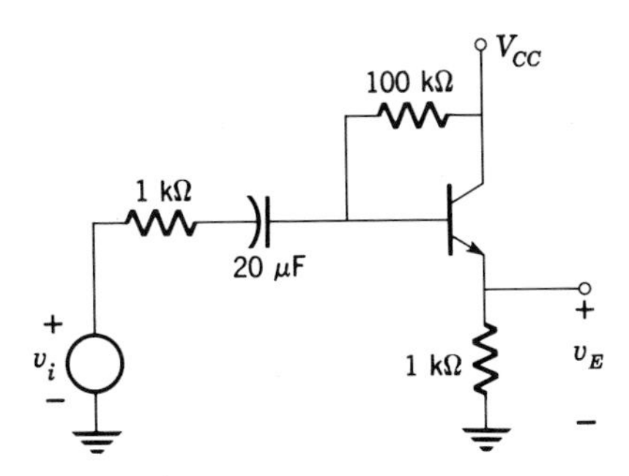

Fig. P13.7

13.8. (*a*) Verify (13.1-13*b*) by analyzing the circuit of Fig. 13.1-6*c* *without* using the feedback formulation.

(*b*) For the circuit of Fig. 13.1-6*c*, show that an exact expression for the output admittance seen by R_L is

$$Y_0 = \left[\frac{j\omega(1 + g_m R_{b'e})C_{b'c}}{1 + j\omega R_{b'e}(C_{b'e} + C_{b'c})}\right]\left[1 + j\omega C_{b'e}\left(\frac{R_{b'e}}{1 + g_m R_{b'e}}\right)\right]$$

Do not use the feedback formulation. Under what conditions does this expression reduce to (13.1-14*a*)? Why does the feedback formulation give a different answer?

13.9. Verify (13.1-20).

13.10. Verify (13.1-21*a*).

13.11. Verify (13.1-22).

13.12. Find and sketch Z_o and Z_i as a function of frequency, for the EF of Fig. P13.7.

13.13. Verify (13.1-26*c*).

13.14. Find the step-function response of the circuit shown in Fig. P13.1. Show that the "rise time" is

$$t_r \triangleq t_{90} - t_{10} = \frac{2.2}{\omega_h}$$

where t_{90} and t_{10} are the times at which the response is 90 and 10 percent of the final value.

SECTION 13.2

13.15. The 2N4223 JFET (see Appendix III for specifications) is used in the circuit of Fig. 13.2-1, with $R_d = 10$ kΩ.

(*a*) Draw the high-frequency equivalent circuit, including the Miller effect.

(*b*) Find the upper 3-dB frequency if $r_i = 50$ Ω.

(*c*) Repeat (*b*) if $r_i = 10$ kΩ.

13.16. The 2N4223 JFET is to be used in the circuit of Fig. 13.2-1. The signal source has a 600-Ω impedance, and the load consists of a 50-kΩ resistance in parallel with 20 pF. Design the stage so that $A_{vm} \geq 20$ dB and $f_h \geq 50$ kHz.

13.17. Verify (13.2-6).

13.18. Verify (13.2-8).

13.19. Verify (13.2-9*a*).

13.20. Verify (13.2-12*b*).

13.21. Design a source follower using the 2N3796 IGFET to operate from a 60-Ω source into a 600-Ω load. Find and sketch $|A_v(j\omega)|$ and $|Z_o|$ for your final design.

13.22. For the source follower shown in Fig. P13.22, plot

(*a*) $|Z_i|$ versus ω

(*b*) $|Z_o|$ versus ω

(*c*) $|A_v| = \left|\dfrac{v_L}{v_i}\right|$ versus ω

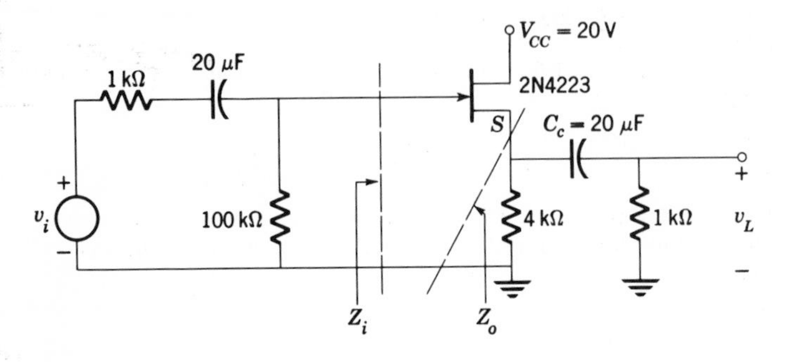

Fig. P13.22

13.23. Calculate and sketch $|Y_o|$ for the source follower shown in Fig. P13.23.

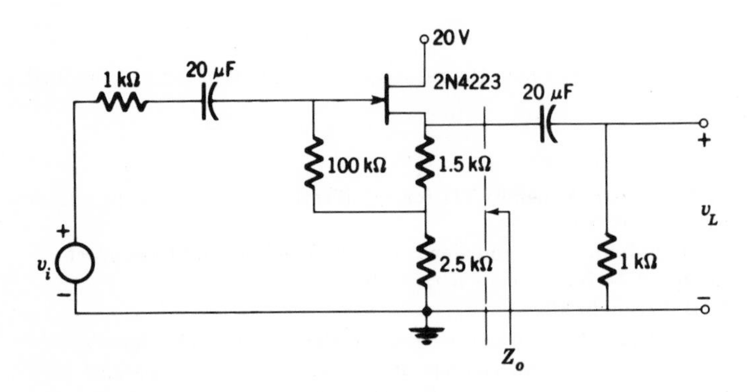

Fig. P13.23

SECTION 13.3

13.24. Verify Eq. (13.3-1).

13.25. For the triode shown in Fig. P13.25, $\mu = 20$, $r_p = 10$ kΩ, $C_{gk} = 10$ pF, $C_{gp} = 5$ pF, and $C_{pk} = 5$ pF.

(*a*) Find the high-frequency response, assuming the Miller impedance is a pure capacitance.

(*b*) Repeat (*a*) without making the assumption that the Miller impedance is a pure capacitance.

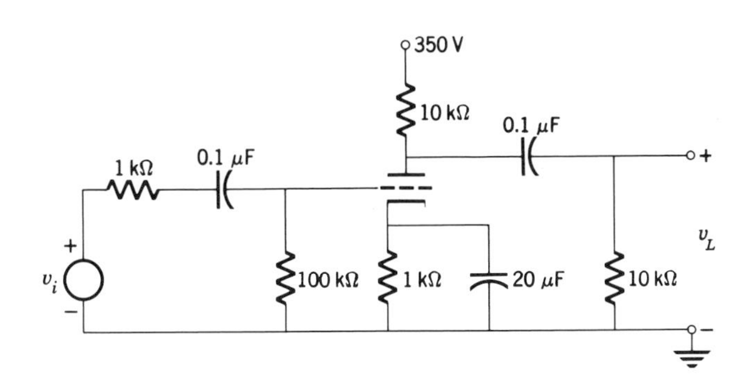

Fig. P13.25

13.26. A pentode is used in the circuit of Fig. P13.25. The pentode parameters are $g_m = 0.003$ mho, $r_p = 200$ kΩ, $C_{gk} = 10$ pF, $C_{gp} = 0.01$ pF, and $C_{pk} = 5$ pF. Find the high-frequency response.

13.27. Derive expressions for the high-frequency voltage gain and the input and output impedances of the cathode follower shown in Fig. 13.3-3.

SECTION 13.4

13.28. Two CE stages, each identical with that shown in Fig. P13.2, are cascaded. The bias network for the second stage then becomes the load of the first stage. Find the midband gain and the 3-db bandwidth.

13.29. Two 2N4223 JFETs are to be cascaded as shown in Fig. 13.4-6*a*. Component values in the circuit are $r_i = 1$ kΩ, $R_g = 1$ MΩ, $R_d = 10$ kΩ, and $R_L = 10$ kΩ. Find the midband gain and the 3-dB bandwidth.

13.30. Verify (13.4-5*a*).

13.31. Verify (13.4-7*a*).

13.32. Two 6SN7 triodes are cascaded as shown in Fig. P13.32. The quiescent point of each triode is $V_{GKQ} = -6$ V and $V_{PKQ} = 250$ V.

(a) Find R_p and R_k
(b) Calculate the midband voltage gain v_L/v_i.
(c) Find the 3-dB frequency f_h.

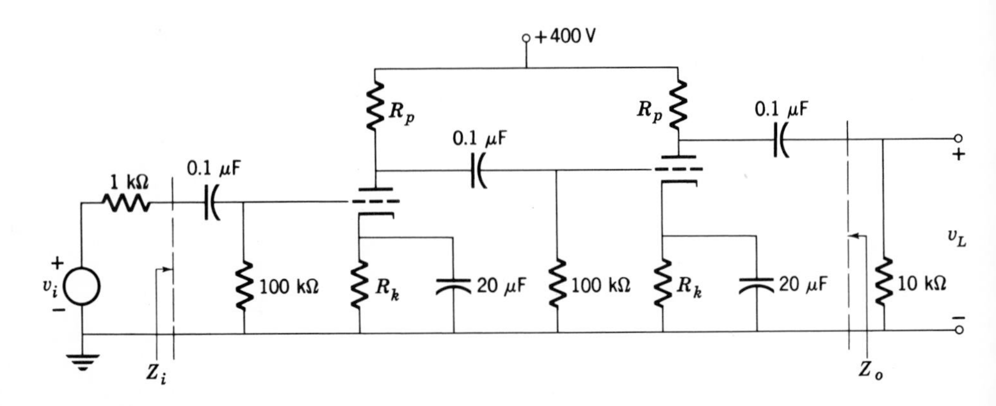

Fig. P13.32

SECTION 13.5

13.33. For the transistors shown, $r_{b'e} = 1$ kΩ, $C_{b'e} = 1000$ pF, $C_{b'c} = 10$ pF, and $g_m = 0.05$ mho.

Determine the GBW product of each configuration.

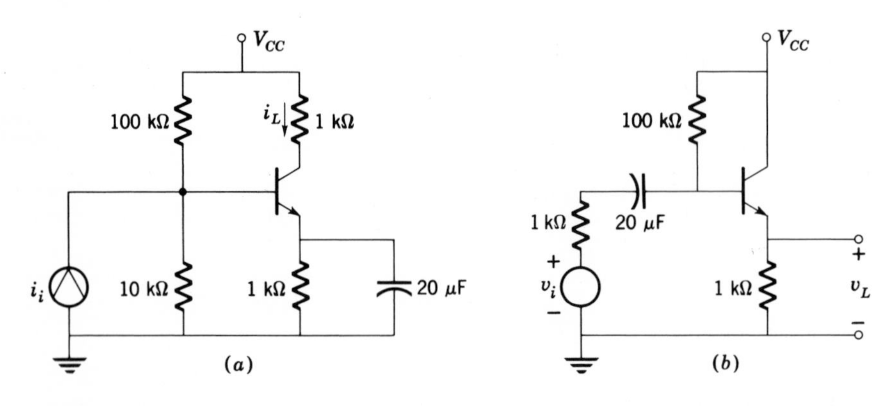

Fig. P13.33

13.34. The transistors are identical with those of Prob. 13.33. Find the GBW product of the cascaded pair.

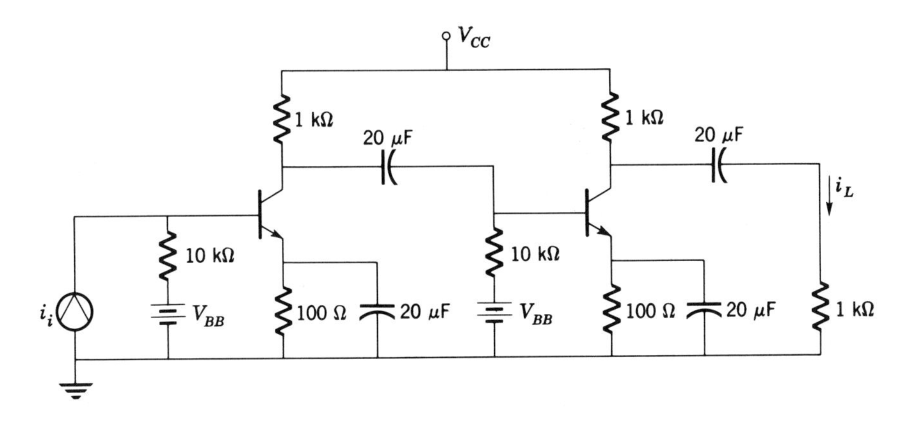

Fig. P13.34

13.35. For the triode shown, $\mu = 20$, $r_p = 10$ kΩ, $C_{gp} = 5$ pF, $C_{gk} = 10$ pF, and $C_{pk} = 0.5$ pF.

Find the GBW product of the cascaded pair.

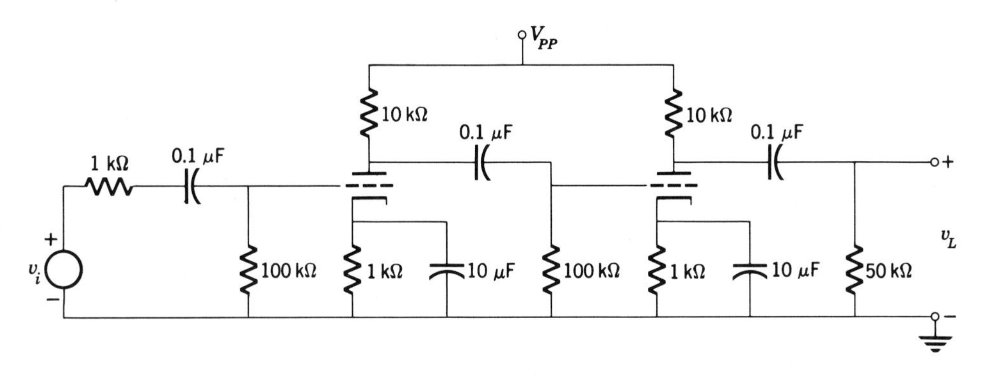

Fig. P13.35

13.36. Repeat Prob. 13.35 if the tubes are pentodes, with $g_m = 0.003$ mho, $r_p = 200$ kΩ, $C_{gk} = 10$ pF, $C_{gp} = 0.01$ pF, and $C_{pk} = 5$ pF.

13.37. A cascade of three identical pentode stages has a 3-dB frequency of 350 kHz. Find the 3-dB frequency of each stage.

13.38. An amplifier is to be designed for a midband current gain of 80 dB and a 3-dB frequency of 350 kHz. Transistors having $h_{fe} = 120$ and $f_T = 90$ MHz are to be used. Neglecting the Miller effect and assuming identical stages, find the number of stages required and the midband gain per stage.

13.39. Repeat Prob. 13.38, taking into account the Miller effect.

13.40. A one-stage FET amplifier uses a low-frequency JFET having the parameters $g_m = 3 \times 10^{-3}$ mho, $r_{ds} = 50$ kΩ, $R_d = 10$ kΩ, $R_g = 500$ kΩ, $C_{gs} = 30$ pF, and $C_{gd} =$ 5pF. Find the GBW product. Assume $R_L = r_i = 10$ kΩ.

SECTION 13.6

13.41. Verify (13.6-2*a*) and (13.6-2*b*).

13.42. Verify (13.6-4*b*) and (13.6-4*c*).

13.43. (*a*) Sketch $v_c(t)$.
(*b*) Calculate t_d, t_r, t_s, and t_f.

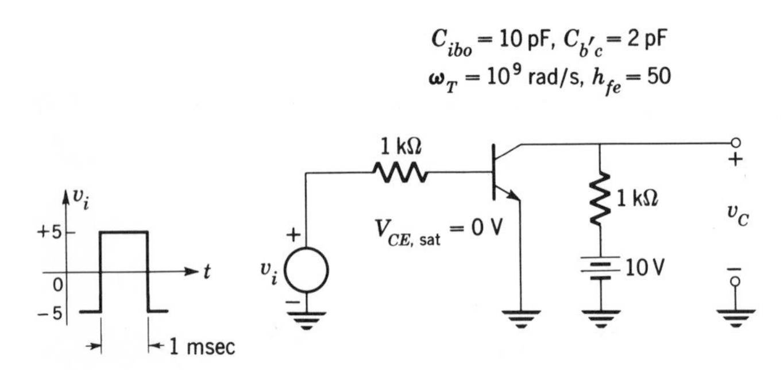

Fig. P13.43

13.44. The transistor used in Fig. P13.43 is operated as shown in Fig. P13.44. The parallel *RC* impedance in the input is used to decrease the delay time and the storage time. Explain.

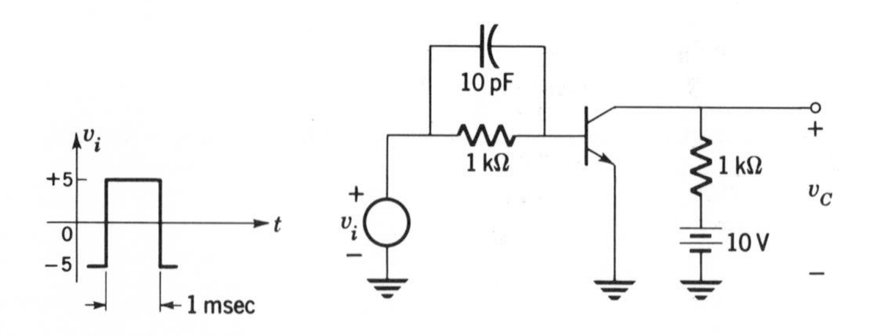

Fig. P13.44

13.45. It is poor practice to attempt to put a negative pulse into the emitter follower shown in Fig. P13.45. Explain why by sketching v_E.

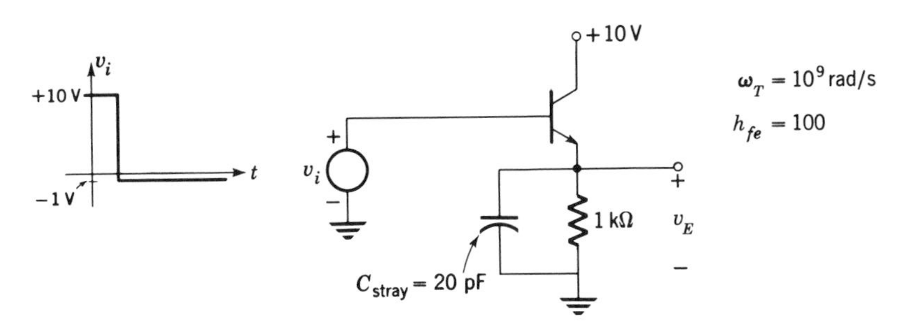

Fig. P13.45

REFERENCES

1. Middlebrook, R.: "An Introduction to Junction Transistor Theory," John Wiley & Sons, Inc., New York, 1958.
2. Motorola Inc. Engineering Staff: "Integrated Circuits," McGraw-Hill Book Company, New York, 1965.
3. Grove, A. S.: "Physics and Technology of Semiconductor Devices," John Wiley & Sons, Inc., New York, 1967.
   Motorola Inc. Engineering Staff: "Analysis and Design of Integrated Circuits," McGraw-Hill Book Company, New York, 1967.
4. Harris, J. N., et al.: "Digital Transistor Circuits," vol. 6, John Wiley & Sons, Inc., New York, 1966.
5. Millman, J., and H. Taub: "Pulse, Digital, and Switching Waveforms," chap. 20, McGraw-Hill Book Company, New York, 1965.
6. Roehr, W. D. (ed.): Switching Transistor Handbook, 2d ed., chap. 5, Motorola Inc., Phoenix, Ariz., 1963.
7. Gibbons, J. F.: "Semiconductor Electronics," chap. 15, McGraw-Hill Book Company, New York, 1966.

14
Tuned Amplifiers

INTRODUCTION

In Chaps. 12 and 13 we discussed the gain and the upper and lower 3-dB frequencies, ω_h and ω_L, of RC-coupled amplifiers. In this chapter we discuss techniques which utilize inductive tuning to extend the upper break frequency ω_h. One technique, called *shunt peaking*, is capable of increasing the gain-bandwidth product by approximately 50 percent. Another technique uses "distributed" amplification, and is capable of delivering high power over a frequency band from 100 kHz to 500 MHz.

Most of this chapter, however, is concerned with a discussion of the amplification of signals within a narrow frequency band centered around a frequency ω_0. These tuned amplifiers are designed to reject all frequencies below a lower break frequency ω_L and above an upper break frequency ω_h.

Tuned circuits are used extensively in almost all communications equipment. An example with which we are all familiar is the radio receiver. When we tune a radio receiver, we are varying ω_0 while keeping

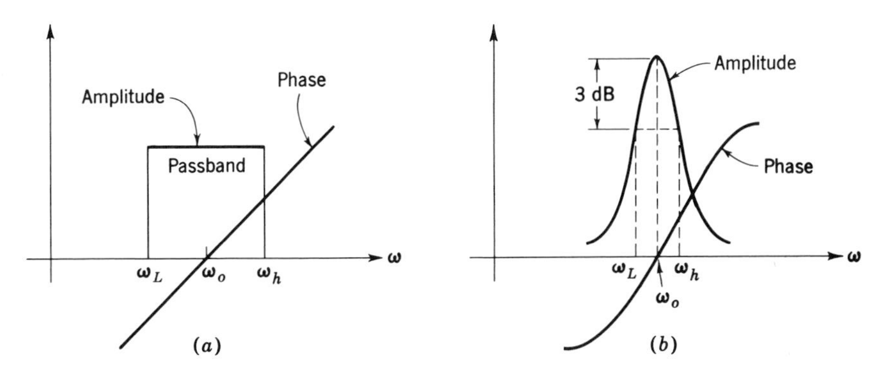

Fig. 14.1 Response of tuned amplifiers. (*a*) Ideal response; (*b*) actual response.

the passband $(\omega_h - \omega_L)$ constant. The particular value of ω_0 at which we set the dial corresponds to the carrier frequency of the broadcast station whose signal we wish to receive, and $\omega_h - \omega_L$ corresponds to the bandwidth required to receive the signal information without significant distortion.

To receive AM and FM signals without significant distortion, and to keep the receiver noise power as small as possible, the ideal bandpass amplifier can be shown to have an amplitude and phase response as in Fig. 14.1*a*. Consider the radio receiver. This characteristic would pass the signal from the station of interest only, rejecting completely signals emanating from stations on adjacent (and all other) channels. A typical response, achieved using the circuits to be described in this chapter, is shown in Fig. 14.1*b*. This characteristic represents an approximation to the ideal response obtained with a single-tuned stage. Better approximations will be considered later in the chapter.

14.1 SINGLE-TUNED AMPLIFIERS

The ordinary common-emitter amplifier is converted to a tuned bandpass amplifier by including a parallel-tuned circuit as shown in Fig. 14.1-1. All bias components have been omitted for simplicity. Let us determine the gain, center frequency, bandwidth, and GBW product of this amplifier.

Before proceeding with these calculations, several practical simplifying assumptions are to be made. First, let us assume that

$$R_L \ll R_c \tag{14.1-1a}$$

and

$$r_{bb'} = 0 \tag{14.1-1b}$$

The effect of $r_{bb'}$ on the response is taken into account in Sec. 14.1-1.

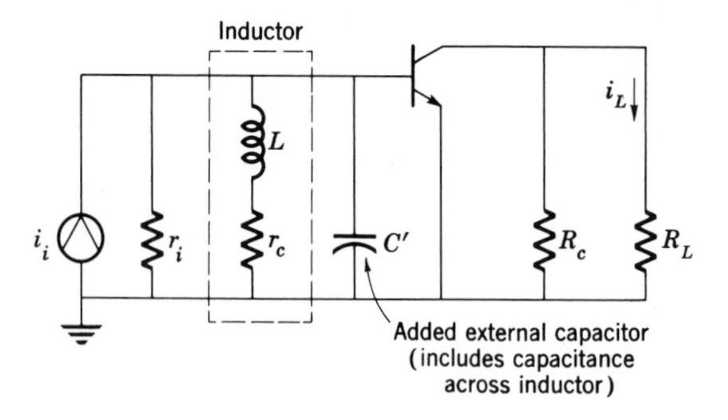

Fig. 14.1-1 A single-tuned amplifier.

The simplified equivalent circuit of this amplifier is shown in Fig. 14.1-2*a*, where

$$C = C' + C_{b'e} + (1 + g_m R_L)C_{b'c} \tag{14.1-2}$$

C' is an added external capacitor, used to tune the circuit and/or help adjust the circuit bandwidth, and $(1 + g_m R_L)C_{b'c}$ is the Miller capacitance C_M. In the series RL circuit (Fig. 14.1-1), which is used as a model for the actual inductor, the resistance r_c represents the losses in the coil. The parallel RL circuit of Fig. 14.1-2*b* is equivalent to the series circuit over the frequency band of interest if the coil losses are low, i.e., if the coil has a high Q_c.

$$Q_c \equiv \frac{\omega L}{r_c} \gg 1 \tag{14.1-3a}$$

The conditions for equivalence are most easily established by equating the admittances of the two circuits shown in Fig. 14.1-2*b* and making use of (14.1-3*a*) as follows:

$$Y_c = \frac{1}{r_c + j\omega L} = \frac{r_c - j\omega L}{r_c{}^2 + \omega^2 L^2} \approx \left(\frac{1}{r_c}\right)\left(\frac{r_c}{\omega L}\right)^2 + \frac{1}{j\omega L} = \frac{1}{R_p} + \frac{1}{j\omega L} \tag{14.1-3b}$$

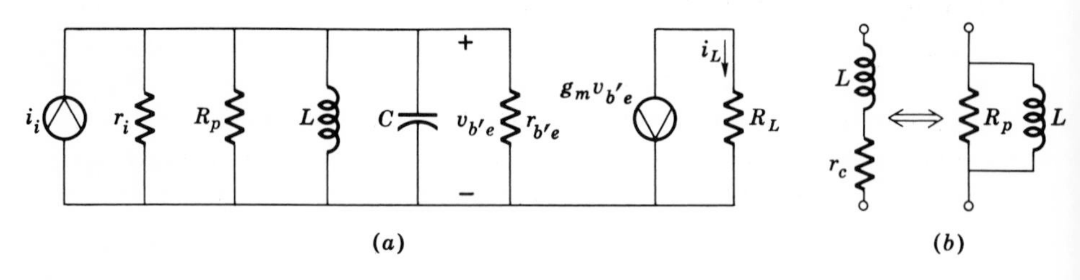

Fig. 14.1-2 (*a*) Equivalent circuit of single-tuned amplifier; (*b*) coil.

and

$$R_p = r_c Q_c^2 = \omega L Q_c \tag{14.1-4}$$

Note that R_p is a function of ω^2 if r_c and L are constant. Representing an inductor by a simple series RL circuit neglects the fact that every inductor has capacitance in parallel with it. In the analyses to follow, this capacitance is assumed to be a part of C'.

Refer to Fig. 14.1-2a. Let

$$R \equiv r_i \| R_p \| r_{b'e} \tag{14.1-5}$$

The current gain of the amplifier is then

$$A_i = \frac{i_L}{i_i} = \left(\frac{i_L}{v_{b'}}\right)\left(\frac{v_{b'}}{i_i}\right) = \frac{-g_m}{1/R + sC + 1/sL} = -\frac{g_m R}{1 + sRC + R/sL} \tag{14.1-6a}$$

Letting $s = j\omega$, this becomes

$$A_i = \frac{-g_m R}{1 + j(\omega RC - R/\omega L)} = \frac{-g_m R}{1 + j\omega_0 RC[(\omega/\omega_0) - (\omega_0/\omega)]} \tag{14.1-6b}$$

where

$$\omega_0^2 = \frac{1}{LC} \tag{14.1-6c}$$

We define the Q of the tuned input circuit at the resonant frequency ω_0 to be

$$Q_i = \frac{R}{\omega_0 L} = \omega_0 RC \tag{14.1-7a}$$

The circuit analysis and the design problems discussed below assume that Q_i and Q_c are greater than 5. This is called the *high-Q* approximation. Using (14.1-7a), (14.1-6b) becomes

$$A_i = \frac{-g_m R}{1 + jQ_i[(\omega/\omega_0) - (\omega_0/\omega)]} \tag{14.1-7b}$$

The gain is a maximum at $\omega = \omega_0$, and is

$$A_{im} = -g_m R \tag{14.1-8}$$

Figure 14.1-3 shows the variation of the magnitude of the gain as a function of frequency, for a single-tuned amplifier. The bandwidth of the amplifier is found by setting

$$|A_i| = \frac{g_m R}{\sqrt{2}} \tag{14.1-9a}$$

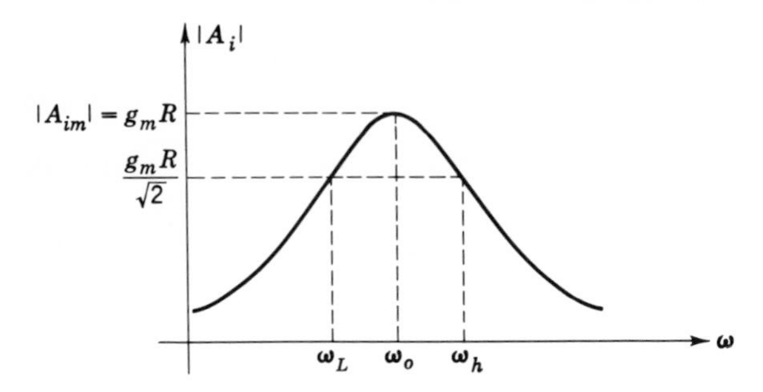

Fig. 14.1-3 Gain vs. frequency for single-tuned amplifier.

in (14.1-7*b*), and solving the resulting equation,

$$1 + Q_i^2 \left(\frac{\omega}{\omega_0} - \frac{\omega_0}{\omega} \right)^2 = 2 \tag{14.1-9b}$$

This equation is quadratic in ω^2 and has two positive solutions, ω_h and ω_L. The 3-dB bandwidth can be shown to be

$$\text{BW} = f_h - f_L = \frac{\omega_0}{2\pi Q_i} = \frac{1}{2\pi RC} \tag{14.1-10}$$

The GBW can be found by combining (14.1-8) and (14.1-10).

$$\text{GBW} = |A_{im}|\text{BW} = \frac{g_m}{2\pi C} \tag{14.1-11}$$

Comparing (14.1-11) with (13.5-3), we see that the gain-bandwidth product is the same as for the *RC*-coupled stage, and depends only on the g_m of the transistor and the total input-circuit capacitance. Thus the addition of the high-*Q* tuning coil has effectively translated the frequency-response curve of the *RC*-coupled amplifier (for the same gain) along the frequency axis without reducing the width of the curve as measured in Hertz.

Example 14.1-1 Design a single-tuned amplifier to operate at a center frequency of 455 kHz with a bandwidth of 10 kHz. The transistor has the parameters $g_m = 0.04$ mho, $h_{fe} = 100$, $C_{b'e} = 1000$ pF, and $C_{b'c} = 10$ pF. The bias network and the input resistance are adjusted so that $r_i = 5$ kΩ and $R_L = 500$ Ω.

Solution In order to obtain a bandwidth of 10 kHz, the *RC* product is [Eq. (14.1-10)]

$$RC = \frac{1}{2\pi \text{BW}} = \frac{1}{2\pi 10^4}$$

where, from (14.1-5),

$$R = r_i \| R_p \| r_{b'e}$$

The input resistance

$$r_i = 5 \text{ k}\Omega$$

$$r_{b'e} = \frac{h_{fe}}{g_m} = 2500 \ \Omega$$

$$R_p = Q_c \omega_0 L = \frac{Q_c}{\omega_0 C}$$

Therefore

$$R = (5 \times 10^3) \| (2.5 \times 10^3) \| \frac{Q_c}{\omega_0 C}$$

and

$$C = \frac{1}{2\pi 10^4 R} = \left(\frac{10^{-4}}{2\pi}\right) \left[\frac{1}{5000} + \frac{1}{2500} + \frac{2\pi(455 \times 10^3)C}{Q_c}\right]$$

Solving for C yields

$$C \approx \frac{0.95 \times 10^{-8}}{1 - 45.5/Q_c}$$

The total input capacitance is

$$C = C' + C_{b'e} + (1 + g_m R_L) C_{b'c} = C' + 1200 \text{ pF}$$

Therefore

$$C' + 1200 \times 10^{-12} \approx \frac{0.95 \times 10^{-8}}{1 - 45.5/Q_c}$$

The choice of a value of Q_c to satisfy this equation is not unique. We know that Q_c must be greater than 45.5 for the capacitance to be positive. The question to be answered by the design engineer is, how large should Q_c be? If, for example, Q_c were chosen to be 45.5, C' would be infinite, C would be infinite, and $L \to 0$. This is not a practical solution! At 455 kHz, a typical range of practical values of Q_c lies between 10 and 150. Let us choose

$$Q_c = 100$$

Then

$$C' \approx 0.016 \ \mu\text{F}$$

and

$$C \approx 0.018 \ \mu\text{F}$$

Note that the input capacitance $C_{b'} = C_{b'e} + C_M$ is negligible.

The inductance required is

$$L = \frac{1}{\omega_0^2 C} \approx 6.9\ \mu\text{H}$$

We can now calculate R_p.

$$R_p = Q_c \omega_0 L \approx 2\ \text{k}\Omega$$

Hence

$$R = r_i \| R_p \| r_{b'e} = 910\ \Omega$$

The resulting midfrequency gain

$$A_{im} = -g_m R = (-0.04)(910) \approx -36.4$$

If a 6.9-μH inductor with a Q_c of 100 is available, the design is complete. If the required Q_c cannot be readily obtained, a transformer may be used in order to transform the input impedances to levels which will allow the specifications to be met. This technique is discussed in Sec. 14.1-2.

The analysis-and-design problem considered above could have employed a FET or VT instead of a transistor. When using a FET or VT, the parallel *RLC* circuit can be formed in the output circuit. Other differences between the transistor and the FET or VT are the element values of the input impedance presented by each device and the presence of the transistor base-spreading resistance $r_{bb'}$. The effect of this resistance on the response of a single-tuned amplifier is discussed in the next section.

14.1-1 THE EFFECT OF $r_{bb'}$ ON THE RESPONSE OF A SINGLE-TUNED AMPLIFIER

The equivalent circuit of the single-tuned amplifier, including $r_{bb'}$, is shown in Fig. 14.1-4. To study the effect of $r_{bb'}$ on the gain and bandwidth, we consider two different operating conditions.

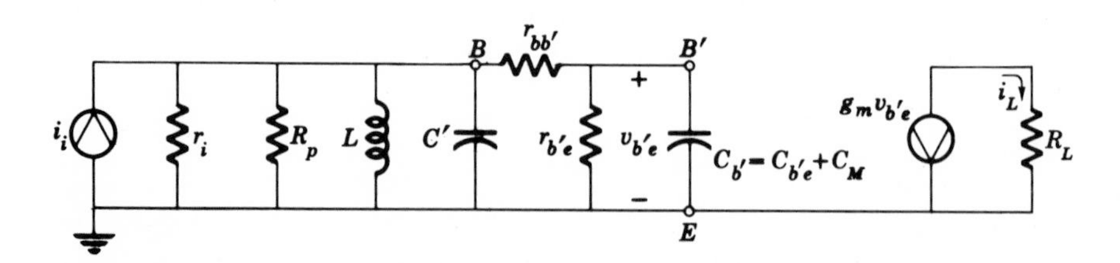

Fig. 14.1-4 Single-tuned amplifier, including $r_{bb'}$.

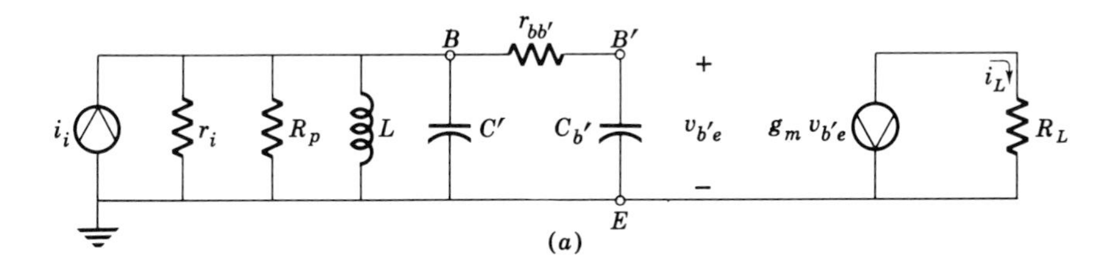

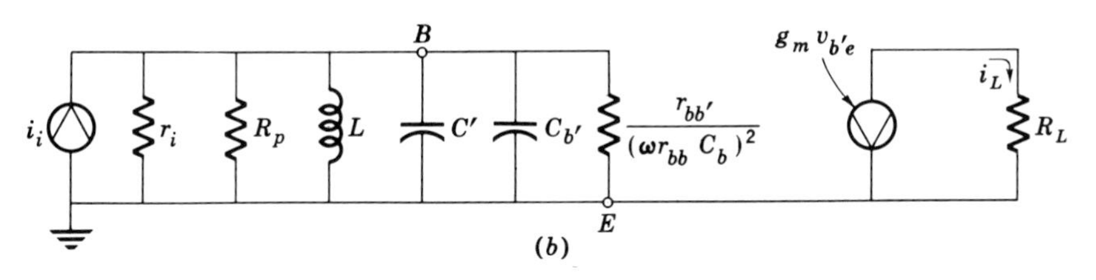

Fig. 14.1-5 Single-tuned amplifier. (a) $\omega_h \gg 1/r_{b'e}C_{b'}$; (b) $r_{bb'} < 1/\omega_h C_{b'}$.

Case 1: $\omega_h \ll 1/r_{b'e}C_{b'}$ When this condition holds, the transistor is operating in the midfrequency range, and internal capacitances can be neglected. This is a common condition when using a device having a GBW larger than that needed. For example, if a transistor has a GBW = 500 MHz and is used in an amplifier tuned to operate at 10 MHz with a gain of 10, all internal capacitances can be neglected. The gain of this amplifier is then

$$A_i = \frac{-g_m R[r_{b'e}/(r_{bb'} + r_{b'e})]}{1 + jQ_i[(\omega/\omega_0) - (\omega_0/\omega)]} \tag{14.1-12a}$$

where

$$R = r_i \| R_p \| (r_{b'e} + r_{bb'}) \tag{14.1-12b}$$

$$C \approx C' \tag{14.1-12c}$$

Note that the midfrequency gain is slightly reduced, and the bandwidth is decreased as compared with the case when $r_{bb'} = 0$.

Case 2: $\omega_h \gg 1/r_{b'e}C_{b'}$ In this region $r_{b'e}$ can be neglected. The equivalent circuit shown in Fig. 14.1-4 can be redrawn as shown in Fig. 14.1-5*a*. If

$$r_{bb'} \gg \frac{1}{\omega_h C_{b'}} \tag{14.1-13}$$

the capacitance $C_{b'}$ can be neglected. However, since $r_{bb'}$ is small, the bandwidth of the circuit will be large and the gain will be small. This region (14.1-13) is *not* a practical operating region.

If

$$r_{bb'} \ll \frac{1}{\omega_h C_{b'}} \tag{14.1-14}$$

we use the technique employed to transform the inductor into a parallel circuit, to reduce the equivalent circuit of Fig. 14.1-5*a* to that of Fig. 14.1-5*b* [Eqs. (14.1-3*a*) and (14.1-3*b*)]. The center frequency is

$$\omega_0 = \frac{1}{\sqrt{LC}} \tag{14.1-15a}$$

where

$$C = C' + C_{b'} \tag{14.1-15b}$$

If the bandwidth is small compared with ω_0, we can let

$$\frac{r_{bb'}}{(\omega r_{bb'} C_{b'})^2} \approx \frac{r_{bb'}}{(\omega_0 r_{bb'} C_{b'})^2} = (Q_{b'})^2 r_{bb'} \tag{14.1-16a}$$

where [Eq. (14.1-14)]

$$Q_{b'} \equiv \frac{1}{\omega_0 r_{bb'} C_{b'}} > 1 \tag{14.1-16b}$$

Then, letting

$$R = r_i \| R_p \| Q_{b'}^2 r_{bb'} \tag{14.1-17a}$$

the bandwidth becomes

$$\text{BW} = \frac{1}{2\pi RC} \tag{14.1-17b}$$

and the gain is

$$A_i = \left\{ \frac{-g_m R}{1 + jQ_i[(\omega/\omega_0) - (\omega_0/\omega)]} \right\} \left(\frac{1}{1 + j\omega/\omega_0 Q_{b'}} \right)$$

$$\approx - \frac{g_m R}{1 + jQ_i[(\omega/\omega_0) - (\omega_0/\omega)]} \tag{14.1-18a}$$

since [Eq. (14.1-14)]

$$\frac{\omega}{\omega_0 Q_{b'}} \leq \frac{\omega_h}{\omega_0 Q_{b'}} \ll 1 \tag{14.1-18b}$$

14.1-2 IMPEDANCE MATCHING TO IMPROVE GAIN

The circuit of Fig. 14.1-1 often yields impractical element values and low gain because of the low effective resistance in the base circuit of the transistor. Since we are dealing with parallel-tuned circuits, low resistance

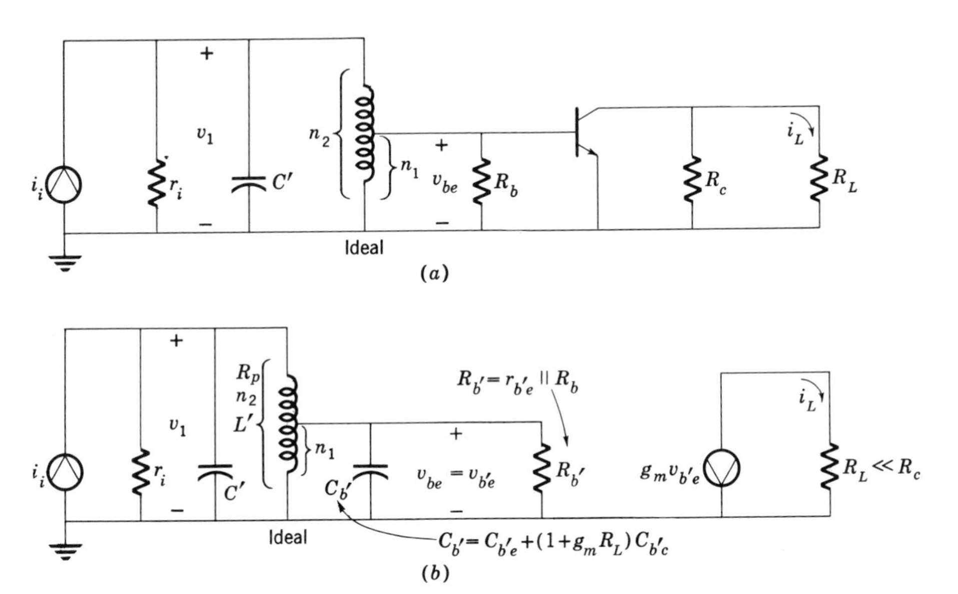

Fig. 14.1-6 Tuned amplifier using an autotransformer ($r_{bb'} = 0$). (*a*) Circuit (bias components omitted); (*b*) equivalent circuit.

implies low Q_i, and consequent difficulty in achieving narrow bandwidth. One method frequently used to circumvent this problem involves the use of a *tapped* inductor as an autotransformer. This serves effectively to transform the low resistance to a more reasonable value. A typical circuit is shown in Fig. 14.1-6.

The turns ratio of the ideal autotransformer is

$$a = \frac{n_1}{n_2} = \frac{v_{b'}}{v_1} < 1 \tag{14.1-19}$$

The impedance of the transistor circuit, which consists of $R_{b'}$ in parallel with $C_{b'}$, is reflected into the input. The resulting parallel RLC circuit is shown in Fig. 14.1-7.

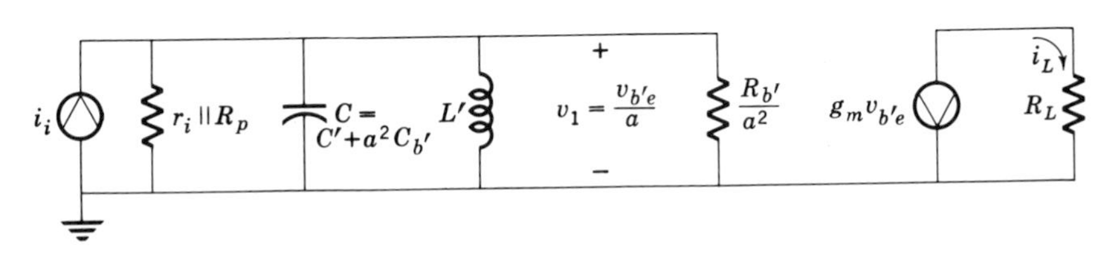

Fig. 14.1-7 Tuned amplifier with reflected impedances.

The gain, center frequency, and bandwidth of the amplifier can now be found as in Sec. 14.1-1. The results are

$$A_i = \frac{i_L}{i_i} = -ag_mR\left\{\frac{1}{1 + jQ_i[(\omega/\omega_0) - (\omega_0/\omega)]}\right\} \tag{14.1-20}$$

where

$$Q_i = \omega_0 RC$$

$$C = C' + a^2C_{b'} \tag{14.1-21}$$

and

$$R = r_i \| R_p \| \frac{R_{b'}}{a^2}$$

The center frequency is

$$\omega_0^2 = \frac{1}{L'C} = \frac{1}{L'(C' + a^2C_{b'})} \tag{14.1-22}$$

If, as is often the case, $C' \gg a^2C_{b'}$,

$$\omega_0^2 \approx \frac{1}{L'C'} \tag{14.1-23}$$

The 3-dB bandwidth, the midfrequency gain, and the GBW can be determined using (14.1-20).

3-dB bandwidth: $$\text{BW} = f_h - f_L = \frac{1}{2\pi RC} \tag{14.1-24}$$

Midfrequency gain: $$A_{im} = -ag_mR \tag{14.1-25}$$

GBW product: $$\text{GBW} = \frac{g_m a}{2\pi C} \tag{14.1-26}$$

The transformer enables us to achieve high gain and a *narrow* bandwidth. However, it decreases the GBW product.

Example 14.1-2 A single-tuned amplifier operates at $f_0 = 455$ kHz and is to have a bandwidth of 10 kHz, $L' = 6.9\ \mu$H, $r_{b'e} = R_{b'} = 1$ kΩ, $r_i = 5$ kΩ, $R_p = 2$ kΩ, $C_{b'e} = 1000$ pF, $g_m = 0.1$ mho, $R_L = 500\ \Omega$, and $C_{b'c} = 4$ pF. Find the turns ratio a and the midfrequency current gain.

Solution We begin by calculating C and R.

$$C = C' + a^2C_1 = C' + 1200 \times 10^{-12}a^2$$

$$\frac{1}{R} = \frac{1}{5000} + \frac{1}{2000} + \frac{a^2}{1000} = (10^{-4})(7 + 10a^2)$$

Then, from Fig. 14.1-7,

$$\omega_0^2 = 4\pi^2(455)^2(10^6) = \frac{1}{L'C} = \frac{1}{(6.9 \times 10^{-6})(C' + a^2\, 1200 \times 10^{-12})}$$

Thus

$$C' + 12a^2 \times 10^{-10} = \frac{10^{-6}}{57.2} \approx 0.017 \times 10^{-6}$$

Since $a \leq 1$,

$$C' \approx 0.017\ \mu\text{F}$$

Using the bandwidth relation (14.1-24) to find R, we have

$$\text{BW} = 10^4 = \frac{1}{2\pi RC} \approx \frac{1}{2\pi RC'} \approx \frac{1}{2\pi R(0.017 \times 10^{-6})}$$

Then $R \approx 930\ \Omega$, and

$$\frac{1}{R} = \frac{1}{930} = (10^{-4})(7 + 10a^2)$$

Solving,

$$a^2 \approx 0.4$$
$$a \approx 0.63$$

The midfrequency gain is

$$A_{im} = -ag_mR \approx -(0.63)(0.1)(930) = -59$$

We now see the advantage of using the autotransformer. In Example 14.1-1, $r_{b'e} = 2500\ \Omega$ and $g_m = 0.04$ ($h_{fe} = 100$). The midfrequency gain was found to be 36.4 (BW = 10 kHz). In this example we let $r_{b'e} = 1\ \text{k}\Omega$ and $g_m = 0.1$ ($h_{fe} = 100$). The transformer multiplied $r_{b'e}$ by $1/a^2 = 2.5$, making it look like 2500 Ω. This resulted in a net current gain of $59/36.4 \approx 1.6$, with the *same* 3-dB bandwidth of 10 kHz.

Note that if the transformer were not used and $r_{b'e} = 1000\ \Omega$ ($g_m = 0.1$ mho), the midfrequency gain would be

$$A_{im} = -g_m(r_{b'e} \| R_p \| r_i) = (-0.1)(5000 \| 2000 \| 1000) \approx -59$$

However, the bandwidth would be

$$\text{BW} = \frac{1}{2\pi RC} = \frac{1}{2\pi(590)(0.017 \times 10^{-6})} \approx 16\ \text{kHz} > 10\ \text{kHz}$$

Thus we see that high gain can be achieved without the transformer, but at the expense of an increased bandwidth.

Example 14.1-3 Redesign the amplifier of Example 14.1-2, using a double-tapped inductance.

Solution An amplifier with a double-tapped autotransformer is shown in Fig. 14.1-8*a*. A simplified equivalent circuit, in which $r_{bb'}$ is assumed to be zero, is shown in Fig. 14.1-8*b*. Note that the effect of the extra tap on the autotransformer is to transform C'' to C'.

$$C' = \left(\frac{n}{n_2}\right)^2 C''$$

Thus, while in Example 14.1-2 $C' = 0.017\ \mu\text{F}$, by employing the second tap with, for example,

$$\frac{n}{n_2} = 2$$

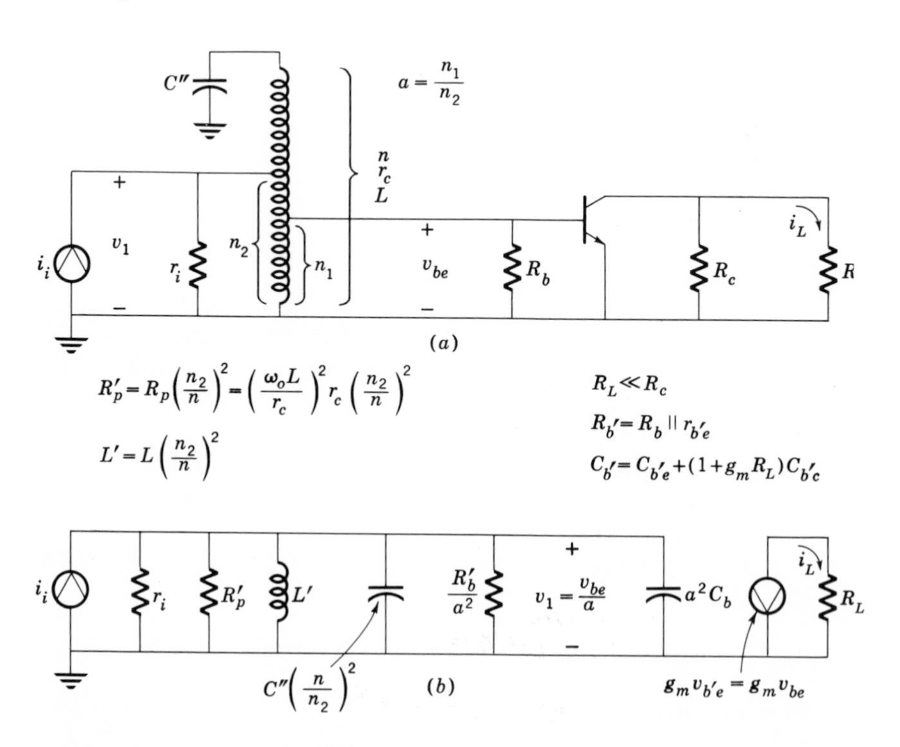

Fig. 14.1-8 Tuned amplifier with double-tapped autotransformer. (*a*) Circuit; (*b*) circuit with reflected impedances.

we can use a smaller external capacitor ($C'' \approx 0.004\ \mu F$). In addition, the inductance L will be larger, and often less difficult to construct.

The second tap is therefore used to adjust impedance levels. Using the result of Example 14.1-2,

$$C''\left(\frac{n}{n_2}\right)^2 = C' \approx 0.017\ \mu F$$

$$L\left(\frac{n_2}{n}\right)^2 = L' = 6.9\ \mu H$$

$$R_p\left(\frac{n_2}{n}\right)^2 = R'_p = 2\ k\Omega$$

and letting

$$n = 2n_2$$

we have

$$C'' \approx 0.004\ \mu F$$

$$L \approx 28\ \mu H$$

and

$$R_p \approx 8\ k\Omega$$

These are more practical element values.

SERIES-RESONANT CIRCUITS

At very high frequencies ($f_0 > 50$ MHz), parallel tuning as used in Examples 14.1-1 to 14.1-3 results in very low Q circuits, and hence wide bandwidths. The reason for this can be seen as follows: If C' were not employed and if r_i, R_p, and R_b were infinite, the circuit Q_i would be, approximately,

$$Q_i \approx \omega_0 r_{b'e} C_{b'}$$

If the Miller capacitance were negligible, $C_{b'} \approx C_{b'e}$, and

$$Q_i \approx \frac{\omega_0}{\omega_\beta}$$

which is less than unity if $\omega_0 < \omega_\beta$. We increase the Q of the circuit by adding C'. This increases the circuit capacitance, but results in a reduction in the required parallel inductance. Very low values of L' are not always obtainable.

At very high frequencies a series-resonant circuit can be employed to provide a very high circuit Q with reasonable inductance values. This technique is illustrated in the following example.

Example 14.1-4 The amplifier shown in Fig. 14.1-9a is to have a 3-dB bandwidth of 2 MHz and a resonant frequency of 100 MHz [the circuit $Q_c = 10^8/(2 \times 10^6) = 50$]. The transistor employed has the parameters $r_{b'e} = 50\ \Omega$, $g_m = 0.1$ mho, $C_{b'e} = 10$ pF, and $C_{b'c} = 1$ pF. The input circuit consists of a 50-Ω resistance ($r_i = 50\ \Omega$) in parallel with a 4-pF capacitor ($C' = 4$ pF). The load resistance R_L is 50 Ω.

(a) Describe the circuit operation.

(b) Find L', L_p, L_c, C_c, and the turns ratio a.

Solution (a) CIRCUIT OPERATION This amplifier is designed so that the circuit Q is determined by the series-resonant circuit. The parallel RLC circuits at the input and the base are each designed to have a

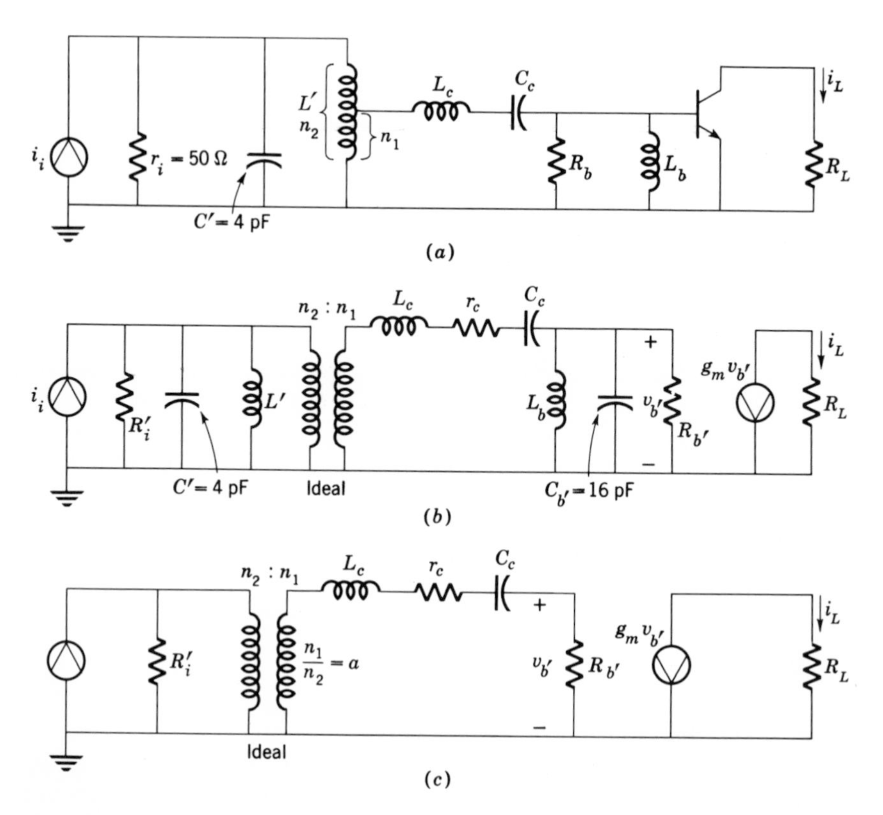

Fig. 14.1-9 Amplifier using a series-resonant circuit. (a) Circuit; (b) equivalent circuit; (c) equivalent circuit without low-Q tuned circuits.

low Q. Quite often, in practice, the two parallel circuits are not even carefully tuned. Figure 14.1-9*b* shows an equivalent circuit, where

$$R_i' = r_i \| \text{ effective parallel resistance of } L'(R_p')$$

$$R_{b'} = R_b \| r_{b'e} \| \text{ effective parallel resistance of } L_b(R_p)$$

$$C_{b'} = C_{b'e} + C_M$$

and

$$\omega_0{}^2 = \frac{1}{L'C'} = \frac{1}{L_cC_c} = \frac{1}{L_bC_{b'}}$$

The circuit Q is determined from the simplified equivalent circuit shown in Fig. 14.1-9*c*. This equivalent circuit assumes that the Q's of the input and base circuits are sufficiently small so that

$$\frac{1}{R_i'} \gg \omega C' - \frac{1}{\omega L'}$$

and

$$\frac{1}{R_{b'}} \gg \omega C_{b'} - \frac{1}{\omega L_b}$$

for ω between ω_L and ω_h. The circuit Q is then essentially the same as the Q of the series-resonant circuit.

$$Q_c = \frac{\omega_0 L_c}{R_{b'} + r_c + a^2 R_i'}$$

(*b*) CIRCUIT DESIGN We begin the design procedure by finding L' and L_b. To resonate the 4-pF input capacitor C' requires that

$$L' = \frac{1}{4\pi^2 f_0{}^2 C'} \approx 0.65\ \mu\text{H}$$

A Q' of 100 at 100 MHz is easily obtained. Assuming that the transformer has this Q', we find R_p'.

$$R_p' = Q'(\omega_0 L') = (100)(2\pi \times 10^8)(0.65 \times 10^{-6}) \approx 41\ \text{k}\Omega$$

Then, since $r_i = 50\ \Omega$,

$$R_i' = r_i \| R_p' \approx 50\ \Omega$$

Let us now consider the base circuit. To resonate $C_{b'} = 16$ pF requires

$$L_b \approx 0.17\ \mu\text{H}$$

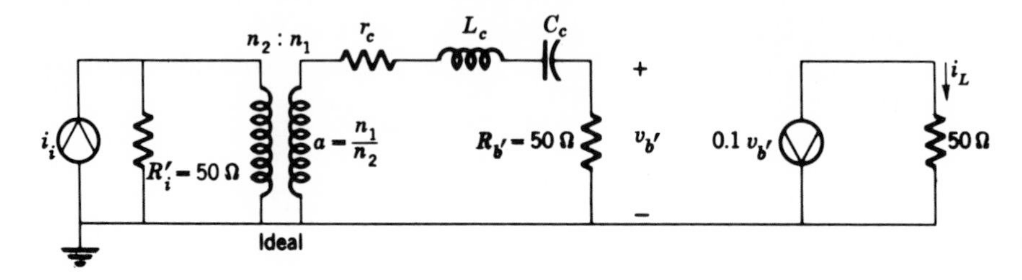

Fig. 14.1-10 Simplified circuit for Example 14.1-4.

Assuming that $Q_b = 100$,

$$R_p = (100)(2\pi \times 10^8)(0.17 \times 10^{-6}) \approx 11\ \text{k}\Omega$$

and since $r_{b'e} = 50\ \Omega$,

$$R_{b'} = R_p \| R_b \| r_{b'e} \approx 50\ \Omega$$

(We assume that $R_b \gg r_{b'e} = 50\ \Omega$.)

Note that the *circuit* Q's are

$$Q_i \approx \omega_0 R_i' C' \approx 0.12$$

and

$$Q_{b'} \approx \omega_0 R_{b'} C_{b'} \approx 0.5$$

The input and base circuit Q's are each much smaller than the required circuit Q of 50. Thus, over the passband of 100 ± 1 MHz, we assume that the equivalent circuit can be represented by Fig. 14.1-10.

To obtain the circuit Q_c of 50 at 100 MHz requires that

$$Q_c = 50 = \frac{1}{\omega_0 C_c(50 + r_c + 50a^2)} = \frac{\omega_0 L_c}{50 + r_c + 50a^2}$$

Note that the Q of the inductor L_c, $\omega_0 L_c/r_c$, must be greater than 50 for the overall circuit Q_c to equal 50. A Q of 250 for the inductor L_c is achievable at 100 MHz. Let us design assuming an inductance with this Q. Thus

$$\frac{\omega_0 L_c}{r_c} = 250$$

Solving for L_c yields

$$\omega_0 L_c = (1.25)(50^2)(1 + a^2)$$

Let

$$a^2 \approx 0.1$$

Then

$$L_c \approx 5.5\ \mu\text{H}$$

and

$$C_c = \frac{1}{4\pi^2 f_0{}^2 L_c} \approx 0.45\ \text{pF}$$

The circuit is tuned by using a variable capacitor for C_c.

14.2 THE CASCODE AMPLIFIER

In the last section we discussed the single-tuned common-emitter bandpass amplifier (the techniques presented are also valid for the FET and VT). We found that its design was complicated by several problems. The first of these was due to the low impedance level in the base circuit of the transistor. The result was that large capacitance values were required to achieve narrow bandwidth. The inductance required to tune this large capacitance then turned out to be quite small. In practice, inductors having a low value of inductance also have a small equivalent parallel resistance ($R_p = \omega_0 LQ$), which leads to an undesired increase in bandwidth. To circumvent this problem, an impedance-transforming device was introduced in the circuit.

A second problem which arises is due to the Miller capacitance. As we found in Chap. 13, the Miller effect reflects a pure capacitance into the input circuit only when the load is a pure resistance. If the load is a complex impedance such as a tuned circuit, or the input impedance of another transistor, it is possible for the Miller effect to produce a negative-resistance component in the input impedance. When this happens, oscillations may occur. In practice, when tuned transformers or tapped inductors are used for coupling between stages, external feedback is often added to cancel the feedback introduced by the collector-base capacitance. This process is called neutralization, and is discussed in Sec. 14.3.

In this section we discuss another solution to the problem, the cascode amplifier (see also Sec. 7.4). In this configuration the load on the common-emitter stage which furnishes the gain is a common-base stage, presenting a very low input impedance. Thus the Miller effect is considerably reduced, and neutralization of the collector-base capacitance is unnecessary. This configuration lends itself nicely to integrated-circuit techniques, and is often used by circuit designers.

Figure 14.2-1 shows the cascode amplifier. For simplicity, we assume that T_1 and T_2 are identical and that the h_{ib} of T_2 is much less than $R_{c1} \| R_{e2}$. In addition, we assume operation below the α cutoff frequency f_α, so that we can neglect all the internal capacitances in T_2.

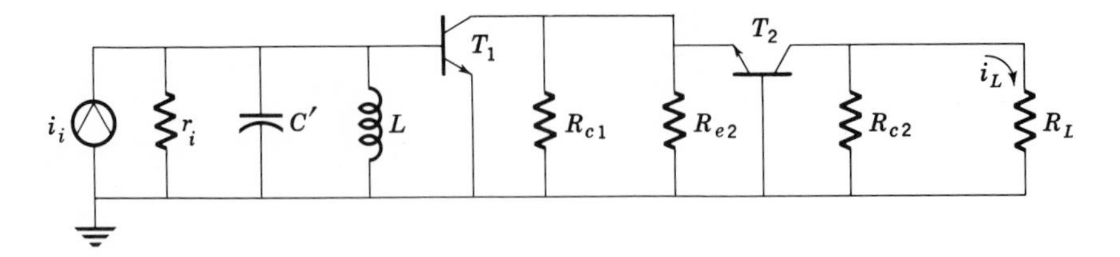

Fig. 14.2-1 Cascode amplifier, ac circuit.

The Miller capacitance reflected into the base of T_1 is

$$C_M = C_{b'c1}(1 + g_{m1}R_{L1}) \tag{14.2-1}$$

In the cascode amplifier the load resistance R_{L1} is the parallel combination of R_{c1}, R_{e2}, and h_{ib2}. Thus

$$C_M \approx C_{b'c1}(1 + g_{m1}h_{ib2}) \tag{14.2-2}$$

But

$$g_{m1} \approx \frac{1}{h_{ib1}} \approx \frac{1}{h_{ib2}}$$

if both T_1 and T_2 are identical transistors operating at the same quiescent point. Hence

$$C_M \approx 2C_{b'c1} \tag{14.2-3}$$

Since $C_{b'c}$ for a high-frequency transistor is of the order of 1 pF, the Miller capacitance is a mere 2 pF. Thus, when using a cascode amplifier, the Miller capacitor can usually be neglected.

These results can be obtained more rigorously by obtaining the equivalent circuit for Fig. 14.2-1, as shown in Fig. 14.2-2. The amplifier of Fig. 14.2-2 is tuned to resonate at $\omega_0 < \omega_\alpha$. The input circuit of T_1 is shown in Fig. 14.2-3, where $r_{bb'}$ is assumed to be equal to zero.

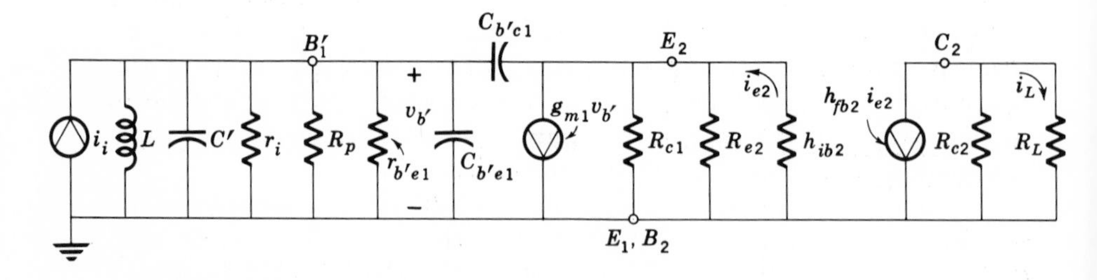

Fig. 14.2-2 Equivalent circuit of cascode amplifier.

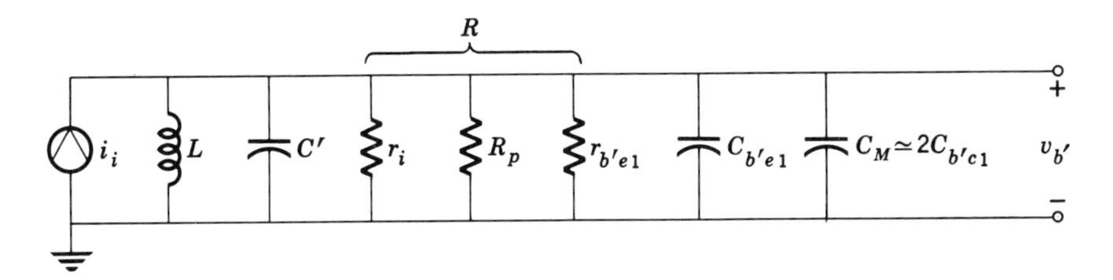

Fig. 14.2-3 Input circuit of cascode amplifier.

The current gain A_i is easily calculated.

$$A_i = \frac{i_L}{i_i} = \left(\frac{i_L}{i_{e2}}\right)\left(\frac{i_{e2}}{v_{b'}}\right)\left(\frac{v_{b'}}{i_i}\right)$$

$$\approx -\left(\frac{h_{fb2}R_{c2}}{R_{c2}+R_L}\right)\left\{\frac{g_{m1}R}{1+jQ_i[(\omega/\omega_0)-(\omega_0/\omega)]}\right\}$$

$$\approx -\frac{g_{m1}R}{1+jQ_i[(\omega/\omega_0)-(\omega_0/\omega)]} \qquad R_L \ll R_{c2} \tag{14.2-4}$$

where

$$R = r_i \| R_p \| r_{b'e1} \tag{14.2-5a}$$

$$C = C' + C_{b'e1} + 2C_{b'c1} \tag{14.2-5b}$$

$$\omega_0{}^2 = \frac{1}{LC} \tag{14.2-5c}$$

$$Q_i = \omega_0 RC \tag{14.2-5d}$$

The GBW product is

$$\text{GBW} = g_{m1}R\left(\frac{1}{2\pi RC}\right) = \frac{g_{m1}}{2\pi(C' + C_{b'e1} + 2C_{b'c1})} \tag{14.2-6}$$

Letting C' be zero, we see that since $C_{b'e1} \gg 2C_{b'c1}$, the GBW product approaches $g_m/2\pi C_{b'e1} = f_T$.

We note that the gain of the cascode amplifier is the same as that obtained using a single common-emitter stage, as given by (14.1-7*b*). The advantage of the cascode amplifier is the reduction in Miller capacitance, and hence isolation between input and output. The disadvantage of using the common-base amplifier is that it does not provide current gain. Thus the cascode amplifier yields the current gain of the grounded-emitter amplifier, along with the isolation of the common-base amplifier. The reduction of the Miller capacitance results in a gain-bandwidth product approximately equal to f_T.

Example 14.2-1 Calculate the gain-bandwidth product of the cascode amplifier shown in Fig. 14.2-4*a*.

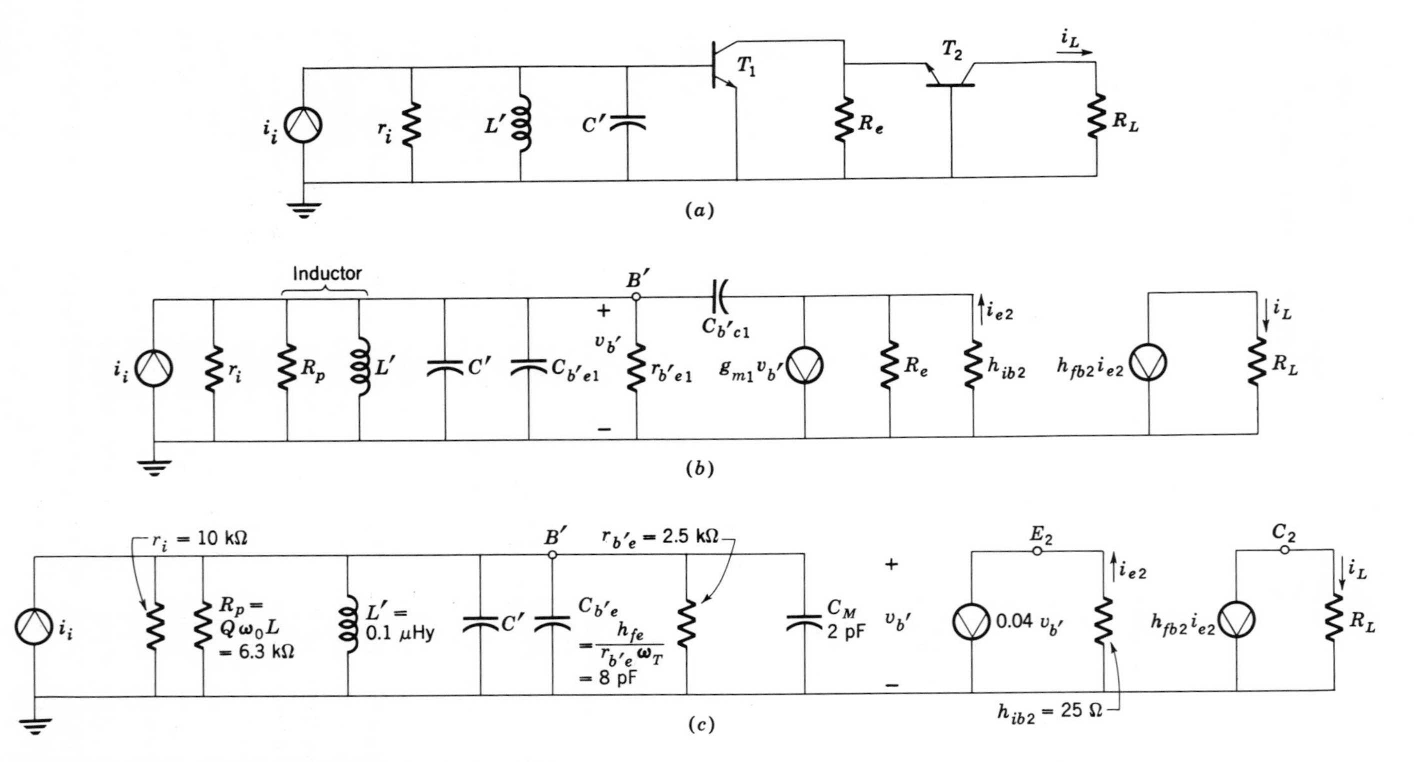

Fig. 14.2-4 (a) Cascode amplifier circuit for Example 14.2-1; (b) equivalent circuit; (c) reduced equivalent circuit.

In this circuit, T_1 and T_2 are identical, $h_{fe1} = h_{fe2} = 100$, $I_{EQ1} = I_{EQ2} = 1$ mA, $C_{b'c} = 1$ pF, $\omega_T = 5$ Grad/s, $L' = 0.1\ \mu$H ($Q = 100$ at 100 MHz), and $r_i = 10$ kΩ. The center frequency of the amplifier is $f_0 = 100$ MHz. We neglect $r_{bb'}$, for simplicity.

Solution The equivalent circuit for the amplifier is shown in Fig. 14.2-4*b*, and can be reduced to the circuit shown in Fig. 14.2-4*c*. Using Fig. 14.2-4*c*, we first determine C'. Thus

$$C' + C_{b'e} + C_M = C' + (8 + 2) \times 10^{-12} = \frac{1}{\omega_0{}^2 L'} = 25 \times 10^{-12}$$

and

$$C' \approx 15 \text{ pF}$$

The 3-dB bandwidth of the circuit is

$$\text{BW} = \frac{1}{2\pi RC}$$

where

$$C = C' + C_{b'e} + C_M = 25 \text{ pF}$$

and

$$\frac{1}{R} = \frac{1}{r_i} + \frac{1}{R_p} + \frac{1}{r_{b'e}} = \frac{1}{10^4} + \frac{1}{0.63 \times 10^4} + \frac{1}{0.25 \times 10^4} = 6.6 \times 10^{-4}$$

Hence

$$\text{BW} = \frac{6.6 \times 10^{-4}}{2\pi(25)(10^{-12})} = 4.2 \text{ MHz}$$

The midfrequency gain of the amplifier is

$$A_{im} = \frac{i_L}{i_i} = \left(\frac{i_L}{i_{e2}}\right)\left(\frac{i_{e2}}{v_{b'}}\right)\left(\frac{v_{b'}}{i_i}\right) = -h_{fb}\,(0.04)\left(\frac{1}{6.6 \times 10^{-4}}\right) \approx -60$$

The GBW product is 252 MHz.

14.3 NEUTRALIZATION

The cascode amplifier discussed in the last section is attractive from several points of view: it reduces the Miller effect to the point where it is almost negligible; and if several stages are required, there is no interaction between them because of the isolation provided by the common-base stage. Thus the alignment (tuning) of the amplifier in production is comparatively easy.

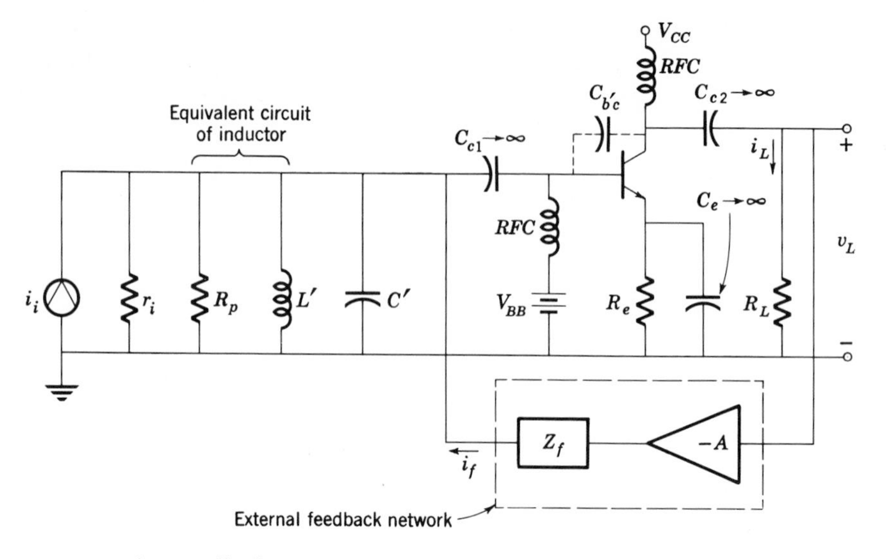

Fig. 14.3-1 A neutralized common-emitter stage.

A disadvantage of the cascode amplifier is that it is wasteful of transistors, since it uses two transistors to obtain the gain usually obtained from one. When integrated circuits are available, this is not a major consideration. However, in some other cases it is important, and cascaded-tuned common-emitter stages must be used. For this type of amplifier, the Miller effect can lead to the possibility of oscillations unless some form of "neutralization" is used to reduce the feedback through $C_{b'c}$. In this section, we discuss the principle of neutralization of a common-emitter amplifier. The analysis presented below is also valid for the FET and triode.

A neutralized amplifier is shown in Fig. 14.3-1. Its operation can be explained qualitatively as follows: The base voltage depends on the output voltage v_L because of the current fed back through $C_{b'c}$ (the Miller effect). To neutralize this collector-to-base feedback, we sample the output voltage and feed back an additional current i_f through the external feedback network. The feedback network is adjusted so that i_f is equal in magnitude but opposite in phase to the current through $C_{b'c}$. This forces the loop gain T to be zero. When this is done, the two feedback currents cancel, and the base voltage is independent of v_L. The stage is then said to be "neutralized."

To determine the required feedback network, consider the equivalent circuit shown in Fig. 14.3-2. The output impedance R_L is omitted because only the two feedback paths are of interest. This circuit can be simplified by obtaining the Thévenin equivalent circuit looking into $B'E$. The

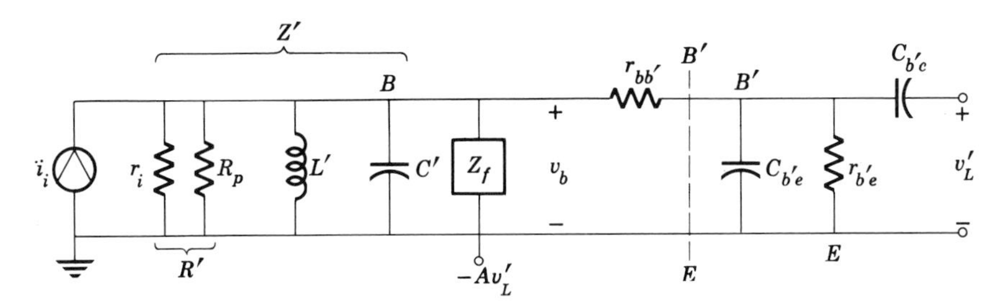

Fig. 14.3-2 Equivalent circuit of neutralized amplifier.

Thévenin voltage is

$$v_t = \left(\frac{Z_{b'e}}{Z_{b'e} + 1/sC_{b'c}}\right) v'_L \tag{14.3-1a}$$

and the Thévenin impedance is

$$Z_t = r_{b'e} \| \frac{1}{s(C_{b'e} + C_{b'c})} \tag{14.3-1b}$$

Figure 14.3-2 can be reduced to the simplified equivalent circuit shown in Fig. 14.3-3, with the aid of (14.3-1*a*) and (14.3-1*b*). We now determine Z_f and A so that v_b is independent of the load voltage v_L.

$$T = \left(\frac{v_L}{v_b}\right)\left(\frac{v_b}{v'_L}\right)\Bigg|_{i_i=0} = 0 \tag{14.3-2a}$$

and since $v_L/v_b \neq 0$,

$$\frac{v_b}{v'_L}\Bigg|_{i_i=0} = 0 \tag{14.3-2b}$$

The equation relating v_b and v'_L is obtained using KVL.

$$v_b\Big|_{i_i=0} = \left(\frac{v_t}{r_{bb'} + Z_t} - \frac{Av'_L}{Z_f}\right)\left(\frac{1}{Z'} + \frac{1}{Z_f} + \frac{1}{r_{bb'} + Z_t}\right) \tag{14.3-3a}$$

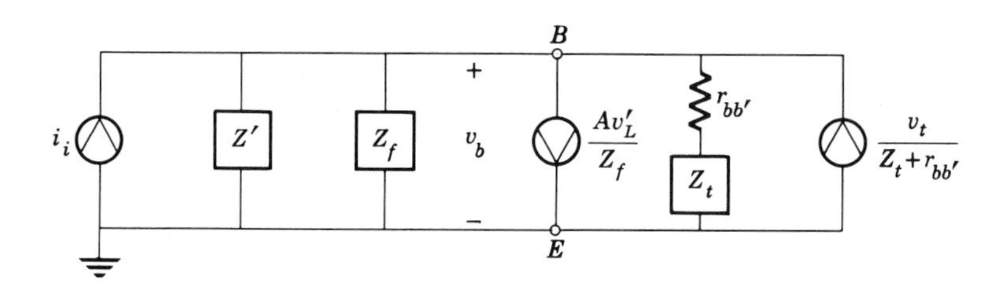

Fig. 14.3-3 Simplified equivalent circuit.

Equation (14.3-2) is satisfied when

$$\frac{v_t}{r_{bb'} + Z_t} = \frac{Av'_L}{Z_f} \tag{14.3-3b}$$

This equation can be simplified using (14.3-1a).

$$A(r_{bb'} + Z_t) = \frac{Z_{b'e}Z_f}{Z_{b'e} + 1/sC_{b'c}} \tag{14.3-4}$$

After some algebra this reduces to

$$Z_f = Ar_{bb'}\left(1 + \frac{C_{b'e}}{C_{b'c}}\right) + \left(\frac{A}{sC_{b'c}}\right)\left(1 + \frac{r_{bb'}}{r_{b'e}}\right) \tag{14.3-5}$$

We see that Z_f consists of a series RC branch with

$$R_f = Ar_{bb'}\left(1 + \frac{C_{b'e}}{C_{b'c}}\right) \tag{14.3-6a}$$

and

$$C_f = \frac{C_{b'c}}{A(1 + r_{bb'}/r_{b'e})} \tag{14.3-6b}$$

The feedback gain A is usually obtained by employing a transformer in the output circuit, as illustrated in the following example.

Example 14.3-1 In the amplifier of Fig. 14.3-1, $h_{fe} = 100$, $r_{bb'} = 10\ \Omega$, $r_{b'e} = 2\ \text{k}\Omega$, $C_{b'e} = 5$ pF, and $C_{b'c} = 1$ pF. Calculate C_f, R_f, and A.

Solution

$$C_f = \frac{C_{b'c}}{A(1 + r_{bb'}/r_{b'e})} = \frac{10^{-12}}{A(1 + 1/200)} \approx \frac{1}{A} \qquad \text{pF}$$

and

$$R_f = Ar_{bb'}\left(1 + \frac{C_{b'e}}{C_{b'c}}\right) = 60A$$

What value of A should be employed? The value of A is chosen to yield a feedback impedance which does not load R_L. If, for example, $r_{bb'} = 0$, the feedback resistor R_f would be zero. In order that C_f not load R_L, A is chosen so that

$$\frac{1}{\omega_0 C_f} \gg R_L$$

If $R_L = 100\ \Omega$, and $\omega_0 = 10^9$ rad/s, then

$$A \gg 0.1$$

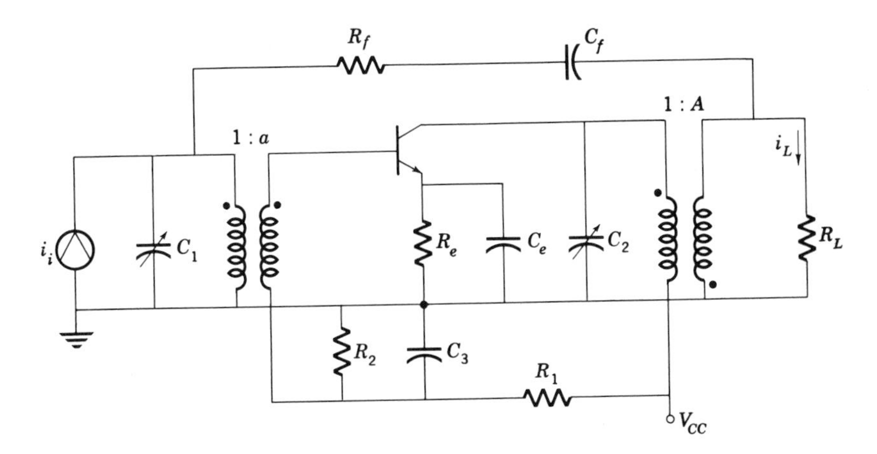

Fig. 14.3-4 Practical neutralized amplifier.

It should also be noted [Eqs. (14.3-6*a*) and (14.3-6*b*)] that

$$R_f C_f \approx r_{bb'} C_{b'e} \ll \frac{1}{\omega_0} \tag{14.3-7}$$

This assumption is usually valid (Sec. 14.1-1). Hence

$$R_f \ll \frac{1}{\omega_0 C_f} \tag{14.3-8}$$

and R_f is often omitted from the neutralizing circuit, even though the resulting amplifier is not completely neutralized.

In practice, A is the turns ratio of a transformer in the output circuit, which also provides the required phase reversal, as shown in Fig. 14.3-4.

Example 14.3-2 Calculate the gain at resonance and the 3-dB bandwidth of the neutralized amplifier of Example 14.3-1. Let $r_i = 10$ kΩ, $R_p = 10$ kΩ, $C' = 4$ pF, and $A = 0.5$. Select L' so that $\omega_0 = 10^9$ rad/s.

Solution Refer to Fig. 14.3-2. Since the amplifier is neutralized, $T = 0$. Thus the gain is the same with or without feedback. To calculate the transfer gain v_b/i_i, we let $v'_L = 0$. The input impedance of the neutralized amplifier is shown in Fig. 14.3-5.

The following analysis is greatly simplified by assuming that at ω_h

$$\frac{1}{\omega_h C_f} \gg R_f$$

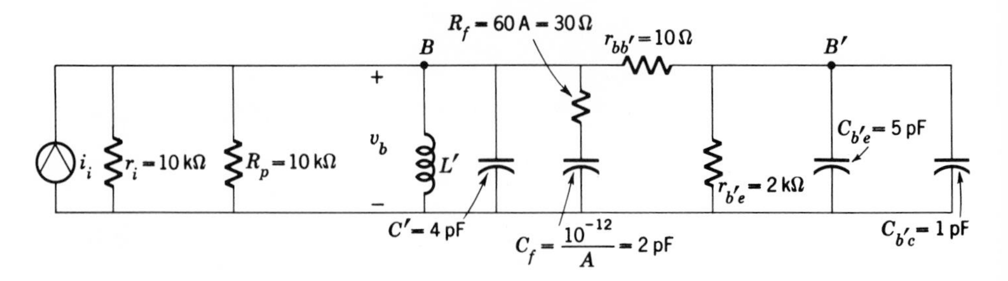

Fig. 14.3-5 Input circuit of neutralized amplifier of Example 14.3-2.

This assumption will be verified subsequently. In addition, we note that $r_{bb'}$ is much less than $r_{b'e}$, and therefore can be neglected. The resulting simplified equivalent circuit is shown in Fig. 14.3-6.

The transfer gain is therefore

$$\frac{v_b}{i_i} \approx \frac{v_{b'}}{i_i} = \frac{1.4 \times 10^3}{1 + jQ_i[(\omega/\omega_0) - (\omega_0/\omega)]}$$

where

$$Q_i = \omega_0 R_i C_i \approx (10^9)(14 \times 10^2)(12 \times 10^{-12}) \approx 17$$

and

$$L' = \frac{1}{\omega_0{}^2 C_i} = \frac{1}{(10^{18})(12 \times 10^{-12})} \approx 0.08\ \mu\text{H}$$

The 3-dB bandwidth of the amplifier is

$$\text{BW} = \frac{\omega_0}{Q_i} = \frac{10^9}{17} \approx 58 \text{ Mrad/s}$$

Note that $\omega_h \approx 10^9$ rad/s. Thus $1/\omega_h C_f = 500\ \Omega$, which is much greater than R_f.

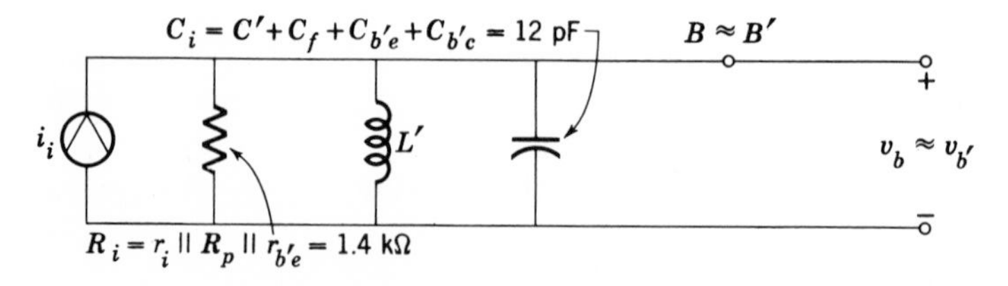

Fig. 14.3-6 Simplified input circuit of the neutralized amplifier of Example 14.3-2 (with R_f neglected).

The current gain of the neutralized amplifier is

$$A_i = \left(\frac{v_{b'}}{i_i}\right)\left(\frac{i_L}{v_{b'}}\right) = \left\{\frac{-1.4 \times 10^3}{1 + jQ_i[(\omega/\omega_0) - (\omega_0/\omega)]}\right\} g_m$$

Thus the midband current gain is

$$A_{im} = -g_m(1.4 \times 10^3) = \frac{-1.4}{20} \times 10^3 = -70$$

14.4 THE SYNCHRONOUSLY TUNED AMPLIFIER

In this section, we discuss the cascading of tuned amplifiers in order to achieve high gain. All amplifier stages are assumed to be identical and to be tuned to the same frequency, ω_0. This is called *synchronous tuning*, and the resulting amplifier has increased gain, and a bandwidth which is narrower than the bandwidth of each of the stages.

In the next section, we discuss the possibility of stagger-tuning the amplifier sections to obtain high gain and a bandwidth greater than that provided by each stage.

To illustrate the effect of cascading N synchronously tuned stages, we first determine the gain and bandwidth of the single-tuned FET amplifier shown in Fig. 14.4-1*a*. The equivalent circuit is given in Fig. 14.4-1*b*,

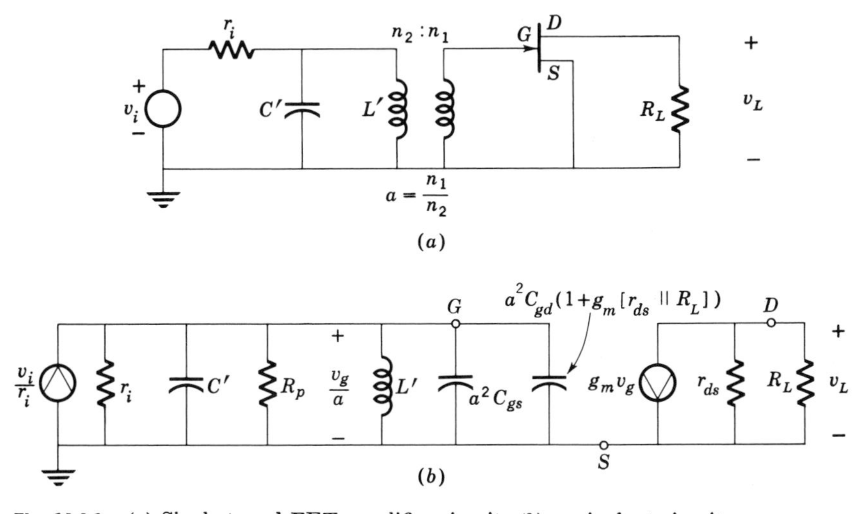

Fig. 14.4-1 (*a*) Single-tuned FET amplifier circuit; (*b*) equivalent circuit.

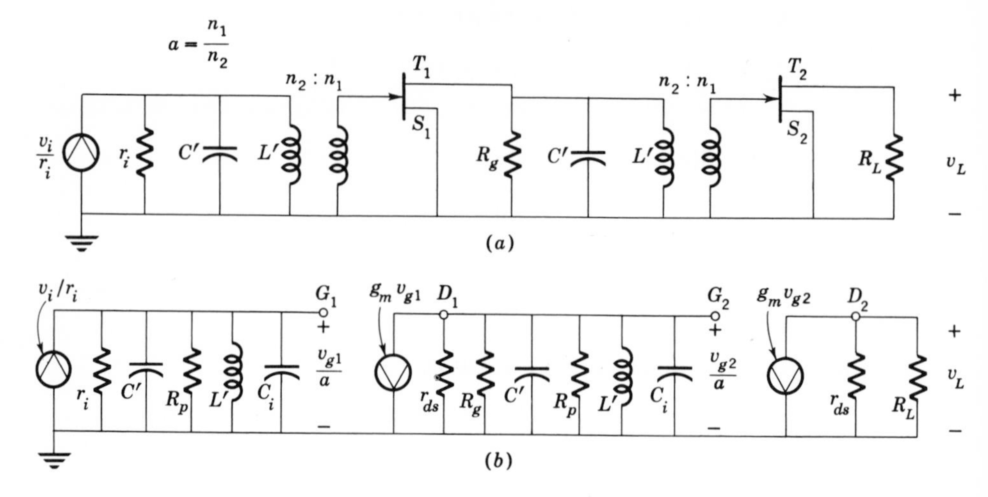

Fig. 14.4-2 (*a*) Synchronously tuned FET amplifier circuit; (*b*) equivalent circuit.

and the voltage gain is (Sec. 14.1)

$$A_v = \frac{-ag_m(r_{ds}\|R_L)[(r_i\|R_p)/r_i]}{1 + jQ_i[(\omega/\omega_0) - (\omega_0/\omega)]} \tag{14.4-1a}$$

where

$$C_i = a^2\{C_{gs} + C_{gd}[1 + g_m(r_{ds}\|R_L)]\}$$

$$Q_i = \omega_0(r_i\|R_p)(C' + C_i) \tag{14.4-1b}$$

$$\omega_0{}^2 = \frac{1}{L(C' + C_i)} \tag{14.4-1c}$$

The center-frequency gain ($\omega = \omega_0$) is

$$A_{vm} = -ag_m(r_{ds}\|R_L)\left(\frac{R_p}{r_i + R_p}\right) \tag{14.4-2a}$$

The 3-dB bandwidth is

$$\text{BW} = \frac{1}{2\pi(r_i\|R_p)(C' + C_i)} \tag{14.4-2b}$$

and the GBW is

$$\text{GBW} = a\left[\frac{g_m(r_{ds}\|R_L)}{2\pi r_i}\right]\left(\frac{1}{C' + C_i}\right) \tag{14.4-2c}$$

Let us now cascade two identical synchronously tuned FET amplifiers, as shown in Fig. 14.4-2*a*. The equivalent circuit is shown in Fig.

14.4-2*b*. In a synchronously tuned amplifier each resonant circuit is tuned to the same frequency and has the same bandwidth. Hence

$$\omega_0^2 = \frac{1}{L'(C' + C_i)} \tag{14.4-3a}$$

and

$$R \equiv r_i \| R_p = r_{ds} \| R_g \| R_p \tag{14.4-3b}$$

The voltage gain of the amplifier is

$$A_v = \frac{v_L}{v_i} = \frac{-ag_m R(R/r_i)[-ag_m(r_{ds}\|R_L)]}{\{1 + jQ_i[(\omega/\omega_0) - (\omega_0/\omega)]\}^2} \tag{14.4-4a}$$

where

$$Q_i = \omega_0 R(C' + C_i) \tag{14.4-4b}$$

The center-frequency gain is

$$A_{vm} = -ag_m R[-ag_m(r_{ds}\|R_L)]\left(\frac{R_p}{r_i + R_p}\right) \tag{14.4-5}$$

The 3-dB bandwidth of the two-stage amplifier is obtained from [Eq. (14.4-4*a*)]

$$\left[1 + Q_i^2\left(\frac{\omega}{\omega_0} - \frac{\omega_0}{\omega}\right)^2\right]^2 = 2 \tag{14.4-6a}$$

Hence

$$\mathrm{BW} = \left(\frac{\omega_0}{2\pi Q_i}\right)\sqrt{2^{1/2} - 1} = 0.643\left(\frac{f_0}{Q_i}\right) = 0.643\left[\frac{1}{2\pi(C' + C_i)R}\right] \tag{14.4-6b}$$

The GBW product is

$$\mathrm{GBW} = A_{vm}\mathrm{BW} = \frac{0.643\ ag_m R[ag_m(r_{ds}\|R_L)]}{2\pi r_i(C' + C_i)} \tag{14.4-7}$$

The result of cascading two synchronously tuned stages is to increase the voltage gain [compare (14.4-5) and (14.4-2*a*)] and to decrease the bandwidth [compare (14.4-6*b*) and (14.4-2*b*)]. The gain-bandwidth product can increase or decrease, as seen from (14.4-7) and (14.4-2*c*).

Example 14.4-1 Let $g_m = 5 \times 10^{-3}$ mho, $r_i = r_{ds} = 10$ kΩ, $R_p = R_L = R_g = 100$ kΩ, $a = \frac{1}{2}$, $C' = 0$, $C_{gs} = 10$ pF, $C_{gd} = 0.1$ pF, and $L' = 0.25$ μH.

Calculate ω_0, A_{vm}, BW, and the GBW product, assuming a single-tuned stage (Fig. 14.4-1) or two synchronously tuned stages (Fig. 14.4-2*a*).

Solution The resonant frequency ω_0 is

$$\omega_0 = \left(\frac{1}{L'(a^2)\{C_{gs} + C_{gd}[1 + g_m(r_{ds}\|R_L)]\}}\right)^{1/2}$$

$$= \left[\frac{1}{(0.25 \times 10^{-6})(\frac{1}{4})(10 \times 10^{-12} + 5 \times 10^{-12})}\right]^{1/2} \approx 10^9 \text{ rad/s}$$

The center-frequency gain for one stage is, from (14.4-2*a*),

$$(A_{vm})_1 \approx -(5 \times 10^{-3})(10^4)(\tfrac{1}{2}) = -25$$

With two stages, from (14.4-5),

$$(A_{vm})_2 = (-5 \times 10^{-3} \times 10^4 \times \tfrac{1}{2})(-5 \times 10^{-3} \times 10^4 \times \tfrac{1}{2}) = 625$$

The 3-dB bandwidth with one stage is, from (14.4-2*b*),

$$(\text{BW})_1 = \frac{1}{2\pi 10^4(\frac{1}{4})(15 \times 10^{-12})} \approx 4.1 \text{ MHz}$$

and for two stages, from (14.4-6*b*),

$$(\text{BW})_2 = 0.643\left[\frac{1}{2\pi(10^4)(\frac{1}{4})(15 \times 10^{-12})}\right] \approx 2.6 \text{ MHz}$$

The GBW in each case is

$$(\text{GBW})_1 \approx 100 \text{ MHz}$$

and

$$(\text{GBW})_2 \approx 1.6 \text{ GHz}$$

Thus, in this example, using two synchronously tuned amplifiers resulted in a significant increase in the GBW product.

N-CASCADED SYNCHRONOUSLY TUNED STAGES

If N transistors are cascaded in a synchronously tuned manner, the voltage gain becomes

$$A_v \approx \left(\frac{(-ag_{m1}R_1)(-ag_{m2}R_2) \cdots (-ag_{mN}R_N)}{\{1 + jQ_i[(\omega/\omega_0) - (\omega_0/\omega)]\}^N}\right)\left(\frac{R_p}{r_i + R_p}\right) \tag{14.4-8}$$

At midband ($\omega = \omega_0$), assuming $r_i \ll R_p$,

$$A_{vm} \approx \prod_{j=1}^{N} (-a)^N g_{mj}R_j \tag{14.4-9}*$$

The bandwidth is found by solving

$$\left| \left[1 + jQ_i \left(\frac{\omega}{\omega_0} - \frac{\omega_0}{\omega} \right) \right]^N \right|^2 = 2 \tag{14.4-10a}$$

Then

$$\text{BW} = \left(\frac{\omega_0}{2\pi Q_i} \right) (2^{1/N} - 1)^{1/2} \tag{14.4-10b}$$

and the GBW product is

$$\text{GBW} = \prod_{j=1}^{N} a^N g_{mj}R_j \left(\frac{f_o}{Q_i} \right) \sqrt{2^{1/N} - 1} \tag{14.4-11}$$

14.5 THE STAGGER- AND DOUBLE-TUNED AMPLIFIERS

In the preceding section we saw that a synchronously tuned multistage amplifier yielded an increased GBW product. However, while the gain increased with N, BW decreased with N. Stagger-tuned and double-tuned amplifiers also increase the GBW product. Both the gain and the resulting bandwidth are larger than the gain and bandwidth of the single-tuned amplifier.

14.5-1 THE STAGGER-TUNED AMPLIFIER

A stagger-tuned amplifier is shown in Fig. 14.5-1. The common-base configuration is used to eliminate coupling between the input and the output circuits. The transistor is chosen so that it is operating at frequencies well below f_α. Capacitors C_1 and C_2 are chosen large enough so that all internal capacitances may be neglected. The equivalent circuit of the amplifier is shown in Fig. 14.5-2.

In a stagger-tuned amplifier each tuned circuit is designed to have the *same* Q at the amplifier center frequency ω_0. However, each circuit resonates at a different frequency, the input at ω_1 and the output at ω_2.

* The symbol $\prod_{j=1}^{N}$ represents a product of terms. Thus

$$\prod_{j=1}^{N} x_j = x_1 \cdot x_2 \cdot x_3 \cdot \cdot \cdot x_N$$

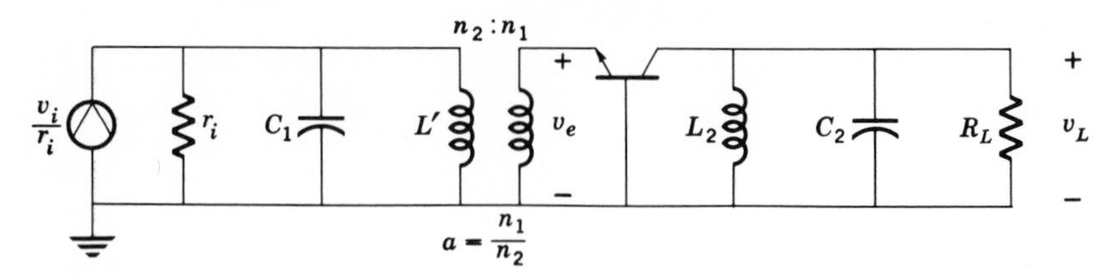

Fig. 14.5-1 Stagger-tuned common-base amplifier (bias components omitted).

Thus

$$\omega_1^2 = \frac{1}{L_1 C_1} \tag{14.5-1}$$

and

$$\omega_2^2 = \frac{1}{L_2 C_2} \tag{14.5-2}$$

Since the Q of each circuit is the same with respect to the center frequency ω_0,

$$Q_0 = Q_1 = \omega_0 R_1 C_1 = Q_2 = \omega_0 R_2 C_2 \tag{14.5-3a}$$

where

$$R_1 = r_i \| R_{p1} \| \frac{h_{ib}}{a^2} \qquad R_2 = R_L \| R_{p2} \tag{14.5-3b}$$

Hence

$$R_1 C_1 = R_2 C_2 = \tau \tag{14.5-3c}$$

The voltage gain of the amplifier is

$$A_v = \frac{v_L}{v_i} = \left(\frac{v_L}{i_e}\right)\left(\frac{i_e}{v_e}\right)\left(\frac{v_e}{v_i}\right)$$

$$\approx \frac{g_m R_1 R_2 (a/r_i)}{\{1 + j\omega\tau[1 - (\omega_2^2/\omega^2)]\}\{1 + j\omega\tau[1 - (\omega_1^2/\omega^2)]\}} \tag{14.5-4}$$

In order to simplify this expression so that we can obtain useful design information, we assume that the overall bandwidth is small com-

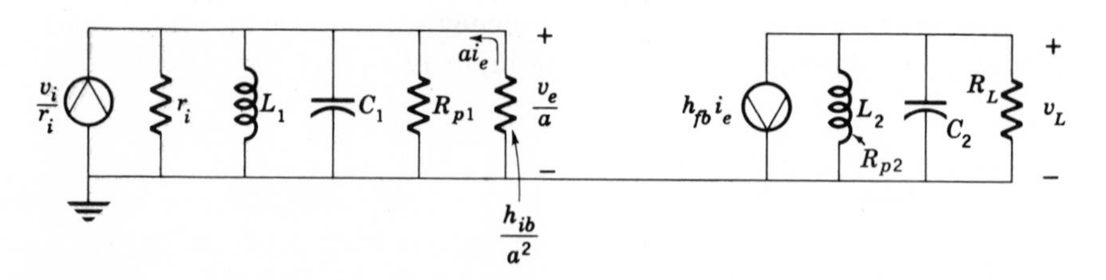

Fig. 14.5-2 Equivalent circuit of stagger-tuned amplifier.

pared with ω_0, which lies somewhere between ω_1 and ω_2. Then we focus our attention on the frequency-dependent (imaginary) part of the denominator, which has two similar terms of the form

$$\tau\left(\frac{\omega^2 - \omega_2{}^2}{\omega}\right) = \tau\left[\frac{(\omega + \omega_2)(\omega - \omega_2)}{\omega}\right] \tag{14.5-5a}$$

Since we restrict our attention to frequencies not far removed from ω_0 and ω_2, and since the bandwidth is small, the frequency variation of (14.5-5*a*) is mainly due to the factor $\omega - \omega_2$. The factor $\omega + \omega_2$ is approximately 2ω within the passband. Thus we assume that

$$\tau\left[\frac{(\omega + \omega_2)(\omega - \omega_2)}{\omega}\right] \approx \tau\left[\frac{2\omega(\omega - \omega_2)}{\omega}\right] = 2\tau(\omega - \omega_2) \tag{14.5-5b}$$

In other words, we are assuming that, within the amplifier passband,

$$\frac{\omega + \omega_2}{\omega} \approx 2 \tag{14.5-5c}$$

Using (14.5-5*b*) in (14.5-4), we obtain

$$A_v \approx \left(\frac{-g_m R_1 R_2 a}{r_i}\right)\left\{\frac{1}{[1 + j2\tau(\omega - \omega_2)][1 + j2\tau(\omega - \omega_1)]}\right\} \tag{14.5-6}$$

It is instructive at this time to digress and sketch the pole-zero portrait of A_v. The diagrams of the poles and zeros of (14.5-4) and (14.5-6) are shown in Fig. 14.5-3. Note that the double zero at the origin and the poles near $j\omega_1$ and $j\omega_2$ are missing in Fig. 14.5-3*b*. Thus the approximation made by using (14.5-5) can be interpreted in the sense that the double zero cancels the two negative poles over the passband of the stagger-tuned amplifier.

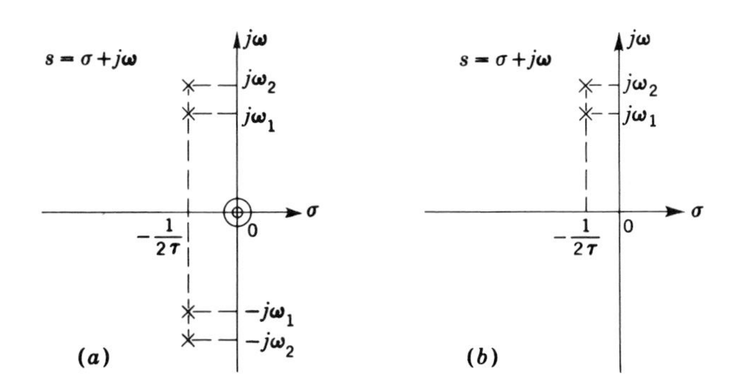

Fig. 14.5-3 Pole-zero patterns for stagger-tuning. (*a*) Complete transfer function (14.5-4); (*b*) narrowband approximation (14.5-6).

Let us now return to (14.5-6) and define ω_0 as

$$\omega_0 = \frac{\omega_1 + \omega_2}{2} \tag{14.5-7a}$$

and let

$$\omega_2 - \omega_1 = 2\Delta \tag{14.5-7b}$$

Then

$$\omega - \omega_2 = \omega - (\omega_0 + \Delta) = (\omega - \omega_0) - \Delta \tag{14.5-8a}$$

and

$$\omega - \omega_1 = \omega - (\omega_0 - \Delta) = (\omega - \omega_0) + \Delta \tag{14.5-8b}$$

Substituting (14.5-8) into (14.5-6) yields

$$A_v = \left(\frac{-g_m R_1 R_2 a}{r_i}\right)\left\{\frac{1}{1 - 4\tau^2[(\omega - \omega_0)^2 - \Delta^2] + j4\tau(\omega - \omega_0)}\right\} \tag{14.5-9}$$

The magnitude of the gain $|A_v|$ is

$$|A_v| = \left(\frac{g_m R_1 R_2 a}{r_i}\right)\sqrt{\frac{1}{16\tau^4(\omega - \omega_0)^4 + (8\tau^2 - 32\tau^4\Delta^2)(\omega - \omega_0)^2 + (1 + 8\Delta^2\tau^2 + 16\tau^4\Delta^4)}}$$

$$= \left(\frac{g_m R_1 R_2 a}{4r_i\tau^2}\right)\sqrt{\frac{1}{(\omega - \omega_0)^4 + \left(\frac{1 - 4\Delta^2\tau^2}{2\tau^2}\right)(\omega - \omega_0)^2 + \left(\frac{1 + 4\Delta^2\tau^2}{4\tau^2}\right)^2}} \tag{14.5-10}$$

The circuit behavior can now be explained using (14.5-10). At the center frequency ω_0,

$$A_{vm} = \left(\frac{-g_m R_1 R_2 a}{r_i}\right)\left(\frac{1}{1 + 4\Delta^2\tau^2}\right) \tag{14.5-11}$$

As $(\omega - \omega_0)^2$ increases, $|A_v|$ decreases or increases, depending on the sign of the coefficient of the $(\omega - \omega_0)^2$ term, $1 - 4\Delta^2\tau^2$. This is shown in Fig. 14.5-4. It is seen from this figure that the widest possible bandwidth without having two *peaks* occurs when $1 - 4\Delta^2\tau^2$ is set equal to zero. This is called the *maximally flat* case. An *underdamped* system displays two peaks as seen from the figure. A study of the phase of A_v reveals that these peaks are accompanied by a highly nonlinear phase characteristic. This is undesirable in phase-sensitive systems such as television or data transmission. Practical systems are usually designed to give maximally flat response or have peaks less than 1 dB above A_{vm}. Although the latter response yields some phase distortion, it results in a wider bandwidth, and so a compromise between bandwidth and fidelity yields this commonly employed +1-dB criterion.

In this section we design for maximal flatness,

$$1 = 4\Delta^2\tau^2 \tag{14.5-12}$$

or, using (14.5-7*b*) and (14.5-3),

$$1 = 4\left(\frac{\omega_2 - \omega_1}{2}\right)^2\left(\frac{Q_0}{\omega_0}\right)^2 \tag{14.5-13}$$

and hence

$$1 = 4\left(\frac{\omega_2 - \omega_1}{\omega_2 + \omega_1}\right)^2 Q_0^2 \tag{14.5-14}$$

Then (14.5-10) becomes

$$|A_v| = \left(\frac{g_m R_1 R_2 a}{4r_i\tau^2}\right)\sqrt{\frac{1}{(\omega - \omega_0)^4 + [(1 + 4\Delta^2\tau^2)/4\tau^2]^2}} \tag{14.5-15}$$

The half-power frequencies occur when $|A_v/A_{vm}| = 1/\sqrt{2}$. The BW is then found from the equation

$$\omega_{3\,\text{dB}} - \omega_0 = \pm\sqrt{\frac{1 + 4\Delta^2\tau^2}{4\tau^2}} = \pm\frac{\sqrt{1 + 4\Delta^2\tau^2}}{2\tau} \tag{14.5-16a}$$

Using (14.5-12), this becomes

$$\omega_{3\,\text{dB}} = \omega_0 \pm \frac{\sqrt{2}}{2\tau} = \omega_0\left(1 \pm \frac{\sqrt{2}}{2Q_0}\right) \tag{14.5-16b}$$

Therefore the 3-dB bandwidth for the maximally flat case is

$$\text{BW} = \frac{\sqrt{2}}{2\pi}\left(\frac{\omega_0}{Q_0}\right) = \frac{\sqrt{2}}{2\pi\tau} \qquad \text{Hz} \tag{14.5-17}$$

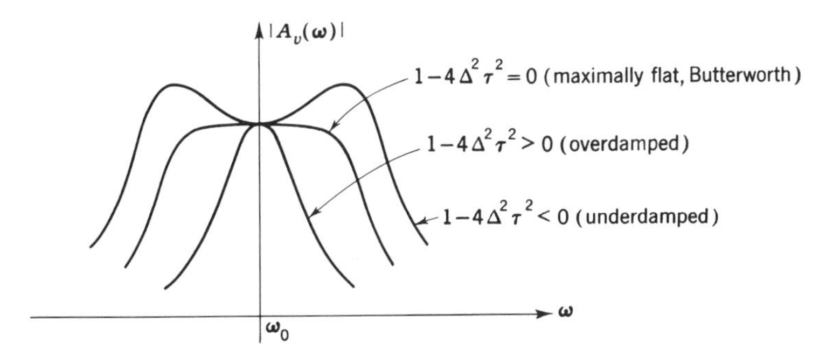

Fig. 14.5-4 Response of one stagger-tuned stage.

The bandwidth of the single-stage stagger-tuned amplifier is *increased* by $\sqrt{2}$ over the bandwidth obtained from a single-tuned amplifier stage. However, the center-frequency gain A_{vm}, given in (14.5-11), is reduced by a factor of 2, when $a = 1$. The GBW product of the maximally flat amplifier is

$$\text{GBW} = -\frac{g_m R_1 a}{2\pi \sqrt{2}\, r_i C_2} \tag{14.5-18}$$

Note that if $r_i = R_1$, the GBW product of the stagger-tuned stage is larger than the GBW product of the single-tuned stage, if $a > \sqrt{2}$.

Example 14.5-1 Design an amplifier with a maximally flat bandwidth of 10 MHz centered at 50 MHz.

Solution Using (14.5-17),

$$Q_0 = \sqrt{2}\left(\frac{50 \times 10^6}{10 \times 10^6}\right) \approx 7$$

and

$$\tau = R_1C_1 = R_2C_2 = \frac{7}{50 \times 10^6} = 1.4 \times 10^{-7}$$

Using (14.5-12),

$$1 = 4\Delta^2\tau^2$$

and

$$\Delta = 3.54 \times 10^6 \text{ rad/s}$$

Therefore (14.5-8)

$$f_1 \approx 46.5 \text{ MHz}$$

and

$$f_2 \approx 53.5 \text{ MHz}$$

Hence

$$L_1C_1 = \frac{10^{-12}}{4\pi^2(46.5)^2} = 0.12 \times 10^{-16}$$

and

$$L_2C_2 = \frac{10^{-12}}{4\pi^2[(53.5)^2]} = 0.87 \times 10^{-17}$$

The values of L_1, L_2, C_1, C_2, R_1, and R_2 can now be selected. This is left as an exercise.

14.5-2 THE DOUBLE-TUNED AMPLIFIER

One form of the double-tuned amplifier uses the mutual inductance present in a transformer to obtain a transfer function similar to that obtained in stagger-tuning. A typical circuit is shown in Fig. 14.5-5*a*. Equivalent circuits are shown in Fig. 14.5-5*b* and *c*. We define the admittance of the parallel combination of R_1, C_1, L_1 as Y, and (Fig. 14.5-5*c*) the parallel combination of L_2, $r_{b'e}$, $C_{b'e} + C_M$ as aY. The current gain can now be obtained.

$$A_i = \frac{i_L}{i_i} = \left(\frac{i_L}{v_{b'}}\right)\left(\frac{v_{b'}}{i_i}\right) = -g_m\left(\frac{1/aY^2}{1/Y + 1/aY + j\omega M}\right)$$

$$= \frac{-g_m}{Y(1 + a + j\omega MY)} = \frac{-g_m/j\omega M}{Y[Y + (1 + a)/j\omega M]}$$

$$= -\frac{g_m/j\omega M}{YY_2} \qquad (14.5\text{-}19a)$$

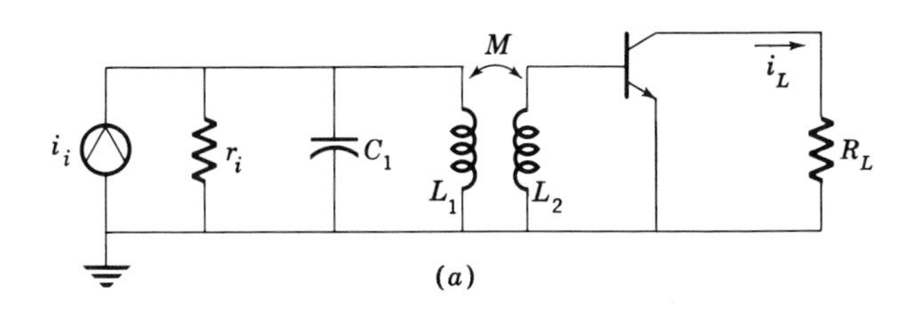

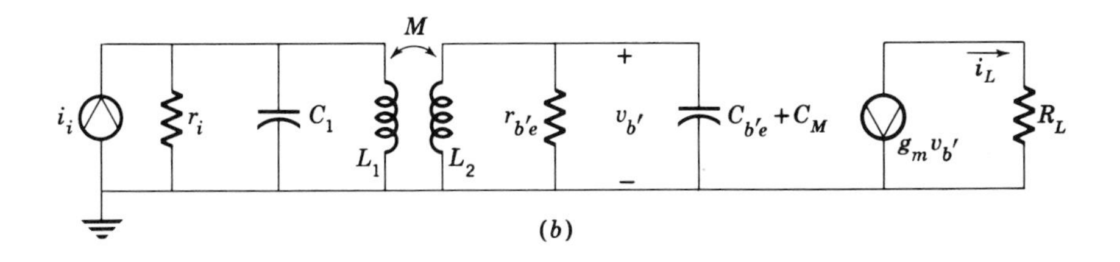

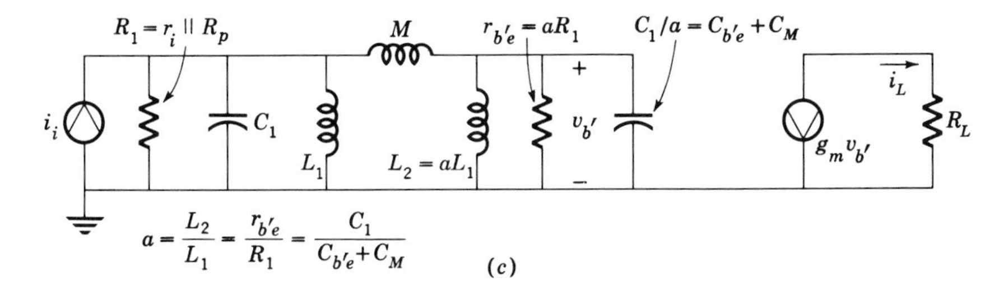

Fig. 14.5-5 (*a*) Double-tuned amplifier circuit (bias components omitted); (*b*) equivalent circuit; (*c*) reduced equivalent circuit.

where

$$Y = \frac{1}{R_1} + j\omega C_1 + \frac{1}{j\omega L_1} = \left(\frac{1}{R_1}\right)\left[1 + j\left(\frac{\omega Q_0}{\omega_0}\right)\left(1 - \frac{\omega_1^2}{\omega^2}\right)\right] \tag{14.5-19b}$$

$$Y_2 = Y + \frac{1}{j\omega M/(1+a)} = \frac{1}{R_1} + j\omega C_1 + \left(\frac{1}{j\omega}\right)\left(\frac{1}{L_1} + \frac{1+a}{M}\right)$$

$$= \left(\frac{1}{R_1}\right)\left[1 + j\left(\frac{\omega}{\omega_0}\right)Q_0\left(1 - \frac{\omega_2^2}{\omega^2}\right)\right] \tag{14.5-19c}$$

$$\omega_1^2 = \frac{1}{C_1 L_1} \tag{14.5-19d}$$

and

$$\omega_2^2 = \frac{1}{C_2[L_1 M/(M + (1+a)L_1)]} \tag{14.5-19e}$$

Proceeding as in (14.5-5), we find that

$$A_i \approx \frac{-g_m R_1^2/j\omega_0 M}{[1 + j2\tau(\omega - \omega_1)][1 + j2\tau(\omega - \omega_2)]} \tag{14.5-20}$$

where, in the numerator, we have assumed that

$$\frac{1}{j\omega M} \approx \frac{1}{j\omega_0 M}$$

over the passband.

Equation (14.5-20) is of the same form as (14.5-6), with

$$\frac{g_m R_1^2}{\omega_0 M} = \frac{g_m R_1 R_2 a}{r_i}$$

Thus the analysis of the double-tuned amplifier, from this point on, is identical with the analysis of the stagger-tuned amplifier, and both amplifiers have similar gain characteristics.

Practical problems A stagger-tuned amplifier is easily designed when using a pentode (Prob. 14.26) or a common-base amplifier (Fig. 14.5-1), since the input and output circuits are isolated. However, stagger-tuning is not generally employed with a common-emitter transistor amplifier, a triode, or a FET because of the Miller effect.

In practice, difficulty is encountered in varying the mutual-inductance coupling of the double-tuned amplifier shown in Fig. 14.5-5. Thus experimental tuning of the amplifier is complex. An alternate configuration is shown in Fig. 14.5-6. The mutual coupling between the two tuned circuits is provided by the capacitor C_m. Since C_m is easily varied, this type of tuned amplifier is often used. Analysis of this circuit is left as an exercise for the reader (Probs. 14.28 and 14.29).

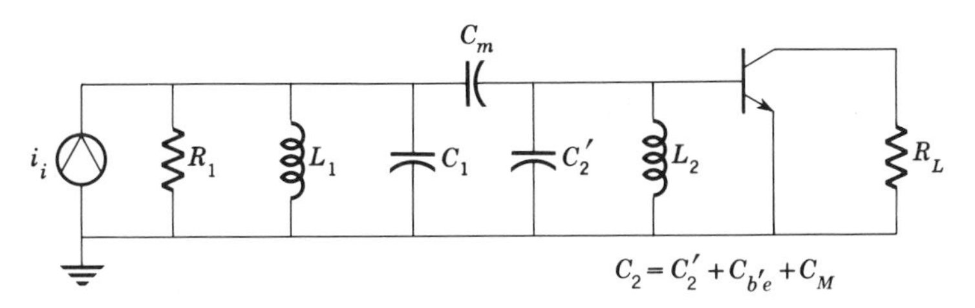

Fig. 14.5-6 Capacitively coupled double-tuned amplifier.

14.6 SHUNT PEAKING

In the preceding sections of this chapter we discussed tuned amplifiers that provided gain over a relatively narrow band of frequencies. Tuned circuits are also used to extend the useful bandwidth of wideband low-pass amplifiers.

Consider the circuit shown in Fig. 14.6-1. This amplifier employs a commonly used form of *shunt peaking* to increase the 3-dB bandwidth. Note that the biasing resistance is in series with an inductor L. The source impedance is assumed to be very large, and is not shown in the figure, and $r_{bb'}$ is assumed to be zero. The equivalent circuit of Fig. 14.6-1 is shown in Fig. 14.6-2, where the capacitor C_i is defined as

$$C_i = C_{b'e} + C_{b'c}(1 + g_m R_L)$$

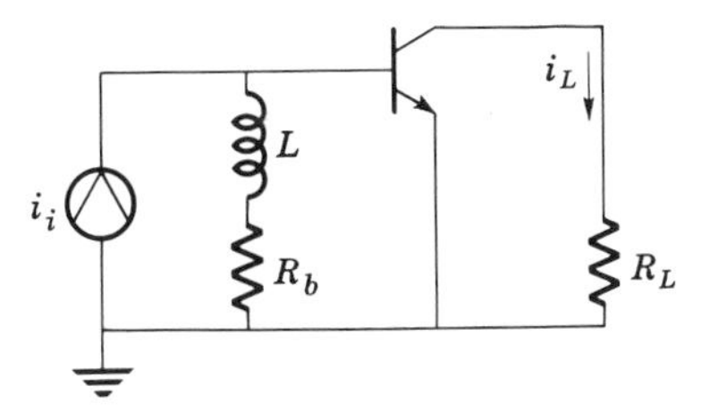

Fig. 14.6-1 Shunt-peaked amplifier.

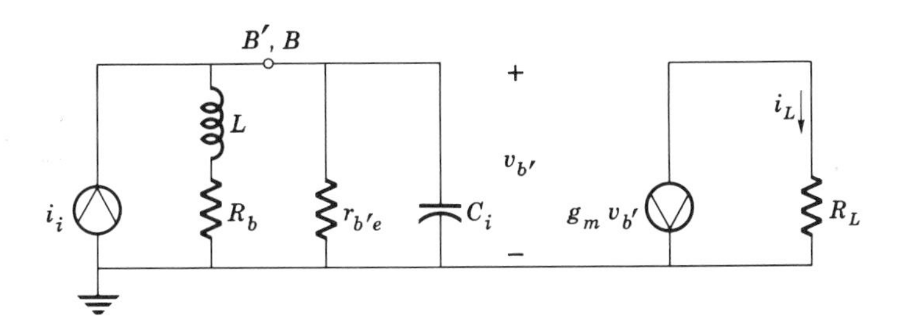

Fig. 14.6-2 Equivalent circuit for shunt peaking.

The current gain is

$$A_i = \frac{i_L}{i_i} = \left(\frac{i_L}{v_{b'}}\right)\left(\frac{v_{b'}}{i_i}\right) = -g_m\left[\frac{R_b + j\omega L}{(R_b + j\omega L)(1/r_{b'e} + j\omega C_i) + 1}\right] \tag{14.6-1}$$

This reduces to

$$A_i = -\frac{h_{fe}(1 + j\omega/\omega_c)}{(1 + r_{b'e}/R_b) + j\omega(1/\omega_c + 1/\omega_i) - \omega^2/\omega_c\omega_i} \tag{14.6-2a}$$

where

$$\omega_c = \frac{R_b}{L} \tag{14.6-2b}$$

$$\omega_i = \frac{1}{r_{b'e}C_i} \tag{14.6-2c}$$

Note that ω_i is the bandwidth without shunt peaking (with R_b and L removed).

We are interested in determining $|A_i|$ as a function of ω. Hence (14.6-2a) can be rewritten

$$|A_i|^2 = h_{fe}^2\left[\frac{1 + \left(\dfrac{\omega}{\omega_c}\right)^2}{\left(1 + \dfrac{r_{b'e}}{R_b} - \dfrac{\omega^2}{\omega_c\omega_i}\right)^2 + \omega^2\left(\dfrac{1}{\omega_c} + \dfrac{1}{\omega_i}\right)^2}\right]$$

$$= h_{fe}^2\left\{\frac{1 + \left(\dfrac{\omega}{\omega_c}\right)^2}{\left(1 + \dfrac{r_{b'e}}{R_b}\right)^2 + \omega^2\left[\left(\dfrac{1}{\omega_c} + \dfrac{1}{\omega_i}\right)^2 - \left(\dfrac{2}{\omega_c\omega_i}\right)\left(1 + \dfrac{r_{b'e}}{R_b}\right)\right] + \left(\dfrac{\omega^4}{\omega_c^2\omega_i^2}\right)}\right\} \tag{14.6-3}$$

The relation between ω_i and ω_c determines the shape of the gain-frequency characteristic and the GBW product. This is seen from the following example.

Example 14.6-1 The gain-bandwidth product of an amplifier can be extended 45 percent, by using shunt peaking and designing the shunt-peaking circuit, so that

$$R_b = r_{b'e}$$

and

$$\omega_c = 3\omega_i$$

Find the current gain, the 3-dB bandwidth, and the GBW product of this amplifier. It is shown in Prob. 14.32 that this design leads to *maximal flatness.*

Solution Substituting $R_b = r_{b'e}$ and $\omega_c = 3\omega_i$ into (14.6-3) yields

$$|A_i|^2 = \left(\frac{h_{fe}}{2}\right)^2 \frac{1 + (\omega/\omega_c)^2}{1 + (\omega/\omega_c)^2 + \frac{9}{4}(\omega/\omega_c)^4}$$

When $\omega/\omega_c \ll \frac{2}{3}$,

$$|A_i| = \frac{h_{fe}}{2}$$

Note that this implies that the midfrequency region extends to

$$\omega \ll \tfrac{2}{3}\omega_c = 2\omega_i$$

The gain at midfrequencies and low frequencies is $h_{fe}/2$. The 3-dB bandwidth f_h is found from the equation

$$\frac{1 + (\omega_h/\omega_c)^2}{1 + (\omega_h/\omega_c)^2 + \frac{9}{4}(\omega_h/\omega_c)^4} = \frac{1}{2}$$

Solving yields

$$\frac{\omega_h}{\omega_c} = \frac{f_h}{f_c} = 0.965$$

Hence

$$f_h = 2.9f_i$$

The GBW product is therefore

$$\text{GBW} \approx 1.45h_{fe}f_i$$

It is interesting to note that different choices of ω_c and ω_i lead, of course, to different forms of response. For example, by properly choosing ω_c and ω_i, a 65 percent increase in GBW can be obtained without any overshoot.[1]

EMITTER DEGENERATION

An alternative technique for shunt peaking, which does not require the use of an inductance, is shown in Fig. 14.6-3*a*. This procedure utilizes the emitter degeneration resulting from an imperfectly bypassed emitter resistance, to widen the bandwidth of the amplifier and to increase the GBW. The use of a capacitor C_e, rather than an inductor, makes the amplifier configuration amenable to IC design.

The circuit operation of the device can be explained using Fig. 14.6-3*a*

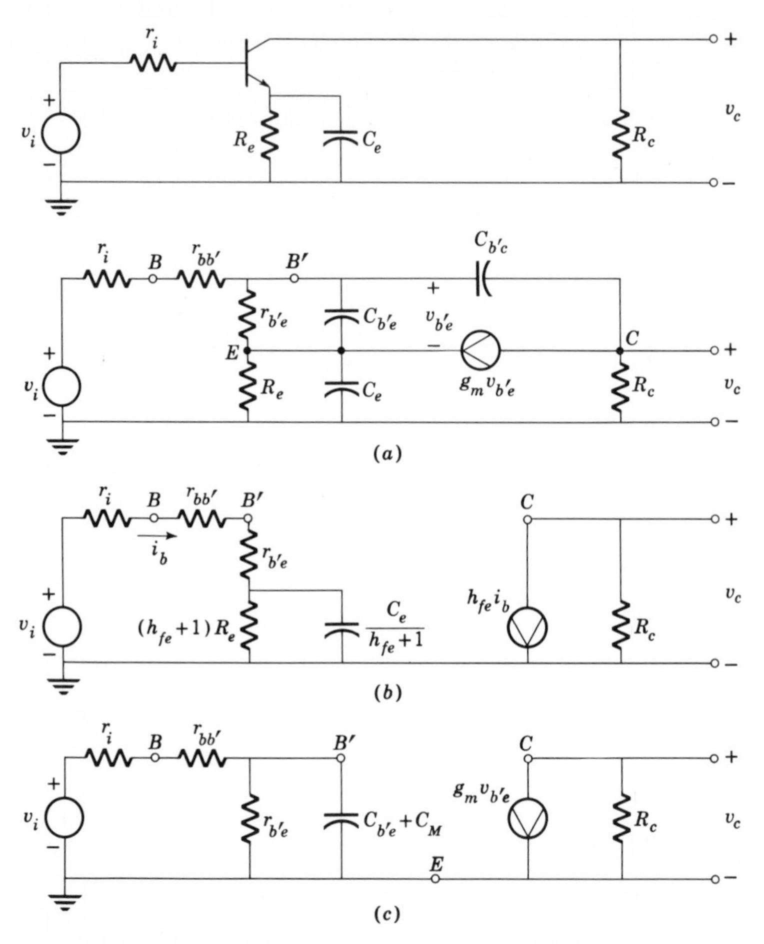

Fig. 14.6-3 (*a*) Shunt peaking using emitter degeneration; (*b*) equivalent circuit at low frequencies; (*c*) equivalent circuit at high frequencies.

and assuming that the low- and high-frequency poles, caused by C_e and the internal capacitance $C_{b'e} + C_M$, respectively, do not interact. The low- and high-frequency equivalent circuits are shown in Fig. 14.6-3*b* and *c*. The magnitude of the gain, A_v, is sketched in Fig. 14.6-4. Notice that the magnitude of the gain, A_v, increases as ω increases beyond $\omega_e = 1/R_eC_e$, and then decreases at high frequencies.

If C_e is decreased, the low-frequency pole and zero increase until interaction between the low- and high-frequency circuits results. It is possible, under these circumstances, to adjust R_e and C_e to achieve an increased bandwidth (Prob. 14-33).

The voltage gain A_v of this amplifier can be calculated using Fig. 14.6-3a, and is shown to be (Prob. 14.33)

$$A_v = \frac{-h_{fe}R_c}{r_{b'e} + r_i'} \left[\frac{1 + j\omega/\omega_e}{\left(\dfrac{r_{b'e} + r_i' + h_{fe}Re}{r_{b'e} + r_i'}\right) + j\omega\left(\dfrac{1}{\omega_1} + \dfrac{1}{\omega_e} + \dfrac{R_e r_{b'e} C_{b'e}}{r_{b'e} + r_i'}\right) - \dfrac{\omega^2}{\omega_e \omega_1}} \right] \tag{14.6-4}$$

where $r_i' = r_i + r_{bb'}$ and $\omega_1 = 1/(r_{b'e} \| r_i')(C_{b'e} + C_M)$

It is interesting to note that the gain in the low-frequency region, obtained using Fig. 14.6-3b, can be calculated from (14.6-4) by letting $\omega_1 \rightarrow \infty$. The gain in the high-frequency region, obtained using Fig. 14.6-3c, can also be determined from (14.6-4), by letting $\omega_e \rightarrow 0$. As we might expect, (14.6-4) is similar to the expression obtained for inductive shunt peaking, which is given in (14.6-2a).

14.7 THE DISTRIBUTED AMPLIFIER*[2]

The distributed amplifier is a wideband device which utilizes transmission lines to achieve increased bandwidth. Its operation can be explained qualitatively using the three-stage amplifier shown in Fig. 14.7-1.

Notice that the LC networks in the collector and emitter circuits comprise lumped-circuit equivalents for transmission lines, which are terminated in their characteristic impedance R_L, to avoid reflections. Refer to the collector circuit. The transistor capacitance $C_{b'c}$ is in parallel with C, and hence forms part of the collector-circuit transmission line. In the emitter circuit, the resistance h_{ib} represents the major component of the emitter impedance, as long as the amplifier is operated below the

* This section should be omitted if the reader does not have some previous background in transmission-line theory.

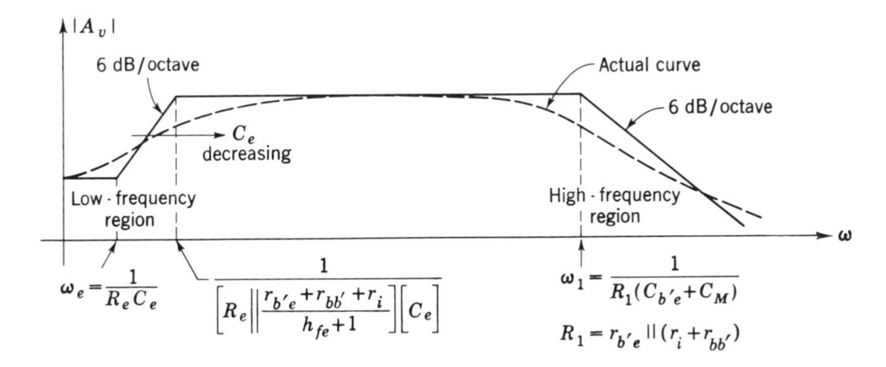

Fig. 14.6-4 Voltage gain, showing the effect of varying C_e.

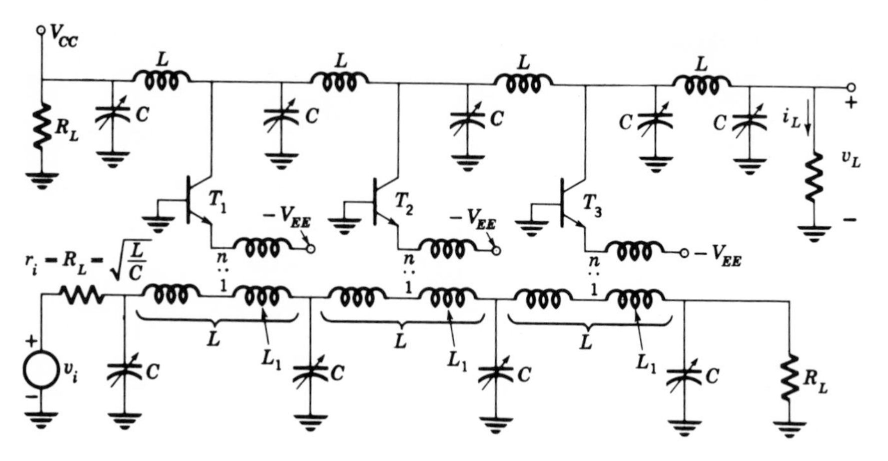

Fig. 14.7-1 Distributed amplifier.

α cutoff frequency. If $h_{ib} \ll \omega L$, its effect on the emitter-circuit transmission line is almost negligible. (Actually, h_{ib} results in a decrease in signal level as v_i "moves down the line." The turns ratios of the transformers are adjusted to reduce this effect.)

Let us look at the input signal v_i. It propagates through the transmission line with a velocity proportional to $1/\sqrt{LC}$ (note that the velocity of propagation is not equal to the speed of light). As v_i reaches each emitter, the transformer induces emitter current, which flows through the collector circuit. The collector current then divides in half, because the collector-circuit transmission line is terminated by R_L at each end. Hence the collector "sees" two equal parallel impedances, each with impedance R_L. The velocity in the collector circuit must be adjusted to be the same as in the emitter circuit so that the collector currents add *in phase*. Thus the load current i_L is

$$i_L = \tfrac{3}{2} h_{fb} i_e \tag{14.7-1}$$

If there were N transistors connected together in this manner, the current gain would be

$$A_i = \frac{i_L}{i_e} = \frac{N h_{fb}}{2} \tag{14.7-2}$$

The voltage gain A_v is (for $n = 1$)

$$A_v = \frac{v_L}{v_i} = \frac{i_L R_L}{i_e(2R_L)} = \frac{A_i}{2} = \frac{N h_{fb}}{4} \tag{14.7-3}$$

Thus, to obtain voltage gain, more than four transistors are required. Current gain can be achieved with three transistors.

The simple circuit shown in Fig. 14.7-1 does not illustrate the many possibilities available with this amplifier. For example, the impedance level in each "line" need not be the same:

$$\sqrt{\frac{L_c}{C_c}} \neq \sqrt{\frac{L_e}{C_e}} \qquad \text{in general}$$

The only restriction is that the signal velocity in the emitter and collector transmission lines be the same:

$$\text{Signal velocity} = \frac{1}{\sqrt{L_c C_c}} = \frac{1}{\sqrt{L_e C_e}} \tag{14.7-4}$$

In addition, a large gain-bandwidth product can often be achieved by using a combination of distributed amplifiers cascaded together. For example, if N transistors form a distributed amplifier and M of these amplifiers are cascaded, the total current gain is

$$A_i = \left(\frac{N h_{fb}}{2}\right)^M \tag{14.7-5}$$

The total number of transistors is then

$$T = NM \tag{14.7-6}$$

To achieve a given gain A_i requires that

$$T = MN = \left(\frac{2M}{h_{fb}}\right) A_i^{1/M} = M\left(\frac{2}{h_{fb}}\right) \epsilon^{(1/M)\ln A_i} \tag{14.7-7}$$

The minimum T for a given A_i can be obtained by differentiating T with respect to M. Thus

$$\frac{\partial T}{\partial M} = \left(\frac{2}{h_{fb}}\right) \epsilon^{(1/M)\ln A_i} + \left(\frac{2M}{h_{fb}}\right)\left(-\frac{1}{M^2}\right) \ln A_i \, \epsilon^{(1/M)\ln A_i} = 0 \tag{14.7-8a}$$

Simplifying yields

$$M = \ln A_i \tag{14.7-8b}$$

as the condition for minimum T. Hence, substituting in (14.7-7),

$$T = \left(\frac{2\epsilon}{h_{fb}}\right) \ln A_i \tag{14.7-9}$$

and

$$N = \frac{T}{M} = \frac{2\epsilon}{h_{fb}} \approx \frac{5.4}{h_{fb}} \tag{14.7-10}$$

Note that this simple derivation states that five or six transistors should be employed in each section before cascading sections. As an example, consider that 20 dB of gain is required and $h_{fb} = 0.9$ (a common value at very high frequencies when high power is required). Then the minimum number of transistors required is

$$T = \left(\frac{2\epsilon}{0.9}\right) \ln 10 = 14$$

and

$$M = \ln A_i = 2.3$$

Then

$$N = 6.05$$

Since N and M must be integers, we might try to let $N = 7$ and $M = 2$. The overall gain is then, from (14.7-5),

$$A_i = \left[\frac{7(0.9)}{2}\right]^2 = 9.9 \ (= 19.4 \text{ dB})$$

Note that we *cannot* obtain the required gain with 14 transistors.

Instead of cascading two equal sections, we use $N_1 = 7$ and $N_2 = 8$. Then

$$A_i = \left(\frac{7h_{fb}}{2}\right)\left(\frac{8h_{fb}}{2}\right) = 11.35 \ (\approx 21 \text{ dB})$$

Thus 15 transistors are required to meet the gain specification.

We must now turn our attention to the design of the transmission line. The design equations for a distributed-element line are well known.[3] They are

$$R_L = \sqrt{\frac{L}{C}} \tag{14.7-11}$$

$$L = \frac{R_L}{\pi f_c} \tag{14.7-12}$$

$$C = \frac{1}{\pi R_L f_c} \tag{14.7-13}$$

The frequency f_c is the cutoff frequency of the transmission line.

The voltage-standing-wave ratio[3] is

$$\text{VSWR} = \frac{1}{\sqrt{1 - (f/f_c)^2}} \tag{14.7-14}$$

The first three expressions are probably familiar. The last expression for the voltage-standing-wave ratio relates the cutoff frequency to the "ripple" in $|A_i|$ caused by the transmission line. For example, if we allow a ± 0.5-dB ripple, the VSWR is 1.8. Thus, for this case,

$$1.8 = \frac{1}{\sqrt{1 - (f_h/f_c)^2}}$$

and

$$f_c = 1.2 f_h$$

Example 14.7-1 The amplifier in Fig. 14.7-1 is to have a 3-dB bandwidth of 100 MHz with a VSWR = 1.8 and an equal load and source termination of 50 Ω. Find L and C (assume a 1:1 transformer).

Solution Substituting $f = 100$ MHz into (14.7-14) with a VSWR = 1.8 yields $f_c = 120$ MHz. Since $R_L = 50\ \Omega$, we have $L = 0.13\ \mu$H and $C = 53$ pF. The value of C inserted in the collector and emitter circuit is 53 pF minus $C_{b'c}$ and $C_{b'e}$, respectively.

PROBLEMS

SECTION 14.1

14.1. Show that the center frequency of the tuned circuit in Fig. P14.1 is $1/\sqrt{LC}$ and that the 3-dB bandwidth is $1/2\pi RC$ [thereby verifying (14.1-10)].

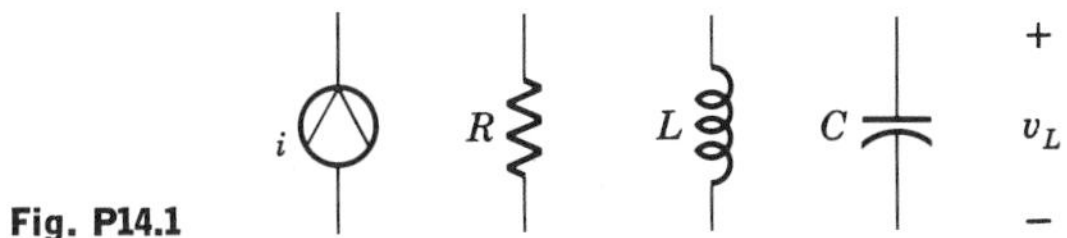

Fig. P14.1

14.2. Find the resonant frequency ω_0 and the bandwidth BW of the tuned circuit shown in Fig. P14.2. Compare the results obtained with the results of Prob. 14.1, where

$$R = r_c \left(\frac{\omega_0 L}{r_c}\right)^2$$

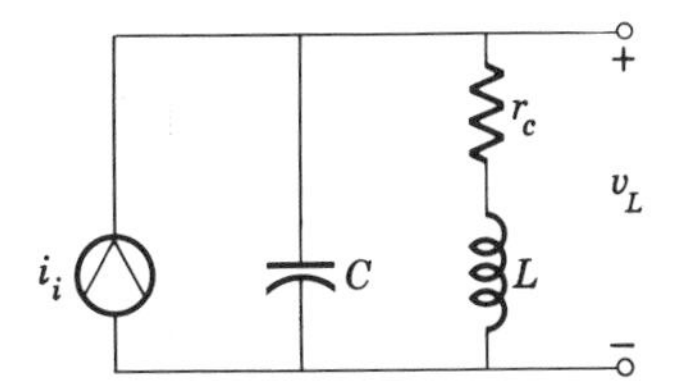

Fig. P14.2

14.3. A tuned-circuit amplifier is shown in Fig. P14.3.

(*a*) Find L so that the circuit resonates at 30 MHz.

(*b*) What is the bandwidth of the amplifier?

(*c*) Calculate the current gain.

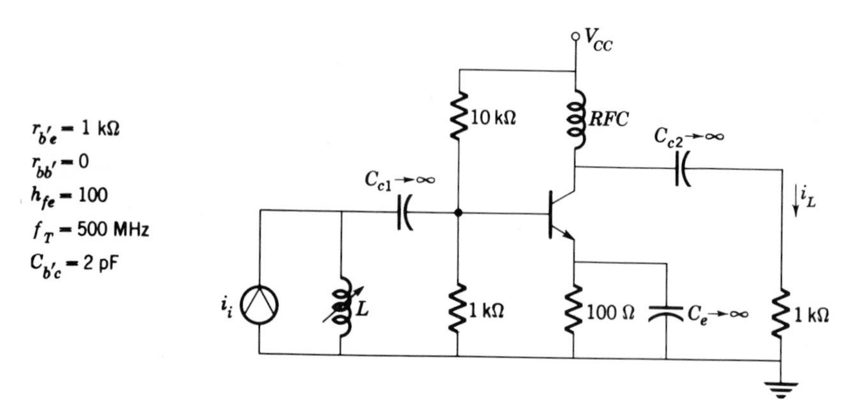

Fig. P14.3

14.4. Design a single-tuned amplifier to have a current gain of 10 dB at 40 MHz. The bandwidth is to be 1 MHz. The source impedance is 1 kΩ, and the load impedance is 1 kΩ. A 10-V supply is available. Assume that the Q of the inductor used is 50.

(*a*) Use only one transistor. Select it from a transistor manual. Note that f_T must exceed the GBW product of the amplifier. Why?

(*b*) Determine the pertinent transistor parameters.

(*c*) Calculate L.

(*d*) Check the design by finding the current gain.

14.5. Design a single-tuned amplifier to have maximum current gain at 30 MHz. Use the circuit shown in Fig. P14.5. Assume that the transformers have extremely high Q. Specify n, m, $A_{i,\max}$, and the bandwidth.

(*a*) What are the input and output admittances of the FET at 30 MHz?

(*b*) What are the forward and reverse transfer admittances of the FET at 30 MHz?

(*c*) Select n and m for maximum current gain.

(*d*) Select L_1 and L_2.

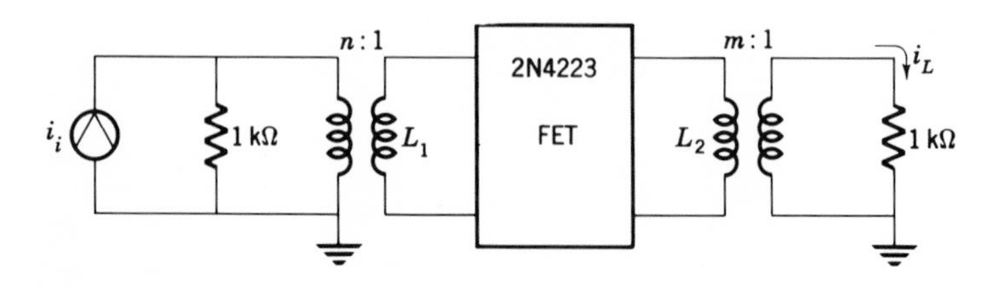

Fig. P14.5

14.6. The admittance Y, seen looking into the RC circuit of Fig. P14.6, can be represented by a resistance in parallel with a capacitance. Sketch the variation of this equivalent R_{eq} and C_{eq} with frequency.

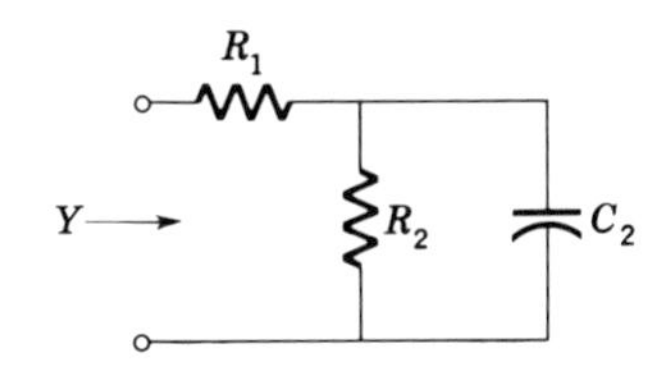

Fig. P14.6

14.7. Show that if, in Fig. 14.1-4,

$$LC' \equiv \frac{1}{\omega_0^2}$$

and if

$$\omega_h = \omega_0 + \frac{1}{4\pi C'(R_p \| r_i)} \ll \frac{1}{r_{b'e}C_{b'}}$$

then (14.1-12a) is valid.

14.8. Refer to Fig. 14.1-5. Let $r_{bb'} \gg 1/C_{b'}\omega_0$.

(a) Transform the series $r_{bb'}$-$C_{b'}$ circuit into a parallel RC equivalent circuit.

(b) If $C_{b'} = 1000$ pF, $r_{bb'} = 50\ \Omega$, $C' = 50$ pF, and $f_0 = 30$ MHz, find the effective parallel RC circuit seen looking into C'.

(c) Comment on the size of the effective parallel resistance as compared with R_p and r_i.

14.9. (a) Show that when (14.1-14) is valid, the series $r_{bb'}$-$C_{b'}$ circuit transforms into a parallel equivalent circuit, and that the effective capacitor is approximately $C_{b'}$, and the effective resistance is given by (14.1-16).

(b) If $r_{bb'} = 50\ \Omega$, $C_{b'} = 100$ pF, and $f_0 = 10$ MHz, calculate the value of the effective parallel resistor.

14.10. Verify (14.1-18).

14.11. Verify (14.1-20).

14.12. The circuit shown in Fig. P14.12 is to resonate at 10 MHz and have a 3-dB BW of 1 MHz.

(a) Find n_2/n_1 for maximum voltage gain at ω_0. What is the gain?

(b) Find n/n_1 to achieve the required bandwidth.

(c) Calculate L for resonance.

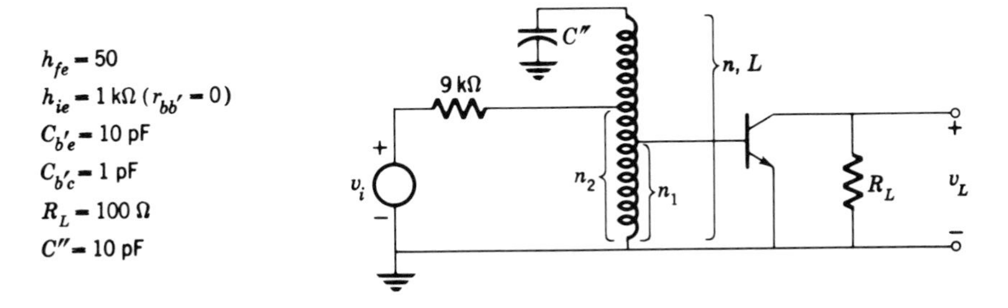

Fig. P14.12

SECTION 14.2

14.13. The capacitor C_1 is chosen so that the circuit resonates at 10 MHz.

(*a*) Calculate the voltage gain.

(*b*) Determine the 3-dB bandwidth.

(*c*) Can detuning (i.e., tuning C_1 so that the circuit resonates at a frequency other than 10 MHz) increase the center-frequency gain?

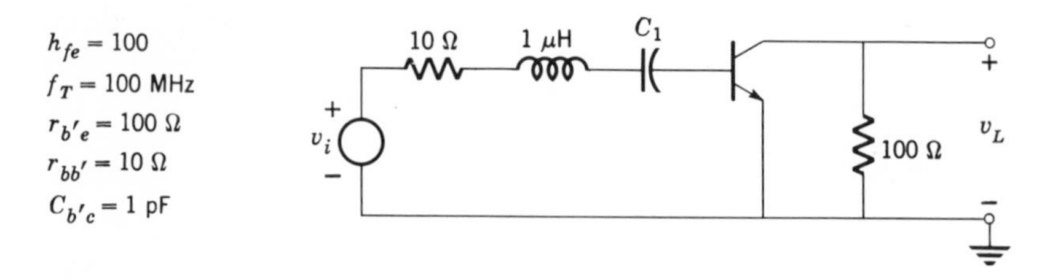

Fig. P14.13

14.14. Verify (14.2-4).

14.15. The FET cascode amplifier is to be tuned at 10 MHz.

(*a*) Find L.

(*b*) Calculate the bandwidth.

(*c*) Calculate the voltage gain.

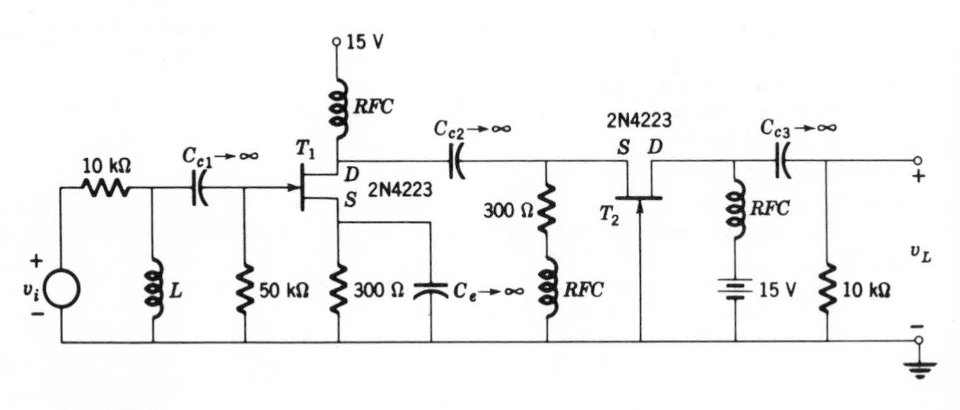

Fig. P14.15

SECTION 14.3

14.16. Verify (14.3-3*a*) and (14.3-3*b*).

14.17. Verify (14.3-5).

14.18. The FET circuit shown is to be neutralized at 100 MHz.

(*a*) Explain the operation of this circuit. Refer to Figs. 14.3-5 and 14.3-1.

(*b*) Find the Q point.

(*c*) Determine all the FET characteristics.

(*d*) Find C_f.

(*e*) Determine C_1, C_2, and L. The answer will not be unique. Discuss your choice of values.

(*f*) Determine L_0, C_0, and R_L. The answer, again, is not unique.

(*g*) Find n.

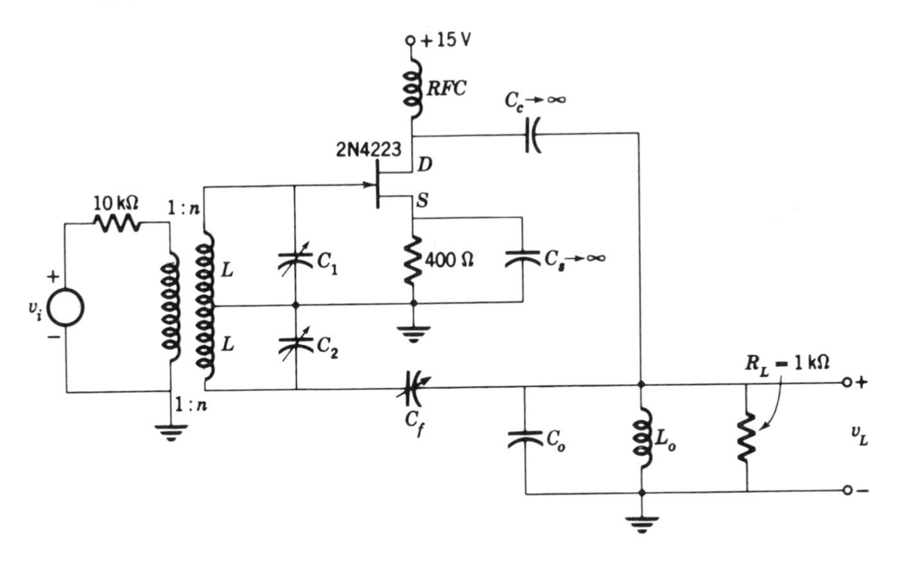

Fig. P14.18

SECTION 14.4

14.19. Verify (14.4-1) and (14.4-2).

14.20. Verify (14.4-4) and (14.4-6*b*).

14.21. Verify (14.4-10*b*).

14.22. The synchronously tuned amplifier is designed to resonate at 100 kHz and have a bandwidth of 2 kHz.

(*a*) Find C_1, L_1, C_2, L_2, and R_1.

(*b*) Calculate the gain v_L/i_i.

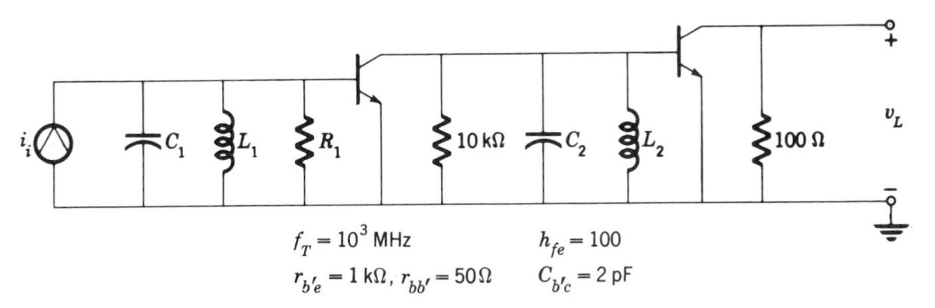

Fig. P14.22

SECTION 14.5

14.23. Verify (14.5-4).

14.24. In connection with the development leading to (14.5-6), let

$$x = \frac{\omega^2 - \omega_2^2}{\omega}$$

and

$$y = 2(\omega - \omega_2)$$

(a) Plot x and y versus ω, when ω is in the range

$$0.5\omega_2 \leq \omega \leq 1.5\omega_2$$

(b) Comment on the range of validity of the approximation made in (14.5-5b).

14.25. Verify (14.5-10).

14.26. Design a maximally flat stagger-tuned vacuum-tube amplifier having a center frequency at 10 MHz and a bandwidth of 3 MHz.

(a) Determine r_p, g_m, C_{gk}, and C_{pk}.
(b) Determine L_1, C_1 and L_2, C_2.
(c) Calculate v_L/v_i.

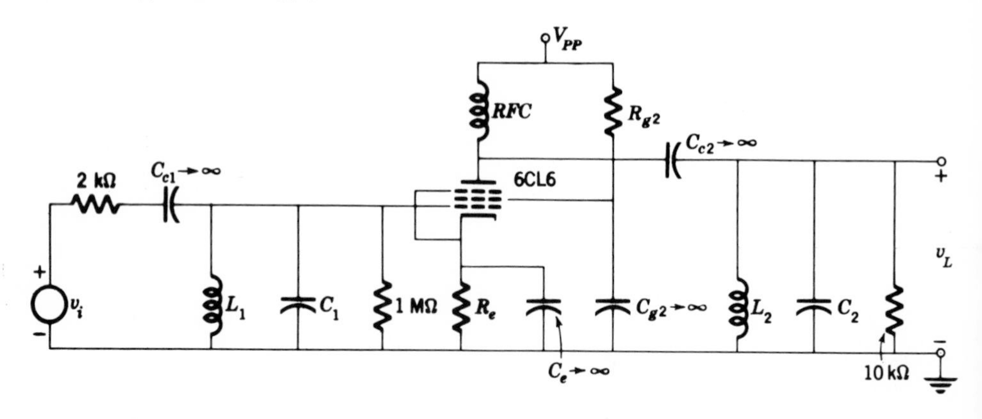

Fig. P14.26

14.27. Verify (14.5-19) and (14.5-20).

14.28. Refer to Fig. 14.5-6. Show that the current gain takes the same form as in (14.5-19) and (14.5-20), where C_m acts like M. (*Hint:* $\tau = R_1/\omega_0^2 L_1$.)

14.29. Using the circuit of Fig. 14.5-6, design a maximally flat double-tuned amplifier having a center frequency of 30 MHz and a 3-dB bandwidth of 10 MHz. The transistor parameters are $f_T = 1$ GHz, $r_{b'e} = 1$ kΩ, $r_{bb'} = 0$, $h_{fe} = 40$, $C_{b'c} = 0.4$ pF, and $R_L = 1$ kΩ.

(a) Find R_1, L_1, C_1, C_2, L_2, and C_m.
(b) Calculate the current gain.

SECTION 14.6

14.30. Verify (14.6-1) and (14.6-2).

14.31. Verify (14.6-3).

14.32. Using (14.6-3) find the conditions required for maximal flatness. Use the following procedure:

(*a*) Write

$$|A_i|^2 = K\left(\frac{1 + x^2}{A + Bx^2 + x^4}\right)$$

where $x = \omega/\omega_c$. Find K, A, and B.

(*b*) Differentiate this equation, and determine the maximum value of gain A_{im} and the value of $x = x_m$.

(*c*) Select A and B so that the gain *does not increase* as x changes from x_m.

14.33. (*a*) Find C_e to provide increased bandwidth and maximal flatness. What is the 3-dB bandwidth and the low-frequency current gain?

(*b*) If $C_e \to \infty$, find the 3-dB bandwidth and the low-frequency current gain.

(*c*) Compare (*a*) and (*b*).

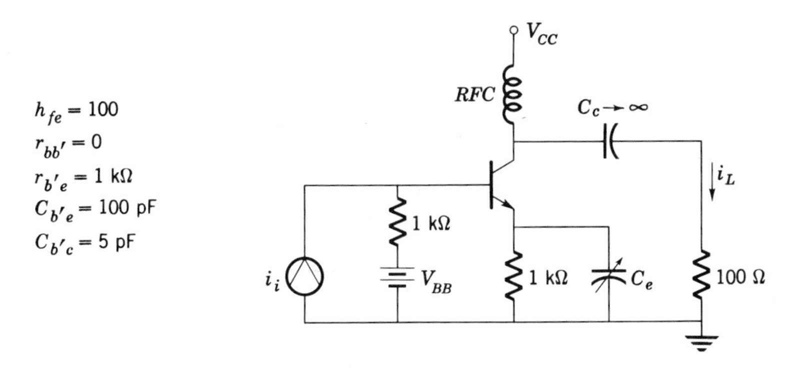

Fig. P14.33

REFERENCES

1. Joyce, M., and K. Clarke: "Transistor Circuit Analysis," p. 260, Addison-Wesley Publishing Company, Inc., Reading, Mass., 1961.
2. Millman, J., and H. Taub: "Pulse, Digital, and Switching Waveforms," McGraw-Hill Book Company, New York, 1965.
3. Johnson, W. C.: "Transmission Lines and Networks," McGraw-Hill Book Company, New York, 1950.

15

Frequency Response of Feedback Amplifiers

INTRODUCTION

The discussion of feedback amplifiers in Chap. 8 centered around the effects of feedback on gain, impedance levels, and sensitivity. In that chapter we found that the effectiveness of the feedback in reducing undesirable internal variations, output impedance, etc., was proportional to the loop gain T. In other words, large values of T are required to make the feedback effective.

In Chap. 8, we assumed that A_i, A_v, and T were independent of frequency. We also assumed that negative feedback was employed, so that T was a negative number, and $1 - T$ was always a positive number. The response of a real amplifier does, of course, depend on frequency, so that while T might be negative at low frequencies, its phase will generally increase at high frequencies. Suppose that, at some frequency ω_0, the magnitude of T is unity and the phase shift reaches 2π rad. Then at ω_0, $T(\omega_0) = +1$ and

$$A_{vf} = \frac{A_v}{1 - T} \rightarrow \infty$$

What does this result mean in terms of the physical system? Qualitatively, let us see what happens as T approaches $+1$. Clearly, A_{vf} increases. Thus, if we continually decrease v_i so as to maintain v_L at a fixed level as T approaches $+1$, we find in the limit that no signal is required to obtain an output when $T = +1$. When this happens, the amplifier is said to be unstable, and it may break into oscillation without any external excitation.

As will be seen, it is not necessary for T to be exactly $1\angle 0°$ or $1\angle 360°$ for instability to occur. When an amplifier is turned on, the loop gain will increase from zero to its nominal value in the time it takes for the amplifier to reach steady-state conditions. When the loop gain passes through the $1\angle 2n\pi$ point, during this transient period, it will break into oscillation.

An alternative way to define stability is in terms of the transient response of the amplifier. Thus an amplifier is stable if its impulse response contains no modes of free vibration which persist or increase indefinitely with time.[1] In terms of the Laplace transform, an equivalent statement is that the transfer function $A_{vf}(s)$ must have *no* poles in the right half of the $s = \sigma + j\omega$ plane, or on the imaginary axis, because such poles have $\sigma \geq 0$, and thus lead to persisting or indefinitely increasing terms in the transient response.

One of the prices we must pay for the benefits of feedback is the need to contend with the stability problem just described. In this chapter we discuss methods of predicting and avoiding high-frequency instability. In Sec. 15.7 we discuss oscillators, where instability is intentionally designed into the circuit.

15.1 BANDWIDTH AND GAIN-BANDWIDTH PRODUCT

In order to determine the frequency response of a typical feedback amplifier, we now consider the voltage-feedback–current-error circuit shown in Fig. 15.1-1.

In Chap. 8 we found that the overall current gain for this circuit could be written in the form

$$A_{if} = \frac{A_i}{1 - T} = \frac{A_i}{1 - A_i\beta_i} \tag{15.1-1}$$

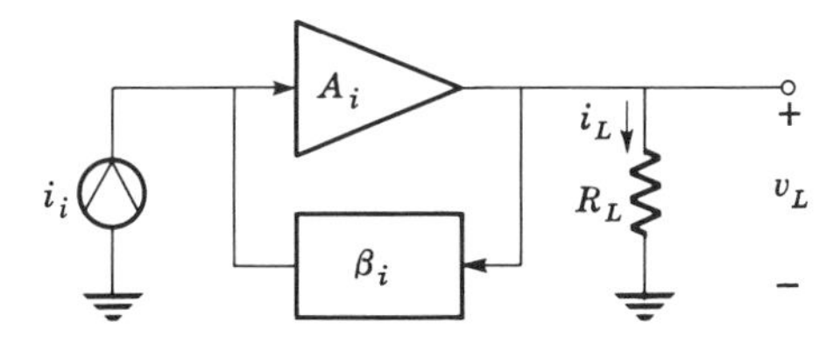

Fig. 15.1-1 Basic feedback circuit.

where A_i and β_i were assumed to be constants. We now drop this assumption so that A_i and β_i will, in general, be functions of the complex frequency s.

SINGLE-POLE AMPLIFIER

Let us assume that the amplifier gain function has a single negative real pole at $s = -\omega_1$, so that

$$A_i(s) = \frac{A_{io}}{1 + s/\omega_1} \tag{15.1-2}$$

If the feedback network is frequency-independent, then $\beta_i(s) = \beta_{io}$, and the overall gain becomes

$$A_{if}(s) = \frac{A_i(s)}{1 - A_i(s)\beta_i(s)}$$
$$= \frac{A_{io}}{1 - A_{io}\beta_{io} + s/\omega_1} \tag{15.1-3}$$

The overall gain, then, has a pole at

$$s = -\omega_1(1 - A_{io}\beta_{io}) \tag{15.1-4}$$

and the 3-dB bandwidth has increased by the factor $1 - A_{io}\beta_{io}$ because of the addition of the feedback. This is the same factor by which the gain is reduced. Thus the gain-bandwidth product is $|A_{io}|\omega_1/2\pi$ for the basic amplifier with or without feedback. Figure 15.1-2*b* is the gain-frequency plot for this amplifier with and without feedback.

Without feedback, the pole (15.1-4) is at $-\omega_1$. As the amount of feedback is increased from zero, the pole moves farther into the left half-plane along the negative real axis as shown in Fig. 15.1-2*a*.* This curve is called a *root locus*. It displays the manner in which the poles of the overall transfer function vary in terms of the variation of a parameter (in this case, the loop gain).

It is clear from (15.1-3) and Fig. 15.1-2 that no stability problem exists here because the only pole of $A_{if}(s)$ lies on the negative real axis as long as $-A_{io}\beta_{io}$ is positive.

DOUBLE-POLE AMPLIFIER

Now let us assume a more complicated case, an amplifier gain function with a pair of coincident real poles.

$$A_i(s) = \frac{A_{io}}{(1 + s/\omega_1)^2} \tag{15.1-5}$$

* Note that this is true *only* if $A_{io}\beta_{io}$ is a negative number. If $A_{io}\beta_{io}$ were positive, the pole would move into the right half-plane and the amplifier would become unstable.

Again assuming $\beta_i(s) = \beta_{io}$, we find for the overall gain

$$A_{if}(s) = \frac{A_{io}}{1 - A_{io}\beta_{io} + 2s/\omega_1 + s^2/\omega_1^2} \tag{15.1-6}$$

Rearranging terms yields

$$A_{if}(s) = \left(\frac{A_{io}}{1 - A_{io}\beta_{io}}\right)\left(\frac{1}{1 + 2\zeta s/\omega_n + s^2/\omega_n^2}\right) \tag{15.1-7}$$

where

$$\omega_n = \omega_1 \sqrt{1 - A_{io}\beta_{io}} \tag{15.1-8a}$$

and

$$\zeta = \frac{1}{\sqrt{1 - A_{io}\beta_{io}}} \tag{15.1-8b}$$

The current gain given in (15.1-7) shows that the system is of second order, with poles at

$$s = -\omega_n(\zeta \pm \sqrt{\zeta^2 - 1}) \tag{15.1-9}$$

The poles of the overall transfer function are negative, real, and coincident when there is no feedback ($\beta_{io} = 0$ and $\zeta = 1$), but become complex as

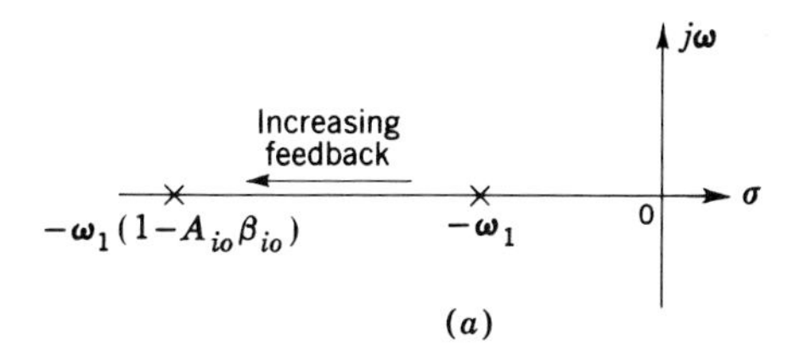

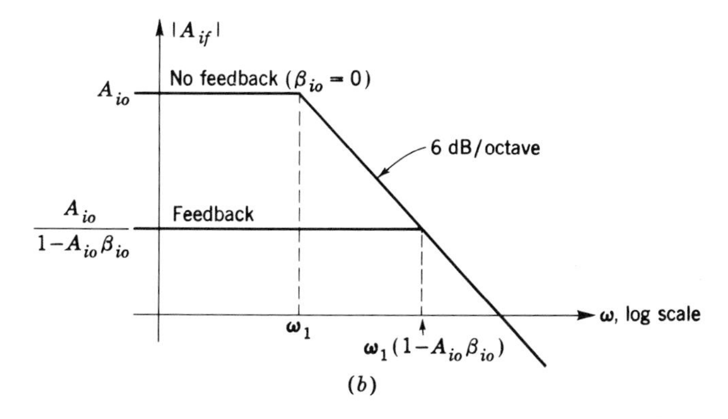

Fig. 15.1-2 Characteristics of single-pole feedback amplifier. (*a*) Locus of pole motion; (*b*) gain vs. frequency.

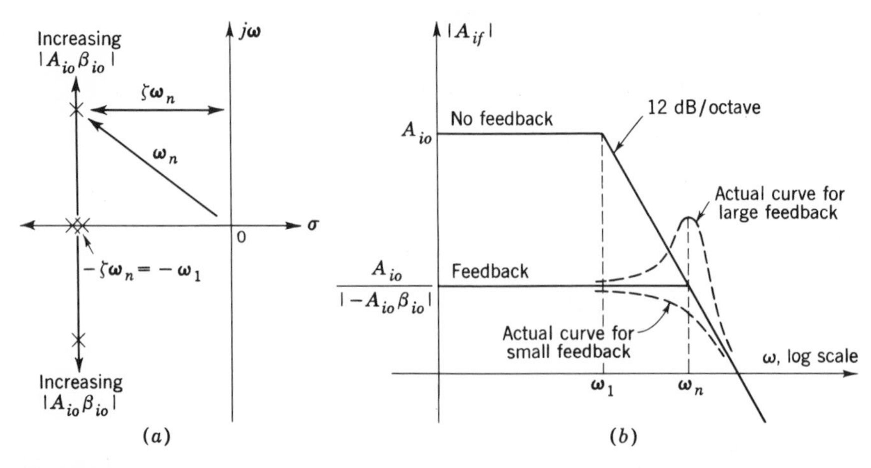

Fig. 15.1-3 Characteristics of two-pole feedback amplifier. (*a*) Root locus; (*b*) gain vs. frequency.

soon as $-A_{io}\beta_{io} > 0$. The loci of these poles for increasing $|A_{io}\beta_{io}|$ are simply two vertical straight lines, as shown in Fig. 15.1-3*a*.

The asymptotic and actual frequency responses with and without feedback are shown in Fig. 15.1-3*b*. The poles of $A_{if}(s)$ are in the left half-plane for all values of feedback; so there is no stability problem in the strict sense. However, when the feedback is large, the frequency response exhibits a sharp peak. This leads to damped oscillations in the transient response of the circuit, which are usually undesirable. The magnitude of the peak depends on the damping ratio ζ, and increases with increasing feedback. Normalized plots of the magnitude and phase of the frequency response are shown in Fig. 15.1-4, for various values of ζ. The graphs can be interpreted in terms of the circuit Q by noting that $Q = 1/2\zeta$ (Prob. 15.6).

The assumptions that the forward amplifier had a simple single pole or a double pole, and that the β network was not frequency-dependent, resulted in stable overall gain with feedback. In the next section we consider an example where the poles can move into the right half-plane.

15.2 THE PROBLEM OF STABILITY, A THREE-POLE AMPLIFIER

To illustrate the stability problem in a multistage feedback amplifier, we consider the three-stage CE cascaded amplifier shown in simplified form in Fig. 15.2-1.

To simplify the analysis, we assume identical transistors, $R \gg h_{ie}$, $R_f \gg R_L$, and $R_f \gg h_{ie}$. The individual stages all have approximately the

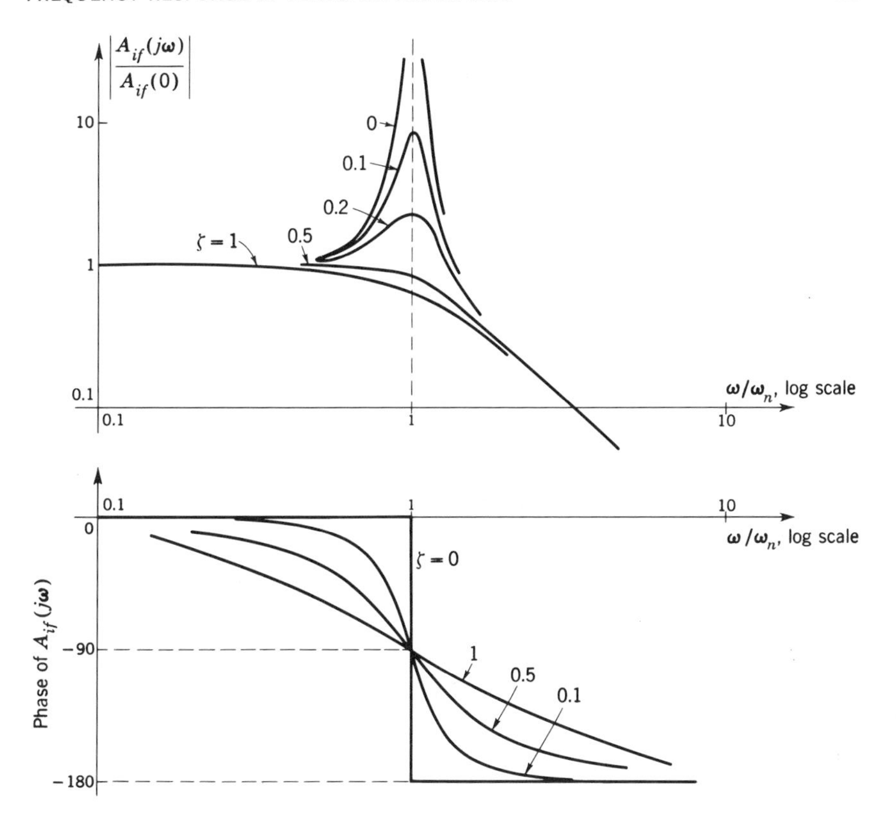

Fig. 15.1-4 Response of second-order system.

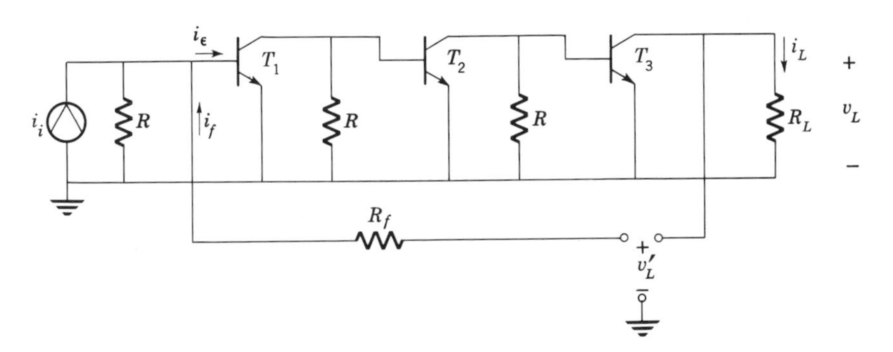

Fig. 15.2-1 Three-stage feedback amplifier.

same gain:

$$A_i \approx \frac{-h_{fe}}{1 + s/\omega_1} \tag{15.2-1}$$

where

$$\omega_1 = \frac{1}{r_{b'e}C}$$

and C represents the total capacitance appearing at the base of each transistor. The gain of the amplifier without feedback is then

$$A_i = \frac{i_L}{i_i}\bigg|_{v'_L=0} \approx \frac{-h_{fe}{}^3}{(1 + s/\omega_1)^3} \tag{15.2-2}$$

The loop gain T is found using the techniques developed in Chap. 8. Then

$$T(s) = \frac{v_L}{v'_L}\bigg|_{i_i=0} \approx A_i(s)\left(\frac{R_L}{R_f}\right) \tag{15.2-3}$$

The gain with feedback is, from (8.2-9),

$$A_{if}(s) = \frac{A_i(s)}{1 - T(s)}$$

$$= \frac{-(h_{fe})^3}{(1 + s/\omega_1)^3 - T(0)} \tag{15.2-4}$$

where

$$T(0) = -(h_{fe})^3\left(\frac{R_L}{R_f}\right)$$

For this case, the loci of the poles of the overall transfer function start from the triple pole at $s = -\omega_1$ when there is no feedback ($R_f = \infty$). As the feedback is increased, one pole moves to the left along the negative real axis, while the other two move along the lines shown in Fig. 15.2-2*a*. When $-T(0) = 8$, the complex poles lie on the $j\omega$ axis at $\omega = \sqrt{3}\,\omega_1$. Thus, for $-T(0) \geq 8$, the amplifier is unstable. Referring to Fig. 15.2-2*b*, we see that as $-T(0)$ increases, the gain with feedback A_{if} decreases. When $-T(0)$ approaches 8, the frequency response of A_{if} exhibits sharp peaks because of the proximity of the complex poles to the $j\omega$ axis.

Since the factor $1 - T(0)$ determines the improvement in sensitivity and the other advantages of feedback, we are not going to be able to make good use of feedback in this circuit unless we can modify it, so as to realize a value greater than 8 for the magnitude of the loop gain $T(0)$. Fortunately, it is usually a relatively easy matter to modify the basic amplifier so that larger values of $|T(0)|$ can be obtained with freedom from oscillations.

In the sections to follow, we present methods for analyzing feedback

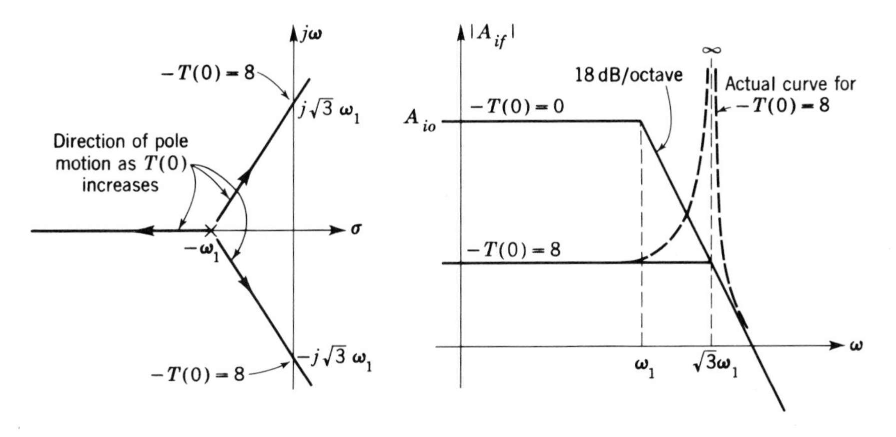

Fig. 15.2-2 Characteristics of three-pole amplifier. (*a*) Root locus; (*b*) gain vs. frequency.

amplifiers with a view toward determining stability. Design methods for ensuring stability within prescribed safety margins are presented as well.

15.3 THE NYQUIST STABILITY CRITERION—BODE PLOTS

In the preceding sections we indicated how the stability problem arises in feedback amplifiers when the loop gain has more than two real poles. The root-locus plots used there provided a visual picture of the effect of feedback on the poles of the overall transfer function. A complete discussion of the root-locus technique is beyond the scope of this text, and the interested reader should consult the References.[1]

The Nyquist criterion which we discuss in this section forms the basis of a steady-state method of determining whether or not an amplifier is stable. We illustrate this method by considering the basic feedback equation

$$A_{if}(s) = \frac{A_i(s)}{1 - T(s)} \tag{15.3-1}$$

The amplifier gain $A_{if}(s)$ is defined as being stable if it has no poles with positive or zero real parts. Thus, to determine whether or not a given amplifier is stable, one need only determine the loop gain $T(s)$, form the function $1 - T(s)$, factor the numerator polynomial,* and inspect the roots for positive or zero real parts. This is usually a tedious task which

* The Routh-Hurwitz criterion (see References) determines stability directly from the coefficients of the polynomial without the need for factoring.

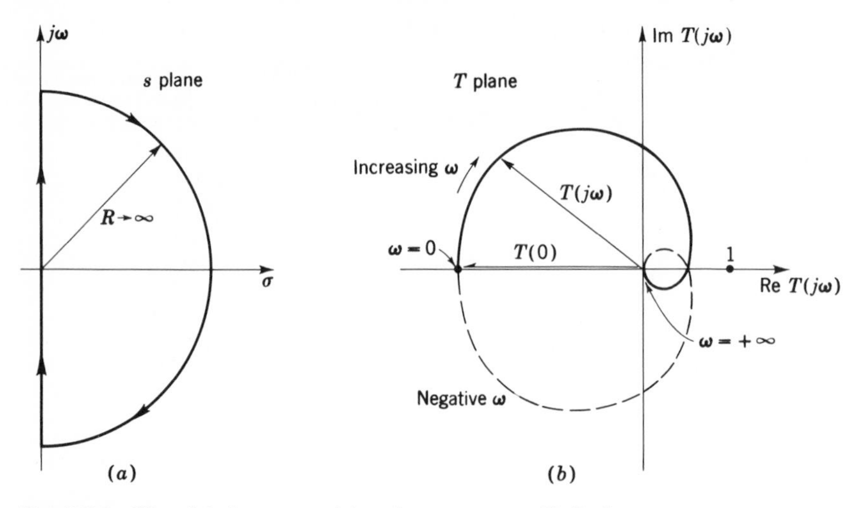

Fig. 15.3-1 Nyquist diagram. (*a*) s-plane contour; (*b*) T-plane contour.

provides little or no design information. If we examine (15.3-1) more closely, a considerably simpler method, which also supplies design information, emerges.

In what follows, we assume that $A_i(s)$ and $T(s)$ are stable transfer functions. With this assumption, the only way for $A_{if}(s)$ to have poles in the right half of the s plane (RHP) is for the denominator $1 - T(s)$ to have zeros in the RHP. Thus our problem is to investigate the zeros of $1 - T(s)$. This is equivalent to finding whether there are values of s with positive or zero real parts at which $T(s) = 1$. In order to do this we plot a *Nyquist diagram* for the loop gain. The Nyquist diagram is simply a polar plot of $T(j\omega)$ for $-\infty < \omega < \infty$. A typical diagram for $T(j\omega)$ having three identical poles as in (15.2-3) is shown in Fig. 15.3-1*b*.

It is shown in the References that the Nyquist diagram maps the right half of the s plane shown in Fig. 15.3-1*a* into the interior of the contour in the T plane (Fig. 15.3-1*b*). If there are any zeros of poles of $1 - T(s)$ in the RHP, the T-plane contour will enclose the point $1\underline{/0}$, which is called the *critical point*. The number of times that the T-plane contour encircles this critical point in a clockwise direction is equal to the number of zeros of $1 - T(s)$ with positive real parts.

The contour of Fig. 15.3-1 applies to the amplifier of Fig. 15.2-1. As shown, there are no encirclements of the critical point; so the amplifier is stable. However, the diagram is sketched for $|T(0)| < 8$. If $|T(0)| = 8$, the contour would pass through the critical point. For $|T(0)| > 8$, the critical point would be encircled twice (once for the

$0 < \omega < \infty$ range and once for $-\infty < \omega < 0$), indicating the presence of a pair of RHP zeros of $1 - T(s)$, as borne out by the root locus of Fig. 15.2-2*a*.

Now that we have established the idea of the critical point in the T plane, we note that it is not necessary to plot the Nyquist diagram to ascertain that the critical point is not encircled. This can be determined from Bode plots of the amplitude and phase of $T(s)$, because they contain all the information in the Nyquist diagram. Bode plots corresponding to the Nyquist diagram of Fig. 15.3-1 are shown in Fig. 15.3-2.

From a comparison of the Bode plots* and the Nyquist diagram, we see that the criterion for stability is that $T(\omega)$ cross the 0-dB (unity-gain) axis at a lower frequency than that required for the phase shift to reach 180°. If this criterion is met, the amplifier will be stable. However, we have no real knowledge of the degree of stability since we do not know the position of the poles of A_{if}, as in the root-locus analysis. We do know that when the 0-db crossover and 180° phase frequencies are very close, the poles of the overall transfer function are very close to the $j\omega$ axis, and the amplifier is on the threshold of instability.

* The negative sign in $T(s)$ is not included in the Bode plot; thus 180° must be added to the phase shift indicated by the curve in order to obtain the Nyquist diagram.

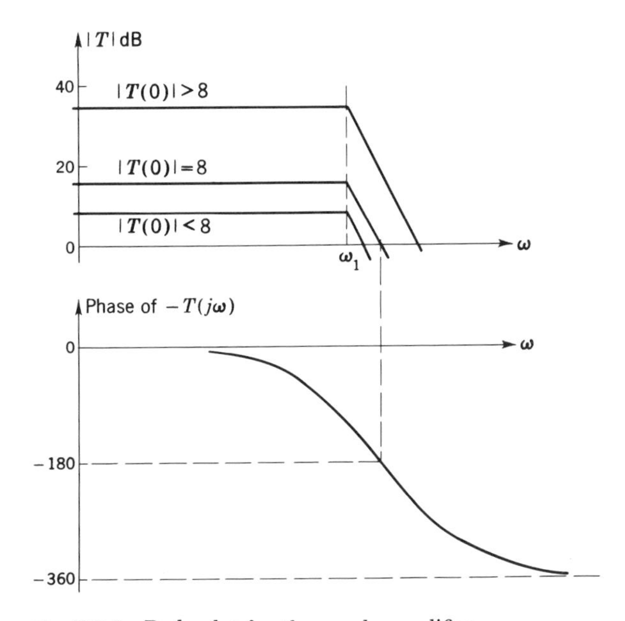

Fig. 15.3-2 Bode plot for three-pole amplifier.

15.4 STABILIZING NETWORKS

In the previous sections we saw that a feedback amplifier can be unstable if its loop gain T is too large and the number of poles in the forward-gain transfer function is greater than 2. Almost *all* integrated circuit amplifiers have more than two poles in the forward gain. When these IC amplifiers are used with feedback, they have a tendency to oscillate unless properly compensated. Various compensation schemes can be employed, all having the same objective, namely, to shape the loop-gain frequency-response characteristic so that the phase shift *is less than* 180° when the magnitude of the loop gain has decreased to unity. Careful design is required in order to ensure that the amplifier will be stable, with an adequate margin of stability over the expected range of variation of transistor h_{fe}, temperature, etc.

In this section we discuss the maximum possible feedback that can be used with the various compensation techniques available.

15.4-1 NO FREQUENCY COMPENSATION

Let us first briefly review several basic principles. To do this we consider a specific example, that of an amplifier with open-loop forward gain,

$$A_v = \frac{-10^4}{\left(1 + \dfrac{s}{2\pi \times 10^6}\right)\left(1 + \dfrac{s}{2\pi 10 \times 10^6}\right)\left(1 + \dfrac{s}{2\pi 30 \times 10^6}\right)} \tag{15.4-1}$$

Equation (15.4-1) is typical of a three-stage transistor amplifier. The equation is plotted in Fig. 15.4-1. We see that the amplifier may be unstable if used in a feedback amplifier with resistive feedback and without compensation, since the magnitude of the gain when the phase shift is 180° is greater than unity (0 dB). To ensure stability, the feedback employed must reduce the magnitude of the loop gain T to a value less than unity when the phase is 180°.

If sufficient feedback is applied to make $|T| = 1$ at 14 MHz (the phase of T at 14 MHz is 180°—see Fig. 15.4-1), the amplifier is said to be *marginally stable*, since there is no margin of safety. If, however, sufficient feedback is applied to make $|T| = 1$ at a frequency less than 14 MHz, the phase at $|T| = 1$ is less than 180°. The difference between 180° and the actual phase at $|T| = 1$ is called the *phase margin*. Designers often employ a value of 45° phase margin as a compromise between acceptable transient response and adequate stability. Thus, using this criterion in the above example, we should design the feedback network so that $|T| = 1$ at $\varphi = 135°$. From the phase curve shown in Fig. 15.4-1, we see that this occurs at approximately 10 MHz. The frequency at which $|T| = 1$ is usually referred to as the *crossover frequency.**

If we assume that the feedback network is not frequency-dependent, we can write

$$T = \beta A_v$$

where β is a constant representing the amount of feedback. Then, if the phase margin is to be 45°, we must have

$$|T| = 1 = |\beta|\,|A_v(10\text{ MHz})| \approx \beta\left(\frac{10^3}{\sqrt{2}}\right)$$

Thus

$$|\beta| = \sqrt{2}\,(10^{-3}) \quad (= -57\text{ dB}) \tag{15.4-2}$$

Let us now look at T. At low frequencies

$$|T| = |\beta|\,|A_v| = \sqrt{2}\,(10^{-3})(10^4) \approx 14\ (= 23\text{ dB})$$

* Strictly speaking, it is the *gain* crossover frequency.

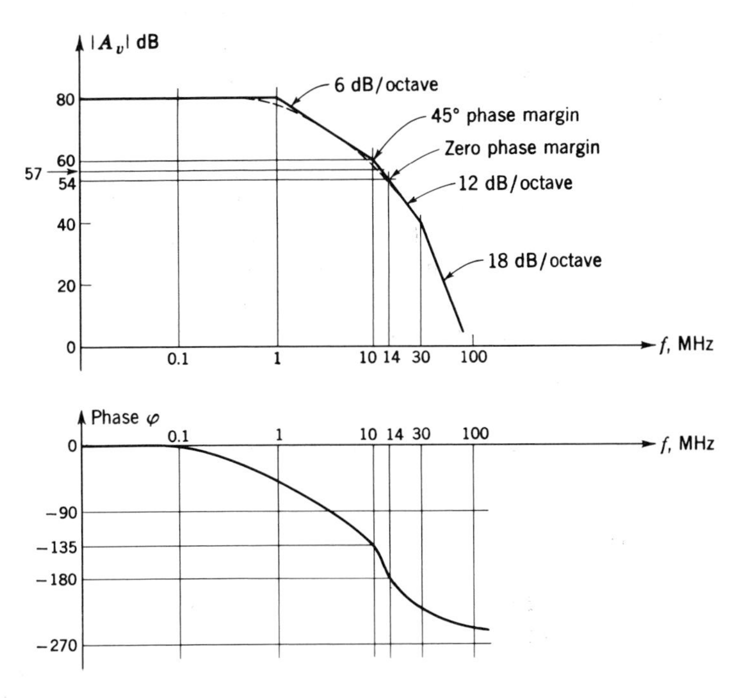

Fig. 15.4-1 Bode plot for amplifier of (15.4-1) (curves neglect interaction of poles).

If $|T|$ at low frequencies were to exceed 23 dB, the phase margin would drop below 45°, and the amplifier might become unstable. Note that a magnitude-frequency plot of T is identical with that of $|A_v|$, with the exception that amplitude values are reduced by 57 dB (for 45° phase margin).

Summarizing, we have found that the amount of feedback necessary to achieve a 45° phase margin is

$$|\beta| = \frac{1}{|A_v\ (\omega \text{ corresponding to } 135°)|} \tag{15.4-3}$$

and the maximum loop gain for the amplifier of (15.4-1) is

$$|T_{\max}| = |\beta|\,|A_{v,\max}| = \frac{A_{v,\max}}{|A_v\ (\omega \text{ corresponding to } 135°)|} \tag{15.4-4}$$

We noted in the above example that at low frequencies the closed-loop gain A_{vf} is

$$A_{vf} = \frac{A_v}{1 - T} = \frac{A_v}{1 - \beta A_v} \approx \frac{-1}{\beta} = -707\ (= 57 \text{ dB})$$

Thus the loss in gain due to the feedback is 23 dB (80 − 57 dB), and the maximum permissible loop gain $|T_{\max}|$ is 23 dB for a phase margin of 45°. A larger loop gain reduces the phase margin below the acceptable level of 45°.

For many applications a loop gain of 23 dB is not adequate to satisfy requirements on sensitivity or impedance level. Thus this example points up the need for compensation schemes which will allow use of larger loop gains along with sufficient phase margin. In succeeding sections we discuss several commonly used compensation techniques.

15.4-2 SIMPLE LAG COMPENSATION

In this section we discuss the simple lag network, which is designed to introduce an additional negative real pole in the transfer function of the basic amplifier gain A_v. When this network is added, the forward gain becomes

$$A_{v1} = \frac{A_v}{1 + s/\alpha} \tag{15.4-5}$$

The pole α is adjusted so that $|T|$ drops to 0 dB at a frequency where the poles of A_v contribute negligible phase shift. Using the example of Sec. 15.4-1, shown in Fig. 15.4-1, we note that with $|T_{\max}| = 23$ dB, a phase margin of 45° is achieved with no compensation. If we sketch the Bode plot for $|T_{\max}| = 26$ dB, as shown in Fig. 15.4-2a, we see that the phase

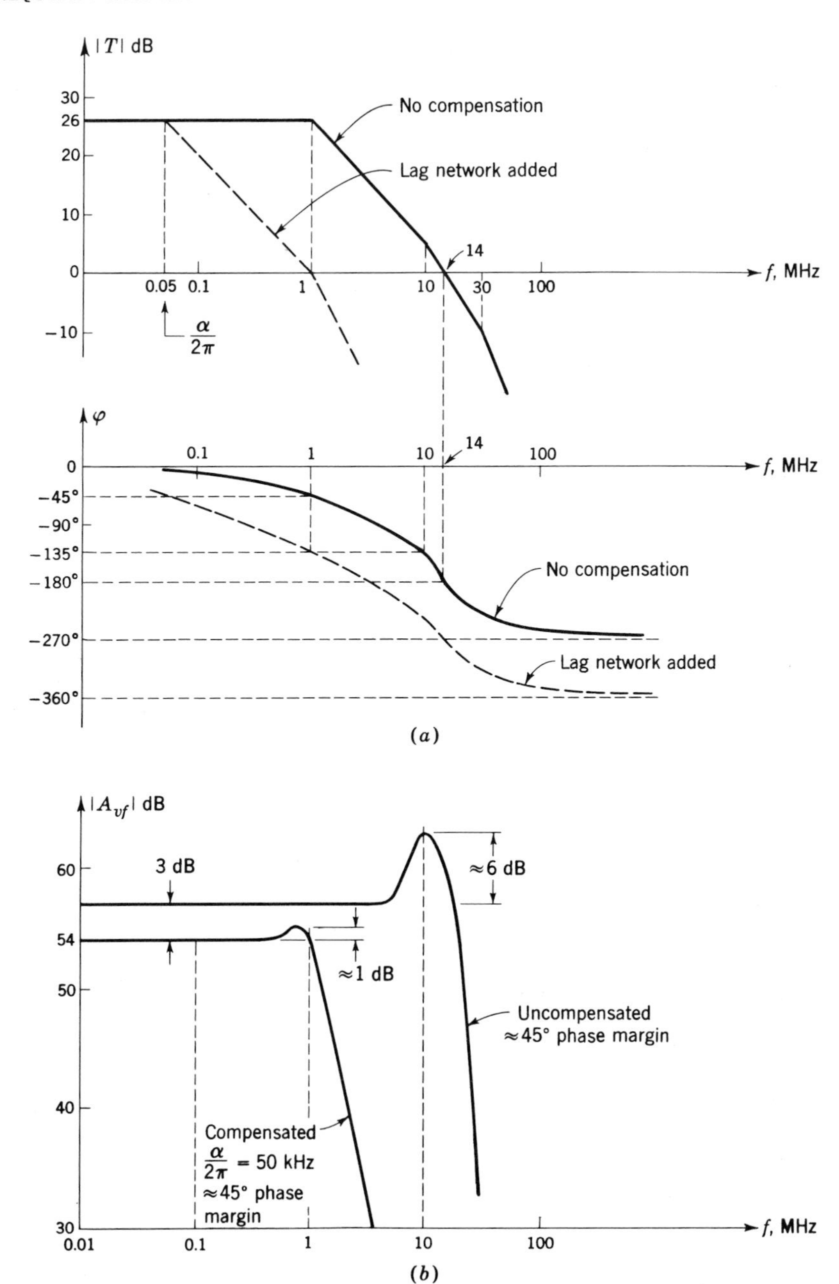

Fig. 15.4-2 The effect of lag compensation. (*a*) Loop gain with simple lag compensation; (*b*) amplifier frequency-response curves.

margin *without compensation* is reduced to zero, and the crossover frequency is 14 MHz.

Now the question is, can we achieve a 45° phase margin by adding a lag network to modify the forward gain as in (15.4-5)? If we set $\alpha/2\pi = 50$ kHz, the Bode plot of $|T|$ takes the form shown in dashed lines in Fig. 15.4-2*a*. The phase margin is now 45°, but the crossover frequency has been reduced to approximately 1 MHz. Thus addition of the lag network has allowed us to increase the loop gain T by a factor of $\sqrt{2}$ (3 dB). However, the crossover frequency has been reduced by a factor of approximately 10.

The gain with feedback for the uncompensated and the compensated feedback amplifiers discussed above [Eqs. (15.4-1) and (15.4-5)] are plotted in Fig. 15.4-2*b* when both are adjusted to have a phase margin of approximately 45°. Note that the uncompensated feedback amplifier has a 6-dB peak in the response. This might be undesirable for some applications. The compensated feedback amplifier has 3 dB less low-frequency gain and a significantly narrower bandwidth than the uncompensated amplifier. However, the loop gain of the compensated amplifier is 3 dB larger, and the peak in the frequency-response curve is only slightly more than 1 dB.

A glance at Figs. 15.4-1 and 15.4-2*a* indicates that, with the lag network added, we can increase the loop gain even further at low frequencies, provided that we lower the break frequency α of the lag network so that the crossover frequency occurs below 1 MHz. This ensures a minimum of 45° phase margin. The loop gain for various values of α is shown in Fig. 15.4-3. From this figure we see that the loop gain can be increased to the full amplifier gain of 80 dB and the phase margin will remain at 45°. The overall 3-dB bandwidth of the feedback amplifier would be about the same for all the loop-gain curves of Fig. 15.4-3. In a given design problem, we may not have this much flexibility, because of simultaneous specifications on overall gain, sensitivity, etc., and trade-offs may have to be made.

Stability for all T Another interesting possibility exists here. If the break frequency of the lag network is set at 100 Hz, the phase margin is 45° when $T = 80$ dB. If now T is *decreased*, the phase margin will increase. Thus, with $\alpha/2\pi = 100$ Hz, the amplifier is stable for *all* possible values of feedback. Clearly, this will result in considerable loss in bandwidth when T is much less than 80 dB, but it may be acceptable in some cases.

Design of the lag network The design of the actual lag network is easily accomplished once a satisfactory value of break frequency α has been

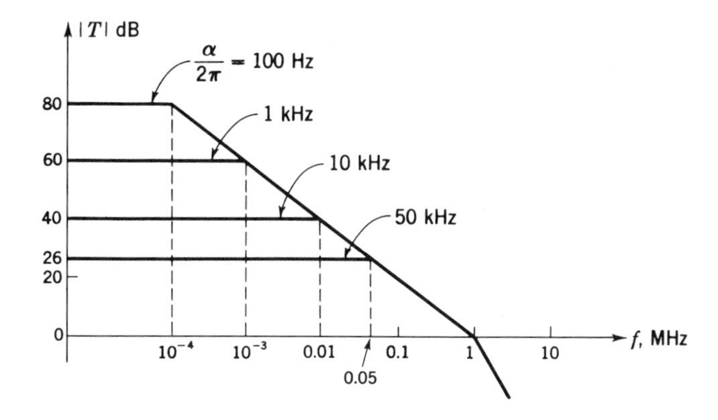

Fig. 15.4-3 Loop gain for 45° phase margin with lag network ($\beta \leq 1$).

determined. Usually, a simple RC filter placed in the forward amplifier, as shown in Fig. 15.4-4, will suffice.

In integrated-circuit amplifiers several terminals are usually provided for the connection of compensation networks. These terminals are connected to suitable high-impedance points within the amplifier. The reason for this is that, at higher impedance levels, smaller capacitance values will be required, and no external resistance R is needed. Examples are presented in Sec. 15.5 to illustrate this point.

15.4-3 MORE COMPLICATED LAG COMPENSATION

In the preceding sections we discussed methods for modifying the loop transmission so as to achieve a 45° phase margin at crossover. The solutions presented were designed to bring the loop gain down to 0 dB before the phase shift due to the basic amplifier became excessive. As a result, a considerable portion of the "available" bandwidth was wasted. In many cases, the specifications will call for maximum bandwidth and fixed closed-loop gain. A lag network with a pole and a zero will usually

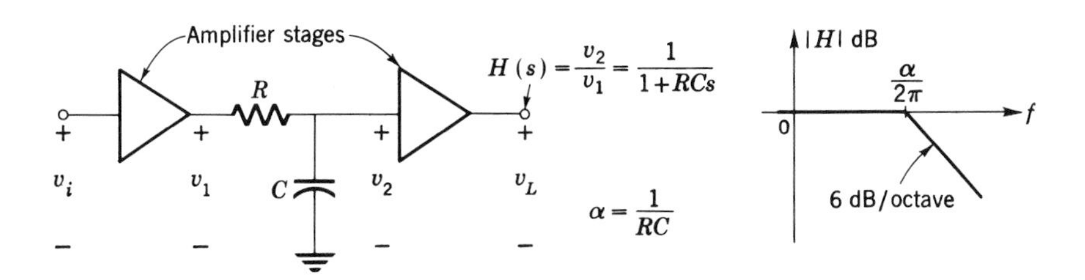

Fig. 15.4-4 Basic lag network.

provide a much wider bandwidth than the single-pole network. The transfer function of this new lag network is

$$H(s) = \frac{1 + s/\alpha_2}{1 + s/\alpha_1} \qquad \alpha_1 < \alpha_2 \tag{15.4-6}$$

The forward gain is then

$$A_{v2} = H(s)A_v \tag{15.4-7}$$

Let us assume that A_v is characterized by three simple poles as in (15.4-1). Then

$$A_v = \frac{A_o}{(1 + s/\gamma_1)(1 + s/\gamma_2)(1 + s/\gamma_3)} \tag{15.4-8}$$

Now

$$A_{v2} = \frac{A_o(1 + s/\alpha_2)}{(1 + s/\alpha_1)(1 + s/\gamma_1)(1 + s/\gamma_2)(1 + s/\gamma_3)} \tag{15.4-9}$$

From this expression we see that we can effectively cancel the smallest pole of A_v by setting the zero of the lag network at the same point; i.e.,

$$\alpha_2 = \gamma_1 \tag{15.4-10}$$

The forward voltage gain becomes

$$A_{v2} = \frac{A_o}{(1 + s/\alpha_1)(1 + s/\gamma_2)(1 + s/\gamma_3)} \tag{15.4-11}$$

In our example [Eq. (15.4-1)]

$$A_o = -10^4 \ (= 80 \text{ dB})$$

$$\frac{\alpha_2}{2\pi} = \frac{\gamma_1}{2\pi} = 1 \text{ MHz}$$

$$\frac{\gamma_2}{2\pi} = 10 \text{ MHz}$$

$$\frac{\gamma_3}{2\pi} = 30 \text{ MHz}$$

$$\frac{\alpha_1}{2\pi} < \frac{\alpha_2}{2\pi}$$

$$T = \beta A_{v2} \tag{15.4-12}$$

The pole of $H(s)$, which occurs at α_1, is chosen so that the amplifier is stable, with a 45° phase margin independent of the value of $T(0)$. In this example,

$$T(\omega) = \frac{A_o\beta}{(1 + j\omega/\alpha_1)(1 + j\omega/\gamma_2)(1 + j\omega/\gamma_3)} \tag{15.4-13}$$

Therefore, at low frequencies ($\omega \to 0$),

$$T(0) = A_o\beta \tag{15.4-14}$$

and at $\omega = \gamma_2$,

$$-T(\gamma_2) = 1\angle{-135°} = \frac{-A_o\beta}{(1 + j\gamma_2/\alpha_1)(1 + j)(1 + j\gamma_2/\gamma_3)}$$
$$\approx \frac{-A_o\beta}{(j\gamma_2/\alpha_1)(1 + j)} = \frac{-T(0)}{(\sqrt{2}\,\gamma_2/\alpha_1)\angle{135°}} \tag{15.4-15a}$$

since

$$\alpha_1 < \alpha_2 = \gamma_1 < \gamma_2 < \gamma_3 \tag{15.4-15b}$$

Hence

$$-T(0) = \sqrt{2}\left(\frac{\gamma_2}{\alpha_1}\right) \tag{15.4-16a}$$

and

$$\alpha_1 = \frac{\sqrt{2}\,\gamma_2}{-T(0)} \tag{15.4-16b}$$

Thus, to achieve a loop gain of 40 dB in the above example requires that [Eq. (15.4-14)]

$$\beta = \frac{T(0)}{A_{v2}(0)} = \frac{-10^2}{-10^4} = 10^{-2} \; (= -40 \text{ dB})$$

and from (15.4-16b)

$$\frac{\alpha_1}{2\pi} = \frac{\sqrt{2}\,(10^7)}{10^2} \approx 140 \text{ kHz}$$

These results are sketched in Fig. 15.4-5. It should be noted that the simplifications used in (15.4-15a) are valid only when $\gamma_3 \gg \gamma_2$. For the values in the example, exact calculation using (15.4-15a) with $\alpha_1/2\pi = 140$ kHz yields $-T(\gamma_2) \approx 0.95\angle{-154°}$. Thus the actual crossover frequency is slightly lower than γ_2 and the actual phase margin is $180° - 154° = 26°$.

Comparing Fig. 15.4-5 with Fig. 15.4-3, we see that the introduction of the zero in the lag network results in an increase in the gain crossover frequency by a factor of 10. The bandwidth of this feedback amplifier can be shown to be approximately 15 MHz.

A network having the transfer function of (15.4-6) is shown in Fig. 15.4-6, and the design equations are given in the figure.

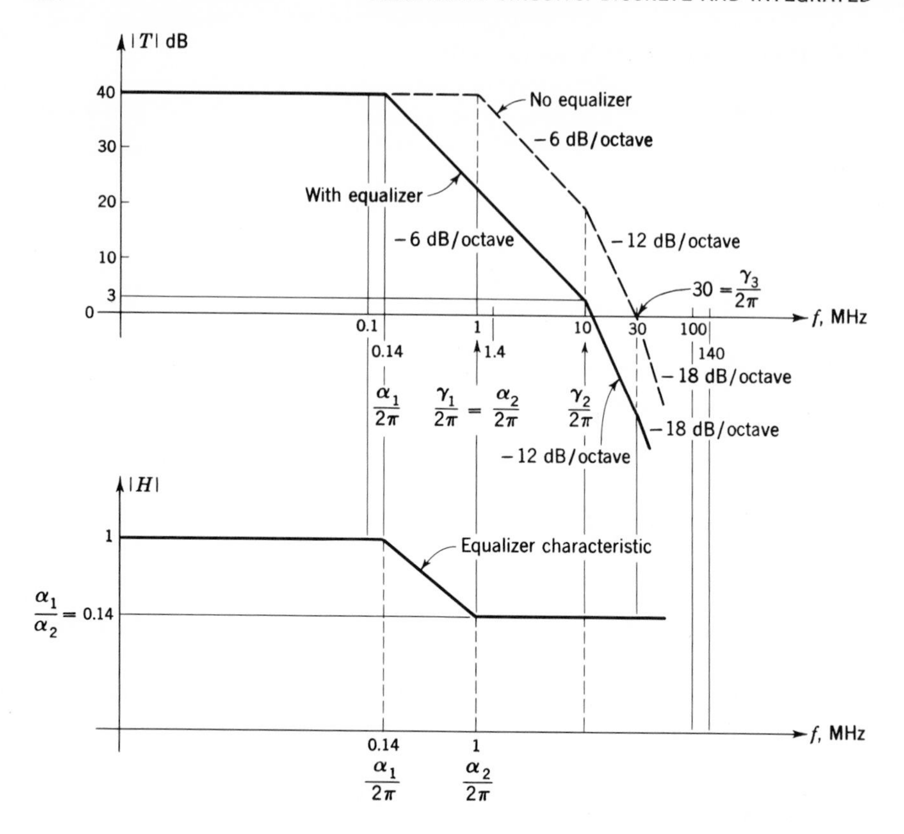

Fig. 15.4-5 Loop gain and equalizer frequency characteristic with pole-zero lag network.

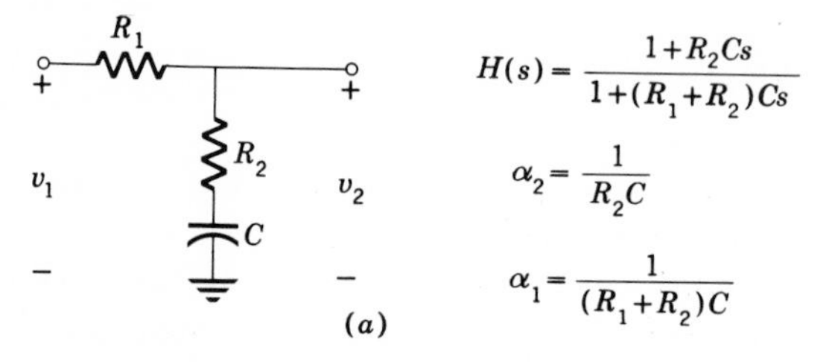

$$H(s) = \frac{1 + R_2Cs}{1 + (R_1 + R_2)Cs}$$

$$\alpha_2 = \frac{1}{R_2C}$$

$$\alpha_1 = \frac{1}{(R_1 + R_2)C}$$

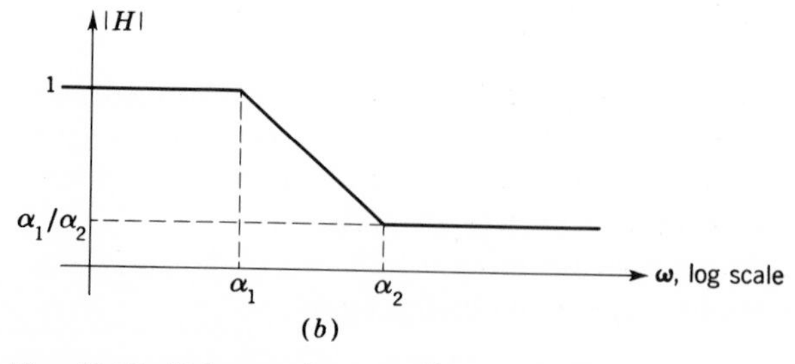

Fig. 15.4-6 Pole-zero lag equalizer. (*a*) Circuit; (*b*) frequency characteristic.

15.4-4 LEAD COMPENSATION

We have seen that to stabilize a feedback amplifier, the loop-gain frequency characteristic must be shaped so that the phase shift at the gain crossover frequency is removed from the critical value of 180° by the required phase margin. We have studied the use of the lag network to achieve this, and found that the lag network could reduce the crossover frequency (and the bandwidth) without any reduction in low-frequency loop gain.

The low-frequency loop gain is often fixed by specifications on A_{vf} and sensitivity. If a lag network is employed which results in a bandwidth that is too narrow, we must seek an alternative solution. One possibility immediately suggests itself: since stability depends only on the phase at crossover, we might try an equalizer which would introduce phase *lead* at this point. The simplest of such networks has the transfer function

$$H(s) = \frac{s + \delta_1}{s + \delta_2} \qquad \delta_2 > \delta_1 \tag{15.4-17a}$$

Note that this transfer function is similar to (15.4-6), except that here the pole occurs at a higher frequency than the zero.

The Bode diagram for $H(s)$ is shown in Fig. 15.4-7, along with a practical network which realizes this transfer function. This should be compared with Fig. 15.4-6. Here we come across a situation which did not arise with the lag network. The lead network is seen from Fig. 15.4-7*b* to introduce a low-frequency attenuation.

$$H(0) = \frac{R_2}{R_1 + R_2} = \frac{\delta_1}{\delta_2} \tag{15.4-17b}$$

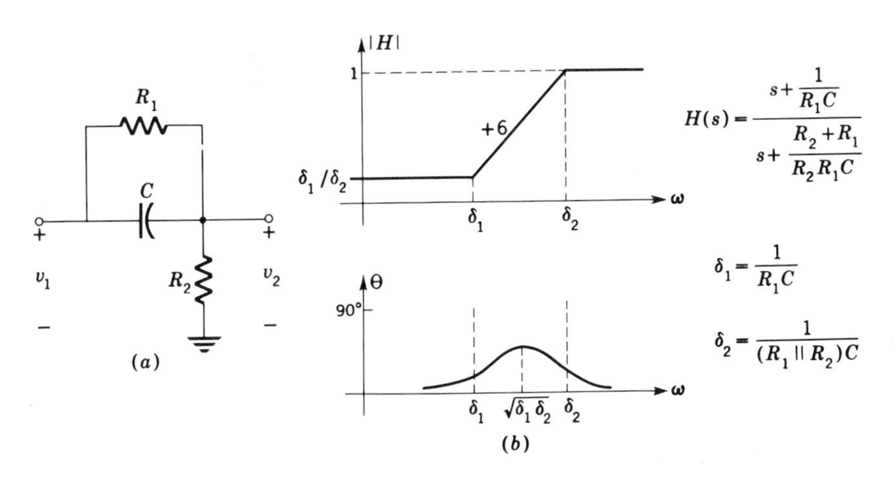

Fig. 15.4-7 Lead network. (*a*) Circuit; (*b*) amplitude and phase.

This will have to be taken into account because it directly affects the low-frequency gain.

Let us consider how this network can be used to stabilize the amplifier of the preceding examples. The loop gain with the lead equalizer is

$$-T(s) = -\beta H(0)A_o \frac{(1 + s/\delta_1)}{\left(1+\frac{s}{\delta_2}\right)\left(1+\frac{s}{\gamma_1}\right)\left(1+\frac{s}{\gamma_2}\right)\left(1+\frac{s}{\gamma_3}\right)} \tag{15.4-18a}$$

where $A_o = -10^4$, $\gamma_1/2\pi = 1$ MHz, $\gamma_2/2\pi = 10$ MHz, $\gamma_3/2\pi = 30$ MHz, and $\delta_1 < \delta_2$. Now we modify the procedure employed when using the lag network, and set

$$\delta_1 = \gamma_2 \tag{15.4-18b}$$

so that the lead-network *zero* cancels the *second-lowest* amplifier pole. We design the lead network to ensure that δ_2 is large enough to exert no significant effect at the gain crossover frequency. Then

$$-T(s) = -\beta H(0)A_o \frac{1}{\left(1+\frac{s}{\gamma_1}\right)\left(1+\frac{s}{\gamma_3}\right)\left(1+\frac{s}{\delta_2}\right)} \tag{15.4-19}$$

Equation (15.4-19) indicates that if

$$\gamma_1 < \gamma_3 \ll \delta_2 \tag{15.4-20a}$$

then the phase shift is $-135°$ at $\omega \approx \gamma_3$, rather than at $\omega \approx \gamma_2$. Thus the gain crossover frequency has been extended by the use of a lead network.

The loop gain at the desired crossover frequency γ_3 is [Eq. (15.4-19)]

$$-T\left(\frac{\gamma_3}{2\pi}\right) = 1\underline{/-135°} \approx \frac{-\beta H(0)A_o}{(j\gamma_3/\gamma_1)(1+j)} = \frac{-\beta H(0)A_o}{\sqrt{2}\,\gamma_3/\gamma_1}\underline{/-135°} \tag{15.4-20b}$$

Since the low-frequency loop gain is

$$T(0) = \beta H(0)A_o \tag{15.4-20c}$$

we have

$$\sqrt{2}\left(\frac{\gamma_3}{\gamma_1}\right) = -T(0) \tag{15.4-21}$$

In this example, $\gamma_1/2\pi = 1$ MHz and $\gamma_3/2\pi = 30$ MHz; hence

$$-T(0) = 30\sqrt{2} \approx 42.5 \ (\approx 32 \text{ dB})$$

From (15.4-17b) and (15.4-20c)

$$\beta H(0) = \beta\left(\frac{\delta_1}{\delta_2}\right) = \frac{T(0)}{A_o} = \frac{-42.5}{-10^4} = 42.5 \times 10^{-4} \ (\approx -48 \text{ dB})$$

To ensure that $\delta_2 \gg \gamma_3$, we might choose

$$\frac{\delta_2}{2\pi} = 300 \text{ MHz}$$

Since [Eq. (15.4-18b)]

$$\frac{\delta_1}{2\pi} = \frac{\gamma_2}{2\pi} = 10 \text{ MHz}$$

$$\beta = \left[\frac{T(0)}{A_o}\right]\left(\frac{\delta_2}{\delta_1}\right) = (42.5)(30)(10^{-4}) \approx 0.13 \ (\approx -18 \text{ dB})$$

The loop gain is plotted in Fig. 15.4-8 as a function of frequency. Note that a maximum low-frequency loop gain of 32 dB results. This should be compared with Fig. 15.4-5. In Fig. 15.4-5, $|T(0)|$ is larger, but is down 3 dB at 140 kHz rather than at 1 MHz.

The low-frequency gain with feedback of the lead-compensated amplifier is, for this example,

$$A_{vf}(0) = \frac{H(0)A_o}{1 - T(0)} = \frac{H(0)A_o}{1 - \beta H(0)A_o} \approx \frac{1}{-\beta} \approx 18 \text{ dB}$$

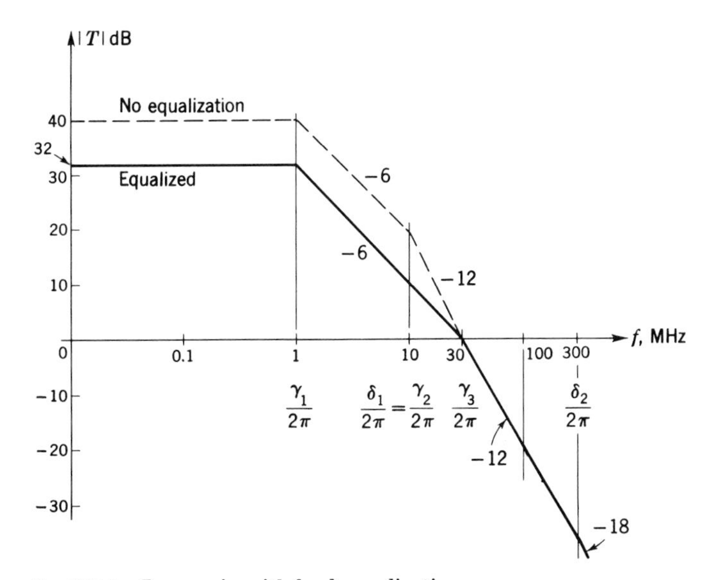

Fig. 15.4-8 Loop gain with lead equalization.

The low-frequency gain with feedback of the lag-compensated amplifier was 40 dB. The bandwidth of the lead-compensated amplifier can be shown to be approximately 45 MHz, compared with the 15-MHz bandwidth obtained using lag compensation. These examples illustrate the trade-off that the designer can make between loop gain T and the bandwidth of the feedback amplifier.

An important disadvantage of lead compensation is that, if the phase near the gain crossover frequency increases rapidly with frequency, it may not be possible to obtain sufficient phase lead to produce effective stabilization. Combination lead-lag equalizers are often used in this situation. A second disadvantage of lead compensation is the low-frequency loss introduced in the lead network, which results in a decreased loop gain.

It should be pointed out that the simple pole-canceling technique used in this section to design the lead network may not yield optimum results. Often a trial-and-error procedure is used to place the break frequencies for optimum compensation.

15.5 EXAMPLES

The compensation networks described in preceding sections can, in theory, be inserted at any convenient point within the basic amplifier. With integrated circuits, however, unless specific terminals are supplied, it is necessary to connect the compensation networks to the external terminals, i.e., across the input or the output, or even to make them part of the feedback network. In this section we present several examples of lead and lag compensation applied to IC amplifiers. The theory is directly applicable to discrete-component amplifiers as well.

15.5-1 LAG-COMPENSATED FEEDBACK AMPLIFIERS

The operational amplifier of Fig. 15.5-1 is shown with a lag-compensation network connected at the output terminals. It is instructive to rederive the gain formula, assuming that $R_f \gg r_o$. Then

$$v_i' = \frac{v_i R_i \| R_f}{r_i + (R_i \| R_f)} + \frac{v_L R_i \| r_i}{R_f + (R_i \| r_i)} \tag{15.5-1}$$

and

$$v_L \approx \frac{A_v' v_i' (R_L + 1/sC_L)}{r_o + R_L + 1/sC_L}$$

$$= v_i' A_v' \left[\frac{1 + sC_L R_L}{1 + sC_L(r_o + R_L)}\right] \tag{15.5-2}$$

Combining (15.5-1) and (15.5-2) to eliminate v_i' yields

$$v_L = A_v' \left[\frac{1 + sC_L R_L}{1 + sC_L(r_o + R_L)}\right] \left(\frac{v_i R_i R_f + v_L R_i r_i}{R_i r_i + R_i R_f + r_i R_f}\right) \tag{15.5-3}$$

Transposing and dividing, we obtain

$$A_{vf} = \frac{A_v}{1 - T}$$

$$= \frac{A'_v \left[\frac{1 + sC_L R_L}{1 + sC_L(r_o + R_L)} \right] \left(\frac{R_i R_f}{R_i r_i + R_i R_f + r_i R_f} \right)}{1 - \left(\frac{r_i}{R_f} \right) A'_v \left[\frac{1 + sC_L R_L}{1 + sC_L(r_o + R_L)} \right] \left(\frac{R_i R_f}{R_i r_i + R_i R_f + r_i R_f} \right)} \tag{15.5-4}$$

The reader will note that this expression can also be derived using the feedback approach described in Chap. 8.

Equation (15.5-4) can be put in the form of (15.4-9) and (15.4-12) by noting that

$$\beta = \frac{r_i}{R_f} \tag{15.5-5a}$$

$$A_o = A'_{vo} \left(\frac{R_i R_f}{R_i r_i + R_i R_f + r_i R_f} \right) \tag{15.5-5b}$$

$$A_{v2} = \left[\frac{A'_v(s)}{A'_v(0)} \right] \left[\frac{1 + sR_L C_L}{1 + s(R_L + r_o) C_L} \right] A_o \tag{15.5-5c}$$

and

$$T = \beta A_{v2} \tag{15.5-5d}$$

It must be noted that in this example β is positive and $A_{v2}(0)$ is negative. Thus $T(0) < 0$. Comparing (15.5-5c) with (15.4-9),

$$\frac{1}{R_L C_L} = \alpha_2 \tag{15.5-6a}$$

and

$$\frac{1}{(R_L + r_o) C_L} = \alpha_1 < \alpha_2 \tag{15.5-6b}$$

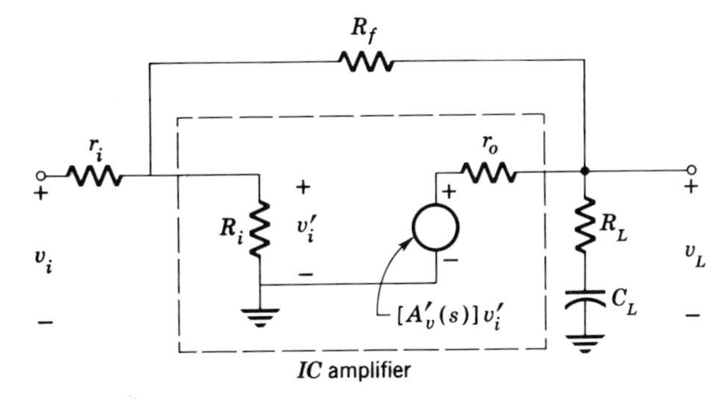

Fig. 15.5-1 Applying lag compensation to IC amplifiers.

The zero α_2 is set equal to the lowest pole of $A'_v(s)$. The pole α_1 is adjusted as in (15.4-16b), so that

$$\alpha_1 = \frac{\sqrt{2}\,\gamma_2}{-T(0)} \tag{15.5-7a}$$

where γ_2 is the second-lowest pole in A'_v and $T(0)$ is

$$T(0) = A'_{vo}\left(\frac{r_i}{R_f}\right)\left(\frac{R_iR_f}{R_ir_i + R_iR_f + r_iR_f}\right) \tag{15.5-7b}$$

When a suitable value of $T(0)$ has been chosen, α_1 is determined from (15.5-7a). Usually, r_o is known; then R_L and C_L can be determined from (15.5-6).

15.5-2 A LAG-COMPENSATED IC AMPLIFIER[3]

The Fairchild μA 702A IC amplifier is shown in Fig. 15.5-2. In this amplifier, terminals are provided in the emitter circuit of T_6 for external connection of either lead or lag compensation. The low-frequency char-

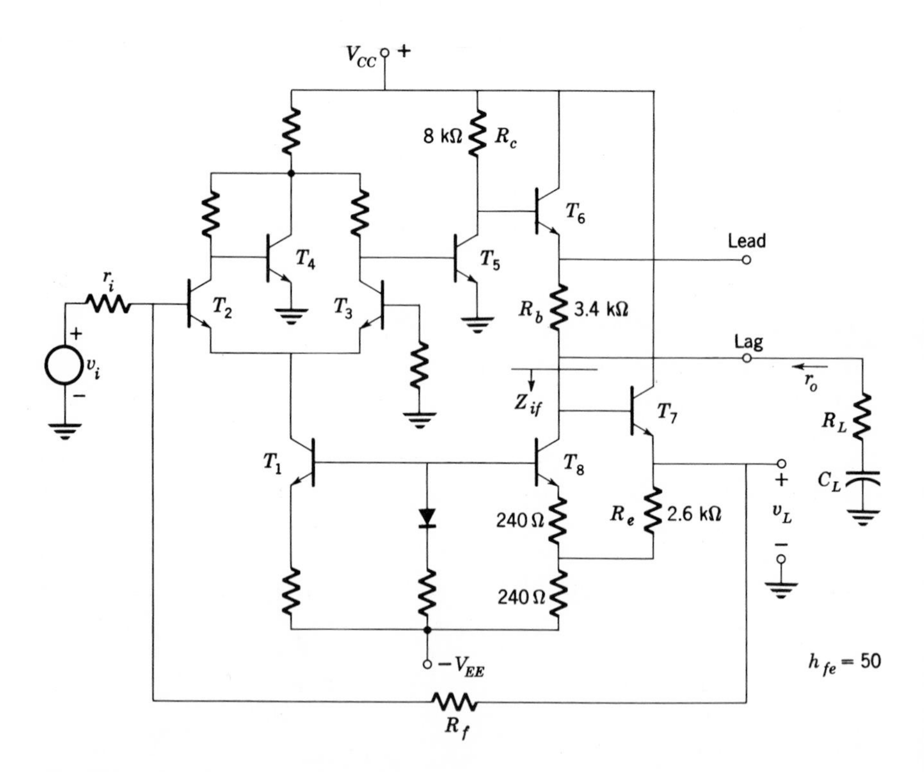

Fig. 15.5-2 Integrated-circuit amplifier with compensation terminals.

acteristics of this amplifier were studied in Sec. 9.8. In (9.8-18) the impedance Z_{if} was found to be approximately -6 kΩ. The impedance seen looking into the *lag* terminal is then $r_0 = (3600)\|(-6000) = 9$ kΩ. Thus we see that the lag-compensation terminal is at a relatively high impedance point within the amplifier. This will result in a smaller value of capacitance being required than if the lag network were connected to the output terminal. It also provides isolation, so that external load changes will not affect the break frequencies.

Manufacturers' data for the μA 702A indicates that the first two amplifier poles occur at 1 and 4 MHz. Then, using (15.5-6) and (15.5-7*a*)

$$\frac{\alpha_2}{2\pi} = \frac{1}{2\pi R_L C_L} = 10^6$$

and

$$\frac{\alpha_1}{2\pi} = \frac{\sqrt{2}\,\gamma_2/2\pi}{-T(0)} = \frac{\sqrt{2}\,(4)(10^6)}{-T(0)}$$

If $T(0) = -100$, as in Sec. 15.4.3, then

$$\frac{\alpha_1}{2\pi} = \frac{1}{2\pi(R_L + r_o)C_L} = 4\sqrt{2} \times 10^4$$

Since $r_o = 9000\ \Omega$,

$$R_L \approx 540\ \Omega \qquad \text{(we should use a standard 560-}\Omega\text{ resistor)}$$
$$C_L \approx 300\text{ pF}$$

In practice, the compensation would be adjusted to the desired degree by varying R_L.

15.5-3 LEAD COMPENSATION

When lead compensation is indicated, the network of Fig. 15.4-7 is often used as an interstage circuit in the forward amplifier. This type of compensation can be applied to the IC amplifier of Fig. 15.5-2 by connecting a capacitor between the lead and lag terminals. The circuit is redrawn in Fig. 15.5-3, where only that part of the IC amplifier of interest has been included. In this circuit C_c is a coupling capacitor. It is sufficiently large so that it does not affect the frequency response of the compensating network. R_1, C_1, and R_2 then correspond, approximately, to the components of the lead network of Fig. 15.4-7. The addition of R_2 allows us to control the frequency range of the 6 dB/octave portion of the lead-network response, and thus the placement of the lead-network pole. The reader will note that the low-frequency gain of the basic amplifier is reduced by a factor of approximately $R_2/(R_1 + R_2)$ because of the addition of the lead network.

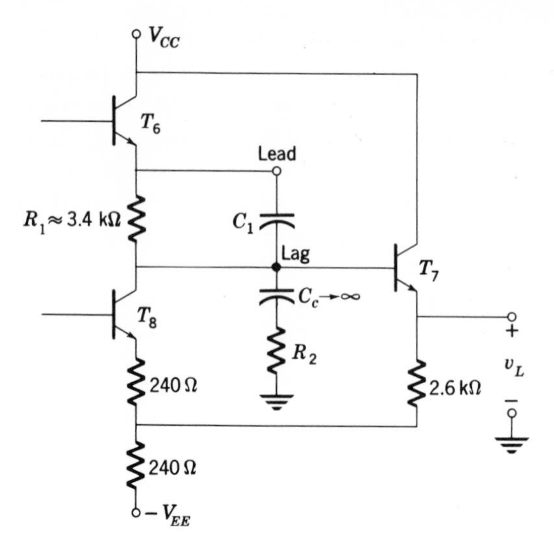

Fig. 15.5-3 IC amplifier with lead compensation.

Another method of applying lead compensation consists in connecting a suitable capacitance C_f across the feedback resistance R_f. This scheme is shown in Fig. 15.5-4. Analysis of this circuit indicates that the zero of the loop gain T, introduced by the presence of C_f, can be used to cancel the *second* pole of A_v, as is done by the conventional lead network of Fig. 15.4-7. Detailed analysis of this circuit is left as an exercise for the reader (Prob. 15-22).

Example 15.5-1 The all-pass lag network The feedback amplifier shown in Fig. 15.5-5*a* is an *all-pass lag network*. Show that the

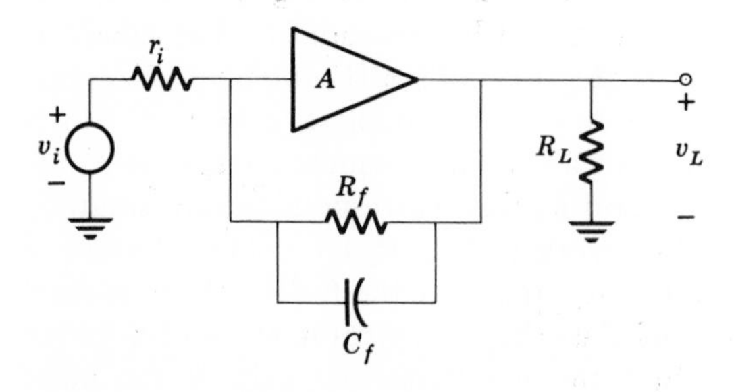

Fig. 15.5-4 Alternative method for applying lead compensation.

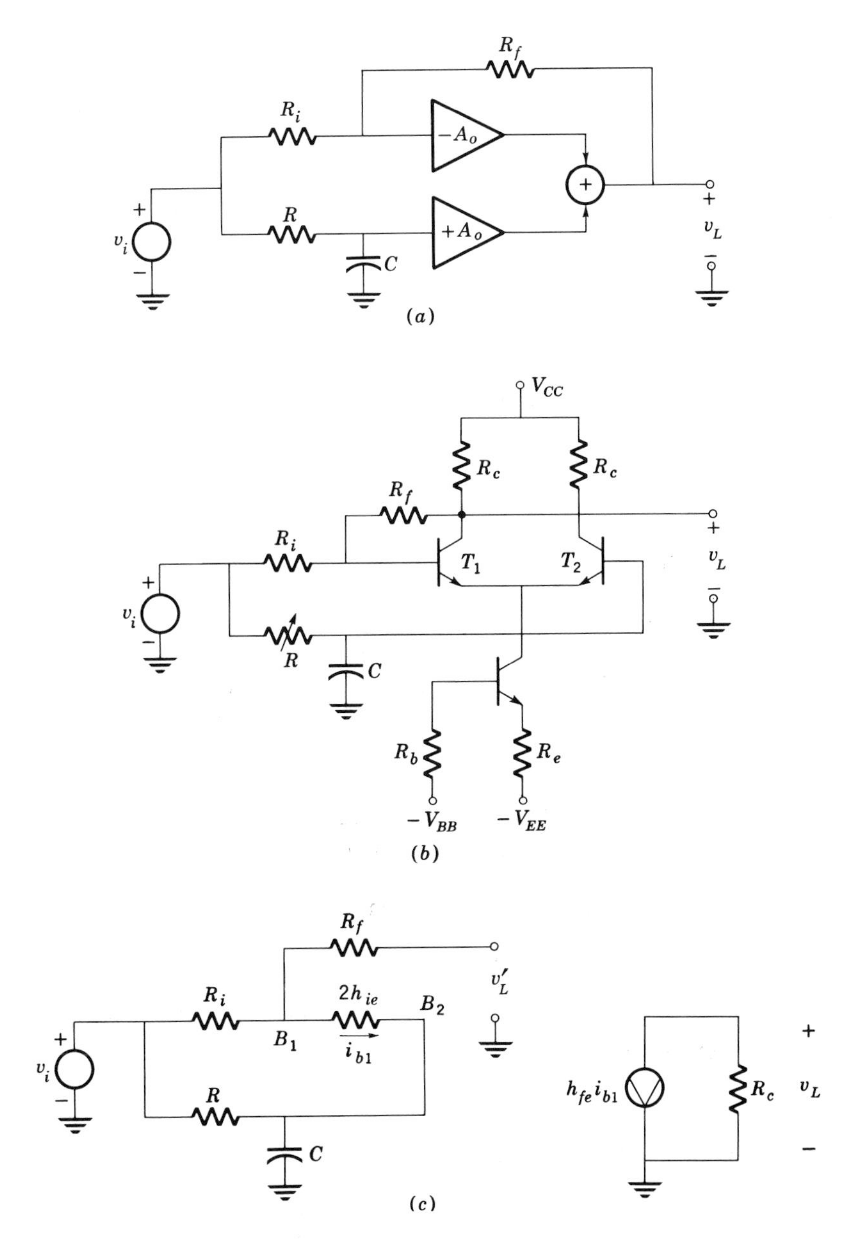

Fig. 15.5-5 All-pass network. (*a*) Circuit; (*b*) difference amplifier; (*c*) equivalent circuit of the difference-amplifier all-pass network.

transfer function of this amplifier is

$$\frac{v_L}{v_i} = \frac{1 - s/\omega_a}{1 + s/\omega_a}$$

where

$$\omega_a = \frac{1}{RC}$$

Solution The forward gain without feedback is

$$A_v = \left.\frac{v_L}{v_i}\right|_{v'_L=0} = \left(\frac{1}{1+sRC} - \frac{R_f}{R_i + R_f}\right) A_o$$

The loop gain T, assuming that the amplifier has an infinite input impedance, is

$$T = \left.\frac{v_L}{v'_L}\right|_{v_i=0} = -\frac{R_i A_o}{R_i + R_f}$$

The gain with feedback becomes

$$A_{vf} = \frac{A_v}{1-T} = \frac{v_L}{v_i} = \frac{[1/(1+sRC) - R_f/(R_i+R_f)]A_o}{1 + R_i A_o/(R_i+R_f)}$$

$$= \frac{A_o\{[1 - R_f/(R_i+R_f)] - s(R_f/R_i + R_f)RC\}}{[1 + R_i A_o/(R_i+R_f)](1+sRC)}$$

If $R_i = R_f$ and $A_o/2 \gg 1$,

$$A_{vf} = \frac{(A_o/2)(1-sRC)}{(1+A_o/2)(1+sRC)} \approx \frac{1-sRC}{1+sRC}$$

This is the desired result. When $s = j\omega$,

$$\angle A_{vf} \approx -2 \tan^{-1} \omega RC$$

When the difference amplifier of Fig. 15.5-5*b* is employed, the resistors are chosen so that

$$R_f \gg R_c \qquad 2h_{ie} \gg R_i \qquad 2h_{ie} \gg R$$

The current i_{b1} is then

$$2h_{ie}i_{b1} \approx \frac{-v_i}{1+sRC} + \frac{v_i R_f}{R_i + R_f} + \frac{v'_L R_i}{R_i + R_f}$$

Letting $v'_L = v_L$, the gain with feedback is

$$A_{vf} = \frac{v_L}{v_i} = \frac{(h_{fe}R_c/2h_{ie})[1/(1+sRC) - R_f/(R_i+R_f)]}{1 + (R_c/2h_{ib})[R_i/(R_i+R_f)]} \qquad (15.5\text{-}8a)$$

If we now let $R_f = R_i$ and if $4h_{ib} \ll R_c$,

$$A_{vf} = \frac{v_L}{v_i} \approx \left(\frac{1 - sRC}{1 + sRC}\right) \tag{15.5-8b}$$

15.6 ACTIVE FILTERS USING FEEDBACK

In this section we discuss active-filter synthesis using feedback. A complete discussion of this topic is beyond the scope of this book. The examples shown, however, demonstrate the principles involved, and the interested reader may refer to the extensive literature available on the subject. Simply stated, any transfer function which can be synthesized theoretically with R, L, and C passive elements can be synthesized using only R and C elements and amplifiers.

15.6-1 THE INTEGRATOR

Perhaps the simplest and most often used circuit is the *integrator*, which is shown in Fig. 15.6-1. We recall from Chap. 8 that the action of an operational amplifier such as this is to reduce the error current i_ϵ. If the gain A_i' is very large, the error current, and hence the voltage v_i', both approach zero. When these conditions are valid, the following relations hold:

$$i_\epsilon = i_1 - i_c \approx 0 \tag{15.6-1a}$$

Therefore

$$i_1 \approx i_c \tag{15.6-1b}$$

and

$$i_c = C\frac{d}{dt}(v_i' - v_L) \approx -C\left(\frac{dv_L}{dt}\right) \tag{15.6-2}$$

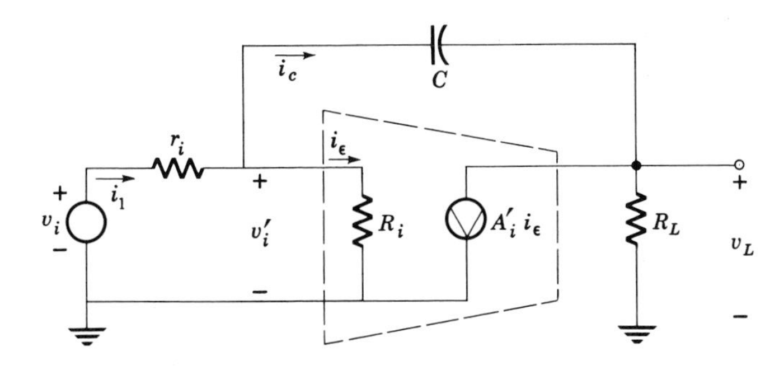

Fig. 15.6-1 Integrator.

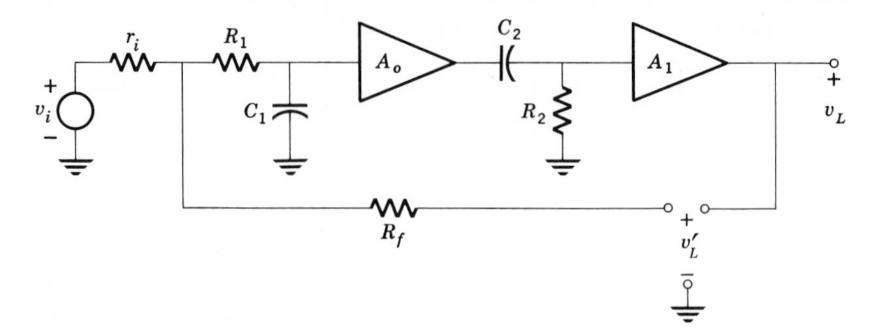

Fig. 15.6-2 Q multiplier.

But, with $v'_i \approx 0$,

$$i_1 \approx \frac{v_i}{r_i} \tag{15.6-3}$$

Combining (15.6-1) to (15.6-3) yields

$$v_i \approx -r_i C \left(\frac{dv_L}{dt}\right) \tag{15.6-4}$$

Integrating both sides, we obtain

$$v_L(t) \approx -\left(\frac{1}{r_i C}\right) \int_0^t v_i(\lambda)\, d\lambda \tag{15.6-5}$$

Thus, over the range of t for which this is valid, the output voltage is proportional to the integral of the input voltage.

This form of operational amplifier is used as a major component in most analog computers. The student can show that the degree to which (15.6-5) approximates the ideal integrator depends upon A'_i. Commercial units with gains of 100 dB or more are available.

15.6-2 THE Q MULTIPLIER

As an example of a narrowband active filter, consider cascading two simple RC filters as shown in Fig. 15.6-2. A_0 and A_1 represent amplifiers with high-input and low-output impedances.

The forward open-loop gain A_v of this system is

$$A_v = \frac{v_L}{v_i}\bigg|_{v'_L=0} = \left(\frac{R_f}{r_i + R_f}\right)\left(\frac{1/R_1C_1}{s + 1/R_1C_1}\right) A_0 \left(\frac{s}{s + 1/R_2C_2}\right) A_1 \tag{15.6-6}$$

The loop gain T is

$$T = \frac{v_L}{v'_L}\bigg|_{v_i=0} = \left(\frac{r_i}{r_i + R_f}\right)\left(\frac{1/R_1C_1}{s + 1/R_1C_1}\right) A_0 \left(\frac{s}{s + 1/R_2C_2}\right) A_1 \tag{15.6-7}$$

Note that if the product A_0A_1 is positive, we have positive feedback.

The closed-loop gain A_{vf} is

$$A_{vf} = \frac{A_v}{1 - T} = \frac{\left(\frac{R_f}{r_i + R_f}\right)\left[\frac{\alpha_1 s}{(s + \alpha_1)(s + \alpha_2)}\right] A_{01}}{1 - \left(\frac{r_i}{r_i + R_f}\right)\left[\frac{\alpha_1 s}{(s + \alpha_1)(s + \alpha_2)}\right] A_{01}} \tag{15.6-8}$$

where

$$\alpha_1 = \frac{1}{R_1C_1} \qquad \alpha_2 = \frac{1}{R_2C_2} \qquad A_{01} = A_0A_1$$

Simplifying (15.6-8) yields

$$A_{vf} = \frac{R_f\alpha_1 s A_{01}/(r_i + R_f)}{s^2 + (\alpha_1 + \alpha_2)s - [r_i/(r_i + R_f)]\alpha_1 s A_{01} + \alpha_1\alpha_2} \tag{15.6-9}$$

Equation (15.6-9) has the same form as the transfer function of the RLC filter circuit shown in Fig. 15.6-3.

For the RLC circuit,

$$\frac{v_L}{v_i} = \frac{R}{R + 1/sC + sL} = \frac{sRC}{s^2LC + sRC + 1}$$

$$= \frac{sR/L}{s^2 + sR/L + 1/LC} \tag{15.6-10}$$

where the resonant frequency ω_0 is

$$\omega_0^2 = \frac{1}{LC} \tag{15.6-11}$$

and the circuit Q is

$$Q = \frac{\omega_0 L}{R} \tag{15.6-12}$$

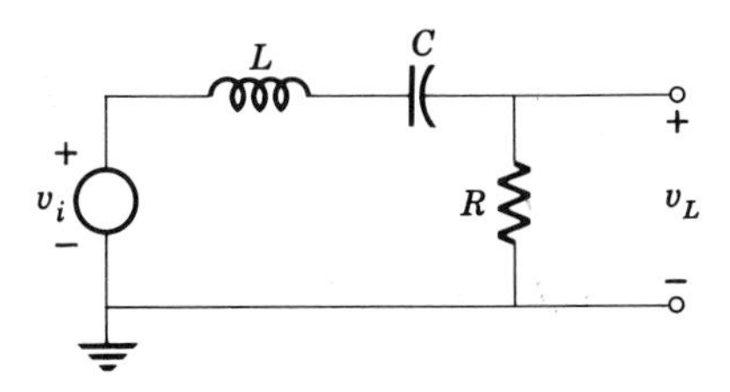

Fig. 15.6-3 RLC filter.

Analogously, we note that in (15.6-9)

$$\omega_0^2 = \alpha_1 \alpha_2 \tag{15.6-13}$$

and the circuit Q is

$$Q = \frac{\omega_0}{\alpha_1 + \alpha_2 - [r_i \alpha_i A_{01}/(r_i + R_f)]} \tag{15.6-14}$$

which can be written

$$Q = \frac{\omega_0 \alpha_2}{\omega_0^2 \{1 - [r_i A_{01}/(r_i + R_f)]\} + \alpha_2^2} \tag{15.6-15}$$

Equation (15.6-15) indicates that the Q can be made as large as desired by adjusting R_f.

Referring to (15.6-8), we see that this is a stable amplifier, even though it has positive feedback, because the magnitude of the loop gain is always less than unity. (In Sec. 9.8 positive feedback was used in the output stage of an IC amplifier to increase the gain of the stage from 1 to 2.5. The loop gain was found to be approximately equal to 0.6.) In this amplifier we use positive feedback to increase the amplifier gain in a selected frequency range. This is called Q multiplication. The larger the desired Q, the closer $T(\omega)$ approaches unity. If $Q \to \infty$, $T(\omega) \to 1$, and the amplifier will oscillate. Q's of several thousand can be obtained using this approach.

The circuit analyzed above demonstrates two important principles. The first is, of course, that the Q of a circuit can be increased using positive feedback. The second is that RC circuits and amplifiers can be employed to obtain RLC circuit characteristics.

Example 15.6-1 Design an RC-tuned circuit to resonate at 1 MHz and have a Q of 1000.

Solution Use the circuit of Fig. 15.6-2. A_0 and A_1 are two IC amplifiers (A_1 is not necessary unless very high gain, i.e., very high Q, is required). Let $R_1C_1 = R_2C_2$.

Using (15.6-13),

$$\alpha_1 \alpha_2 = \alpha_1^2 = \frac{1}{R_1^2 C_1^2} = \frac{1}{R_2^2 C_2^2} = \omega_0^2 = 4\pi^2(10^{12})$$

and

$$R_1 C_1 \approx 16 \times 10^{-8}$$

Then, from (15.6-14),

$$Q = \frac{\omega_0}{\omega_0 + \omega_0 - \left(\dfrac{r_i}{r_i + R_f}\right) A_{01}\omega_0} = \frac{1}{2 - \left(\dfrac{r_i}{r_i + R_f}\right) A_{01}}$$

Since $Q = 1000$,

$$2 - \left(\frac{r_i}{r_i + R_f}\right) A_{01} = 10^{-3}$$

If $A_{01} = +10$ and $r_i = 100\ \Omega$,

$$R_f = r_i \left(\frac{A_{01}}{2 - 10^{-3}} - 1\right) \approx 400\ \Omega$$

Note that $R_f = 400\ \Omega$ places the amplifier on the verge of oscillation. In practice, great care must be taken in the adjustment of R_f.

15.7 OSCILLATORS

In preceding sections we saw that when the loop gain $T = 1\angle 0$, the feedback amplifier became unstable. To prevent this from occurring we employed lead- and lag-compensation networks. In Sec. 15.6 we saw that the Q of the closed-loop gain, A_{vf}, could be increased almost indefinitely using *positive* feedback. Now consider the possibility of increasing the Q so that $T = 1\angle 0$ at a single frequency, ω_0, the magnitude of T being less than unity at all other frequencies. Then $Q(\omega_0)$ is infinite, and the feedback amplifier will be unstable at only one frequency, ω_0. Physically, this means that an output is possible with no input present at the one frequency, ω_0. Thus the output must be sinusoidal, and such a device is a sinusoidal oscillator.

15.7-1 THE PHASE-SHIFT OSCILLATOR

One of the simplest oscillators to design and construct at low frequencies is the phase-shift oscillator shown in Fig. 15.7-1. To determine the conditions for oscillation we must calculate $T(\omega)$ and set it equal to $1\angle 0$.

Before analyzing this circuit, let us attempt to anticipate the results to be obtained. The transistor will provide a phase shift of 180°. Thus, if T is equal to $1\angle 0$, an additional 180° must be provided by the three RC circuits [the third $R = R' + (h_{ie} \| R_b)$]. If each RC circuit could act independently, it could be adjusted to provide a 60° phase shift at ω_0. Under these conditions the transfer function for each section

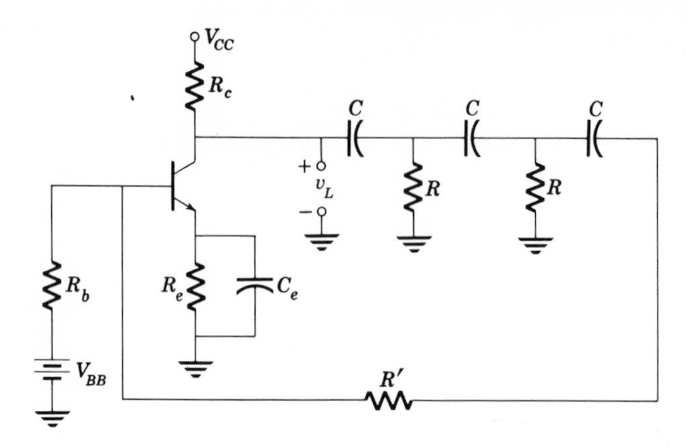

Fig. 15.7-1 Phase-shift oscillator.

would be

$$\frac{s}{s+\alpha} = \left(\frac{\omega}{\sqrt{\alpha^2+\omega^2}}\right) \epsilon^{j[\pi/2 - \tan^{-1}(\omega/\alpha)]} \qquad \alpha = \frac{1}{RC} \tag{15.7-1}$$

To obtain a 60° phase shift at ω_0,

$$\frac{\pi}{2} - \left(\tan^{-1}\frac{\omega_0}{\alpha}\right) = \frac{\pi}{3} \tag{15.7-2}$$

or

$$\frac{\pi}{6} = \tan^{-1}\frac{\omega_0}{\alpha} \tag{15.7-3}$$

and

$$\omega_0 = \frac{\alpha}{\sqrt{3}} = \frac{1}{\sqrt{3}\,RC} \tag{15.7-4}$$

This is the condition for oscillation, assuming independent RC sections. The loop gain T is adjusted to unity at ω_0 by adjusting the attenuation in the circuit. This oscillator works well as long as $\omega_0 \ll \omega_\beta$, since near ω_β, the transistor input appears capacitive.

In the circuit of Fig. 15.7-1, the RC circuits are not independent, and we must proceed with an analysis of the equivalent circuit shown in Fig. 15.7-2.

The loop gain is defined as

$$T = \frac{i_b}{i_b'} \tag{15.7-5}$$

Solving the circuit of Fig. 15.7-2 yields

$$-h_{fe}i_b \approx i_b'\left(3 + \frac{4}{sRC} + \frac{1}{s^2R^2C^2} + \frac{R}{R_c} + \frac{6}{sR_cC} + \frac{5}{s^2C^2RR_c} + \frac{1}{s^3R^2R_cC^3}\right) \tag{15.7-6}$$

Since $i_b = i_b'$ for $T = 1\underline{/0°}$, at ω_0, we can equate real and imaginary parts on both sides of this equation to determine the conditions for oscillation. From the imaginary terms,

$$\frac{1}{\omega_0^2RR_cC^2} = 4 + 6\left(\frac{R}{R_c}\right)$$

Thus

$$\omega_0 = \left(\frac{1}{RC}\right)\left[\frac{1}{\sqrt{6 + 4(R_c/R)}}\right] \tag{15.7-7}$$

Setting the real parts equal on both sides and using (15.7-7),

$$-h_{fe} = 3 + \frac{R}{R_c} - \frac{4 + 6R/R_c}{R/R_c} - 5\left(4 + 6\frac{R}{R_c}\right) \tag{15.7-8}$$

Solving for R/R_c in terms of h_{fe} yields

$$\frac{R}{R_c} = \left(\frac{h_{fe} - 23}{58}\right) + \sqrt{\left(\frac{h_{fe} - 23}{58}\right)^2 - \frac{4}{29}} \tag{15.7-9}$$

Hence, given h_{fe} and ω_0, we can determine R/R_c from (15.7-9), and then RC from (15.7-7).

In order to have the term under the square root positive, we must have

$$h_{fe} > 23 + 21.6 = 44.6 \tag{15.7-10}$$

If h_{fe} is less than this value, the circuit will not oscillate, since $T < 1\underline{/0°}$.

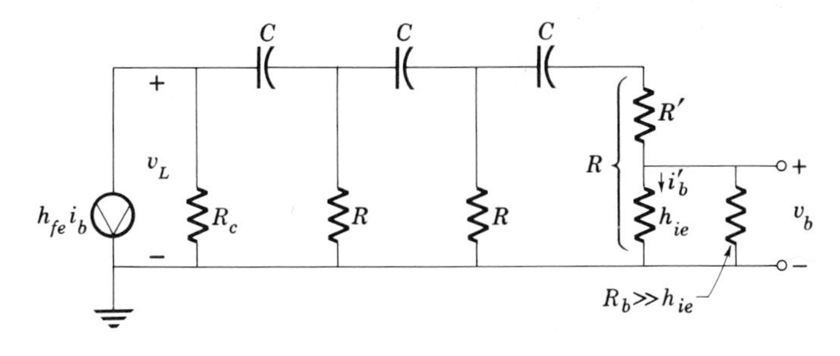

Fig. 15.7-2 Equivalent circuit of phase-shift oscillator.

At this value of h_{fe}, from (15.7-9),

$$\frac{R}{R_c} = 0.375$$

If this ratio is increased, the oscillator waveform will be distorted. Thus, when constructing this type of oscillator, h_{fe} is not known exactly, and R_c is usually a variable resistor, which is adjusted to eliminate the distortion.

Experimental results The oscillator frequency ω_0 depends on the ratio R/R_c and therefore on h_{fe}. Thus, interchanging transistors will usually result in a change in the frequency of oscillation.

The oscillator was constructed and tested using a 2N3904 *npn* transistor with $h_{fe} \approx 150$. The system parameters were $R = 1$ kΩ, $C = 0.0068$ μF, $R_e = 1$ kΩ, $R' = 0$, and $R_b = 10$ kΩ; R_c was a 1 kΩ variable resistance. The dc collector supply voltage was 9 V, and the quiescent collector current was approximately 4 mA.

Nine different 2N3904 transistors were used to determine the frequency variability. The frequency of oscillation varied between 8.6 and 9.2 kHz.

15.7-2 THE WIEN BRIDGE OSCILLATOR

An IC Wien bridge oscillator is shown in Fig. 15.7-3. The operation of the oscillator can be explained using Fig. 15.7-3*a*. The loop gain is

$$T = \frac{v_L}{v_L'} = \left\{ \frac{R/(1 + sRC)}{[R/(1 + sRC)] + R + 1/sC} - \frac{R_i}{R_i + R_f} \right\} A_o$$

$$= \left[\frac{s/\omega_0}{s/\omega_0 + (1 + s/\omega_0)^2} - \frac{R_i}{R_i + R_f} \right] A_o \qquad (15.7\text{-}11a)$$

where

$$\omega_0 = \frac{1}{RC} \qquad (15.7\text{-}11b)$$

The condition for oscillation is $T = 1\underline{/0}$. Solving (15.7-11*b*), with $T = 1\underline{/0}$ and $s = j\omega$, yields

$$1 - \left(\frac{\omega}{\omega_0}\right)^2 + j\frac{\omega}{\omega_0}\left[3 - \frac{A_o(R_i + R_f)}{R_i(1 + A_o) + R_f}\right] = 0 \qquad (15.7\text{-}12a)$$

Equating real and imaginary parts to zero, we find that

$$\omega = \omega_0 = \frac{1}{RC} \qquad (15.7\text{-}12b)$$

and

$$\frac{R_f}{R_i} = \frac{2A_o + 3}{A_o - 3} \qquad (15.7\text{-}12c)$$

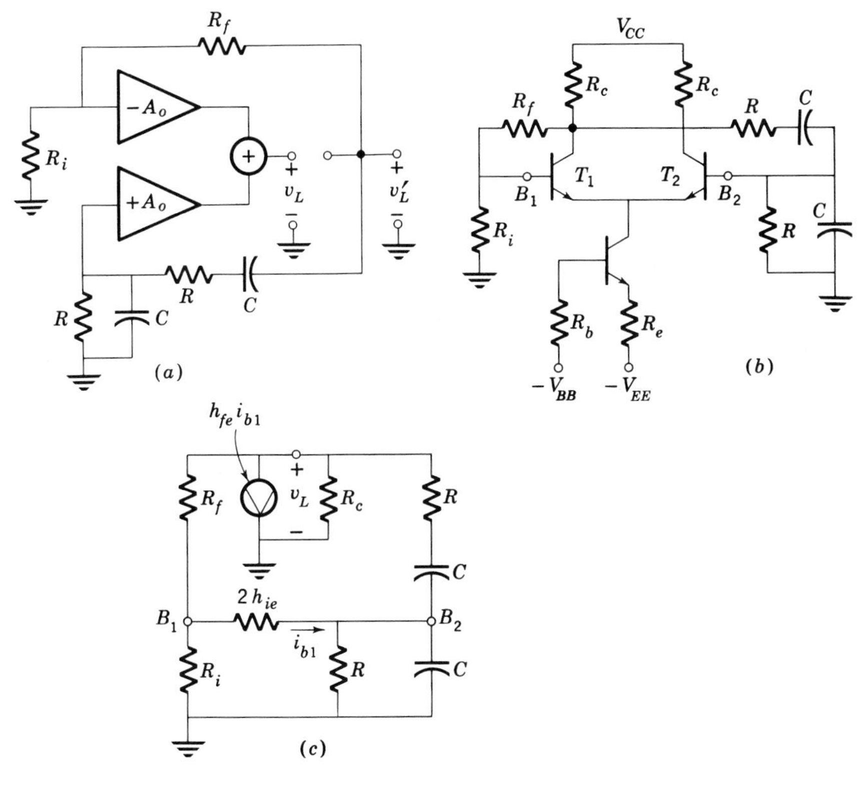

Fig. 15.7-3 Wien bridge oscillator. (*a*) Circuit; (*b*) difference amplifier; (*c*) equivalent circuit.

If A_o is very large (note that A_o must be greater than 3),

$$\frac{R_f}{R_i} \to 2$$

Equation (15.7-12*b*) shows that the resonant frequency is independent of the amplifier gain A_o.

Consider using the difference amplifier shown in Fig. 15.7-3*b* and *c* to synthesize the Wien bridge oscillator. From the equivalent circuit it can be shown that, at ω_0, the bridge is almost balanced, and v_{b1} is *almost* equal to v_{b2}. The gain of the amplifier is

$$A_o = \frac{v_L}{v_{b1} - v_{b2}} = \left(\frac{v_L}{i_{b1}}\right)\left(\frac{i_{b1}}{v_{b1} - v_{b2}}\right) \approx -h_{fe}R_c\left(\frac{1}{2h_{ie}}\right) = -\frac{R_c}{2h_{ib}} \tag{15.7-13a}$$

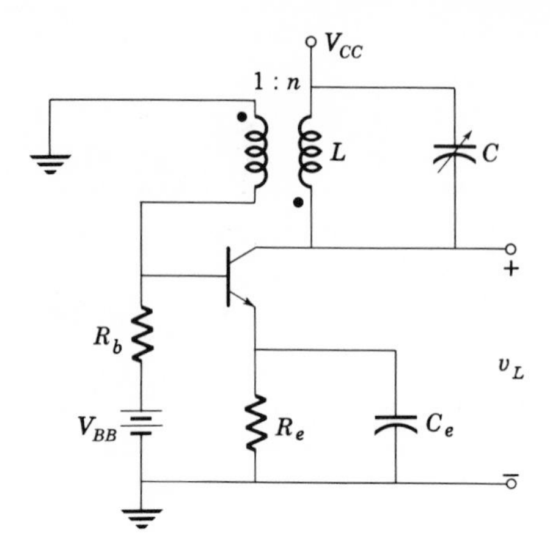

Fig. 15.7-4 Tuned-circuit oscillator.

where we have assumed that

$$R_f \gg R_c \tag{15.7-13b}$$

and

$$R \gg R_c \tag{15.7-13c}$$

In order for the circuit of Fig. 15.7-3*b* to be described by (15.7-11*a*), it is also necessary that

$$2h_{ie} \gg R_i \tag{15.7-13d}$$

and

$$2h_{ie} \gg R \tag{15.7-13e}$$

The above inequalities represent sufficient, rather than necessary, conditions for oscillation to occur. In addition, they tend to simplify the analysis.

It is interesting to note that ω_0 and the ratio R_f/R_i are almost independent of the transistor parameters.

15.7-3 THE TUNED-CIRCUIT OSCILLATOR

A simple tuned-circuit oscillator is shown in Fig. 15.7-4. The operation of the circuit can be explained by assuming that a small collector current flows at the frequency $\omega_0 \approx 1/\sqrt{LC}$. Then a voltage at the frequency ω_0 will appear at the collector, part of which is fed back to the base through the transformer. The polarity of the transformer is adjusted so that the feedback is positive, and the base current (also at ω_0) therefore tends to increase. Eventually, of course, the nonlinearities of the transistor limit the current swing.

Now, to put this on a quantitative basis, let us study the equivalent circuit shown in Fig. 15.7-5. Figure 15.7-5*a* shows the basic equivalent circuit, where $T = v_b/v_b'$. Note that the voltage nv_b is reversed because of the phase reversal in the transformer. It is this phase reversal which provides the extra 180° phase shift needed for positive feedback.

Figure 15.7-5*b* is the final equivalent circuit with the base-circuit impedance reflected through the transformer into the collector. The loop gain T is now

$$T = \frac{v_b}{v_b'} = \frac{g_m/n}{1/n^2h_{ie} + 1/j\omega L + j\omega[C + (C_{in}/n^2)]} \tag{15.7-14}$$

The criterion for oscillation is that T must be $1\underline{/0}$. This leads to the relations

$$\omega_0{}^2 = \frac{1}{L[C + (C_{in}/n^2)]} \tag{15.7-15}$$

and

$$nh_{fe} = 1 \tag{15.7-16}$$

Note that the frequency of oscillation is dependent on C_{in}. Since

$$C_{in} = C_{b'e} + C_M$$

the frequency ω_0 depends on the quiescent point. Note also that

$$n = \frac{1}{h_{fe}}$$

Since h_{fe} is very large, n can be small. If nh_{fe} exceeds unity, the amplitude of oscillation will be limited by the transistor nonlinearities. However, the tuned collector circuit tends to filter out the resulting harmonics present in the collector current. The collector voltage is therefore almost sinusoidal.

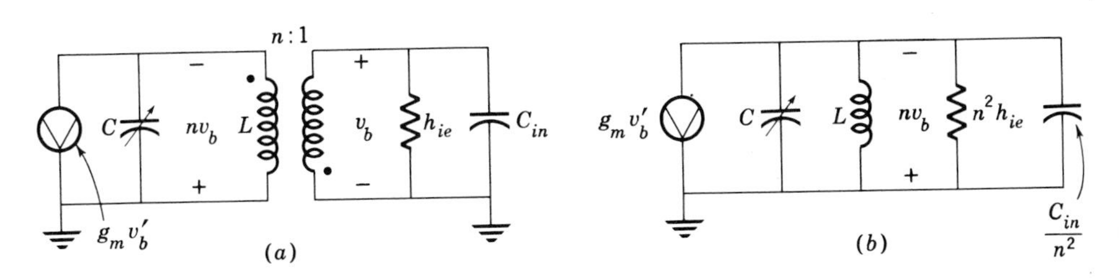

Fig. 15.7-5 Small-signal equivalent circuit of tuned-circuit oscillator ($R_b \gg r_{b'e}$). (*a*) Transformer included; (*b*) transformer eliminated.

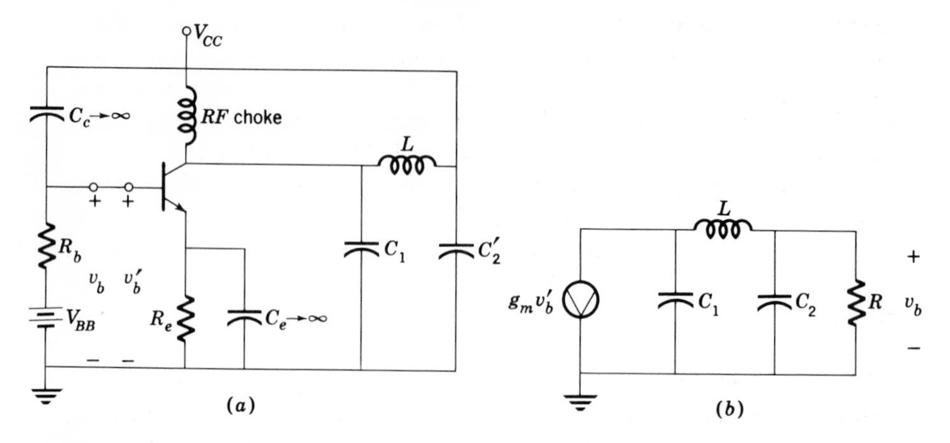

Fig. 15.7-6 (a) The Colpitts oscillator circuit; (b) small-signal equivalent circuit.

15.7-4 THE COLPITTS OSCILLATOR

A Colpitts oscillator, which is used at radio frequencies, is shown in Fig. 15.7-6a. The equivalent circuit is shown in Fig. 15.7-6b, where

$$C_2 = C_2' + C_{b'e} + C_M \tag{15.7-17a}$$

and

$$R = R_b \| r_{b'e} \tag{15.7-17b}$$

The loop gain T is

$$T = \frac{v_b}{v_b'} = 1\angle 0 = \frac{-g_m R}{sRC_1 + (1 + sRC_2)(1 + s^2LC_1)} \tag{15.7-18}$$

Solving (15.7-18), we obtain

$$\omega^2 = \omega_0{}^2 = \frac{1}{L[C_1C_2/(C_1 + C_2)]} \tag{15.7-19a}$$

and

$$\omega_0{}^2 LC_1 = (1 + g_m R) \tag{15.7-19b}$$

Dividing (15.7-19a) by (15.7-19b) yields

$$1 + g_m R = 1 + \frac{C_1}{C_2} \tag{15.7-20a}$$

If $R \approx r_{b'e}$ this yields

$$h_{fe} \approx \frac{C_1}{C_2} \tag{14.7-20b}$$

Thus the conditions for oscillation are strongly dependent on the h_{fe} of the transistor.

Experimental results The inductor L in the feedback network of the Colpitts oscillator can be replaced by a piezoelectric crystal, as shown in Fig. 15.7-7*a*. The equivalent circuit of the crystal is shown in Fig. 15.7-7*b*. A typical crystal will have an extremely high Q (several thousand), and thus tends to stabilize the oscillator and prevent frequency variation with transistor replacement. The circuit shown in Fig. 15.7-7*a* was constructed. Nine different 2N3904 transistors were tested, and the frequency of oscillation was found to vary from 99.924 to 99.925 kHz.

The analysis of the circuit of Fig. 15.7-7*a* is left as an exercise for the reader.

15.7-5 THE HARTLEY OSCILLATOR

The final oscillator circuit to be considered here is the Hartley oscillator, shown in Fig. 15.7-8*a*. This oscillator employs the common-base configuration, and is generally used at very high frequencies. Positive feedback is achieved by returning a portion of the output to the input through a transformer. In practice, an autotransformer is often used, rather than the transformer shown. The loop gain of unity is obtained by adjusting the turns ratio. R_e is used as a bias resistance, and is much greater than h_{ib}.

The loop gain T can be found from the equivalent circuit shown in Fig. 15.7-8*c*.

$$T = 1\underline{/0^\circ} = \frac{(n_2/n_1)h_{fb}(L/C)}{(L/C) + (n_2/n_1)^2 h_{ib}(j\omega L + 1/j\omega C)} \tag{15.7-21}$$

The conditions for oscillation, which are obtained by solving (15.7-21), are

$$\omega_0{}^2 = \frac{1}{LC} \tag{15.7-22a}$$

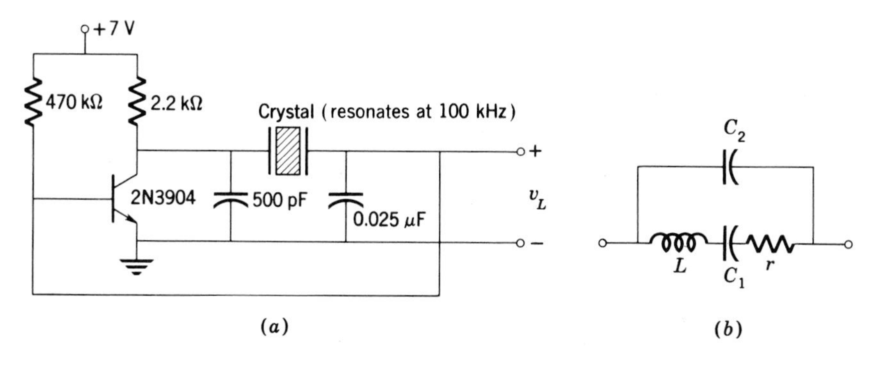

Fig. 15.7-7 (*a*) A Colpitts crystal oscillator; (*b*) equivalent circuit of crystal.

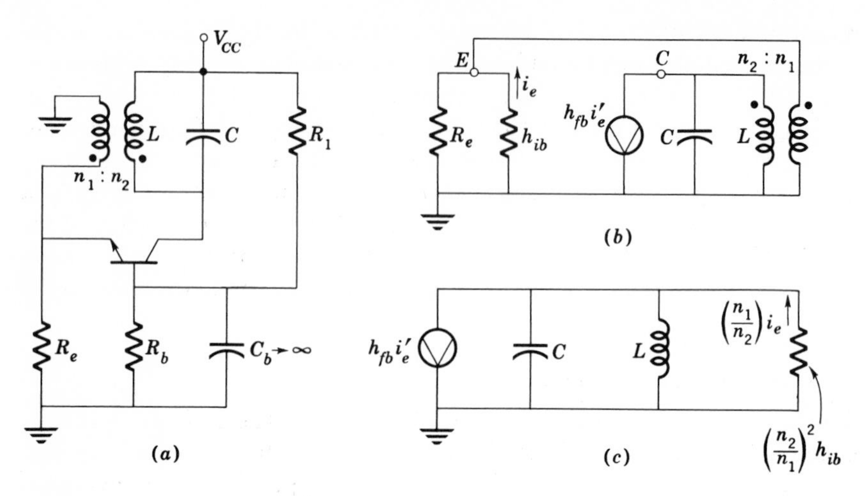

Fig. 15.7-8 Hartley oscillator. (a) Circuit; (b) equivalent circuit; (c) transformer eliminated ($R_e \gg h_{ib}$).

and

$$\frac{n_1}{n_2} = h_{fb} \tag{15.7-22b}$$

Thus the turns ratio is adjusted to make up the loss in current gain present in the common-base configuration.

Experimental results The circuit shown in Fig. 15.7-9 is a Hartley oscillator which operates in the common-collector mode instead of the common-base mode. The emitter voltage is stepped up by the autotransformer and returned to the base. The frequency of oscillation was found to vary from 1.01 to 1.02 MHz with transistor replacement.

The analysis of this circuit is left as an exercise (Prob. 15.36).

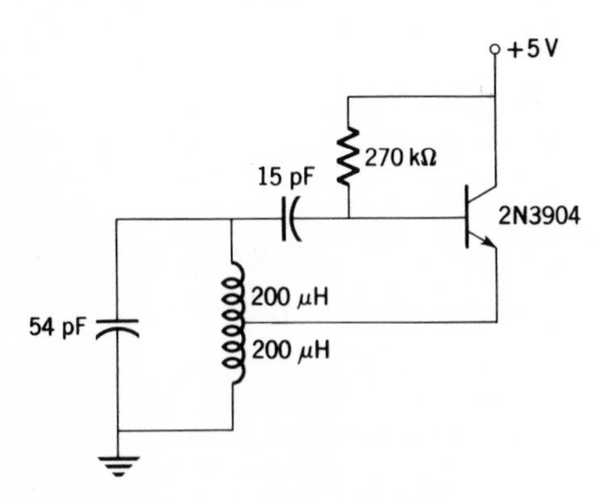

Fig. 15.7-9 A common-collector Hartley oscillator.

SUMMARY

In all the circuits analyzed in this section, the criteria that the magnitude of the loop gain must be unity and the net loop phase shift be zero were used to find the frequency of oscillation and the circuit-parameter relations for sustained oscillations. In the practical design of oscillator circuits these relations are used to establish a preliminary design. Important performance criteria, such as buildup time and frequency and amplitude stabilities, are then evaluated, using a combination of the experimental and analytical approaches.

PROBLEMS

SECTION 15.1

15.1. For the transistor, $r_{b'e} = 1\ \text{k}\Omega$, $r_{bb'} = 0$, $C_{b'e} + C_M = 0.01\ \mu\text{F}$, and $g_m = 0.05$ mho.

(a) Find A_v, Z_i, and Z_o in the mid- and high-frequency range if the switch is at point B.

(b) Find A_{vf}, Z_{if}, and Z_{of} if the switch is at point A. Compare.

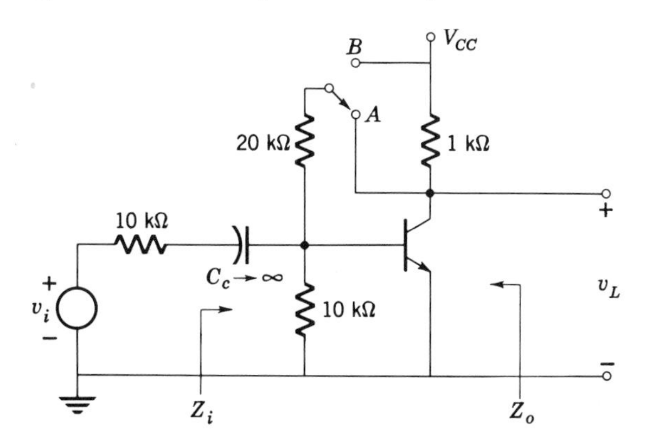

Fig. P15.1

15.2. Find and sketch the asymptotic gain and phase for T and A_{vf} if $g_m = 0.01$, 1, 100. Find ζ and ω_n for each value of g_m. Estimate the 3-dB bandwidth in each case.

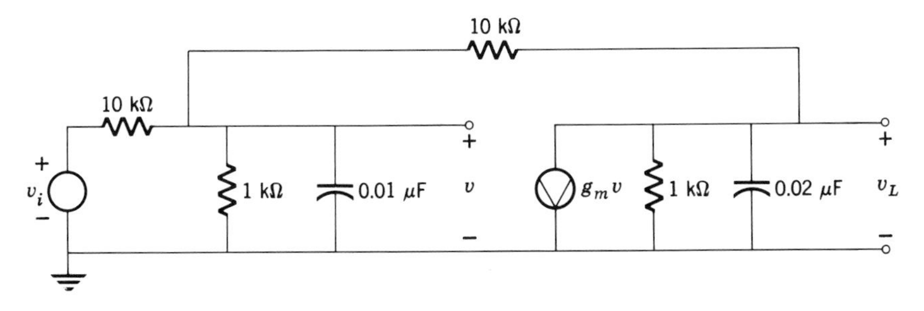

Fig. P15.2

15.3. In the circuit of Fig. P8.4 the transistors have $C_{b'e} = 0.002\ \mu F$.

(*a*) Find and sketch the asymptotic current gain, $A_{if} = i_L/i_i$.

(*b*) Find and sketch the step-function response.

(*c*) Determine the 3-dB bandwidth.

15.4. Low-frequency instability often occurs in multistage ac-coupled feedback amplifiers. Consider a two-stage amplifier with two coupling capacitors adjusted so that

$$A_i(s) = \frac{A_{im}s^2}{(s + \omega_L)^2}$$

Assuming $\beta_i(s) = \beta_{io}$, find $A_{if}(s)$ and comment on the stability.

15.5. In Prob. 15.1, the g_m of the transistor can vary from 0.025 to 0.1 mho. Find and sketch (asymptotic plots) upper and lower limits on A_v and A_{vf} for switch positions A and B.

15.6. The transfer function of (15.1-7) can be synthesized using an *RLC* circuit and and ideal amplifier.

(*a*) Find the circuit.

(*b*) Show that the Q of the circuit is $1/2\zeta$.

SECTIONS 15.2 AND 15.3

15.7. The three transistors shown are identical, with parameters $r_{b'e} = 1\ k\Omega$, $r_{bb'} = 0$, $C_{b'e} = 0.01\ \mu F$, $C_{b'c} = 0$, $h_{fe} = 200$, and $C_c \to \infty$.

Find and sketch the magnitude and the phase of T. Is the amplifier stable?

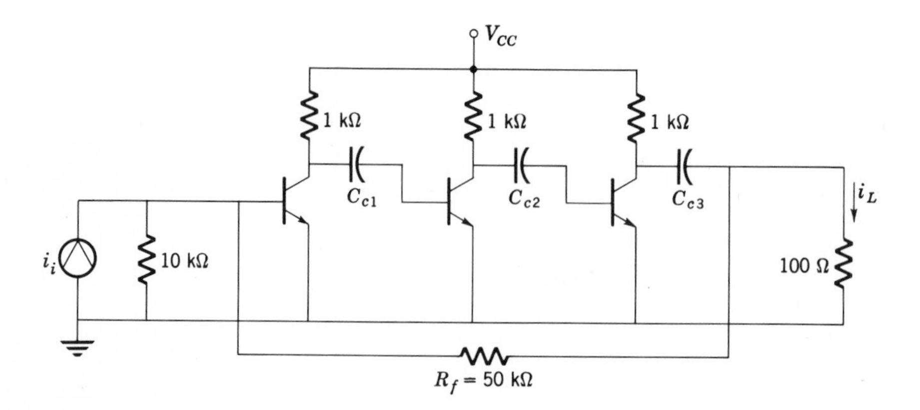

Fig. P15.7

15.8. In the circuit of Fig. P15.7, $C_{b'e} = 0$, but $C_c = 10\ \mu F$. Find and sketch the magnitude and the phase of T. Is the amplifier stable?

15.9. Sketch the Nyquist diagrams for Probs. 15.7 and 15.8 and compare.

15.10. In the feedback amplifier of Prob. 15.7, find the value of R_f for which the phase margin is between 45° and 60°. Comment on the result.

15.11. Determine the gain with feedback for Prob. 15.10. Plot the magnitude and the phase of A_{if}.

15.12. In Prob. 15.8, $C_{c1} = 0.1\ \mu F$, $C_{c2} = 1000\ \mu F$, and $C_{c3} = 10{,}000\ \mu F$. Investigate the stability of the amplifier.

SECTION 15.4

15.13. Stabilize the amplifier of Fig. P15.7 by connecting a capacitor across the signal path at the base of the first stage. Design for a phase margin of approximately 45°. Find and plot $|A_{if}|$ as a function of frequency and estimate the bandwidth.

15.14. Repeat Prob. 15.13 using the lag equalizer of Fig. 15.4-6 in order to increase the overall bandwidth.

15.15. Try to stabilize the amplifier of Prob. 15.7 by using the lead network of Fig. 15.4-7. $T(0) \geq 40$ dB, and $A_{if} = 500$. Plot $|A_{if}|$ as a function of frequency, and estimate the bandwidth.

15.16. The amplifier of Fig. P15.7 can be stabilized by "broadbanding" (increasing the 3-dB frequency) of one stage so that the high-frequency response appears to be that of a two-stage amplifier. The third stage can be broadbanded by applying local feedback so that its 3-dB frequency is increased by a factor of 10 and its midband gain is reduced by this same factor. Does this result in a stable amplifier? If not, indicate additional steps to insure stability.

15.17. Verify Fig. 15.4-2*b*.

15.18. Plot $|A_{vf}|$ versus frequency for the compensated feedback amplifier of Sec. 15.4-3. Verify that the 3-dB bandwidth is approximately 15 MHz.

15.19. Plot $|A_{vf}|$ as a function of frequency for the lead-compensated amplifier of Sec. 15.4-4. Show that the 3-dB bandwidth is approximately 45 MHz.

SECTION 15.5

15.20. Verify (15.5-4) using the feedback technique of Chap. 8.

15.21. In the circuit of Fig. 15.5-1, $r_i = 2\ \text{k}\Omega$, $R_f = 10\ \text{k}\Omega$, $R_i = 100\ \text{k}\Omega$, $r_o = 100\ \Omega$, and $A'_v(0) = -10^4$. The dominant poles of the amplifier are at 1 and 4 MHz.

Find R_L and C_L for a phase margin of about 45°.

15.22. Analyze the circuit of Fig. 15.5-4, assuming that the amplifier has the parameters given in Prob. 15.21. Find C_f to maximize the phase margin. Let $R_L = 100\ \Omega$.

15.23. (*a*) Show that the gain with feedback A_{vf} of the amplifier in Fig. 15.5-5*b* is given by (15.5-8*a*) and (15.5-8*b*).

(*b*) Plot the magnitude and phase of A_{vf} as a function of frequency.

(*c*) Find the response of the network to a step function.

SECTION 15.6

15.24. In the integrator circuit of Fig. 15.6-1, the parameters are $r_i = 1\ \text{M}\Omega$, $C = 1\ \mu F$, $R_i = 100\ \text{k}\Omega$, $R_L = 1\ \text{k}\Omega$, and $A'_i = 80$ dB.

Find and plot v_L when v_i is a unit step function. Compare with the output of an ideal integrator. Up to what point does the actual output depart from the ideal by less than 1 percent? 10 percent?

15.25. Rederive (15.6-4) without assuming that $v'_i = 0$. However, assume that $|i_\epsilon| \ll |i_c|$. Show that the resulting equation is of the form

$$\frac{dv_L}{dt} + Bv_L = Kv_i$$

Find B and K. Show that as $A'_i \to \infty$, $B \to 0$.

15.26. Find the sensitivity of Q to changes in R_f and A_{01} for (15.6-15). Discuss the result.

15.27. In Example 15.6-1 the tuned circuit is to have a Q which lies between 100 and 200. Design the circuit, specifying tolerances on A_0, A_1, and R_f so that the Q lies between the specified limits.

SECTION 15.7

15.28. Verify (15.7-6) and (15.7-9).

15.29. Find the frequency of oscillation for the nominal value of $h_{fe} = 150$. By what percent will the frequency change if h_{fe} changes by ± 50 percent? Discuss your results in relation to the experimental results presented in Sec. 15.7-1.

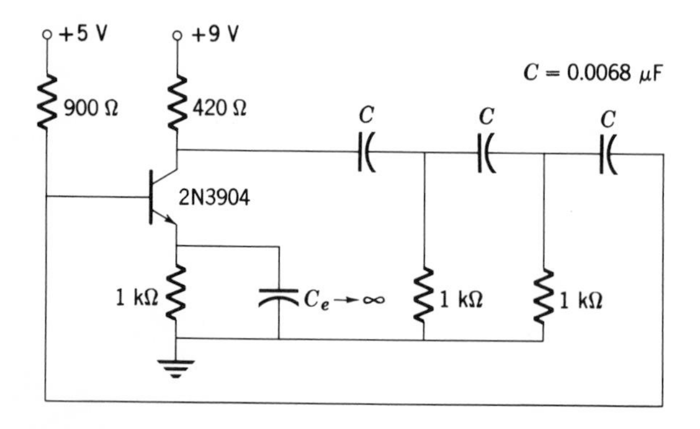

Fig. P15.29

15.30. Verify (15.7-11*a*).

15.31. Show that when the inequalities of (15.7-13) are satisfied, Fig. 15.7-3*b* is a Wien bridge oscillator.

15.32. Verify (15.7-14) to (15.7-16).

15.33. Verify (15.7-18) to (15.7-20*b*).

15.34. Find the condition under which the circuit shown in Fig. 15.7-7*a* will oscillate.

15.35. Verify (15.7-21) to (15.7-22*b*).

15.36. Find the conditions under which the circuit shown in Fig. 15.7-9 will oscillate.

REFERENCES

1. Bower, J. L., and P. Schultheiss: "Introduction to the Design of Servomechanisms," John Wiley & Sons, Inc., New York, 1958.
2. Thornton, R. D., et al.: "Multistage Transistor Circuits," vol. 5, John Wiley & Sons, Inc., New York, 1965.
3. Giles, J. N.: "Fairchild Semiconductor Linear Integrated Circuit Applications Handbook," Fairchild Semiconductor Corp., Mountainview, Calif., 1967.

appendix I

Gain Expressed in Logarithmic Units—The Decibel (dB)

It is often convenient to express the magnitude of the gains of individual stages in logarithmic units. When this is done, the overall gain is found by the simple addition of the individual stage gains. This logarithmic unit also simplifies plotting of frequency response (Chaps. 12 and 13). For most purposes, the most convenient unit is the decibel (dB), which, when used to express the power gain of an amplifier, is defined as

$$A_p = 10\left(\log \frac{P_2}{P_1}\right) \qquad \text{dB} \tag{I.1}$$

where log means $\log_{10}$.

This definition is used in two different ways: first, to express a power ratio in logarithmic units, and second, to express power level with respect to a fixed reference power. A common reference is one milliwatt, for which the units are abbreviated dBm. Thus, for a power level in decibels with respect to one milliwatt, we have

$$A_p = 10\left(\log \frac{P_2}{10^{-3}}\right) = 10 \log (P_2 \times 10^3) \qquad \text{dBm} \qquad P_2 \text{ in watts} \tag{I.2}$$

The decibel was originally defined in terms of power gain as in (I.1). Because of its usefulness, it is generally applied directly to voltage and current gain in the following way: Referring to (I.1), if P_2 and P_1 are dissipated in identical resistances R, then

$$P_2 = \frac{V_2^2}{R} = I_2^2R \qquad \text{and} \qquad P_1 = \frac{V_1^2}{R} = I_1^2R$$

Substituting in (I.1),

$$A_v = 10\left(\log\frac{V_2^2}{V_1^2}\right) = 20\left(\log\frac{V_2}{V_1}\right) \quad \text{dB} \tag{I.3}$$

or

$$A_i = 10\left(\log\frac{I_2^2}{I_1^2}\right) = 20\left(\log\frac{I_2}{I_1}\right) \quad \text{dB} \tag{I.4}$$

Thus voltage and current gains in decibels are equal to power gain only if the resistances, in which the powers are dissipated, are equal. However, by convention, (I.3) and (I.4) are used to express the voltage and current gains in decibels independent of the resistance level.

Example I.1 Measurements on a certain amplifier yield the following data: $R_{in} = 1\ \text{k}\Omega$; $R_L = 100\ \Omega$.

When the input voltage is 1 mV peak, the load voltage is 10 V peak. Find, in decibels, the voltage, current, and power gains and the output power level in decibels above 1 mW.

Solution From (I.3)

$$A_v = 20\left(\log\frac{V_2}{V_1}\right) = 20\left(\log\frac{10}{10^{-3}}\right) = 80\ \text{dB}$$

To find A_i, note that $I_{in} = 1\ \mu\text{A}$ and $I_2 = 0.1$ A so that

$$A_i = 20\left(\log\frac{I_2}{I_1}\right) = 20\left(\log\frac{10^{-1}}{10^{-6}}\right) = 100\ \text{dB}$$

The powers are

$$P_1 = \frac{10^{-6}}{(2)(10^3)} = \tfrac{1}{2} \times 10^{-9}\ \text{W} \qquad P_2 = \frac{10^2}{(2)(100)} = \tfrac{1}{2}\text{W}$$

so

$$A_p = 10\left(\log\frac{\tfrac{1}{2}}{\tfrac{1}{2} \times 10^{-9}}\right) = 90\ \text{dB}$$

Relative to 1 mW, the output power level is

$$P_2 = 10\left(\log\frac{\tfrac{1}{2}}{10^{-3}}\right) = 10\ (3 \log 10 - \log 2) = 27\ \text{dBm}$$

---

appendix II

Standard Values of Resistance and Capacitance

The following lists of standard components are included for use in conjunction with the design problems. These lists are typical and, especially in the case of capacitors, are subject to some variation from one manufacturer to another.

RESISTORS

Carbon resistors of 10 percent tolerance are available in power ratings of $\frac{1}{4}$, $\frac{1}{2}$, 1, and 2 W in the following range:

$$\left.\begin{matrix} 2.7 \\ 3.3 \\ 3.9 \\ 4.7 \\ 5.6 \\ 6.8 \\ 8.2 \\ 10 \\ 12 \\ 15 \\ 18 \\ 22 \end{matrix}\right\} \text{all} \times 10^n, \text{ where } n = 0, 1, 2, 3, 4, 5, 6$$

CAPACITORS

Typical ranges of capacitor values available from one manufacturer are as follows:

CERAMIC-DISK CAPACITORS, 10 PERCENT TOLERANCE, pF

3.3	30	200	560	2200
5	39	220	600	2500
6	47	240	680	2700
6.8	50	250	750	3000
7.5	51	270	800	3300
8	56	300	820	3900
10	68	330	910	4000
12	75	350	1000	4300
15	82	360	1200	4700
18	91	390	1300	5000
20	100	400	1500	5600
22	120	470	1600	6800
24	130	500	1800	7500
25	150	510	2000	8200
27	180			

TANTALUM CAPACITORS, 10 PERCENT TOLERANCE, μF

0.0047
0.0056
0.0068
0.0082
0.010
0.012
0.015
0.018
0.022
0.027
0.033
0.039

(all of the above) all $\times 10^n$, where $n = 0, 1, 2, 3, 4, 5$ (to 330 μF)

ELECTROLYTIC CAPACITORS FOR BYPASS APPLICATION, μF

250	2000
500	3000
1000	4000
1500	5000

appendix III

Device Characteristics

III.1 ZENER DIODE SPECIFICATIONS

Figure III.1-1 shows the variation of Zener test voltage V_{ZT} with Zener resistance r_{ZT} for several typical power levels P_Z (see Sec. 2.10-2). Note that the test current I_{ZT} is measured at 25 percent of the maximum power rating,

$$I_{ZT} = \frac{1}{4}\frac{P_Z}{V_{ZT}}$$

In addition, it should be noted that the current at the knee of the characteristic I_{Zk} is approximately constant for a specified P_Z and is independent of V_{ZT}.

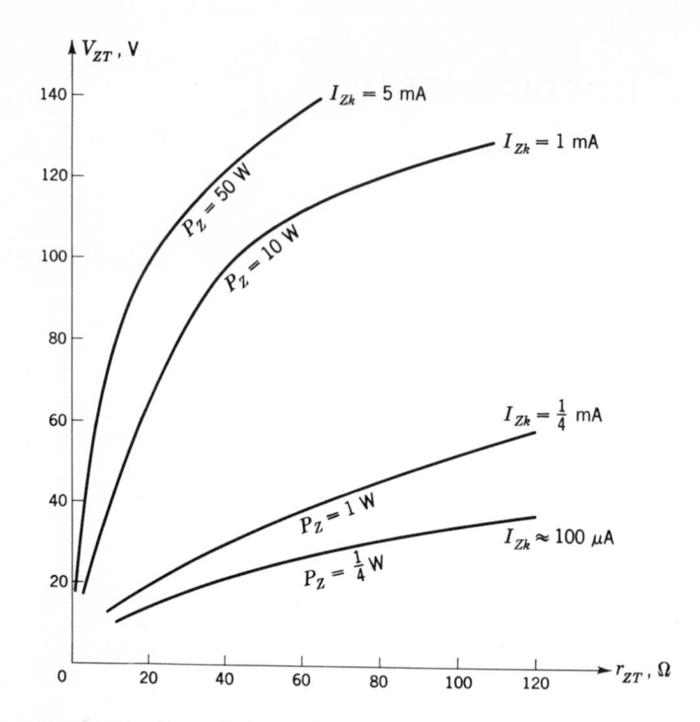

Fig. III.1-1 Zener diode specifications ($I_{ZT} = \frac{1}{4} P_Z / V_{ZT}$).

III.2 TRANSISTOR CHARACTERISTICS

MAXIMUM RATINGS

Characteristic	Symbol	Rating	Unit
Collector-Base Voltage	V_{CB}	60	Vdc
Collector-Emitter Voltage	V_{CEO}	40	Vdc
Emitter-Base Voltage	V_{EB}	6	Vdc
Collector Current	I_C	200	mAdc
Total Device Dissipation @ T_A = 60°C	P_D	210	mW
Total Device Dissipation @ T_A = 25°C Derate above 25°C	P_D	310 2.81	mW mW/°C
Thermal Resistance, Junction to Ambient	θ_{JA}	0.357	°C/mW
Junction Operating Temperature	T_J	135	°C
Storage Temperature Range	T_{stg}	-55 to +135	°C

Fig. III.2-1 Characteristics of the 2N3903 and 2N3904 *npn* silicon transistors, designed for general-purpose switching and amplifier applications and for complementary circuitry with types 2N3905 and 2N3906. *(Courtesy of Motorola, Inc.)*

ELECTRICAL CHARACTERISTICS ($T_A = 25°C$ unless otherwise noted)

Characteristic		Fig. No.	Symbol	Min	Max	Unit
OFF CHARACTERISTICS						
Collector-Base Breakdown Voltage (I_C = 10 μAdc, I_E = 0)			BV_{CBO}	60	-	Vdc
Collector-Emitter Breakdown Voltage* (I_C = 1.0 mAdc, I_B = 0)			BV_{CEO}*	40	-	Vdc
Emitter-Base Breakdown Voltage (I_E = 10 μAdc, I_C = 0)			BV_{EBO}	6.0	-	Vdc
Collector Cutoff Current (V_{CE} = 30 Vdc, $V_{EB(off)}$ = 3.0 Vdc)			I_{CEX}	-	50	nAdc
Base Cutoff Current (V_{CE} = 30 Vdc, $V_{EB(off)}$ = 3.0 Vdc)			I_{BL}	-	50	nAdc
ON CHARACTERISTICS						
DC Current Gain*		15	h_{FE}*			-
(I_C = 0.1 mAdc, V_{CE} = 1.0 Vdc)	2N3903			20	-	
	2N3904			40	-	
(I_C = 1.0 mAdc, V_{CE} = 1.0 Vdc)	2N3903			35	-	
	2N3904			70	-	
(I_C = 10 mAdc, V_{CE} = 1.0 Vdc)	2N3903			50	150	
	2N3904			100	300	
(I_C = 50 mAdc, V_{CE} = 1.0 Vdc)	2N3903			30	-	
	2N3904			60	-	
(I_C = 100 mAdc, V_{CE} = 1.0 Vdc)	2N3903			15	-	
	2N3904			30	-	
Collector-Emitter Saturation Voltage*		16, 17	$V_{CE(sat)}$*			Vdc
(I_C = 10 mAdc, I_B = 1.0 mAdc)				-	0.2	
(I_C = 50 mAdc, I_B = 5.0 mAdc)				-	0.3	
Base-Emitter Saturation Voltage*		17	$V_{BE(sat)}$*			Vdc
(I_C = 10 mAdc, I_B = 1.0 mAdc)				0.65	0.85	
(I_C = 50 mAdc, I_B = 5.0 mAdc)				-	0.95	

SMALL-SIGNAL CHARACTERISTICS

Characteristic	Device	Fig. No.	Symbol	Min	Max	Unit
Current-Gain—Bandwidth Product (I_C = 10 mAdc, V_{CE} = 20 Vdc, f = 100 MHz)	2N3903 2N3904		f_T	250 300	- -	MHz
Output Capacitance (V_{CB} = 5.0 Vdc, I_E = 0, f = 100 kHz)		3	C_{ob}	-	4.0	pF
Input Capacitance (V_{BE} = 0.5 Vdc, I_C = 0, f = 100 kHz)		3	C_{ib}	-	8.0	pF
Input Impedance (I_C = 1.0 mAdc, V_{CE} = 10 Vdc, f = 1.0 kHz)	2N3903 2N3904	13	h_{ie}	0.5 1.0	8.0 10	k ohms
Voltage Feedback Ratio (I_C = 1.0 mAdc, V_{CE} = 10 Vdc, f = 1.0 kHz)	2N3903 2N3904	14	h_{re}	0.1 0.5	5.0 8.0	X 10-4
Small-Signal Current Gain (I_C = 1.0 mAdc, V_{CE} = 10 Vdc, f = 1.0 kHz)	2N3903 2N3904	11	h_{fe}	50 100	200 400	-
Output Admittance (I_C = 1.0 mAdc, V_{CE} = 10 Vdc, f = 1.0 kHz)		12	h_{oe}	1.0	40	μmhos
Noise Figure (I_C = 100 μAdc, V_{CE} = 5.0 Vdc, R_S = 1.0 k ohms, f = 10 Hz to 15.7 kHz)	2N3903 2N3904	9, 10	NF	- -	6.0 5.0	dB

SWITCHING CHARACTERISTICS

Characteristic	Conditions	Device	Fig. No.	Symbol	Min	Max	Unit
Delay Time	(V_{CC} = 3.0 Vdc, $V_{BE(off)}$ = 0.5 Vdc, I_C = 10 mAdc, I_{B1} = 1.0 mAdc)		1, 5	t_d	-	35	ns
Rise Time			1, 5, 6	t_r	-	35	ns
Storage Time	(V_{CC} = 3.0 Vdc, I_C = 10 mAdc, I_{B1} = I_{B2} = 1.0 mAdc)	2N3903 2N3904	2, 7	t_s	- -	175 200	ns
Fall Time			2, 8	t_f	-	50	ns

* Pulse Test: Pulse Width = 300 μs, Duty Cycle = 2.0%.

Fig. III.2-1 *(Continued)*

FIGURE 1 — DELAY AND RISE TIME EQUIVALENT TEST CIRCUIT

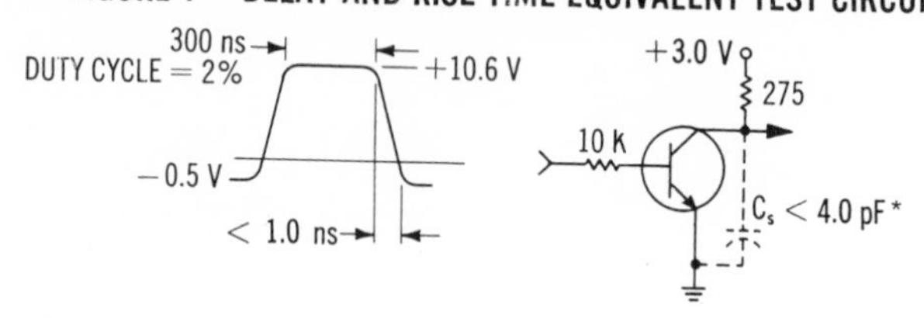

FIGURE 2 — STORAGE AND FALL TIME EQUIVALENT TEST CIRCUIT

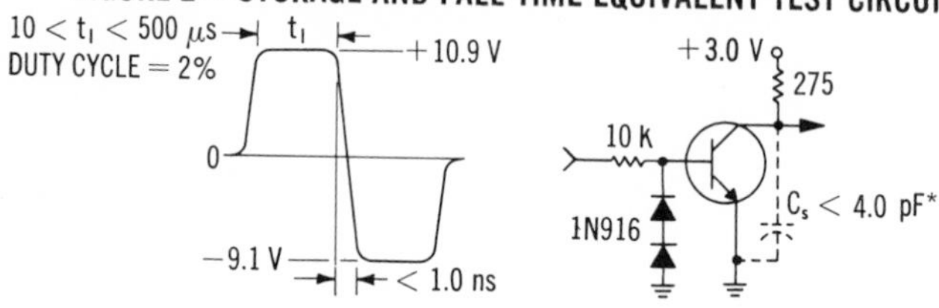

*Total shunt capacitance of test jig and connectors

Fig. III.2-1 *(Continued)*

TRANSIENT CHARACTERISTICS

—— $T_J = 25°C$ – – $T_J = 125°C$

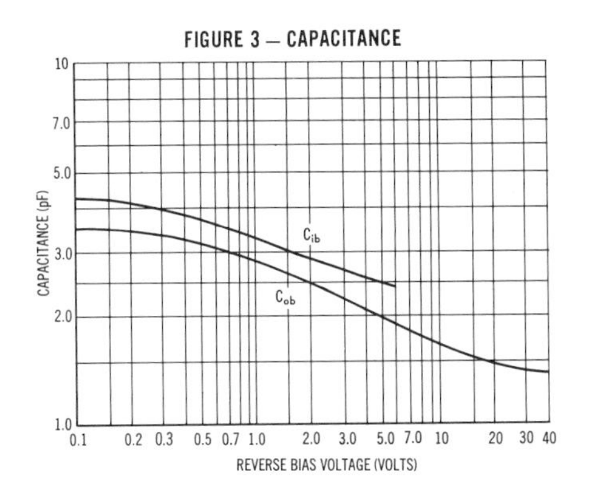

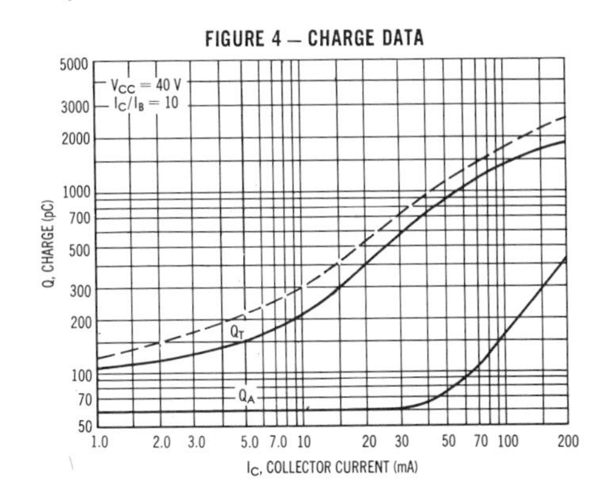

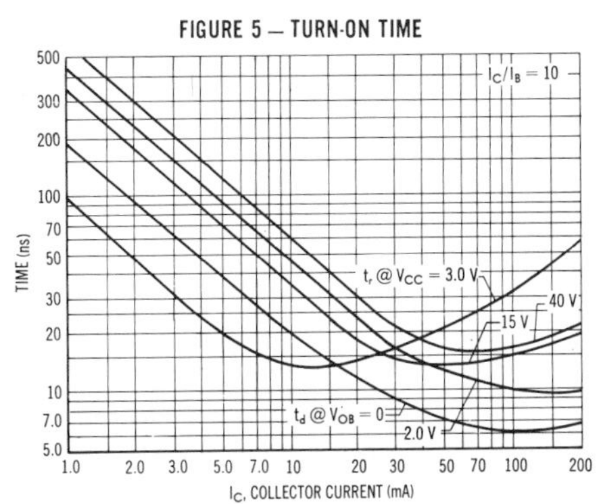

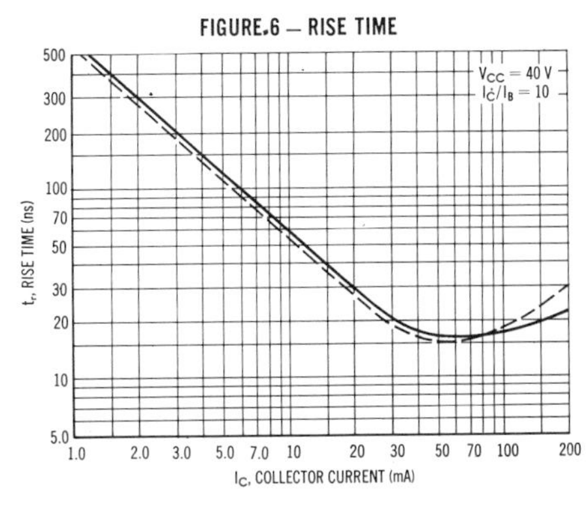

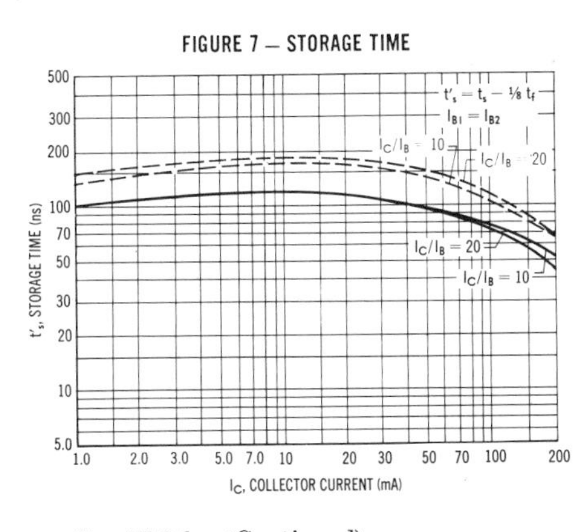

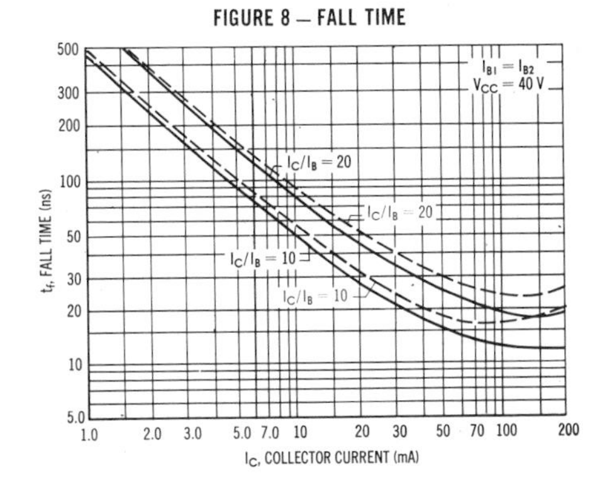

Fig. III.2-1 *(Continued)*

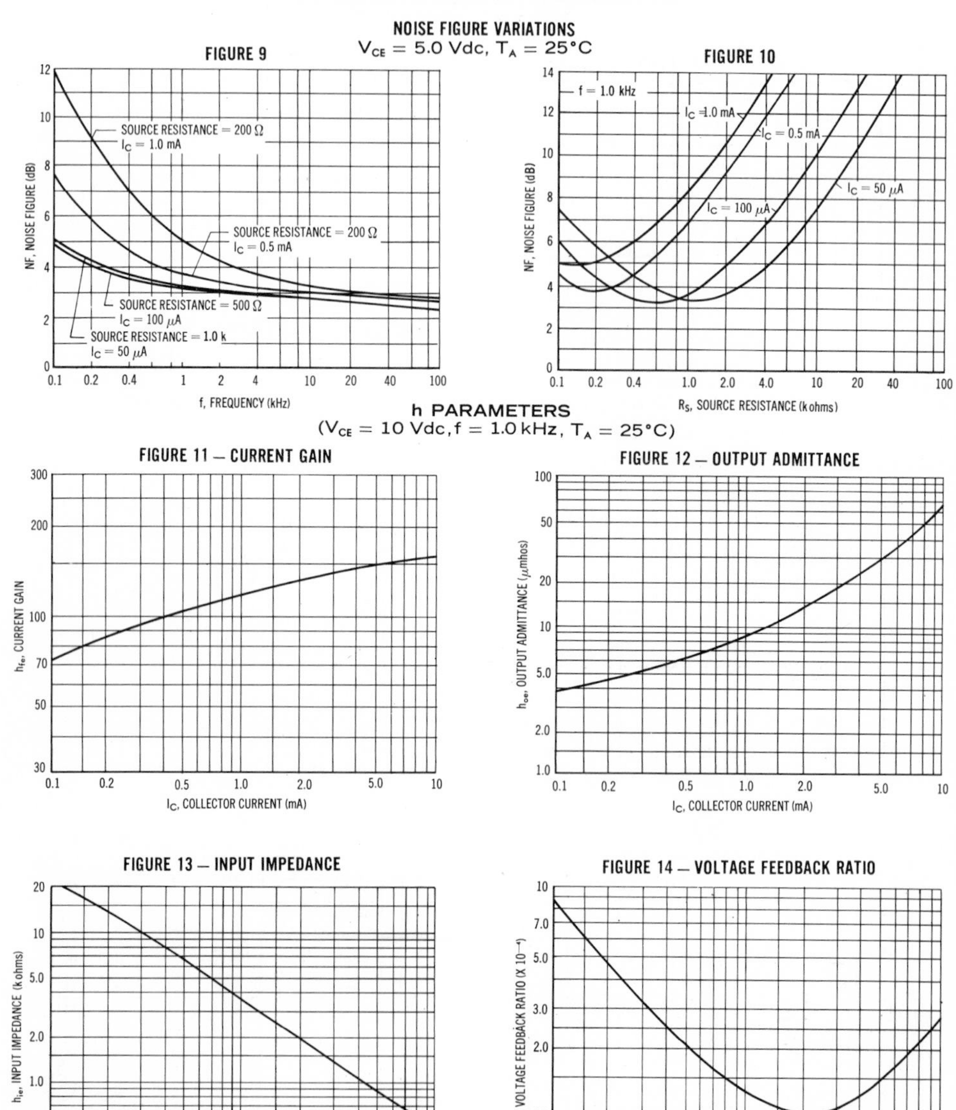

Fig. III.2-1 *(Continued)*

STATIC CHARACTERISTICS

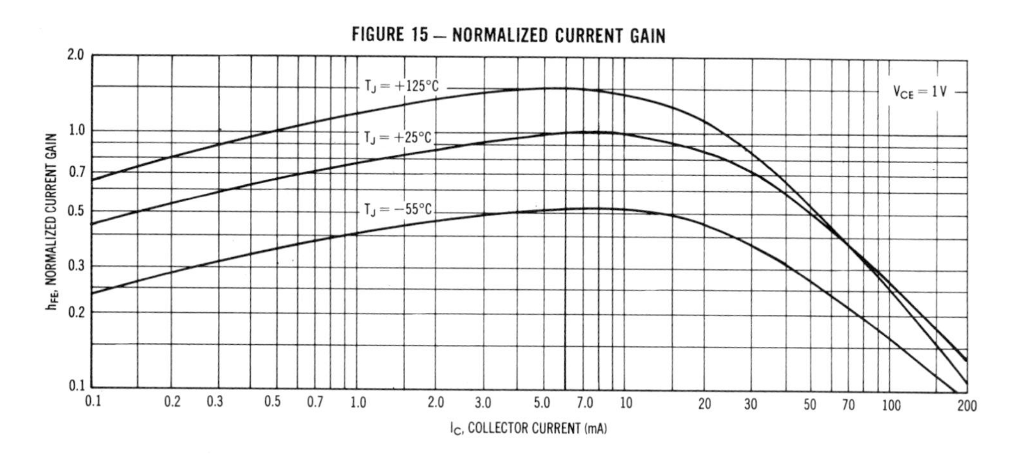

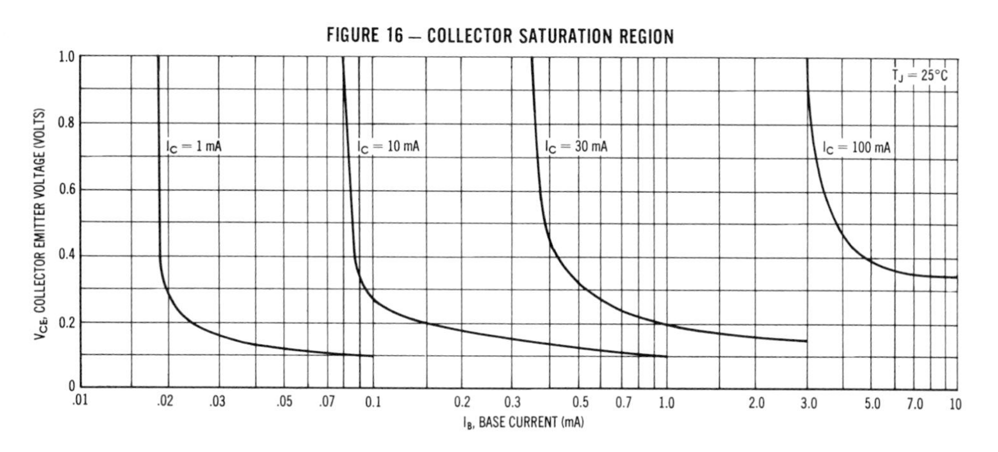

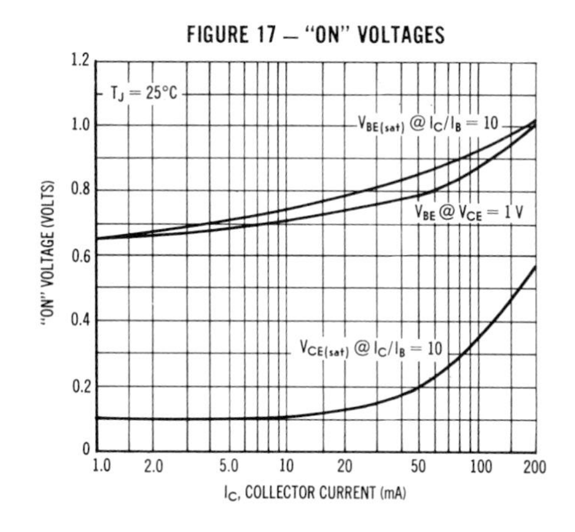

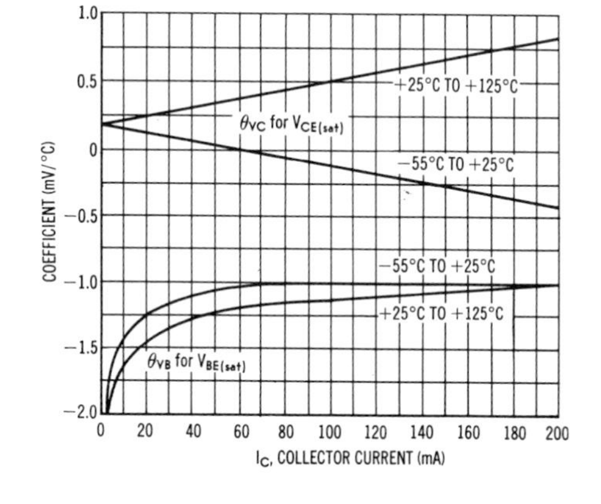

Fig. III.2-1 *(Continued)*

MAXIMUM RATINGS

Characteristic	Symbol	Rating	Unit
Collector-Base Voltage	V_{CB}	30	Volts
Collector-Emitter Voltage	V_{CEO}	20	Volts
Emitter-Base Voltage	V_{EB}	3	Volts
Collector Current	I_C	100	mA
Total Device Dissipation @ $T_A = 60°C$ @ $T_A = 25°C$	P_D	210 310	mW
Thermal Resistance, Junction to Ambient	θ_{JA}	0.357	°C/mW
Junction Temperature	T_J	135	°C

Fig. III.2-2 Characteristics of the MPS6507 *npn* silicon transistor, designed as a VHF mixer in TV applications. (*Courtesy of Motorola, Inc.*)

ELECTRICAL CHARACTERISTICS ($T_A = 25°C$ unless otherwise noted)

Characteristic	Symbol	Min	Typ	Max	Unit
Collector-Emitter Breakdown Voltage ($I_C = 1$ mAdc, $I_B = 0$)	BV_{CEO}	20	—	—	Vdc
Collector-Emitter Breakdown Voltage* ($I_C = 10$ mAdc, $V_{EB} = 0$)	BV_{CES}^*	30	—	—	Vdc
Collector Cutoff Current	I_{CBO}				μAdc
($V_{CB} = 15$ Vdc, $I_E = 0$)		—	—	0.05	
($V_{CB} = 15$ Vdc, $I_E = 0$, $T_A = 60$ °C)		—	—	1.0	
DC Current Gain ($I_C = 2$ mAdc, $V_{CE} = 10$ Vdc)	h_{FE}	25	—	—	—
High Frequency Current Gain ($I_C = 2$ mAdc, $V_{CE} = 10$ Vdc, f = 44 mc)	$\|h_{fe}\|$	20	—	—	db
Output Capacitance ($V_{CB} = 10$ Vdc, $I_E = 0$, f = 100 kc)	C_{ob}	—	—	2.5	pf
Current-Gain — Bandwidth Product ($I_C = 10$ mAdc, $V_{CE} = 10$ Vdc)	f_T	700	—	—	mc

* Pulse Test: Pulse Width ≤ 300 μsec, Duty Cycle ≤ 2%

Fig. III.2-2 *(Continued)*

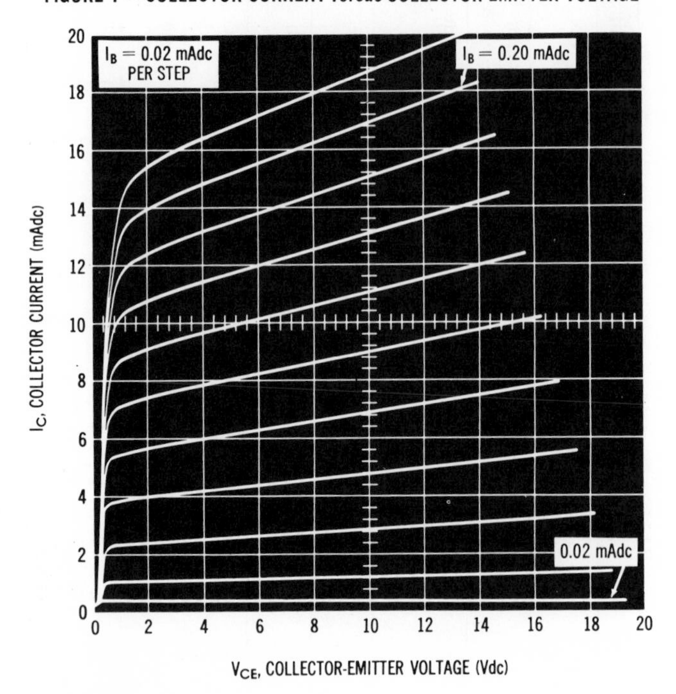

Fig. III.2-2 *(Continued)*

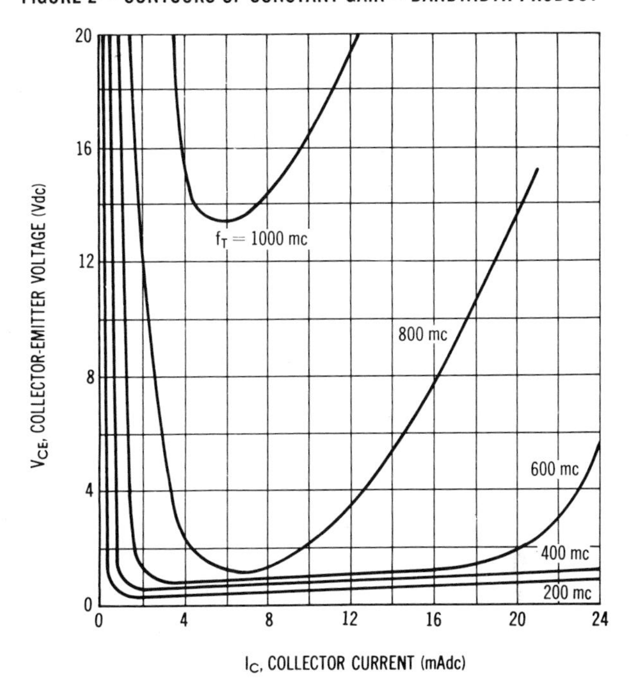

Fig. III.2-2 *(Continued)*

y PARAMETER VARIATIONS

(V_{CE} = 10 Vdc, I_C = 3 mAdc, T_A = 25°C)

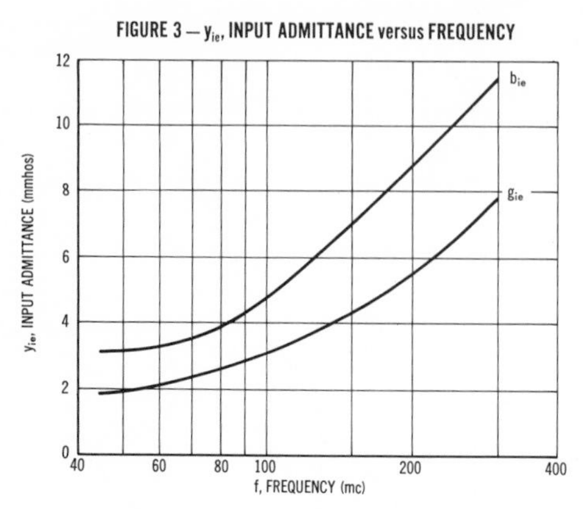

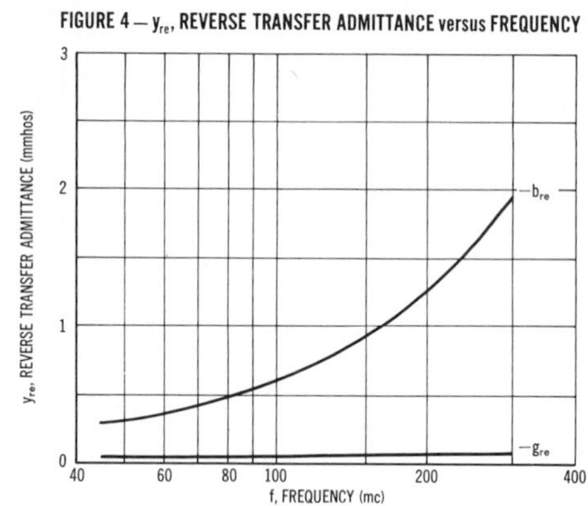

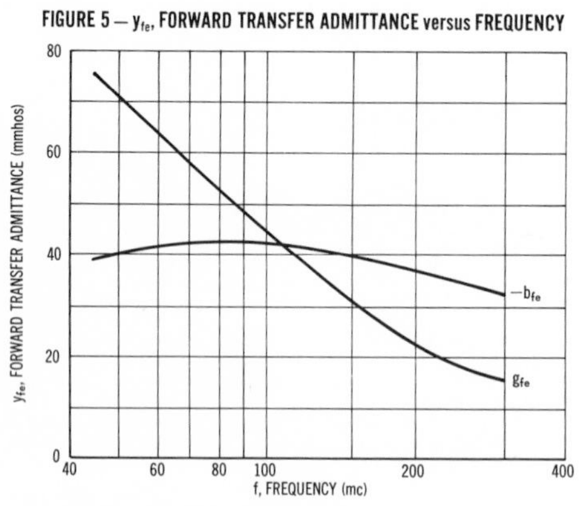

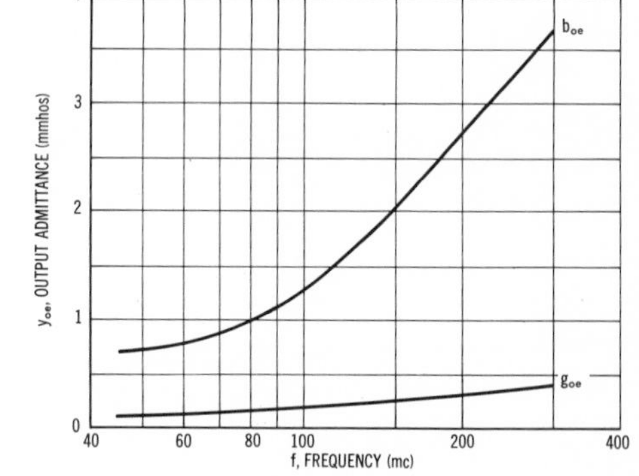

Fig. III.2-2 *(Continued)*

III.3 FET CHARACTERISTICS

MAXIMUM RATINGS ($T_A = 25°C$)

Characteristic	Symbol	Rating	Unit
Drain-Source Voltage	V_{DS}	30	Vdc
Drain-Gate Voltage	V_{DG}	30	Vdc
Gate-Source Voltage	V_{GS}	-30	Vdc
Drain Current	I_D	20	mAdc
Power Dissipation Derate above 25°C	P_D	300 2	mW mW/°C
Operating Junction Temperature	T_J	175	°C
Storage Temperature Range	T_{stg}	-65 to +200	°C

Fig. III.3-1 Characteristics of the 2N4223 and 2N4224 silicon *n*-channel junction FETs, designed for VHF amplifier and mixer applications. (*Courtesy of Motorola, Inc.*)

ELECTRICAL CHARACTERISTICS (T_A = 25°C unless otherwise noted)

Characteristic		Symbol	Min	Max	Unit
OFF CHARACTERISTICS					
Gate-Source Breakdown Voltage (I_G = -10 μAdc, V_{DS} = 0)		$V_{(BR)GSS}$	-30	-	Vdc
Gate Reverse Current (V_{GS} = -20 Vdc, V_{DS} = 0)	2N4223	I_{GSS}	-	-0.25	nAdc
	2N4224		-	-0.50	
(V_{GS} = -20 Vdc, V_{DS} = 0, T_A = 100°C)	2N4223		-	-250	
	2N4224		-	-500	
Gate-Source Cutoff Voltage (I_D = 0.25 nAdc, V_{DS} = 15 Vdc)	2N4223	$V_{GS(off)}$	-	-8	Vdc
(I_D = 0.50 nAdc, V_{DS} = 15 Vdc)	2N4224		-	-8	
Gate-Source Voltage (I_D = 0.3 mAdc, V_{DS} = 15 Vdc)	2N4223	V_{GS}	-1.0	-7.0	Vdc
(I_D = 0.2 mAdc, V_{DS} = 15 Vdc)	2N4224		-1.0	-7.5	
ON CHARACTERISTICS					
Zero-Gate-Voltage Drain Current* (V_{DS} = 15 Vdc, V_{GS} = 0)	2N4223	I_{DSS}*	3	18	mAdc
	2N4224		2	20	

DYNAMIC CHARACTERISTICS

Characteristic	Device	Symbol	Min	Max	Unit
Forward Transfer Admittance (V_{DS} = 15 Vdc, V_{GS} = 0, f = 1 kHz)*	2N4223 2N4224	$\|y_{fs}\|$	3000 2000	7000 7500	μmhos
(V_{DS} = 15 Vdc, V_{GS} = 0, f = 200 MHz)	2N4223 2N4224		2700 1700	- -	
Input Conductance (V_{DS} = 15 Vdc, V_{GS} = 0, f = 200 MHz)		$Re(y_{is})$	-	800	μmhos
Output Conductance (V_{DS} = 15 Vdc, V_{GS} = 0, f = 200 MHz)		$Re(y_{os})$	-	200	μmhos
Input Capacitance (V_{DS} = 15 Vdc, V_{GS} = 0, f = 1 MHz)		C_{iss}	-	6	pF
Reverse Transfer Capacitance (V_{DS} = 15 Vdc, V_{GS} = 0, f = 1 MHz)		C_{rss}	-	2	pF
Noise Figure (V_{DS} = 15 Vdc, V_{GS} = 0, R_S = 1 kohm, f = 200 MHz)	2N4223	NF	-	5	dB
Small-Signal Power Gain (V_{DS} = 15 Vdc, V_{GS} = 0, f = 200 MHz)	2N4223	G_{ps}	10	-	dB

*Pulse Test: Pulse Width ≦ 630 ms, Duty Cycle ≦ 10%

Fig. III.3-1 *(Continued)*

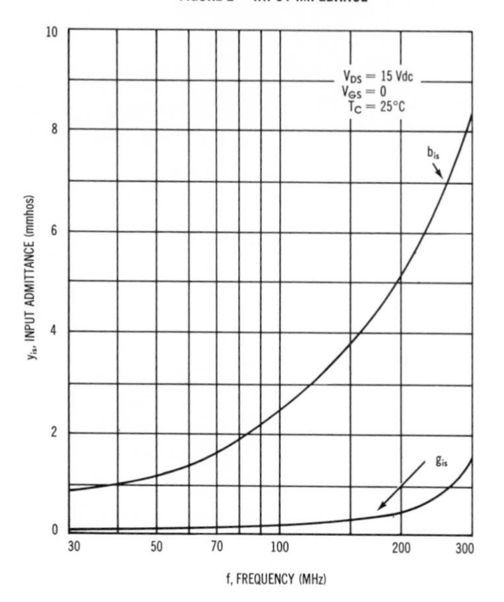

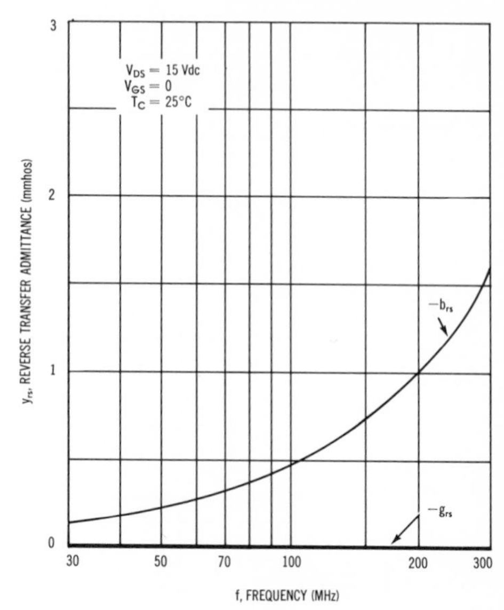

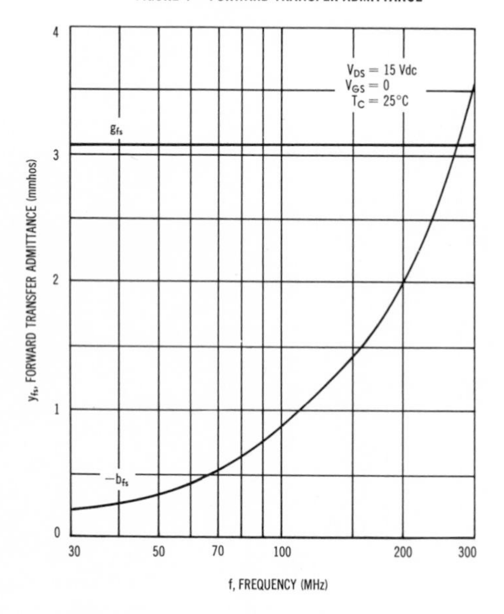

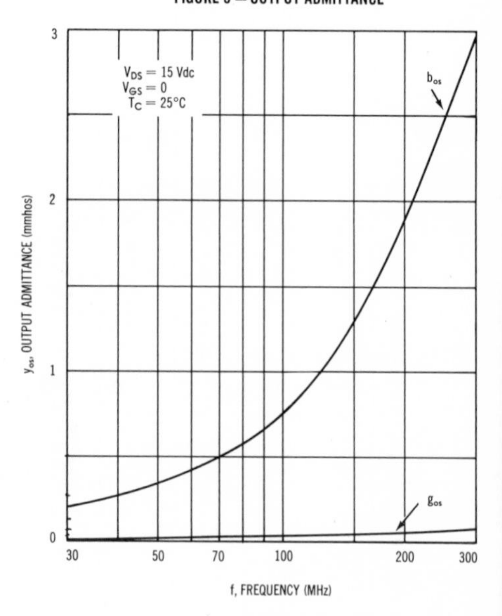

Fig. III.3-1 *(Continued)*

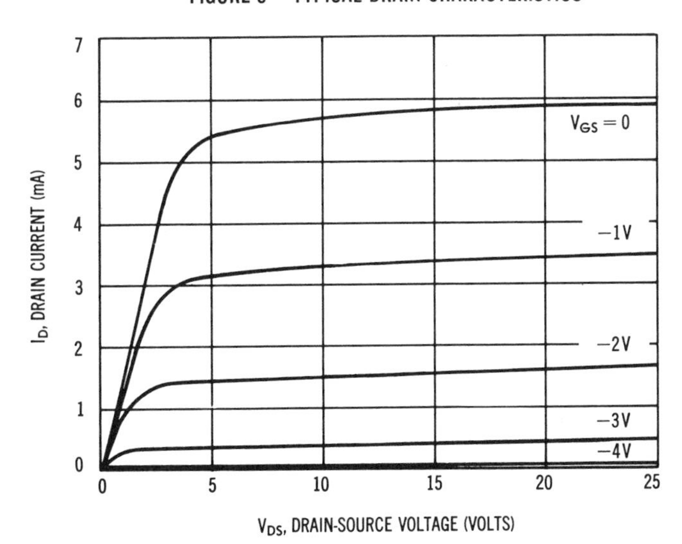

Fig. III.3-1 (*Continued*)

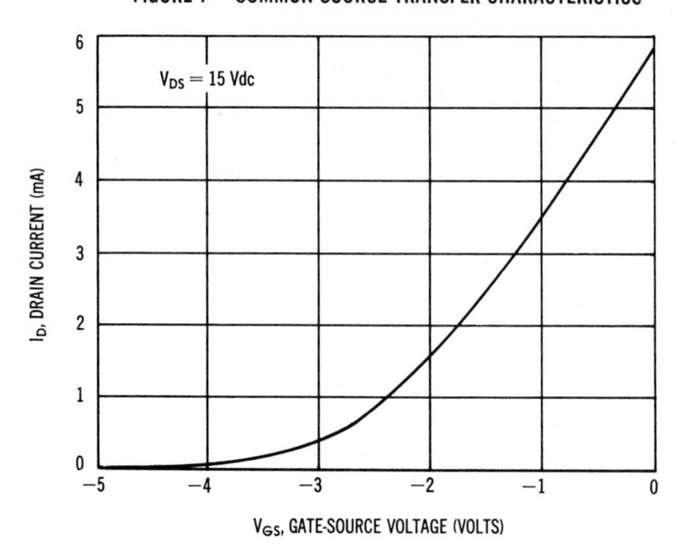

Fig. III.3-1 *(Continued)*

MAXIMUM RATINGS ($T_A = 25°C$ unless otherwise noted)

Rating	Symbol	Value	Unit
Drain-Source Voltage 2N3796 2N3797	V_{DS}	 25 20	Vdc
Gate-Source Voltage	V_{GS}	±10	Vdc
Drain Current	I_D	20	mAdc
Power Dissipation at $T_A = 25^oC$ Derate above 25^oC	P_D	200 1.14	mW $mW/^oC$
Operating Junction Temperature	T_J	+200	^oC
Storage Temperature	T_{stg}	-65 to +200	^oC

Fig. III.3-2 Characteristics of the 2N3796 and 2N3797 silicon n-channel MOSFETs, designed for low-power applications in the audio-frequency range. *(Courtesy of Motorola, Inc.)*

ELECTRICAL CHARACTERISTICS ($T_A = 25°C$ unless otherwise noted)

Characteristic		Symbol	Min	Typ	Max	Unit
Drain-Source Breakdown Voltage		BV_{DSX}				Vdc
(V_{GS} = -4.0 V, I_D = 5.0 μA)	2N3796		25	30	—	
(V_{GS} = -7.0 V, I_D = 5.0 μA)	2N3797		20	25	—	
Zero-Gate-Voltage Drain Current		I_{DSS}				mAdc
(V_{DS} = 10 V, V_{GS} = 0)	2N3796		0.5	1.5	3.0	
	2N3797		2.0	2.9	6.0	
Gate-Source Voltage Cutoff		$V_{GS(off)}$				Vdc
(I_D = 0.5 μA, V_{DS} = 10 V)	2N3796		—	-3.0	-4.0	
(I_D = 2.0 μA, V_{DS} = 10 V)	2N3797		—	-5.0	-7.0	
"On" Drain Current		$I_{D(on)}$				mAdc
(V_{DS} = 10 V, V_{GS} = +3.5 V)	2N3796		7.0	8.3	14	
	2N3797		9.0	14	18	
Drain-Gate Reverse Current *		I_{DGO} *				pAdc
(V_{DG} = 10 V, I_S = 0)			—	—	1.0	
Gate-Reverse Current *		I_{GSS} *				pAdc
(V_{GS} = -10 V, V_{DS} = 0)			—	—	1.0	
(V_{GS} = -10 V, V_{DS} = 0, T_A = 150°C)			—	—	200	

Characteristic		Symbol	Min	Typ	Max	Unit
Small-Signal, Common-Source Forward Transfer Admittance (V_{DS} = 10 V, V_{GS} = 0, f = 1.0 kHz)	2N3796	$\|y_{fs}\|$	900	1200	1800	μmhos
	2N3797		1500	2300	3000	
(V_{DS} = 10 V, V_{GS} = 0, f = 1.0 MHz)	2N3796		900	—	—	
	2N3797		1500	—	—	
Small-Signal, Common-Source, Output Admittance (V_{DS} = 10 V, V_{GS} = 0, f = 1.0 kHz)	2N3796	$\|y_{os}\|$	—	12	25	μmhos
	2N3797		—	27	60	
Small-Signal, Common-Source, Input Capacitance (V_{DS} = 10 V, V_{GS} = 0, f = 1.0 MHz)	2N3796	C_{iss}	—	5.0	7.0	pF
	2N3797		—	6.0	8.0	
Small-Signal, Common-Source, Reverse Transfer Capacitance (V_{DS} = 10 V, V_{GS} = 0, f = 1.0 MHz)		C_{rss}	—	0.5	0.8	pF
Noise Figure (V_{DS} = 10 V, V_{GS} = 0, f = 1.0 kHz, R_S = 3 megohms)		NF	—	3.8	—	dB

* This value of current includes both the FET leakage current as well as the leakage current associated with the test socket and fixture when measured under best attainable conditions.

Fig. III.3-2 *(Continued)*

TYPICAL DRAIN CHARACTERISTICS

FIGURE 1 — 2N3796

FIGURE 2 — 2N3797

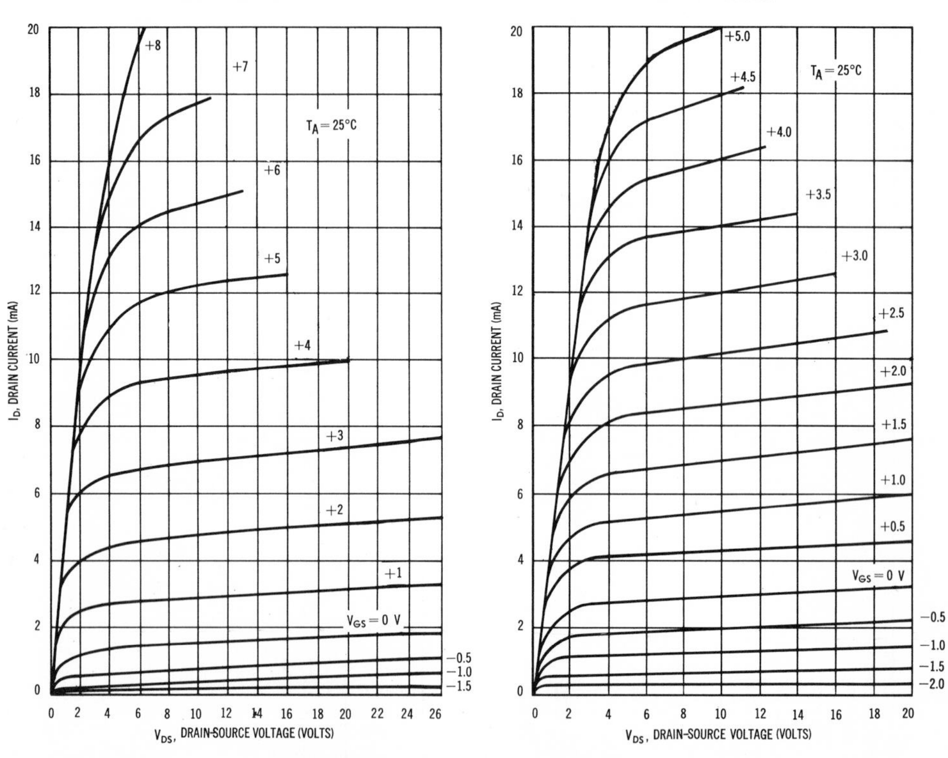

COMMON SOURCE TRANSFER CHARACTERISTICS

FIGURE 3 — 2N3796

FIGURE 4 — 2N3797

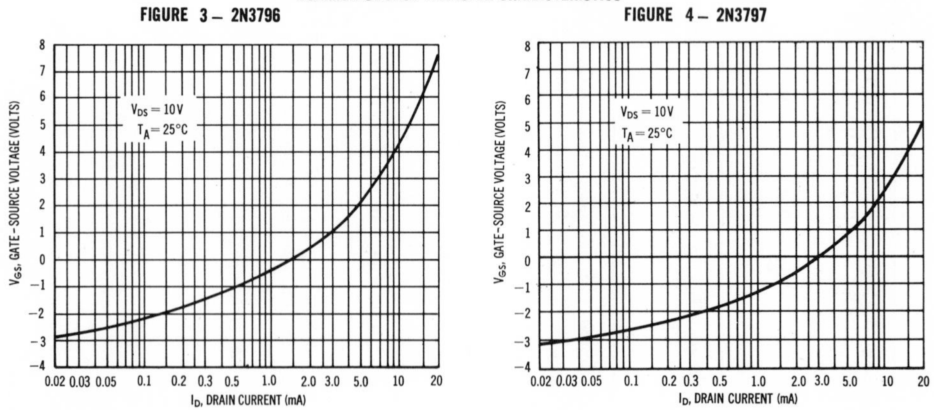

Fig. III.3-2 *(Continued)*

FIGURE 5 — FORWARD TRANSFER ADMITTANCE

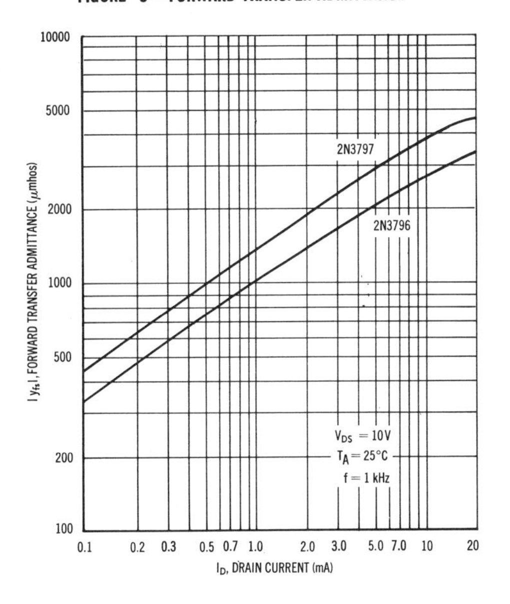

FIGURE 6 — AMPLIFICATION FACTOR

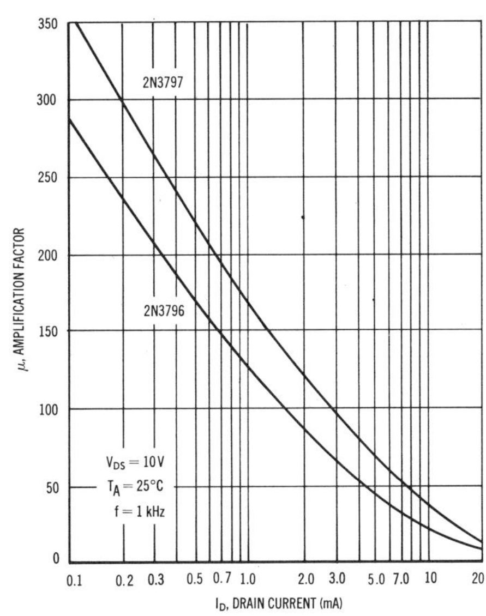

FIGURE 7 — OUTPUT ADMITTANCE

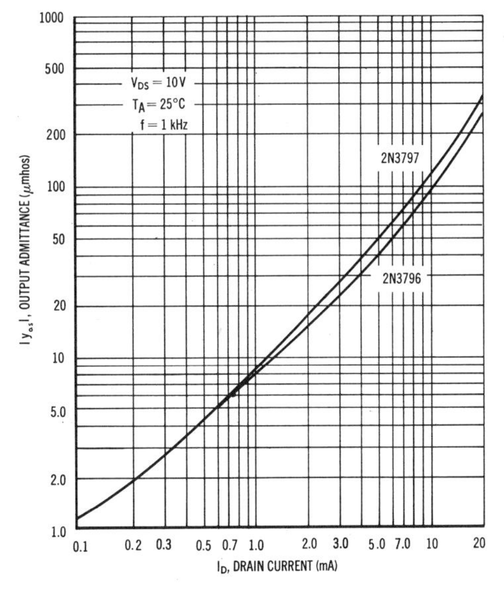

FIGURE 8 — NOISE FIGURE

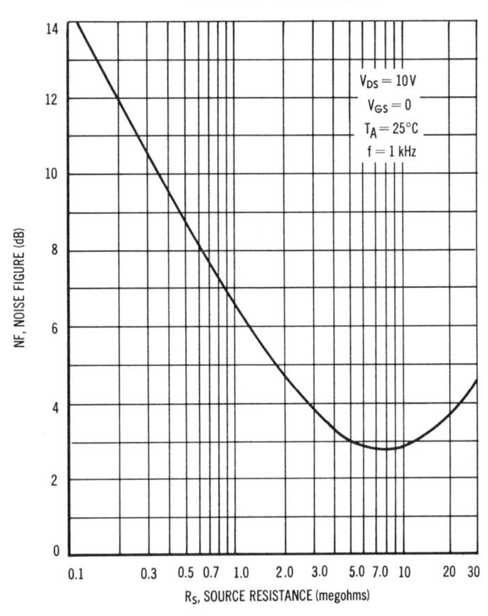

Fig. III.3-2 *(Continued)*

III.4 VACUUM-TUBE CHARACTERISTICS

GENERAL DATA

Electrical:

Heater, for Unipotential Cathodes:

Voltage	6.3	ac or dc volts
Current	0.6	amp

Direct Interelectrode Capacitances (With no external shield):

	Unit No. 1	*Unit No. 2*	
Grid to plate	4	3.8	μμf
Grid to cathode and heater	2.2	2.6	μμf
Plate to cathode and heater	0.7	0.7	μμf

Characteristics, Class A_1 Amplifier (Each Unit):

Plate Voltage	90	250	volts
Grid Voltage	0	−8	volts
Amplification Factor	20	20	volts
Plate Resistance (Approx.)	6700	7700	ohms
Transconductance	3000	2600	μmhos
Plate Current	10	9	ma
Plate Current for grid voltage of −12.5 volts	–	1.3	ma
Grid Voltage (Approx.) for plate current of 10 μamp	−7	−18	volts

AMPLIFIER - Class A_1

Values are for Each Unit

Maximum Ratings, *Design-Center Values:*

PLATE VOLTAGE	450 max.	volts
CATHODE CURRENT	20 max.	ma
PLATE DISSIPATION:		
Either plate	5 max.	watts
Both plates (Both units operating)	7.5 max.	watts
PEAK HEATER-CATHODE VOLTAGE:		
Heater negative with respect to cathode	200 max.	volts
Heater positive with respect to cathode	200 max.	volts

Maximum Circuit Values:

Grid-Circuit Resistance:

For fixed-bias operation	1 max.	megohm

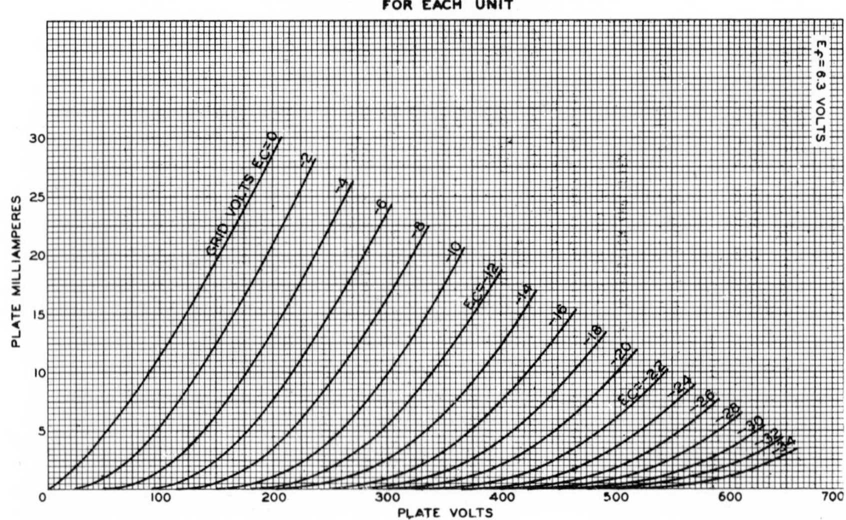

Fig. III.4-1 Characteristics of the 6SN7 twin medium-mu triode. (*Courtesy of Radio Corporation of America.*)

GENERAL DATA

Electrical:

Heater, for Unipotential Cathodes:

Heater Arrangement	*Series*	*Parallel*	
Voltage	12.6	6.3	ac or dc volts
Current	0.15	0.3	amp

Direct Interelectrode Capacitances (Approx.)°:

	Unit No. 1	*Unit No. 2*	
Grid-Drive Operation:			
Grid to Plate	1.5	1.5	μμf
Grid to Cathode	2.2	2.2	μμf
Plate to Cathode	0.5	0.4	μμf
Heater to Cathode	2.4	2.4	μμf
Cathode-Drive Operation:			
Plate to Cathode	0.2	0.2	μμf
Grid & Heater to Cathode	4.6	4.6	μμf
Grid & Heater to Plate	1.8	1.8	μμf
Grid to Grid	0.005 max.		μμf
Plate to Plate	0.4 max.		μμf

° With no external shield.

AMPLIFIER - Class A_1

Values are for each unit

Maximum Ratings, *Design-Center Values:*

PLATE VOLTAGE	300 max.	volts
GRID VOLTAGE:		
Negative Bias Value	−50 max.	volts
PLATE DISSIPATION	2.5 max.	watts
PEAK HEATER-CATHODE VOLTAGE:		
Heater negative with respect to cathode	90 max.	volts
Heater positive with respect to cathode	90 max.	volts

Characteristics:

Plate Supply Voltage	100	250	volts
Cathode-Bias Resistor	270	200	ohms
Amplification Factor	60	60	
Plate Resistance (Approx.)	15000	10900	ohms
Transconductance	4000	5500	μmhos
Grid Voltage (Approx.) for plate current of 10 μamp	−5	−12	volts
Plate Current	3.7	10	ma

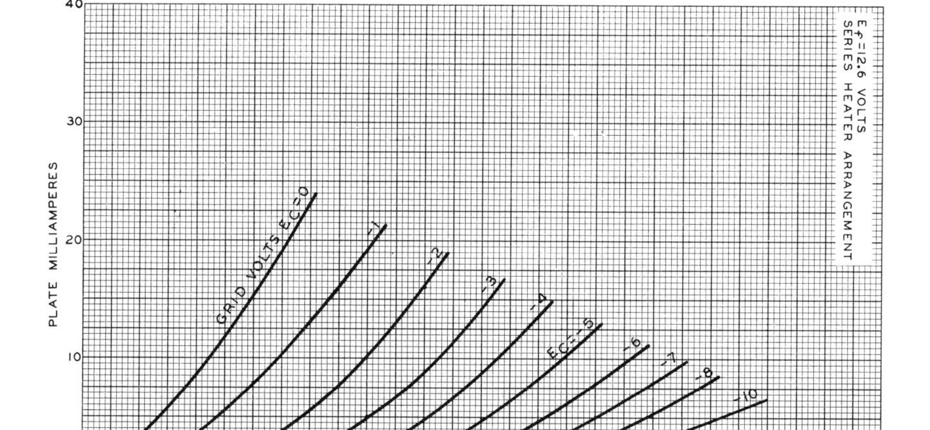

Fig. III.4-2 Characteristics of the 12AT7 twin high-mu triode. *(Courtesy of Radio Corporation of America.)*

GENERAL DATA

Electrical:

Heater, for Unipotential Cathode:		
Voltage (AC or DC)	6.3 ± 10%	volts
Current	0.2	amp
Direct Interelectrode Capacitances (Approx.):[o]		
Grid No.1 to plate	0.03	$\mu\mu$f
Grid No.1 to cathode & internal shield, grid No.2, and heater	4.5	$\mu\mu$f
Plate to cathode & internal shield, grid No.2, and heater	3	$\mu\mu$f

Characteristics, Class A_1 Amplifier:

Plate Voltage	125	volts
Grid-No.2 Voltage	80	volts
Grid-No.1 Voltage	−1	volt
Plate Resistance (Approx.)	0.1	megohm
Transconductance	8000	μmhos
Plate Current	10	ma
Grid-No.2 Current	1.5	ma
Grid-No.1 Voltage (Approx.) for plate μa = 20	−6	volts

AMPLIFIER — Class A_1

Maximum Ratings, *Design-Maximum Values:*

PLATE VOLTAGE	180 max.	volts
GRID-No.2 (SCREEN-GRID) SUPPLY VOLTAGE	180 max.	volts
GRID-No.1 (CONTROL-GRID) VOLTAGE:		
Positive-bias value	0 max.	volts
CATHODE CURRENT	20 max.	ma
GRID-No.2 INPUT:		
For grid-No.2 voltages up to 90 volts	0.5 max.	watt
PLATE DISSIPATION	2 max.	watts
PEAK HEATER-CATHODE VOLTAGE:		
Heater negative with respect to cathode	100 max.	volts
Heater positive with respect to cathode	100 max.	volts

Maximum Circuit Values:

Grid-No.1-Circuit Resistance	0.5 max.	megohm

[o] With external shield JEDEC No.316 connected to cathode.

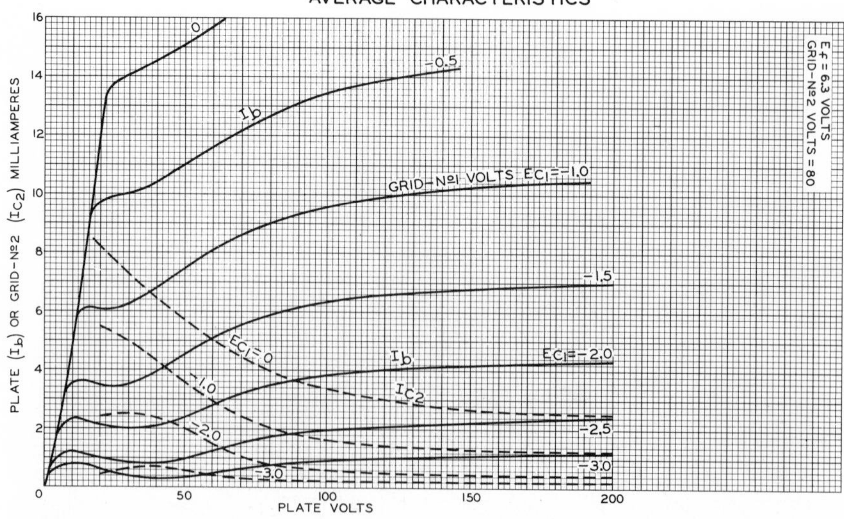

Fig. III.4-3 Characteristics of the 6CY5 sharp-cutoff tetrode. (*Courtesy of Radio Corporation of America.*)

GENERAL DATA

Electrical:

Heater, for Unipotential Cathode:

Voltage	6.3	ac or dc volts
Current	0.65	amp

Direct Interelectrode Capacitances (without external shield):

Grid No.1 to Plate	0.120	μμf
Input	11	μμf
Output	5.5	μμf

Characteristics, Amplifier Class A_1:

Plate Voltage	250	volts
Grid No.3	Connected to cathode at socket	
Grid-No.2 Voltage	150	volts
Grid-No.1 Voltage	-3	volts
Peak AF Grid-No.1 Signal Voltage	3	volts
Zero-Signal DC Plate Current	30	ma
Max.-Signal DC Plate Current	31	ma
Zero-Signal DC Grid-No.2 Current	7	ma
Max.-Signal DC Grid-No.2 Current	7.2	ma
Plate Resistance (Approx.)	0.15	megohm
Transconductance	11000	μmhos
Grid-No.1 Voltage (Approx.) for plate current of 10 μamp	-14	volts
Load Resistance	7500	ohms
Total Harmonic Distortion	8	per cent
Max.-Signal Power Output	2.8	watts

AMPLIFIER - Class A_1

Maximum Ratings, ***Design-Center Values:***

PLATE VOLTAGE	300 max.	volts
PLATE SUPPLY VOLTAGE	300 max.	volts
GRID-No.3 (SUPPRESSOR) VOLTAGE	0 max.	volts
GRID-No.2 SUPPLY VOLTAGE	300 max.	volts
GRID-No.1 (CONTROL-GRID) VOLTAGE:		
Negative bias value	50 max.	volts
Positive bias value	0 max.	volts
PLATE DISSIPATION	7.5 max.	watts
GRID-No.2 INPUT	1.7 max.	watts
PEAK HEATER-CATHODE VOLTAGE:		
Heater negative with respect to cathode	90 max.	volts
Heater positive with respect to cathode	90 max.	volts
BULB TEMPERATURE (At hottest point on bulb surface)	200 max.	°C

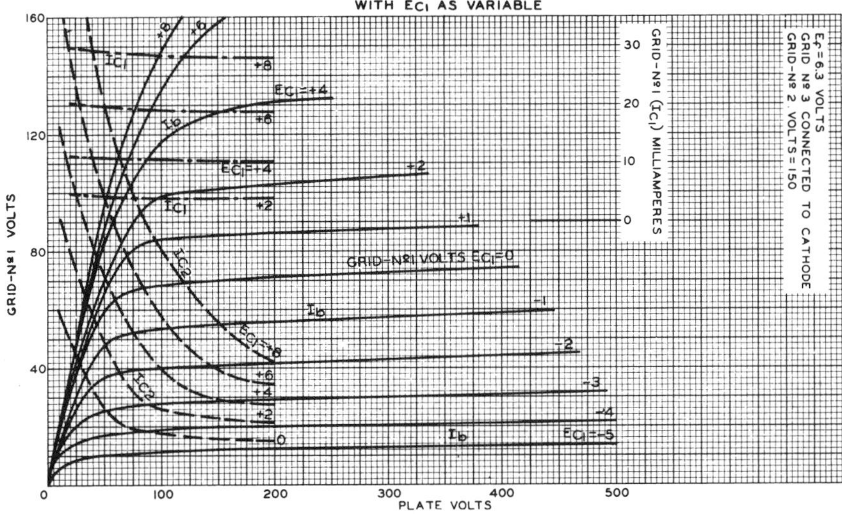

Fig. III.4-4 Characteristics of the 6CL6 power pentode. *(Courtesy of Radio Corporation of America.)*

Index

Index

Ac load line, 31, 32, 36, 102, 108, 111, 112, 134, 159
Acceptor material, 18
Active filters, 605, 609
  integrator, 605–606
  Q multiplier, 606–609
    using IC amplifier, 609
All-pass network, 602, 605
  using IC amplifier, 602
Ambient temperature, 51–53, 56, 143
Amplification factor, 376, 401, 416
Amplifiers, binary, 2
  capacitively-coupled double-tuned, 561
  cascading stages in, 221–233
    design of cascaded amplifier, 229–233
  complementary symmetry, 177–179
  distortion, 2, 372–374, 410–413
  distributed, 565–569
  double-tuned, 559–560
  gain-bandwidth [*see* Gain-bandwidth (GBW) products]
  lag-compensated feedback, 600–602, 608
  neutralized, 549
  push-pull, 152, 212
  single-tuned, 523–539
    design of, 526
  stagger-tuned, 553–558
  unstable (*see* Feedback amplifier; Oscillations; Stability of feedback amplifier)
Amplitude modulation (AM), 2–7, 302, 303
Analog computers, 42, 606
Arsenic, 329
Automatic gain control (AGC) [*see* Automatic volume control (AVC)]
Automatic volume control (AVC), 6, 302–309
Automatic volume control (AVC), amplifier, equivalent circuits of, 305
  amplitude modulation, 302, 303
  audio output, 303
  circuit, experimental, 306
  receiver, 302
  rectifier, 303
  sensitivity function, 306
  transfer characteristic, 308
  use of diode equation, 350
  waveforms, 304
Autotransformer, 531
  double-tapped, 534
Avalanche breakdown, 20, 21, 85
  current multiplication, 146

Bandwidth, 465–504, 526
  of feedback amplifiers, 577–580, 589
Base-collector capacitance, C_{ob}, $C_{b'c}$, 466
Base-emitter capacitance, $C_{b'e}$, 466
Base-injection bias, 113–115
Base-spreading resistance, $r_{bb'}$, 464
  effect of, on response of single-tuned amplifier, 528
β variation, 92, 95, 105, 113
Bias, 88
  network, 112
  stability, 123–146, 228–233
  turn on, 169
Biasing, cathode follower, 406
    grid-biased, 404, 406
    self-biased, 406
  IGFET, 369–374
  JFET, 368–369, 404–410
  pentode, 408
    screen-current, 408
    screen-grid resistance, 410

Biasing, transistor, 123–146
stability, 123, 142
triode, 404
Binary amplifier, 2
Bode plots, 433–439, 583–586
break frequency, 434
logarithmic scales, 433
phase-angle variation, 436
Boltzmann's constant, 20
"Bootstrap" circuit, 269
Boron, 18
Breakdown, 367
regions, 87
Breakdown voltage, 20, 145, 146, 212–213, 364
Butterworth filter, 59, 61
Bypass capacitor, combined effect of, with coupling capacitor, 446–447
(*See also* Emitter bypass capacitor)

Capacitance, standard values, 626
Capacitive tuning, 560
Cascaded amplifier, bias stability, 228–233
feedback in, 352
FET's, 495
integrated-circuit amplifiers, 351–354
Miller capacitance, 491
Miller impedances, 490, 495
N stages, 222
oscillations in, 354
pentodes, 495
RC amplifiers, 490–498
3-dB bandwidth, 494
transistors, 221–233
vacuum tubes, 495
Cascode amplifier, 250–255, 344, 539–543
dc analysis, 252
gain-bandwidth product, 541
level shifting in direct-coupled amplifier, 253–255
Miller capacitance, 540
small-signal analysis, 251
Case temperature, 51, 53, 57, 143, 144
Cathode, 399
Cathode follower, at high frequencies, 406, 490
oscillations in, 490
output impedance, 418
self-bias, 406, 418–419
voltage gain, 419
Channel, of the IGFET, 365
of the JFET, 361
Clamping circuit, 60
Class A operation, 152, 164
ac load line, 159
collector dissipation, 157
efficiency, 157
figure of merit, 157
inductor-coupled power-amplifier, 153
maximum-dissipation hyperbola, 158–164
maximum symmetrical swing, 154, 159
power-amplifier load lines, 153
power transferred to load, 156
Q point placement, 153
saturation, 162
supplied power, 156
Class A power amplifier, design of, 158–162
Class B push-pull power amplifiers, 168–176
crossover distortion, 169, 170
design, 174–176
efficiency, 173
figure of merit, 173
load line determination, 171
maximum efficiency, 168
power calculations, 171–174
power dissipated in the collector, 172
power transferred to load, 172
supplied power, 172
turn-on bias, 169
CLC filters, 13
Clipping circuit, 10
Collector-base junction, 78–80, 83
capacitance, 463
circuit, 80
cutoff, 85
dissipation, 145, 157, 166, 212–213

Collector-base junction, impedance, 74
Colpitts oscillator, 616–617
  conditions for oscillation, 616
  piezoelectric crystal, 617
  variation of frequency of oscillation, 617
Common-base amplifiers, 77–81
  cutoff frequency, 212–213
  equivalent circuit, 198
  h parameters, 198–202
  input impedance, 202
  output admittance, h_{ob}, 199
  output impedance, 202
  short-circuit current gain, h_{fb}, 199
  voltage gain, 202
Common-collector amplifier, 110, 202–212
  [*See also* Emitter-follower (common-collector) configuration]
Common collector (CC) hybrid parameters, 202
Common-drain amplifier, 380–383
Common-emitter (CE) amplifier, 81
  cutoff frequency, 145
  output characteristics, 84
  short-circuit 3-dB frequency, f_β, 464
  vi characteristic, 83
Common-gate amplifier, 386–387
Common mode (*see* Difference amplifier)
Common-mode current, 236
Common-mode current gain, 238
Common-mode rejection ratio, 239–240
Common-source amplifier, 377–380
  input impedance, 377
  output impedance, 377
  voltage gain, 377
Complementary-symmetry amplifier, 177–179
Compound amplifier [*see* Darlington configuration (compound amplifier)]
Conditions for oscillation (*see* Oscillators)
Conduction band, 17, 18
Constant-emitter current source, 240
Control grid, 400, 402
Controlled switches, 504
Coupling capacitor, 106–110, 440–446
  ac load line, 107, 108
  combined effect of, with bypass capacitor, 446–447
  current gain, 443
  design, 445
  maximum symmetrical swing, 107, 108
  Q point, 107, 108
  3-dB frequency, 441, 445, 446
Crossover distortion, 169, 170
Current amplification in the transistor, 80–87
  α, β, definition of, 80
  avalanche breakdown, 85
  base current, 85
  basic amplifier, 82
  breakdown regions, 87
  common-base configuration, 81
  common-emitter configuration, 81
  common-emitter output characteristics, 84
  current-source model, 87
  saturation, 84–87
  small-signal current-amplification factor h_{fe}, 81, 82
  transistor equivalent circuit, 86
Current amplification factor, α, 75, 80
  β, 75
  small-signal, 81, 82
Current error in feedback amplifier, 265, 284–287
  (*See also* Feedback)
Current feedback, 265, 370
  (*See also* Feedback)
Current gain, 190, 192, 194, 195
  bandwidth, 443, 471
  with feedback, 268

Darlington configuration (compound amplifier), 245–250
  load lines, 246
  quiescent point, 246
  short-circuit current gain, 245–250
  small-signal operation, 247

Darlington emitter-follower, 289
Dc amplifiers, 220
Decibel (dB), 623–624
one milliwatt (dBm), 623
Delay time, 505, 507
Depletion region, 361, 363, 462
Derating curve, 54, 57, 143
Derating factor, 212–213
Design, of feedback amplifiers, 298–302
gain, 298
input impedance, 298
operational amplifier, 300
output impedance, 298
ripple reduction, 301
sensitivity, 298
(*See also* Feedback)
of a simple integrated circuit, 341–342
photomicrograph of an integrated differential-amplifier circuit, 341
of thin film resistor, 340
Difference amplifier, 233–245, 276, 344, 613
balance control, 241
common-mode current, 236
common-mode current gain, 238
common-mode rejection ratio, 239–240
constant-emitter current source, 240–245
differential-mode current, 236
differential-mode current gain A_d, 239
quiescent-point analysis, 234–235
small-signal analysis, 235–239, 241
Difference circuit, 263
Differential-mode current, 236
Diffusion, 326, 329
Diode characteristics, 21, 54, 55
equation, 20
equivalent circuits, 42
large signal analysis, 33–37
ac load lines, 36
distortion, 33, 34
graphical solution, 35
Diode characteristics, large signal analysis, iterative graphical procedure, 34
piecewise linear, 33
Q-point shift, 33
manufacturer's specifications, 54–58
breakdown voltage, 55
case temperature, 57
derating curve, 54, 57
half-wave rectifier, 55
junction temperature, 54, 56, 57
peak inverse voltage, 54, 55
rectifier, 54
reverse current, 54, 55
temperature coefficient, 56, 57
nonlinear properties, clipping circuit, 10
Fourier series, 12
frequency multiplication, 17
full-wave-rectifier, 25
harmonics, 13
ideal diode, 9–17
power supplies, 15
power supply filters, 12
rectifier, 10, 14–17
regulation, 16
ripple voltage, 12
switch, 9
piecewise linear analysis, 42–50
equivalent circuits, 42, 44
nonlinear differential equations, 42, 47, 49
scaling variables, 46
small signal analysis, 31–32
ac load line, 31, 32
dc load line, 30–32
distortion, 28
dynamic resistance, 25–31
equivalent circuits, 29
graphical analysis, 26, 27
linear circuit, 25
Q point, 30, 31
RC filter, 29
reactive elements, 29
Taylor series expansion, 27, 28
temperature effects in, 50–54
ambient temperature, 51–53

Diode characteristics, temperature effects in, analog, 50, 52
case temperature, 51, 53
germanium diodes, 52
heat dissipation, 50
heat sink, 50, 51, 53
junction temperature, 50–53
mica insulator, 53
power dissipation, 53
thermal resistance, 50–53
theory of, 17–22
acceptor material, 18
avalanche breakdown, 20, 21
breakdown voltage, 20
characteristics, 21
conduction band, 17, 18
current rating, 22
dc load line, 22–25
doping, 18
half-wave rectifier, 22
ideal diode, 22
junction diode, 19, 22
quiescent point, 24
reverse-biased, 19
reverse-saturation current, 20
silicon diode, 22
surface leakage, 21, 22
Zener breakdown, 20
(*See also* Zener diode)
vacuum, cathode, 399
filament, 398
Langmuir-Childs law, 400
perveance, 400
plate, 399
Zener diode, 37–42
dc load line, 39
dynamic resistance, 39
output-voltage variation, 39
power dissipation, 39
reverse characteristic, 37
ripple reduction, 40, 42
shunt regulator, 38
voltage regulator, 38, 40
Distortion, 2, 5, 28, 33, 34, 152, 283, 410, 413
harmonic, 410
second, 413
Distortion, harmonic, third, 413
in the FET, 372–374
second-harmonic, 373
Distributed amplifier, 565–569
cutoff-frequency of transmission line, 568
design equations, 568
signal velocity, 567
transmission lines, 565
voltage gain, 566
voltage-standing wave ratio, 568
Doping, 18
Double-tapped autotransformer, 534
Double-tuned amplifier, 559, 560
capacitive tuning, 560
capacitively-coupled, 561
mutual coupling, 560
practical problems, 560
Drain, IGFET, 365
JFET, 360
Drain coupling capacitor, 450–451
3-dB frequency, 450
voltage gain, 450
Drain current, 364
Drain-source resistance, 376
Drift, 280
Dynamic parameters, 88
Dynamic resistance, 26, 29, 39

Efficiency, 99, 101, 157, 166, 173
Electron, 17
Electron charge, 20
Electron motion, 18
Emitter-base circuit, 76, 79
Emitter-base junction, 74–80, 83
break voltage, 76
piecewise linear equivalent, 77
vi characteristic, 76
Emitter bypass capacitor, 101–106
ac load line, 102
effect on low-frequency response, 431, 433
maximum symmetrical swing, 102, 103
Q point, 102–105
3-dB frequency, 433
Emitter current, 75

Emitter degeneration, 563–565
Emitter-follower (common-collector) configuration, 110–115, 202–212
ac load line, 111, 112
base-injection bias, 113–115
β variation, 113
bias network, 112
dc load line, 112
design, 208–211
high frequencies, 474–480
input impedance, 474
oscillations, 479
output capacitance, 467, 473
voltage gain, 474
hybrid parameters, 202
impedance reflection applied to, 205–207
input impedance, 115, 203, 207, 474
maximum symmetrical swing, 111
output impedance, 204, 205, 207, 475
phase inverter (splitter), 211–212
push-pull amplifier, 212
Q point, 111, 112, 209
regulator, 309
variation of EF parameters, 208
voltage gain, 207, 474
Enhancement mode, 365
Epitaxial layer, 329
Equivalent circuit, of integrated transistor, 331–332
saturation voltage, 331
series-collector resistance r_{sc}, 331
transmission lines, 331
(*See also* Integrated circuits)
of a junction resistor, 338
of transistor amplifier at high frequencies, 462–480
collector-base (output) capacitance, 463, 466, 469
depletion region, 462
hybrid-pi equivalent circuit, 463, 467
base-emitter capacitance, 466
base-spreading resistance, 464
common-emitter short circuit 3-dB frequency, f_β, 464
gain-bandwidth product, 465
Fabrication of an integrated circuit transistor, 328–331
chip, 330
diffusion, 329
disk, 330
epitaxial layer, 329
h_{fe}, 330
photolithographic isolation masking, 328
substrate, 328, 330
tolerances in, 330
Fairchild μA702 (*see* Integrated circuits)
Fall time, 511–512
Feedback, 6, 352
basic concepts of, 263–266
current, 265, 370
current error, 265, 284–287
difference circuit, 263
gain, 263
negative, 265
network, 263
summing network, 263
voltage error, 265
voltage, 265
(*See also* Voltage feedback)
Feedback amplifier, 262–316
bandwidth, 577–580
design, 298, 302
frequency response, 576–619
gain-bandwidth product, 577–580
lag compensated IC, 600–601
lead compensation, 601–602
stability, 580–583
Field-effect transistor, characteristics of, 2N3796, 647–651
2N4224, 641–646
drain-coupling capacitor, 450–451
gain-bandwidth product, 499
gate-coupling capacitor, 451
gate resistance, 450
high frequencies, 386, 387, 480–487
low-frequency response, 447–453
mid frequencies, 368, 387
Miller capacitance, 481
source bypass capacitor, 448–450

Field-effect transistor, source follower at high frequency, 483–487
input impedance of, 483
output impedance of, 485
voltage gain of, 486
[*See also* Insulated-gate field-effect transistor (IGFET); Junction field-effect transistor (JFET)]
Figure of merit of a power amplifier, 157, 166, 173
Filament, 398
Fourier series, 12, 59
Frequency of oscillation (*see* Oscillators)
Frequency compensation, 586, 600–602
(*See also* Feedback amplifier)
Frequency multiplication, 17

Gain-bandwidth (GBW) product, 498–504, 532, 541, 551, 553, 558, 562
of a cascaded amplifier, 501–504
design using, 503
of feedback amplifiers, 577–580
of FET amplifier, 499
of a transistor amplifier, 498
of a vacuum tube amplifier, 499
Gate, 365
(*See also* Field-effect transistor)
Gate-source resistance, 374
(*See also* Field-effect transistor)
Germanium, 17, 142
Germanium diode, 22
Graphical analysis, 26, 27, 368–374, 404–410
IGFET, 369–372
current feedback, 370
self-bias, 370
temperature variations, sensitivity to, 371
JFET, 368–369
self-bias, 369
transistor circuits, 87–95
β variation, 92, 95
cutoff in a transistor, 90
dc load line, 90, 93
Graphical analysis, transistor circuits, dc (quiescent) operating conditions, 90
maximum symmetrical swing, 91, 93, 96
Q point, 92–94
saturation, 90, 91, 93, 95
variations in α, 89
vacuum tube, 404–410
Grid-bias equation, 404, 406
Grid-plate capacitance, 402, 403
Grounded-cathode amplifier, 416–418
input impedance, 417
output impedance, 417
voltage gain, 417
(*See also* Vacuum tube at high frequencies)
Grounded-grid amplifier, 419–420
input impedance, 419
output impedance, 419
voltage gain, 420
(*See also* Vacuum tube at high frequencies)

h parameters, 75, 213–214
h_{FB} (α), 75
h_{fe}, 81, 330
h_{FE} (β), 81, 86
h_{ie}, 190
h_{oe}, effect of, 195
h_{re}, effect of, 194
Harmonic distortion (*see* Distortion)
Hartley oscillators, 617–618
(*See also* Oscillators)
Heat dissipation, 50
Heat sink, 50, 51, 53, 143–145
infinite, 143
with insulator, 145
Heterodyning, 5
High-frequency oscillations, 479
High-frequency response, of FET amplifier, 480–487
of transistor amplifier, 462–480
of vacuum-tube amplifier, 487–489
High-Q approximation, 525
(*See also* Single-tuned amplifier)

Hole, 18
Hybrid parameters, 186–188
  definitions of, 187
Hybrid-pi equivalent circuit, 463, 467
  (*See also* Equivalent circuit, of transistor amplifier at high frequencies)

I_{CBO}, 75, 145
Ideal diode, 22, 42
IF amplifier, 6
  (*See also* Double-tuned amplifier; Single-tuned amplifier; Stagger-tuned amplifier)
IF frequency, 5
Impedance matching, 530
Impedance reflection, in the FET, 380
  in the transistor, 205–207
Inductor-coupled power amplifier, 153
Input admittance, 374, 470
Input impedance, 115, 192, 193, 202
  common-source amplifier, 377, 381, 417, 419
  effect of loop gain on, 286
  emitter follower, 203, 283–288
    at high frequencies 351, 474–480
  without feedback, 285
  negative feedback, 284
  seen by current source in feedback amplifier, 286
  seen by voltage source in feedback amplifier, 286, 292, 298
  source follower, 483
Insulated-gate field-effect transistor (IGFET), 364–368
  breakdown, 367
  channel, 365
  drain, 365
  enhancement mode, 365
  gate, 365
  pinch-off voltage, 366
  saturation, 367
  source, 365
  substrate, 365
  (*See also* Field-effect transistor)
Integrated capacitors, 335–337
  junction capacitor, 335–336
    equivalent circuit of, 336
  thin-film capacitor, 336–337
Integrated-circuit amplifiers, biasing, 137–142
  cascading, 351–354
  design of, 341, 342
  fabrication, 326–341
  lead compensation, 601, 602, 608
  oscillators (*see* Oscillators)
  typical integrated-circuit amplifier (Fairchild μA702), 342–351
Integrated circuits, 326–360
  diffusion, 326
  hybrid, 326, 327
  monolithic, 326, 327
  oxidized wafers, 328
  thin-film techniques, 326
Integrated diode, 333–336
Integrated inductor, 341
Integrated resistors, 337–341
  junction resistor, 337–339
    equivalent circuit of, 338
    temperature effects, 339
  thin-film resistor, design of, 340
    materials used in manufacture, 339
    power rating, 340
    sheet resistance, 339
Integrator, 605–606
Interelectrode capacitance, 420, 422
Internal disturbances, reduction of, 280–283

Junction field-effect transistor (JFET), 360, 364
  breakdown voltage, 364
  channel, 361
  depletion region, 361, 363
  drain, 360
  gate, 360
  pinch-off voltage, 363
  saturation region, 364
  source, 360
  temperature dependence, 364
  (*See also* Field-effect transistor)

Lag-compensated feedback amplifiers, 598–601
IC amplifiers, 599–601
lag network design, 590
Lag compensation, 588–594
Langmuir-Childs law, 400
LC filters, 13
Lead compensation of feedback amplifiers, 595–598, 601–602
Limiter, 63
Load line, 23, 31
Load-resistance variation, 311, 314
Local oscillator, 5, 6
Logarithmic scales, 433
Loop gain, 268, 271, 272, 276
effect of, 286, 289, 314, 469
(*See also* Feedback)
Loudspeaker, 303
Low-frequency response, of FET amplifier, 447–453
of transistor amplifier, 431–447
of vacuum tube, 453–455
Low-power transistor circuits, 99
Low-power transistors, 84

Manufacturer's specifications, diode, 54–56
FET, 387–388
transistors, high-power, 145–146
low-power, 213–214
vacuum tubes, 420–422
Zener diode, 56–58
Maximal flatness, 556, 557, 563
(*See also* Stagger-tuned amplifier)
Maximum average power, 142
collector dissipation, 101
dissipation hyperbola, 158–164
efficiency, 168
(*See also* Power amplifier; Power-amplifier calculations)
Maximum power dissipated, 88
collector-current swing, 91
ratings, 213–214, 387, 420, 422
symmetrical swing, 93, 102, 103, 107, 108, 111, 159, 165
(*See also* Power amplifier; Power-amplifier calculations)
Metal-oxide-semiconductor field-effect transistor (MOSFET) [*see* Insulated-gate field-effect transistor (IGFET)]
Mica insulator, 53
Miller capacitance, 467–473, 540
$C_{b'c}$, 469
current gain and bandwidth, 471, 481, 491
input admittance, 470
loop gain, 470
Miller effect, 471
Norton's equivalent circuit, 471
output admittance, 470
short-circuit current gain, 471
Mixing circuit, 64
Modulation index, 3
Monolithic, 326, 327
MOSFET [*see* Insulated-gate field-effect transistor (IGFET)]
Multiple transistor circuits, 220–255
dc amplifiers, 220
Mutual conductance, 414
coupling, 560

Negative feedback (*see* Feedback)
Neutralization, 543–549
gain of neutralized amplifier, 549
3-dB bandwidth, 548
Noise, 5
Nonlinear amplifiers, 1
Nonlinear differential equation, 42, 49
Nonlinear equation, 47
Nonlinear properties–the ideal diode, 9
Nonlinear terms, 43
Nyquist stability criterion, 583–586
bode plots, 579
critical point, 584
root locus, 585
Routh-Hurwitz criterion, 583
stability, 585

Operational amplifier, 300
Oscillations, 269, 354, 490

Oscillators, 609–618
  Colpitts, 616–617
    conditions for oscillation, 616
    piezoelectric crystal, 617
    variation of frequency of oscillation, 617
  Hartley, 617–618
    conditions for oscillation, 617
    variation of frequency of oscillation, 618
  IC Wien bridge, 612
  phase-shift, 609–612
    conditions for oscillation, 610
    frequency variability of, 612
  tuned-circuit, 614–616
    criterion for oscillation, 615
    frequency of oscillation, 615
  Wien bridge, 612–614
    conditions for oscillation, 612
Output impedance, 192, 193, 202, 207, 288–289, 374, 470
  common-source amplifier, 377, 381, 417–419
  emitter follower, 204, 205, 283, 288–289, 475
  without feedback, 290, 292, 298
  source follower, 351, 485
  voltage feedback-current error, 288
  voltage feedback-voltage error, 288–289
Oxidized wafers, 328

Peak-inverse voltage, 54–55
Pentode, 403–404
  amplifier, 497
  grid-to-plate capacitance, 403
  6CL6 pentode characteristics, 655
  suppressor grid, 403
Perveance, 400
Phase-angle variation, 436–439
Phase inverter (splitter), 211–212, 383–385
Phase margin, 586, 588
Phase-shift oscillator, 609–612
  (*See also* Oscillators)
Phase-splitting circuit [*see* Phase inverter (splitter)]
Photolithographic isolation masking, 328
Photomicrograph of integrated differential-amplifier circuit, 431
Piecewise-linear curves, 22
Piezoelectric crystal, 617
Pinch-off voltage, 363, 366
Plate, 399
Plate resistance (r_p), 413, 414, 416
Positive feedback, 266, 268, 344
  (*See also* Feedback)
Power amplifier, 4, 151–179
  Class A, AB, B, C, definition of, 152
  complementary symmetry, 177–179
  figure of merit of, 157, 166, 173
  load lines, 154
  low distortion, 152
  push-pull amplifier, 152, 168–176
  Q-point placement, 153
Power-amplifier calculations, 96–101, 155–158, 165–167, 171, 174
  efficiency, 99, 101
  low-power transistor circuits, 99
  maximum collector dissipation, 101
    dissipation hyperbola, 39
  power delivered by the supply, 98
  power dissipated, in the collector, 98, 172
    in the load, 97, 98, 156, 166, 172
Power supply, 6, 15
  drift, 281
  variation, 280
Power supply filters, 12
Push-pull amplifier, 152, 168–176, 212
  (*See also* Class B push-pull amplifiers)

Q multiplier, 606–609
Q-point placement, in a FET, 368–374
  in a transistor, 88–95
  in a vacuum tube, 404–410
Quiescent collector current, variation in, 137

Quiescent-point variation in a transistor, due to temperature variation, 128–130
collector-current change, 129
due to uncertainty in β, 124–128
design procedure to minimize, 126
due to variation of I_{CBO} with temperature, 128
due to variation of V_{BE} with temperature, 128

RC filter, 15, 29
Rectifier, 16, 17, 54, 303
full-wave, 14, 15
half-wave, 10, 15, 22, 55
harmonics in, 13
Regulated power supply, 6, 309–316
emitter-follower regulator, 309
input-voltage variation, 310, 313
load-resistance variation, 311, 314
series regulator, 309
series-shunt regulator, 312
two-transistor feedback regulator, 312
Zener diode, 309, 312
Resistance, standard values, 625
Reverse-biased diode, 19
Reverse-saturation current, 20, 54
Ripple reduction, 40, 301
Ripple voltage, 12, 14, 40–42
Rise time, 507–509
Root locus, 578, 580, 583, 585

Saturation, 84–87, 91, 93, 95, 134, 367, 504
region, 364
voltage, 331
Scaling variables, 46
Screen, 403
current, 408
(*See also* Pentode; Tetrode)
grid, 402
(*See also* Pentode; Tetrode)
resistance, 410
secondary electrons, 403
Second harmonic, 373
Second-harmonic distortion (*see* Distortion)
Self-biasing, of the IGFET, 370, 406
of the JFET, 369
vacuum tube, 404–410
Sensitivity analysis (*see* Stability-factor analysis)
Sensitivity function, 279–283, 306
gain variations, 279–280
power-supply drift, 281
power-supply variation, 280
reduction of internal disturbances, 280–283
temperature effects, 280
Series-collector resistance r_{sc}, 31
Series regulator, 309
Series-shunt regulator, 312
Sheet resistance, 339
Shunt peaking, 561–565
emitter degeneration, 563–565
gain-bandwidth product, 562
maximal flatness, 563
Shunt regulator, 38
Single-tuned amplifier, 523–539
autotransformer, 531
bandwidth, 525
circuit design, 537
design, 526
double-tapped autotransformer, 534
GBW product, 532
high-Q approximation, 525
impedance matching, 530
$r_{bb'}$, effect of, on response of, 528
Small-signal analysis, diode, 25–32
FET, 374–387
transistor, 185–212, 221–258
vacuum tube 413–420
Source, of IGFET, 365
of the JFET, 360
Source bypass capacitor, 448–450
3-dB frequency, 449
voltage gain, 448
Source follower, high frequency, 483–487
input impedance, 381, 483

Source follower, output impedance, 381, 485
  voltage gain, 380–383, 486
Stability of feedback amplifier, 580–583, 585
  (*See also* Bode plots; Nyquist stability criterion)
Stability-factor analysis, 130–137
  ac load line, 134
  dc load line, 134
  design to minimize ΔI_{CQ}, 132
  Q-point variation, 132, 134
  saturation, 134
  stability factors, definition of, 131
  worst possible Q-point shift, 137
Stabilizing networks, 354, 586–598
  gain crossover frequency, 587
  lag compensation, 588–594
    design of lag network, 590
  lead compensation, 595–598
  marginally-stable amplifier, 586
  phase margin, 586, 588
Stagger-tuned amplifier, 553–558
  design, 558
  GBW product, 558
  maximally flat, 556, 557
  practical problems, 560
  3-dB bandwidth 557
  underdamped, 556
  voltage gain, 554

Temperature compensation using diode biasing, 137–142
  practical diode biasing circuit for integrated circuits, 139
  temperature stabilization, 139
  temperature variations, 138
  variation in quiescent collector current, 137
Temperature dependence, 364
Temperature effects in transistors, 50–54, 280
Temperature stabilization, 139
Temperature variations, sensitivity to, 371
Tetrode, 402–403
  characteristics, 6CY5, 654
  control grid, 402
  grid-to-plate capacitance, 402
  screen, 402, 403
  secondary electrons, 403
Thermal analog, 50
Thermal considerations in transistor amplifiers, 142–145
  ambient temperature, 143
  case temperature, 143, 144
  derating curves, 143
  heat sink, 143, 144
    infinite, 143
    with insulator, 145
  junction temperature, 143, 144
  manufacturer's ratings, 142
  maximum average power, 142
  practical design, 142
  silicon, 142
  thermal ratings, 143
  thermal resistances, 142, 144
Thermal ratings, 143
Thermal resistance, 50–53, 142, 144, 145
Thermal system, 52
Thin-film capacitor (*see* Integrated capacitors)
Thin-film resistor (*see* Integrated resistors)
Transconductance, 375, 416
Transfer characteristic, 387
Transformer, 15
Transformer-coupled amplifier, 164–168
  Class A operation, 164
  collector dissipation, 166
  design, 167–168
  efficiency, 166
  figure of merit, 166
  maximum symmetrical swing, 165
  power calculations, 165–167
  power transferred to load, 166
  supplied power, 166
Transistor amplifier, equivalent circuit at high frequencies, 462–480

Transistor characteristics, MPS6507
npn silicon transistor, 636–640
2N3796, 2N3797 FET, 647–651
2N3903, 2N3904, 629–640
2N4223, 2N4224 FET, 641–646
Transistor current-flow mechanism, 73–75
collector-base impedance, 74
current amplification factor (α), 75
current amplification factor (β), 75
emitter-base junction, 74
emitter current, 75
I_{CBO}, 75
Transistor switch, 504–512
delay time, 505–507
fall time, 511–512
rise time, 507–509
saturation, 504
storage time, 509–511
turn-off time, 509
turn-on time, 509
Transistor theory, 73–83
Transmission lines, 331, 565, 568
Transmitter, 2, 4
Triode, 400–402
amplification factor, 401
control grid, 400
Triode characteristic, 6SN7, 12AT7, 652, 653
Tuned-amplifier (*see* Double-tuned amplifier; Single-tuned amplifier; Stagger-tuned amplifier)
Tuned-circuit oscillator, 614–616
(*See also* Oscillators)
Turn-off time, 509
(*See also* Transistor switch)
Turn-on time, 509
(*See also* Transistor switch)
Two-transistor feedback regulator, 312
(*See also* Regulated power supply)

Unstable amplifier, 582
(*See also* Oscillators)

Vacuum tube at high frequencies, 487–489
Miller impedance, 488
3-dB frequency of, 489
voltage gain, 489
Vacuum-tube characteristics, 652–655
Valence bands, 17, 18
Variability analysis (*see* Stability-factor analysis)
Variation, of frequency of oscillation (*see* Oscillators)
of I_{CBO} with temperature, 128
of V_{BE} with temperature, 128
Voltage feedback, with current error, 266–275, 288
with voltage error, 275, 287–289
Voltage-feedback amplifiers, bootstrap circuit, 269
current gain with feedback, 268, 276
design, 266
difference amplifier, 276
gain, 266–279
loop gain, 268, 271, 272, 276
negative feedback, 266, 268
oscillations, 269
positive feedback, 266, 268
Voltage regulator, 38
(*See also* Regulated power supply)
Voltage regulation, 40
(*See also* Regulated power supply)
Voltage-standing wave ratio, 568

Wien bridge oscillator, 612–614
(*See also* Oscillators)

Zener diode, 37–42, 53, 56–58, 309, 312
breakdown, 20
specifications, 627–628
voltage, 37
(*See also* Diode characteristics)